STATISTICAL DIGITAL SIGNAL PROCESSING AND MODELING

MONSON H. HAYES

Georgia Institute of Technology

JOHN WILEY & SONS, INC.

Acquisitions Editor Steven Elliot
Marketing Manager Debra Riegert
Senior Production Editor Tony VenGraitis
Designer Laura Ierardi
Manufacturing Manager Mark Cirillo
Production Service Ingrao Associates

This book was set in Times Roman by
Techsetters, Inc. and printed and bound
by Hamilton Printing. The cover was
printed by New England Book Components.

Recognizing the importance of preserving
what has been written, it is a policy of
John Wiley & Sons, Inc. to have books
of enduring value published in the United States
printed on acid-free paper, and we exert our best
efforts to that end.

The paper in this book was manufactured by a
mill whose forest management programs include
sustained yield harvesting of its timberlands.
Sustained yield harvesting principles ensure that
the number of trees cut each year does not exceed
the amount of new growth.

To order books or for customer service please, call 1(800)-CALL-WILEY (225-5945).
ISBN 0-471 59431-8

10 9

This book is dedicated to my beautiful wife, Sandy, for all of her love, patience, and support, and for all of the sacrifices that she has made so that this book could become a reality.

It is also dedicated to Michael, Kimberly, and Nicole, for the joy that they have brought into my life, and to my parents for all of their love and support of me in all of my endeavors.

CONTENTS

PREFACE

This book is the culmination of a project that began as a set of notes for a graduate level course that is offered at Georgia Tech. In writing this book, there have been many challenges. One of these was the selection of an appropriate title for the book. Although the title that was selected is *Statistical Signal Processing and Modeling*, any one of a number of other titles could equally well have been chosen. For example, if the title of a book is to capture its central theme, then the title perhaps could have been *Least Squares Theory in Signal Processing*. If, on the other hand, the title should reflect the role of the book within the context of a course curriculum, then the title should have been *A Second Course in Discrete-Time Signal Processing*. Whatever the title, the goal of this book remains the same: to provide a comprehensive treatment of signal processing algorithms for modeling discrete-time signals, designing optimum digital filters, estimating the power spectrum of a random process, and designing and implementing adaptive filters.

In looking through the Table of Contents, the reader may wonder what the reasons were in choosing the collection of topics in this book. There are two. The first is that each topic that has been selected is not only important, in its own right, but is also important in a wide variety of applications such as speech and audio signal processing, image processing, array processing, and digital communications. The second is that, as the reader will soon discover, there is a remarkable relationship that exists between these topics that tie together a number of seemingly unrelated problems and applications. For example, in Chapter 4 we consider the problem of modeling a signal as the unit sample response of an all-pole filter. Then, in Chapter 7, we find that all-pole signal modeling is equivalent to the problem of designing an optimum (Wiener) filter for linear prediction. Since both problems require finding the solution to a set of Toeplitz linear equations, the Levinson recursion that is derived in Chapter 5 may be used to solve both problems, and the properties that are shown to apply to one problem may be applied to the other. Later, in Chapter 8, we find that an all-pole model performs a maximum entropy extrapolation of a partial autocorrelation sequence and leads, therefore, to the maximum entropy method of spectrum estimation.

This book possesses some unique features that set it apart from other treatments of statistical signal processing and modeling. First, each chapter contains numerous examples that illustrate the algorithms and techniques presented in the text. These examples play an important role in the learning process. However, of equal or greater importance is the working of new problems by the student. Therefore, at the end of each chapter, the reader will find numerous problems that range in difficulty from relatively simple exercises to more involved problems. Since many of the problems introduce extensions, generalizations, or applications of the material in each chapter, the reader is encouraged to read through the problems that are not worked. In addition to working problems, another important step in the learning process is to experiment with signal processing algorithms on a computer using either real or synthetic data. Therefore, throughout the book the reader will find MATLAB programs that have been written for most of the algorithms and techniques that are presented in the book and, at the end of each chapter, the reader will find a variety of computer exercises that use these programs. In these exercises, the student will study the performance of signal processing algorithms, look at ways to generalize the algorithms or make them more efficient, and write new programs.

Another feature that is somewhat unique to this book concerns the treatment of complex signals. Since a choice had to be made about how to deal with complex signals, the easy thing to have done would have been to consider only real-valued signals, leaving the generalization to complex signals to the exercises. Another possibility would have been to derive all results assuming that signals are real-valued, and to state how the results would change for complex-valued signals. Instead, however, the approach that was taken was to assume, from the beginning, that all signals are complex-valued. This not only saves time and space, and avoids having to jump back and forth between real-valued and complex-valued signals, but it also allows the reader or instructor to easily treat real-valued signals as a special case.

This book consists of nine chapters and an appendix and is suitable for a one-quarter or one-semester course at the Senior or Graduate level. It is intended to be used in a course that follows an introductory course in discrete-time signal processing. Although the prerequisites for this book are modest, each is important. First, it is assumed that the reader is familiar with the fundamentals of discrete-time signal processing including difference equations, discrete convolution and linear filtering, the discrete-time Fourier transform, and the z-transform. This material is reviewed in Chapter 2 and is covered in most standard textbooks on discrete-time signal processing [3,6]. The second prerequisite is a familiarity with linear algebra. Although an extensive background in linear algebra is not required, it will be necessary for the reader to be able to perform a variety of matrix operations such as matrix multiplication, finding the inverse of a matrix, and evaluating its determinant. It will also be necessary for the reader to be able to find the solution to a set of linear equations and to be familiar with eigenvalues and eigenvectors. This is standard material that may be found in any one of a number of excellent textbooks on linear algebra [2,5], and is reviewed in Chapter 2. Also in Chapter 2 is a section of particular importance that may not typically be in a student's background. This is the material covered in Section 2.3.10, which is concerned with optimization theory. The specific problem of interest is the minimization of a quadratic function of one or more *complex* variables. Although the minimization of a quadratic function of one or more *real* variables is fairly straightforward and only requires setting the derivative of the function with respect to each variable equal to zero, there are some subtleties that arise when the variables are complex. For example, although we know that the quadratic function $f(z) = |z|^2$ has a unique minimum that occurs at $z = 0$, it is not clear how to formally demonstrate this since this function is not differentiable. The last prerequisite is a course in basic probability theory. Specifically, it is necessary for the reader to be able to compute ensemble averages such as the mean and variance, to be familiar with jointly distributed random variables, and to know the meaning of terms such as statistical independence, orthogonality, and correlation. This material is reviewed in Chapter 3 and may be found in many textbooks [1,4].

This book is structured to allow a fair amount of flexibility in the order in which the topics may be taught. The first three chapters stand by themselves and, depending upon the background of the student, may either be treated as optional reading, reviewed quickly, or used as a reference. Chapter 2, for example, reviews the basic principles of discrete-time signal processing and introduces the principles and techniques from linear algebra that will be used in the book. Chapter 3, on the other hand, reviews basic principles of probability theory and provides an introduction to discrete-time random processes. Since it is not assumed that the reader has had a course in random processes, this chapter develops all of the necessary tools and techniques that are necessary for our treatment of stochastic signal modeling, optimum linear filters, spectrum estimation, and adaptive filtering.

In Chapter 4, we begin our treatment of statistical signal processing and modeling with the development of a number of techniques for modeling a signal as the output of a linear shift-invariant filter. Most of this chapter is concerned with models for deterministic signals, which include the methods of Padé, Prony, and Shanks, along with the autocorrelation and covariance methods. Due to its similarity with the problem of signal modeling, we also look at the design of a least-squares inverse filter. Finally, in Section 4.7, we look at models for discrete-time random processes, and briefly explore the important application of spectrum estimation. Depending on the interest of the reader or the structure of a course, this section may be postponed or omitted without any loss in continuity.

The initial motivation for the material in Chapter 5 is to derive efficient algorithms for the solution to a set of Toeplitz linear equations. However, this chapter accomplishes much more than this. Beginning with the derivation of a special case of the Levinson recursion known as the Levinson-Durbin recursion, this chapter then proceeds to establish a number of very important properties and results that may be derived from this recursion. These include the introduction of the lattice filter structure, the proof of the stability of the all-pole model that is formed using the autocorrelation method, the derivation of the Schur-Cohn stability test for digital filters, a discussion of the autocorrelation extension problem, the Cholesky decomposition of a Toeplitz matrix, and a procedure for recursively computing the inverse of a Toeplitz matrix that may be used to derive the general Levinson recursion and establish an interesting relationship that exists between the spectrum estimates produced using the minimum variance method and the maximum entropy method. The chapter concludes with the derivation of the Levinson and the split Levinson recursions.

The focus of Chapter 6 is on lattice filters and on how they may be used for signal modeling. The chapter begins with the derivation of the FIR lattice filter, and then proceeds to develop other lattice filter structures, which include the all-pole and allpass lattice filters, lattice filters that have both poles and zeros, and the split lattice filter. Then, we look at lattice methods for all-pole signal modeling. These methods include the forward covariance method, the backward covariance method, and the Burg algorithm. Finally, the chapter concludes by looking at how lattice filters may be used in modeling discrete-time random processes.

In Chapter 7 we turn our attention to the design of optimum linear filters for estimating one process from another. The chapter begins with the design of an FIR Wiener filter and explores how this filter may be used in such problems as smoothing, linear prediction, and noise cancellation. We then look at the problem of designing IIR Wiener filters. Although a noncausal Wiener filter is easy to design, when a causality constraint in imposed on the filter structure, the design requires a spectral factorization. One of the limitations of the Wiener filter, however, is the underlying assumption that the processes that are being filtered are wide-sense stationary. As a result, the Wiener filter is linear and shift-invariant. The chapter concludes with an introduction to the discrete Kalman filter which, unlike the Wiener filter, may be used for stationary as well as nonstationary processes.

Chapter 8 considers the problem of estimating the power spectrum of a discrete-time random process. Beginning with the classical approaches to spectrum estimation, which involve taking the discrete-time Fourier transform of an estimated autocorrelation sequence, we will examine the performance of these methods and will find that they are limited in resolution when the data records are short. Therefore, we then look at some modern approaches to spectrum estimation, which include the minimum variance method, the maximum entropy method, and parametric methods that are based on developing a model for a random process and using this model to estimate the power spectrum. Finally, we look at eigenvector methods for estimating the frequencies of a harmonic process. These methods include the Pisarenko harmonic decomposition, MUSIC, the eigenvector method, the minimum norm algorithm, and methods that are based on a principle components analysis of the autocorrelation matrix.

Finally, Chapter 9 provides an introduction to the design, implementation, and analysis of adaptive filters. The focus of this chapter is on the LMS algorithm and its variations, and the recursive least squares algorithm. There are many applications in which adaptive filters have played an important role such as linear prediction, echo cancellation, channel equalization, interference cancellation, adaptive notch filtering, adaptive control, system identification, and array processing. Therefore, included in this chapter are examples of some of these applications.

Following Chapter 9, the reader will find an Appendix that contains some documentation on how to use the MATLAB m-files that are found in the book. As noted in the appendix, these m-files are available by anonymous ftp from

ftp.eedsp.gatech.edu/pub/users/mhayes/stat_dsp

and may be accessed from the Web server for this book at

http://www.ece.gatech.edu/users/mhayes/stat_dsp

The reader may also wish to browse this Web site for additional information such as new problems and m-files, notes and errata, and reader comments and feedback.

A typical one quarter course for students who have not been exposed to discrete-time random processes might cover, in depth, Sections 3.3–3.7 of Chapter 3; Chapter 4; Sections 5.1–5.3 of Chapter 5; Sections 6.1, 6.2, and 6.4–6.7 of Chapter 6; Chapter 7; and Chapter 8. For students that have had a formal course in random processes, Chapter 3 may be quickly reviewed or omitted and replaced with Sections 9.1 and 9.2 of Chapter 9. Alternatively, the instructor may wish to introduce adaptive approaches to signal modeling at the end of Chapter 4, introduce the adaptive Wiener filter after Chapter 7, and discuss techniques for adaptive spectrum estimation in Chapter 8. For a semester course, on the other hand, Chapter 9 could be covered in its entirety. Starred (*) sections contain material that is a bit more advanced, and these sections may be easily omitted without any loss of continuity.

In teaching from this book, I have found it best to review linear algebra only as it is needed, and to begin the course by spending a couple of lectures on some of the more advanced material from Sections 3.3–3.7 of Chapter 3, such as ergodicity and the mean ergodic theorems, the relationship between the power spectrum and the maximum and minimum eigenvalues of an autocorrelation matrix, spectral factorization, and the Yule-Walker equations. I have also found it best to restrict attention to the case of real-valued signals, since treating the complex case requires concepts that may be less familiar to the student. This typically only requires replacing derivatives that are taken with respect to z^* with derivatives that are taken with respect to z.

Few textbooks are ever written in isolation, and this book is no exception. It has been shaped by many discussions and interactions over the years with many people. I have been fortunate in being able to benefit from the collective experience of my colleagues at Georgia Tech: Tom Barnwell, Mark Clements, Jim McClellan, Vijay Madisetti, Francois Malassenet, Petros Maragos, Russ Mersereau, Ron Schafer, Mark Smith, and Doug Williams. Among these people, I would like to give special thanks to Russ Mersereau, my friend and colleague, who has followed the development of this book from its beginning, and who will probably never believe that this project is finally done. Many of the ways in which the topics in this book have been developed and presented are attributed to Jim McClellan, who first exposed me to this material during my graduate work at M.I.T. His way of looking at the problem of signal modeling has had a significant influence in much of the material in this book. I would also like to acknowledge Jae Lim and Alan Oppenheim, whose leadership and guidance played a significant and important role in the development of my professional career. Other people I have been fortunate to be able to interact with over the years include my doctoral students Mitch Wilkes, Erlandur Karlsson, Wooshik Kim, David Mazel, Greg Vines, Ayhan Sakarya, Armin Kittel, Sam Liu, Baldine-Brunel Paul, Halük Aydinoglu, Antai Peng, and Qin Jiang. In some way, each of these people has influenced the final form of this book. I would also like to acknowledge all of the feedback and comments given to me by all of the students who have studied this material at Georgia Tech using early drafts of this book. Special thanks goes to Ali Adibi, Abdelnaser Adas, David Anderson, Mohamed-Slim Alouini, Osama Al-Sheikh, Rahmi Hezar, Steven Kogan, and Jeff Schodorf for the extra effort that they took in proof-reading and suggesting ways to improve the presentation of the material.

I would like to express thanks and appreciation to my dear friends Dave and Annie, Pete and Linda, and Olivier and Isabelle for all of the wonderful times that we have shared together while taking a break from our work to relax and enjoy each other's company.

Finally, we thank the following reviewers for their suggestions and encouragement throughout the development of this text: Tom Alexander, North Carolina State University; Jan P. Allenbach, Purdue University; Andreas Antoniou, University of Victoria; Ernest G. Baxa, Clemson University; Takis Kasparis, University of Central Florida; JoAnn B. Koskol, Widener University; and Charles W. Therrien, Naval Postgraduate School.

References

1. H. L. Larson and B. O. Shubert, *Probabilistic Models in Engineering Sciences: Vol. I, Random Variables and Stochastic Processes*, John Wiley & Sons, New York, 1979.
2. B. Nobel and J. W. Daniel, *Applied Linear Algebra*, Prentice-Hall, Englewood Cliffs, NJ, 1977.
3. A. V. Oppenheim and R. W. Schafer, *Discrete-Time Signal Processing*, Prentice-Hall, Englewood Cliffs, NJ, 1989.
4. A. Papoulis, *Probability, Random Variables, and Stochastic Processes*, 3rd Ed, McGraw-Hill, New York, 1991.
5. G. Strang, *Linear Algebra and its Applications*, Academic Press, New York, 1980.
6. R. D. Strum and D. E. Kirk, *First Principles of Discrete Systems and Digital Signal Processing*, Addison-Wesley, MA, 1988.

INTRODUCTION 1

Our ability to communicate is key to our society. Communication involves the exchange of information, and this exchange may be over short distances, as when two people engage in a face-to-face conversation, or it may occur over large distances through a telephone line or satellite link. The entity that carries this information from one point to another is a *signal*. A signal may assume a variety of different forms and may carry many different types of information. For example, an acoustic wave generated by the vocal tract of a human carries speech, whereas electromagnetic waves carry audio and video information from a radio or television transmitter to a receiver. A signal may be a function of a single continuous variable, such as time, it may be a function of two or more continuous variables, such as (x, y, t) where x and y are spatial coordinates, or it may be a function of one or more discrete variables.

Signal processing is concerned with the representation, manipulation, and transformation of signals and the information that they carry. For example, we may wish to enhance a signal by reducing the noise or some other interference, or we may wish to process the signal to extract some information such as the words contained in a speech signal, the identity of a person in a photograph, or the classification of a target in a radar signal.

Digital signal processing (*DSP*) is concerned with the processing of information that is represented in digital form. Although DSP, as we know it today, began to bloom in the 1960s, some of the important and powerful processing techniques that are in use today may be traced back to numerical algorithms that were proposed and studied centuries ago. In fact, one of the key figures whose work plays an important role in laying the groundwork for much of the material in this book, and whose name is attached to a number of terms and techniques, is the mathematician Karl Friedrick Gauss. Although Gauss is usually associated with terms such as the Gaussian density function, Gaussian elimination, and the Gauss-Seidel method, he is perhaps best known for his work in least squares estimation.[1] It should be interesting for the reader to keep a note of some of the other personalities appearing in this book and the dates in which they made their contributions. We will find, for example, that Prony's work on modeling a signal as a sum of exponentials was published in 1795, the work of Padé on signal matching was published in 1892, and Schuster's work

[1]Gauss is less well known for his work on fast algorithms for computing the Fourier series coefficients of a sampled signal. Specifically, he has been attributed recently with the derivation of the radix-2 decimation-in-time Fast Fourier Transform (FFT) algorithm [6].

on the periodogram appeared in 1898. Two of the more recent pioneers whose names we will encounter and that have become well known in the engineering community are those of N. Wiener (1949) and R. E. Kalman (1960).

Since the early 1970s when the first DSP chips were introduced, the field of Digital Signal Processing (DSP) has evolved dramatically. With a tremendously rapid increase in the speed of DSP processors along with a corresponding increase in their sophistication and computational power, digital signal processing has become an integral part of many products and applications. Coupled with the development of efficient algorithms for performing complex signal processing tasks, digital signal processing is radically changing the way that signal processing is done and is becoming a commonplace term.

The purpose of this book is to introduce and explore the relationships between four very important signal processing problems: signal modeling, optimum filtering, spectrum estimation, and adaptive filtering. Although these problems may initially seem to be unrelated, as we progress through this book we will find a number of different problems reappearing in different forms and we will often be using solutions to previous problems to solve new ones. For example, we will find that one way to derive a high-resolution spectrum estimate is to use the tools and techniques derived for modeling a signal and then use the model to estimate the spectrum.

The prerequisites necessary for this book are modest, consisting of an introduction to the basic principles of digital signal processing, an introduction to basic probability theory, and a general familiarity with linear algebra. For purposes of review as well as to introduce the notation that will be used throughout this book, Chapter 2 provides an overview of the fundamentals of DSP and discusses the concepts from linear algebra that will be useful in our representation and manipulation of signals. The following paragraphs describe, in general terms, the topics that are covered in this book and overview the importance of these topics in applications.

DISCRETE-TIME RANDOM PROCESSES

An introductory course in digital signal processing is concerned with the analysis and design of systems for processing deterministic discrete-time signals. A deterministic signal may be defined as one that can be described by a mathematical expression or that can be reproduced repeatedly. Simple deterministic signals include the unit sample, a complex exponential, and the response of a digital filter to a given input. In almost any application, however, it becomes necessary to consider a more general type of signal known as a random process. Unlike a deterministic signal, a random process is an ensemble or collection of signals that is defined in terms of the statistical properties that the ensemble possesses. Although a sinusoid is a deterministic signal, it may also be used as the basis for the random process consisting of the ensemble of all possible sinusoids having a given amplitude and frequency. The randomness or uncertainty in this process is contained in the phase of the sinusoid. A more common random process that we will be frequently concerned with is noise. Noise is pervasive and occurs in many different places and in many forms. For example, noise may be quantization errors that occur in an A/D converter, it may be round-off noise injected into the output of a fixed point digital filter, or it may be an unwanted disturbance such as engine noise in the cockpit of an airplane or the random disturbances picked up by a sonar array on the bottom of the ocean floor. As the name implies, a random process is distinguished from a deterministic signal because it is random or nondeterministic. Thus, until a particular signal from the ensemble is selected or observed, the exact values of the process are generally unknown. As a result of this randomness it is necessary to use a different language to represent and describe these signals.

A discrete-time random process is an indexed sequence of random variables. Therefore, what is needed is a framework for describing these random variables and the relationships between them. What will be of primary interest will be the notions of an autocorrelation sequence and a power spectrum. The autocorrelation is a second-order statistical characterization of a random process that provides information about how much linear dependence there is between signal values. This dependence may be exploited in the design of systems for predicting a signal. These predictors, in turn, are important in a number of applications such as speech compression, systems for detecting signals in noise, and filters for reducing interference. The power spectrum, on the other hand, is the Fourier transform of the autocorrelation sequence and provides another representation for the second-order statistics of the process. As is the case for deterministic signals, the frequency domain description of random processes provides a different window through which we may view the process and, in some applications, is so important that we will spend an entire chapter on methods for estimating the power spectrum.

SIGNAL MODELING

The efficient representation of signals is at the heart of many signal processing problems and applications. For example, with the explosion of information that is transmitted and stored in digital form, there is an increasing need to compress these signals so that they may be more efficiently transmitted or stored. Although this book does not address the problem of signal compression directly, it is concerned with the representation of discrete-time signals and with ways of processing these signals to enhance them or to extract some information from them. We will find that once it is possible to accurately model a signal, it then becomes possible to perform important signal processing tasks such as extrapolation and interpolation, and we will be able to use the model to classify signals or to extract certain features or characteristics from them.

One approach that may be used to compress or code a discrete-time signal is to find a model that is able to provide an accurate representation for the signal. For example, we may consider modeling a signal as a sum of sinusoids. The model or code would consist of the amplitudes, frequencies, and phases of the sinusoidal components. There is, of course, a plethora of different models that may be used to represent discrete-time signals, ranging from simple harmonic decompositions to fractal representations [2]. The approach that is used depends upon a number of different factors including the type of signal that is to be compressed and the level of fidelity that is required in the decoded or uncompressed signal. In this book, our focus will be on how we may most accurately model a signal as the unit sample response of a linear shift-invariant filter. What we will discover is that there are many different ways in which to formulate such a signal modeling problem, and that each formulation leads to a solution having different properties. We will also find that the techniques used to solve the signal modeling problem, as well as the solutions themselves, will be useful in our finding solutions to other problems. For example, the FIR Wiener filtering problem may be solved almost by inspection from the solutions that we derive for all-pole signal modeling, and many of the approaches to the problem of spectrum estimation are based on signal modeling techniques.

THE LEVINSON AND RELATED RECURSIONS

In spite of the ever-increasing power and speed of digital signal processors, there will always be a need to develop *fast algorithms* for performing a specific task or executing a particular algorithm. Many of the problems that we will be solving in this book will require that we find the solution to a set of linear equations. Many different approaches have been developed for solving these equations, such as Gaussian elimination and, in some cases, efficient algorithms exist for finding the solution. As we will see in our development of different approaches to signal modeling and as we will see in our discussions of Wiener filtering, linear equations having a Toeplitz form will appear often. Due to the tremendous amount of structure in these Toeplitz linear equations, we will find that the number of computations required to solve these equations may be reduced from order n^3 required for Gaussian elimation to order n^2. The key in this reduction is the Levinson and the more specialized Levinson-Durbin recursions. Interesting in their own right, what is perhaps even more important are the properties and relationships that are hidden within these recursions. We will see, for example, that embedded in the Levinson-Durbin recursion is a filter structure known as the lattice filter that has many important properties that make them attractive in signal processing applications. We will also find that the Levinson-Durbin recursion forms the basis for a remarkably simple stability test for digital filters.

LATTICE FILTERS

As mentioned in the previous paragraph, the lattice filter is a structure that emerges from the Levinson-Durbin recursion. Although most students who have taken a first course in digital signal processing are introduced to a number of different filter structures such as the direct form, cascade, and parallel structures, the lattice filter is typically not introduced. Nevertheless, there are a number of advantages that a lattice filter enjoys over these other filter structures that often make it a popular structure to use in signal processing applications. The first is the modularity of the filter. It is this modularity and the stage-by-stage optimality of the lattice filter for linear prediction and all-pole signal modeling that allows for the order of the lattice filter to be easily increased or decreased. Another advantage of these filters is that it is trivial to determine whether or not the filter is minimum phase (all of the roots inside the unit circle). Specifically, all that is required is to check that all of the filter coefficients (reflection coefficients) are less than 1 in magnitude. Finally, compared to other filter structures, the lattice filter tends to be less sensitive to parameter quantization effects.

WIENER AND KALMAN FILTERING

There is always a desire or the need to design the optimum filter—the one that will perform a given task or function better than any other filter. For example, in a first course in DSP one learns how to design a linear phase FIR filter that is optimum in the Chebyshev sense of minimizing the maximum error between the frequency response of the filter and the response of the ideal filter [7]. The Wiener and Kalman filters are also optimum filters. Unlike a typical linear shift-invariant filter, a Wiener filter is designed to process a given signal $x(n)$, the input, and form the best estimate in the mean square sense of a related signal $d(n)$, called the desired signal. Since $x(n)$ and $d(n)$ are not known in advance and may be described only in a statistical sense, it is not possible to simply use $H(e^{j\omega}) = D(e^{j\omega})/X(e^{j\omega})$ as the frequency response of the filter. The Kalman filter may be used similarly to recursively find the best estimate. As we will see, these filters are very general and may be used to solve a variety of different problems such as prediction, interpolation, deconvolution, and smoothing.

SPECTRUM ESTIMATION

As mentioned earlier in this chapter, the frequency domain provides a different window through which one may view a discrete-time signal or random process. The power spectrum is the Fourier transform of the autocorrelation sequence of a stationary process. In a number of applications it is necessary that the power spectrum of a process be known. For example, the IIR Wiener filter is defined in terms of the power spectral densities of two processes, the input to the filter and the desired output. Without prior knowledge of these power spectral densities it becomes necessary to estimate them from observations of the processes. The power spectrum also plays an important role in the detection and classification of periodic or narrowband processes buried in noise.

In developing techniques for estimating the power spectrum of a random process, we will find that the simple approach of Fourier transforming the data from a sample realization does not provide a statistically reliable or high-resolution estimate of the underlying spectrum. However, if we are able to find a model for the process, then this model may be used to estimate the spectrum. Thus, we will find that many of the techniques and results developed for signal modeling will prove useful in understanding and solving the spectrum estimation problem.

ADAPTIVE FILTERS

The final topic considered in this book is adaptive filtering. Throughout most of the discussions of signal modeling, Wiener filtering, and spectrum estimation, it is assumed that the signals that are being processed or analyzed are stationary; that is, their statistical properties are not varying in time. In the real world, however, this will never be the case. Therefore, these problems are reconsidered within the context of nonstationary processes. Beginning with a general FIR Wiener filter, it is shown how a gradient descent algorithm may be used to solve the Wiener-Hopf equations and design the Wiener filter. Although this algorithm is well behaved in terms of its convergence properties, it is not generally used in practice since it requires knowledge of the process statistics, which are generally unknown. Another approach, which has been used successfully in many applications, is the stochastic gradient algorithm known as LMS. Using a simple gradient estimate in place of the true gradient in a gradient descent algorithm, LMS is efficient and well understood in terms of its convergence properties. A variation of the LMS algorithm is the perceptron algorithm used in pattern recognition and is the starting point for the design of a neural network. Finally, while the LMS algorithm is designed to solve the Wiener filtering problem by minimizing a mean square error, a deterministic least squares approach leads to the development of the RLS algorithm. Although computationally much more involved than the stochastic gradient algorithms, RLS enjoys a significant performance advantage.

There are many excellent textbooks that deal in much more detail and rigor with the subject of adaptive filtering and adaptive signal processing [1, 3, 4, 5, 8]. Here, our goal is to simply provide the bridge to that literature, illustrating how the problems that we have solved in earlier chapters may be adapted to work in nonstationary environments. It is here, too, that we are able to look at some important applications such as linear prediction, channel equalization, interference cancelation, and system identification.

CLOSING

The problems that are considered in the following chapters are fundamental and important. The similarities and relationships between these problems are striking and remarkable. Many of the problem solving techniques presented in this text are powerful and general and may be successfully applied to a variety of other problems in digital signal processing as well as in other disciplines. It is hoped that the reader will share in the excitement of unfolding these relationships and embarking on a journey into the world of statistical signal processing.

References

1. S. T. Alexander, *Adaptive Signal Processing*, Springer-Verlag, New York, 1986.
2. M. Barnsley, *Fractals Everywhere*, Academic Press, New York, 1988.
3. M. G. Bellanger, *Adaptive Digital Filters and Signal Analysis*, Marcel Dekker, Inc., New York, 1987.
4. P. M. Clarkson, *Optimal and Adaptive Signal Processing*, CRC Press, Boca Raton, FL, 1993.
5. S. Haykin, *Adaptive Filter Theory*, Prentice-Hall, Englewood-Cliffs, NJ, 1986.
6. M. T. Heideman, D. H. Johnson, and C. S. Burrus, "Gauss and the history of the Fast Fourier Transform," *IEEE ASSP Magazine*, vol. 1, no. 4, pp. 14–21, October 1984.
7. A. V. Oppenheim and R. W. Schafer, *Discrete-Time Signal Processing*, Prentice-Hall, Englewood Cliffs, NJ, 1989.
8. B. Widrow and S. Stearns, *Adaptive Signal Processing*, Prentice-Hall, Englewood Cliffs, NJ, 1985.

BACKGROUND 2

2.1 INTRODUCTION

There are two goals of this chapter. The first is to provide the reader with a summary as well as a reference for the fundamentals of discrete-time signal processing and the basic vector and matrix operations of linear algebra. The second is to introduce the notation and terminology that will be used throughout the book. Although it would be possible to move directly to Chapter 3 and only refer to this chapter as a reference as needed, the reader should, at a minimum, peruse the topics presented in this chapter in order to become familiar with what is expected in terms of background and to become acquainted with the terminology and notational conventions that will be used.

The organization of this chapter is as follows. The first part contains a review of the fundamentals of discrete-time signal processing. This review includes a summary of important topics in discrete-time signal manipulation and representation such as linear filtering, difference equations, digital networks, discrete-time Fourier transforms, z-transforms, and the DFT. Since this review is not intended to be a comprehensive treatment of discrete-time signal processing, the reader wishing a more in-depth study may consult any one of a number of excellent texts including [7, 8, 11]. The second part of this chapter summarizes the basic ideas and techniques from linear algebra that will be used in later chapters. The topics that are covered include vector and matrix manipulations, the solution of linear equations, eigenvalues and eigenvectors, and the minimization of quadratic forms. For a more detailed treatment of linear algebra, the reader may consult any one of a number of standard textbooks such as [6, 10].

2.2 DISCRETE-TIME SIGNAL PROCESSING

In this brief overview of the fundamentals of discrete-time signal processing, we focus primarily on the specification and characterization of discrete-time signals and systems. Of particular importance are the topics of discrete-time filters, the transform analysis of discrete-time signals and systems, and digital filter flowgraphs. Since the goal of this section is only to review these topics while introducing notation and terminology, the discussions will be cursory with most of the mathematical details being left to the references.

2.2.1 Discrete-Time Signals

A discrete-time signal is an indexed sequence of real or complex numbers.[1] Thus, a discrete-time signal is a function of an integer-valued variable, n, that is denoted by $x(n)$. Although the independent variable n need not necessarily represent "time" (n may, for example, correspond to a spatial coordinate or distance), $x(n)$ is generally referred to as a function of time. Since a discrete-time signal is undefined for noninteger values of n, the graphical representation of $x(n)$ will be in the form of a *lollipop* plot, as shown in Fig. 2.1.

Discrete-time signals may arise as a result of sampling a continuous-time signal, such as speech, with an analog-to-digital (A/D) converter. For example, if a continuous-time signal $x_a(t)$ is sampled at a rate of $f_s = 1/T_s$ samples per second, then the sampled signal $x(n)$ is related to $x_a(t)$ as follows

$$x(n) = x_a(nT_s)$$

Not all discrete-time signals, however, are obtained in this manner. In particular, some signals may be considered to be naturally occurring discrete-time sequences since there is no physical analog-to-digital converter that is converting an analog signal into a discrete-time signal. Examples of signals that fall into this category include daily stock market prices, population statistics, warehouse inventories, and the Wolfer sunspot numbers.

Although most information-bearing signals of practical interest are complicated functions of time, there are three simple yet important discrete-time signals that are frequently used in the representation and description of more complicated signals. These are the unit sample, the unit step, and the complex exponential. The *unit sample*, denoted by $\delta(n)$, is defined by

$$\delta(n) = \begin{cases} 1 & ; \quad n = 0 \\ 0 & ; \quad \text{otherwise} \end{cases}$$

and plays the same role in discrete-time signal processing that the unit impulse plays in continuous-time signal processing. The unit sample may be used to decompose an arbitrary signal $x(n)$ into a sum of weighted (scaled) and shifted unit samples as follows

$$x(n) = \sum_{k=-\infty}^{\infty} x(k)\delta(n-k)$$

This decomposition is the discrete version of the *sifting property* for continuous-time signals. The *unit step*, denoted by $u(n)$, is defined by

$$u(n) = \begin{cases} 1 & ; \quad n \geq 0 \\ 0 & ; \quad \text{otherwise} \end{cases}$$

and is related to the unit sample by

$$u(n) = \sum_{k=-\infty}^{n} \delta(k)$$

Finally, a *complex exponential* is the periodic signal

$$e^{jn\omega_0} = \cos(n\omega_0) + j\sin(n\omega_0)$$

[1] Discrete-time signals may be either deterministic or random (stochastic). Here it is assumed that the signals are deterministic. In Chapter 3 we will consider the characterization of discrete-time random processes.

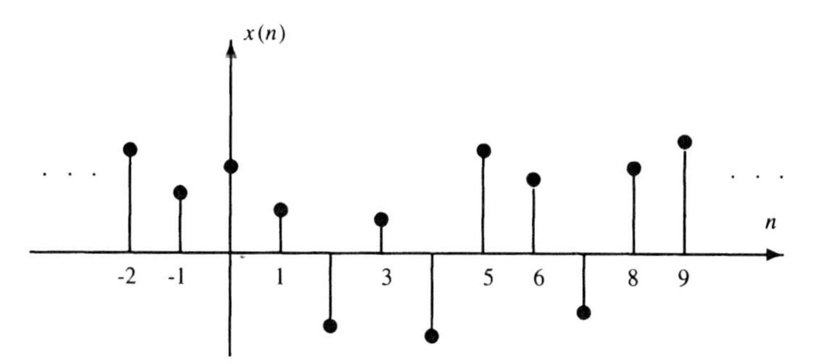

Figure 2.1 *The graphical representation of a discrete-time signal $x(n)$.*

where ω_0 is some real constant. Complex exponentials are useful in the Fourier decomposition of signals as described in Section 2.2.4.

Discrete-time signals may be conveniently classified in terms of their duration or extent. For example, a discrete-time sequence is said to be a *finite length sequence* if it is equal to zero for all values of n outside a finite interval $[N_1, N_2]$. Signals that are not finite in length, such as the unit step and the complex exponential, are said to be *infinite length sequences.* Since the unit step is equal to zero for $n < 0$, it is said to be a *right-sided sequence.* In general, a right-sided sequence is any infinite-length sequence that is equal to zero for all values of $n < n_0$ for some integer n_0. Similarly, an infinite-length sequence $x(n)$ is said to be *left sided* if, for some integer n_0, $x(n) = 0$ for all $n > n_0$. An example of a left-sided sequence is

$$x(n) = u(n_0 - n) = \begin{cases} 1 & ; \quad n \le n_0 \\ 0 & ; \quad n > n_0 \end{cases}$$

which is a time-reversed and delayed unit step. An infinite-length signal that is neither right sided nor left sided, such as the complex exponential, is referred to as a *two-sided sequence.*

2.2.2 Discrete-time Systems

A discrete-time system is a mathematical operator or mapping that transforms one signal (the input) into another signal (the output) by means of a fixed set of rules or functions. The notation $T[-]$ will be used to represent a general system, such as the one shown in Fig. 2.2, in which an input signal $x(n)$ is transformed into an output signal $y(n)$ through the transformation $T[-]$. The input/output properties of such a system may be specified in any one of a number of different ways. The relationship between the input and output, for example, may be expressed in terms of a concise mathematical rule or function such as

$$y(n) = x^2(n)$$

or

$$y(n) = 0.5y(n-1) + x(n)$$

It is also possible, however, to describe a system in terms of an algorithm that provides a sequence of instructions or operations that is to be applied to the input signal values. For

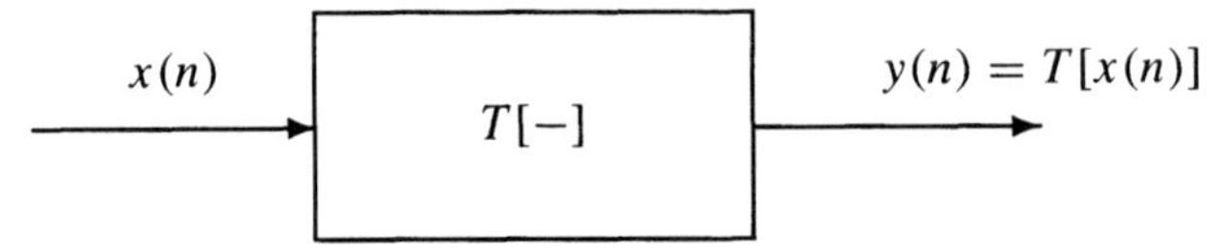

Figure 2.2 *The representation of a discrete-time system as a transformation $T[-]$ that maps an input signal $x(n)$ into an output signal $y(n)$.*

example,

$$\begin{aligned} y_1(n) &= 0.5y_1(n-1) + 0.25x(n) \\ y_2(n) &= 0.25y_2(n-1) + 0.5x(n) \\ y_3(n) &= 0.4y_3(n-1) + 0.5x(n) \\ y(n) &= y_1(n) + y_2(n) + y_3(n) \end{aligned}$$

is a sequence of instructions that defines a third-order recursive digital filter in parallel form. In some cases, a system may conveniently be specified in terms of a table that defines the set of all possible input/output signal pairs of interest. In the Texas Instruments *Speak and Spell*™, for example, pushing a specific button (the input signal) results in the synthesis of a given spoken letter or word (the system output).

Discrete-time systems are generally classified in terms of the properties that they possess. The most common properties include linearity, shift-invariance, causality, stability, and invertibility, which are described in the following sections.

Linearity and Shift-Invariance. The two system properties that are of the greatest importance for simplifying the analysis and design of discrete-time systems are *linearity* and *shift-invariance*. A system $T[-]$ is said to be *linear* if, for any two inputs $x_1(n)$ and $x_2(n)$ and for any two (complex-valued) constants a and b,

$$T[ax_1(n) + bx_2(n)] = aT[x_1(n)] + bT[x_2(n)]$$

In other words, the response of a linear system to a sum of two signals is the sum of the two responses, and scaling the input by a constant results in the output being scaled by the same constant. The importance of this property is evidenced in the observation that if the input is decomposed into a superposition of weighted and shifted unit samples,

$$x(n) = \sum_{k=-\infty}^{\infty} x(k)\delta(n-k)$$

then the output is

$$y(n) = \sum_{k=-\infty}^{\infty} x(k)T[\delta(n-k)] = \sum_{k=-\infty}^{\infty} x(k)h_k(n) \qquad (2.1)$$

where $h_k(n) = T[\delta(n-k)]$ is the response of the system to the delayed unit sample $\delta(n-k)$. Equation (2.1) is referred to as the *superposition sum* and it illustrates that a linear system is completely specified once the signals $h_k(n)$ are known.

The second important property is *shift-invariance*.[2] A system is said to be shift-invariant if a shift in the input by n_0 results in a shift in the output by n_0. Thus, if $y(n)$ is the response

[2]Some authors use the term *time-invariance* instead of shift-invariance. However, since the independent variable, n, may represent something other than time, such as distance, we prefer to use the term *shift-invariance*.

of a shift-invariant system to an input $x(n)$, then for any shift in the input, $x(n-n_0)$, the response of the system will be $y(n-n_0)$. In effect, shift-invariance means that the properties of the system do not change with time.

A system that is both linear and shift-invariant is called a *linear shift-invariant* (LSI) system. For a shift-invariant system, if $h(n) = T[\delta(n)]$ is the response to a unit sample $\delta(n)$, then the response to $\delta(n-k)$ is $h(n-k)$. Therefore, for an LSI system the superposition sum given in Eq. (2.1) becomes

$$y(n) = \sum_{k=-\infty}^{\infty} x(k)h(n-k) \tag{2.2}$$

which is the *convolution sum*. In order to simplify notation, the convolution sum is often expressed as

$$y(n) = x(n) * h(n)$$

Since the convolution sum allows one to evaluate the response of an LSI system to an arbitrary input $x(n)$, LSI systems are uniquely characterized by their response, $h(n)$, to a unit sample. This signal is referred to as the *unit sample response* of the system.

Causality. A system property that is important for real-time applications is *causality*. A system is said to be *causal* if, for any n_0, the response of the system at time n_0 depends only upon the values of the input for $n \leq n_0$. For a causal system it is not possible for changes in the output to precede changes in the input. The system described by the equation $y(n) = x(n) + x(n-1)$, for example, is causal since the value of the output at any time $n = n_0$ depends only on the input $x(n)$ at time n_0 and at time $n_0 - 1$. The system described by $y(n) = x(n) + x(n+1)$, on the other hand, is noncausal since the output at time $n = n_0$ depends on the value of the input at time $n_0 + 1$. If a system is linear and shift-invariant then it will be causal if and only if $h(n) = 0$ for $n < 0$.

Stability. In many applications, it is important for a system to have a response, $y(n)$, that is bounded in amplitude whenever the input is bounded. A system with this property is said to be *stable* in the Bounded-Input Bounded-Output (BIBO) sense. More specifically, a system is BIBO stable if, for any bounded input, $|x(n)| \leq A < \infty$, the output is bounded, $|y(n)| \leq B < \infty$. In the case of a linear shift-invariant system, stability is guaranteed whenever the unit sample response is absolutely summable

$$\sum_{n=-\infty}^{\infty} |h(n)| < \infty \tag{2.3}$$

For example, an LSI system with $h(n) = a^n u(n)$ is stable whenever $|a| < 1$.

Invertibility. A system property that is important in applications such as channel equalization and deconvolution is *invertibility* (see Section 4.4.5 for the design of an FIR least squares inverse filter). A system is said to be *invertible* if the input to the system may be uniquely determined by observing the output. In order for a system to be invertible, it is necessary for distinct inputs to produce distinct outputs. In other words, given any two inputs $x_1(n)$ and $x_2(n)$ with $x_1(n) \neq x_2(n)$, it must be true that $y_1(n) \neq y_2(n)$. For example, the system defined by

$$y(n) = x(n)g(n)$$

is invertible if and only if $g(n) \neq 0$ for all n. Specifically, if $g(n)$ is nonzero for all n, then, given $y(n)$, the input may be reconstructed by $x(n) = y(n)/g(n)$.

2.2.3 Time-Domain Descriptions of LSI Filters

An important class of linear shift-invariant systems are those for which the input $x(n)$ and output $y(n)$ are related by a linear constant coefficient difference equation (LCCDE) of the form

$$y(n) + \sum_{k=1}^{p} a(k)y(n-k) = \sum_{k=0}^{q} b(k)x(n-k) \tag{2.4}$$

In this difference equation, p and q are integers that determine the *order* of the system and $a(1), \ldots, a(p)$ and $b(0), \ldots, b(q)$ are the *filter coefficients* that define the system. The difference equation is often written in the form

$$\boxed{y(n) = \sum_{k=0}^{q} b(k)x(n-k) - \sum_{k=1}^{p} a(k)y(n-k)} \tag{2.5}$$

which clearly shows that the output $y(n)$ is equal to a linear combination of past output values, $y(n-k)$ for $k = 1, 2, \ldots, p$, along with past and present input values, $x(n-k)$ for $k = 0, 1, \ldots, q$. For the special case of $p = 0$, the difference equation becomes

$$y(n) = \sum_{k=0}^{q} b(k)x(n-k) \tag{2.6}$$

and the output is simply a weighted sum of the current and past input values. As a result, the unit sample response is finite in length

$$h(n) = \sum_{k=0}^{q} b(k)\delta(n-k)$$

and the system is referred to as a *Finite length Impulse Response* (*FIR*) system. However, if $p \neq 0$ then the unit sample response is, in general, infinite in length and the system is referred to as an *Infinite length Impulse Response* (*IIR*) system. For example, if

$$y(n) = ay(n-1) + x(n)$$

then the unit sample response is $h(n) = a^n u(n)$.

2.2.4 The Discrete-Time Fourier Transform

Frequency analysis of discrete-time signals and systems provides an important analysis and design tool and often provides insights into the solution of problems that would not otherwise be possible. Of central importance in the frequency analysis of discrete-time signals is the Discrete-Time Fourier Transform (DTFT). The DTFT of a signal $x(n)$ is the complex-valued function of the continuous (frequency) variable, ω, defined by

$$\boxed{X(e^{j\omega}) = \sum_{n=-\infty}^{\infty} x(n)e^{-jn\omega}} \tag{2.7}$$

In order for the DTFT of a signal to be defined, however, the sum in Eq. (2.7) must converge. A sufficient condition for the sum to converge uniformly to a continuous function of ω is that $x(n)$ be absolutely summable

$$\sum_{n=-\infty}^{\infty} |x(n)| < \infty \tag{2.8}$$

Although most signals of interest will have a DTFT, signals such as the unit step and the complex exponential are not absolutely summable and do not have a DTFT. However, if we allow the DTFT to contain generalized functions then the DTFT of a complex exponential is an impulse

$$x(n) = e^{jn\omega_0} \longrightarrow X(e^{j\omega}) = 2\pi u_0(\omega - \omega_0) \quad ; \quad |\omega| < \pi$$

where $u_0(\omega - \omega_0)$ is used to denote an impulse at frequency $\omega = \omega_0$. Similarly, the DTFT of a unit step is

$$u(n) \longrightarrow U(e^{j\omega}) = \frac{1}{1 - e^{-j\omega}} + \pi u_0(\omega) \quad ; \quad |\omega| < \pi$$

The DTFT possesses some symmetry properties of interest. For example, since $e^{-jn\omega}$ is periodic in ω with a period of 2π, it follows that $X(e^{j\omega})$ is also periodic with a period of 2π. In addition, if $x(n)$ is real-valued then $X(e^{j\omega})$ will be conjugate symmetric

$$X(e^{j\omega}) = X^*(e^{-j\omega})$$

The DTFT is, in general, a complex-valued function of ω. Therefore, it is normally represented in polar form in terms of its magnitude and phase

$$X(e^{j\omega}) = \left|X(e^{j\omega})\right| e^{j\phi_x(\omega)}$$

For real signals, the conjugate symmetry of $X(e^{j\omega})$ implies that the magnitude is an even function, $|X(e^{j\omega})| = |X(e^{-j\omega})|$ and that the phase is an odd function, $\phi_x(\omega) = -\phi_x(\omega)$.

A discrete-time Fourier transform of special importance is the DTFT of the unit sample response of a linear shift-invariant system,

$$\boxed{H(e^{j\omega}) = \sum_{n=-\infty}^{\infty} h(n)e^{-jn\omega}} \tag{2.9}$$

This DTFT is called the *frequency response* of the filter, and it defines how a complex exponential is changed in amplitude and phase by the system. Note that the condition for the existence of the DTFT given in Eq. (2.8) is the same as the condition for BIBO stability of an LSI system. Therefore, it follows that the DTFT of $h(n)$ exists for BIBO stable systems.

The DTFT is an invertible transformation in the sense that, given the DTFT $X(e^{j\omega})$ of a signal $x(n)$, the signal may be recovered using the Inverse DTFT (IDTFT) as follows

$$\boxed{x(n) = \frac{1}{2\pi}\int_{-\pi}^{\pi} X(e^{j\omega})e^{jn\omega}d\omega} \tag{2.10}$$

In addition to providing a method for computing $x(n)$ from $X(e^{j\omega})$, the IDTFT may also be viewed as a decomposition of $x(n)$ into a linear combination of complex exponentials.

There are a number of useful and important properties of the DTFT. Perhaps the most important of these is the *convolution theorem*, which states that the DTFT of a convolution of two signals

$$y(n) = x(n) * h(n)$$

is equal to the product of the transforms,

$$Y(e^{j\omega}) = X(e^{j\omega})H(e^{j\omega})$$

Another useful property is *Parseval's theorem*, which states that the sum of the squares of a signal, $x(n)$, is equal to the integral of the square of its DTFT,

$$\boxed{\sum_{n=-\infty}^{\infty} |x(n)|^2 = \frac{1}{2\pi}\int_{-\pi}^{\pi} |X(e^{j\omega})|^2 d\omega} \tag{2.11}$$

Some additional properties of the DTFT are listed in Table 2.1. Derivations and applications of these properties may be found in references [7, 8, 11].

2.2.5 The z-Transform

The z-transform is a generalization of the discrete-time Fourier transform that allows many signals not having a DTFT to be described using transform techniques. The z-transform of a discrete-time signal $x(n)$ is defined as

$$\boxed{X(z) = \sum_{n=-\infty}^{\infty} x(n)z^{-n}} \tag{2.12}$$

where $z = re^{j\omega}$ is a complex variable. Note that when $r = 1$, or $z = e^{j\omega}$, the z-transform

Table 2.1 ***Properties of the DTFT***

Property	Sequence	Transform
	$x(n)$	$X(e^{j\omega})$
Delay	$x(n - n_0)$	$e^{-j\omega n_0}X(e^{j\omega})$
Modulation	$e^{j\omega_0 n}x(n)$	$X(e^{j(\omega-\omega_0)})$
Conjugation	$x^*(n)$	$X^*(e^{-j\omega})$
Time reversal	$x(-n)$	$X(e^{-j\omega})$
Convolution	$x(n) * y(n)$	$X(e^{j\omega})Y(e^{j\omega})$
Multiplication	$x(n)y(n)$	$\frac{1}{2\pi}\int_{-\pi}^{\pi} X(e^{j\theta})Y(e^{j(\omega-\theta)})d\theta$
Multiplication by n	$nx(n)$	$j\frac{d}{d\omega}X(e^{j\omega})$
Parseval	$\sum_{n=-\infty}^{\infty} x(n)y^*(n)$	$\frac{1}{2\pi}\int_{-\pi}^{\pi} X(e^{j\omega})Y^*(e^{j\omega})d\omega$

becomes the discrete-time Fourier transform,

$$X(e^{j\omega}) = X(z)\big|_{z=e^{j\omega}} = \sum_{n=-\infty}^{\infty} x(n)e^{-jn\omega}$$

As with the DTFT, the z-transform is only defined when the sum in Eq. (2.12) converges. Since the sum generally does not converge for all values of z, associated with each z-transform is a *region of convergence* that defines those values of z for which the sum converges. For a finite length sequence, the sum in Eq. (2.12) contains only a finite number of terms. Therefore, the z-transform of a finite length sequence is a polynomial in z and the region of convergence will include all values of z (except possibly $z = 0$ or $z = \infty$). For right-sided sequences, on the other hand, the region of convergence is the exterior of a circle, $|z| > R_-$, and for left-sided sequences it is the interior of a circle, $|z| < R_+$, where R_- and R_+ are positive numbers. For two-sided sequences, the region of convergence is an annulus

$$R_- < |z| < R_+$$

Just as for the discrete-time Fourier transform, the z-transform has a number of important and useful properties, some of which are summarized in Table 2.2. In addition, a symmetry condition that will be of interest in our discussions of the power spectrum is the following. If $x(n)$ is a conjugate symmetric sequence,

$$x(n) = x^*(-n)$$

then it z-transform satisfies the relation

$$\boxed{X(z) = X^*(1/z^*)} \tag{2.13}$$

This property follows by combining the conjugation and time-reversal properties listed in Table 2.2. Finally, given in Table 2.3 are some closed-form expressions for some commonly found summations. These are often useful in evaluating the z-transform given in Eq. (2.12).

A z-transform of special importance in the design and analysis of linear shift-invariant systems is the *system function*, which is the z-transform of the unit sample response,

$$H(z) = \sum_{n=-\infty}^{\infty} h(n)z^{-n}$$

Table 2.2 ***Properties of the z-Transform***

Property	Sequence	Transform
	$x(n)$	$X(z)$
Delay	$x(n - n_0)$	$z^{-n_0}X(z)$
Multiplication by α^n	$\alpha^n x(n)$	$X(z/\alpha)$
Conjugation	$x^*(n)$	$X^*(z^*)$
Time reversal	$x(-n)$	$X(z^{-1})$
Convolution	$x(n) * h(n)$	$X(z)H(z)$
Multiplication by n	$nx(n)$	$-z\dfrac{d}{dz}X(z)$

Table 2.3 ***Closed-form Expressions for Some Commonly Encountered Series***

$\sum_{n=0}^{N-1} a^n = \frac{1-a^N}{1-a}$	$\sum_{n=0}^{\infty} a^n = \frac{1}{1-a} \quad ; \quad \lvert a\rvert < 1$
$\sum_{n=0}^{N-1} na^n = \frac{(N-1)a^{N+1} - Na^N + a}{(1-a)^2}$	$\sum_{n=0}^{\infty} na^n = \frac{a}{(1-a)^2} \quad ; \quad \lvert a\rvert < 1$
$\sum_{n=0}^{N-1} n = \frac{1}{2}N(N-1)$	$\sum_{n=0}^{N-1} n^2 = \frac{1}{6}N(N-1)(2N-1)$

For an FIR filter described by a LCCDE of the form given in Eq. (2.6), the system function is a polynomial in z^{-1},

$$H(z) = \sum_{k=0}^{q} b(k)z^{-k} = b(0)\prod_{k=1}^{q}(1 - z_k z^{-1}) \tag{2.14}$$

and the roots of this polynomial, z_k, are called the *zeros* of the filter. Due to the form of $H(z)$ for FIR filters, they are often referred to as *all-zero* filters. For an IIR filter described by the general difference equation given in Eq. (2.5), the system function is a ratio of two polynomials in z^{-1},

$$H(z) = \frac{\sum_{k=0}^{q} b(k)z^{-k}}{1 + \sum_{k=1}^{p} a(k)z^{-k}} = b(0)\frac{\prod_{k=1}^{q}(1 - z_k z^{-1})}{\prod_{k=1}^{p}(1 - p_k z^{-1})} \tag{2.15}$$

The roots of the numerator polynomial, z_k, are the zeros of $H(z)$ and the roots of the denominator polynomial, p_k, are called the *poles*. If the order of the numerator polynomial is zero, $q = 0$, then

$$H(z) = \frac{b(0)}{1 + \sum_{k=1}^{p} a(k)z^{-k}} = \frac{b(0)}{\prod_{k=1}^{p}(1 - p_k z^{-1})} \tag{2.16}$$

and $H(z)$ is called an *all-pole* filter.

If the coefficients $a(k)$ and $b(k)$ in the system function are real-valued (equivalently, if $h(n)$ is real-valued) then

$$H(z) = H^*(z^*)$$

and the poles and zeros of $H(z)$ will occur in conjugate pairs. That is, if $H(z)$ has a pole (zero) at $z = a$, then $H(z)$ will have a pole (zero) at $z = a^*$. Some useful z-transform pairs may be found in Table 2.4.

2.2.6 Special Classes of Filters

There are several special classes of filters that we will be encountering in the following chapters. The first are filters that have *linear phase*. These filters are important in applications

Table 2.4 Some Useful z-Transform Pairs

Sequence	Transform	Region of Convergence
$\delta(n)$	1	All z
$\alpha^n\left[u(n) - u(n-N)\right]$	$\dfrac{1-\alpha^N z^{-N}}{1-\alpha z^{-1}}$	$\|z\| > 0$
$\alpha^n u(n)$	$\dfrac{1}{1-\alpha z^{-1}}$	$\|z\| > \alpha$
$-\alpha^n u(-n-1)$	$\dfrac{1}{1-\alpha z^{-1}}$	$\|z\| < \alpha$
$\alpha^{\|n\|}$	$\dfrac{1-\alpha^2}{(1-\alpha z^{-1})(1-\alpha z)}$	$\alpha < \|z\| < 1/\alpha$

such as speech and image processing and have a frequency response of the form

$$H(e^{j\omega}) = A(e^{j\omega})e^{j(\beta - \alpha\omega)} \tag{2.17}$$

where $A(e^{j\omega})$ is a real-valued function of ω and α and β are constants.[3] In order for a *causal* filter to have linear phase and be realizable using a *finite-order* linear constant coefficient difference equation, the filter must be FIR [2, 7]. In addition, the linear phase condition places the constraint that the unit sample response $h(n)$ be either conjugate symmetric (Hermitian),

$$h^*(n) = h(N-1-n)$$

or conjugate antisymmetric (anti-Hermitian)

$$h^*(n) = -h(N-1-n)$$

These constraints, in turn, impose the constraint that the zeros of the system function $H(z)$ occur in conjugate reciprocal pairs,

$$H^*(z^*) = \pm z^{N-1} H(1/z)$$

In other words, if $H(z)$ has a zero at $z = z_0$, then $H(z)$ must also have a zero at $z = 1/z_0^*$.

Another filter having a special form is the *allpass filter*. Allpass filters are useful in applications such as phase equalization and have a frequency response with a constant magnitude

$$|H(e^{j\omega})| = A$$

For an allpass filter having a rational system function, $H(z)$ must be of the form

$$H(z) = z^{-n_0} A \prod_{k=1}^{N} \frac{z^{-1} - \alpha_k^*}{1 - \alpha_k z^{-1}}$$

Thus, if $H(z)$ has a zero (pole) at $z = \alpha_k$, then $H(z)$ must also have a pole (zero) at $z = 1/\alpha_k^*$.

[3]The term linear phase is often reserved for the special case in which $\beta = 0$ and $A(e^{j\omega})$ is nonnegative. Filters of the form given in Eq. (2.17) are then said to have *generalized linear phase*. Here we will not be concerned about this distinction.

Finally, a stable and causal filter that has a rational system function with all of its poles and zeros *inside* the unit circle is said to be a *minimum phase filter*. Thus, for a minimum phase filter, $|z_k| < 1$ and $|p_k| < 1$ in Eq. (2.15). Note that a minimum phase filter will have a stable and causal inverse, $1/H(z)$, which will also be minimum phase. This property will be useful in the development of the spectral factorization theorem in Section 3.5 and in the derivation of the causal Wiener filter in Chapter 7.

2.2.7 Filter Flowgraphs

A LCCDE defines the relationship between the filter input, $x(n)$, and the filter output, $y(n)$. In the implementation of a filter, either in hardware or software, there are many different ways in which the computations necessary to compute the value of the output $y(n)$ at time n may be ordered. For example, the difference equation

$$y(n) = 0.2y(n-1) + x(n) + 0.5x(n-1)$$

and the pair of coupled difference equations

$$\begin{aligned} w(n) &= 0.2w(n-1) + x(n) \\ y(n) &= w(n) + 0.5w(n-1) \end{aligned}$$

are two implementations of a first-order IIR filter having a system function

$$H(z) = \frac{1 + 0.5z^{-1}}{1 - 0.2z^{-1}}$$

In describing these different filter implementations it is convenient to use *flowgraphs* or *digital networks* that show how a given system is realized in terms of interconnections of the basic computational elements of the filter, i.e., adders, multipliers, and delays. The notation that will be used to represent these computational elements is shown in Fig. 2.3. As an example, shown in Fig. 2.4*a* is a fourth-order IIR filter implemented in *direct form* and in Fig. 2.4*b* is an Nth-order FIR filter, referred to as a *tapped delay line*.

One structure of particular importance that will be developed in Chapter 6 is the *lattice filter*. As we will see, this structure has a number of useful and important properties compared to other filter structures.

2.2.8 The DFT and FFT

The discrete-time Fourier transform provides a Fourier representation of a discrete-time signal that is useful as a design and analysis tool. However, since the DTFT is a function of a continuous variable, ω, it is not directly amenable to digital computation. For finite length sequences, there is another representation called the *Discrete Fourier Transform* (*DFT*) that is a function of an integer variable, k, and is therefore easily evaluated with a digital computer. For a finite-length sequence $x(n)$ of length N that is equal to zero outside the interval $[0, N-1]$, the *N-point DFT* is

$$X(k) = \sum_{n=0}^{N-1} x(n) e^{-j2\pi kn/N} \tag{2.18}$$

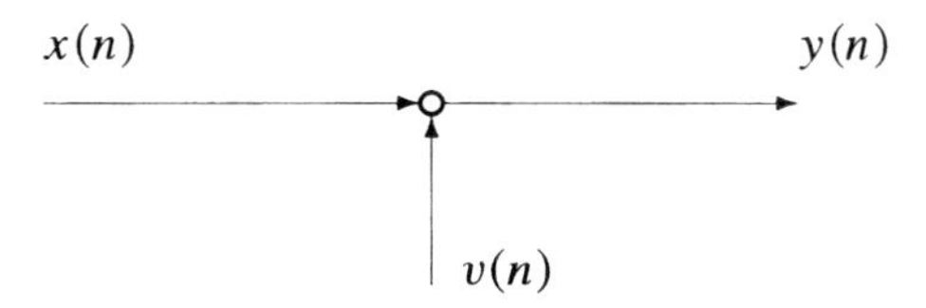

(a) An adder.

$x(n)$ a $ax(n)$

(b) Multiplication by a constant.

$x(n)$ z^{-1} $x(n-1)$

(c) A unit delay.

Figure 2.3 *Flowgraph notation for digital networks.*

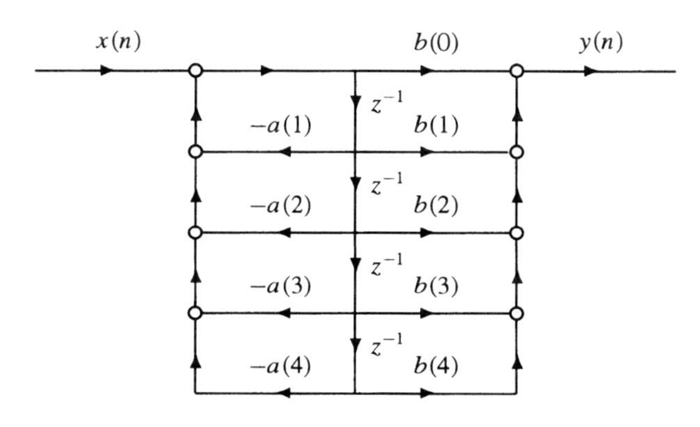

(a) Direct form II filter structure for a 4th order IIR filter.

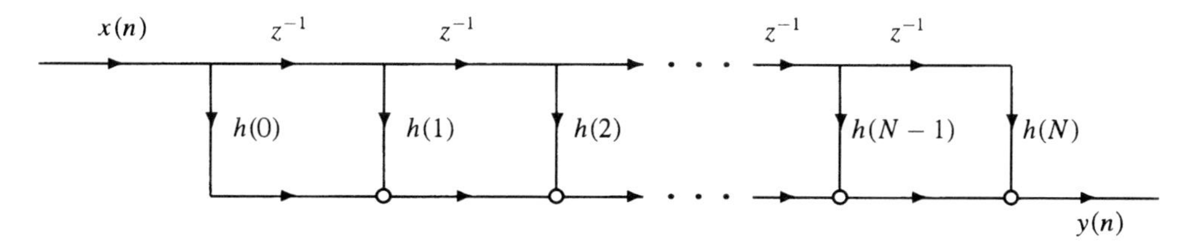

(a) Tapped delay line filter structure for an FIR filter.

Figure 2.4 *Filter flowgraphs for a fourth-order IIR and an Nth-order FIR filter.*

Note that the DFT of $x(n)$ is equal to the discrete-time Fourier transform sampled at N frequencies that are equally spaced between 0 and 2π,

$$X(k) = X(e^{j\omega})|_{\omega=2\pi k/N} \tag{2.19}$$

The Inverse Discrete Fourier Transform (IDFT)

$$\boxed{x(n) = \frac{1}{N}\sum_{k=0}^{N-1} X(k)e^{j2\pi kn/N}} \tag{2.20}$$

provides the relationship required to determine $x(n)$ from the DFT coefficients $X(k)$.

Recall that the product of the discrete-time Fourier transforms of two signals corresponds, in the time domain, to the (linear) convolution of the two signals. For the DFT, however, if $H(k)$ and $X(k)$ are the N-point DFTs of $h(n)$ and $x(n)$, respectively, and if $Y(k) = X(k)H(k)$, then

$$y(n) = \sum_{k=0}^{N} x((k))_N h((n-k))_N$$

which is the N-point *circular convolution* of $x(n)$ and $h(n)$ where

$$x((n))_N \equiv x(n \bmod N)$$

In general, circular convolution of two sequences is not the same as linear convolution. However, there is a special and important case in which the two are the same. Specifically, if $x(n)$ is a finite-length sequence of length N_1 and $h(n)$ is a finite-length sequence of length N_2, then the linear convolution of $x(n)$ and $h(n)$ is of length $L = N_1 + N_2 - 1$. In this case, the N-point circular convolution of $x(n)$ and $h(n)$ will be equal to the linear convolution provided $N \geq L$.

In using the DFT to perform convolutions or as a method for evaluating samples of the discrete-time Fourier transform in real-time applications, it is useful to know what the computational requirements would be to find the DFT. If the DFT is evaluated directly using Eq. (2.19), then N complex multiplications and $N - 1$ complex additions would be required for each value of k. Therefore, an N-point DFT requires a total of N^2 complex multiplications and $N^2 - N$ complex additions. [4] This number may be reduced significantly, however, by employing any one of a number of *Fast Fourier Transform* (*FFT*) algorithms [7]. For example, if N is a power of 2, $N = 2^{\mu}$, a radix-2 decimation-in-time FFT algorithm requires approximately $\frac{N}{2}\log_2 N$ complex multiplications and $N\log_2 N$ complex additions. For large N, this represents a significant reduction.

2.3 LINEAR ALGEBRA

In many of the mathematical developments that will be encountered in later chapters, it will be convenient to use vector and matrix notation for the representation of signals and the operations that are performed on them. Such a representation will greatly simplify many of the mathematical expressions from a notational point of view and allow us to draw upon

[4] Since we are not taking into account the fact that some of these multiplications are by ± 1, the actual number is a bit smaller than this.

many useful results from linear algebra to simplify or solve these expressions. Although it will not be necessary to delve too deeply into the theory of linear algebra, it will be important to become familiar with the basic tools of vector and matrix analysis. Therefore, in this section we summarize the results from linear algebra that will prove to be useful in this book.

2.3.1 Vectors

A vector is an array of real-valued or complex-valued numbers or functions. Vectors will be denoted by lowercase bold letters, such as $\mathbf{x}$, $\mathbf{a}$, and $\mathbf{v}$ and, in all cases, these vectors will be assumed to be column vectors. For example,

$$\mathbf{x} = \begin{bmatrix} x_1 \\ x_2 \\ \vdots \\ x_N \end{bmatrix}$$

is a column vector containing N scalars. If the elements of $\mathbf{x}$ are real, then $\mathbf{x}$ it is said to be a *real vector*, whereas if they are complex, then $\mathbf{x}$ is said to be a *complex vector*. A vector having N elements is referred to as an N-dimensional vector. The transpose of a vector, $\mathbf{x}^T$, is a row vector

$$\mathbf{x}^T = \left[x_1,\ x_2, \ldots,\ x_N\right]$$

and the *Hermitian transpose*, $\mathbf{x}^H$, is the complex conjugate of the transpose of $\mathbf{x}$

$$\mathbf{x}^H = \left(\mathbf{x}^T\right)^* = \left[x_1^*,\ x_2^*, \ldots,\ x_N^*\right]$$

Vectors are useful for representing the values of a discrete-time signal in a concise way. For example, a finite length sequence $x(n)$ that is equal to zero outside the interval $[0, N-1]$ may be represented in vector form as

$$\mathbf{x} = \begin{bmatrix} x(0) \\ x(1) \\ \vdots \\ x(N-1) \end{bmatrix} \tag{2.21}$$

It is also convenient, in some cases, to consider the set of vectors, $\mathbf{x}(n)$, that contain the signal values $x(n), x(n-1), \ldots, x(n-N+1)$,

$$\mathbf{x}(n) = \begin{bmatrix} x(n) \\ x(n-1) \\ \vdots \\ x(n-N+1) \end{bmatrix} \tag{2.22}$$

Thus, $\mathbf{x}(n)$ is a vector of N elements that is parameterized by the time index n.

Many of the operations that we will be performing on vectors will involve finding the *magnitude* of a vector. In order to find the magnitude, however, it is necessary to define a *norm* or *distance metric*. The *Euclidean* or *$\mathcal{L}_2$ Norm* is one such measure that, for a vector

$\mathbf{x}$ of dimension N, is

$$\|\mathbf{x}\|_2 = \left\{\sum_{i=1}^{N} |x_i|^2\right\}^{1/2} \tag{2.23}$$

Other useful norms include the $\mathcal{L}_1$ norm

$$\|\mathbf{x}\|_1 = \sum_{i=1}^{N} |x_i|$$

and the $\mathcal{L}_\infty$ norm

$$\|\mathbf{x}\|_\infty = \max_i |x_i|$$

In this book, since the $\mathcal{L}_2$ norm will be used almost exclusively, the vector norm will be denoted simply by $\|\mathbf{x}\|$ and will be interpreted as the $\mathcal{L}_2$ norm unless indicated otherwise.

A vector may be *normalized* to have unit magnitude by dividing by its norm. For example, assuming that $\|\mathbf{x}\| \neq 0$, then

$$\mathbf{v}_x = \frac{\mathbf{x}}{\|\mathbf{x}\|}$$

is a *unit norm* vector that lies in the same *direction* as $\mathbf{x}$. For a vector whose elements are signal values, $x(n)$, the norm squared represents the energy in the signal. For example, the squared norm of the vector $\mathbf{x}$ in Eq. (2.21) is

$$\|\mathbf{x}\|^2 = \sum_{n=0}^{N-1} |x(n)|^2$$

In addition to providing a metric for the *length* of a vector, the norm may also be used to measure the distance between two vectors

$$d(\mathbf{x}, \mathbf{y}) = \|\mathbf{x} - \mathbf{y}\| = \left\{\sum_{i=1}^{N} |x_i - y_i|^2\right\}^{1/2}$$

Given two complex vectors $\mathbf{a} = \left[a_1, \ldots, a_N\right]^T$ and $\mathbf{b} = \left[b_1, \ldots, b_N\right]^T$, the *inner product* is the scalar defined by

$$\langle \mathbf{a}, \mathbf{b} \rangle = \mathbf{a}^H \mathbf{b} = \sum_{i=1}^{N} a_i^* b_i$$

For real vectors the inner product becomes

$$\langle \mathbf{a}, \mathbf{b} \rangle = \mathbf{a}^T \mathbf{b} = \sum_{i=1}^{N} a_i b_i$$

The inner product defines the geometrical relationship between two vectors. This relationship is given by

$$\langle \mathbf{a}, \mathbf{b} \rangle = \|\mathbf{a}\| \, \|\mathbf{b}\| \cos\theta \tag{2.24}$$

where θ is the angle between the two vectors. Thus, two nonzero vectors $\mathbf{a}$ and $\mathbf{b}$ are said to be *orthogonal* if their inner product is zero

$$\langle \mathbf{a}, \mathbf{b} \rangle = 0$$

Two vectors that are orthogonal and have unit norm are said to be *orthonormal.*

Example 2.3.1 *Inner Product*

Consider the two unit norm vectors

$$\mathbf{v}_1 = \begin{bmatrix} 1 \\ 0 \end{bmatrix} \quad ; \quad \mathbf{v}_2 = \frac{1}{\sqrt{2}} \begin{bmatrix} 1 \\ 1 \end{bmatrix}$$

The inner product between $\mathbf{v}_1$ and $\mathbf{v}_2$ is

$$< \mathbf{v}_1, \mathbf{v}_2 > = \frac{1}{\sqrt{2}} = \cos\theta$$

where $\theta = \pi/4$. Therefore, as we know from Euclidean geometry, $\mathbf{v}_1$ and $\mathbf{v}_2$ form an angle of 45 degrees with respect to each other. If, on the other hand,

$$\mathbf{v}_1 = \begin{bmatrix} 1 \\ 0 \end{bmatrix} \quad ; \quad \mathbf{v}_2 = \begin{bmatrix} 0 \\ 1 \end{bmatrix}$$

then the inner product is zero and $\mathbf{v}_1$ and $\mathbf{v}_2$ are orthogonal.

Note that since $|\cos\theta| \leq 1$, the inner product between two vectors is bounded by the product of their magnitudes

$$\boxed{|\langle \mathbf{a}, \mathbf{b} \rangle| \leq \|\mathbf{a}\| \, \|\mathbf{b}\|} \tag{2.25}$$

where equality holds if and only if $\mathbf{a}$ and $\mathbf{b}$ are colinear, i.e., $\mathbf{a} = \alpha\mathbf{b}$ for some constant α. Equation (2.25) is known as the *Cauchy-Schwarz inequality*. Another useful inequality is the following

$$\boxed{2|\langle \mathbf{a}, \mathbf{b} \rangle| \leq \|\mathbf{a}\|^2 + \|\mathbf{b}\|^2} \tag{2.26}$$

with equality holding if and only if $|\mathbf{a}| = |\mathbf{b}|$. This inequality may be established by noting that, for any two vectors $\mathbf{a}$ and $\mathbf{b}$,

$$\|\mathbf{a} \pm \mathbf{b}\|^2 \geq 0$$

Expanding the norm we have

$$\|\mathbf{a} \pm \mathbf{b}\|^2 = \|\mathbf{a}\|^2 \pm 2\langle \mathbf{a}, \mathbf{b} \rangle + \|\mathbf{b}\|^2 \geq 0$$

from which the inequality easily follows.

One of the uses of the inner product is to represent, in a concise way, the output of a linear shift-invariant filter. For example, let $h(n)$ be the unit sample response of an FIR filter of order $N - 1$ and let $x(n)$ be the filter input. The filter output is the convolution of $h(n)$ and $x(n)$,

$$y(n) = \sum_{k=0}^{N-1} h(k)x(n-k)$$

Therefore, expressing $x(n)$ in vector form using Eq. (2.22) and writing $h(n)$ in vector form as

$$\mathbf{h} = \begin{bmatrix} h(0) \\ h(1) \\ \vdots \\ h(N-1) \end{bmatrix}$$

then $y(n)$ may be written as the inner product

$$y(n) = \mathbf{h}^T \mathbf{x}(n)$$

2.3.2 Linear Independence, Vector Spaces, and Basis Vectors

Linear dependence and linear independence of a set of vectors are extremely important concepts in linear algebra. A set of n vectors, $\mathbf{v}_1, \mathbf{v}_2, \ldots, \mathbf{v}_n$ is said to be *linearly independent* if

$$\alpha_1 \mathbf{v}_1 + \alpha_2 \mathbf{v}_2 + \cdots + \alpha_n \mathbf{v}_n = 0 \tag{2.27}$$

implies that $\alpha_i = 0$ for all i. If a set of nonzero α_i can be found so that Eq. (2.27) holds, then the set is said to be *linearly dependent*. If $\mathbf{v}_1, \mathbf{v}_2, \ldots, \mathbf{v}_n$ is a set of linearly dependent vectors, then it follows that at least one of the vectors, say $\mathbf{v}_1$, may be expressed as a linear combination of the remaining vectors

$$\mathbf{v}_1 = \beta_2 \mathbf{v}_2 + \beta_3 \mathbf{v}_3 + \cdots + \beta_n \mathbf{v}_n$$

for some set of scalars, β_i. For vectors of dimension N, no more than N vectors may be linearly independent, i.e., any set containing more than N vectors will always be linearly dependent.

Example 2.3.2 *Linear Independence*

Given the following pair of vectors

$$\mathbf{v}_1 = \begin{bmatrix} 1 \\ 2 \\ 1 \end{bmatrix} \qquad \mathbf{v}_2 = \begin{bmatrix} 1 \\ 0 \\ 1 \end{bmatrix}$$

it is easily shown that $\mathbf{v}_1$ and $\mathbf{v}_2$ are linearly independent. Specifically, note that if we are to find values for the scalars α and β such that

$$\alpha_1 \mathbf{v}_1 + \alpha_2 \mathbf{v}_2 = \alpha_1 \begin{bmatrix} 1 \\ 2 \\ 1 \end{bmatrix} + \alpha_2 \begin{bmatrix} 1 \\ 0 \\ 1 \end{bmatrix} = \begin{bmatrix} 0 \\ 0 \\ 0 \end{bmatrix}$$

then it follows that

$$\begin{aligned} \alpha_1 + \alpha_2 &= 0 \\ 2\alpha_1 &= 0 \end{aligned}$$

The only solution to these equations is $\alpha_1 = \alpha_2 = 0$. Therefore, the vectors are linearly independent. However, adding the vector

$$\mathbf{v}_3 = \begin{bmatrix} 0 \\ 1 \\ 0 \end{bmatrix}$$

to the set results in a set of linearly *dependent* vectors since

$$\mathbf{v}_1 = \mathbf{v}_2 + 2\mathbf{v}_3$$

Given a set of N vectors,

$$\mathbf{V} = \left\{ \mathbf{v}_1, \mathbf{v}_2, \ldots, \mathbf{v}_N \right\}$$

consider the set of all vectors $\mathcal{V}$ that may be formed from a linear combination of the vectors $\mathbf{v}_i$,

$$\mathbf{v} = \sum_{i=1}^{N} \alpha_i \mathbf{v}_i$$

This set forms a *vector space*, and the vectors $\mathbf{v}_i$ are said to *span* the space $\mathcal{V}$. Furthermore, if the vectors $\mathbf{v}_i$ are linearly independent, then they are said to form a *basis* for the space $\mathcal{V}$, and the number of vectors in the basis, N, is referred to as the *dimension* of the space. For example, the set of all real vectors of the form $\mathbf{x} = [x_1, x_2, \ldots, x_N]^T$ forms an N-dimensional vector space, denoted by R^N, that is spanned by the basis vectors

$$\begin{aligned} \mathbf{u}_1 &= [1, 0, 0, \ldots, 0]^T \\ \mathbf{u}_2 &= [0, 1, 0, \ldots, 0]^T \\ &\vdots \\ \mathbf{u}_N &= [0, 0, 0, \ldots, 1]^T \end{aligned}$$

In terms of this basis, any vector $\mathbf{v} = \left[v_1, v_2, \ldots, v_N \right]^T$ in R^N may be uniquely decomposed as follows

$$\mathbf{v} = \sum_{i=1}^{N} v_i \mathbf{u}_i$$

It should be pointed out, however, that the basis for a vector space is not unique.

2.3.3 Matrices

An $n \times m$ matrix is an array of numbers (real or complex) or functions having n rows and m columns. For example,

$$\mathbf{A} = \left\{ a_{ij} \right\} = \begin{bmatrix} a_{11} & a_{12} & a_{13} & \cdots & a_{1m} \\ a_{21} & a_{22} & a_{23} & \cdots & a_{2m} \\ a_{31} & a_{32} & a_{33} & \cdots & a_{3m} \\ \vdots & \vdots & \vdots & & \vdots \\ a_{n1} & a_{n2} & a_{n3} & \cdots & a_{nm} \end{bmatrix}$$

is an $n \times m$ matrix of numbers a_{ij} and

$$\mathbf{A}(z) = \{a_{ij}(z)\} = \begin{bmatrix} a_{11}(z) & a_{12}(z) & a_{13}(z) & \cdots & a_{1m}(z) \\ a_{21}(z) & a_{22}(z) & a_{23}(z) & \cdots & a_{2m}(z) \\ a_{31}(z) & a_{32}(z) & a_{33}(z) & \cdots & a_{3m}(z) \\ \vdots & \vdots & \vdots & & \vdots \\ a_{n1}(z) & a_{n2}(z) & a_{n3}(z) & \cdots & a_{nm}(z) \end{bmatrix}$$

is an $n \times m$ matrix of functions $a_{ij}(z)$. If $n = m$, then $\mathbf{A}$ is a *square* matrix of n rows and n columns. In some cases, we will have a need to consider matrices having an infinite number of rows or columns. For example, recall that the output of an FIR linear shift-invariant filter with a unit sample response $h(n)$ may be written in vector form as follows

$$y(n) = \mathbf{h}^T\mathbf{x}(n) = \mathbf{x}^T(n)\mathbf{h}$$

If $x(n) = 0$ for $n < 0$, then we may express $y(n)$ for $n \geq 0$ as

$$\mathbf{X}_0\mathbf{h} = \mathbf{y}$$

where $\mathbf{X}_0$ is a *convolution matrix* defined by[5]

$$\mathbf{X}_0 = \begin{bmatrix} x(0) & 0 & 0 & \cdots & 0 \\ x(1) & x(0) & 0 & \cdots & 0 \\ x(2) & x(1) & x(0) & \cdots & 0 \\ \vdots & \vdots & \vdots & & \\ x(N-1) & x(N-2) & x(N-3) & \cdots & x(0) \\ \vdots & \vdots & \vdots & & \vdots \end{bmatrix}$$

and $\mathbf{y} = \left[y(0),\ y(1),\ y(2),\ \ldots\right]^T$. Note that, in addition to its structure of having equal values along each of the diagonals, $\mathbf{X}_0$ has $N - 1$ columns and an infinite number of rows.

Matrices will be denoted by bold upper case letters, e.g., $\mathbf{A}$, $\mathbf{B}$, and $\mathbf{H}(z)$. On occasion it will be convenient to represent an $n \times m$ matrix $\mathbf{A}$ as a set of m column vectors,

$$\mathbf{A} = \left[\mathbf{c}_1,\ \mathbf{c}_2, \ldots, \mathbf{c}_m\right]$$

or a set of n row vectors

$$\mathbf{A} = \begin{bmatrix} \mathbf{r}_1^H \\ \mathbf{r}_2^H \\ \vdots \\ \mathbf{r}_n^H \end{bmatrix}$$

A matrix may also be partitioned into submatrices. For example, an $n \times m$ matrix $\mathbf{A}$ may be partitioned as

$$\mathbf{A} = \begin{bmatrix} \mathbf{A}_{11} & \mathbf{A}_{12} \\ \mathbf{A}_{21} & \mathbf{A}_{22} \end{bmatrix}$$

where $\mathbf{A}_{11}$ is $p \times q$, $\mathbf{A}_{12}$ is $p \times (m - q)$, $\mathbf{A}_{21}$ is $(n - p) \times q$, and $\mathbf{A}_{22}$ is $(n - p) \times (m - q)$.

[5]The subscript on $\mathbf{X}_0$ is used to indicate the value of the index of $x(n)$ in the first entry of the convolution matrix.

Example 2.3.3 *Partitioning Matrices*

Consider the 3 × 3 matrix

$$\mathbf{A} = \begin{bmatrix} 1 & 0 & 0 \\ 0 & 2 & 1 \\ 0 & 1 & 2 \end{bmatrix} = \left[\begin{array}{c|cc} 1 & 0 & 0 \\ \hline 0 & 2 & 1 \\ 0 & 1 & 2 \end{array}\right]$$

This matrix may be partitioned as

$$\mathbf{A} = \left[\begin{array}{c|c} 1 & \mathbf{0}^T \\ \hline \mathbf{0} & \mathbf{A}_{22} \end{array}\right]$$

where $\mathbf{0} = [0,\ 0]^T$ is a zero vector of length 2 and

$$\mathbf{A}_{22} = \begin{bmatrix} 2 & 1 \\ 1 & 2 \end{bmatrix}$$

is a 2 × 2 matrix.

If $\mathbf{A}$ is an $n \times m$ matrix, then the *transpose*, denoted by $\mathbf{A}^T$, is the $m \times n$ matrix that is formed by interchanging the rows and columns of $\mathbf{A}$. Thus, the $(i,\ j)$th element becomes the (j, i)th element and vice versa. If the matrix is square, then the transpose is formed by simply reflecting the elements of $\mathbf{A}$ about the diagonal. For a square matrix, if $\mathbf{A}$ is equal to its transpose,

$$\mathbf{A} = \mathbf{A}^T$$

then $\mathbf{A}$ is said to be a *symmetric* matrix. For complex matrices, the *Hermitian transpose* is the complex conjugate of the transpose of $\mathbf{A}$ and is denoted by $\mathbf{A}^H$. Thus,

$$\mathbf{A}^H = (\mathbf{A}^*)^T = (\mathbf{A}^T)^*$$

If a square complex-valued matrix is equal to its Hermitian transpose

$$\mathbf{A} = \mathbf{A}^H$$

then the matrix is said to be *Hermitian.* A few properties of the Hermitian transpose are

1. $(\mathbf{A} + \mathbf{B})^H = \mathbf{A}^H + \mathbf{B}^H$
2. $(\mathbf{A}^H)^H = \mathbf{A}$
3. $(\mathbf{AB})^H = \mathbf{B}^H\mathbf{A}^H$

Equivalent properties for the transpose may be obtained by replacing the Hermitian transpose H with the transpose T.

2.3.4 Matrix Inverse

Let $\mathbf{A}$ be an $n \times m$ matrix that is partitioned in terms of its m column vectors

$$\mathbf{A} = \left[\mathbf{c}_1, \mathbf{c}_2, \ldots, \mathbf{c}_m\right]$$

The *rank* of **A**, $\rho(\mathbf{A})$, is defined to be the number of linearly independent columns in **A**, i.e., the number of linearly independent vectors in the set $\{\mathbf{c}_1, \mathbf{c}_2, \ldots, \mathbf{c}_m\}$. One of the properties of $\rho(\mathbf{A})$ is that the rank of a matrix is equal to the rank of its Hermitian transpose, $\rho(\mathbf{A}) = \rho(\mathbf{A}^H)$. Therefore, if **A** is partitioned in terms of its n row vectors,

$$\mathbf{A} = \begin{bmatrix} \mathbf{r}_1^H \\ \mathbf{r}_2^H \\ \vdots \\ \mathbf{r}_n^H \end{bmatrix}$$

then the rank of **A** is equivalently equal to the number of linearly independent row vectors, i.e., the number of linearly independent vectors in the set $\{\mathbf{r}_1, \mathbf{r}_2, \ldots, \mathbf{r}_n\}$. A useful property of the rank of a matrix is the following:

Property. The rank of **A** is equal to the rank of $\mathbf{A}\mathbf{A}^H$ and $\mathbf{A}^H\mathbf{A}$,

$$\rho(\mathbf{A}) = \rho(\mathbf{A}\mathbf{A}^H) = \rho(\mathbf{A}^H\mathbf{A})$$

Since the rank of a matrix is equal to the number of linearly independent rows and the number of linearly independent columns, it follows that if **A** is an $m \times n$ matrix then

$$\rho(\mathbf{A}) \leq \min(m, n)$$

If **A** is an $m \times n$ matrix and $\rho(\mathbf{A}) = \min(m, n)$ then **A** is said to be of *full rank*. If **A** is a square matrix of full rank, then there exists a unique matrix $\mathbf{A}^{-1}$, called the inverse of **A**, such that

$$\mathbf{A}^{-1}\mathbf{A} = \mathbf{A}\mathbf{A}^{-1} = \mathbf{I}$$

where

$$\mathbf{I} = \begin{bmatrix} 1 & 0 & 0 & \cdots & 0 \\ 0 & 1 & 0 & \cdots & 0 \\ 0 & 0 & 1 & \cdots & 0 \\ \vdots & \vdots & \vdots & & \vdots \\ 0 & 0 & 0 & \cdots & 1 \end{bmatrix}$$

is the *identity matrix*, which has ones along the main diagonal and zeros everywhere else. In this case **A** is said to be *invertible* or *nonsingular*. If **A** is not of full rank, $\rho(\mathbf{A}) < n$, then it is said to be *noninvertible* or *singular* and **A** does not have an inverse. Some properties of the matrix inverse are as follows. First, if **A** and **B** are invertible, then the inverse of their product is

$$(\mathbf{A}\mathbf{B})^{-1} = \mathbf{B}^{-1}\mathbf{A}^{-1}$$

Second, the Hermitian transpose of the inverse is equal to the inverse of the Hermitian transpose

$$(\mathbf{A}^H)^{-1} = (\mathbf{A}^{-1})^H$$

Finally, a formula that will be useful for inverting matrices that arise in adaptive filtering algorithms is the *Matrix Inversion lemma* [3]

$$(\mathbf{A}+\mathbf{BCD})^{-1} = \mathbf{A}^{-1} - \mathbf{A}^{-1}\mathbf{B}(\mathbf{C}^{-1}+\mathbf{DA}^{-1}\mathbf{B})^{-1}\mathbf{DA}^{-1} \tag{2.28}$$

where it is assumed that $\mathbf{A}$ is $n \times n$, $\mathbf{B}$ is $n \times m$, $\mathbf{C}$ is $m \times m$, and $\mathbf{D}$ is $m \times n$ with $\mathbf{A}$ and $\mathbf{C}$ nonsingular matrices. A special case of this lemma occurs when $\mathbf{C} = 1$, $\mathbf{B} = \mathbf{u}$, and $\mathbf{D} = \mathbf{v}^H$ where $\mathbf{u}$ and $\mathbf{v}$ are n-dimensional vectors. In this case the lemma becomes

$$(\mathbf{A}+\mathbf{uv}^H)^{-1} = \mathbf{A}^{-1} - \frac{\mathbf{A}^{-1}\mathbf{uv}^H\mathbf{A}^{-1}}{1+\mathbf{v}^H\mathbf{A}^{-1}\mathbf{u}} \tag{2.29}$$

which is sometimes referred to as *Woodbury's Identity* [5]. As a special case, note that for $\mathbf{A} = \mathbf{I}$, Eq. (2.29) becomes

$$(\mathbf{I}+\mathbf{uv}^H)^{-1} = \mathbf{I} - \frac{1}{1+\mathbf{v}^H\mathbf{u}}\mathbf{uv}^H \tag{2.30}$$

2.3.5 The Determinant and the Trace

If $\mathbf{A} = a_{11}$ is a 1×1 matrix, then its *determinant* is defined to be $\det(\mathbf{A}) = a_{11}$. The determinant of an $n \times n$ matrix is defined recursively in terms of the determinants of $(n-1) \times (n-1)$ matrices as follows. For any j

$$\det(\mathbf{A}) = \sum_{i=1}^{n}(-1)^{i+j}a_{ij}\det(\mathbf{A}_{ij})$$

where $\mathbf{A}_{ij}$ is the $(n-1) \times (n-1)$ matrix that is formed by deleting the ith row and the jth column of $\mathbf{A}$.

Example 2.3.4 *The Determinant*

For the 2×2 matrix

$$\mathbf{A} = \begin{bmatrix} a_{11} & a_{12} \\ a_{21} & a_{22} \end{bmatrix}$$

the determinant is

$$\det(\mathbf{A}) = a_{11}a_{22} - a_{12}a_{21}$$

and for the 3×3 matrix

$$\mathbf{A} = \begin{bmatrix} a_{11} & a_{12} & a_{13} \\ a_{21} & a_{22} & a_{23} \\ a_{31} & a_{32} & a_{33} \end{bmatrix}$$

the determinant is

$$\begin{aligned}\det(\mathbf{A}) &= a_{11}\det\begin{bmatrix} a_{22} & a_{23} \\ a_{32} & a_{33} \end{bmatrix} - a_{12}\det\begin{bmatrix} a_{21} & a_{23} \\ a_{31} & a_{33} \end{bmatrix} + a_{13}\det\begin{bmatrix} a_{21} & a_{22} \\ a_{31} & a_{32} \end{bmatrix} \\ &= a_{11}\left[a_{22}a_{33} - a_{23}a_{32}\right] - a_{12}\left[a_{21}a_{33} - a_{31}a_{23}\right] + a_{13}\left[a_{21}a_{32} - a_{31}a_{22}\right]\end{aligned}$$

The determinant may be used to determine whether or not a matrix is invertible. Specifically,

> ***Property.*** An $n \times n$ matrix $\mathbf{A}$ is invertible if and only if its determinant is nonzero,
>
> $$\det(\mathbf{A}) \neq 0$$

Some additional properties of the determinant are listed below. It is assumed that $\mathbf{A}$ and $\mathbf{B}$ are $n \times n$ matrices.

1. $\det(\mathbf{AB}) = \det(\mathbf{A})\det(\mathbf{B})$.
2. $\det(\mathbf{A}^T) = \det(\mathbf{A})$
3. $\det(\alpha\mathbf{A}) = \alpha^n \det(\mathbf{A})$, where α is a constant.
4. $\det(\mathbf{A}^{-1}) = \dfrac{1}{\det(\mathbf{A})}$, assuming that $\mathbf{A}$ is invertible.

Another function of a matrix that will occasionally be useful is the *trace*. Given an $n \times n$ matrix, $\mathbf{A}$, the *trace* is the sum of the terms along the diagonal,

$$\operatorname{tr}(\mathbf{A}) = \sum_{i=1}^{n} a_{ii}$$

2.3.6 Linear Equations

Many problems addressed in later chapters, such as signal modeling, Wiener filtering, and spectrum estimation, require finding the solution or solutions to a set of linear equations. Many techniques are available for solving linear equations and, depending on the form of the equations, there may be "fast algorithms" for efficiently finding the solution. In addition to solving the linear equations, however, it is often important to characterize the form of the solution in terms of existence and uniqueness. Therefore, this section briefly summarizes the conditions under which a unique solution exists, discusses how constraints might be imposed if multiple solutions exist, and indicates how an approximate solution may be obtained if no solution exists.

Consider the following set of n linear equations in the m unknowns x_i, $i = 1, 2, \ldots, m$,

$$\begin{aligned}
a_{11}x_1 + a_{12}x_2 + \cdots + a_{1m}x_m &= b_1 \\
a_{21}x_1 + a_{22}x_2 + \cdots + a_{2m}x_m &= b_2 \\
&\vdots \\
a_{n1}x_1 + a_{n2}x_2 + \cdots + a_{nm}x_m &= b_n
\end{aligned}$$

These equations may be written concisely in matrix form as follows

$$\mathbf{A}\mathbf{x} = \mathbf{b} \tag{2.31}$$

where $\mathbf{A}$ is an $n \times m$ matrix with entries a_{ij}, $\mathbf{x}$ is an m-dimensional vector containing the unknowns x_i, and $\mathbf{b}$ is an n-dimensional vector with elements b_i. A convenient way to view Eq. (2.31) is as an expansion of the vector $\mathbf{b}$ in terms of a linear combination of the column vectors $\mathbf{a}_i$ of the matrix $\mathbf{A}$, i.e,

$$\mathbf{b} = \sum_{i=1}^{m} x_i \mathbf{a}_i \tag{2.32}$$

As discussed below, solving Eq. (2.31) depends upon a number of factors including the relative size of m and n, the rank of the matrix $\mathbf{A}$, and the elements in the vector $\mathbf{b}$. The case of a square matrix ($m = n$) is considered first.

Square Matrix: $m = n$. If $\mathbf{A}$ is a square $n \times n$ matrix then the nature of the solution to the linear equations $\mathbf{Ax} = \mathbf{b}$ depends upon whether or not $\mathbf{A}$ is singular. If $\mathbf{A}$ is nonsingular, then the inverse matrix $\mathbf{A}^{-1}$ exists and the solution is uniquely defined by

$$\mathbf{x} = \mathbf{A}^{-1}\mathbf{b}$$

However, if $\mathbf{A}$ is singular, then there may either be no solution (the equations are inconsistent) or many solutions.

Example 2.3.5 *Linear Equations—Singular Case*

Consider the following pair of equations in two unknowns x_1 and x_2,

$$\begin{aligned} x_1 + x_2 &= 1 \\ x_1 + x_2 &= 2 \end{aligned}$$

In matrix form, these equations are

$$\begin{bmatrix} 1 & 1 \\ 1 & 1 \end{bmatrix} \begin{bmatrix} x_1 \\ x_2 \end{bmatrix} = \begin{bmatrix} 1 \\ 2 \end{bmatrix}$$

Clearly, the matrix $\mathbf{A}$ is singular, $\det(\mathbf{A}) = 0$, and no solution exists. However, if the second equation is modified so that

$$\begin{bmatrix} 1 & 1 \\ 1 & 1 \end{bmatrix} \begin{bmatrix} x_1 \\ x_2 \end{bmatrix} = \begin{bmatrix} 1 \\ 1 \end{bmatrix}$$

then there are many solutions. Specifically, note that for any constant α, the vector

$$\mathbf{x} = \begin{bmatrix} 1 \\ 0 \end{bmatrix} + \alpha \begin{bmatrix} 1 \\ -1 \end{bmatrix}$$

will satisfy these equations.

For the case in which $\mathbf{A}$ is singular, the columns of $\mathbf{A}$ are linearly dependent and there exists nonzero solutions to the homogeneous equations

$$\mathbf{Az} = \mathbf{0} \tag{2.33}$$

In fact, there will be $k = n - \rho(\mathbf{A})$ linearly independent solutions to the homogeneous equations. Therefore, if there is *at least one* vector, $\mathbf{x}_0$, that solves Eq. (2.31), then any vector of the form

$$\mathbf{x} = \mathbf{x}_0 + \alpha_1 \mathbf{z}_1 + \cdots + \alpha_k \mathbf{z}_k$$

will also be a solution where $\mathbf{z}_i$, $i = 1, 2, \ldots, k$ are linearly independent solutions of Eq. (2.33).

Rectangular Matrix: $n < m$**.** If $n < m$, then there are fewer equations than unknowns. Therefore, provided the equations are not inconsistent, there are many vectors that satisfy the equations, i.e., the solution is *underdetermined* or *incompletely specified*. One approach that is often used to define a unique solution is to find the vector satisfying the equations that has the minimum norm, i.e.,

$$\min \|\mathbf{x}\| \quad \text{such that} \quad \mathbf{A}\mathbf{x} = \mathbf{b}$$

If the rank of $\mathbf{A}$ is n (the rows of $\mathbf{A}$ are linearly independent), then the $n \times n$ matrix $\mathbf{A}\mathbf{A}^H$ is invertible and the *minimum norm solution* is [10]

$$\boxed{\mathbf{x}_0 = \mathbf{A}^H(\mathbf{A}\mathbf{A}^H)^{-1}\mathbf{b}} \tag{2.34}$$

The matrix

$$\mathbf{A}^+ = \mathbf{A}^H(\mathbf{A}\mathbf{A}^H)^{-1}$$

is known as the *pseudo-inverse* of the matrix $\mathbf{A}$ for the underdetermined problem. The following example illustrates how the pseudoinverse is used to solve an underdetermined set of linear equations.

Example 2.3.6 ***Linear Equations—Underdetermined Case***

Consider the following equation in the four unknowns x_1, x_2, x_3, x_4,

$$x_1 - x_2 + x_3 - x_4 = 1 \tag{2.35}$$

This equation may be written in matrix form as

$$\mathbf{A}\mathbf{x} = [1, \ -1, \ 1, \ -1]\mathbf{x} = \mathbf{b}$$

where $\mathbf{b} = 1$ and $\mathbf{x}$ is the vector containing the unknowns x_i. Clearly, the solution is incompletely specified since there are many solutions that satisfy this equation. However, the minimum norm solution is unique and given by Eq. (2.34). Specifically, since

$$\mathbf{A}\mathbf{A}^H = 4$$

and $(\mathbf{A}\mathbf{A}^H)^{-1} = \frac{1}{4}$ then the minimum norm solution is

$$\mathbf{x} = \frac{1}{4}\begin{bmatrix} 1 \\ -1 \\ 1 \\ -1 \end{bmatrix}$$

If the following equation is added to Eq. (2.35)

$$x_1 + x_2 + x_3 + x_4 = 1$$

then there are two equations in four unknowns with

$$\mathbf{A} = \begin{bmatrix} 1 & -1 & 1 & -1 \\ 1 & 1 & 1 & 1 \end{bmatrix}$$

and $\mathbf{b} = \begin{bmatrix} 1, & 1 \end{bmatrix}^T$. Again, it is easy to see that the solution is incompletely specified. Since

$$\mathbf{A}\mathbf{A}^H = \begin{bmatrix} 4 & 0 \\ 0 & 4 \end{bmatrix}$$

then $(\mathbf{A}\mathbf{A}^H)^{-1} = \frac{1}{4}\mathbf{I}$ and the minimum norm solution is

$$\mathbf{x} = \mathbf{A}^H(\mathbf{A}\mathbf{A}^H)^{-1}\mathbf{b} = \frac{1}{4}\mathbf{A}^H\mathbf{b} = \frac{1}{2}\begin{bmatrix} 1 \\ 0 \\ 1 \\ 0 \end{bmatrix}$$

Rectangular Matrix: $m < n$. If $m < n$, then there are more equations than unknowns and, in general, no solution exists. In this case, the equations are *inconsistent* and the solution is said to be *overdetermined*. The geometry of this problem is illustrated in Fig. 2.5 for the case of three equations in two unknowns. Since an arbitrary vector $\mathbf{b}$ cannot be represented in terms of a linear combination of the columns of $\mathbf{A}$ as given in Eq. (2.32), the goal is to find the coefficients x_i that produce the best approximation to $\mathbf{b}$,

$$\hat{\mathbf{b}} = \sum_{i=1}^{m} x_i \mathbf{a}_i$$

The approach that is commonly used in this situation is to find the *least squares solution*, i.e., the vector $\mathbf{x}$ that minimizes the norm of the error

$$\|\mathbf{e}\|^2 = \|\mathbf{b} - \mathbf{A}\mathbf{x}\|^2 \tag{2.36}$$

As illustrated in Fig. 2.5, the least squares solution has the property that the error,

$$\mathbf{e} = \mathbf{b} - \mathbf{A}\mathbf{x}$$

is *orthogonal* to each of the vectors that are used in the approximation for $\mathbf{b}$, i.e., the column vectors of $\mathbf{A}$. This orthogonality implies that

$$\mathbf{A}^H\mathbf{e} = 0 \tag{2.37}$$

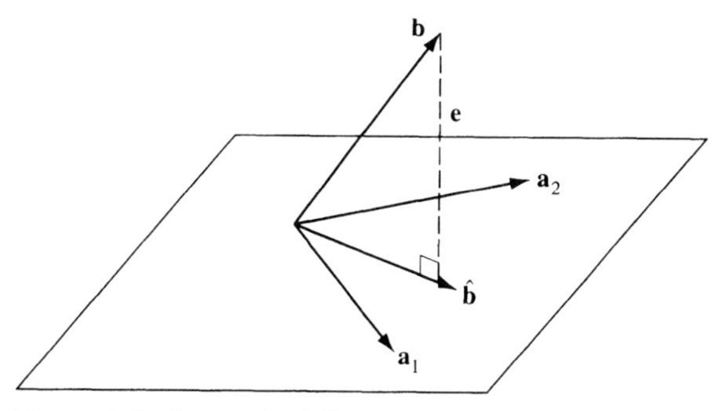

Figure 2.5 *Geometrical illustration of the least squares solution to an overdetermined set of linear equations. The best approximation to* $\mathbf{b}$ *is formed when the error* $\mathbf{e}$ *is orthogonal to the vectors* $\mathbf{a}_1$ *and* $\mathbf{a}_2$.

or,

$$\mathbf{A}^H\mathbf{A}\mathbf{x} = \mathbf{A}^H\mathbf{b} \tag{2.38}$$

which are known as the *normal equations* [10]. If the columns of **A** are linearly independent (**A** has full rank), then the matrix $\mathbf{A}^H\mathbf{A}$ is invertible and the *least squares solution* is

$$\mathbf{x}_0 = (\mathbf{A}^H\mathbf{A})^{-1}\mathbf{A}^H\mathbf{b} \tag{2.39}$$

or,

$$\mathbf{x}_0 = \mathbf{A}^+\mathbf{b}$$

where the matrix

$$\mathbf{A}^+ = (\mathbf{A}^H\mathbf{A})^{-1}\mathbf{A}^H$$

is the *pseudo-inverse* of the matrix **A** for the overdetermined problem. Furthermore, the *best approximation* $\hat{\mathbf{b}}$ to **b** is given by the *projection* of the vector **b** onto the subspace spanned by the vectors $\mathbf{a}_i$,

$$\hat{\mathbf{b}} = \mathbf{A}\mathbf{x}_0 = \mathbf{A}(\mathbf{A}^H\mathbf{A})^{-1}\mathbf{A}^H\mathbf{b} \tag{2.40}$$

or

$$\hat{\mathbf{b}} = \mathbf{P}_A\,\mathbf{b}$$

where

$$\mathbf{P}_A = \mathbf{A}(\mathbf{A}^H\mathbf{A})^{-1}\mathbf{A}^H \tag{2.41}$$

is called the *projection matrix*. Finally, expanding the square in Eq. (2.36) and using the *orthogonality condition* given in Eq. (2.37) it follows that the minimum least squares error is

$$\min \|\mathbf{e}\|^2 = \mathbf{b}^H\mathbf{e} = \mathbf{b}^H\mathbf{b} - \mathbf{b}^H\mathbf{A}\mathbf{x}_0 \tag{2.42}$$

The following example illustrates the use of the pseudoinverse to solve an overdetermined set of linear equations.

Example 2.3.7 ***Linear Equations—Overdetermined Case***

Consider the set of three equations in two unknowns

$$\begin{bmatrix} 2 & 1 \\ 1 & 2 \\ 1 & 1 \end{bmatrix} \begin{bmatrix} x_1 \\ x_2 \end{bmatrix} = \begin{bmatrix} 1 \\ 1 \\ 1 \end{bmatrix}$$

Since the rank of **A** is two, the least squares solution is unique. With

$$\mathbf{A}^H\mathbf{A} = \begin{bmatrix} 6 & 5 \\ 5 & 6 \end{bmatrix}$$

and

$$(\mathbf{A}^H\mathbf{A})^{-1} = \frac{1}{11}\begin{bmatrix} 6 & -5 \\ -5 & 6 \end{bmatrix}$$

the least squares solution is

$$\mathbf{x}_0 = \frac{1}{11}\begin{bmatrix} 6 & -5 \\ -5 & 6 \end{bmatrix}\begin{bmatrix} 2 & 1 & 1 \\ 1 & 2 & 1 \end{bmatrix}\begin{bmatrix} 1 \\ 1 \\ 1 \end{bmatrix} = \frac{4}{11}\begin{bmatrix} 1 \\ 1 \end{bmatrix}$$

The error, $\mathbf{e}$, is

$$\mathbf{e} = \mathbf{b} - \mathbf{A}\mathbf{x}_0 = \frac{1}{11}[-1,\ -1,\ 3]^T$$

and the least square error is

$$\|\mathbf{e}\|^2 = \mathbf{b}^H\mathbf{e} = \frac{1}{11}$$

A general treatment of the problem of solving m linear equations in n unknowns may be found in references [6, 10], including the cases for which $\mathbf{A}\mathbf{A}^H$ and $\mathbf{A}^H\mathbf{A}$ are not invertible in the minimum norm and least squares solutions, respectively.

2.3.7 Special Matrix Forms

In this section, we describe some of the special types of matrices and symmetries that will be encountered in the following chapters. The first is a *diagonal matrix*, which is a square matrix that has all of its entries equal to zero except possibly those along the main diagonal. Thus, a diagonal matrix has the form

$$\mathbf{A} = \begin{bmatrix} a_{11} & 0 & 0 & \cdots & 0 \\ 0 & a_{22} & 0 & \cdots & 0 \\ 0 & 0 & a_{33} & \cdots & 0 \\ \vdots & \vdots & \vdots & & \vdots \\ 0 & 0 & 0 & \cdots & a_{nn} \end{bmatrix} \tag{2.43}$$

and may be written concisely as

$$\mathbf{A} = \text{diag}\,\{a_{11}, a_{22}, \ldots, a_{nn}\}$$

A diagonal matrix with ones along the diagonal is referred to as the *identity matrix* and will be represented by $\mathbf{I}$,

$$\mathbf{I} = \text{diag}\,\{1, 1, \ldots, 1\} = \begin{bmatrix} 1 & 0 & 0 & \cdots & 0 \\ 0 & 1 & 0 & \cdots & 0 \\ 0 & 0 & 1 & \cdots & 0 \\ \vdots & \vdots & \vdots & & \vdots \\ 0 & 0 & 0 & \cdots & 1 \end{bmatrix}$$

If the entries along the diagonal in Eq. (2.43) are replaced with matrices,

$$\mathbf{A} = \begin{bmatrix} \mathbf{A}_{11} & 0 & \cdots & 0 \\ 0 & \mathbf{A}_{22} & \cdots & 0 \\ \vdots & \vdots & & 0 \\ 0 & 0 & \cdots & \mathbf{A}_{kk} \end{bmatrix}$$

then $\mathbf{A}$ is said to be a *block diagonal matrix*. The matrix in Example 2.3.3 is an example of a 3×3 block diagonal matrix.

Another matrix that will be useful is the *exchange matrix*

$$\mathbf{J} = \begin{bmatrix} 0 & \cdots & 0 & 0 & 1 \\ 0 & \cdots & 0 & 1 & 0 \\ 0 & \cdots & 1 & 0 & 0 \\ \vdots & & \vdots & \vdots & \vdots \\ 1 & \cdots & 0 & 0 & 0 \end{bmatrix}$$

which is symmetric and has ones along the *cross-diagonal* and zeros everywhere else. Note that since $\mathbf{J}^2 = \mathbf{I}$ then $\mathbf{J}$ is its own inverse. The effect of multiplying a vector $\mathbf{v}$ by the exchange matrix is to reverse the order of the entries, i.e.,

$$\mathbf{J} \begin{bmatrix} v_1 \\ v_2 \\ \vdots \\ v_n \end{bmatrix} = \begin{bmatrix} v_n \\ v_{n-1} \\ \vdots \\ v_1 \end{bmatrix}$$

Similarly, if a matrix $\mathbf{A}$ is multiplied on the left by the exchange matrix, the effect is to reverse the order of each column. For example, if

$$\mathbf{A} = \begin{bmatrix} a_{11} & a_{12} & a_{13} \\ a_{21} & a_{22} & a_{23} \\ a_{31} & a_{32} & a_{33} \end{bmatrix}$$

then

$$\mathbf{J}^T\mathbf{A} = \begin{bmatrix} a_{31} & a_{32} & a_{33} \\ a_{21} & a_{22} & a_{23} \\ a_{11} & a_{12} & a_{13} \end{bmatrix}$$

Similarly, if $\mathbf{A}$ is multiplied on the right by $\mathbf{J}$, then the order of the entries in each row is reversed,

$$\mathbf{AJ} = \begin{bmatrix} a_{13} & a_{12} & a_{11} \\ a_{23} & a_{22} & a_{21} \\ a_{33} & a_{32} & a_{31} \end{bmatrix}$$

Finally, the effect of forming the product $\mathbf{J}^T\mathbf{AJ}$ is to reverse the order of each row and column,

$$\mathbf{J}^T\mathbf{AJ} = \begin{bmatrix} a_{33} & a_{32} & a_{31} \\ a_{23} & a_{22} & a_{21} \\ a_{13} & a_{12} & a_{11} \end{bmatrix}$$

thereby *reflecting* each element of $\mathbf{A}$ about the central element.

An *upper triangular* matrix is a square matrix in which all of the terms below the diagonal are equal to zero, i.e., with $\mathbf{A} = \{a_{ij}\}$ then $a_{ij} = 0$ for $i > j$. The following is an example of a 4×4 upper triangular matrix

$$\mathbf{A} = \begin{bmatrix} a_{11} & a_{12} & a_{13} & a_{14} \\ 0 & a_{22} & a_{23} & a_{24} \\ 0 & 0 & a_{33} & a_{34} \\ 0 & 0 & 0 & a_{44} \end{bmatrix}$$

A *lower triangular* matrix, on the other hand, is a square matrix in which all of the entries above the diagonal are equal to zero, i.e., $a_{ij} = 0$ for $i < j$. Clearly, the transpose of a lower triangular matrix is upper triangular and vice versa. Some additional properties of lower and upper triangular matrices are as follows.

1. The determinant of a lower triangular matrix or an upper triangular matrix is equal to the product of the terms along the diagonal

$$\det(\mathbf{A}) = \prod_{i=1}^{n} a_{ii}$$

2. The inverse of an upper (lower) triangular matrix is upper (lower) triangular.
3. The product of two upper (lower) triangular matrices is upper (lower) triangular.

A matrix having a tremendous amount of structure that will play a prominent role in many of our discussions throughout this book is one that is referred to as a *Toeplitz* matrix. An $n \times n$ matrix $\mathbf{A}$ is said to be Toeplitz if all of the elements along each of the diagonals have the same value, i.e.,

$$a_{ij} = a_{i+1,j+1} \quad ; \qquad \text{for all } i < n \text{ and } j < n$$

An example of a 4×4 Toeplitz matrix is

$$\mathbf{A} = \begin{bmatrix} 1 & 3 & 5 & 7 \\ 2 & 1 & 3 & 5 \\ 4 & 2 & 1 & 3 \\ 6 & 4 & 2 & 1 \end{bmatrix}$$

Note that all of the entries in a Toeplitz matrix are completely defined once the first column and the first row have been specified. A convolution matrix is an example of a Toeplitz matrix. A matrix with a similar property is a *Hankel matrix*, which has equal elements along the diagonals that are perpendicular to the main diagonal, i.e.,

$$a_{ij} = a_{i+1,j-1} \quad ; \qquad \text{for all } i < n \text{ and } j \le n$$

An example of a 4×4 Hankel matrix is

$$\mathbf{A} = \begin{bmatrix} 1 & 3 & 5 & 7 \\ 3 & 5 & 7 & 4 \\ 5 & 7 & 4 & 2 \\ 7 & 4 & 2 & 1 \end{bmatrix}$$

Another example of a Hankel matrix is the exchange matrix $\mathbf{J}$.

Toeplitz matrices are a special case of a larger class of matrices known as *persymmetric* matrices. A persymmetric matrix is symmetric about the cross-diagonal, $a_{ij} = a_{n-j+1,n-i+1}$.

An example of a persymmetric but non-Toeplitz matrix is

$$\mathbf{A} = \begin{bmatrix} 1 & 3 & 5 & 7 \\ 2 & 2 & 4 & 5 \\ 4 & 4 & 2 & 3 \\ 6 & 4 & 2 & 1 \end{bmatrix}$$

If a Toeplitz matrix is symmetric, or Hermitian in the case of a complex matrix, then all of the elements of the matrix are completely determined by either the first row or the first column of the matrix. An example of a symmetric Toeplitz matrix is

$$\mathbf{A} = \begin{bmatrix} 1 & 3 & 5 & 7 \\ 3 & 1 & 3 & 5 \\ 5 & 3 & 1 & 3 \\ 7 & 5 & 3 & 1 \end{bmatrix}$$

Since we will be dealing frequently with symmetric Toeplitz and Hermitian Toeplitz matrices it will be convenient to adopt the notation

$$\mathbf{A} = \text{Toep}\{a(0),\ a(1), \ldots, a(p)\}$$

for the Hermitian Toeplitz matrix that has the elements $a(0),\ a(1), \ldots, a(p)$ in the first *column*. For example,

$$\mathbf{A} = \text{Toep}\{1,\ j,\ 1-j\} = \begin{bmatrix} 1 & -j & 1+j \\ j & 1 & -j \\ 1-j & j & 1 \end{bmatrix}$$

Symmetric Toeplitz matrices are a special case of a larger class of matrices known as *centrosymmetric* matrices. A centrosymmetric matrix is both symmetric and persymmetric. An example of a centrosymmetric matrix that is not Toeplitz is

$$\mathbf{A} = \begin{bmatrix} 1 & 3 & 5 & 6 \\ 3 & 2 & 4 & 5 \\ 5 & 4 & 2 & 3 \\ 6 & 5 & 3 & 1 \end{bmatrix}$$

There are many interesting and useful properties of Toeplitz, persymmetric, and centrosymmetric matrices. For example, if $\mathbf{A}$ is a symmetric Toeplitz matrix, then

$$\mathbf{J}^T\mathbf{A}\mathbf{J} = \mathbf{A}$$

whereas if $\mathbf{A}$ is a Hermitian Toeplitz matrix, then

$$\mathbf{J}^T\mathbf{A}\mathbf{J} = \mathbf{A}^*$$

Of particular interest will be the following properties that are concerned with the inverse.

- **Property 1.** The inverse of a symmetric matrix is symmetric.
- **Property 2.** The inverse of a persymmetric matrix is persymmetric.
- **Property 3.** The inverse of a Toeplitz matrix is not, in general, Toeplitz. However, since a Toeplitz matrix is persymmetric, the inverse will always be persymmetric. Furthermore, the inverse of a symmetric Toeplitz matrix will be centrosymmetric.

These symmetry properties of matrix inverses along with those previously mentioned are summarized in Table 2.5.

Table 2.5 ***The Relationship between the Symmetries of a Matrix and Its Inverse***

Matrix	Inverse
Symmetric	Symmetric
Hermitian	Hermitian
Persymmetric	Persymmetric
Centrosymmetric	Centrosymmetric
Toeplitz	Persymmetric
Hankel	Symmetric
Triangular	Triangular

Finally, we conclude with the definition of orthogonal and unitary matrices. A real $n \times n$ matrix is said to be *orthogonal* if the columns (and rows) are orthonormal. Thus, if the columns of $\mathbf{A}$ are $\mathbf{a}_i$,

$$\mathbf{A} = \left[\mathbf{a}_1, \mathbf{a}_2, \ldots, \mathbf{a}_n\right]$$

and

$$\mathbf{a}_i^T \mathbf{a}_j = \begin{cases} 1 & \text{for } i = j \\ 0 & \text{for } i \neq j \end{cases}$$

then $\mathbf{A}$ is orthogonal. Note that if $\mathbf{A}$ is orthogonal, then

$$\mathbf{A}^T \mathbf{A} = \mathbf{I}$$

Therefore, the inverse of an orthogonal matrix is equal to its transpose

$$\mathbf{A}^{-1} = \mathbf{A}^T$$

The exchange matrix is an example of an orthogonal matrix since $\mathbf{J}^T \mathbf{J} = \mathbf{J}^2 = \mathbf{I}$.

In the case of a complex $n \times n$ matrix, if the columns (rows) are orthogonal,

$$\mathbf{a}_i^H \mathbf{a}_j = \begin{cases} 1 & \text{for } i = j \\ 0 & \text{for } i \neq j \end{cases}$$

then

$$\mathbf{A}^H \mathbf{A} = \mathbf{I}$$

and $\mathbf{A}$ is said to be a *unitary* matrix. The inverse of a unitary matrix is equal to its Hermitian transpose

$$\mathbf{A}^{-1} = \mathbf{A}^H$$

2.3.8 Quadratic and Hermitian Forms

The *quadratic form* of a real symmetric $n \times n$ matrix $\mathbf{A}$ is the scalar defined by

$$Q_A(\mathbf{x}) = \mathbf{x}^T \mathbf{A} \mathbf{x} = \sum_{i=1}^{n} \sum_{j=1}^{n} x_i a_{ij} x_j$$

where $\mathbf{x}^T = [x_1, x_2, \ldots, x_n]$ is a vector of n real variables. Note that the quadratic form is a quadratic function in the n variables $x_1, x_2, \ldots, x_n$. For example, the quadratic form of

$$\mathbf{A} = \begin{bmatrix} 3 & 1 \\ 1 & 2 \end{bmatrix}$$

is

$$Q_A(\mathbf{x}) = \mathbf{x}^T \mathbf{A} \mathbf{x} = 3x_1^2 + 2x_1x_2 + 2x_2^2$$

In a similar fashion, for a Hermitian matrix, the Hermitian form is defined as

$$Q_A(\mathbf{x}) = \mathbf{x}^H \mathbf{A} \mathbf{x} = \sum_{i=1}^{n} \sum_{j=1}^{n} x_i^* a_{ij} x_j$$

If the quadratic form of a matrix $\mathbf{A}$ is positive for all nonzero vectors $\mathbf{x}$,

$$Q_A(\mathbf{x}) > 0$$

then $\mathbf{A}$ is said to be *positive definite* and we write $\mathbf{A} > 0$. For example, the matrix

$$\mathbf{A} = \begin{bmatrix} 2 & 0 \\ 0 & 3 \end{bmatrix}$$

which has the quadratic form

$$Q_A(\mathbf{x}) = \begin{bmatrix} x_1, x_2 \end{bmatrix} \begin{bmatrix} 2 & 0 \\ 0 & 3 \end{bmatrix} \begin{bmatrix} x_1 \\ x_2 \end{bmatrix} = 2x_1^2 + 3x_2^2$$

is positive definite since $Q_A(\mathbf{x}) > 0$ for all $\mathbf{x} \neq \mathbf{0}$.

If the quadratic form is nonnegative for all nonzero vectors $\mathbf{x}$

$$Q_A(\mathbf{x}) \geq 0$$

then $\mathbf{A}$ is said to be *positive semidefinite*. For example,

$$\mathbf{A} = \begin{bmatrix} 2 & 0 \\ 0 & 0 \end{bmatrix}$$

is positive semidefinite since $Q_A(x) = 2x_1^2 \geq 0$, but it is not positive definite since $Q_A(\mathbf{x}) = 0$ for any vector $\mathbf{x}$ of the form $\mathbf{x} = [0, x_2]^T$. A test for positive definiteness is given in Section 2.3.9 (see Property 3 on p. 43).

In a similar fashion, a matrix is said to be *negative definite* if $Q_A(\mathbf{x}) < 0$ for all nonzero $\mathbf{x}$, whereas it is said to be *negative semidefinite* if $Q_A(\mathbf{x}) \leq 0$ for all nonzero $\mathbf{x}$. A matrix that is none of the above is said to be *indefinite*.

For any $n \times n$ matrix $\mathbf{A}$ and for any $n \times m$ matrix $\mathbf{B}$ having full rank, the definiteness of $\mathbf{A}$ and $\mathbf{B}^H\mathbf{A}\mathbf{B}$ will be the same. For example, if $\mathbf{A} > 0$ and $\mathbf{B}$ is full rank, then $\mathbf{B}^H\mathbf{A}\mathbf{B} > 0$. This follows by noting that for any vector $\mathbf{x}$,

$$\mathbf{x}^H \left(\mathbf{B}^H \mathbf{A} \mathbf{B}\right) \mathbf{x} = (\mathbf{B}\mathbf{x})^H \mathbf{A} (\mathbf{B}\mathbf{x}) = \mathbf{v}^H \mathbf{A} \mathbf{v}$$

where $\mathbf{v} = \mathbf{B}\mathbf{x}$. Therefore, if $\mathbf{A} > 0$, then $\mathbf{v}^H\mathbf{A}\mathbf{v} > 0$ and $\mathbf{B}^H\mathbf{A}\mathbf{B}$ is positive definite (The constraint that $\mathbf{B}$ have full rank ensures that $\mathbf{v} = \mathbf{B}\mathbf{x}$ is nonzero for any nonzero vector $\mathbf{x}$).

2.3.9 Eigenvalues and Eigenvectors

Eigenvalues and eigenvectors provide useful and important information about a matrix. For example, given the eigenvalues it is possible to determine whether or not a matrix is positive definite. The eigenvalues may also be used to determine whether or not a matrix is invertible as well as to indicate how sensitive the determination of the inverse will be to numerical errors. Eigenvalues and eigenvectors also provide an important representation for matrices known as the *eigenvalue decomposition*. This decomposition, developed below, will be useful in our discussions of spectrum estimation as well as in our study of the convergence properties of adaptive filtering algorithms. This section begins with the definition of the eigenvalues and eigenvectors of a matrix and then proceeds to review some properties of eigenvalues and eigenvectors that will be useful in later chapters.

Let $\mathbf{A}$ be an $n \times n$ matrix and consider the following set of linear equations,

$$\mathbf{A}\mathbf{v} = \lambda \mathbf{v} \tag{2.44}$$

where λ is a constant. Equation (2.44) may equivalently be expressed as a set of homogeneous linear equations of the form

$$(\mathbf{A} - \lambda \mathbf{I})\, \mathbf{v} = 0 \tag{2.45}$$

In order for a nonzero vector to be a solution to this equation, it is necessary for the matrix $\mathbf{A} - \lambda\mathbf{I}$ to be singular. Therefore, the determinant of $\mathbf{A} - \lambda\mathbf{I}$ must be zero,

$$p(\lambda) = \det\big(\mathbf{A} - \lambda \mathbf{I}\big) = 0 \tag{2.46}$$

Note that $p(\lambda)$ is an nth-order polynomial in λ. This polynomial is called the characteristic polynomial of the matrix $\mathbf{A}$ and the n roots, λ_i for $i = 1, 2, \ldots, n$, are the *eigenvalues* of $\mathbf{A}$. For each eigenvalue, λ_i, the matrix $(\mathbf{A} - \lambda_i \mathbf{I})$ will be singular and there will be at least one nonzero vector, $\mathbf{v}_i$, that solves Eq. (2.44), i.e.,

$$\mathbf{A}\mathbf{v}_i = \lambda_i \mathbf{v}_i \tag{2.47}$$

These vectors, $\mathbf{v}_i$, are called the *eigenvectors* of $\mathbf{A}$. For any eigenvector $\mathbf{v}_i$, it is clear that $\alpha \mathbf{v}_i$ will also be an eigenvector for any constant α. Therefore, eigenvectors are often normalized so that they have unit norm, $\|\mathbf{v}_i\| = 1$. The following property establishes the linear independence of eigenvectors that correspond to distinct eigenvalues.

Property 1. The nonzero eigenvectors $\mathbf{v}_1, \mathbf{v}_2, \ldots, \mathbf{v}_n$ corresponding to distinct eigenvalues $\lambda_1, \lambda_2, \ldots, \lambda_n$ are linearly independent.

If $\mathbf{A}$ is an $n \times n$ singular matrix, then there are nonzero solutions to the homogeneous equation

$$\mathbf{A}\mathbf{v}_i = \mathbf{0} \tag{2.48}$$

and it follows that $\lambda = 0$ is an eigenvalue of $\mathbf{A}$. Furthermore, if the rank of $\mathbf{A}$ is $\rho(\mathbf{A})$, then there will be $k = n - \rho(\mathbf{A})$ linearly independent solutions to Eq. (2.48). Therefore, it follows that $\mathbf{A}$ will have $\rho(\mathbf{A})$ nonzero eigenvalues and $n - \rho(\mathbf{A})$ eigenvalues that are equal to zero.

Example 2.3.8 *Eigenvalues of a 2 x 2 Symmetric Matrix*

Consider the following 2×2 symmetric matrix

$$\mathbf{A} = \begin{bmatrix} 4 & 1 \\ 1 & 4 \end{bmatrix}$$

The characteristic polynomial is

$$\det(\mathbf{A} - \lambda\mathbf{I}) = \det\begin{bmatrix} 4-\lambda & 1 \\ 1 & 4-\lambda \end{bmatrix} = (4-\lambda)(4-\lambda) - 1 = \lambda^2 - 8\lambda + 15$$

Therefore, the eigenvalues of **A** are the roots of

$$\lambda^2 - 8\lambda + 15 = 0$$

which are

$$\lambda_1 = 5 \qquad \lambda_2 = 3$$

The eigenvectors are found by solving Eq. (2.47) with $\lambda_1 = 5$ and $\lambda_2 = 3$. Thus, for $\lambda_1 = 5$, Eq. (2.47) becomes

$$\begin{bmatrix} 4 & 1 \\ 1 & 4 \end{bmatrix}\begin{bmatrix} v_{11} \\ v_{12} \end{bmatrix} = 5\begin{bmatrix} v_{11} \\ v_{12} \end{bmatrix}$$

and it follows that

$$4v_{11} + v_{12} = 5v_{11}$$

or

$$v_{11} = v_{12}$$

Similarly, for $\lambda_2 = 3$ we find that

$$v_{11} = -v_{12}$$

Thus, the normalized eigenvectors of **A** are

$$\mathbf{v}_1 = \frac{1}{\sqrt{2}}\begin{bmatrix} 1 \\ 1 \end{bmatrix} \qquad \mathbf{v}_2 = \frac{1}{\sqrt{2}}\begin{bmatrix} 1 \\ -1 \end{bmatrix}$$

In general, it is not possible to say much about the eigenvalues and eigenvectors of a matrix without knowing something about its properties. However, in the case of symmetric or Hermitian matrices, the eigenvalues and eigenvectors have several useful and important properties. For example,

> ***Property* 2.** The eigenvalues of a Hermitian matrix are real.

This property is easily established as follows. Let **A** be a Hermitian matrix with eigenvalue λ_i and eigenvector $\mathbf{v}_i$,

$$\mathbf{A}\mathbf{v}_i = \lambda_i \mathbf{v}_i$$

Multiplying this equation on the left by $\mathbf{v}_i^H$

$$\mathbf{v}_i^H \mathbf{A}\mathbf{v}_i = \lambda_i \mathbf{v}_i^H \mathbf{v}_i \tag{2.49}$$

and taking the Hermitian transpose gives

$$\mathbf{v}_i^H \mathbf{A}^H \mathbf{v}_i = \lambda_i^* \mathbf{v}_i^H \mathbf{v}_i \tag{2.50}$$

Since $\mathbf{A}$ is Hermitian, then Eq. (2.50) becomes

$$\mathbf{v}_i^H \mathbf{A} \mathbf{v}_i = \lambda_i^* \mathbf{v}_i^H \mathbf{v}_i \tag{2.51}$$

Comparing Eq. (2.49) with Eq. (2.51) it follows that λ_i must be real, i.e., $\lambda_i = \lambda_i^*$. ∎

The next property establishes a fundamental result linking the positive definiteness of a matrix to the positivity of its eigenvalues.

Property 3. A Hermitian matrix is positive definite, $\mathbf{A} > 0$, if and only if the eigenvalues of $\mathbf{A}$ are positive, $\lambda_k > 0$.

Similar properties hold for positive semidefinite, negative definite, and negative semidefinite matrices. For example, if $\mathbf{A}$ is Hermitian and $\mathbf{A} \geq 0$ then $\lambda_k \geq 0$.

The determinant of a matrix is related to its eigenvalues by the following relationship,

$$\det(\mathbf{A}) = \prod_{i=1}^{n} \lambda_i \tag{2.52}$$

Therefore, a matrix is nonsingular (invertible) if and only if all of its eigenvalues are nonzero. Conversely, if a matrix has one or more zero eigenvalues then it will be singular (noninvertible). As a result, it follows that any positive definite matrix is nonsingular.

In Property 2 we saw that the eigenvalues of a Hermitian matrix are real. The next property establishes the orthogonality of the eigenvectors corresponding to distinct eigenvalues.

Property 4. The eigenvectors of a Hermitian matrix corresponding to distinct eigenvalues are orthogonal, i.e., if $\lambda_i \neq \lambda_j$, then $< \mathbf{v}_i, \mathbf{v}_j >= 0$.

To establish this property, let λ_i and λ_j be two distinct eigenvalues of a Hermitian matrix with eigenvectors $\mathbf{v}_i$ and $\mathbf{v}_j$, respectively,

$$\begin{aligned} \mathbf{A}\mathbf{v}_i &= \lambda_i \mathbf{v}_i \\ \mathbf{A}\mathbf{v}_j &= \lambda_j \mathbf{v}_j \end{aligned} \tag{2.53}$$

Multiplying the first equation on the left by $\mathbf{v}_j^H$ and the second by $\mathbf{v}_i^H$ gives

$$\mathbf{v}_j^H \mathbf{A} \mathbf{v}_i = \lambda_i \mathbf{v}_j^H \mathbf{v}_i \tag{2.54}$$

$$\mathbf{v}_i^H \mathbf{A} \mathbf{v}_j = \lambda_j \mathbf{v}_i^H \mathbf{v}_j \tag{2.55}$$

Taking the Hermitian transpose of Eq. (2.55) yields

$$\mathbf{v}_j^H \mathbf{A}^H \mathbf{v}_i = \lambda_j^* \mathbf{v}_j^H \mathbf{v}_i \tag{2.56}$$

If $\mathbf{A}$ is Hermitian, then $\mathbf{A}^H = \mathbf{A}$ and it follows from Property 2 that $\lambda_j^* = \lambda_j$. Therefore, Eq. (2.56) becomes

$$\mathbf{v}_j^H \mathbf{A} \mathbf{v}_i = \lambda_j \mathbf{v}_j^H \mathbf{v}_i \tag{2.57}$$

Subtracting Eq. (2.57) from (2.54) leads to

$$(\lambda_i - \lambda_j)\mathbf{v}_j^H \mathbf{v}_i = 0$$

Therefore, if $\lambda_i \neq \lambda_j$, then $\mathbf{v}_j^H \mathbf{v}_i = 0$ and it follows that $\mathbf{v}_i$ and $\mathbf{v}_j$ are orthogonal. ∎

Although stated and verified only for the case of distinct eigenvalues, it is also true that for any $n \times n$ Hermitian matrix there exists a set of n orthonormal eigenvectors [10]. For example, consider the 2×2 identity matrix,

$$\mathbf{I} = \begin{bmatrix} 1 & 0 \\ 0 & 1 \end{bmatrix}$$

which has two eigenvalues that are equal to one. Since any vector is an eigenvector of $\mathbf{I}$, then $\mathbf{v}_1 = \begin{bmatrix} 1, & 0 \end{bmatrix}^T$ and $\mathbf{v}_2 = \begin{bmatrix} 0, & 1 \end{bmatrix}^T$ is one possible set of orthonormal eigenvectors.

For any $n \times n$ matrix $\mathbf{A}$ having a set of n linearly independent eigenvectors we may perform an *eigenvalue decomposition* that expresses $\mathbf{A}$ in the form

$$\boxed{\mathbf{A} = \mathbf{V}\mathbf{\Lambda}\mathbf{V}^{-1}} \tag{2.58}$$

where $\mathbf{V}$ is a matrix that contains the eigenvectors of $\mathbf{A}$ and $\mathbf{\Lambda}$ is a diagonal matrix that contains the eigenvalues. This decomposition is performed as follows. Let $\mathbf{A}$ be an $n \times n$ matrix with eigenvalues λ_k and eigenvectors $\mathbf{v}_k$,

$$\mathbf{A}\mathbf{v}_k = \lambda_k \mathbf{v}_k \quad ; \quad k = 1, 2, \ldots, n$$

These n equations may be written in matrix form as

$$\mathbf{A}\begin{bmatrix} \mathbf{v}_1, & \mathbf{v}_2, & \ldots, & \mathbf{v}_n \end{bmatrix} = \begin{bmatrix} \lambda_1 \mathbf{v}_1, & \lambda_2 \mathbf{v}_2, & \ldots, & \lambda_n \mathbf{v}_n \end{bmatrix} \tag{2.59}$$

Therefore, with

$$\mathbf{V} = \begin{bmatrix} \mathbf{v}_1, & \mathbf{v}_2, & \ldots, & \mathbf{v}_n \end{bmatrix}$$

and

$$\mathbf{\Lambda} = \text{diag}\{\lambda_1, \ \lambda_2, \ \ldots, \ \lambda_n\}$$

Eq. (2.59) may be written as

$$\mathbf{A}\mathbf{V} = \mathbf{V}\mathbf{\Lambda} \tag{2.60}$$

If the eigenvectors $\mathbf{v}_i$ are independent, then $\mathbf{V}$ is invertible and the decomposition follows by multiplying both sides of Eq. (2.60) on the right by $\mathbf{V}^{-1}$.

For a Hermitian matrix, the harmonic decomposition assumes a special form. In particular, recall that for a Hermitian matrix we may always find an orthonormal set of eigenvectors. Therefore, if $\mathbf{A}$ is Hermitian, then $\mathbf{V}$ is unitary and the eigenvalue decomposition becomes

$$\mathbf{A} = \mathbf{V}\mathbf{\Lambda}\mathbf{V}^H = \sum_{i=1}^{n} \lambda_i \mathbf{v}_i \mathbf{v}_i^H \tag{2.61}$$

This result is known as the *spectral theorem*.

Spectral Theorem. Any Hermitian matrix $\mathbf{A}$ may be decomposed as

$$\mathbf{A} = \mathbf{V}\mathbf{\Lambda}\mathbf{V}^H = \lambda_1 \mathbf{v}_1 \mathbf{v}_1^H + \lambda_2 \mathbf{v}_2 \mathbf{v}_2^H + \cdots + \lambda_n \mathbf{v}_n \mathbf{v}_n^H$$

where λ_i are the eigenvalues of $\mathbf{A}$ and $\mathbf{v}_i$ are a set of orthonormal eigenvectors.

As an application of the spectral theorem, suppose that $\mathbf{A}$ is a nonsingular Hermitian matrix. Using the spectral theorem we may find the inverse of $\mathbf{A}$ as follows,

$$\mathbf{A}^{-1} = \left(\mathbf{V}\mathbf{\Lambda}\mathbf{V}^H\right)^{-1} = \left(\mathbf{V}^H\right)^{-1}\mathbf{\Lambda}^{-1}\mathbf{V}^{-1} = \mathbf{V}\mathbf{\Lambda}^{-1}\mathbf{V}^H = \sum_{i=1}^{n}\frac{1}{\lambda_i}\mathbf{v}_i\mathbf{v}_i^H$$

Note that the invertibility of $\mathbf{A}$ guarantees that $\lambda_i \neq 0$ so this sum is always well defined.

In many signal processing applications one finds that a matrix $\mathbf{B}$ is singular or ill conditioned (one or more eigenvalues are close to zero). In these cases, we may sometimes stabilize the problem by adding a constant to each term along the diagonal,

$$\mathbf{A} = \mathbf{B} + \alpha\mathbf{I}$$

The effect of this operation is to leave the eigenvectors of $\mathbf{B}$ unchanged while changing the eigenvalues of $\mathbf{B}$ from λ_k to $\lambda_k + \alpha$. To see this, note that if $\mathbf{v}_k$ is an eigenvector of $\mathbf{B}$ with eigenvalue λ_k, then

$$\mathbf{A}\mathbf{v}_k = \mathbf{B}\mathbf{v}_k + \alpha\mathbf{v}_k = \left(\lambda_k + \alpha\right)\mathbf{v}_k$$

Therefore, $\mathbf{v}_k$ is also an eigenvector of $\mathbf{A}$ with eigenvalue $\lambda_k + \alpha$. This result is summarized in the following property for future reference.

Property 5. Let $\mathbf{B}$ be an $n \times n$ matrix with eigenvalues λ_i and let $\mathbf{A}$ be a matrix that is related to $\mathbf{B}$ as follows

$$\mathbf{A} = \mathbf{B} + \alpha\mathbf{I}$$

Then $\mathbf{A}$ and $\mathbf{B}$ have the same eigenvectors and the eigenvalues of $\mathbf{A}$ are $\lambda_i + \alpha$.

The following example considers finding the inverse of a matrix of the form $\mathbf{A} = \mathbf{B} + \alpha\mathbf{I}$ where $\mathbf{B}$ is a noninvertible Hermitian matrix of rank one.

Example 2.3.9 *Using the Spectral Theorem to Invert a Matrix*

Let $\mathbf{A}$ be an $n \times n$ Hermitian matrix of the form

$$\mathbf{A} = \mathbf{B} + \alpha\mathbf{I}$$

where α is a nonzero constant and where

$$\mathbf{B} = \lambda\mathbf{u}_1\mathbf{u}_1^H$$

with $\mathbf{u}_1$ a unit norm vector, $\mathbf{u}_1^H\mathbf{u}_1 = 1$, and $\lambda \neq 0$.[6] Since $\mathbf{B}$ is an $n \times n$ matrix of rank one, it has only one nonzero eigenvalue, the rest are equal to zero. Since

$$\mathbf{B}\mathbf{u}_1 = \lambda\left(\mathbf{u}_1\mathbf{u}_1^H\right)\mathbf{u}_1 = \lambda\mathbf{u}_1\left(\mathbf{u}_1^H\mathbf{u}_1\right) = \lambda\mathbf{u}_1$$

we see that $\mathbf{u}_1$ is an eigenvector of $\mathbf{B}$ with eigenvalue λ. This vector is also an eigenvector of $\mathbf{A}$ with eigenvalue $\lambda + \alpha$. Since $\mathbf{A}$ is a Hermitian matrix, there exists an orthonormal set of eigenvectors, call them $\mathbf{v}_i$. One of these eigenvectors is $\mathbf{u}_1$, so we will set $\mathbf{v}_1 = \mathbf{u}_1$. The eigenvalues corresponding to the eigenvectors $\mathbf{v}_i$ for $i \geq 2$ are equal to α. Using the

[6]We will encounter matrices of this form in Chapter 8 when we consider the autocorrelation matrix of a random process consisting of a complex exponential in white noise.

spectral theorem, we find for the inverse of **A**,

$$\mathbf{A}^{-1} = \frac{1}{\alpha + \lambda}\mathbf{u}_1\mathbf{u}_1^H + \sum_{i=2}^{n} \frac{1}{\alpha}\mathbf{v}_i\mathbf{v}_i^H \tag{2.62}$$

Since the n orthonormal eigenvectors of **A** also form an orthonormal set of eigenvectors for **I**, then the spectral theorem applied to **I** gives

$$\mathbf{I} = \sum_{i=1}^{n} \mathbf{v}_i\mathbf{v}_i^H$$

Therefore, the second term in Eq. (2.62) may be written as

$$\sum_{i=2}^{n} \frac{1}{\alpha}\mathbf{v}_i\mathbf{v}_i^H = \frac{1}{\alpha}\mathbf{I} - \frac{1}{\alpha}\mathbf{u}_1\mathbf{u}_1^H$$

and the inverse of **A** becomes

$$\mathbf{A}^{-1} = \frac{1}{\alpha + \lambda}\mathbf{u}_1\mathbf{u}_1^H + \frac{1}{\alpha}\mathbf{I} - \frac{1}{\alpha}\mathbf{u}_1\mathbf{u}_1^H = \frac{1}{\alpha}\mathbf{I} - \frac{\lambda}{\alpha(\alpha + \lambda)}\mathbf{u}_1\mathbf{u}_1^H$$

which is the desired solution.

It will be useful in our development of adaptive filtering algorithms in Chapter 9 to be able to characterize the geometry of the surface described by the equation

$$\mathbf{x}^T\mathbf{A}\mathbf{x} = 1 \tag{2.63}$$

where **A** is symmetric and positive definite. Using the spectral theorem this becomes

$$\mathbf{x}^T(\mathbf{V}\mathbf{\Lambda}\mathbf{V}^T)\mathbf{x} = 1 \tag{2.64}$$

With the change of variables

$$\mathbf{y} = \begin{bmatrix} y_1 \\ y_2 \\ \vdots \\ y_n \end{bmatrix} = \begin{bmatrix} \mathbf{v}_1^T\mathbf{x} \\ \mathbf{v}_2^T\mathbf{x} \\ \vdots \\ \mathbf{v}_n^T\mathbf{x} \end{bmatrix} = \mathbf{V}^T\mathbf{x} \tag{2.65}$$

Eq. (2.64) simplifies to

$$\mathbf{y}^T\mathbf{\Lambda}\mathbf{y} = \lambda_1 y_1^2 + \lambda_2 y_2^2 + \cdots + \lambda_n y_n^2 = 1 \tag{2.66}$$

which is an equation for an ellipse in n dimensions that is centered at the origin. Note that Eq. (2.65) performs a rotation of the coordinate system with the new axes being aligned with the eigenvectors of **A**. Assuming that the eigenvalues have been ordered so that $\lambda_1 \leq \lambda_2 \leq \cdots \leq \lambda_n$ we see that $\mathbf{y} = \left[1/\sqrt{\lambda_1},\ 0, \ldots, 0\right]^T$ is the point on the ellipse that is the farthest from the origin, lying at the end of the major axis of the ellipse (see Fig. 2.6). Furthermore, this point lies along the direction of the eigenvector $\mathbf{v}_1$, i.e.,

$$\mathbf{x} = \frac{1}{\sqrt{\lambda_1}}\mathbf{v}_1$$

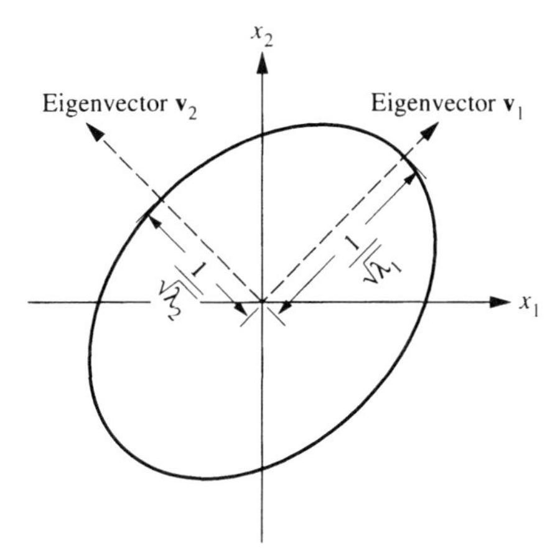

Figure 2.6 *The locus of points described by the equation* $\mathbf{x}^T\mathbf{A}\mathbf{x} = 1$ *where* $\mathbf{A}$ *is a positive definite symmetric matrix with eigenvalues* λ_1 *and* λ_2 *and eigenvectors* $\mathbf{v}_1$ *and* $\mathbf{v}_2$.

Similarly, the point closest to the origin is located at

$$\mathbf{x} = \frac{1}{\sqrt{\lambda_n}}\mathbf{v}_n$$

which is at the end of the minor axis of the ellipse (recall that since $\mathbf{A}$ is symmetric, then the eigenvectors $\mathbf{v}_i$ form an orthonormal set and, since $\mathbf{A} > 0$, then the eigenvalues λ_i are positive). The intermediate eigenvectors correspond to the intermediate axes. Thus, we have the following property of the surface described by Eq. (2.63).

Property 6. For a symmetric positive definite matrix $\mathbf{A}$, the equation

$$\mathbf{x}^T\mathbf{A}\mathbf{x} = 1$$

defines an ellipse in n dimensions whose axes are in the direction of the eigenvectors $\mathbf{v}_j$ of $\mathbf{A}$ with the half-length of these axes equal to $1/\sqrt{\lambda_j}$.

In our treatment of adaptive filtering algorithms in Chapter 9, it will be useful to be able to place an upper bound on the largest eigenvalue of a matrix. A loose upper bound follows from the following property of the eigenvalues of a matrix,

$$\mathrm{tr}(\mathbf{A}) = \sum_{i=1}^{n} \lambda_i$$

Denoting the maximum eigenvalue of a matrix $\mathbf{A}$ by $\lambda_{\max}$, if $\mathbf{A}$ is positive semi-definite then we have the following upper bound

$$\lambda_{\max} \leq \sum_{i=1}^{n} \lambda_i = \mathrm{tr}(\mathbf{A})$$

A tighter bound that follows from *Gershgorin's Circle theorem* [3, 10] is the following

Property 7. The largest eigenvalue of an $n \times n$ matrix $\mathbf{A} = \{a_{ij}\}$ is upper bounded by

$$|\lambda_{\max}| \leq \max_i \sum_{j=1}^{n} |a_{ij}|$$

Thus, the maximum eigenvalue is upper bounded by the maximum *row sum* of the matrix $\mathbf{A}$. Since the eigenvalues of a matrix and its transpose are the same, a similar statement may be made for *column sums*.

Finally, we conclude with a truly remarkable result known as the *Bordering theorem*, which is an interlacing theorem for the eigenvalues of a matrix [3].

Bordering Theorem. Let $\mathbf{A}$ be an $n \times n$ Hermitian matrix with ordered eigenvalues $\lambda_1 \leq \lambda_2 \leq \cdots \leq \lambda_n$. Let $\mathbf{y}$ be an arbitrary complex vector and a an arbitrary complex number. If $\widetilde{\mathbf{A}}$ is the $(n+1) \times (n+1)$ Hermitian matrix formed from $\mathbf{A}$, $\mathbf{y}$, and a as follows,

$$\widetilde{\mathbf{A}} = \left[\begin{array}{c|c} \mathbf{A} & \mathbf{y} \\ \hline \mathbf{y}^H & a \end{array}\right]$$

then the ordered eigenvalues of $\widetilde{\mathbf{A}}$, denoted by $\widetilde{\lambda}_1 \leq \widetilde{\lambda}_2 \leq \cdots \leq \widetilde{\lambda}_{n+1}$ are interlaced with those of $\mathbf{A}$ as follows

$$\widetilde{\lambda}_1 \leq \lambda_1 \leq \widetilde{\lambda}_2 \leq \lambda_2 \leq \cdots \leq \widetilde{\lambda}_n \leq \lambda_n \leq \widetilde{\lambda}_{n+1}$$

As a consequence of this theorem we see that, as a matrix is enlarged, the maximum eigenvalue does not decrease and the minimum eigenvalue does not increase. This result is useful in describing how the eigenvalues of an autocorrelation matrix vary as the dimension is increased.

2.3.10 Optimization Theory

Many of the problems that we will be formulating in the following chapters will involve the minimization of a function of one or more variables. Such problems fall in the domain of *optimization theory* [9]. The simplest form of optimization problem is that of finding the minimum of a scalar function $f(x)$ of a single real variable x. Assuming that this *objective function* $f(x)$ is differentiable, the stationary points of $f(x)$, the local and global minima, must satisfy the following conditions,

$$\frac{d\,f(x)}{dx} = 0 \tag{2.67}$$

$$\frac{d^2\,f(x)}{dx^2} > 0 \tag{2.68}$$

If $f(x)$ is a strictly convex function[7] then there is only one solution to Eq. (2.67) and this solution corresponds to the *global minimum* of $f(x)$. If the objective function is not convex, then each stationary point must be checked to see if it is the global minimum.

When the objective function depends upon a complex variable z and is differentiable (analytic), finding the stationary points of $f(z)$ is the same as in the real case. Unfortunately, however, many of the functions that we want to minimize are not differentiable. For example, consider the simple function $f(z) = |z|^2$. Although it is clear that $f(z)$ has a unique minimum that occurs at $z = 0$, this function is not differentiable [3]. The problem is that $f(z) = |z|^2$ is a function of z and z^* and any function that depends upon the complex conjugate z^* is not differentiable with respect to z. There are two ways that we may get around this problem [4]. The first is to express $f(z)$ in terms of its real and imaginary parts, $z = x + jy$, and minimize $f(x, y)$ with respect to these two variables. A more elegant solution is to treat z and z^* as independent variables and minimize $f(z, z^*)$ with respect to both z and z^* [1]. For example, if we write $f(z) = |z|^2$ as $f(z, z^*) = zz^*$ and treat $f(z, z^*)$ as a function of z with z^* constant, then

$$\frac{d}{dz}|z|^2 = z^* \tag{2.69}$$

whereas if we treat $f(z, z^*)$ as a function of z^* with z constant, then we have

$$\frac{d}{dz^*}|z|^2 = z \tag{2.70}$$

Setting both derivatives equal to zero and solving this pair of equations we see that the solution is $z = 0$. Minimizing a real-valued function of z and z^* is actually a bit easier than this as stated in the following theorem [1].

Theorem. If $f(z, z^*)$ is a *real-valued* function of z and z^* and if $f(z, z^*)$ is analytic with respect to both z and z^*, then the stationary points of $f(z, z^*)$ may be found by setting the derivative of $f(z, z^*)$ with respect to either z or z^* equal to zero and solving for z.

Thus, for $f(z, z^*) = |z|^2$, it is sufficient to set either Eq. (2.69) or (2.70) equal to zero and solve for z.

Another problem that we will be concerned with in this book is finding the minimum of a function of two or more variables. For a scalar function of n real variables, $f(\mathbf{x}) = f(x_1, x_2, \ldots, x_n)$, this involves computing the gradient, which is the vector of partial derivatives defined by

$$\nabla_x f(\mathbf{x}) = \frac{d}{d\mathbf{x}} f(\mathbf{x}) = \begin{bmatrix} \dfrac{\partial}{\partial x_1} f(\mathbf{x}) \\ \vdots \\ \dfrac{\partial}{\partial x_n} f(\mathbf{x}) \end{bmatrix}$$

The gradient points in the direction of the maximum rate of change of $f(\mathbf{x})$ and is equal to zero at the stationary points of $f(\mathbf{x})$. Thus, a necessary condition for a point $\mathbf{x}$ to be a

[7] A function $f(x)$ is strictly convex over a closed interval $[a, b]$ if, for any two point x_1 and x_2 in $[a, b]$, and for any scalar α such that $0 \le \alpha \le 1$, then $f\big(\alpha x_1 + (1-\alpha)x_2\big) < \alpha f(x_1) + (1-\alpha) f(x_2)$.

stationary point of $f(\mathbf{x})$ is

$$\boxed{\nabla_x f(\mathbf{x}) = 0} \tag{2.71}$$

In order for this stationary point to be a minimum the Hessian matrix, $\mathbf{H}_x$, must be positive definite, i.e.,

$$\mathbf{H}_x > 0$$

where $\mathbf{H}_x$ is the $n \times n$ matrix of second-order partial derivatives with the (i, j)th element given by

$$\{\mathbf{H}_x\}_{i,j} = \frac{\partial^2 f(\mathbf{x})}{\partial x_i \partial x_j}$$

As with a function of a single real variable, if $f(\mathbf{x})$ is strictly convex, then the solution to Eq. (2.71) is unique and is equal to the global minimum of $f(\mathbf{x})$.

When the function to be minimized is a real-valued function of the complex vectors $\mathbf{z}$ and $\mathbf{z}^*$, finding the minimum of $f(\mathbf{z}, \mathbf{z}^*)$ is complicated by the fact that $f(\mathbf{z}, \mathbf{z}^*)$ is not differentiable. However, as in the scalar case, treating $\mathbf{z}$ and $\mathbf{z}^*$ as independent variables we may solve for the stationary points using the following theorem [1].

Theorem. If $f(\mathbf{z}, \mathbf{z}^*)$ is a *real-valued* function of the complex vectors $\mathbf{z}$ and $\mathbf{z}^*$, then the vector pointing in the direction of the maximum rate of change of $f(\mathbf{z}, \mathbf{z}^*)$ is $\nabla_{\mathbf{z}^*} f(\mathbf{z}, \mathbf{z}^*)$, which is the derivative of $f(\mathbf{z}, \mathbf{z}^*)$ with respect to $\mathbf{z}^*$.

As a result of this theorem it follows that the stationary points of $f(\mathbf{z}, \mathbf{z}^*)$ are solutions to the equation

$$\boxed{\nabla_{\mathbf{z}^*} f(\mathbf{z}, \mathbf{z}^*) = 0} \tag{2.72}$$

Some examples of gradient vectors are given in Table 2.6.

Finally, we consider a special type of constrained minimization problem. Let $\mathbf{z} = [z_1, z_2, \ldots, z_n]^T$ be a complex vector and $\mathbf{R}$ a positive-definite Hermitian matrix. Suppose that we want to find the vector $\mathbf{z}$ that minimizes the quadratic form $\mathbf{z}^H\mathbf{R}\mathbf{z}$ subject to the constraint that the solution must satisfy the linear equality $\mathbf{z}^H\mathbf{a} = 1$, where $\mathbf{a}$ is a given complex vector. This is a problem that arises in array processing [4] and, as we will see in Chapter 8, is the problem that must be solved to find the minimum variance spectrum estimate of a random process. Geometrically, this constrained minimization problem is as illustrated in Fig. 2.7 for the case of a two-dimensional vector. The solution may be derived in a number of different ways [9]. One approach, presented here, is to introduce a Lagrange

Table 2.6 ***The Complex Gradient of a Complex Function****

	$\mathbf{a}^H\mathbf{z}$	$\mathbf{z}^H\mathbf{a}$	$\mathbf{z}^H\mathbf{A}\mathbf{z}$
$\nabla_{\mathbf{z}}$	$\mathbf{a}^*$	0	$(\mathbf{A}\mathbf{z})^*$
$\nabla_{\mathbf{z}^*}$	0	$\mathbf{a}$	$\mathbf{A}\mathbf{z}$

*$\mathbf{A}$ is assumed to be Hermitian.

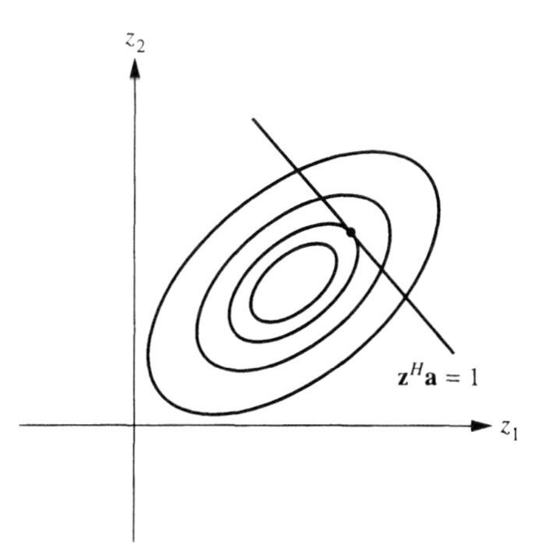

Figure 2.7 *Geometric illustration of the minimization of $\mathbf{z}^H\mathbf{R}\mathbf{z}$ subject to the constraint $\mathbf{z}^H\mathbf{a} = 1$. Each ellipse represents the locus of points for which the quadratic form is constant, $\mathbf{z}^H\mathbf{R}\mathbf{z} = c$, for some value of c. The solution occurs at the point where the line $\mathbf{z}^H\mathbf{a} = 1$ is tangent to a contour.*

multiplier, λ, and minimize the unconstrained objective function

$$Q_R(\mathbf{z}, \lambda) = \mathbf{z}^H\mathbf{R}\mathbf{z} + \lambda(1 - \mathbf{z}^H\mathbf{a}) \tag{2.73}$$

As we have seen, minimizing $Q_R(\mathbf{z}, \lambda)$ may be accomplished by setting the gradient of $Q_R(\mathbf{z}, \lambda)$ with respect to $\mathbf{z}^*$ equal to zero. Since

$$\nabla_{\mathbf{z}^*} Q_R(\mathbf{z}, \lambda) = \mathbf{R}\mathbf{z} - \lambda\mathbf{a}$$

then

$$\mathbf{z} = \lambda\mathbf{R}^{-1}\mathbf{a} \tag{2.74}$$

The value of the Lagrange multiplier is then found by setting the derivative of $Q_R(\mathbf{z}, \lambda)$ with respect to λ equal to zero as follows

$$\frac{\partial Q_R(\mathbf{z}, \lambda)}{\partial \lambda} = 1 - \mathbf{z}^H\mathbf{a} = 0 \tag{2.75}$$

Substituting Eq. (2.74) for $\mathbf{z}$ into Eq. (2.75) and solving for λ gives

$$\lambda = \frac{1}{\mathbf{a}^H\mathbf{R}^{-1}\mathbf{a}} \tag{2.76}$$

Finally, substituting Eq. (2.76) into Eq. (2.74) yields

$$\boxed{\mathbf{z} = \frac{\mathbf{R}^{-1}\mathbf{a}}{\mathbf{a}^H\mathbf{R}^{-1}\mathbf{a}}} \tag{2.77}$$

as the vector $\mathbf{z}$ that minimizes the objective function $Q_R(\mathbf{z}, \lambda)$.[8] Substituting this vector into the quadratic form $Q_R(\mathbf{z}) = \mathbf{z}^H\mathbf{R}\mathbf{z}$ gives, for the minimum value of $\mathbf{z}^H\mathbf{R}\mathbf{z}$,

$$\min_{\mathbf{z}}\left\{\mathbf{z}^H\mathbf{R}\mathbf{z}\right\} = \frac{\mathbf{z}^H\mathbf{a}}{\mathbf{a}^H\mathbf{R}^{-1}\mathbf{a}} = \frac{1}{\mathbf{a}^H\mathbf{R}^{-1}\mathbf{a}}$$

where the last equality follows from Eq. (2.75).

2.4 SUMMARY

In this chapter we have reviewed the fundamentals of digital signal processing and established the notation that will be used to describe discrete-time signals and systems. In addition, we have introduced the key concepts from linear algebra that will be used throughout this book. Since many of the problems that we will be solving require that we find the solution to a set of linear equations, of particular importance will be the discussions in Section 2.3.6 related to underdetermined and overdetermined sets of linear equations. For a more complete and rigorous treatment of linear equations, the reader may consult any one of a number of excellent references [3, 6, 10].

References

1. D. H. Brandwood, "A complex gradient operator and its application in adaptive array theory," *IEE Proc., Parts F and H*, vol. 130, pp. 11–16, February 1983.
2. M. A. Clements and J. Pease, "On causal linear phase IIR digital filters", *IEEE Trans. Acoust., Speech, Sig. Proc.*, vol. ASSP-3, no. 4, pp. 479–484, March 1989.
3. R. A. Horn and C. R. Johnson, *Matrix Analysis*, Cambridge University Press, Cambridge, 1985.
4. D. H. Johnson and D. E. Dudgeon, *Array Processing*, Prentice-Hall, Englewood Cliffs, NJ, 1993.
5. S. Kay, *Modern Spectrum Estimation: Theory and Applications*, Prentice-Hall, Englewood Cliffs, NJ, 1988.
6. B. Nobel and J. W. Daniel, *Applied Linear Algebra*, Prentice-Hall, Englewood Cliffs, NJ, 1977.
7. A. V. Oppenheim and R. W. Schafer, *Discrete-Time Signal Processing*, Prentice-Hall, Englewood Cliffs, NJ, 1989.
8. A. V. Oppenheim and A. S. Willsky, *Signals and Systems*, Prentice-Hall, Englewood Cliffs, NJ, 1983.
9. D. A. Pierce, *Optimization Theory with Applications*, John Wiley & Sons, New York, 1969.
10. G. Strang, *Linear Algebra and its Applications*, Academic Press, New York, 1980.
11. R. D. Strum and D. E. Kirk, *First Principles of Discrete Systems and Digital Signal Processing*, Addison-Wesley, MA, 1988.

2.5 PROBLEMS

2.1. The DFT of a sequence $x(n)$ of length N may be expressed in matrix form as follows

$$\mathbf{X} = \mathbf{W}\mathbf{x}$$

where $\mathbf{x} = \left[x(0),\ x(1),\ \ldots, x(N-1)\right]^T$ is a vector containing the signal values and $\mathbf{X}$ is a vector containing the DFT coefficients $X(k)$,

[8]The assumption that $\mathbf{R}$ is positive definite guarantees that the matrix inverse exists and that this solution is the global minimum.

(a) Find the matrix $\mathbf{W}$.

(b) What properties does the matrix $\mathbf{W}$ have?

(c) What is the inverse of $\mathbf{W}$?

2.2. Prove or disprove each of the following statements:

(a) The product of two upper triangular matrices is upper triangular.

(b) The product of two Toeplitz matrices is Toeplitz.

(c) The product of two centrosymmetric matrices is centrosymmetric.

2.3. Find the minimum norm solution to the following set of underdetermined linear equations,

$$\begin{bmatrix} 1 & 0 & 2 & -1 \\ -1 & 1 & 0 & 1 \end{bmatrix} \begin{bmatrix} x_1 \\ x_2 \\ x_3 \\ x_4 \end{bmatrix} = \begin{bmatrix} 1 \\ 1 \end{bmatrix}$$

2.4. Consider the set of inconsistent linear equations $\mathbf{Ax} = \mathbf{b}$ given by

$$\begin{bmatrix} 1 & 0 \\ 0 & 1 \\ 1 & 1 \end{bmatrix} \begin{bmatrix} x_1 \\ x_2 \end{bmatrix} = \begin{bmatrix} 1 \\ 1 \\ 0 \end{bmatrix}$$

(a) Find the least squares solution to these equations.

(b) Find the projection matrix $\mathbf{P}_A$.

(c) Find the best approximation $\hat{\mathbf{b}} = \mathbf{P}_A \mathbf{b}$ to $\mathbf{b}$.

(d) Consider the matrix

$$\mathbf{P}_{\mathbf{A}}^{\perp} = \mathbf{I} - \mathbf{P}_A$$

Find the vector $\mathbf{b}^{\perp} = \mathbf{P}_{\mathbf{A}}^{\perp}\mathbf{b}$ and show that it is orthogonal to $\hat{\mathbf{b}}$. What does the matrix $\mathbf{P}_A^{\perp}$ represent?

2.5. Consider the problem of trying to model a sequence $x(n)$ as the sum of a constant plus a complex exponential of frequency ω_0 [5]

$$\hat{x}(n) = c + ae^{jn\omega_0} \quad ; \quad n = 0, 1, \ldots, N-1$$

where c and a are unknown. We may express the problem of finding the values for c and a as one of solving a set of overdetermined linear equations

$$\begin{bmatrix} 1 & 1 \\ 1 & e^{j\omega_0} \\ \vdots & \vdots \\ 1 & e^{j(N-1)\omega_0} \end{bmatrix} \begin{bmatrix} c \\ a \end{bmatrix} = \begin{bmatrix} x(0) \\ x(1) \\ \vdots \\ x(N-1) \end{bmatrix}$$

(a) Find the least squares solution for c and a.

(b) If N is even and $\omega_0 = 2\pi k/N$ for some integer k, find the least squares solution for c and a.

2.6. It is known that the sum of the squares of n from $n = 1$ to $N - 1$ has a closed form solution of the following form

$$\sum_{n=0}^{N-1} n^2 = a_0 + a_1 N + a_2 N^2 + a_3 N^3$$

Given that a third-order polynomial is uniquely determined in terms of the values of the polynomial at four distinct points, derive a closed form expression for this sum by setting up a set of linear equations and solving these equations for a_0, a_1, a_2, a_3. Compare your solution to that given in Table 2.3.

2.7. Show that a projection matrix $\mathbf{P}_A$ has the following two properties,

1. It is *idempotent*, $\mathbf{P}_A^2 = \mathbf{P}_A$.
2. It is Hermitian.

2.8. Let $\mathbf{A} > 0$ and $\mathbf{B} > 0$ be positive definite matrices. Prove or disprove the following statements.

(a) $\mathbf{A}^2 > 0$.

(b) $\mathbf{A}^{-1} > 0$.

(c) $\mathbf{A} + \mathbf{B} > 0$.

2.9. (a) Prove that each eigenvector of a symmetric Toeplitz matrix is either symmetric or antisymmetric, i.e., $\mathbf{v}_k = \pm \mathbf{J}\mathbf{v}_k$.

(b) What property can you state about the eigenvalues of a Hermitian Toeplitz matrix?

2.10. (a) Find the eigenvalues and eigenvectors of the 2×2 symmetric Toeplitz matrix

$$\mathbf{A} = \begin{bmatrix} a & b \\ b & a \end{bmatrix}$$

(b) Find the eigenvalues and eigenvectors of the 2×2 Hermitian matrix

$$\mathbf{A} = \begin{bmatrix} a & b^* \\ b & a \end{bmatrix}$$

2.11. Establish Property 5 on p. 45.

2.12. A necessary and sufficient condition for a Hermitian matrix $\mathbf{A}$ to be positive definite is that there exists a nonsingular matrix $\mathbf{W}$ such that

$$\mathbf{A} = \mathbf{W}^H \mathbf{W}$$

(a) Prove this result.

(b) Find a factorization of the form $\mathbf{A} = \mathbf{W}^H \mathbf{W}$ for the matrix

$$\mathbf{A} = \begin{bmatrix} 2 & -1 \\ -1 & 2 \end{bmatrix}$$

2.13. Consider the 2×2 matrix

$$\mathbf{A} = \begin{bmatrix} 0 & 1 \\ 1 & 0 \end{bmatrix}$$

(a) Find the eigenvalues and eigenvectors of **A**.

(b) Are the eigenvectors unique? Are they linearly independent? Are they orthogonal?

(c) Diagonalize **A**, i.e., find **V** and **D** such that

$$\mathbf{V}^H\mathbf{A}\mathbf{V} = \mathbf{D}$$

where **D** is a diagonal matrix.

2.14. Find the eigenvalues and eigenvectors of the matrix

$$\mathbf{A} = \begin{bmatrix} 1 & -1 \\ 2 & 4 \end{bmatrix}$$

2.15. Consider the following 3×3 symmetric matrix

$$\mathbf{A} = \begin{bmatrix} 1 & -1 & 0 \\ -1 & 2 & -1 \\ 0 & -1 & 1 \end{bmatrix}$$

(a) Find the eigenvalues and eigenvectors of **A**.

(b) Find the determinant of **A**.

(c) Find the spectral decomposition of **A**.

(d) What are the eigenvalues of $\mathbf{A} + \mathbf{I}$ and how are the eigenvectors related to those of **A**?

2.16. Suppose that an $n \times n$ matrix **A** has eigenvalues $\lambda_1, \ldots, \lambda_n$ and eigenvectors $\mathbf{v}_1, \ldots, \mathbf{v}_n$.

(a) What are the eigenvalues and eigenvectors of $\mathbf{A}^2$?

(b) What are the eigenvalues and eigenvectors of $\mathbf{A}^{-1}$?

2.17. Find a matrix whose eigenvalues are $\lambda_1 = 1$ and $\lambda_2 = 4$ with eigenvectors $\mathbf{v}_1 = [3, 1]^T$ and $\mathbf{v}_2 = [2, 1]^T$.

2.18. Gerschgorin's circle theorem states that every eigenvalue of a matrix **A** lies in at least one of the circles $C_1, \ldots, C_N$ in the complex plane where C_i has center at the diagonal entry a_{ii} and its radius is $r_i = \sum_{j \neq i} |a_{ij}|$.

(a) Prove this theorem by using the eigenvalue equation $\mathbf{Ax} = \lambda\mathbf{x}$ to write

$$(\lambda - a_{ii})x_i = \sum_{j \neq i} a_{ij}x_j$$

and then use the triangular inequality,

$$\left| \sum_{j \neq i} a_{ij}x_j \right| \leq \sum_{j \neq i} |a_{ij}x_j|$$

(b) Use this theorem to establish the bound on $\lambda_{\max}$ given in Property 7.

(c) The matrix

$$\begin{bmatrix} 4 & 1 & 2 \\ 2 & 3 & 0 \\ 3 & 2 & 6 \end{bmatrix}$$

is said to be *diagonally dominant* since $a_{ii} > r_i$. Use Gerschgorin's circle theorem to show that this matrix is nonsingular.

2.19. Consider the following quadratic function of two real-valued variables z_1 and z_2,

$$f(z_1, z_2) = 3z_1^2 + 3z_2^2 + 4z_1z_2 + 8$$

Find the values of z_1 and z_2 that minimize $f(z_1, z_2)$ subject to the constraint that $z_1 + z_2 = 1$ and determine the minimum value of $f(z_1, z_2)$.

DISCRETE-TIME RANDOM PROCESSES 3

3.1 INTRODUCTION

Signals may be broadly classified into one of two types—deterministic and random. A *deterministic* signal is one that may be reproduced exactly with repeated measurements. The unit sample response of a linear shift-invariant filter is an example of a deterministic signal. A *random* signal, or random process, on the other hand, is a signal that is not repeatable in a predictable manner. Quantization noise produced by an A/D converter or a fixed-point digital filter is an example of a random process. Other examples include tape hiss or turntable rumble in audio signals, background clutter in radar images, speckle noise in synthetic aperture radar (SAR) images, and engine noise in speech transmissions from the cockpit of an airplane. Some signals may be considered to be either deterministic or random, depending upon the application. Speech, for example, may be considered to be a deterministic signal if a specific speech waveform is to be processed or analyzed. However, speech may also be viewed as a random process if one is considering the ensemble of all possible speech waveforms in order to design a system that will optimally process speech signals, in general.

In this book, many of the signals that we will be characterizing and analyzing will be random processes. Since a random process may only be described probabilistically or in terms of its average behavior, in this chapter we develop the background that is necessary to understand how a random process may be described and how its statistical properties are affected by a linear shift-invariant system. Although it is assumed that the reader is familiar with the basic theory of random variables, the chapter begins with a brief review of random variables in Section 3.2, focusing primarily on the concepts of statistical averages such as the mean, variance, and correlation. Discrete-time random processes are introduced in Section 3.3 as an indexed sequence of random variables. The characterization of a random process in terms of its ensemble averages such as the mean, autocorrelation, and autocovariance are defined, and properties of these averages are explored. The notion of stationarity and the concept of ergodicity are introduced next. Then, the power spectral density, or power spectrum, of a discrete random process is defined. The power spectrum, which is the Fourier transform of the autocorrelation function, provides an important characterization of a random process and will appear frequently in this book. In Chapter 8, for example, we consider techniques for estimating the power spectrum of a random process from a single observation of the process. In Section 3.4, we consider the effect of filtering a random process with a linear shift-invariant filter. Specifically, the relationship between the autocorrelation of the input process and the autocorrelation of the output process is established. Finally, Section 3.6 provides a description of a useful and important class of random processes: those that may be generated by filtering white noise with a linear shift-invariant filter. These include autoregressive (AR), moving average (MA), and autoregressive moving average (ARMA) random processes.

3.2 RANDOM VARIABLES

In this section, we introduce the concept of a *random variable* and develop some basic properties and characteristics of random variables. These results are important for two reasons. First, since random variables are pervasive and may be found in almost any practical application or natural setting ranging from the theory of games to detection and estimation theory and the restoration and enhancement of distorted signals, it is important to be able to understand and manipulate random variables. The second reason is that the characterization of random variables will serve as the starting point for our development of random processes in Section 3.3.

3.2.1 Definitions

The concept of a random variable may best be illustrated by means of the following simple example. Given a fair coin that is equally likely to result in *Heads* or *Tails* when flipped, one would expect that if the coin were to be flipped repeatedly, then the outcome would be *Heads* approximately half of the time and *Tails* the other half. For example, if flipping the coin N_T times results in n_H *Heads* and n_T *Tails*, then, for N_T large, we should find that

$$\frac{n_H}{N_T} \approx 0.5 \quad ; \quad \frac{n_T}{N_T} \approx 0.5$$

Therefore, with a fair coin there is an equal probability of getting either *Heads* or *Tails* and the following *probability assignment* is made for the two possible experimental outcomes $H = \{\text{Heads}\}$ and $T = \{\text{Tails}\}$,

$$\Pr\{H\} = 0.5 \quad ; \quad \Pr\{T\} = 0.5 \tag{3.1}$$

The set of all possible experimental outcomes is called the *sample space* or the *certain event* and the sample space, denoted by Ω, is always assigned a probability of one

$$\Pr\{\Omega\} = 1$$

For the coin flipping experiment

$$\Omega = \{H, T\}$$

and

$$\Pr\{H, T\} = 1$$

Subsets of the sample space are called *events*, and events consisting of a single element are called *elementary events*. For the coin tossing experiment there are only two elementary events,

$$\omega_1 = \{H\} \quad ; \quad \omega_2 = \{T\}$$

Given the coin flipping experiment, suppose that a real-valued variable, x, is defined in such a way that a value of 1 is assigned to x if the outcome of a coin flip is *Heads* and a value of -1 is assigned if the outcome is *Tails*. With this definition, a mapping has been defined between the set of experimental outcomes, $\omega_i \in \Omega$, and the set of real numbers, R, i.e., $f : \Omega \longrightarrow R$. This mapping is given by

$$\begin{aligned} \omega_1 = \{H\} &\Longrightarrow x = 1 \\ \omega_2 = \{T\} &\Longrightarrow x = -1 \end{aligned}$$

Based on the probability assignments defined for the elementary events *Heads* and *Tails* in Eq. (3.1), it follows that there is an equal probability that the variable x will be assigned a value of 1 or -1

$$\begin{aligned} \Pr\{x = 1\} &= 0.5 \\ \Pr\{x = -1\} &= 0.5 \end{aligned} \tag{3.2}$$

Since the only two possible outcomes of the coin flip are *Heads* and *Tails*, then the only possible values that may be assigned to the variable x are $x = 1$ and $x = -1$. Therefore, for any number α that is different from 1 or -1, the probability that $x = \alpha$ is equal to zero

$$\Pr\{x = \alpha\} = 0 \quad \text{if} \quad \alpha \neq \pm 1$$

and the probability that x assumes one of the two values, $x = 1$ or $x = -1$, is equal to one

$$\Pr\{\Omega\} = \Pr\{x = \pm 1\} = 1$$

With this probability assignment on each of the elementary events in the sample space Ω, and with the mapping of each elementary event $\omega_i \in \Omega$ to a value of x, a *random variable* has been defined that is specified in terms of the likelihood (probability) that it assumes a particular value. This random variable has an equal probability of having a value of 1 or -1. More generally, a random variable x that may assume only one of two values, $x = 1$ and $x = -1$, with

$$\Pr\{x = 1\} = p \qquad \text{and} \qquad \Pr\{x = -1\} = 1 - p$$

is referred to as a *Bernoulli random variable*.

A slightly more complicated random variable may be defined by replacing the coin tossing experiment with the roll of a fair die. Specifically, if the number that is obtained with the roll of a fair die is assigned to the variable x, then x will be a random variable that is equally likely to take on any integer value from one to six. In a similar fashion, a *complex random variable* may be defined by assigning complex numbers to elementary events in the sample space. For example, with an experiment consisting of a roll of two fair die, one black and one white, a complex random variable z may be defined with the assignment

$$z = m + jn$$

where m is the number showing on the black die and n is the number showing on the white die.

Each of the random variables considered so far are examples of *discrete random variables* since the sample space Ω consists of a discrete set of events, ω_i. Unlike a discrete random variable, a random variable of the *continuous type* may assume a continuum of values. An example of a random variable of the *continuous type* is the following. Consider an infinite resolution roulette wheel for which any number in the interval [0, 1] may result from the spin of the wheel. The sample space for this experiment is

$$\Omega = \{\omega : 0 \le \omega \le 1\}$$

If the wheel is fair so that any number in the interval [0, 1] is equally likely to occur, then a probability assignment on Ω may be made as follows. For any interval $I = (\alpha_1, \alpha_2]$ that is a subset of [0, 1], define the probability of the event $\omega \in I$ as follows,

$$\Pr\{\omega \in I\} = \Pr\{\alpha_1 < \omega \le \alpha_2\} = \alpha_2 - \alpha_1$$

In addition, suppose we postulate that, for any two *disjoint intervals*, I_1 and I_2, the probability that an outcome will lie either in I_1 or in I_2 is equal to the sum

$$\Pr\{\omega \in I_1 \text{ or } \omega \in I_2\} = \Pr\{\omega \in I_1\} + \Pr\{\omega \in I_2\}.$$

With this postulate or axiom for the probability of the sum of disjoint events, we have established a probability assignment on the sample space that may be used to determine the probability of any event defined on Ω. Finally, if the value that results from the spin of the wheel is assigned to the variable x, then we have specified a random variable of the continuous type.

Looking back at the random variables that have been considered so far, we see that for each random variable there is an underlying experiment—the flip of a coin, the roll of a die, or the spin of a roulette wheel. In addition, for each experimental outcome ω_i in the sample space Ω, a real or complex number is assigned to the variable x. Thus, as illustrated in Fig. 3.1, a random variable is a function in which the domain is the sample space Ω and the range is a subset of the real or complex numbers. Furthermore, note that the characterization of a random variable x is given statistically in terms of a probability assignment, or *probability law*, that is defined on events in the sample space Ω. This probability law is a rule that assigns a number, called the probability, to each event A in the sample space. In order to be a valid probability assignment, the following three axioms must be satisfied:

1. $\Pr(A) \ge 0$ for every event $A \in \Omega$.
2. $\Pr(\Omega) = 1$ for the certain event Ω.
3. For any two mutually exclusive events A_1 and A_2,

$$\Pr(A_1 \cup A_2) = \Pr(A_1) + \Pr(A_2)$$

Once a probability assignment is defined on events in the sample space, it is possible to develop a probabilistic description for the random variable x. This description is often expressed in terms of the probability that x assumes a value within a given range of values such as

$$\Pr\{x = 1\} \quad \text{or} \quad \Pr\{x \le 0\} \quad \text{or} \quad \Pr\{0 < x \le 1\}$$

In signal processing applications, it is the probabilistic description of the random variable, rather than the statistical characterization of events in the sample space, that is generally of interest. Therefore, it is more convenient to have a probability law that is assigned directly to the random variable itself rather than to events in the sample space. For a real-valued random variable, x, one such statistical characterization is the *probability distribution function*,

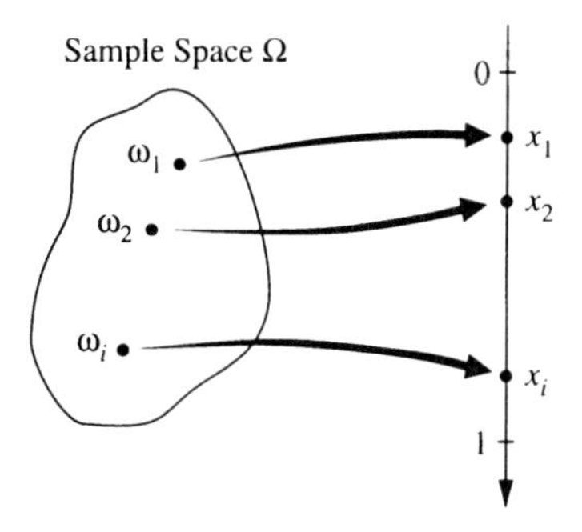

Figure 3.1 *The definition of a random variable in terms of a mapping from elementary events in a sample space Ω to points on the real line.*

$F_x(\alpha)$, given by

$$\boxed{F_x(\alpha) = \Pr\{x \leq \alpha\}} \tag{3.3}$$

For the random variable defined on the coin tossing experiment, the probability distribution function is

$$F_x(\alpha) = \begin{cases} 0 & ; \quad \alpha < -1 \\ 0.5 & ; \quad -1 \leq \alpha < 1 \\ 1 & ; \quad 1 \leq \alpha \end{cases} \tag{3.4}$$

which is plotted in Fig. 3.2*a*. Note that there are two step changes in $F_x(\alpha)$, one at $x = -1$ and the other at $x = 1$. These discontinuities in $F_x(\alpha)$ are due to the discrete *probability masses* at these points.

Another useful statistical characterization of a random variable is the *probability density function*, $f_x(\alpha)$, which is the derivative of the distribution function

$$\boxed{f_x(\alpha) = \frac{d}{d\alpha} F_x(\alpha)} \tag{3.5}$$

For the random variable having the distribution function given in Eq. (3.4), the probability density function, plotted in Fig. 3.2*b*, is

$$f_x(\alpha) = \frac{1}{2} u_0(\alpha + 1) + \frac{1}{2} u_0(\alpha - 1) \tag{3.6}$$

where $u_0(\alpha)$ is the unit impulse function. Density functions containing impulses are characteristic of random variables of the discrete type.

For the continuous random variable defined on the experiment of spinning a roulette wheel, the probability distribution function is

$$F_x(\alpha) = \begin{cases} 0 & ; \quad \alpha < 0 \\ \alpha & ; \quad 0 \leq \alpha < 1 \\ 1 & ; \quad 1 \leq \alpha \end{cases} \tag{3.7}$$

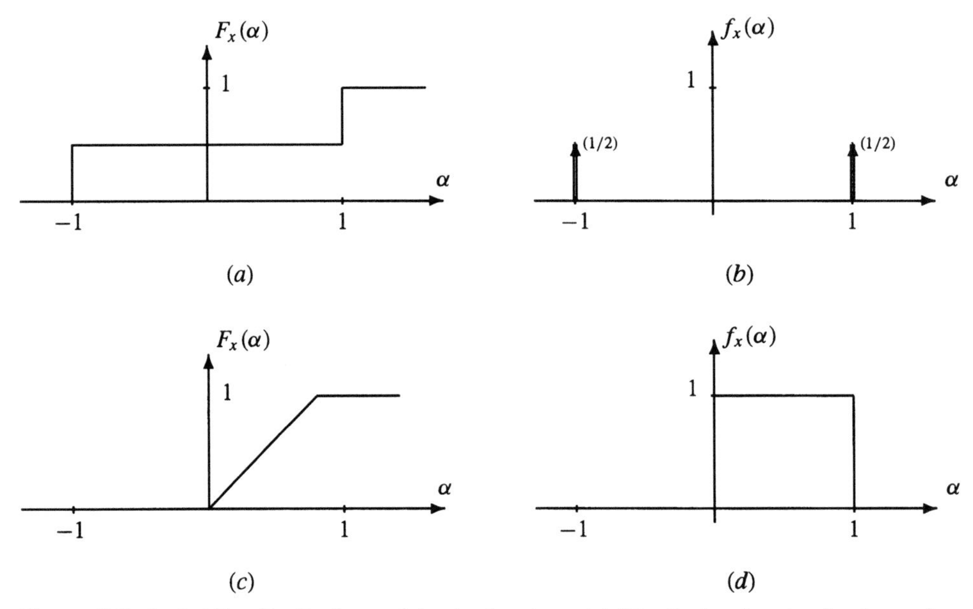

Figure 3.2 *Probability distribution and density functions: (a) Distribution function for the random variable defined on the coin tossing experiment and (b) the density function. (c) Distribution function for the random variable defined on the roulette wheel experiment and (d) the density function.*

and the probability density function is

$$f_x(\alpha) = \begin{cases} 1 & ; \quad 0 \le \alpha \le 1 \\ 0 & ; \quad \text{otherwise} \end{cases}$$

These distribution and density functions are shown in Figs. 3.2*c* and 3.2*d*, respectively. Note that, in contrast to a discrete random variable, the probability distribution function of a random variable of the continuous type is a continuous function of α and the density function is piece-wise continuous.

For complex random variables, the distribution function as defined in Eq. (3.3) is not appropriate since the inequality $z \le \alpha$ is meaningless when z is a complex number. Since a complex random variable

$$z = x + jy$$

is, in essence, a pair of random variables, x and y, the probability assignments on z are made in terms of the joint distribution and joint density functions of x and y, as discussed in Section 3.2.3.

3.2.2 Ensemble Averages

A complete statistical characterization of a random variable requires that it be possible to determine the probability of any event that may be defined on the sample space. In many applications, however, a complete statistical characterization of a random variable may not be necessary if the average behavior of the random variable is known. For example, in making the decision whether or not to play blackjack, it is the expected rate of return on a hand that is of interest rather than a complete statistical characterization of the game. Since the primary focus of this book will be with statistical averages rather than with probabilistic descriptions of random variables, in this section we review what is meant by the *expected value* of a random variable and the expected value of a function of a random variable.

Let x be the random variable that was defined on the die experiment. Suppose that the die is rolled N_T times and that the number k appears n_k times. Computing the average value that is rolled by summing all of the numbers and dividing by the total number of times that the die is rolled yields the *sample mean*

$$\langle x \rangle_{N_T} = \frac{n_1 + 2n_2 + 3n_3 + 4n_4 + 5n_5 + 6n_6}{N_T} \tag{3.8}$$

If $N_T \gg 1$, then $n_k/N_T \approx \Pr\{x = k\}$ and Eq. (3.8) becomes

$$\langle x \rangle_{N_T} \approx \sum_{k=1}^{6} k \Pr\{x = k\} \tag{3.9}$$

The mean or *expected value* of a discrete random variable x that assumes a value of α_k with probability $\Pr\{x = \alpha_k\}$ is therefore defined as

$$\boxed{E\{x\} = \sum_k \alpha_k \Pr\{x = \alpha_k\}} \tag{3.10}$$

In terms of the probability density function $f_x(\alpha)$, this expectation may be written as

$$\boxed{E\{x\} = \int_{-\infty}^{\infty} \alpha\, f_x(\alpha) d\alpha} \tag{3.11}$$

For random variables of the continuous type, Eq. (3.11) is also used as the definition for the expected value.

Example 3.2.1 ***Computing the Mean of a Random Variable***

For the random variable defined on the coin flipping experiment, the expected value is

$$E\{x\} = \Pr\{x = 1\} - \Pr\{x = -1\} = 0$$

For the random variable defined on the die experiment, on the other hand,

$$E\{x\} = \sum_{k=1}^{6} k \Pr\{x = k\} = \frac{1}{6} \sum_{k=1}^{6} k = 3.5$$

Finally, for the continuous random variable defined in terms of the roulette wheel experiment,

$$E\{x\} = \int_0^1 \alpha f_x(\alpha) d\alpha = \int_0^1 \alpha\, d\alpha = 0.5$$

There are many examples for which it is necessary to be able to compute the expected value of a function of a random variable. For example, in finding the average power that is dissipated in a one ohm resistor when the voltage, x, is a random variable, it is necessary to find the expected value of $y = x^2$. If x is a random variable with a probability density function $f_x(\alpha)$ and if $y = g(x)$, then the expected value of y is

$$E\{y\} = E\{g(x)\} = \int_{-\infty}^{\infty} g(\alpha) f_x(\alpha) d\alpha \tag{3.12}$$

For example, the expected value of $y = x^2$ is

$$E\{x^2\} = \int_{-\infty}^{\infty} \alpha^2 f_x(\alpha) d\alpha \tag{3.13}$$

and the expected value of $y = |x|$ is

$$E\{|x|\} = \int_{-\infty}^{\infty} |\alpha| f_x(\alpha) d\alpha = \int_{0}^{\infty} \alpha\big[f_x(\alpha) + f_x(-\alpha)\big] d\alpha$$

In order to motivate Eq. (3.12), consider the problem of computing the expected value of x^2 where x is the random variable defined in terms of the roll of a die. Assuming that the die is rolled N_T times and that the number k is rolled n_k times, it follows that the average squared value of the number rolled is approximately equal to the sample mean-square value,

$$\langle x^2 \rangle_{N_T} = \sum_{k=1}^{6} k^2 \frac{n_k}{N_T} \approx \sum_{k=1}^{6} k^2 \Pr\{x = k\} = E\{x^2\} \tag{3.14}$$

In terms of the probability density function, Eq. (3.14) becomes

$$\langle x^2 \rangle_{N_T} \approx \int_{-\infty}^{\infty} \alpha^2 f_x(\alpha) d\alpha$$

which is the same as Eq. (3.13).

The expected value of x^2 is an important statistical average referred to as the *mean-square value*. The mean-square value is frequently used as a measure for the quality of an estimate (e.g., see Section 3.2.6). For example, in developing an estimator for a random variable x, it is common to find the estimator $\hat{x}$ that minimizes the mean-square error,

$$\xi = E\big\{[x - \hat{x}]^2\big\}$$

A related statistical average is the *variance* which is the mean-square value of the random variable $y = x - E\{x\}$. Denoted either by $\mathrm{Var}\{x\}$ or σ_x^2, the variance is

$$\sigma_x^2 = E\Big\{\big[x - E\{x\}\big]^2\Big\} = \int_{-\infty}^{\infty} \big[\alpha - E\{x\}\big]^2 f_x(\alpha) d\alpha \tag{3.15}$$

The square root of the variance, σ_x, is the *standard deviation*. For complex random variables, the mean-square value is $E\{zz^*\} = E\{|z|^2\}$ and the variance is

$$\sigma_z^2 = E\Big\{\big[z - E\{z\}\big]\big[z^* - E\{z^*\}\big]\Big\} = E\big\{|z - E\{z\}|^2\big\}$$

It is clear from Eq. (3.12) that the expectation is a linear operator since, for any two random variables x and y and for any constants a and b,

$$E\{ax + by\} = aE\{x\} + bE\{y\}$$

Using the linearity of the expectation operator, the variance may be expressed as

$$\mathrm{Var}\{x\} = E\left\{[x - E\{x\}]^2\right\} = E\{x^2\} - E^2\{x\} \tag{3.16}$$

Therefore, it follows that if the mean of x is zero, then $\mathrm{Var}\{x\} = E\left\{x^2\right\}$.

3.2.3 Jointly Distributed Random Variables

Having considered the basic properties of a single random variable, we now consider what is involved in working with two or more random variables. Unlike the case for a single random variable, with two or more random variables it is necessary to consider the statistical dependencies that may exist between the random variables. For example, consider once again the simple experiment of flipping a fair coin. Suppose that the random variable x is assigned a value of one if the flip results in *Heads* and a value of minus one if the flip results is *Tails*. If two fair coins are tossed and the outcome of the toss of the first coin determines the value of a random variable, $x(1)$, and the outcome of the toss of the second coin determines the value of a second random variable, $x(2)$, then $\{x(1), x(2)\}$ represents a *pair* of random variables with

$$\begin{aligned} \Pr\{x(i) = 1\} &= 0.5 \\ \Pr\{x(i) = -1\} &= 0.5 \end{aligned}$$

for $i = 1, 2$. If the outcome of the first coin toss does not affect the outcome of the second, then any one of the four possible pairings $\{0, 0\}$, $\{0, 1\}$, $\{1, 0\}$, and $\{1, 1\}$ are equally likely to occur. Compare this, however, with the following experiment. Suppose that we have three coins: one coin which is fair, another coin which is unfair and is much more likely to result in *Heads*, and a third coin that is also unfair but much more likely to result in *Tails*. Just as in the first experiment, let $x(1)$ be a random variable whose value is determined by the flip of the fair coin. Let $x(2)$ be a second random variable whose value is also determined by the flip of a coin, but suppose that the coin that is used depends upon the outcome of the flip of the fair coin. Specifically, if the flip of the fair coin results in *Heads*, then the unfair coin that is more likely to show *Heads* is flipped. Otherwise the coin that is more likely to show *Tails* is flipped. With this definition of $x(1)$ and $x(2)$, we observe that there is a dependence between the two random variables in the sense that it is more likely for $x(1)$ and $x(2)$ to have the same value than it is for them to have different values. As a result, observing the value of $x(1)$ increases the probability of being able to correctly "predict" the value of $x(2)$. This example may be carried to the extreme by using a two-headed coin for the second coin and a two-tailed coin for the third coin. In this case, $x(1)$ and $x(2)$ will always have the same value. What this example illustrates is that the probabilistic description necessary to completely characterize two or more random variables requires that the statistical relationship between them be specified. This relationship is contained in the joint probability distribution and joint probability density functions, which are defined as follows. Given two random variables, $x(1)$ and $x(2)$, the *joint distribution function* is

$$\boxed{F_{x(1),x(2)}(\alpha_1, \alpha_2) = \Pr\big\{x(1) \le \alpha_1,\ x(2) \le \alpha_2\big\}} \tag{3.17}$$

which defines the probability that $x(1)$ is less than α_1 <u>and</u> $x(2)$ is less than α_2. As with the marginal density function, the *joint density function* for two random variables is the derivative of the distribution function

$$\boxed{f_{x(1),x(2)}(\alpha_1, \alpha_2) = \frac{\partial^2}{\partial\alpha_1 \partial\alpha_2} F_{x(1),x(2)}(\alpha_1,\ \alpha_2)} \tag{3.18}$$

The joint distribution and density functions are also used for the statistical characterization of complex random variables. For example, if $z = x + jy$ is a complex random variable and $\alpha = a + jb$ is a complex number, then the distribution function for z is given by the joint distribution function

$$F_z(\alpha) = \Pr\{x \le a, y \le b\} \tag{3.19}$$

For more than two random variables, the joint distribution and joint density functions are defined in a similar manner. For example, the joint distribution function for n random variables $x(1), x(2), \ldots, x(n)$ is

$$F_{x(1),\ldots,x(n)}(\alpha_1, \ldots, \alpha_n) = \Pr\{x(1) \le \alpha_1, \ldots, x(n) \le \alpha_n\}$$

and the joint density function is

$$f_{x(1),\ldots,x(n)}(\alpha_1, \ldots, \alpha_n) = \frac{\partial^n}{\partial\alpha_1 \cdots \partial\alpha_n} F_{x(1),\ldots,x(n)}(\alpha_1, \ldots, \alpha_n)$$

3.2.4 Joint Moments

Just as with a single random variable, ensemble averages provide an important and useful characterization of jointly distributed random variables. The two averages of primary importance are the *correlation* and the *covariance*. The *correlation*, denoted by r_{xy}, is the second-order joint moment

$$\boxed{r_{xy} = E\left\{xy^*\right\}} \tag{3.20}$$

where, in the case of complex random variables, y^* is used to denote the complex conjugate of y. An ensemble average related to the correlation is the covariance, c_{xy}, which is

$$\boxed{c_{xy} = \mathrm{Cov}(x, y) = E\Big\{\big[x - m_x\big]\big[y - m_y\big]^*\Big\} = E\left\{xy^*\right\} - m_x m_y^*} \tag{3.21}$$

where $m_x = E\{x\}$ and $m_y = E\{y\}$ are the means (expected values) of the random variables x and y, respectively. Clearly, if either x or y have zero mean, then the covariance is equal to the correlation.

Frequently, it is useful to normalize the covariance so that it is invariant to scaling. Such a normalized covariance is the *correlation coefficient*, which is

$$\boxed{\rho_{xy} = \frac{E\Big\{[x - m_x][y - m_y]^*\Big\}}{\sigma_x\sigma_y} = \frac{E\{xy^*\} - m_x m_y^*}{\sigma_x\sigma_y}} \tag{3.22}$$

For zero mean random variables the correlation coefficient becomes

$$\rho_{xy} = \frac{E\{xy^*\}}{\sigma_x\sigma_y}$$

Due to the normalization by $\sigma_x\sigma_y$, the correlation coefficient is bounded by one in magnitude,

$$\boxed{|\rho_{xy}| \le 1} \tag{3.23}$$

This property may be easily shown for the case of real random variables as follows. Assuming, without any loss in generality, that x and y have zero mean, it follows that if a is any real number, then $(ax - y)^2 > 0$ and,

$$E\{[ax - y]^2\} = a^2 E\{x^2\} - 2aE\{xy\} + E\{y^2\} \geq 0 \tag{3.24}$$

Since Eq. (3.24) is a quadratic that is nonnegative for all values of a, then the roots of this equation must either be complex or, if they are real, they must be equal to each other. In other words, the discriminant must be nonpositive

$$4E^2\{xy\} - 4E\{x^2\}E\{y^2\} \leq 0$$

or

$$E^2\{xy\} \leq E\{x^2\}E\{y^2\} \tag{3.25}$$

which is the *cosine inequality* [5]. Therefore, it follows that $\rho_{xy}^2 \leq 1$. Note that if there is a value of a for which Eq. (3.24) holds with equality, then $E\{[ax - y]^2\} = 0$ which implies that $y = ax$ (with probability one).

3.2.5 Independent, Uncorrelated, and Orthogonal Random Variables

There are many examples of random variables that arise in applications in which the value of one random variable does not depend on the value of another. For example, if $x(1)$ is a random variable that is defined to be equal to the number of sunspots that occur on the last day of the year and if $x(2)$ is the random variable that is defined to be equal to the amount of the U.S. national debt (in dollars) on the same day, then, in accordance with traditional economic theory, there should not be any relationship or dependence between the values of $x(1)$ and $x(2)$. As a result, the random variables are said to be statistically independent. A more precise formulation of the meaning of statistical independence is given in the following definition.

Definition. Two random variables x and y are said to be *statistically independent* if the joint probability density function is separable,

$$f_{x,y}(\alpha, \beta) = f_x(\alpha) f_y(\beta)$$

A weaker form of independence occurs when the joint second-order moment $E\{xy^*\}$ is separable,

$$E\{xy^*\} = E\{x\} E\{y^*\} \tag{3.26}$$

or

$$r_{xy} = m_x m_y^*$$

Two random variables that satisfy Eq. (3.26) are said to be *uncorrelated.* Note that since

$$c_{xy} = r_{xy} - m_x m_y^*$$

then two random variables x and y will be uncorrelated if their covariance is zero, $c_{xy} = 0$. Clearly, statistically independent random variables are always uncorrelated. The converse, however, is not true in general [5]. A useful property of uncorrelated random variables is the following.

Property. The variance of a sum of uncorrelated random variables is equal to the sum of the variances,

$$\text{Var}\{x + y\} = \text{Var}\{x\} + \text{Var}\{y\}$$

The correlation between random variables provides an important characterization of the statistical dependence that exists between them and it will play an important role in our study of random processes as well as in our discussions of signal modeling, Wiener filtering, and spectrum estimation. As a result, it will be important to understand what the correlation means, beyond simply how it is defined. In the following section we will establish, for example, the relationship between correlation and *linear predictability* when we look at the problem of linear mean-square estimation.

A property related to uncorrelatedness is *orthogonality*. Specifically, two random variables are said to be orthogonal if their correlation is zero,

$$r_{xy} = 0$$

Although orthogonal random variables need not necessarily be uncorrelated, zero mean random variables that are uncorrelated will always be orthogonal.

Since the primary interest in this book is with statistical averages rather than with probabilistic descriptions of random variables, our concern will be primarily with whether or not random variables are uncorrelated rather than whether or not they are statistically independent.

3.2.6 Linear Mean-Square Estimation

In this section we consider briefly the problem of estimating a random variable y in terms of an observation of another random variable x. This problem generally arises when y cannot be directly measured or observed so a related random variable is measured and used to estimate y. For example, we may wish to estimate the IQ (Intelligence Quotient) of an individual in terms of his or her performance on a standardized test. If we represent the IQ by the random variable y and the test performance by the random variable x, assuming that there is some relation (correlation) between x and y, the goal is to find the best estimate of y in terms of x. In mean-square estimation, an estimate $\hat{y}$ is to be found that minimizes the mean-square error

$$\xi = E\left\{(y - \hat{y})^2\right\}$$

Although the solution to this problem generally leads to a nonlinear estimator,[1] in many cases a linear estimator is preferred. In linear mean-square estimation, the estimator is constrained to be of the form

$$\hat{y} = ax + b$$

and the goal is to find the values for a and b that minimize the mean-square error

$$\xi = E\left\{(y - \hat{y})^2\right\} = E\left\{(y - ax - b)^2\right\} \tag{3.27}$$

[1]The optimum estimate is the conditional mean, $\hat{y} = E\left\{y|x\right\}$.

There are several advantages to using a linear estimator. The first is that the parameters a and b depend only on the second-order moments of x and y and not on the joint density functions. Second, the equations that must be solved for a and b are linear. Finally, for Gaussian random variables (see Section 3.2.7 for a discussion of Gaussian random variables) the optimum nonlinear mean-square estimate is linear. As we will see in later chapters, this problem is a special case of a more general mean-square estimation problem that arises in many different signal processing applications.

Solving the linear mean-square estimation problem may be accomplished by differentiating ξ with respect to a and b and setting the derivatives equal to zero as follows,

$$\frac{\partial \xi}{\partial a} = -2E\big\{(y - ax - b)x\big\} = -2E\,\{xy\} + 2aE\{x^2\} + 2bm_x = 0 \tag{3.28}$$

$$\frac{\partial \xi}{\partial b} = -2E\big\{y - ax - b\big\} = -2m_y + 2am_x + 2b = 0. \tag{3.29}$$

Before solving these equations for a and b, note that Eq. (3.28) says that

$$\boxed{E\big\{(y - \hat{y})x\big\} = E\{ex\} = 0} \tag{3.30}$$

where $e = y - \hat{y}$ is the *estimation error*. This relationship, known as the *orthogonality principle*, states that for the optimum linear predictor the estimation error will be orthogonal to the data x. The orthogonality principle is fundamental in mean-square estimation problems and will reappear many times in our discussions of signal modeling and Wiener filtering.

Solving Eqs. (3.28) and (3.29) for a and b we find

$$a = \frac{E\big\{xy\big\} - m_x m_y}{\sigma_x^2} \tag{3.31}$$

$$b = \frac{E\big\{x^2\big\}m_y - E\big\{xy\big\}m_x}{\sigma_x^2} \tag{3.32}$$

From Eq. (3.31) it follows that

$$E\{xy\} = a\sigma_x^2 + m_x m_y \tag{3.33}$$

which, when substituted into Eq. (3.32), gives the following expression for b

$$b = \frac{E\,\big\{x^2\big\}\,m_y - \big[a\sigma_x^2 + m_x m_y\big]\,m_x}{\sigma_x^2} = m_y - am_x$$

where we have used the relation $\sigma_x^2 = E\big\{x^2\big\} - m_x^2$. As a result, the estimate for y may be written as

$$\hat{y} = ax + (m_y - am_x) = a(x - m_x) + m_y \tag{3.34}$$

where

$$a = \frac{E\{xy\} - m_x m_y}{\sigma_x^2} = \frac{\rho_{xy}\sigma_x\sigma_y}{\sigma_x^2} = \rho_{xy}\frac{\sigma_y}{\sigma_x} \tag{3.35}$$

Substituting Eq. (3.35) into Eq. (3.34) we have

$$\boxed{\hat{y} = \rho_{xy}\frac{\sigma_y}{\sigma_x}\big(x - m_x\big) + m_y} \tag{3.36}$$

Having found the optimum linear estimator for y, we may now evaluate the minimum mean-square error. With

$$\begin{aligned} E\{[y-\hat{y}]^2\} &= E\{[(y-m_y)-a(x-m_x)]^2\} \\ &= \sigma_y^2 + a^2\sigma_x^2 - 2aE\{(x-m_x)(y-m_y)\} \\ &= \sigma_y^2 + a^2\sigma_x^2 - 2a\big[E\{xy\} - m_x m_y\big] \end{aligned} \tag{3.37}$$

substituting Eq. (3.33) into Eq. (3.37) we have

$$E\Big\{[y-\hat{y}]^2\Big\} = \sigma_y^2 - a^2\sigma_x^2$$

Finally, using expression (3.35) for a, the minimum mean-square error becomes

$$\boxed{\xi_{\min} = \sigma_y^2(1-\rho_{xy}^2)} \tag{3.38}$$

which is plotted in Fig. 3.3 as a function of ρ_{xy}. Note that since the minimum mean-square error $\xi_{\min}$ must be nonnegative, Eq. (3.38) establishes again the property that the correlation coefficient can be no larger than one in absolute value.

We now look at some special cases of the linear mean-square estimation problem. First note that if x and y are uncorrelated, then $a = 0$ and $b = E\{y\}$. Therefore, the estimate for y is

$$\hat{y} = E\{y\} \tag{3.39}$$

and the minimum mean-square error is

$$\xi_{\min} = E\Big\{(y-\hat{y})^2\Big\} = E\Big\{(y-E(y))^2\Big\} = \sigma_y^2$$

Thus, x is not used in the estimation of y, so that knowing the value of the random variable x does not improve the accuracy of the estimate of y. Another special case occurs when $|\rho_{xy}| = 1$. In this case, the minimum mean-square error is zero,

$$\xi_{\min} = E\Big\{(y-\hat{y})^2\Big\} = 0$$

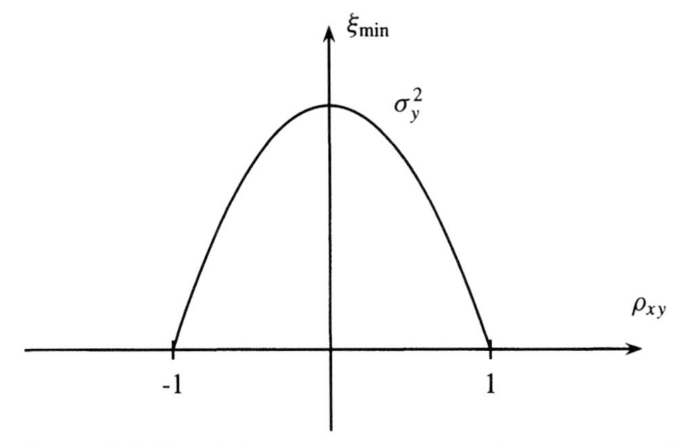

Figure 3.3 *The minimum mean-square error in the estimate of y from x, plotted as a function of the correlation coefficient, ρ_{xy}.*

and it follows that[2]

$$y = ax + b$$

Thus, when the magnitude of the correlation coefficient is equal to one, the random variables x and y are linearly related to each other. Based on these observations we see that the correlation coefficient provides a measure of the *linear predictability* between random variables. The closer $|\rho_{xy}|$ is to 1, the smaller the mean-square error in the estimation of y using a linear estimator. This property will be seen again in later chapters when we consider the more general problem of estimating a random variable, y, in terms of a linear combination of two or more random variables.

3.2.7 Gaussian Random Variables

Gaussian random variables play a central role in probability theory. A random variable x is said to be Gaussian if its density function is of the form

$$f_x(\alpha) = \frac{1}{\sigma_x\sqrt{2\pi}} \exp\left\{-\frac{(\alpha - m_x)^2}{2\sigma_x^2}\right\}$$

where m_x and σ_x^2 are the mean and the variance of x, respectively. Note that the density function of a Gaussian random variable is completely defined once the mean and the variance are specified. Two random variables x and y are said to be *jointly Gaussian* if their joint density function is

$$f_{x,y}(\alpha, \beta) = A \exp\left\{-\frac{1}{2(1 - \rho_{xy}^2)}\left[\frac{(\alpha - m_x)^2}{\sigma_x^2} - 2\rho_{xy}\frac{(\alpha - m_x)(\beta - m_y)}{\sigma_x\sigma_y} + \frac{(\beta - m_y)^2}{\sigma_y^2}\right]\right\}$$

where

$$A = \frac{1}{2\pi\sigma_x\sigma_y\sqrt{1 - \rho_{xy}^2}}$$

Here, m_x and m_y are the means and σ_x^2 and σ_y^2 the variances of the Gaussian random variables x and y, respectively. The correlation between x and y is ρ_{xy}. As with a single Gaussian random variable, the joint density function is completely specified once the means, variances, and correlation are known.

Gaussian random variables have a number of interesting and important properties. A few of these properties are as follows.

Property 1. If x and y are jointly Gaussian, then for any constants a and b the random variable

$$z = ax + by$$

is Gaussian with mean

$$m_z = am_x + bm_y$$

[2]Strictly speaking, y is said to be equal to $ax + b$ in a *mean-square sense* since y may be different from $ax + b$ as long as the probability of this happening is equal to zero [5].

and variance

$$\sigma_z^2 = a^2\sigma_x^2 + b^2\sigma_y^2 + 2ab\sigma_x\sigma_y\rho_{xy}$$

Property 2. If two jointly Gaussian random variables are uncorrelated, $\rho_{xy} = 0$, then they are statistically independent, $f_{xy}(\alpha, \beta) = f_x(\alpha)f_y(\beta)$.

Property 3. If x and y are jointly Gaussian random variables then the optimum nonlinear estimator for y

$$\hat{y} = g(x)$$

that minimizes the mean-square error

$$\xi = E\left\{[y - g(x)]^2\right\}$$

is a linear estimator

$$\hat{y} = ax + b$$

Property 4. If x is Gaussian with zero mean then

$$E\{x^n\} = \begin{cases} 1 \times 3 \times 5 \times \cdots \times (n-1)\sigma_x^n & ; \quad n \text{ even} \\ 0 & ; \quad n \text{ odd} \end{cases}$$

For other interesting and useful properties of Gaussian random variable, the reader may consult any one of a number of standard probability textbooks [3, 5].

3.2.8 Parameter Estimation: Bias and Consistency

There are many examples in signal processing and other scientific fields where it is necessary to estimate the value of an unknown parameter from a set of observations of a random variable. For example, given a set of observations from a Gaussian distribution, estimating the mean or variance from these observations is a problem of parameter estimation. As a specific application, recall that in linear mean-square estimation, estimating the value of a random variable y from an observation of a related random variable, x, the coefficients a and b in the estimator $\hat{y} = ax + b$ depend upon the mean and variance of x and y as well as on the correlation. If these statistical averages are unknown, then it is necessary to estimate these parameters from a set of observations of x and y. Since any estimate will be a function of the observations, the estimates themselves will be random variables. Therefore, in classifying the effectiveness of a particular estimator, it is important to be able to characterize its statistical properties. The statistical properties of interest include the bias and the variance which are described as follows.

Consider the problem of estimating the value of a parameter, θ, from a sequence of random variables, x_n, for $n = 1, 2, \ldots, N$. Since the estimate is a function of N random variables, we will denote it by $\hat{\theta}_N$. In general, we would like the estimate to be equal, *on the average*, to the true value. The difference between the expected value of the estimate and the actual value, θ, is called the *bias* and will be denoted by B,

$$B = \theta - E\{\hat{\theta}_N\}$$

If the bias is zero, then the expected value of the estimate is equal to the true value

$$E\{\hat{\theta}_N\} = \theta$$

and the estimate is said to be *unbiased*. If $B \neq 0$, then $\hat{\theta}$ is said to be *biased*. If an estimate is biased but the bias goes to zero as the number of observations, N, goes to infinity,

$$\lim_{N\to\infty} E\{\hat{\theta}_N\} = \theta$$

then the estimate is said to be *asymptotically unbiased*. In general, it is desirable that an estimator be either unbiased or asymptotically unbiased. However, as the following example illustrates, the bias is not the only statistical measure of importance.

Example 3.2.2 *An Unbiased Estimator*

Let x be the random variable defined on the coin flipping experiment, with $x = 1$ if the outcome is heads and $x = -1$ if the outcome is tails. If the coin is unfair so that the probability of flipping *Heads* is p and the probability of flipping *Tails* is $1 - p$, then the mean of x is

$$m_x = E\{x\} = p - (1 - p) = 2p - 1$$

However, suppose the value for p is unknown and that the mean of x is to be estimated. Flipping the coin N times and denoting the resulting values for x by x_i, consider the following estimator for m_x,

$$\hat{m}_x = x_N$$

Since the expected value of $\hat{m}_x$ is

$$E\{\hat{m}_x\} = E\{x_N\} = 2p - 1$$

then this estimator is *unbiased*. However, it is clear that $\hat{m}_x = x_N$ is not a very good estimator of the mean. The reason is that $\hat{m}_x$ will either be equal to one, with a probability of p, or it will be equal to minus one, with a probability of $1 - p$. Therefore, the accuracy of the estimate does not improve as the number of observations N increases. In fact, note that the variance of the estimate,

$$\text{Var}\{\hat{m}_x\} = \text{Var}\{x_N\} = 4p(1 - p)$$

does not decrease with N.

In order for the estimate of a parameter to converge, in some sense, to its true value, it is necessary that the variance of the estimate go to zero as the number of observations goes to infinity,

$$\lim_{N\to\infty} \text{Var}\{\hat{\theta}_N\} = \lim_{N\to\infty} E\left\{\left|\hat{\theta}_N - E\{\hat{\theta}_N\}\right|^2\right\} = 0 \tag{3.40}$$

If $\hat{\theta}_N$ is unbiased, $E\{\hat{\theta}_N\} = \theta$, it follows from the Tchebycheff inequality [5] that, for any $\epsilon > 0$,

$$\Pr\left\{|\hat{\theta}_N - \theta| \geq \epsilon\right\} \leq \frac{\text{Var}\{\hat{\theta}_N\}}{\epsilon^2}$$

Therefore, if the variance goes to zero as $N \to \infty$, then the probability that $\hat{\theta}_N$ differs by more than ϵ from the true value will go to zero. In this case, $\hat{\theta}_N$ is said to *converge to θ with probability one*.

Another form of convergence, stronger than convergence with probability one, is *mean-square convergence*. An estimate $\hat{\theta}_N$ is said to converge to θ in the mean-square sense if

$$\lim_{N\to\infty} E\left\{\left|\hat{\theta}_N - \theta\right|^2\right\} = 0 \tag{3.41}$$

Note that for an unbiased estimator this is equivalent to the condition given in Eq. (3.40) that the variance of the estimate goes to zero.

Finally, an estimate is said to be *consistent* if it converges, in some sense, to the true value of the parameter. Depending upon the form of convergence that is used one has different definitions of consistency [8]. Here we will say that an estimate is consistent if it is asymptotically unbiased and has a variance that goes to zero as N goes to infinity.

Example 3.2.3 ***The Sample Mean***

Let x be a random variable with a mean m_x and variance σ_x^2. Given N uncorrelated observations of x, denoted by x_n, suppose that an estimator for m_x is formed as follows,

$$\hat{m}_x = \frac{1}{N}\sum_{n=1}^{N} x_n$$

This estimate, known as the *sample mean*, has an expected value of

$$E\{\hat{m}_x\} = \frac{1}{N}\sum_{n=1}^{N} E\{x_n\} = m_x$$

Therefore, the sample mean is an *unbiased* estimator. In addition, the variance of the estimate is

$$\text{Var}\{\hat{m}_x\} = \frac{1}{N^2}\sum_{n=1}^{N} \text{Var}\{x\} = \frac{\sigma_x^2}{N}$$

Since the variance goes to zero as $N \to \infty$, the sample mean is a *consistent* estimator.

3.3 RANDOM PROCESSES

In this section, we consider the characterization and analysis of *discrete-time random processes*. Since a discrete-time random process is simply an indexed sequence of random variables, with a basic understanding of random variables the extension of the concepts from Section 3.2 to discrete-time random processes is straightforward.

3.3.1 Definitions

Just as a random variable may be thought of as a mapping from a sample space of a given experiment into the set of real or complex numbers, a discrete-time random process is a mapping from the sample space Ω into the set of discrete-time signals $x(n)$. Thus, a discrete-time random process is a collection, or *ensemble*, of discrete-time signals. A simple example of a discrete-time random process is the following. Consider the experiment of rolling a fair die and let the outcome of a roll of the die be assigned to the random variable A. Thus, A is allowed to assume any integer value between one and six, each with equal probability. With

$$x(n) = A\cos(n\omega_0) \tag{3.42}$$

a random process has been created that consists of an ensemble of six different and equally probable discrete-time signals. A more complicated process may be constructed by considering the experiment of repeatedly flipping a fair coin. By setting the value of $x(n)$ at time n equal to one if *Heads* is flipped at time n and setting $x(n)$ equal to minus one if *Tails* is flipped, then $x(n)$ becomes a discrete-time random process consisting of a random sequence of 1's and -1's. If the flip of the coin at time n in no way affects the outcome of the flip of the coin at any other time (the coin flips are independent), then $x(n)$ is known as a *Bernoulli process*. A sample function of a Bernoulli process is shown in Fig. 3.4*a*.

Given a random process $x(n)$, other processes may be generated by transforming $x(n)$ by means of some mathematical operation. A particularly important and useful transformation is that of linear filtering. For example, given a Bernoulli process $x(n)$, a new process may be generated by filtering $x(n)$ with the first-order recursive filter defined by the difference equation

$$y(n) = 0.5y(n-1) + x(n)$$

An example of a Bernoulli process filtered in this way is shown in Fig. 3.4*b*.

As a final example of a discrete-time random process, consider the experiment of spinning an infinite resolution roulette wheel for which any number in the interval [0, 1] is equally likely to occur. If the number obtained with the spin of the wheel is assigned to the random variable x, let the infinite binary expansion of x be

$$x = \sum_{n=0}^{\infty} x(n)2^{-n}$$

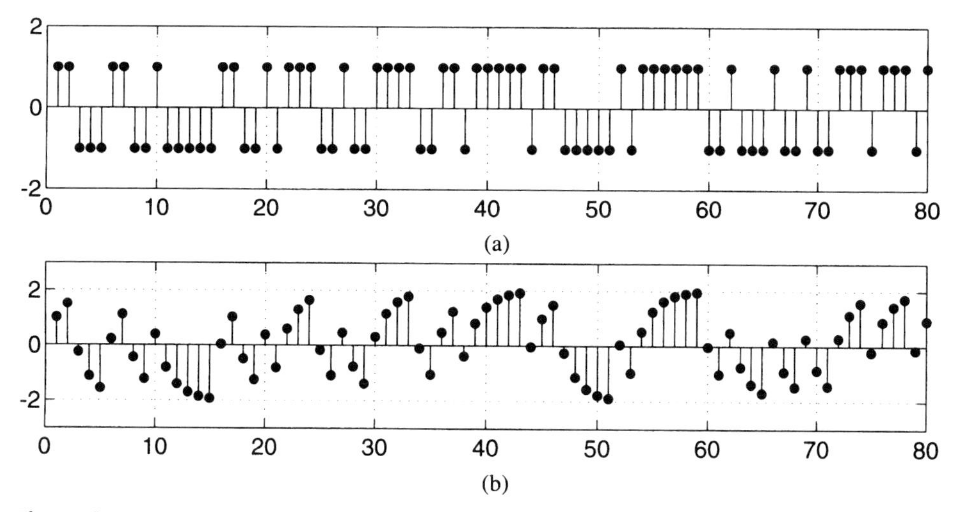

Figure 3.4 *Sample realization of two different random processes: (a) a Bernoulli process $x(n)$ with* $\Pr\{x(n) = 1\} = 0.5$ *and (b) a filtered Bernoulli process $y(n)$ generated by the difference equation* $y(n) = 0.5y(n-1) + x(n)$.

where $x(n)$ is either equal to zero or one for all $n \geq 0$. The sequence of binary coefficients, $x(n)$, forms a discrete-time random process.

As with random variables, a discrete-time random process is a mapping from a sample space of experimental outcomes to a set (ensemble) of discrete-time signals. For example, as shown in Fig. 3.5, to each experimental outcome ω_i in the sample space Ω, there is a corresponding discrete-time signal, $x_i(n)$. However, in the description and analysis of random processes, another viewpoint is often more convenient. Specifically, note that for a particular value of n, say $n = n_0$, the signal value $x(n_0)$ is a random variable that is defined on the sample space Ω. That is to say, for each $\omega \in \Omega$ there is a corresponding value of $x(n_0)$. Therefore, a random process may also be viewed as an indexed sequence of random variables,

$$\ldots, x(-2), x(-1), x(0), x(1), x(2), \ldots$$

where each random variable in the sequence has an underlying probability distribution function

$$F_{x(n)}(\alpha) = \Pr\{x(n) \leq \alpha\}$$

and probability density function

$$f_{x(n)}(\alpha) = \frac{d}{d\alpha} F_{x(n)}(\alpha) \tag{3.43}$$

However, in order to form a complete statistical characterization of a random process, in addition to the first-order density function, the joint probability density or distribution functions that define how collections of random variables are related to each other must be defined. Specifically, what is required are the joint distribution functions

$$F_{x(n_1),\ldots,x(n_k)}(\alpha_1, \ldots, \alpha_k) = \Pr\big\{x(n_1) \leq \alpha_1, \ldots, x(n_k) \leq \alpha_k\big\}$$

(or the corresponding joint density functions) for any collection of random variables $x(n_i)$. Depending upon the specific form of these joint distribution functions, significantly different

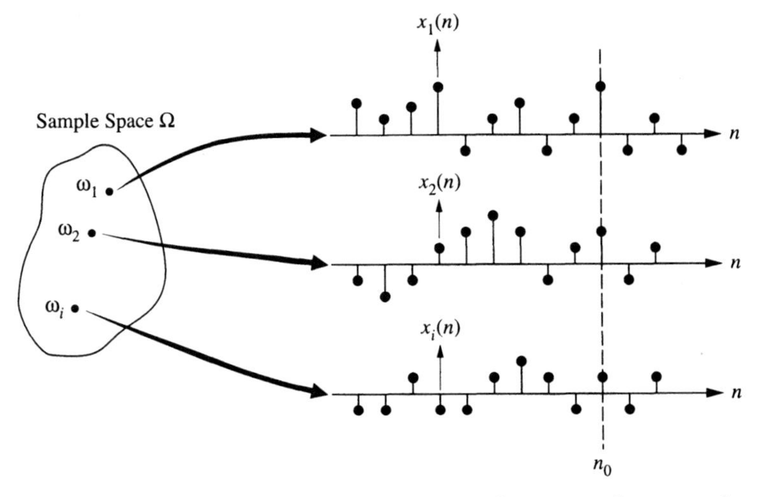

Figure 3.5 *The generation of a random process from a sample space point of view.*

types of random process may be generated. Consider, for example, a random process formed from a sequence of Gaussian random variables $x(n)$. If the random variables $x(n)$ are uncorrelated, then the process, known as white Gaussian noise, is a very random, noiselike sequence. If, on the other hand, $x(n) = \alpha$ where α is a Gaussian random variable, then each random process in the ensemble is equal to a constant for all n. Thus, although both processes have the same first-order statistics, they are significantly different as a result of the differences in their higher-order density functions.

3.3.2 Ensemble Averages

Since a discrete-time random process is an indexed sequence of random variables, we may calculate the mean of each of these random variables and generate the (deterministic) sequence

$$\boxed{m_x(n) = E\{x(n)\}} \tag{3.44}$$

known as the *mean* of the process. Similarly, computing the variance of each random variable in the sequence

$$\boxed{\sigma_x^2(n) = E\left\{|x(n) - m_x(n)|^2\right\}} \tag{3.45}$$

defines the *variance* of the process. These first-order statistics represent ensemble averages and both, in general, depend upon n. Whereas the mean defines the average value of the process as a function of n, the variance represents the average squared deviation of the process away from the mean.

Two additional ensemble averages that are important in the study of random processes are the *autocovariance*

$$\boxed{c_x(k,l) = E\big\{[x(k) - m_x(k)]\,[x(l) - m_x(l)]^*\big\}} \tag{3.46}$$

and the *autocorrelation*

$$\boxed{r_x(k,l) = E\big\{x(k)x^*(l)\big\}} \tag{3.47}$$

relating the random variables $x(k)$ and $x(l)$. Note that if $k = l$ then the autocovariance function reduces to the variance,

$$c_x(k,k) = \sigma_x^2(k)$$

Note also that if the product in Eq. (3.46) is expanded, then it follows that the autocovariance and autocorrelation sequences and related by

$$c_x(k,l) = r_x(k,l) - m_x(k)m_x^*(l)$$

Thus, for zero mean random processes the autocovariance and autocorrelation are equal. In this book, for convenience, unless stated otherwise, **random processes will always be assumed to have zero mean** so that the autocovariance and autocorrelation sequences may be used interchangeably. This assumption results in no loss of generality since, for any process $x(n)$ that has a nonzero mean, a zero mean process $y(n)$ may always be formed by subtracting the mean from $x(n)$ as follows

$$y(n) = x(n) - m_x(n)$$

As in the case of random variables, the autocorrelation and autocovariance functions provide information about the degree of linear dependence between two random variables. For example, if $c_x(k, l) = 0$ for $k \neq l$, then the random variables $x(k)$ and $x(l)$ are uncorrelated and knowledge of one does not help in the estimation of the other using a linear estimator.

Example 3.3.1 *The Harmonic Process*

An important random process that is found in applications such as radar and sonar signal processing is the harmonic process. An example of a real-valued harmonic process is the random phase sinusoid, which is defined by

$$x(n) = A \sin(n\omega_0 + \phi)$$

where A and ω_0 are fixed constants and ϕ is a random variable that is uniformly distributed over the interval $-\pi$ to π, i.e., the probability density function for ϕ is

$$f_\phi(\alpha) = \begin{cases} (2\pi)^{-1} & ; \quad -\pi \le \alpha < \pi \\ 0 & ; \quad \text{otherwise} \end{cases}$$

The mean of this process,

$$m_x(n) = E\{x(n)\} = E\{A \sin(n\omega_0 + \phi)\}$$

may be found using Eq. (3.12) as follows

$$m_x(n) = \int_{-\infty}^{\infty} A \sin(n\omega_0 + \alpha) f_\phi(\alpha) d\alpha = \int_{-\pi}^{\pi} \frac{1}{2\pi} A \sin(n\omega_0 + \alpha) d\alpha = 0$$

Thus, $x(n)$ is a zero mean process. The autocorrelation of $x(n)$ may be determined in a similar fashion. Specifically, with

$$r_x(k, l) = E\big\{x(k)x^*(l)\big\} = E\{A \sin(k\omega_0 + \phi) A \sin(l\omega_0 + \phi)\}$$

using the trigonometric identity

$$2 \sin A \sin B = \cos(A - B) - \cos(A + B)$$

gives

$$r_x(k, l) = \frac{1}{2} A^2 E\Big\{\cos\big[(k - l)\omega_0\big]\Big\} - \frac{1}{2} A^2 E\Big\{\cos\big[(k + l)\omega_0 + 2\phi\big]\Big\}$$

Note that the first term is the expected value of a constant and the second term is equal to zero. Therefore,

$$r_x(k, l) = \frac{1}{2} A^2 \cos\big[(k - l)\omega_0\big]$$

As another example, consider the complex harmonic process

$$x(n) = A e^{j(n\omega_0 + \phi)}$$

where, as with the random phase sinusoid, ϕ is a random variable that is uniformly distributed between $-\pi$ and π. The mean of this process is zero

$$m_x(n) = E\big\{A e^{j(n\omega_0 + \phi)}\big\} = 0$$

and the autocorrelation is

$$\begin{aligned} r_x(k,l) &= E\{x(k)x^*(l)\} = E\{Ae^{j(k\omega_0+\phi)}A^*e^{-j(l\omega_0+\phi)}\} \\ &= |A|^2E\{e^{j(k-l)\omega_0}\} = |A|^2e^{j(k-l)\omega_0} \end{aligned}$$

Note that, for both processes, the mean is a constant and the autocorrelation $r_x(k,l)$ is a function only of the difference between k and l

$$r_x(k,l) = r_x(k-l,0)$$

Thus, the first- and second-order statistics do not depend upon an absolute time origin. That is, the mean and autocorrelation do not change if the process is shifted in time. As discussed in Section 3.3.4, the harmonic process is an example of a *wide-sense stationary process.*

The autocovariance and the autocorrelation sequences provide information about the statistical relationship between two random variables that are derived from the same process, e.g., $x(k)$ and $x(l)$. In applications involving more than one random process it is often of interest to determine the covariance or the correlation between a random variable in one process, $x(k)$, and a random variable in another, $y(l)$. Given two random processes, $x(n)$ and $y(n)$, the *cross-covariance* is defined by

$$\boxed{c_{xy}(k,l) = E\Big\{\big[x(k)-m_x(k)\big]\big[y(l)-m_y(l)\big]^*\Big\}} \tag{3.48}$$

and the *cross-correlation* by

$$\boxed{r_{xy}(k,l) = E\{x(k)y^*(l)\}} \tag{3.49}$$

These two functions satisfy the relation

$$c_{xy}(k,l) = r_{xy}(k,l) - m_x(k)m_y^*(l)$$

Just as a pair of random variables are said to be uncorrelated if $c_{xy} = 0$, two random processes $x(n)$ and $y(n)$ are said to be *uncorrelated* if

$$c_{xy}(k,l) = 0$$

for all k and l or, equivalently, if

$$r_{xy}(k,l) = m_x(k)m_y^*(l) \tag{3.50}$$

Two random processes $x(n)$ and $y(n)$ are said to be *orthogonal* if their cross-correlation is zero

$$r_{xy}(k,l) = 0$$

Although orthogonal random processes are not necessarily uncorrelated, zero mean processes that are uncorrelated are orthogonal.

Example 3.3.2 *Cross-correlation*

Consider the pair of processes, $x(n)$ and $y(n)$, where $y(n)$ is related to $x(n)$ as follows

$$y(n) = x(n-1)$$

The cross-correlation between $x(n)$ and $y(n)$ is

$$r_{xy}(k,l) = E\{x(k)y^*(l)\} = E\{x(k)x^*(l-1)\} = r_x(k, l-1)$$

On the other hand, if $y(n)$ is equal to the convolution of $x(n)$ with a deterministic sequence $h(n)$, such as the unit sample response of a linear shift-invariant filter,

$$y(n) = \sum_{m=-\infty}^{\infty} h(m)x(n-m)$$

then the cross-correlation between $x(n)$ and $y(n)$ is

$$r_{xy}(k,l) = E\{x(k)y^*(l)\} = E\left\{x(k)\sum_{m=-\infty}^{\infty} h^*(m)x^*(l-m)\right\} = \sum_{m=-\infty}^{\infty} h^*(m)r_x(k,l-m)$$

In any practical data acquisition device, noise or measurement errors are invariably introduced into the recorded data. In many applications, this noise is modeled as additive so that if $x(n)$ denotes the "signal" and $w(n)$ the "noise," the recorded signal is

$$y(n) = x(n) + w(n)$$

Often, this additive noise is assumed to have zero mean and to be uncorrelated with the signal. In this case, the autocorrelation of the measured data, $y(n)$, is the sum of the autocorrelations of $x(n)$ and $w(n)$. Specifically, note that since

$$\begin{aligned} r_y(k,l) &= E\{y(k)y^*(l)\} = E\Big\{\big[x(k)+w(k)\big]\big[x(l)+w(l)\big]^*\Big\} \\ &= E\{x(k)x^*(l)\} + E\{w(k)w^*(l)\} + E\{x(k)w^*(l)\} + E\{w(k)x^*(l)\} \end{aligned} \tag{3.51}$$

if $x(n)$ and $w(n)$ are uncorrelated, then

$$E\{x(k)w^*(l)\} = E\{w(k)x^*(l)\} = 0$$

and it follows that

$$r_y(k,l) = r_x(k,l) + r_w(k,l)$$

This fundamental result is summarized in the following property.

Property. If two random processes $x(n)$ and $y(n)$ are uncorrelated, then the autocorrelation of the sum

$$z(n) = x(n) + y(n)$$

is equal to the sum of the autocorrelations of $x(n)$ and $y(n)$,

$$r_z(k,l) = r_x(k,l) + r_y(k,l)$$

Example 3.3.3 ***Autocorrelation of a Sum of Processes***

Example 3.3.1 considered a harmonic process consisting of a single sinusoid. Now consider a process consisting of a sum of M sinusoids in additive noise

$$x(n) = \sum_{m=1}^{M} A_m \sin(n\omega_m + \phi_m) + v(n)$$

where A_m and ω_m are constants and ϕ_m are uncorrelated random variables that are uniformly distributed between $-\pi$ and π. Since the random variables ϕ_m are uncorrelated it follows that each sinusoidal process is uncorrelated with the others. Assuming that the noise is uncorrelated with the sinusoids, using the results of Example 3.3.1 it follows that

$$r_x(k,l) = \frac{1}{2}\sum_{m=1}^{M} A_m^2 \cos\big[(k-l)\omega_m\big] + r_v(k,l)$$

where $r_v(k,l)$ is the autocorrelation of the additive noise.

3.3.3 Gaussian Processes

In Section 3.2.7, we defined what it means for a pair of random variables to be jointly Gaussian. This definition may be extended to a collection of n random variables as follows. With $\mathbf{x} = \big[x_1, \ x_2, \dots, x_n\big]^T$ a vector of n real-valued random variables, $\mathbf{x}$ is said to be a Gaussian random vector and the random variables x_i are said to be jointly Gaussian if the joint probability density function of the n random variables x_i is

$$f_x(\mathbf{x}) = \frac{1}{(2\pi)^{n/2}|\mathbf{C}_x|^{1/2}} \exp\Big\{-\frac{1}{2}(\mathbf{x}-\mathbf{m}_x)^T \mathbf{C}_x^{-1}(\mathbf{x}-\mathbf{m}_x)\Big\}$$

where $\mathbf{m}_x = \big[m_1, \ m_2, \dots, m_n\big]^T$ is a vector containing the means of x_i,

$$m_i = E\{x_i\}$$

$\mathbf{C}_x$ is a symmetric positive definite matrix with elements c_{ij} that are the covariances between x_i and x_j,

$$c_{ij} = E\Big\{(x_i - m_i)(x_j - m_j)\Big\}$$

and $|\mathbf{C}_x|$ is the determinant of the covariance matrix.

A discrete-time random process $x(n)$ is said to be *Gaussian* if every finite collection of samples of $x(n)$ are jointly Gaussian. Note that a Gaussian random process is completely defined once the mean vector and covariance matrix are known. Gaussian processes are of great interest both from a theoretical as well as a practical point of view. Many of the processes that are found in applications, for example, are Gaussian, or approximately Gaussian as a result of the Central Limit theorem [5].

3.3.4 Stationary Processes

In signal processing applications, the statistics or ensemble averages of a random process are often independent of time. For example, quantization noise that results from roundoff errors in a fixed point digital signal processor typically has a constant mean and a constant variance whenever the input signal is "sufficiently complex." In addition, it is often assumed that the quantization noise has first- and second-order probability density functions that are independent of time. These conditions are examples of "statistical time-invariance" or *stationarity*. In this section several different types of stationarity are defined. As we will see in Section 3.3.6, a stationarity assumption is important for the estimation of ensemble averages.

If the first-order density function of a random process $x(n)$ is independent of time, i.e.,

$$f_{x(n)}(\alpha) = f_{x(n+k)}(\alpha)$$

for all k, then the process is said to be *first-order stationary*. For a first-order stationary process, the first-order statistics will be independent of time. For example, the mean of the process will be constant

$$m_x(n) = m_x$$

and the same will be true of the variance, $\sigma_x^2(n) = \sigma_x^2$. In a similar manner, a process is said to be *second-order stationary* if the second-order joint density function $f_{x(n_1),x(n_2)}(\alpha_1, \alpha_2)$ depends only on the difference, $n_2 - n_1$, and not on the individual times n_1 and n_2. Equivalently, the process $x(n)$ is second-order stationary if, for any k, the processes $x(n)$ and $x(n+k)$ have the same second-order joint density function

$$f_{x(n_1),x(n_2)}(\alpha_1, \alpha_2) = f_{x(n_1+k),x(n_2+k)}(\alpha_1, \alpha_2)$$

If a process is second-order stationary, then it will also be first-order stationary. In addition, second-order stationary processes have second-order statistics that are invariant to a time shift of the process. For example, it follows that the autocorrelation sequence has the property

$$\begin{aligned} r_x(k,l) &= \int_{-\infty}^{\infty} \alpha\beta f_{x(k),x(l)}(\alpha, \beta) d\alpha\, d\beta \\ &= \int_{-\infty}^{\infty} \alpha\beta f_{x(k+n),x(l+n)}(\alpha, \beta) d\alpha\, d\beta = r_x(k+n, l+n) \end{aligned} \tag{3.52}$$

Therefore, the correlation between the random variables $x(k)$ and $x(l)$ depends only on the difference, $k - l$, separating the two random variables in time

$$r_x(k,l) = r_x(k-l, 0)$$

This difference, $k - l$, is called the *lag*, and with a slight abuse in notation, we will drop the second argument and write $r_x(k,l)$ simply as a function of the lag,

$$r_x(k,l) \equiv r_x(k-l)$$

Continuing to higher-order joint density functions, a process is said to be *stationary of order L* if the processes $x(n)$ and $x(n+k)$ have the same Lth-order joint density functions. Finally, a process that is stationary for all orders $L > 0$ is said to be *stationary in the strict sense*.

Since it is primarily the mean and autocorrelation of a process that will be of interest in this book, and not probability density functions, we will be concerned with another form of stationarity known as *wide-sense stationary* (WSS), which is defined as follows.

Wide Sense Stationarity. A random process $x(n)$ is said to be *wide-sense stationary* if the following three conditions are satisfied:

1. The mean of the process is a constant, $m_x(n) = m_x$.
2. The autocorrelation $r_x(k,l)$ depends only on the difference, $k - l$.
3. The variance of the process is finite, $c_x(0) < \infty$.

Since constraints are placed on ensemble averages rather than on density functions, wide-sense stationarity is a weaker constraint than second-order stationarity. However, in the case of a Gaussian process, wide-sense stationarity is equivalent to strict-sense stationarity.

This is a consequence of the fact that a Gaussian random process is completely defined in terms of the mean and covariance. Some examples of WSS random processes include the Bernoulli process (Fig. 3.4*a*) and the random phase sinusoid considered in Example 3.3.1. An example of a process that is not WSS is the sinusoidal process defined in Eq. (3.42).

In the case of two or more processes, similar definitions exist for *joint stationarity*. For example, two processes $x(n)$ and $y(n)$ are said to be *jointly wide-sense stationary* if $x(n)$ and $y(n)$ are wide-sense stationary and if the cross-correlation $r_{xy}(k,l)$ depends only on the difference, $k-l$,

$$r_{xy}(k,l) = r_{xy}(k+n, l+n)$$

Again, for jointly WSS processes, we will write the cross-correlation as a function only of the lag, $k-l$, as follows

$$r_{xy}(k-l) = E\{x(k)y^*(l)\}$$

The autocorrelation sequence of a WSS process has a number of useful and important properties, some of which are presented as follows.

***Property 1**—Symmetry.* The autocorrelation sequence of a WSS random process is a conjugate symmetric function of k,

$$r_x(k) = r_x^*(-k)$$

For a real process, the autocorrelation sequence is symmetric

$$r_x(k) = r_x(-k)$$

This property follows directly from the definition of the autocorrelation function,

$$r_x(k) = E\left\{x(n+k)x^*(n)\right\} = E\left\{x^*(n)x(n+k)\right\} = r_x^*(-k)$$

The next property relates the value of the autocorrelation sequence at lag $k=0$ to the mean-square value of the process.

***Property 2**—Mean-square value.* The autocorrelation sequence of a WSS process at lag $k=0$ is equal to the mean-square value of the process

$$r_x(0) = E\left\{|x(n)|^2\right\} \geq 0$$

As with property 1, this property follows directly from the definition of the autocorrelation sequence. The next property places an upper bound on the autocorrelation sequence in terms of its value at lag $k=0$.

***Property 3**—Maximum value.* The magnitude of the autocorrelation sequence of a WSS random process at lag k is upper bounded by its value at lag $k=0$,

$$r_x(0) \geq |r_x(k)|$$

This property may be established as follows. Let a be an arbitrary complex number. Since

$$E\left\{\left|x(n+k) - ax(n)\right|^2\right\} \geq 0$$

expanding the square we have

$$r_x(0) - a^* r_x(k) - a r_x(-k) + |a|^2 r_x(0) \geq 0$$

Using the conjugate symmetry of the autocorrelation sequence, $r_x(-k) = r_x^*(k)$, this becomes

$$\left(1 + |a|^2\right) r_x(0) - a^* r_x(k) - a r_x^*(k) \geq 0 \tag{3.53}$$

Since $r_x(k)$ is, in general, complex, let

$$r_x(k) = |r_x(k)| e^{j\phi(k)}$$

where $\phi(k)$ is the phase of $r_x(k)$. If we set $a = e^{j\phi(k)}$, then Eq. (3.53) becomes

$$2r_x(0) - 2|r_x(k)| \geq 0$$

and the property follows. ■

The next property is concerned with periodic random processes.

Property 4—Periodicity. If the autocorrelation sequence of a WSS random process is such that

$$r_x(k_0) = r_x(0)$$

for some k_0, then $r_x(k)$ is periodic with period k_0. Furthermore,

$$E\left\{|x(n) - x(n - k_0)|^2\right\} = 0$$

and $x(n)$ is said to be *mean-square periodic*.

We will establish this property in the case of a real-valued random process using the cosine inequality, Eq. (3.25), as follows. With

$$E^2\Big\{\big[x(n+k+k_0) - x(n+k)\big]\big[x(n)\big]\Big\} \leq E\Big\{\big[x(n+k+k_0) - x(n+k)\big]^2\Big\} E\Big\{x^2(n)\Big\}$$

expanding the products and using the linearity of the expectation we have

$$\big[r_x(k+k_0) - r_x(k)\big]^2 \leq 2\big[r_x(0) - r_x(k_0)\big] r_x(0)$$

for all k. Thus, if $r_x(k_0) = r_x(0)$, then $r_x(k+k_0) = r_x(k)$ and $r_x(k)$ is periodic. To show that $x(n)$ is mean-square periodic, note that

$$E\Big\{\big[x(n) - x(n-k_0)\big]^2\Big\} = 2r_x(0) - 2r_x(k_0)$$

Therefore, if $r_x(k_0) = r_x(0)$, then

$$E\Big\{\big[x(n) - x(n-k_0)\big]^2\Big\} = 0$$

and the result follows. ■

An example of a periodic process is the random phase sinusoid considered in Example 3.3.1. With $x(n) = A\cos(n\omega_0 + \phi)$ the autocorrelation sequence is $r_x(k) = \frac{1}{2}A^2\cos(k\omega_0)$. Therefore, if $\omega_0 = 2\pi/N$, then $r_x(k)$ is periodic with period N and $x(n)$ is mean-square periodic.

3.3.5 The Autocovariance and Autocorrelation Matrices

The autocovariance and autocorrelation sequences are important second-order statistical characterizations of discrete-time random processes that are often represented in matrix form. For example, if

$$\mathbf{x} = \left[x(0), x(1), \ldots, x(p)\right]^T$$

is a vector of $p+1$ values of a process $x(n)$, then the outer product

$$\mathbf{x}\mathbf{x}^H = \begin{bmatrix} x(0)x^*(0) & x(0)x^*(1) & \cdots & x(0)x^*(p) \\ x(1)x^*(0) & x(1)x^*(1) & \cdots & x(1)x^*(p) \\ \vdots & \vdots & & \vdots \\ x(p)x^*(0) & x(p)x^*(1) & \cdots & x(p)x^*(p) \end{bmatrix} \tag{3.54}$$

is a $(p+1) \times (p+1)$ matrix. If $x(n)$ is wide-sense stationary, taking the expected value and using the Hermitian symmetry of the autocorrelation sequence, $r_x(k) = r_x^*(-k)$, leads to the $(p+1) \times (p+1)$ matrix of autocorrelation values

$$\mathbf{R}_x = E\left\{\mathbf{x}\mathbf{x}^H\right\} = \begin{bmatrix} r_x(0) & r_x^*(1) & r_x^*(2) & \cdots & r_x^*(p) \\ r_x(1) & r_x(0) & r_x^*(1) & \cdots & r_x^*(p-1) \\ r_x(2) & r_x(1) & r_x(0) & \cdots & r_x^*(p-2) \\ \vdots & \vdots & \vdots & & \vdots \\ r_x(p) & r_x(p-1) & r_x(p-2) & \cdots & r_x(0) \end{bmatrix} \tag{3.55}$$

referred to as the *autocorrelation matrix*. In a similar fashion, forming an outer product of the vector $(\mathbf{x} - \mathbf{m}_x)$ with itself and taking the expected value leads to a $(p+1) \times (p+1)$ matrix referred to as the *autocovariance matrix*

$$\mathbf{C}_x = E\left\{(\mathbf{x} - \mathbf{m}_x)(\mathbf{x} - \mathbf{m}_x)^H\right\}$$

The relationship between $\mathbf{R}_x$ and $\mathbf{C}_x$ is

$$\mathbf{C}_x = \mathbf{R}_x - \mathbf{m}_x\mathbf{m}_x^H$$

where

$$\mathbf{m}_x = \left[m_x, m_x, \ldots, m_x\right]^T$$

is a vector of length $(p+1)$ containing the mean value of the process.[3] For zero mean processes the autocovariance and autocorrelation matrices are equal. As we stated earlier, it will generally be assumed that all random processes have zero mean. Therefore, only rarely will we encounter the autocovariance matrix. However, since the autocorrelation matrix will arise frequently, in the following paragraphs we explore some of the important properties of this matrix.

The first thing to note about the autocorrelation matrix of a WSS process is that it has a great deal of structure. In addition to being Hermitian, all of the terms along each of the diagonals are equal. Thus, $\mathbf{R}_x$ is a Hermitian Toeplitz matrix. In the case of a real-valued random process, the autocorrelation matrix is a symmetric Toeplitz matrix. Thus, we have the following property of $\mathbf{R}_x$.

[3]Recall that since $x(n)$ is WSS, the mean is a constant.

Property 1. The autocorrelation matrix of a WSS random process $x(n)$ is a Hermitian Toeplitz matrix,

$$\mathbf{R}_x = \text{Toep}\Big\{r_x(0),\ r_x(1), \ldots, r_x(p)\Big\}$$

The converse, however, is not true—not all Hermitian Toeplitz matrices represent a valid autocorrelation matrix. Note, for example, that since $r_x(0) = E\left\{|x(n)|^2\right\}$, then the terms along the main diagonal of $\mathbf{R}_x$ must be nonnegative. Therefore,

$$\mathbf{R}_x = \begin{bmatrix} -2 & 3 \\ 3 & -2 \end{bmatrix}$$

cannot be the autocorrelation matrix of a WSS process. However, positivity of the terms along the main diagonal is not sufficient to guarantee that a Hermitian Toeplitz matrix is a valid autocorrelation matrix. For example,

$$\mathbf{R}_x = \begin{bmatrix} 2 & 3 \\ 3 & 2 \end{bmatrix}$$

does not correspond to a valid autocorrelation matrix. What is required is that $\mathbf{R}_x$ be a nonnegative definite matrix.

Property 2. The autocorrelation matrix of a WSS random process is nonnegative definite, $\mathbf{R}_x > 0$.

To establish this property it must be shown that if $\mathbf{R}_x$ is an autocorrelation matrix, then

$$\mathbf{a}^H\mathbf{R}_x\mathbf{a} \geq 0 \tag{3.56}$$

for any vector $\mathbf{a}$. Since $\mathbf{R}_x = E\left\{\mathbf{x}\mathbf{x}^H\right\}$, we may write Eq. (3.56) as follows

$$\begin{aligned}\mathbf{a}^H\mathbf{R}_x\mathbf{a} &= \mathbf{a}^H E\left\{\mathbf{x}\mathbf{x}^H\right\}\mathbf{a} = E\left\{\mathbf{a}^H\left(\mathbf{x}\mathbf{x}^H\right)\mathbf{a}\right\} \\ &= E\Big\{\left(\mathbf{a}^H\mathbf{x}\right)\left(\mathbf{x}^H\mathbf{a}\right)\Big\}\end{aligned} \tag{3.57}$$

Therefore,

$$\mathbf{a}^H\mathbf{R}_x\mathbf{a} = E\Big\{|\mathbf{a}^H\mathbf{x}|^2\Big\}$$

and, since $|\mathbf{a}^H\mathbf{x}|^2 \geq 0$ for any $\mathbf{a}$, it follows that

$$E\Big\{|\mathbf{a}^H\mathbf{x}|^2\Big\} \geq 0$$

and the property is established. ■

Property 2 provides a necessary condition for a given sequence $r_x(k)$ for $k = 0, 1, \ldots, p$, to represent the autocorrelation values of a WSS random process. A more difficult question concerns how to find a random process having an autocorrelation whose first $p+1$ values match the given sequence. Since the autocorrelation sequence of a random process is, in general, infinite in length, it is necessary to find a valid extrapolation of $r_x(k)$ for all $|k| > p$. Unfortunately, simply extending $r_x(k)$ with zeros will not always produce a valid autocorrelation sequence. This problem will be considered in more detail in Section 5.2.8 of Chapter 5.

The next property is a result of the fact that the autocorrelation matrix is Hermitian and nonnegative definite. Specifically, recall that the eigenvalues of a Hermitian matrix are real and, for a nonnegative definite matrix, they are nonnegative (p. 43). This leads to the following property.

Property 3. The eigenvalues, λ_k, of the autocorrelation matrix of a WSS random process are real-valued and nonnegative.

Example 3.3.4 *Autocorrelation Matrix*

As we saw in Example 3.3.1 (p. 78), the autocorrelation sequence of a random phase sinusoid is

$$r_x(k) = \frac{1}{2}A^2 \cos(k\omega_0)$$

Therefore, the 2×2 autocorrelation matrix is

$$\mathbf{R}_x = \frac{1}{2}A^2 \begin{bmatrix} 1 & \cos\omega_0 \\ \cos\omega_0 & 1 \end{bmatrix}$$

The eigenvalues of $\mathbf{R}_x$ are

$$\begin{aligned} \lambda_1 &= 1 + \cos\omega_0 \geq 0 \\ \lambda_2 &= 1 - \cos\omega_0 \geq 0 \end{aligned}$$

and the determinant is

$$\det\big(\mathbf{R}_x\big) = 1 - \cos^2\omega_0 = \sin^2\omega_0 \geq 0$$

Clearly, $\mathbf{R}_x$ is nonnegative definite, and, if $\omega_0 \neq 0, \pi$, then $\mathbf{R}_x$ is positive definite.

As another example, consider the complex-valued process consisting of a sum of two complex exponentials

$$y(n) = Ae^{j(n\omega_1 + \phi_1)} + Ae^{j(n\omega_2 + \phi_2)}$$

where A, ω_1, and ω_2 are constants and ϕ_1 and ϕ_2 are uncorrelated random variables that are uniformly distributed between $-\pi$ and π. As we saw in Example 3.3.1, the autocorrelation sequence of a single complex exponential $x(n) = Ae^{j(n\omega_0 + \phi)}$ is

$$r_x(k) = |A|^2 e^{jk\omega_0}$$

Since $y(n)$ is the sum of two uncorrelated processes, the autocorrelation sequence of $y(n)$ is

$$r_y(k) = |A|^2 e^{jk\omega_1} + |A|^2 e^{jk\omega_2}$$

and the 2×2 autocorrelation matrix is

$$\mathbf{R}_y = |A|^2 \begin{bmatrix} 2 & e^{-j\omega_1} + e^{-j\omega_2} \\ e^{j\omega_1} + e^{j\omega_2} & 2 \end{bmatrix}$$

The eigenvalues of $\mathbf{R}_y$ are

$$\lambda_1 = 2 + 2\cos\left(\frac{\omega_1 - \omega_2}{2}\right) \geq 0$$

$$\lambda_2 = 2 - 2\cos\left(\frac{\omega_1 - \omega_2}{2}\right) \geq 0$$

Note that if $\omega_2 = -\omega_1$, then this reduces to the random phase sinusoid discussed above.

3.3.6 Ergodicity

The mean and autocorrelation of a random process are examples of ensemble averages that describe the statistical averages of the process over the ensemble of all possible discrete-time signals. Although these ensemble averages are often required in problems such as signal modeling, optimum filtering, and spectrum estimation, they are not generally known a priori. Therefore, being able to estimate these averages from a realization of a discrete-time random process becomes an important problem. In this section, we consider the estimation of the mean and autocorrelation of a random process and present some conditions for which it is possible to estimate these averages using an appropriate *time average*.

Let us begin by considering the problem of estimating the mean, $m_x(n)$, of a random process $x(n)$. If a large number of sample realizations of the process were available, e.g., $x_i(n), i = 1, \ldots, L$, then an average of the form

$$\hat{m}_x(n) = \frac{1}{L}\sum_{i=1}^{L} x_i(n)$$

could be used to estimate the mean $m_x(n)$. In most situations, however, such a collection of sample realizations is not generally available and it is necessary to consider methods for estimating the ensemble average from a *single* realization of the process. Given only a single realization of $x(n)$, we may consider estimating the ensemble average $E\{x(n)\}$ with a *sample mean* that is taken as the average, over time, of $x(n)$ as follows

$$\hat{m}_x(N) = \frac{1}{N}\sum_{n=0}^{N-1} x(n) \tag{3.58}$$

In order for this to be a meaningful estimator for the mean, however, it is necessary that some constraints be imposed on the process. Consider, for example, the random process

$$x(n) = A\cos(n\omega_0)$$

where A is a random variable that is equally likely to assume a value of 1 or 2. The mean of this process is

$$E\{x(n)\} = E\{A\}\cos(n\omega_0) = 1.5\cos(n\omega_0)$$

However, given a single realization of this process, for large N the sample mean will be approximately zero

$$\hat{m}_x(N) \approx 0 \quad ; \quad N \gg 1$$

It should be clear from this example that since the sample mean involves an average over time, the ensemble mean must be a constant, $E\{x(n)\} = m_x$, at least over the interval over which the time average is being taken. Even with a constant mean, however, the sample mean may not converge to m_x as $N \to \infty$ without further restrictions on $x(n)$. An illustration of what may happen is given in the following example.

Example 3.3.5 *The Sample Mean*

Consider the random process

$$x(n) = A$$

where A is a random variable whose value is determined by the flip of a fair coin. Specifically, if the toss of a coin results in *Heads*, then $A = 1$; if the result is *Tails*, then $A = -1$. Since the coin is fair, $\Pr\{A = 1\} = \Pr\{A = -1\} = 0.5$, then the mean of the process is

$$m_x(n) = E\{x(n)\} = E\{A\} = 0$$

However, with the sample mean,

$$\hat{m}_x(N) = 1$$

with a probability of one half and

$$\hat{m}_x(N) = -1$$

with a probability of one half. Therefore, $\hat{m}_x(N)$ will not converge, in any sense, to the true mean, $m_x = 0$, as $N \to \infty$. The difficulty is that the random variables $x(n)$ are perfectly correlated, $r_x(k) = 1$, and no new information about the process is obtained as more sample values are used.

At the other extreme, consider the Bernoulli process that is generated by repeatedly flipping a fair coin. With $x(n) = 1$ if *Heads* is flipped at time n and $x(n) = -1$ if *Tails* is flipped, then, as in the previous case,

$$m_x(n) = E\{x(n)\} = 0$$

However, given a single realization of the process the sample mean is

$$\hat{m}_x(N) = \frac{1}{N}\sum_{n=0}^{N-1} x(n) = \frac{n_1}{N} - \frac{n_2}{N}$$

where n_1 is the number of times that *Heads* is flipped and n_2 is the number of times that *Tails* is flipped. As $N \to \infty$, since n_1/N converges to $\Pr\{H\} = 1/2$ and n_2/N converges to $\Pr\{T\} = 1/2$, then $\hat{m}_x(N)$ converges to zero, the mean of the process. Unlike the first example, the Bernoulli process consists of a sequence of uncorrelated random variables. Therefore, each new value of $x(n)$ brings additional information about the statistics of the process.

We now consider the convergence of the sample mean to the ensemble mean, m_x, of a WSS process $x(n)$. However, note that since the sample mean is the average of the random variables, $x(0), \ldots, x(N)$, then $\hat{m}_x(N)$ is also a random variable. In fact, viewed as a sequence that is indexed by N, the sample mean is a sequence of random variables. Therefore, in discussing the convergence of the sample mean it is necessary to consider convergence within a statistical framework. Although there are a number of different ways in which to formulate conditions for the convergence of a sequence of random variables (see Section 3.2.8), the condition of interest here is mean-square convergence [3, 5],

$$\lim_{N\to\infty} E\left\{\left|\hat{m}_x(N) - m_x\right|^2\right\} = 0 \tag{3.59}$$

If Eq. (3.59) is satisfied then $x(n)$ is said to be ergodic in the mean.

> ***Definition.*** If the sample mean $\hat{m}_x(N)$ of a wide-sense stationary process converges to m_x in the mean-square sense, then the process is said to be *ergodic in the mean* and we write
>
> $$\lim_{N\to\infty} \hat{m}_x(N) = m_x$$

In order for the sample mean to converge in the mean-square sense it is necessary and sufficient that [2, 5].

1. The sample mean be asymptotically unbiased,

$$\lim_{N\to\infty} E\left\{\hat{m}_x(N)\right\} = m_x \tag{3.60}$$

2. The variance of the estimate go to zero as $N \to \infty$,

$$\lim_{N\to\infty} \mathrm{Var}\left\{\hat{m}_x(N)\right\} = 0 \tag{3.61}$$

From the definition of the sample mean it follows easily that the sample mean is unbiased for any wide-sense stationary process,

$$E\left\{\hat{m}_x(N)\right\} = \frac{1}{N}\sum_{n=0}^{N-1} E\{x(n)\} = m_x$$

In order for the variance to go to zero, however, some constraints must be placed on the process $x(n)$. Evaluating the variance of $\hat{m}_x(N)$ we have

$$\begin{aligned}
\mathrm{Var}\{\hat{m}_x(N)\} &= E\left\{\left|\hat{m}_x(N) - m_x\right|^2\right\} = E\left\{\left|\frac{1}{N}\sum_{n=0}^{N-1}[x(n) - m_x]\right|^2\right\} \\
&= \frac{1}{N^2}\sum_{n=0}^{N-1}\sum_{m=0}^{N-1} E\left\{[x(m) - m_x][x(n) - m_x]^*\right\} \\
&= \frac{1}{N^2}\sum_{n=0}^{N-1}\sum_{m=0}^{N-1} c_x(m-n)
\end{aligned} \tag{3.62}$$

where $c_x(m-n)$ is the autocovariance of $x(n)$. Grouping together common terms we may write the variance as

$$\begin{aligned}
\mathrm{Var}\{\hat{m}_x(N)\} &= \frac{1}{N^2}\sum_{n=0}^{N-1}\sum_{m=0}^{N-1} c_x(m-n) = \frac{1}{N^2}\sum_{k=-N+1}^{N-1} (N - |k|)\, c_x(k) \\
&= \frac{1}{N}\sum_{k=-N+1}^{N-1} \left(1 - \frac{|k|}{N}\right) c_x(k)
\end{aligned} \tag{3.63}$$

Therefore, $x(n)$ will be ergodic in the mean if and only if

$$\lim_{N\to\infty} \frac{1}{N} \sum_{k=-N+1}^{N-1} \left(1 - \frac{|k|}{N}\right) c_x(k) = 0 \tag{3.64}$$

An equivalent condition that is necessary and sufficient for $x(n)$ to be ergodic in the mean is given in the following theorem [5, 8]

> ***Mean Ergodic Theorem 1.*** Let $x(n)$ be a WSS random process with autocovariance sequence $c_x(k)$. A necessary and sufficient condition for $x(n)$ to be ergodic in the mean is
>
> $$\lim_{N\to\infty} \frac{1}{N} \sum_{k=0}^{N-1} c_x(k) = 0 \tag{3.65}$$

Equation (3.65) places a necessary and sufficient constraint on the asymptotic decay of the autocorrelation sequence. A sufficient condition that is much easier to apply is given in the following theorem [5]

> ***Mean Ergodic Theorem 2.*** Let $x(n)$ be a WSS random process with autocovariance sequence $c_x(k)$. Sufficient conditions for $x(n)$ to be ergodic in the mean are that $c_x(0) < \infty$ and
>
> $$\lim_{k\to\infty} c_x(k) = 0 \tag{3.66}$$

Thus, according to Eq. (3.66) a WSS process will be ergodic in the mean if it is asymptotically uncorrelated.

We now establish the sufficiency of (3.66). Due to the conjugate symmetry of $c_x(k)$, the magnitude of $c_x(k)$ is symmetric, $|c_x(k)| = |c_x(-k)|$, and the variance of the sample mean may be upper bounded using Eq.(3.63) as follows:

$$\text{Var}\{\hat{m}_x(N)\} \le \frac{1}{N} c_x(0) + \frac{2}{N} \sum_{k=1}^{N-1} \left(1 - \frac{|k|}{N}\right) |c_x(k)| \tag{3.67}$$

If $c_x(k) \to 0$ as $k \to \infty$ then, for any $\epsilon > 0$, we may find a value of k, say k_0, such that

$$|c_x(k)| < \epsilon \qquad \text{for all } |k| > k_0$$

Thus, for $N > k_0$ we may write Eq. (3.67) as follows

$$\text{Var}\{\hat{m}_x(N)\} \le \frac{1}{N} c_x(0) + \frac{2}{N} \sum_{k=1}^{k_0} \left(1 - \frac{|k|}{N}\right) |c_x(k)| + \frac{2}{N} \sum_{k=k_0+1}^{N-1} \left(1 - \frac{|k|}{N}\right) \epsilon \tag{3.68}$$

Since[4] $|c_x(k)| \le c_x(0)$ and

$$\sum_{k=1}^{k_0} k = \frac{1}{2} k_0(k_0 + 1)$$

[4]This is a straightforward extension of the maximum value property given on p. 83 for the autocorrelation.

we may write (3.68) as

$$\text{Var}\{\hat{m}_x(N)\} \le \frac{1}{N}c_x(0)\Big[1 + 2k_0 - \frac{1}{N}k_0(k_0+1)\Big] + \frac{2}{N}\epsilon\big(N - k_0 - 1)(1 - \frac{k_0}{N})$$

If we fix k_0 and let $N \to \infty$ then

$$\lim_{N\to\infty} \text{Var}\{\hat{m}_x(N)\} \le 2\epsilon$$

Since the value of ϵ is arbitrary, it follows that the variance goes to zero as $N \to \infty$ and the theorem is established. ■

Example 3.3.6 ***Ergodicity in the Mean***

In Example 3.3.5 we considered the random process

$$x(n) = A$$

where A is a random variable with $\Pr\{A = 1\} = 0.5$ and $\Pr\{A = -1\} = 0.5$. Since the variance of A is equal to one then the autocovariance of $x(n)$ is

$$c_x(k) = 1$$

Therefore,

$$\frac{1}{N}\sum_{k=0}^{N-1} c_x(k) = 1$$

and it follows that this process is not ergodic in the mean (the variance of the sample mean is, in fact, equal to one). The Bernoulli process, on the other hand, consisting of a sequence of independent random variables, each with a variance of one, has an autocovariance of

$$c_x(k) = \delta(k)$$

Therefore, by the second mean ergodic theorem, it follows that the Bernoulli process is ergodic in the mean. Finally, consider the random phase sinusoid $x(n) = A\sin(n\omega_0 + \phi)$ which, assuming $\omega_0 \neq 0$, has an autocovariance of

$$c_x(k) = \frac{1}{2}A^2\cos(k\omega_0)$$

With

$$\sum_{k=0}^{N-1}\cos k\omega_0 = \text{Re}\left\{\sum_{k=0}^{N-1} e^{jk\omega_0}\right\} = \text{Re}\left\{\frac{1 - e^{jN\omega_0}}{1 - e^{j\omega_0}}\right\} = \frac{\sin(N\omega_0/2)}{\sin(\omega_0/2)}\cos\big[(N-1)\omega_0/2\big]$$

we have

$$\frac{1}{N}\sum_{k=0}^{N-1} c_x(k) = \frac{A^2}{2N}\frac{\sin(N\omega_0/2)}{\sin(\omega_0/2)}\cos\big[(N-1)\omega_0/2\big]$$

which goes to zero as $N \to \infty$ provided $\omega_0 \neq 0$. For $\omega_0 = 0$

$$x(n) = A\sin\phi$$

which is not ergodic in the mean since the covariance is

$$c_x(k) = \frac{1}{2}A^2$$

Therefore, the random phase sinusoid is ergodic in the mean provided $\omega_0 \neq 0$.

The mean ergodic theorems may be generalized to the estimation of other ensemble averages. For example, consider the estimation of the autocorrelation sequence

$$r_x(k) = E\left\{x(n)x^*(n-k)\right\}$$

from a single realization of a process $x(n)$. Since, for each k, the autocorrelation is the expected value of the process

$$y_k(n) = x(n)x^*(n-k)$$

we may estimate the autocorrelation from the sample mean of $y_k(n)$ as follows:

$$\hat{r}_x(k, N) = \frac{1}{N}\sum_{n=0}^{N-1} y_k(n) = \frac{1}{N}\sum_{n=0}^{N-1} x(n)x^*(n-k)$$

If $\hat{r}_x(k, N)$ converges in the mean-square sense to $r_x(k)$ as $N \to \infty$,

$$\lim_{N\to\infty} E\left\{\left|\hat{r}_x(k, N) - r_x(k)\right|^2\right\} = 0$$

then the process is said to be *autocorrelation ergodic* and we write

$$\lim_{N\to\infty} \hat{r}_x(k, N) = r_x(k)$$

Since $\hat{r}_x(k, N)$ is the sample mean of $y_k(n)$, it follows that $x(n)$ will be autocorrelation ergodic if $y_k(n)$ is ergodic in the mean. Applying the first mean ergodic theorem to the sample mean of $y_k(n)$ places a constraint on the covariance of $y_k(n)$, which is equivalent to a constraint on the fourth-order moment of $x(n)$. Another result of interest that is applicable to Gaussian processes is the following [8].

Autocorrelation Ergodic Theorem. A necessary and sufficient condition for a wide-sense stationary Gaussian process with covariance $c_x(k)$ to be autocorrelation ergodic is

$$\lim_{N\to\infty} \frac{1}{N}\sum_{k=0}^{N-1} c_x^2(k) = 0$$

Clearly, in most applications, determining whether or not a given process is ergodic is not practical. Therefore, whenever the solution to a problem requires knowledge of the mean, the autocorrelation, or some other ensemble average, the process is typically assumed to be ergodic and time averages are used to estimate these ensemble averages. Whether or not this assumption is appropriate will be determined by the performance of the algorithm that uses these estimates.

3.3.7 White Noise

An important and fundamental discrete-time process that will be frequently encountered in our treatment of discrete-time random processes is white noise. A wide-sense stationary process $v(n)$, either real or complex, is said to be *white* if the autocovariance function is zero for all $k \neq 0$,

$$c_v(k) = \sigma_v^2 \delta(k)$$

Thus, white noise is simply a sequence of uncorrelated random variables, each having a variance of σ_v^2. Since white noise is defined only in terms of the form of its second-order moment, there is an infinite variety of white noise random processes. For example, a random process consisting of a sequence of uncorrelated real-valued Gaussian random variables is a white noise process referred to as white Gaussian noise (WGN). A sample realization of zero mean unit variance white Gaussian noise is shown in Fig. 3.6*a*. A significantly different type of white noise process is the *Bernoulli process* shown in Fig. 3.6*b*, which consists of a sequence of uncorrelated Bernoulli random variables.

For complex white noise, note that if

$$v(n) = v_1(n) + jv_2(n)$$

then

$$E\{|v(n)|^2\} = E\{|v_1(n)|^2\} + E\{|v_2(n)|^2\}$$

Therefore, it is important to note that the variance of $v(n)$ is the sum of the variances of the real and imaginary components, $v_1(n)$ and $v_2(n)$, respectively.

As we will see in Section 3.4, a wide variety of different and important random processes may be generated by filtering white noise with a linear shift-invariant filter.

3.3.8 The Power Spectrum

Just as Fourier analysis is an important tool in the description and analysis of deterministic discrete-time signals, it also plays an important role in the study of random processes. However, since a random process is an ensemble of discrete-time signals, we cannot compute the Fourier transform of the process itself. Nevertheless, as we will see below, it is possible to develop a frequency domain representation of the process if we express the Fourier transform in terms of an ensemble average.

Recall that the autocorrelation sequence of a WSS process provides a time domain description of the second-order moment of the process. Since $r_x(k)$ is a deterministic sequence

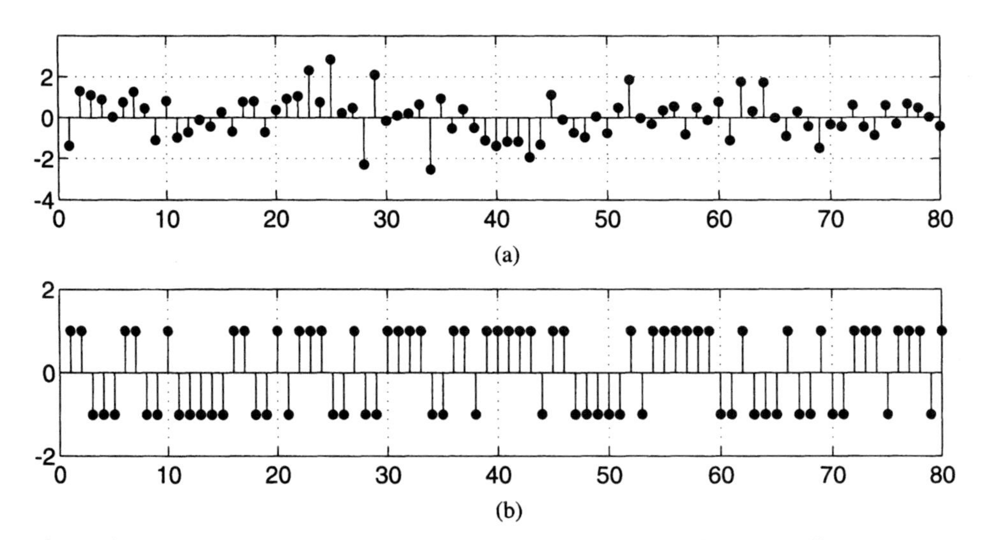

Figure 3.6 *Examples of white noise. (a) White Gaussian noise. (b) White Bernoulli noise.*

we may compute its discrete-time Fourier transform,

$$P_x(e^{j\omega}) = \sum_{k=-\infty}^{\infty} r_x(k)e^{-jk\omega} \tag{3.69}$$

which is called the *power spectrum* or *power spectral density* of the process.[5] Given the power spectrum, the autocorrelation sequence may be determined by taking the inverse discrete-time Fourier transform of $P_x(e^{j\omega})$,

$$r_x(k) = \frac{1}{2\pi}\int_{-\pi}^{\pi} P_x(e^{j\omega})e^{jk\omega}d\omega \tag{3.70}$$

Thus, the power spectrum provides a frequency domain description of the second-order moment of the process. In some cases it may be more convenient to use the z-transform instead of the discrete-time Fourier transform, in which case

$$P_x(z) = \sum_{k=-\infty}^{\infty} r_x(k)z^{-k} \tag{3.71}$$

will also be referred to as the power spectrum of $x(n)$.

Just as with the autocorrelation sequence, there are a number of properties of the power spectrum that will be useful. First of all, since the autocorrelation of a WSS random process is conjugate symmetric, it follows that the power spectrum will be a real-valued function of ω. In the case of a *real-valued* random process, the autocorrelation sequence is real and even, which implies that the power spectrum is real and *even*. Thus, we have the following symmetry property of the power spectrum.

Property 1—Symmetry. The power spectrum of a WSS random process $x(n)$ is real-valued, $P_x(e^{j\omega}) = P_x^*(e^{j\omega})$, and $P_x(z)$ satisfies the symmetry condition

$$P_x(z) = P_x^*(1/z^*)$$

In addition, if $x(n)$ is real then the power spectrum is *even*, $P_x(e^{j\omega}) = P_x(e^{-j\omega})$, which implies that

$$P_x(z) = P_x^*(z^*)$$

In addition to being real-valued, the power spectrum is nonnegative. Specifically,

Property 2—Positivity. The power spectrum of a WSS random process is nonnegative

$$P_x(e^{j\omega}) \geq 0$$

[5] If $x(n)$ does not have zero mean, then the power spectrum as it is defined here will have an impulse at $\omega = 0$. For nonzero mean random processes, the power spectrum is normally defined to be the discrete-time Fourier transform of the autocovariance.

This property follows from the constraint that the autocorrelation matrix is nonnegative definite and will be established in the next section. Finally, a property that relates the average power in a random process to the power spectrum is as follows.

> ***Property 3—Total power.*** The power in a zero mean WSS random process is proportional to the area under the power spectral density curve
>
> $$E\{|x(n)|^2\} = \frac{1}{2\pi}\int_{-\pi}^{\pi} P_x(e^{j\omega})d\omega$$

This property follows from Eq. (3.70) with $k = 0$ and the fact that $r_x(0) = E\{|x(n)|^2\}$.

Example 3.3.7 ***The Power Spectrum***

The autocorrelation sequence of a zero mean white noise process is $r_v(k) = \sigma_v^2\delta(k)$ where σ_v^2 is the variance of the process. The power spectrum, therefore, is equal to a constant

$$P_v(e^{j\omega}) = \sigma_v^2$$

The random phase sinusoid, on the other hand, has an autocorrelation sequence that is sinusoidal,

$$r_x(k) = \frac{1}{2}A^2\cos(k\omega_0)$$

and has a power spectrum given by

$$P_x(e^{j\omega}) = \frac{1}{2}\pi A^2\big[u_0(\omega - \omega_0) + u_0(\omega + \omega_0)\big]$$

As a final example, consider the autocorrelation sequence

$$r_x(k) = \alpha^{|k|}$$

where $|\alpha| < 1$, which, as we will see in Section 3.6.2, corresponds to a first-order autoregressive process. The power spectrum is

$$\begin{aligned} P_x(e^{j\omega}) &= \sum_{k=-\infty}^{\infty} r_x(k)e^{-jk\omega} = \sum_{k=0}^{\infty}\alpha^k e^{-jk\omega} + \sum_{k=0}^{\infty}\alpha^k e^{jk\omega} - 1 \\ &= \frac{1}{1-\alpha e^{-j\omega}} + \frac{1}{1-\alpha e^{j\omega}} - 1 = \frac{1-\alpha^2}{1-2\alpha\cos\omega+\alpha^2} \end{aligned}$$

which is clearly real, even, and nonnegative.

The next property relates the extremal eigenvalues of the autocorrelation matrix of a WSS process to the maximum and minimum values of the power spectrum.

Property 4—Eigenvalue Extremal Property. The eigenvalues of the $n \times n$ autocorrelation matrix of a zero mean WSS random process are upper and lower bounded by the maximum and minimum values, respectively, of the power spectrum,

$$\min_{\omega} P_x(e^{j\omega}) \le \lambda_i \le \max_{\omega} P_x(e^{j\omega})$$

To establish this property, let λ_i and $\mathbf{q}_i$ be the eigenvalues and eigenvectors, respectively, of the $n \times n$ autocorrelation matrix $\mathbf{R}_x$,

$$\mathbf{R}_x \mathbf{q}_i = \lambda_i \mathbf{q}_i \quad ; \quad i = 1, 2, \ldots, n$$

Since

$$\mathbf{q}_i^H \mathbf{R}_x \mathbf{q}_i = \lambda_i \mathbf{q}_i^H \mathbf{q}_i$$

it follows that

$$\lambda_i = \frac{\mathbf{q}_i^H \mathbf{R}_x \mathbf{q}_i}{\mathbf{q}_i^H \mathbf{q}_i} \tag{3.72}$$

Expanding the Hermitian form in the numerator and denoting the coefficients of $\mathbf{q}_i$ by $q_i(k)$,

$$\mathbf{q}_i = \left[q_i(0),\ q_i(1), \ldots,\ q_i(n-1)\right]^T$$

we have

$$\mathbf{q}_i^H \mathbf{R}_x \mathbf{q}_i = \sum_{k=0}^{n-1} \sum_{l=0}^{n-1} q_i^*(k) r_x(k-l) q_i(l) \tag{3.73}$$

With

$$r_x(k-l) = \frac{1}{2\pi} \int_{-\pi}^{\pi} P_x(e^{j\omega}) e^{j\omega(k-l)} d\omega$$

we may rewrite Eq. (3.3.7) as follows

$$\begin{aligned}
\mathbf{q}_i^H \mathbf{R}_x \mathbf{q}_i &= \frac{1}{2\pi} \sum_{k=0}^{n-1} \sum_{l=0}^{n-1} q_i^*(k) q_i(l) \int_{-\pi}^{\pi} P_x(e^{j\omega}) e^{j\omega(k-l)} d\omega \\
&= \frac{1}{2\pi} \int_{-\pi}^{\pi} P_x(e^{j\omega}) \left[\sum_{k=0}^{n-1} q_i^*(k) e^{jk\omega} \sum_{l=0}^{n-1} q_i(l) e^{-jl\omega} \right] d\omega
\end{aligned} \tag{3.74}$$

With

$$Q_i(e^{j\omega}) = \sum_{k=0}^{n-1} q_i(k) e^{-jk\omega}$$

which is the Fourier transform of the eigenvector $\mathbf{q}_i$, Eq. (3.74) may be written as

$$\mathbf{q}_i^H \mathbf{R}_x \mathbf{q}_i = \frac{1}{2\pi} \int_{-\pi}^{\pi} P_x(e^{j\omega}) |Q(e^{j\omega})|^2 d\omega$$

Repeating these steps for the denominator of Eq. (3.72) we find

$$\mathbf{q}_i^H \mathbf{q}_i = \frac{1}{2\pi} \int_{-\pi}^{\pi} |Q(e^{j\omega})|^2 d\omega$$

Therefore,

$$\lambda_i = \frac{\mathbf{q}_i^H \mathbf{R}_x \mathbf{q}_i}{\mathbf{q}_i^H \mathbf{q}_i} = \frac{\frac{1}{2\pi}\int_{-\pi}^{\pi} P_x(e^{j\omega})|Q(e^{j\omega})|^2 d\omega}{\frac{1}{2\pi}\int_{-\pi}^{\pi} |Q(e^{j\omega})|^2 d\omega}$$

Finally, since

$$\frac{1}{2\pi}\int_{-\pi}^{\pi} P_x(e^{j\omega})|Q(e^{j\omega})|^2 d\omega \le \left[\max_{\omega} P_x(e^{j\omega})\right]\frac{1}{2\pi}\int_{-\pi}^{\pi} |Q(e^{j\omega})|^2 d\omega$$

and

$$\frac{1}{2\pi}\int_{-\pi}^{\pi} P_x(e^{j\omega})|Q(e^{j\omega})|^2 d\omega \ge \left[\min_{\omega} P_x(e^{j\omega})\right]\frac{1}{2\pi}\int_{-\pi}^{\pi} |Q(e^{j\omega})|^2 d\omega$$

then it follows that

$$\min_{\omega} P_x(e^{j\omega}) \le \lambda_i \le \max_{\omega} P_x(e^{j\omega})$$

as was to be demonstrated. ■

In addition to providing a frequency domain representation of the second-order moment, the power spectrum may also be related to the ensemble average of the squared Fourier magnitude, $|X(e^{j\omega})|^2$. In particular, consider

$$\begin{aligned} P_N(e^{j\omega}) &= \frac{1}{2N+1}\left|\sum_{n=-N}^{N} x(n)e^{-jn\omega}\right|^2 \\ &= \frac{1}{2N+1}\sum_{n=-N}^{N}\sum_{m=-N}^{N} x(n)x^*(m)e^{-j(n-m)\omega} \end{aligned} \tag{3.75}$$

which is proportional to the squared magnitude of the discrete-time Fourier transform of $2N+1$ samples of a given realization of the random process. Since, for each frequency ω, $P_N(e^{j\omega})$ is a random variable, taking the expected value it follows that

$$E\left\{P_N(e^{j\omega})\right\} = \frac{1}{2N+1}\sum_{n=-N}^{N}\sum_{m=-N}^{N} r_x(n-m)e^{-j(n-m)\omega} \tag{3.76}$$

With the substitution $k = n - m$, Eq. (3.76) becomes, after some rearrangement of terms,

$$\begin{aligned} E\left\{P_N(e^{j\omega})\right\} &= \frac{1}{2N+1}\sum_{k=-2N}^{2N} (2N+1-|k|)r_x(k)e^{-jk\omega} \\ &= \sum_{k=-2N}^{2N}\left(1 - \frac{|k|}{2N+1}\right) r_x(k)e^{-jk\omega} \end{aligned} \tag{3.77}$$

Assuming that the autocorrelation sequence decays to zero fast enough so that

$$\sum_{k=-\infty}^{\infty} |k| r_x(k) < \infty$$

taking the limit of Eq. (3.77) by letting $N \to \infty$ it follows that

$$\lim_{N\to\infty} E\left\{P_N(e^{j\omega})\right\} = \sum_{k=-\infty}^{\infty} r_x(k)e^{-jk\omega} = P_x(e^{j\omega}) \tag{3.78}$$

Therefore, combining Eqs. (3.75) and (3.78) we have

$$P_x(e^{j\omega}) = \lim_{N\to\infty} \frac{1}{2N+1} E\left\{ \left| \sum_{n=-N}^{N} x(n)e^{-jn\omega} \right|^2 \right\} \tag{3.79}$$

Thus, the power spectrum may be viewed as the expected value of $P_N(e^{j\omega})$ in the limit as $N \to \infty$.

3.4 FILTERING RANDOM PROCESSES

Linear shift-invariant filters are frequently used to perform a variety of different signal processing tasks in applications that range from signal detection and estimation, to deconvolution, signal representation and synthesis. Since the inputs to these filters are often random processes, it is important to be able to determine how the statistics of these processes change as a result of filtering. In this section we derive the relationship between the mean and autocorrelation of the input process to the mean and autocorrelation of the output process.

Let $x(n)$ be a WSS random process with mean m_x and autocorrelation $r_x(k)$. If $x(n)$ is filtered with a *stable* linear shift-invariant filter having a unit sample response $h(n)$, then the output, $y(n)$, is a random process that is related to $x(n)$ by the convolution sum

$$y(n) = x(n) * h(n) = \sum_{k=-\infty}^{\infty} h(k)x(n-k) \tag{3.80}$$

The mean of $y(n)$ may be found by taking the expected value of Eq. (3.80) as follows,

$$\begin{aligned} E\{y(n)\} &= E\left\{ \sum_{k=-\infty}^{\infty} h(k)x(n-k) \right\} = \sum_{k=-\infty}^{\infty} h(k)E\{x(n-k)\} \\ &= m_x \sum_{k=-\infty}^{\infty} h(k) = m_x H(e^{j0}) \end{aligned} \tag{3.81}$$

Thus, the mean of $y(n)$ is constant and it is related to the mean of $x(n)$ by a scale factor that is equal to the frequency response of the filter at $\omega = 0$.

We may also use Eq. (3.80) to relate the autocorrelation of $y(n)$ to the autocorrelation of $x(n)$. This is done by first computing the cross-correlation between $x(n)$ and $y(n)$ as follows,

$$\begin{aligned} r_{yx}(n+k, n) &= E\{y(n+k)x^*(n)\} = E\left\{ \sum_{l=-\infty}^{\infty} h(l)x(n+k-l)x^*(n) \right\} \\ &= \sum_{l=-\infty}^{\infty} h(l)E\{x(n+k-l)x^*(n)\} \\ &= \sum_{l=-\infty}^{\infty} h(l)r_x(k-l) \end{aligned} \tag{3.82}$$

Thus, for a wide-sense stationary process $x(n)$, the cross-correlation $r_{yx}(n+k, n)$ depends only on the difference between $n+k$ and n and Eq. (3.82) may be written as

$$r_{yx}(k) = r_x(k) * h(k) \tag{3.83}$$

The autocorrelation of $y(n)$ may now be determined as follows,

$$\begin{aligned} r_y(n+k, n) &= E\{y(n+k)y^*(n)\} = E\left\{y(n+k)\sum_{l=-\infty}^{\infty} x^*(l)h^*(n-l)\right\} \\ &= \sum_{l=-\infty}^{\infty} h^*(n-l)E\{y(n+k)x^*(l)\} \\ &= \sum_{l=-\infty}^{\infty} h^*(n-l)r_{yx}(n+k-l) \end{aligned} \tag{3.84}$$

Changing the index of summation by setting $m = n - l$ we have

$$r_y(n+k, n) = \sum_{m=-\infty}^{\infty} h^*(m)r_{yx}(m+k) = r_{yx}(k) * h^*(-k) \tag{3.85}$$

Therefore, the autocorrelation sequence $r_y(n+k, n)$ depends only on k, the difference between the indices $n+k$ and n, i.e.,

$$r_y(k) = r_{yx}(k) * h^*(-k) \tag{3.86}$$

Combining Eqs. (3.83) and (3.86) we have

$$r_y(k) = \sum_{l=-\infty}^{\infty} \sum_{m=-\infty}^{\infty} h(l)r_x(m-l+k)h^*(m) \tag{3.87}$$

or,

$$\boxed{r_y(k) = r_x(k) * h(k) * h^*(-k)} \tag{3.88}$$

These relationships are illustrated in Fig. 3.7. Thus, it follows from Eqs. (3.81) and (3.88) that if $x(n)$ is WSS, then $y(n)$ will be WSS provided $\sigma_y^2 < \infty$ (this condition requires that the filter be stable). In addition, it follows from Eq. (3.82) that $x(n)$ and $y(n)$ will be jointly WSS.

Another interpretation of Eq. (3.88) is as follows. Defining $r_h(k)$ to be the (deterministic) autocorrelation of the unit sample response, $h(n)$,

$$r_h(k) = h(k) * h^*(-k) = \sum_{n=-\infty}^{\infty} h(n)h^*(n+k)$$

we see that $r_y(k)$ is the convolution of the autocorrelation of the input process with the

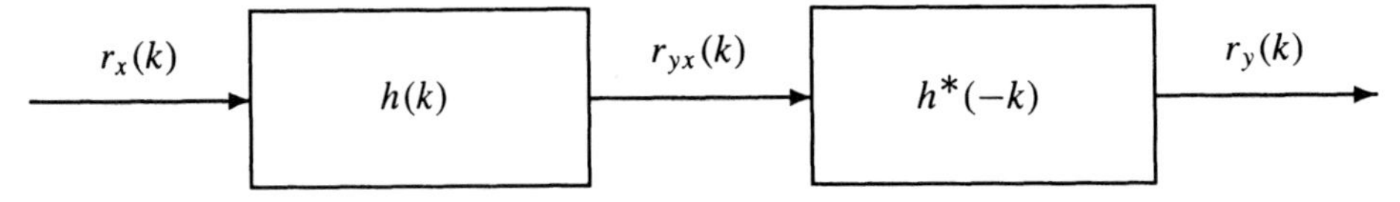

Figure 3.7 *Computing the autocorrelation sequence of a filtered process $y(n) = x(n) * h(n)$.*

deterministic autocorrelation of the filter

$$r_y(k) = r_x(k) * r_h(k) \tag{3.89}$$

For the variance of the output process, $y(n)$, we see from Eq. (3.87) that

$$\sigma_y^2 = r_y(0) = \sum_{l=-\infty}^{\infty} \sum_{m=-\infty}^{\infty} h(l) r_x(m-l) h^*(m)$$

As a special case, note that if $h(n)$ is finite in length and zero outside the interval $[0, N-1]$ then the variance (power) of $y(n)$ may be expressed in terms of the autocorrelation matrix $\mathbf{R}_x$ of $x(n)$ and the vector of filter coefficients $\mathbf{h}$ as follows:

$$\boxed{\sigma_y^2 = E\left\{|y(n)|^2\right\} = \mathbf{h}^H \mathbf{R}_x \mathbf{h}} \tag{3.90}$$

Having found the relationship between the autocorrelation of $x(n)$ and the autocorrelation of $y(n)$, we may relate the power spectrum of $y(n)$ to that of $x(n)$ as follows:

$$\boxed{P_y(e^{j\omega}) = P_x(e^{j\omega})|H(e^{j\omega})|^2} \tag{3.91}$$

Therefore, if a WSS process is filtered with a linear shift-invariant filter, then the power spectrum of the input signal is multiplied by the squared magnitude of the frequency response of the filter. In terms of z-transforms, Eq. (3.88) becomes

$$\boxed{P_y(z) = P_x(z) H(z) H^*(1/z^*)} \tag{3.92}$$

If $h(n)$ is real, then $H(z) = H^*(z^*)$ and Eq. (3.92) becomes

$$P_y(z) = P_x(z) H(z) H(1/z) \tag{3.93}$$

Thus, if $h(n)$ is complex and the system function $H(z)$ has a pole at $z = z_0$, then the power spectrum $P_y(z)$ will have a pole at $z = z_0$ and another at the conjugate reciprocal location $z = 1/z_0^*$.[6] Similarly, if $H(z)$ has a zero at $z = z_0$, then the power spectrum of $y(n)$ will have a zero at $z = z_0$ as well as a zero at $z = 1/z_0^*$. This is a special case of *spectral factorization* as discussed in the following section.

Example 3.4.1 *Filtering White Noise*

Let $x(n)$ be the random process that is generated by filtering white noise $w(n)$ with a first-order linear shift-invariant filter having a system function

$$H(z) = \frac{1}{1 - 0.25z^{-1}}$$

If the variance of the white noise is equal to one, $\sigma_w^2 = 1$, then the power spectrum of $x(n)$ is

$$P_x(z) = \sigma_w^2 H(z) H(z^{-1}) = \frac{1}{(1 - 0.25z^{-1})(1 - 0.25z)}$$

[6]It is assumed that there are no pole/zero cancelations between $P_x(z)$ and $H(z)$.

Therefore, the power spectrum has a pair of poles, one at $z = 0.25$ and one at the reciprocal location $z = 4$.

The autocorrelation of $x(n)$ may be found from $P_x(z)$ as follows. Performing a partial fraction expansion of $P_x(z)$,

$$P_x(z) = \frac{z^{-1}}{(1-0.25z^{-1})(z^{-1}-0.25)} = \frac{16/15}{1-0.25z^{-1}} + \frac{4/15}{z^{-1}-0.25}$$
$$= \frac{16/15}{1-0.25z^{-1}} - \frac{16/15}{1-4z^{-1}}$$

and taking the inverse z-transform yields

$$r_x(k) = \tfrac{16}{15}\left(\tfrac{1}{4}\right)^k u(k) + \tfrac{16}{15}4^k u(-k-1) = \tfrac{16}{15}\left(\tfrac{1}{4}\right)^{|k|}$$

The following example shows that it is possible to work backwards from a rational power spectral density function to find a filter that may be used to generate the process.

Example 3.4.2 *Spectral Shaping Filter*

Suppose that we would like to generate a random process having a power spectrum of the form

$$P_x(e^{j\omega}) = \frac{5+4\cos 2\omega}{10+6\cos\omega} \tag{3.94}$$

by filtering unit variance white noise with a linear shift-invariant filter. Writing $P_x(e^{j\omega})$ in terms of complex exponentials we have

$$P_x(e^{j\omega}) = \frac{5+2e^{j2\omega}+2e^{-j2\omega}}{10+3e^{j\omega}+3e^{-j\omega}}$$

Replacing $e^{j\omega}$ by z gives

$$P_x(z) = \frac{5+2(z^2+z^{-2})}{10+3(z+z^{-1})} = \frac{(2z^2+1)(2z^{-2}+1)}{(3z+1)(3z^{-1}+1)}$$

Performing the factorization

$$P_x(z) = H(z)H(z^{-1}) \tag{3.95}$$

where

$$H(z) = \frac{2z^2+1}{3z+1} = z\frac{2}{3}\frac{1+\frac{1}{2}z^{-2}}{1+\frac{1}{3}z^{-1}}$$

we see that $H(z)$ is a stable filter. Therefore, it follows from Eq. (3.93) that if unit variance white noise is filtered with $H(z)$, then the filtered process will have the power spectrum given in Eq. (3.94). Since introducing a delay into $H(z)$ will not alter the power spectrum of the filtered process, we may equivalently use the filter

$$H(z) = \frac{2}{3}\frac{1+\frac{1}{2}z^{-2}}{1+\frac{1}{3}z^{-1}}$$

which is *causal* and has a unit sample response given by

$$h(n) = \tfrac{2}{3}\left(-\tfrac{1}{3}\right)^n u(n) + \tfrac{1}{3}\left(-\tfrac{1}{3}\right)^{n-2} u(n-2)$$

We will have more to say about factorizations of the form given in Eq. (3.95) in the following section.

Thus far, we have defined the power spectrum to be the Fourier transform of the autocorrelation sequence of a WSS process. Physically, the power spectrum describes how the signal power is distributed as a function of frequency. To see this, let $H(e^{j\omega})$ be a narrow-band bandpass filter with center frequency ω_0 and bandwidth $\Delta\omega$, as shown in Fig. 3.8. If $x(n)$ is a zero mean WSS process that is filtered with $H(e^{j\omega})$, then the output process, $y(n)$, will have a power spectrum given by Eq. (3.91). Since the average power in $y(n)$ is

$$E\{|y(n)|^2\} = r_y(0) = \frac{1}{2\pi}\int_{-\pi}^{\pi} P_y(e^{j\omega})d\omega$$

it follows from Eq. (3.91) that

$$\begin{aligned} E\{|y(n)|^2\} &= \frac{1}{2\pi}\int_{-\pi}^{\pi} |H(e^{j\omega})|^2 P_x(e^{j\omega})d\omega \\ &= \frac{1}{2\pi}\int_{\omega_0-\Delta\omega/2}^{\omega_0+\Delta\omega/2} P_x(e^{j\omega})d\omega \\ &\approx \frac{\Delta\omega}{2\pi} P_x(e^{j\omega_0}) \end{aligned} \tag{3.96}$$

Therefore, $P_x(e^{j\omega})$ may be viewed as a density function that describes how the power in $x(n)$ varies with ω.

Equation (3.96) may be used to show that the power spectrum is a nonnegative function of ω. To do this, note that since $E\{|y(n)|^2\}$ is always nonnegative, then

$$\frac{1}{2\pi}\int_{\omega_0-\Delta\omega/2}^{\omega_0+\Delta\omega/2} P_x(e^{j\omega})d\omega \geq 0 \tag{3.97}$$

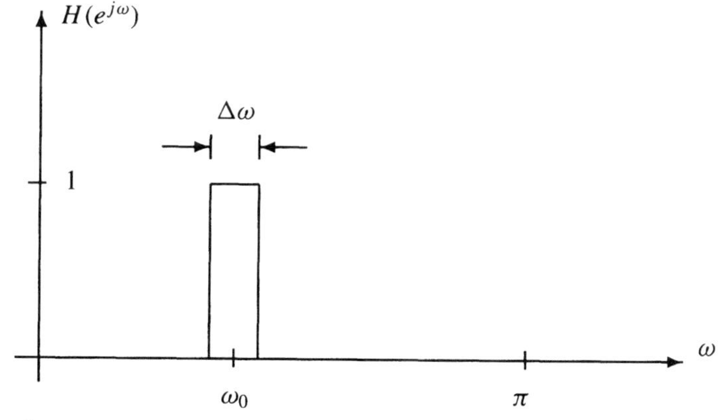

Figure 3.8 *A narrow-band bandpass filter with a center frequency of ω_0 and a passband width of $\Delta\omega$.*

However, the only way that Eq. (3.97) may hold for any ω_0 and for all $\Delta\omega$ is for $P_x(e^{j\omega})$ to be nonnegative.

3.5 SPECTRAL FACTORIZATION

The power spectrum $P_x(e^{j\omega})$ of a wide-sense stationary process is a real-valued, positive, and periodic function of ω. In this section, we show that if $P_x(e^{j\omega})$ is a continuous function of ω, then these constraints imply that $P_x(z)$ may be factored into a product of the form

$$P_x(z) = \sigma_0^2 Q(z) Q^*(1/z^*)$$

This is known as the *spectral factorization* of $P_x(z)$.

Let $x(n)$ be a WSS process with autocorrelation $r_x(k)$ and power spectrum $P_x(e^{j\omega})$. We will assume that $P_x(e^{j\omega})$ is a continuous function of ω, which implies that $x(n)$ contains no periodic components. With

$$P_x(z) = \sum_{k=-\infty}^{\infty} r_x(k) z^{-k}$$

let us further assume that $\ln\big[P_x(z)\big]$ is *analytic* in an annulus $\rho < |z| < 1/\rho$ that contains the unit circle. Analyticity of the logarithm of $P_x(z)$ implies that $\ln\big[P_x(z)\big]$ and all of its derivatives are continuous functions of z and that $\ln[P_x(z)]$ can be expanded in a Laurent series of the form [1]

$$\ln P_x(z) = \sum_{k=-\infty}^{\infty} c(k) z^{-k}$$

where $c(k)$ are the coefficients of the expansion. Thus, $c(k)$ is the sequence that has $\ln\big[P_x(z)\big]$ as its z-transform.[7] Alternatively, evaluating $\ln\big[P_x(z)\big]$ on the unit circle

$$\ln P_x(e^{j\omega}) = \sum_{k=-\infty}^{\infty} c(k) e^{-jk\omega} \tag{3.98}$$

we see that $c(k)$ may also be viewed as the Fourier coefficients of the periodic function $\ln\big[P_x(e^{j\omega})\big]$. Therefore,

$$c(k) = \frac{1}{2\pi} \int_{-\pi}^{\pi} \ln P_x(e^{j\omega}) e^{jk\omega} d\omega$$

Since $P_x(e^{j\omega})$ is real, it follows that the coefficients $c(k)$ are conjugate symmetric, $c(-k) = c^*(k)$. In addition, note that $c(0)$ is proportional to the area under the logarithm of the power spectrum,

$$\boxed{c(0) = \frac{1}{2\pi} \int_{-\pi}^{\pi} \ln P_x(e^{j\omega}) d\omega} \tag{3.99}$$

Using the expansion given in Eq. (3.98), we may write the power spectrum in factored form

[7] The sequence $c(k)$ is the *cepstrum* of the sequence $r_x(k)$ [4].

as follows:

$$
\begin{aligned}
P_x(z) &= \exp\left\{\sum_{k=-\infty}^{\infty} c(k)z^{-k}\right\} \\
&= \exp\left\{c(0)\right\}\exp\left\{\sum_{k=1}^{\infty} c(k)z^{-k}\right\}\exp\left\{\sum_{k=-\infty}^{-1} c(k)z^{-k}\right\}
\end{aligned} \tag{3.100}
$$

Let us define

$$
Q(z) = \exp\left\{\sum_{k=1}^{\infty} c(k)z^{-k}\right\} \quad ; \quad |z| > \rho \tag{3.101}
$$

which is the z-transform of a causal and stable sequence, $q(k)$. Therefore, $Q(z)$ may be expanded in a power series of the form

$$
Q(z) = 1 + q(1)z^{-1} + q(2)z^{-2} + \cdots
$$

where $q(0) = 1$ due to the fact that $Q(\infty) = 1$. Since $Q(z)$ and $\ln\left[Q(z)\right]$ are both analytic for $|z| > \rho$, $Q(z)$ is a *minimum phase* filter. For a rational function of z this implies that $Q(z)$ has no poles or zeros outside the unit circle. As a result, $Q(z)$ has a stable and causal inverse $1/Q(z)$. Using the conjugate symmetry of $c(k)$, we may express the second factor in Eq. (3.100) in terms of $Q(z)$ as follows:

$$
\exp\left\{\sum_{k=-\infty}^{-1} c(k)z^{-k}\right\} = \exp\left\{\sum_{k=1}^{\infty} c^*(k)z^{k}\right\} = \exp\left\{\sum_{k=1}^{\infty} c(k)(1/z^*)^{-k}\right\}^* = Q^*(1/z^*)
$$

This leads to the following *spectral factorization* of the power spectrum $P_x(z)$,

$$
\boxed{P_x(z) = \sigma_0^2\, Q(z)\, Q^*(1/z^*)} \tag{3.102}
$$

where

$$
\boxed{\sigma_0^2 = \exp\left\{c(0)\right\} = \exp\left\{\frac{1}{2\pi}\int_{-\pi}^{\pi} \ln P_x(e^{j\omega})d\omega\right\}} \tag{3.103}
$$

is real and nonnegative. For a real-valued process the spectral factorization takes the form

$$
P_x(z) = \sigma_0^2 Q(z)Q(z^{-1}) \tag{3.104}
$$

Any process that can be factored as in Eq. (3.102) is called a *regular* process [8, 10]. From this factorization we have the following properties of regular processes.

1. Any regular process may be realized as the output of a causal and stable filter that is driven by white noise having a variance of σ_0^2. This is known as the *innovations representation* of the process and is illustrated in Fig. 3.9*a*.
2. The inverse filter $1/H(z)$ is a *whitening filter*. That is, if $x(n)$ is filtered with $1/H(z)$, then the output is white noise with a variance of σ_0^2. The formation of this white noise process, called the *innovations process*, is illustrated in Fig. 3.9*b*,
3. Since $v(n)$ and $x(n)$ are related by an invertible transformation, either process may be derived from the other. Therefore, they both contain the same *information*.

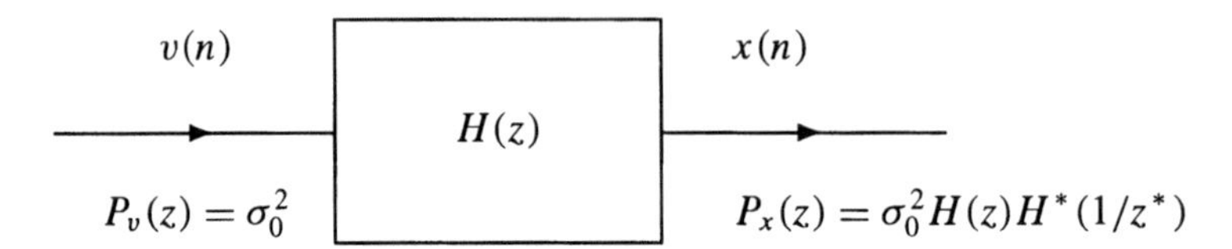

(a) A random process $x(n)$ represented as the response of a minimum phase filter to white noise.

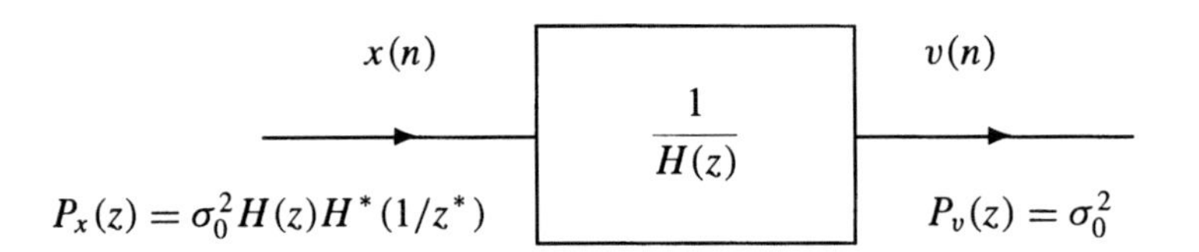

(b) A whitening filter for the process $x(n)$.

Figure 3.9 *Innovations representation of a WSS process.*

For the special case in which $P_x(z)$ is a rational function

$$P_x(z) = \frac{N(z)}{D(z)}$$

the spectral factorization (3.102) states that $P_x(z)$ may be written in factored form as

$$\boxed{P_x(z) = \sigma_0^2 Q(z) Q^*(1/z^*) = \sigma_0^2 \left[\frac{B(z)}{A(z)}\right]\left[\frac{B^*(1/z^*)}{A^*(1/z^*)}\right]} \qquad (3.105)$$

where σ_0^2 is a constant, $B(z)$ is a monic polynomial [8]

$$B(z) = 1 + b(1)z^{-1} + \cdots + b(q)z^{-q}$$

having all of its roots inside the unit circle, and $A(z)$ is a monic polynomial

$$A(z) = 1 + a(1)z^{-1} + \cdots + a(p)z^{-p}$$

with all of its roots inside the unit circle. The fact that any rational power spectrum has a factorization of this form is easy to show. As stated in the symmetry property for the power spectrum (Property 1 on p. 95), since $P_x(e^{j\omega})$ is real, then $P_x(z) = P_x^*(1/z^*)$. Therefore, for each pole (or zero) in $P_x(z)$ there will be a matching pole (or zero) at the conjugate reciprocal location. These symmetry constraints are illustrated in Fig. 3.10 for the zeros of the power spectrum.

Before concluding this discussion, we should point out that the power spectrum given in Eq. (3.102) is not the most general representation for a WSS random process. According to the Wold decomposition theorem, any WSS random process may be decomposed into a sum of two orthogonal processes, a regular process $x_r(n)$ and a predictable process $x_p(n)$ [6, 11]. A process $x_p(n)$ is said to be predictable (deterministic) if there exists a set of coefficients,

[8] A monic polynomial is one for which the coefficient of the zeroth-order term is equal to one. For example, $f(z) = 1 + 2z - 3z^2$ is a second-order monic polynomial whereas $f(z) = 3 + 2z - 3z^2$ is not.

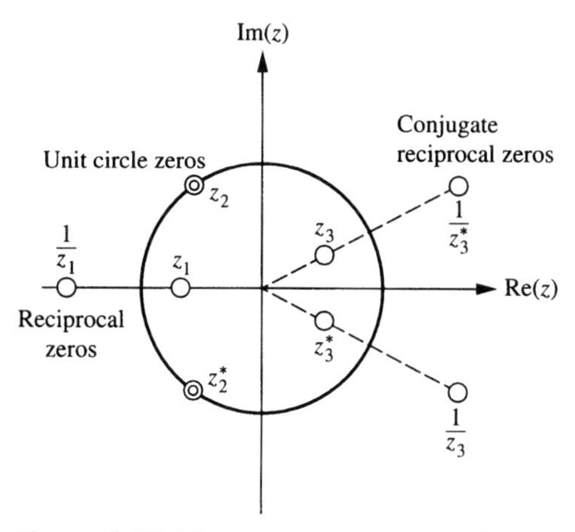

Figure 3.10 *The symmetry constraints that are placed on the zeros of the power spectrum of a real-valued random process when the power spectrum is a rational function of z.*

$a(k)$, such that

$$x_p(n) = \sum_{k=1}^{\infty} a(k) x_p(n-k)$$

In other words, $x_p(n)$ may be predicted without error in terms of a linear combination of its previous values. It may be shown [6] that a process is predictable if and only if its spectrum consists of impulses,

$$P_{x_p}(e^{j\omega}) = \sum_{k=1}^{N} \alpha_k u_0(\omega - \omega_k)$$

Examples of predictable processes include the random phase sinusoid and the complex harmonic process given in Example 3.3.1.

> ***Wold Decomposition Theorem.*** A general random process can be written as a sum of two processes,
>
> $$x(n) = x_p(n) + x_r(n)$$
>
> where $x_r(n)$ is a regular random process and $x_p(n)$ is a predictable process, with $x_r(n)$ being *orthogonal* to $x_p(n)$,
>
> $$E\{x_r(m) x_p^*(n)\} = 0$$

As a corollary to the Wold decomposition theorem, it follows that the general form for the power spectrum of a wide sense stationary process is

$$\boxed{P_x(e^{j\omega}) = P_{x_r}(e^{j\omega}) + \sum_{k=1}^{N} \alpha_k u_0(\omega - \omega_k)} \tag{3.106}$$

where $P_{x_r}(e^{j\omega})$ is the continuous part of the spectrum corresponding to a regular process and where the sum of weighted impulses that give rise to the line spectrum represent the predictable part of the process [6].

3.6 SPECIAL TYPES OF RANDOM PROCESSES

In this section, we look at the characteristics and properties of some of the random processes that we will be encountering in the following chapters. We begin by looking at those processes that may be generated by filtering white noise with a linear shift-invariant filter that has a rational system function. These include the autoregressive, moving average, and autoregressive moving average processes. Of interest will be the form of the autocorrelation sequence and the power spectrum of these processes. We will show, for example, that the autocorrelation sequences of these processes satisfy a set of equations, known as the Yule-Walker equations, relating $r_x(k)$ to the parameters of the filter. In addition to the filtered white noise processes, we will also look at harmonic processes that consist of a sum of random phase sinusoids or complex exponentials. These processes are important in applications where signals have periodic components.

3.6.1 Autoregressive Moving Average Processes

Suppose that we filter white noise $v(n)$ with a causal linear shift-invariant filter having a rational system function with p poles and q zeros

$$H(z) = \frac{B_q(z)}{A_p(z)} = \frac{\sum_{k=0}^{q} b_q(k)z^{-k}}{1 + \sum_{k=1}^{p} a_p(k)z^{-k}} \tag{3.107}$$

Assuming that the filter is stable, the output process $x(n)$ will be wide-sense stationary and, with $P_v(z) = \sigma_v^2$, the power spectrum will be

$$P_x(z) = \sigma_v^2 \frac{B_q(z)B_q^*(1/z^*)}{A_p(z)A_p^*(1/z^*)}$$

or, in terms of ω,

$$\boxed{P_x(e^{j\omega}) = \sigma_v^2 \frac{|B_q(e^{j\omega})|^2}{|A_p(e^{j\omega})|^2}} \tag{3.108}$$

A process having a power spectrum of this form is known as an *autoregressive moving average process* of order (p, q) and is referred to as an ARMA(p, q) process. Note that the power spectrum of an ARMA(p, q) process contains $2p$ poles and $2q$ zeros with the reciprocal symmetry discussed in Section 3.4.

Example 3.6.1 *An ARMA(2,2) Random Process*

The power spectrum of an ARMA(2,2) process is shown in Fig. 3.11. This process is generated by filtering white noise $v(n)$ with a linear shift-invariant system having two poles

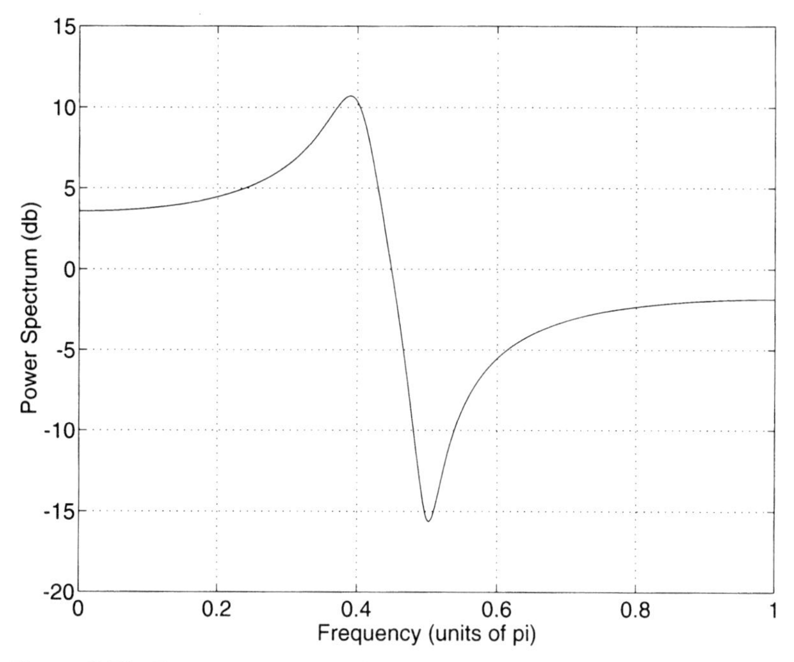

Figure 3.11 *The power spectrum of an ARMA(2,2) random process with zeros at* $z = 0.95e^{\pm j\pi/2}$ *and poles at* $z = 0.9e^{\pm j2\pi/5}$.

and two zeros. The zeros of $H(z)$ are at $z = 0.95e^{\pm j\pi/2}$, and the poles are at $z = 0.9e^{\pm j2\pi/5}$. Thus,

$$H(z) = \frac{1 + 0.9025z^{-2}}{1 - 0.5562z^{-1} + .81z^{-2}}$$

Note that there is a peak in the power spectrum close to $\omega = 0.4\pi$ due to the poles in the denominator of $H(z)$ and a null in the spectrum close to $\omega = 0.5\pi$ due to the zeros of $H(z)$.

Since $x(n)$ and $v(n)$ are related by the linear constant coefficient difference equation

$$x(n) + \sum_{l=1}^{p} a_p(l)x(n-l) = \sum_{l=0}^{q} b_q(l)v(n-l) \tag{3.109}$$

it follows that the autocorrelation of $x(n)$ and the cross-correlation between $x(n)$ and $v(n)$ satisfy the same difference equation. To see this, note that if we multiply both sides of Eq. (3.109) by $x^*(n-k)$ and take the expected value, then we have

$$r_x(k) + \sum_{l=1}^{p} a_p(l)r_x(k-l) = \sum_{l=0}^{q} b_q(l)E\Big\{v(n-l)x^*(n-k)\Big\} \tag{3.110}$$

Since $v(n)$ is WSS, it follows that $v(n)$ and $x(n)$ are jointly WSS (see Section 3.4, p. 100) and

$$E\Big\{v(n-l)x^*(n-k)\Big\} = r_{vx}(k-l)$$

Therefore, Eq. (3.110) becomes

$$r_x(k) + \sum_{l=1}^{p} a_p(l) r_x(k-l) = \sum_{l=0}^{q} b_q(l) r_{vx}(k-l) \tag{3.111}$$

which is a difference equation of the same form as Eq. (3.109). In its present form, this difference equation is not very useful. However, by writing the cross-correlation $r_{vx}(k)$ in terms of autocorrelation $r_x(k)$ and the unit sample response of the filter, we may derive a set of equations relating the autocorrelation of $x(n)$ to the filter. Since

$$x(n) = h(n) * v(n) = \sum_{m=-\infty}^{\infty} v(m) h(n-m)$$

the cross-correlation may be written as

$$\begin{aligned} E\Big\{v(n-l)x^*(n-k)\Big\} &= E\left\{\sum_{m=-\infty}^{\infty} v(n-l)v^*(m)h^*(n-k-m)\right\} \\ &= \sum_{m=-\infty}^{\infty} E\Big\{v(n-l)v^*(m)\Big\}h^*(n-k-m) \\ &= \sigma_v^2 h^*(l-k) \end{aligned} \tag{3.112}$$

where the last equality follows from the fact that $E\{v(n-l)v^*(m)\} = \sigma_v^2\delta(n-l-m)$, i.e., $v(n)$ is white noise. Substituting Eq. (3.112) into (3.111) gives

$$r_x(k) + \sum_{l=1}^{p} a_p(l) r_x(k-l) = \sigma_v^2 \sum_{l=0}^{q} b_q(l) h^*(l-k) \tag{3.113}$$

Assuming that $h(n)$ is causal, the sum on the right side of Eq. (3.113), which we denote by $c_q(k)$, may be written as

$$c_q(k) = \sum_{l=k}^{q} b_q(l) h^*(l-k) = \sum_{l=0}^{q-k} b_q(l+k) h^*(l) \tag{3.114}$$

Since $c_q(k) = 0$ for $k > q$, we may write Eq. (3.113) for $k \geq 0$, as follows

$$r_x(k) + \sum_{l=1}^{p} a_p(l) r_x(k-l) = \begin{cases} \sigma_v^2 c_q(k) & ; \quad 0 \leq k \leq q \\ 0 & ; \quad k > q \end{cases} \tag{3.115}$$

which are the *Yule-Walker equations*. Writing these equations for $k = 0, 1, \ldots, p+q$ in matrix form we have

$$\left[\begin{array}{cccc} r_x(0) & r_x(-1) & \cdots & r_x(-p) \\ r_x(1) & r_x(0) & \cdots & r_x(-p+1) \\ \vdots & \vdots & & \vdots \\ r_x(q) & r_x(q-1) & \cdots & r_x(q-p) \\ \hline r_x(q+1) & r_x(q) & \cdots & r_x(q-p+1) \\ \vdots & \vdots & & \vdots \\ r_x(q+p) & r_x(q+p-1) & \cdots & r_x(q) \end{array}\right] \left[\begin{array}{c} 1 \\ a_p(1) \\ a_p(2) \\ \vdots \\ a_p(p) \end{array}\right] = \sigma_v^2 \left[\begin{array}{c} c_q(0) \\ c_q(1) \\ \vdots \\ c_q(q) \\ \hline 0 \\ \vdots \\ 0 \end{array}\right] \tag{3.116}$$

Note that Eq. (3.115) defines a recursion for the autocorrelation sequence in terms of the filter coefficients, $a_p(k)$ and $b_q(k)$.[9] Therefore, the Yule-Walker equations may be used to extrapolate the autocorrelation sequence from a finite set of values of $r_x(k)$. For example, if $p \geq q$ and if $r_x(0), \ldots, r_x(p-1)$ are known, then $r_x(k)$ for $k \geq p$ may be computed recursively using the difference equation

$$r_x(k) = -\sum_{l=1}^{p} a_p(l) r_x(k-l)$$

Note that the Yule-Walker equations also provide a relationship between the filter coefficients and the autocorrelation sequence. Thus, Eq. (3.115) may be used to estimate the filter coefficients, $a_p(k)$ and $b_q(k)$, from the autocorrelation sequence $r_x(k)$. However, due to the product $h^*(l)b_q(k+l)$ that appears in Eq. (3.114), the Yule-Walker equations are *nonlinear* in the filter coefficients and solving them for the filter coefficients is, in general, difficult. As we will see in later chapters, the Yule-Walker equations are important in problems such as signal modeling and spectrum estimation.

In the next two sections, we look at two special cases of ARMA processes. The first is when $q = 0$ (autoregressive process), and the second is when $p = 0$ (moving average process).

3.6.2 Autoregressive Processes

A special type of ARMA(p, q) process results when $q = 0$ in Eq. (3.107), i.e., when $B_q(z) = b(0)$. In this case $x(n)$ is generated by filtering white noise with an all-pole filter of the form

$$H(z) = \frac{b(0)}{1 + \sum_{k=1}^{p} a_p(k) z^{-k}} \tag{3.117}$$

An ARMA$(p, 0)$ process is called an *autoregressive process* of order p and will be referred to as an AR(p) process. From Eq. (3.92) it follows that if $P_v(z) = \sigma_v^2$, then the power spectrum of $x(n)$ is

$$P_x(z) = \sigma_v^2 \frac{|b(0)|^2}{A_p(z) A_p^*(1/z^*)} \tag{3.118}$$

[9]The unit sample response $h(k)$ may be computed if the filter coefficients $a_p(k)$ and $b_q(k)$ are known.

or, in terms of the frequency variable ω,

$$P_x(e^{j\omega}) = \sigma_v^2 \frac{|b(0)|^2}{\left|A_p(e^{j\omega})\right|^2} \tag{3.119}$$

Thus, the power spectrum of an AR(p) process contains $2p$ poles and no zeros (except those located at $z = 0$ and $z = \infty$). The Yule-Walker equations for an AR(p) process may be found from Eq. (3.115) by setting $q = 0$. With $c_0(0) = b(0)h^*(0) = |b(0)|^2$ these equations are

$$r_x(k) + \sum_{l=1}^{p} a_p(l) r_x(k-l) = \sigma_v^2 |b(0)|^2 \delta(k) \quad ; \quad k \geq 0 \tag{3.120}$$

which, in matrix form become

$$\begin{bmatrix} r_x(0) & r_x(-1) & \cdots & r_x(-p) \\ r_x(1) & r_x(0) & \cdots & r_x(-p+1) \\ \vdots & \vdots & & \vdots \\ r_x(p) & r_x(p-1) & \cdots & r_x(0) \end{bmatrix} \begin{bmatrix} 1 \\ a_p(1) \\ \vdots \\ a_p(p) \end{bmatrix} = \sigma_v^2 |b(0)|^2 \begin{bmatrix} 1 \\ 0 \\ \vdots \\ 0 \end{bmatrix} \tag{3.121}$$

Note that since the Yule-Walker equations are *linear* in the coefficients $a_p(k)$, it is a simple matter to find the coefficients $a_p(k)$ from the autocorrelation sequence $r_x(k)$. For example, suppose that we are given the first two autocorrelation values of a real-valued AR(1) process. Assuming that $\sigma_v^2 = 1$, using the property that $r_x(k) = r_x(-k)$ for real processes the Yule-Walker equations for $k = 0, 1$ are as follows:

$$\begin{aligned} r_x(0) + r_x(1)a(1) &= b^2(0) \\ r_x(0)a(1) &= -r_x(1) \end{aligned}$$

From the second equation, we find for $a(1)$

$$a(1) = -\frac{r_x(1)}{r_x(0)}$$

and, from the first equation, we may solve for $b(0)$,

$$b^2(0) = r_x(0) + r_x(1)a(1) = \frac{r_x^2(0) - r_x^2(1)}{r_x(0)}$$

Expressed in terms of $r_x(0)$ and $r_x(1)$, a first-order filter that generates an AR(1) process with the given autocorrelation values is

$$H(z) = \frac{\sqrt{r_x(0)\left[r_x^2(0) - r_x^2(1)\right]}}{r_x(0) - r_x(1)z^{-1}}$$

In a similar fashion, we may use the Yule-Walker equations to generate the autocorrelation sequence from a given set of filter coefficients. For example, again let $x(n)$ be a first-order autoregressive process. Writing the first two Yule-Walker equations in matrix form in terms of the unknowns $r_x(0)$ and $r_x(1)$ we have

$$\begin{bmatrix} 1 & a(1) \\ a(1) & 1 \end{bmatrix} \begin{bmatrix} r_x(0) \\ r_x(1) \end{bmatrix} = \begin{bmatrix} b^2(0) \\ 0 \end{bmatrix}$$

Solving for $r_x(0)$ and $r_x(1)$ we find

$$r_x(0) = \frac{b^2(0)}{1-a^2(1)}$$

$$r_x(1) = -a(1)r_x(0) = -a(1)\frac{b^2(0)}{1-a^2(1)}$$

Now observe from Eq. (3.120) that for all $k > 0$

$$r_x(k) = -a(1)r_x(k-1).$$

Therefore, the autocorrelation sequence for $k \geq 0$ may be written as

$$r_x(k) = \frac{b^2(0)}{1-a^2(1)}\Big(-a(1)\Big)^k \quad ; \quad k \geq 0$$

Finally, using the symmetry of $r_x(k)$ gives

$$\boxed{r_x(k) = \frac{b^2(0)}{1-a^2(1)}\big[-a(1)\big]^{|k|}} \tag{3.122}$$

We now look at specific examples of AR(1) and AR(2) processes.

Example 3.6.2 ***AR(1) and AR(2) Random Processes***

Consider the real AR(1) process that is formed by filtering unit variance white noise with the first-order all-pole filter

$$H(z) = \frac{b(0)}{1-a(1)z^{-1}}$$

where $b(0)$ and $a(1)$ are real-valued coefficients. The power spectral density of this process is

$$P_x(z) = \frac{b^2(0)}{\big[1-a(1)z^{-1}\big]\big[1-a(1)z\big]} = \frac{b^2(0)}{1+a^2(1)-a(1)\big[z+z^{-1}\big]}$$

or, in terms of Fourier transforms,

$$P_x(e^{j\omega}) = \frac{b^2(0)}{\big[1-a(1)e^{-j\omega}\big]\big[1-a(1)e^{j\omega}\big]} = \frac{b^2(0)}{1+a^2(1)-2a(1)\cos\omega}$$

The power spectrum of this process is shown in Fig. 3.12*a* for $a(1) = 0.5,\ 0.75$, and 0.9 with $b(0)$ chosen so that the peak at $\omega = 0$ is equal to one. Note that the power in $x(n)$ is concentrated in the low frequencies and, as a result, $x(n)$ is referred to as a low-pass process. Also note that as the pole moves closer to the unit circle the process bandwidth decreases. Now consider what happens if $a(1) < 0$. The power spectrum in this case is shown in Fig. 3.12*b* for $a(1) = -0.5,\ -0.75$, and -0.9 with the value of $b(0)$ again chosen so that the peak at $\omega = \pi$ is equal to one. Since the power in $x(n)$ is now concentrated in the high frequencies, $x(n)$ is referred to as a high-pass process.

Now consider the AR(2) process that is generated by filtering white noise with the second-order filter

$$H(z) = \frac{b(0)}{\big[1-a(1)z^{-1}\big]\big[1-a^*(1)z^{-1}\big]}$$

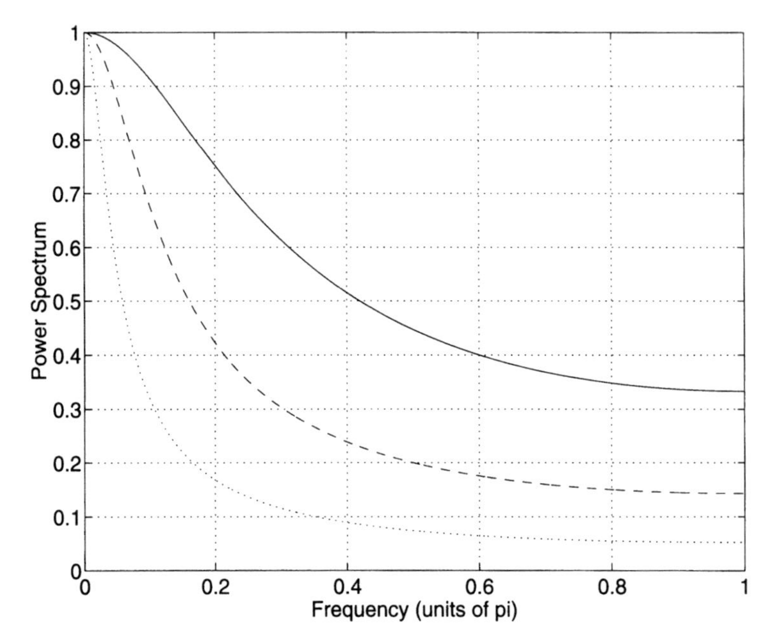

(a) The power spectrum of a low-pass AR(1) random process with a pole at $z = 0.5$ (solid line), a pole at $z = 0.75$ (dashed line), and a pole at $z = 0.9$ (dotted line).

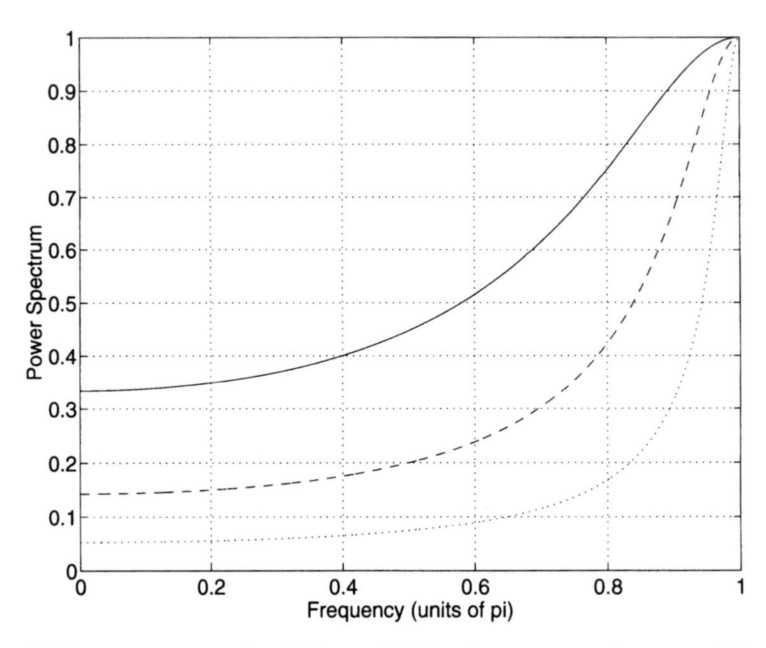

(b) The power spectrum for a high-pass AR(1) random process, pole at $z = -0.5$ (solid line), pole at $z = -0.75$ (dashed line), and pole at $z = -0.9$ (dotted line).

Figure 3.12 *The power spectrum of first-order autoregressive processes. In each case, the numerator coefficient, $b(0)$, is chosen so that the power spectrum is normalized to one at $\omega = 0$.*

With $a(1) = re^{j\omega_0}$ and $\omega_0 = \pi/2$, the power spectrum is shown in Fig. 3.13 for $r = 0.5$, 0.75, and 0.9 with the value of $b(0)$ chosen so that the peak value of the power spectrum is equal to one. Note that the power in $x(n)$ is now concentrated within a band of frequencies centered around ω_0 and that as the pole moves closer to the unit circle the process bandwidth decreases. A process of this form is referred to as a *bandpass* process.

3.6.3 Moving Average Processes

Another special case of ARMA(p, q) process results when $p = 0$. With $p = 0$, $x(n)$ is generated by filtering white noise with an FIR filter that has a system function of the form

$$H(z) = \sum_{k=0}^{q} b_q(k) z^{-k}$$

i.e., $A_p(z) = 1$ in Eq. (3.107). An ARMA($0, q$) process is known as a *moving average process* of order q and will be referred to as an MA(q) process. From Eq. (3.92) note that if $P_v(z) = \sigma_v^2$, then the power spectrum is

$$P_x(z) = \sigma_v^2 B_q(z) B_q^*(1/z^*) \tag{3.123}$$

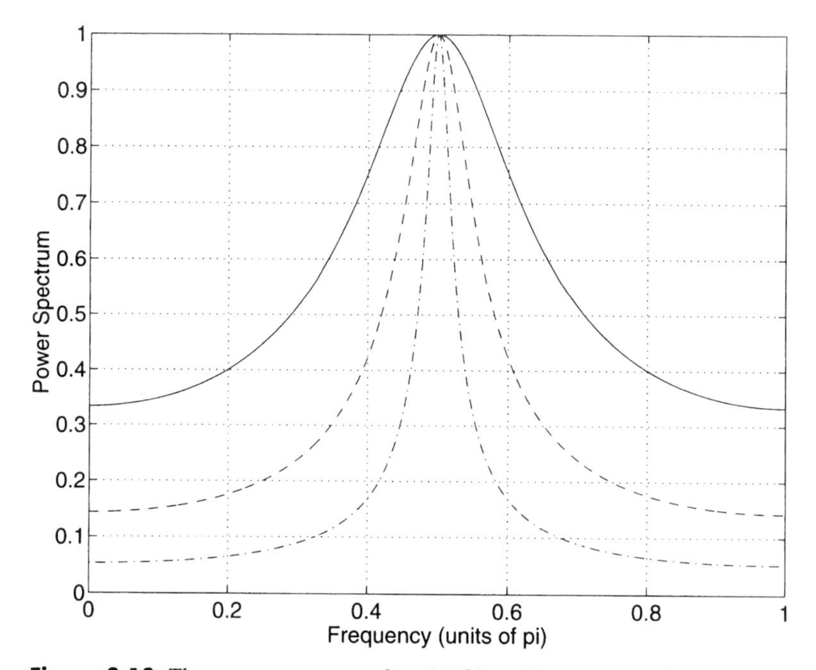

Figure 3.13 *The power spectrum of an AR(2) random process where the filter generating the process has a complex pole pair at $z = re^{\pm j\omega_0}$ where $\omega_0 = \pi/2$ and $r = 0.5$ (solid line), $r = 0.75$ (dashed line), and $r = 0.9$ (dashdot line). The numerator $b(0)$ is chosen so that the peak value of the power spectrum is normalized to one.*

or, in terms of ω,

$$P_x(e^{j\omega}) = \sigma_v^2 \left|B_q(e^{j\omega})\right|^2 \tag{3.124}$$

Therefore, $P_x(z)$ contains $2q$ zeros and no poles (except those that lie at $z = 0$ and $z = \infty$). The Yule-Walker equations for an MA(q) process may be found by simply taking the inverse z-transform of Eq. (3.123). Alternatively, they may be found from Eq. (3.115) by setting $a_p(k) = 0$ and noting that since $h(n) = b(n)$, then

$$c_q(k) = \sum_{l=0}^{q-k} b_q(l+k)b^*(l)$$

In either case we have

$$r_x(k) = \sigma_v^2 b_q(k) * b_q^*(-k) = \sigma_v^2 \sum_{l=0}^{q-|k|} b_q(l+|k|)b_q^*(l) \tag{3.125}$$

where we have used the absolute value on k so that the expression for $r_x(k)$ is valid for all k. Thus, we see that the autocorrelation sequence of an MA(q) process is equal to zero for all values of k that lie outside the interval $[-q, q]$. In addition, note that $r_x(k)$ depends nonlinearly on the moving average parameters, $b_q(k)$. For example, for $q = 2$

$$\begin{aligned} r_x(0) &= |b(0)|^2 + |b(1)|^2 + |b(2)|^2 \\ r_x(1) &= b^*(0)b(1) + b^*(1)b(2) \\ r_x(2) &= b^*(0)b(2) \end{aligned} \tag{3.126}$$

Therefore, unlike the case for an autoregressive process, estimating the moving average parameters for a moving average process is a nontrivial problem.

Moving average processes are characterized by slowly changing functions of frequency that will have sharp nulls in the spectrum if $P_x(z)$ contains zeros that are close to the unit circle. The following example provides an illustration of this behavior for an MA(4) process.

Example 3.6.3 *An MA(4) Process*

Shown in Fig. 3.14 is the power spectrum of an MA(4) process that is generated by filtering white noise with a filter having a pair of complex zeros at $z = e^{\pm j\pi/2}$ and a pair of complex zeros at $z = 0.8e^{\pm j3\pi/4}$. Thus, the autocorrelation sequence is equal to zero for $|k| > 4$. Due to the zeros on the unit circle, the power spectrum goes to zero at $\omega = \pm\pi/2$. In addition, due to the zeros at a radius of 0.8 there is an additional null in the spectrum close to $\omega = 3\pi/4$.

3.6.4 Harmonic Processes

Harmonic processes provide useful representations for random processes that arise in applications such as array processing, when the signals contain periodic components. An example of a wide sense stationary harmonic process is the random phase sinusoid

$$x(n) = A\sin\big(n\omega_0 + \phi\big)$$

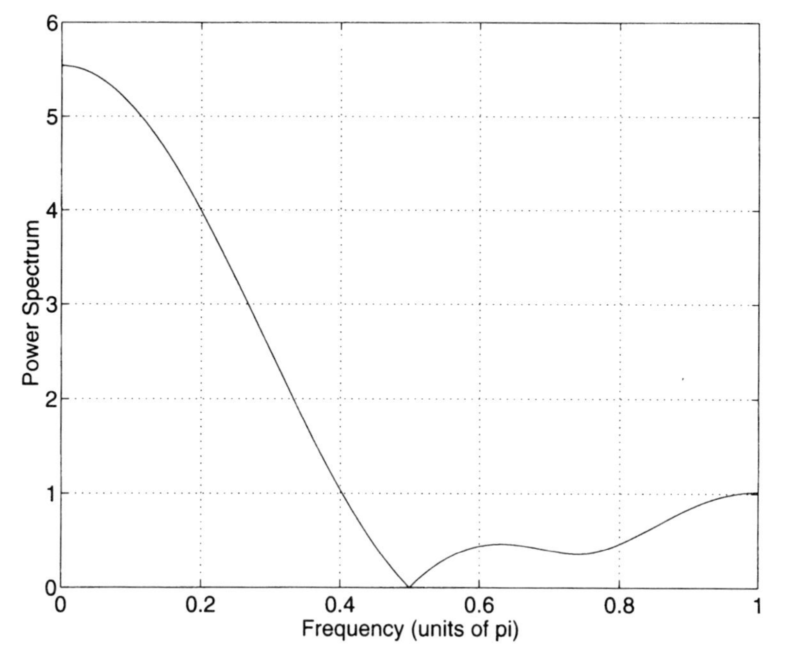

Figure 3.14 *The power spectrum of an MA(4) process having zeros at* $z = e^{\pm j\pi/2}$ *and* $z = 0.8e^{\pm j3\pi/4}$.

where A and ω_0 are constants and ϕ is a random variable that is uniformly distributed between $-\pi$ and π. As shown in Example 3.3.1, the autocorrelation sequence of $x(n)$ is periodic with frequency ω_0,

$$r_x(k) = \tfrac{1}{2}A^2 \cos(k\omega_0) \tag{3.127}$$

Therefore, the power spectrum is

$$P_x(e^{j\omega}) = \frac{\pi}{2}A^2\Big[u_0(\omega - \omega_0) + u_0(\omega + \omega_0)\Big]$$

where $u_0(\omega - \omega_0)$ is an impulse at frequency ω_0 and $u_0(\omega + \omega_0)$ is an impulse at $-\omega_0$. If the amplitude is also a random variable that is uncorrelated with ϕ, then the autocorrelation sequence will be the same as Eq. (3.127) with A^2 replaced with $E\{A^2\}$,

$$r_x(k) = \tfrac{1}{2}E\left\{A^2\right\}\cos(k\omega_0)$$

Higher-order harmonic processes may be formed from a sum of random phase sinusoids as follows:

$$x(n) = \sum_{l=1}^{L} A_l \sin\big(n\omega_l + \phi_l\big)$$

Assuming the random variables ϕ_l and A_l are uncorrelated, the autocorrelation sequence is

$$r_x(k) = \sum_{l=1}^{L} \tfrac{1}{2}E\big\{A_l^2\big\}\cos(k\omega_l)$$

and the power spectrum is

$$P_x(e^{j\omega}) = \sum_{l=1}^{L} \tfrac{1}{2}\pi E\{A_l^2\}\Big[u_0(\omega - \omega_l) + u_0(\omega + \omega_l)\Big]$$

In the case of complex signals, a first order harmonic process has the form

$$x(n) = |A|e^{j(n\omega_0+\phi)}$$

where ϕ, and possibly $|A|$, are random variables. Similar to what we found in the case of a sinusoid, if ϕ and A are uncorrelated random variables, then the autocorrelation sequence is a complex exponential with frequency ω_0,

$$r_x(k) = E\{|A|^2\}\exp\big(jk\omega_0\big)$$

and the power spectrum is

$$P_x(e^{j\omega}) = 2\pi E\{|A|^2\}u_0(\omega - \omega_0)$$

With a sum of L uncorrelated harmonic processes

$$x(n) = \sum_{l=1}^{L} A_l \exp\big[j(n\omega_l + \phi_l)\big]$$

the autocorrelation sequence is

$$r_x(k) = \sum_{l=1}^{L} E\{|A_l|^2\}\exp\big[jk\omega_l\big]$$

and the power spectrum is

$$P_x(e^{j\omega}) = \sum_{l=1}^{L} 2\pi E\{|A_l|^2\}u_0(\omega - \omega_l)$$

3.7 SUMMARY

This chapter provided an introduction to the basic theory of discrete-time random processes. Beginning with a review of random variables, we then introduced a discrete-time random process as an indexed sequence of random variables. Although a complete description of a random process requires the specification of the joint distribution or density functions, our concern in this book is primarily with ensemble averages. Therefore, we proceeded to define the mean, the variance, and the autocorrelation of a discrete-time random process. Next, we introduced the notion of stationarity and defined a process to be wide-sense stationary (WSS) if its mean is constant and autocorrelation $E\big\{x(k)x^*(l)\big\}$ depends only on the difference, $k - l$, between the times of the samples $x(k)$ and $x(l)$. Although it is not generally possible to assume, in practice, that a process is wide-sense stationary, in some cases it is safe to assume that it is wide-sense stationary over a sufficiently short time interval. Throughout most of this book, the processes that we will be considering will be assumed to be wide-sense stationary. However, in Chapter 9 we will take a broader view of statistical signal processing and look at the problem of filtering nonstationary random processes.

In many of the problems that we will be considering in this book it will be necessary to know the first- and second-order statistics of a process. Since these statistics are not generally known a priori, we will find it necessary to estimate these statistics from a single realization of a process. This leads us to the notion of ergodicity, which is concerned with the problem of estimating the statistical properties of a random process from time averages of a single realization. Although a proper treatment of ergodicity requires a deep knowledge of probability theory, we introduced two useful and interesting mean ergodic theorems and an autocorrelation ergodic theorem for Gaussian processes. Instead of concerning ourselves with whether or not the conditions of the ergodic theorems are satisfied (interestingly, these theorems cannot be applied without some a priori knowledge of higher order statistics), in this book it will be assumed that when an ensemble average is required it may be estimated by simply taking an appropriate time average. The ultimate justification for this assumption, in practice, will be in the performance of the algorithm that uses these estimates.

Another important statistical quantity introduced in this chapter is the power spectrum, which is the discrete-time Fourier transform of a wide-sense stationary process. Just as with deterministic signals, the power spectrum is an important tool in the description, modeling, and analysis of wide-sense stationary processes. So important and useful is the power spectrum, in fact, that an entire chapter will be devoted to the problem of power spectrum estimation. We also developed the very important *spectral factorization theorem*, which gives us insight into how to design filters to whiten a wide-sense stationary process or to generate a process with a given rational power spectrum by filtering white noise.

Next we looked at the process of filtering a wide-sense stationary process with a discrete-time filter. Of particular importance is how the mean, autocorrelation, and power spectrum of the input process are modified by the filter. The fundamental result here was that the power spectrum of the input process is multiplied by the magnitude squared of the frequency response of the filter. Finally, we concluded the chapter with a listing of some of the important types of processes that will be encountered in this book. These processes include those that result from filtering white noise with a linear shift-invariant filter. Depending upon whether the filter is FIR, all-pole, or a general pole-zero filter, we saw that it was possible to generate moving average, autoregressive, and autoregressive moving average processes, respectively.

References

1. R. V. Churchill and J. W. Brown, *Introduction to Complex Variables and Applications*, 4th ed., McGraw-Hill, New York, 1984.
2. S. Haykin, *Modern Filters*, MacMillan, New York, 1989.
3. H. L. Larson and B. O. Shubert, *Probabilistic Models in Engineering Sciences: Vol. I, Random Variables and Stochastic Processes*, John Wiley & Sons, New York, 1979.
4. A. V. Oppenheim and R. W. Schafer, *Discrete-Time Signal Processing*, Prentice-Hall, Englewood Cliffs, NJ, 1989.
5. A. Papoulis, *Probability, Random Variables, and Stochastic Processes*, 3rd ed., McGraw-Hill, New York, 1991.
6. A. Papoulis, "Predictable processes and Wold's decomposition: A review," *IEEE Trans. Acoust., Speech, Sig. Proc.*, vol. 33, no. 4, pp. 933–938, August, 1985.
7. B. Picinbono, *Random Signals and Noise*, Prentice-Hall, Englewood Cliffs, NJ, 1993.

8. B. Porat, *Digital Processing of Random Signals*, Prentice-Hall, Englewood Cliffs, NJ, 1994.
9. I. S. Reed, "On a moment factoring theorem for complex Gaussian processes," *IRE Trans. Inf. Theory*, vol. IT-8, pp. 194–195.
10. C. W. Therrien, *Discrete Random Signals and Statistical Signal Processing*, Prentice-Hall, Englewood Cliffs, NJ, 1992.
11. H. Wold, *The Analysis of Stationary Time Series*, 2nd ed., Almquist and Wicksell, Uppsala, Sweden, 1954 (originally published in 1934).

3.8 PROBLEMS

3.1. Let x be a random variable with mean m_x and variance σ_x^2. Let x_i for $i = 1, 2, \ldots, N$ be N independent measurements of the random variable x.

(a) With $\hat{m}_x$ the sample mean defined by

$$\hat{m}_x = \frac{1}{N}\sum_{i=1}^{N} x_i$$

determine whether or not the sample variance

$$\hat{\sigma}_x^2 = \frac{1}{N}\sum_{i=1}^{N}(x_i - \hat{m}_x)^2$$

is unbiased, i.e., is $E\{\hat{\sigma}_x^2\} = \sigma_x^2$?

(b) If x is a Gaussian random variable, find the variance of the sample variance,

$$E\{(\hat{\sigma}_x^2 - E\{\hat{\sigma}_x^2\})^2\}.$$

3.2. Let $x(n)$ be a stationary random process with zero mean and autocorrelation $r_x(k)$. We form the process, $y(n)$, as follows:

$$y(n) = x(n) + f(n)$$

where $f(n)$ is a known deterministic sequence. Find the mean $m_y(n)$ and the autocorrelation $r_y(k, l)$ of the process $y(n)$.

3.3. A discrete-time random process $x(n)$ is generated as follows:

$$x(n) = \sum_{k=1}^{p} a(k)x(n-k) + w(n)$$

where $w(n)$ is a white noise process with variance σ_w^2. Another process, $z(n)$, is formed by adding noise to $x(n)$,

$$z(n) = x(n) + v(n)$$

where $v(n)$ is white noise with a variance of σ_v^2 that is uncorrelated with $w(n)$.

(a) Find the power spectrum of $x(n)$.

(b) Find the power spectrum of $z(n)$.

3.4. Suppose we are given a linear shift-invariant system having a system function

$$H(z) = \frac{1 - \frac{1}{2}z^{-1}}{1 - \frac{1}{3}z^{-1}}$$

that is excited by zero mean exponentially correlated noise $x(n)$ with an autocorrelation sequence

$$r_x(k) = \left(\tfrac{1}{2}\right)^{|k|}$$

Let $y(n)$ be the output process, $y(n) = x(n) * h(n)$.

(a) Find the power spectrum, $P_y(z)$, of $y(n)$.

(b) Find the autocorrelation sequence, $r_y(k)$, of $y(n)$.

(c) Find the cross-correlation, $r_{xy}(k)$, between $x(n)$ and $y(n)$.

(d) Find the *cross-power spectral density*, $P_{xy}(z)$, which is the z transform of the cross-correlation $r_{xy}(k)$.

3.5. Find the power spectrum for each of the following wide-sense stationary random processes that have the given autocorrelation sequences.

(a) $r_x(k) = 2\delta(k) + j\delta(k-1) - j\delta(k+1)$.

(b) $r_x(k) = \delta(k) + 2(0.5)^{|k|}$.

(c) $r_x(k) = 2\delta(k) + \cos(\pi k/4)$.

(d) $r_x(k) = \begin{cases} 10 - |k| & ; \quad |k| < 10 \\ 0 & ; \quad \text{otherwise} \end{cases}$

3.6. Find the autocorrelation sequence corresponding to each of the following power spectral densities.

(a) $P_x(e^{j\omega}) = 3 + 2\cos\omega$.

(b) $P_x(e^{j\omega}) = \dfrac{1}{5 + 3\cos\omega}$.

(c) $P_x(z) = \dfrac{-2z^2 + 5z - 2}{3z^2 + 10z + 3}$.

3.7. Let $x(n)$ be a zero mean WSS process with an $N \times N$ autocorrelation matrix $\mathbf{R}_x$. Determine whether each of the following statements are *True* or *False*.

(a) If the eigenvalues of $\mathbf{R}_x$ are equal, $\lambda_1 = \lambda_2 = \cdots = \lambda_N$, then $r_x(k) = 0$ for $k = 1, 2, \ldots, N$.

(b) If $\lambda_1 > 0$ and $\lambda_k = 0$ for $k = 2, 3, \ldots, N$, then $r_x(k) = Ae^{jk\omega_0}$.

3.8. Consider the random process

$$x(n) = A\cos(n\omega_0 + \phi) + w(n)$$

where $w(n)$ is zero mean white Gaussian noise with a variance σ_w^2. For each of the following cases, find the autocorrelation sequence and if the process is WSS, find the power spectrum.

(a) A is a Gaussian random variable with zero mean and variance σ_A^2 and both ω_0 and ϕ are constants.

(b) ϕ is uniformly distributed over the interval $[-\pi, \pi]$ and both A and ω_0 are constants.

(c) ω_0 is a random variable that is uniformly distributed over the interval $[\omega_0 - \Delta, \omega_0 + \Delta]$ and both A and ϕ are constants.

3.9. Determine whether or not each of the following are valid autocorrelation matrices. If they are not, explain why not.

(a) $\mathbf{R}_1 = \begin{bmatrix} 4 & 1 & 1 \\ -1 & 4 & 1 \\ -1 & -1 & 4 \end{bmatrix}$

(b) $\mathbf{R}_2 = \begin{bmatrix} 2 & 2 \\ 2 & 2 \end{bmatrix}$

(c) $\mathbf{R}_3 = \begin{bmatrix} 1 & 1+j \\ 1-j & 1 \end{bmatrix}$

(d) $\mathbf{R}_4 = \begin{bmatrix} 3 & 2 & 1 \\ 2 & 4 & 2 \\ 1 & 2 & 3 \end{bmatrix}$

(e) $\mathbf{R}_5 = \begin{bmatrix} 2 & j & 1 \\ -j & 4j & -j \\ 1 & j & 2 \end{bmatrix}$

3.10. The input to a linear shift-invariant filter with unit sample response

$$h(n) = \delta(n) + \tfrac{1}{2}\delta(n-1) + \tfrac{1}{4}\delta(n-2)$$

is a zero mean wide-sense stationary process with autocorrelation

$$r_x(k) = \left(\tfrac{1}{2}\right)^{|k|}$$

(a) What is the variance of the output process?

(b) Find the autocorrelation of the output process, $r_y(k)$, for all k.

3.11. Consider a first-order AR process that is generated by the difference equation

$$y(n) = ay(n-1) + w(n)$$

where $|a| < 1$ and $w(n)$ is a zero mean white noise random process with variance σ_w^2.

(a) Find the unit sample response of the filter that generates $y(n)$ from $w(n)$.

(b) Find the autocorrelation of $y(n)$.

(c) Find the power spectrum of $y(n)$.

3.12. Consider an MA(q) process that is generated by the difference equation

$$y(n) = \sum_{k=0}^{q} b(k)w(n-k)$$

where $w(n)$ is zero mean white noise with variance σ_w^2.

(a) Find the unit sample response of the filter that generates $y(n)$ from $w(n)$.

(b) Find the autocorrelation of $y(n)$.

(c) Find the power spectrum of $y(n)$.

3.13. Suppose we are given a zero mean process $x(n)$ with autocorrelation

$$r_x(k) = 10\left(\tfrac{1}{2}\right)^{|k|} + 3\left(\tfrac{1}{2}\right)^{|k-1|} + 3\left(\tfrac{1}{2}\right)^{|k+1|}$$

(a) Find a filter which, when driven by unit variance white noise, will yield a random process with this autocorrelation.

(b) Find a stable and causal filter which, when excited by $x(n)$, will produce zero mean, unit variance, white noise.

3.14. For each of the following, determine whether or not the random process is

i. Wide-sense stationary.

ii. Mean ergodic.

(a) $x(n) = A$ where A is a random variable with probability density function $f_A(\alpha)$.

(b) $x(n) = A\cos(n\omega_0)$ where A is a Gaussian random variable with mean m_A and variance σ_A^2.

(c) $x(n) = A\cos(n\omega_0 + \phi)$ where ϕ is a random variable that is uniformly distributed between $-\pi$ and π.

(d) $x(n) = A\cos(n\omega_0) + B\sin(n\omega_0)$ where A and B are uncorrelated zero mean random variables with variance σ^2.

(e) A Bernoulli process with $\Pr\{x(n) = 1\} = p$ and $\Pr\{x(n) = -1\} = 1 - p$.

(f) $y(n) = x(n) - x(n-1)$ where $x(n)$ is the Bernoulli process defined in part (e).

3.15. Determine which of the following correspond to a valid autocorrelation sequence for a WSS random process. For those that are not valid, state why not. For those that are valid, describe a way for generating a process with the given autocorrelation.

(a) $r_x(k) = \delta(k-1) + \delta(k+1)$

(b) $r_x(k) = 3\delta(k) + 2\delta(k-1) + 2\delta(k+1)$

(c) $r_x(k) = \exp(jk\pi/4)$

(d) $r_x(k) = \begin{cases} 1 & ; \quad |k| < N \\ 0 & ; \quad \text{else} \end{cases}$

(e) $r_x(k) = \begin{cases} \dfrac{N - |k|}{N} & ; \quad |k| < N \\ 0 & ; \quad \text{else} \end{cases}$

(f) $r_x(k) = 2^{-k^2}$

3.16. Show that the cross-correlation, $r_{xy}(k)$, between two jointly wide-sense stationary processes $x(n)$ and $y(n)$ satisfies the following inequalities,

(a) $|r_{xy}(k)| \le \Big[r_x(0)r_y(0)\Big]^{1/2}$

(b) $|r_{xy}(k)| \le \frac{1}{2}\Big[r_x(0) + r_y(0)\Big]$

3.17. Given a wide-sense stationary random process $x(n)$, we would like to design a "linear predictor" that will predict the value of $x(n+1)$ using a linear combination of $x(n)$ and $x(n-1)$. Thus, our predictor for $x(n+1)$ is of the form

$$\hat{x}(n+1) = ax(n) + bx(n-1)$$

where a and b are constants. Assume that the process has zero mean

$$E\{x(n)\} = 0$$

and that we want to minimize the mean-square error

$$\xi = E\left\{\left[x(n+1) - \hat{x}(n+1)\right]^2\right\}$$

(a) With $r_x(k)$ the autocorrelation of $x(n)$, determine the optimum predictor of $x(n)$ by finding the values of a and b that minimize the mean-square error.

(b) What is the minimum mean-square error of the predictor? Express your answer in terms of the autocorrelation $r_x(k)$.

(c) If $x(n+1)$ is uncorrelated with $x(n)$, what form does your predictor take?

(d) If $x(n+1)$ is uncorrelated with both $x(n)$ and $x(n-1)$, what form does your predictor take?

3.18. *True* or *False:* If $x(n)$ is a WSS process and $y(n)$ is the process that is formed by filtering $x(n)$ with a stable, linear shift-invariant filter $h(n)$, then

$$\sigma_y^2 = \sigma_x^2 \sum_{n=-\infty}^{\infty} |h(n)|^2$$

where σ_x^2 and σ_y^2 are the variances of the processes $x(n)$ and $y(n)$, respectively.

3.19. Show that a sufficient condition for a wide-sense stationary process to be ergodic in the mean is that the autocovariance be absolutely summable,

$$\sum_{k=-\infty}^{\infty} |c_x(k)| < \infty$$

3.20. For each of the following, determine whether the statement is *True* or *False*.

(a) All wide-sense stationary moving average processes are ergodic in the mean.

(b) All wide-sense stationary autoregressive processes are ergodic in the mean.

3.21. Let $x(n)$ be a real WSS Gaussian random process with autocovariance function $c_x(k)$. Show that $x(n)$ will be *correlation ergodic* if and only if

$$\lim_{N\to\infty} \frac{1}{N} \sum_{k=0}^{N-1} c_x^2(k) = 0$$

Hint: Use the *moment factoring theorem* for real Gaussian random variables which states that

$$E\left\{x_1x_2x_3x_4\right\} = E\left\{x_1x_2\right\}E\left\{x_3x_4\right\} + E\left\{x_1x_3\right\}E\left\{x_2x_4\right\} + E\left\{x_1x_4\right\}E\left\{x_2x_3\right\}$$

3.22. Ergodicity in the mean depends on the asymptotic behavior of the autocovariance of a process, $c_x(k)$. The asymptotic behavior of $c_x(k)$, however, is related to the behavior of the power spectrum $P_x(e^{j\omega})$ at $\omega = 0$. Show that $x(n)$ is ergodic in the mean if and only if $P_x(e^{j\omega})$ is continuous at the origin, $\omega = 0$. *Hint:* Express Eq. (3.65) as a limit, as $N \to \infty$, of the convolution of $c_x(k)$ with a pulse

$$p_N(k) = \begin{cases} 1/N & ; \quad 0 \le k < N \\ 0 & ; \quad \text{otherwise} \end{cases}$$

with the convolution being evaluated at $k = N$.

3.23. In Section 3.2.4 it was shown that for real-valued zero mean random variables the autocorrelation is bounded by one in magnitude

$$|\rho_{xy}| \leq 1$$

Show that this bound also applies when x and y are complex random variables with nonzero mean. Determine what relationship must hold between x and y if

$$|\rho_{xy}| = 1$$

3.24. Let $P_x(e^{j\omega})$ be the power spectrum of a wide-sense stationary process $x(n)$ and let λ_k be the eigenvalues of the $M \times M$ autocorrelation matrix $\mathbf{R}_x$. *Szegö's theorem* states that if $g(\cdot)$ is a continuous function then

$$\lim_{M\to\infty} \frac{g(\lambda_1) + g(\lambda_2) + \cdots + g(\lambda_M)}{M} = \frac{1}{2\pi}\int_{-\pi}^{\pi} g\big[P_x(e^{j\omega})\big]d\omega$$

Using this theorem, show that

$$\lim_{M\to\infty}\Big[\det \mathbf{R}_x\Big]^{1/M} = \exp\left\{\frac{1}{2\pi}\int_{-\pi}^{\pi} \ln\Big[P_x(e^{j\omega})\Big]d\omega\right\}$$

3.25. In some applications, the data collection process may be flawed so that there are either missing data values or outliers that should be discarded. Suppose that we are given N samples of a WSS process $x(n)$ with one value, $x(n_0)$, missing. Let $\mathbf{x}$ be the vector containing the given sample values,

$$\mathbf{x} = \big[x(0),\ x(1), \ldots, x(n_0-1),\ x(n_0+1), \ldots, x(N)\big]^T$$

(a) Let $\mathbf{R}_x$ be the autocorrelation matrix for the vector $\mathbf{x}$,

$$\mathbf{R}_x = E\{\mathbf{x}\mathbf{x}^H\}$$

Which of the following statements are true:

1. $\mathbf{R}_x$ is Toeplitz.
2. $\mathbf{R}_x$ is Hermitian.
3. $\mathbf{R}_x$ is positive semidefinite.

(b) Given the autocorrelation matrix for $\mathbf{x}$, is it possible to find the autocorrelation matrix for the vector

$$\mathbf{x} = \big[x(0),\ x(1), \ldots, x(N)\big]^T$$

that does not have $x(n_0)$ missing? If so, how would you find it? If not, explain why not.

3.26. The power spectrum of a wide-sense stationary process $x(n)$ is

$$P_x(e^{j\omega}) = \frac{25 - 24\cos\omega}{26 - 10\cos\omega}$$

Find the whitening filter $H(z)$ that produces unit variance white noise when the input is $x(n)$.

3.27. We have seen that the autocorrelation matrix of a WSS process is positive semidefinite,

$$\mathbf{R}_x \geq 0$$

The spectral factorization theorem states that if $P_x(e^{j\omega})$ is continuous then the power spectrum may be factored as

$$P_x(e^{j\omega}) = \sigma_0^2 |Q(e^{j\omega})|^2$$

where $Q(e^{j\omega})$ corresonds to a causal and stable filter.

(a) If $\sigma_0^2 \neq 0$ and $Q(e^{j\omega})$ is nonzero for all ω, show that the autocorrelation matrix is positive definite.

(b) Give an example of a nontrivial process for which $\mathbf{R}_x$ is *not* positive definite.

Computer Exercises

C3.1. If u is a random variable with a uniform probability density function,

$$f_u(\alpha) = \begin{cases} 1 & ; \quad 0 \leq \alpha \leq 1 \\ 0 & ; \quad \text{otherwise} \end{cases}$$

then a random variable x with a distribution function $F_x(\alpha)$ may be formed from u using the transformation [5]

$$x = F_x^{-1}(u)$$

(a) Let u be uniformly distributed between 0 and 1. How would you create an exponentially distributed random variable with a density function

$$f_x(\alpha) = \frac{1}{\mu} e^{-\alpha/\mu} \quad ; \quad \alpha \geq 0$$

(b) Find the mean and variance of the exponentially distributed random variable x given in (a).

(c) How many samples of an exponentially distributed random variable x are required in order for the sample mean

$$\hat{m}_x = \frac{1}{N} \sum_{i=1}^{N} x_i$$

to be within 1% of the true mean with a probability of 0.01, i.e.,

$$\text{Prob}\left\{|\hat{m}_x - m_x| > 0.01 m_x\right\} < 0.01$$

(d) Using the MATLAB m-file `rand`, which generates random numbers that are uniformly distributed between 0 and 1, generate a sufficient number of exponentially distributed random numbers having a density function

$$f_x(\alpha) = \frac{1}{2} e^{-\alpha/2} \quad ; \quad \alpha \geq 0$$

so that the sample mean will be within 1% of the true mean with a probability of 0.01. Find the sample mean and compare it to the true mean. Repeat this experiment for several different sets of random samples.

C3.2. In many applications it is necessary to be able to efficiently estimate the autocorrelation sequence of a random process from a finite number of samples, e.g. $x(n)$ for

$n = 0, 1, \ldots, N-1$. The autocorrelation may be estimated using the sample autocorrelation

$$\hat{r}_x(k) = \frac{1}{N}\sum_{n=0}^{N-1} x(n)x^*(n-k)$$

Since $x(n)$ is only given for values of n within the interval $[0, N-1]$, in evaluating this sum $x(n)$ is assumed to be equal to zero for values of n that are outside the interval $[0, N-1]$. Alternatively, the sample mean may be expressed as

$$\hat{r}_x(k) = \frac{1}{N}\sum_{n=k}^{N-1} x(n)x^*(n-k) = \frac{1}{N}\sum_{n=0}^{N-1-k} x(n+k)x^*(n) \quad ; \quad 0 \le k < N$$

with $\hat{r}_x(k) = \hat{r}_x^*(-k)$ for $k < 0$, and $\hat{r}_x(k) = 0$ for $|k| \ge N$.

(a) Show that $\hat{r}_x(k)$ may be written as a convolution

$$\hat{r}_x(k) = \frac{1}{N}x(k) * x^*(-k)$$

if it is assumed that the values of $x(n)$ outside the interval $[0, N-1]$ are zero.

(b) Show that the discrete-time Fourier transform of $\hat{r}_x(k)$ is equal to the magnitude squared of the discrete-time Fourier transform of $x(n)$,

$$\sum_{k=-N+1}^{N-1} \hat{r}_x(k)e^{-jk\omega} = \frac{1}{N}\left|X(e^{j\omega})\right|^2$$

where

$$X(e^{j\omega}) = \sum_{n=0}^{N-1} x(n)e^{-jn\omega}$$

In other words, the sample autocorrelation may be computed by taking the inverse DTFT of the magnitude squared of the DTFT of $x(n)$, scaled by $1/N$.

(c) With $x(n) = 1$ for $n = 0, 1, \ldots, 7$, plot the sample autocorrelation sequence $\hat{r}_x(k)$. Using MATLAB, find the sample autocorrelation of $x(n)$ using DFT's of length 8, 16, and 32. Comment on what you observe. (Note: the sample autocorrelation may be computed using a single MATLAB command line.)

C3.3. In this exercise we will look at how accurately the sample autocorrelation is in estimating the autocorrelation of white noise.

(a) Generate 1000 samples of zero mean unit variance white Gaussian noise.

(b) Estimate the first 100 lags of the autocorrelation sequence using the sample autocorrelation

$$\hat{r}_x(k) = \frac{1}{1000}\sum_{n=0}^{999} x(n)x(n-k)$$

How close is your estimate to the *true* autocorrelation sequence $r_x(k) = \delta(k)$?

(c) Segment your white noise sequence into ten different white noise sequences each having a length of 100 samples, and estimate the autocorrelation by averaging the sample autocorrelations of each subsequence, i.e.,

$$\hat{r}_x(k) = \frac{1}{1000}\sum_{m=0}^{9}\sum_{n=0}^{99} x(n+100m)x(n-k+100m) \quad ; \quad k=0,1,\ldots,99$$

How does your estimate compare to that in part (b)? How does it compare to the true autocorrelation sequence $r_x(k)$?

(d) Generate 10,000 samples of a zero mean unit variance white Gaussian noise sequence and estimate the first 100 lags of the autocorrelation sequence as in part (b). How does your estimate compare to that in part (b)? What conclusions can you draw from these experiments?

C3.4. Consider the autoregressive random process

$$x(n) = a(1)x(n-1) + a(2)x(n-2) + b(0)v(n)$$

where $v(n)$ is unit variance white noise.

(a) With $a(1) = 0$, $a(2) = -0.81$, and $b(0) = 1$, generate 24 samples of the random process $x(n)$.

(b) Estimate the autocorrelation sequence using the sample autocorrelation and compare it to the true autocorrelation sequence.

(c) Using your estimated autocorrelation sequence, estimate the power spectrum of $x(n)$ by computing the Fourier transform of $\hat{r}_x(k)$. (Hint: this may be done easily using the results derived in Prob. C3.3.)

(d) Using the estimated autocorrelation values found in (b), use the Yule-Walker equations to estimate the values of $a(1)$, $a(2)$, and $b(0)$ and comment on the accuracy of your estimates.

(e) Estimate the power spectrum using your estimates of $a(k)$ and $b(0)$ derived in (d) as follows

$$\hat{P}_x(e^{j\omega}) = \frac{b^2(0)}{\left|1 + a(1)e^{-j\omega} + a(2)e^{-2j\omega}\right|^2}$$

(f) Compare your power spectrum estimates in parts (c) and (e) with the true power spectrum.

SIGNAL MODELING

4

4.1 INTRODUCTION

Signal modeling is an important problem that arises in a variety of applications. In data compression, for example, a waveform $x(n)$ consisting of N data values, $x(0), \ldots, x(N-1)$, is to be transmitted across a communication channel or archived on tape or disk. One method that may be used for transmission or storage is to process the signal on a point-by-point basis, i.e., transmit $x(0)$ followed by $x(1)$ and so on. However, if it is possible to accurately model the signal with a small number of parameters k, where $k \ll N$, then it would be more efficient to transmit or store these parameters instead of the signal values. The reconstruction of the signal from the model parameters would then be performed either by the communications receiver or the data retrieval system. As a more concrete example, consider the signal $x(n) = \alpha \cos(n\omega_0 + \phi)$. If this signal is to be saved on disk, then, clearly, the signal values themselves could be stored. A more efficient approach, however, would be to record the amplitude α, the frequency ω_0, and the phase ϕ. At any later time the signal $x(n)$ could be reconstructed exactly from these parameters. In a more practical setting, suppose that a signal to be stored or transmitted can be accurately modeled as a sum of L sinusoids,

$$x(n) \approx \sum_{k=1}^{L} \alpha_k \cos(n\omega_k + \phi_k) \tag{4.1}$$

Again, either the signal values $x(n)$ or the model parameters $\{\alpha_k, \omega_k, \phi_k : k = 1, \ldots, L\}$ could be recorded or transmitted. The tradeoff between the two representations is one of accuracy versus efficiency.

Another application of signal modeling is in the area of signal prediction (extrapolation) and signal interpolation. In both of these problems a signal $x(n)$ is known over a given interval of time and the goal is to determine $x(n)$ over some other interval. In the prediction problem, for example, $x(n)$ may be known for $n = 0, 1, \ldots, N-1$ and the goal is to predict the next signal value, $x(N)$. In the case of interpolation, one or more consecutive values may be missing or severely distorted over some interval, say $[N_1, N_2]$, and the problem is to use the data outside this interval to recover or estimate the values of $x(n)$ inside the interval. In both cases, if a model can be found that provides an accurate representation for $x(n)$, then the model may be used to estimate the unknown values of $x(n)$. For example, suppose that $x(n)$ may be modeled with a recursion of the form

$$x(n) \approx \sum_{k=1}^{p} a_p(k)x(n-k) \tag{4.2}$$

Given $x(n)$ for $n \in [0, N-1]$, the next signal value, $x(N)$, could be estimated using Eq. (4.2) as follows[1]:

$$\hat{x}(N) = \sum_{k=1}^{p} a_p(k)x(N-k)$$

Signal modeling is concerned with the representation of signals in an efficient manner. In general, there are two steps in the modeling process. The first is to choose an appropriate parametric form for the model. One possibility, as described above, is to model the signal as a sum of sinusoids. Another is to use a sum of damped sinusoids

$$x(n) \approx \sum_{k=1}^{L} \alpha_k \lambda_k^n \cos(n\omega_k + \phi_k)$$

In this chapter, the model that will be used is one that represents a signal as the output of a causal linear shift-invariant filter that has a rational system function of the form

$$H(z) = \frac{B_q(z)}{A_p(z)} = \frac{\sum_{k=0}^{q} b_q(k)z^{-k}}{1 + \sum_{k=1}^{p} a_p(k)z^{-k}} \tag{4.3}$$

as shown in Fig. 4.1. Thus, the *signal model* includes the filter coefficients, $a_p(k)$ and $b_q(k)$, along with a description of the input signal $v(n)$. For the most part, the signals that we will be trying to model will be deterministic. Therefore, in order to keep the model as efficient as possible, the filter input will typically either be a fixed signal, such as a unit sample $\delta(n)$, or it will be a signal that may be represented parametrically using only a few parameters, such as

$$v(n) = \sum_{k=0}^{p} \delta(n - kn_0)$$

which is used in modeling voiced speech where n_0 represents the pitch period [11] or

$$v(n) = \sum_{k=0}^{p} \alpha_k \delta(n - n_k)$$

which is used in multipulse linear predictive coding [1, 4]. In Section 4.7, where we consider models for random processes, the input to the filter will be a stochastic process, which, generally, will be taken to be white noise.

Once the form of the model has been selected, the next step is to find the model parameters that provide the *best* approximation to the given signal. There are, however, many different ways to define what is meant by "*the best*" approximation and, depending upon the definition that is used, there will be different solutions to the modeling problem along

[1] In general, a "hat" above any signal or function will be used to indicate an estimate or approximation to the given signal or function.

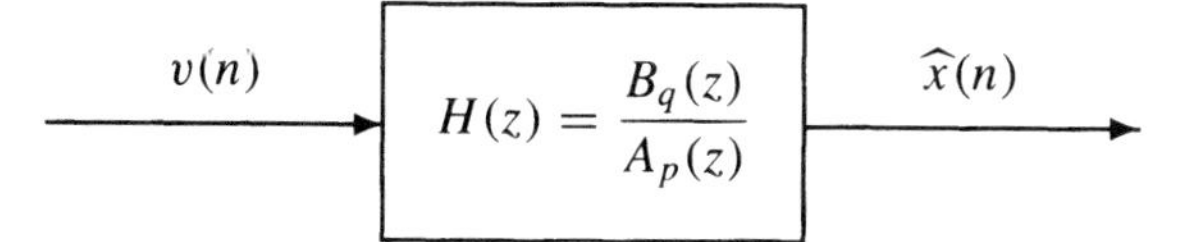

Figure 4.1 *Modeling a signal $x(n)$ as the response of a linear shift-invariant filter to an input $v(n)$. The goal is to find the filter $H(z)$ and the input $v(n)$ that make $\hat{x}(n)$ "as close as possible" to $x(n)$.*

with different techniques for finding the model parameters. Therefore, in developing an approach to signal modeling, it will be important not only to find a model that is useful, i.e., works well, but also one that has a computationally efficient procedure for deriving the model parameters from the given data.

In this chapter, a number of different approaches to signal modeling will be developed. The first, presented in Section 4.2, is a *least squares method* that minimizes the sum of the squares of the error

$$e'(n) = x(n) - \hat{x}(n)$$

which is the difference between the signal $x(n)$ and the output of the filter. As we will see, this approach requires finding the solution to a set of nonlinear equations. Therefore, it is rarely used in practice. In Section 4.3, a more practical solution is developed that forces the model to be exact, $e'(n) = 0$, over a fixed finite interval such as $[0, N-1]$. This approach, known as the Padé approximation method, has the advantage of only requiring the solution to a set of linear equations. However, this method is limited in that there is no guarantee on how accurate the model will be for values of n outside the interval $[0, N-1]$. Therefore, in Section 4.4 we develop Prony's method which is a blend between the least squares approach and the Padé approximation method and has the advantage of only requiring the solution to a set of linear equations, while, at the same time, producing a model that is more accurate, on the average, than the Padé method. We then derive Shanks' method, which modifies the Prony approach for finding the zeros of the system function. The special case of all-pole modeling using Prony's method is then considered, and the relationship between all-pole modeling and linear prediction is discussed. In Section 4.5, we return to the least squares method of Section 4.2 and develop an iterative algorithm for solving the nonlinear modeling equations. In Section 4.6, we develop the autocorrelation and covariance methods of all-pole modeling, which are modifications to the all-pole Prony method for the case in which only a finite length data record is available. Then, in Section 4.7, we consider the problem of modeling random processes. Our discussion of signal modeling will not end, however, with this chapter. In Chapter 6, for example, models that are based on the lattice filter will be considered; in Chapter 9, we will present some methods that may be used to model nonstationary signals.

4.2 THE LEAST SQUARES (DIRECT) METHOD

In this section, we consider the problem of modeling a deterministic signal, $x(n)$, as the unit sample response of a linear shift-invariant filter, $h(n)$, having a rational system function of the form given in Eq. (4.3). Thus, $v(n)$ in Fig. 4.1 is taken to be the unit sample, $\delta(n)$. We will assume, without any loss in generality, that $x(n) = 0$ for $n < 0$ and that the filter $h(n)$

is causal. Denoting the modeling error by $e'(n)$,

$$e'(n) = x(n) - h(n)$$

the problem is to find the filter coefficients, $a_p(k)$ and $b_q(k)$, that make $e'(n)$ *as small as possible*. Before $e'(n)$ may be minimized, however, it is necessary to define what is meant by *small*. For example, does *small* mean that the maximum value of $|e'(n)|$ is to be made as small as possible? Does it mean that we should minimize the mean square value? Or does it mean something else? In the *least squares* or *direct method* of signal modeling, the error measure that is to be minimized is the squared error [6]

$$\mathcal{E}_{LS} = \sum_{n=0}^{\infty} |e'(n)|^2$$

(Note that since $h(n)$ and $x(n)$ are both assumed to be zero for $n < 0$, then $e'(n) = 0$ for $n < 0$ and the summation begins at $n = 0$.) A block diagram of the direct method of signal modeling is shown in Fig. 4.2. A necessary condition for the filter coefficients $a_p(k)$ and $b_q(k)$ to minimize the squared error is that the partial derivative of $\mathcal{E}_{LS}$ with respect to each of the coefficients vanish, i.e.,

$$\frac{\partial \mathcal{E}_{LS}}{\partial a_p^*(k)} = 0 \quad ; \quad k = 1, 2, \ldots, p$$

and

$$\frac{\partial \mathcal{E}_{LS}}{\partial b_q^*(k)} = 0 \quad ; \quad k = 0, 1, \ldots, q$$

Using Parseval's theorem, the least squares error may be written in the frequency domain in terms of $E'(e^{j\omega})$, the Fourier transform of $e'(n)$, as follows:

$$\mathcal{E}_{LS} = \frac{1}{2\pi} \int_{-\pi}^{\pi} |E'(e^{j\omega})|^2 d\omega \tag{4.4}$$

This representation allows us to express $\mathcal{E}_{LS}$ explicitly in terms of the model parameters $a_p(k)$ and $b_q(k)$. Thus, setting the partial derivatives with respect to $a_p^*(k)$ equal to zero we have

$$\frac{\partial \mathcal{E}_{LS}}{\partial a_p^*(k)} = \frac{1}{2\pi} \int_{-\pi}^{\pi} \frac{\partial}{\partial a_p^*(k)} [E'(e^{j\omega}) E'^*(e^{j\omega})] d\omega = 0 \tag{4.5}$$

Since

$$E'(e^{j\omega}) = X(e^{j\omega}) - \frac{B_q(e^{j\omega})}{A_p(e^{j\omega})}$$

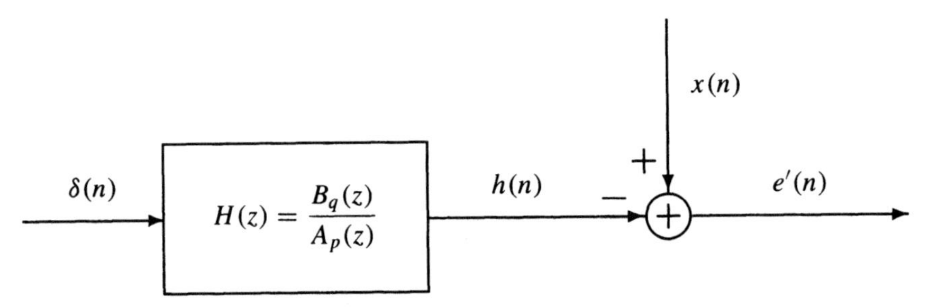

Figure 4.2 *The direct method of signal modeling for approximating a signal $x(n)$ as the unit sample response of a linear shift-invariant system having p poles and q zeros.*

treating $a_p(k)$ and $a_p^*(k)$ as independent variables, as discussed in Section 2.3.10, Eq. (4.5) becomes

$$\frac{\partial \mathcal{E}_{LS}}{\partial a_p^*(k)} = \frac{1}{2\pi}\int_{-\pi}^{\pi}\left[X(e^{j\omega}) - \frac{B_q(e^{j\omega})}{A_p(e^{j\omega})}\right]\frac{B_q^*(e^{j\omega})}{[A_p^*(e^{j\omega})]^2}e^{jk\omega}d\omega = 0 \tag{4.6}$$

for $k = 1, 2, \ldots, p$. Similarly, differentiating Eq. (4.4) with respect to $b_q(k)$ we have

$$\frac{\partial \mathcal{E}_{LS}}{\partial b_q^*(k)} = -\frac{1}{2\pi}\int_{-\pi}^{\pi}\left[X(e^{j\omega}) - \frac{B_q(e^{j\omega})}{A_p(e^{j\omega})}\right]\frac{e^{jk\omega}}{A_p^*(e^{j\omega})}d\omega = 0 \tag{4.7}$$

for $k = 0, 1, \ldots, q$. What is clear from Eqs. (4.6) and (4.7) is that the optimum set of model parameters are defined implicitly in terms of a set of $p + q + 1$ nonlinear equations. Although iterative techniques such as the method of steepest descent, Newton's method, or iterative prefiltering (see Section 4.5) could be used to solve these equations, the least squares approach is not mathematically tractable and not amenable to real-time signal processing applications. It is for this reason that we now turn our attention to some indirect methods of signal modeling. In these methods, the modeling problem is changed slightly so that the model parameters may be found more easily. In the following sections it is important to keep in mind, however, that the problems that are being solved are different from the true least squares solution presented above.

4.3 THE PADÉ APPROXIMATION

Since a signal model that is based on a least squares error criterion leads to a mathematically intractable solution, in this section we will reformulate the signal modeling problem and derive what has become known as the Padé approximation method [6, 9]. Unlike the least squares solution, the Padé approximation only requires solving a set of linear equations. In order to motivate the Padé approximation, let $x(n)$ be a signal that is to be modeled as the unit sample response of a causal linear shift-invariant filter. With a system function having p poles and q zeros as in Eq. (4.3), note that there are $p + q + 1$ *degrees of freedom* in the model, i.e., the p denominator coefficients and the $q + 1$ numerator coefficients. Therefore, with the appropriate choice for the coefficients $a_p(k)$ and $b_q(k)$ it would seem plausible that we could force the filter output, $h(n)$, to be equal to the given signal $x(n)$ for $p + q + 1$ values of n. In order to explore this idea a little further, let us consider a simple example.

Example 4.3.1 *Exact Matching of a Signal*

Let $x(n)$ be a signal that is to be modeled using a causal first-order all-pole filter of the form

$$H(z) = \frac{b(0)}{1 + a(1)z^{-1}}$$

Since the unit sample response of this filter is

$$h(n) = b(0)\big[-a(1)\big]^n u(n)$$

the constraint that $h(n) = x(n)$ for $n = 0, 1$ requires that values for $a(1)$ and $b(0)$ be found such that

$$\begin{aligned} x(0) &= b(0) \\ x(1) &= -b(0)a(1) \end{aligned}$$

Assuming that $x(0) \neq 0$, the solution to these two equations is $b(0) = x(0)$ and $a(1) = -x(1)/x(0)$ and the model is[2]

$$H(z) = \frac{x^2(0)}{x(0) - x(1)z^{-1}}$$

Therefore, with a first-order model it is always possible to match the first two signal values exactly if $x(0) \neq 0$.

If we increase the order of the model to include one pole and one zero,

$$H(z) = \frac{b(0) + b(1)z^{-1}}{1 + a(1)z^{-1}}$$

the problem becomes a bit more complicated. Since the unit sample response in this case is

$$h(n) = b(0)\big[-a(1)\big]^n u(n) + b(1)\big[-a(1)\big]^{n-1} u(n-1)$$

the equations that must be solved so that $h(n) = x(n)$ for $n = 0, 1, 2$ are

$$\begin{aligned} x(0) &= b(0) \\ x(1) &= -b(0)a(1) + b(1) \\ x(2) &= b(0)a^2(1) - b(1)a(1) = a(1)\big[b(0)a(1) - b(1)\big] \end{aligned}$$

Although these equations are nonlinear, if we substitute the second equation into the third, then we have

$$x(2) = -a(1)x(1)$$

Therefore,

$$a(1) = -\frac{x(2)}{x(1)}$$

and it follows from the first two equations that

$$\begin{aligned} b(0) &= x(0) \\ b(1) &= x(1) - x(0)\frac{x(2)}{x(1)} \end{aligned}$$

Thus, if $x(1) \neq 0$, then we may match the first three values of $x(n)$ with a model consisting of one pole and one zero.

In the previous example, it is important to note that, in spite of the fact that we were able to find the filter coefficients that make $h(n) = x(n)$ for $n = 0, 1$ in the first case and $n = 0, 1, 2$ in the second, the equations that we needed to solve were *nonlinear*. Although this will be the case in general, we will see in the following discussion that it is possible to

[2] The form of the model presupposes that $h(0) \neq 0$ and, therefore, that $x(0) \neq 0$. For those cases in which $x(0) = 0$, the signal may simply be shifted until $x(0) \neq 0$.

reformulate the problem in such a way that the filter coefficients may be found by solving a set of *linear* equations. We will also establish that, for a model containing p poles and q zeros, it is almost always possible to find a set of filter coefficients that force $h(n) = x(n)$ for $p + q + 1$ values of n. In the discussions that follow, $x(n)$ may be either a real- or complex-valued sequence.

The development of the Padé method begins by expressing the system function,

$$H(z) = \frac{B_q(z)}{A_p(z)} = \frac{\sum_{k=0}^{q} b_q(k)z^{-k}}{1 + \sum_{k=1}^{p} a_p(k)z^{-k}}$$

as follows:

$$H(z)A_p(z) = B_q(z) \tag{4.8}$$

In the time domain, Eq. (4.8) becomes a convolution

$$h(n) + \sum_{k=1}^{p} a_p(k)h(n-k) = b_q(n) \tag{4.9}$$

where $h(n) = 0$ for $n < 0$ and $b_q(n) = 0$ for $n < 0$ and $n > q$. To find the coefficients $a_p(k)$ and $b_q(k)$ that give an exact fit of the data to the model over the interval $[0, p+q]$ we set $h(n) = x(n)$ for $n = 0, 1, \ldots, p+q$ in Eq. (4.9). This leads to the following set of $p+q+1$ linear equations in $p+q+1$ unknowns,

$$x(n) + \sum_{k=1}^{p} a_p(k)x(n-k) = \begin{cases} b_q(n) & ; \quad n = 0, 1, \ldots, q \\ 0 & ; \quad n = q+1, \ldots, q+p \end{cases} \tag{4.10}$$

In matrix form these equations may be written as follows:

$$\left[\begin{array}{cccc} x(0) & 0 & \cdots & 0 \\ x(1) & x(0) & \cdots & 0 \\ x(2) & x(1) & \cdots & 0 \\ \vdots & \vdots & & \vdots \\ x(q) & x(q-1) & \cdots & x(q-p) \\ \hline x(q+1) & x(q) & \cdots & x(q-p+1) \\ \vdots & \vdots & & \vdots \\ x(q+p) & x(q+p-1) & \cdots & x(q) \end{array}\right] \begin{bmatrix} 1 \\ a_p(1) \\ a_p(2) \\ \vdots \\ a_p(p) \end{bmatrix} = \left[\begin{array}{c} b_q(0) \\ b_q(1) \\ b_q(2) \\ \vdots \\ b_q(q) \\ \hline 0 \\ \vdots \\ 0 \end{array}\right] \tag{4.11}$$

To solve Eq. (4.11) for $a_p(k)$ and $b_q(k)$ we use a two-step approach, first solving for the denominator coefficients $a_p(k)$ and then solving for the numerator coefficients $b_q(k)$.

In the first step of the Padé approximation method, solving for the coefficients $a_p(k)$, we use the last p equations of Eq. (4.11) as indicated by the partitioning, i.e.,

$$\begin{bmatrix} x(q+1) & x(q) & \cdots & x(q-p+1) \\ x(q+2) & x(q+1) & \cdots & x(q-p+2) \\ \vdots & \vdots & & \vdots \\ x(q+p) & x(q+p-1) & \cdots & x(q) \end{bmatrix} \begin{bmatrix} 1 \\ a_p(1) \\ \vdots \\ a_p(p) \end{bmatrix} = \begin{bmatrix} 0 \\ 0 \\ \vdots \\ 0 \end{bmatrix} \tag{4.12}$$

Expanding Eq. (4.12) as follows

$$\begin{bmatrix} x(q+1) \\ x(q+2) \\ \vdots \\ x(q+p) \end{bmatrix} + \begin{bmatrix} x(q) & x(q-1) & \cdots & x(q-p+1) \\ x(q+1) & x(q) & \cdots & x(q-p+2) \\ \vdots & \vdots & & \vdots \\ x(q+p-1) & x(q+p-2) & \cdots & x(q) \end{bmatrix} \begin{bmatrix} a_p(1) \\ a_p(2) \\ \vdots \\ a_p(p) \end{bmatrix} = \begin{bmatrix} 0 \\ 0 \\ \vdots \\ 0 \end{bmatrix}$$

and then bringing the vector on the left to the right-hand side leads to the following set of p equations in the p unknowns $a_p(1), a_p(2), \ldots, a_p(p)$,

$$\begin{bmatrix} x(q) & x(q-1) & \cdots & x(q-p+1) \\ x(q+1) & x(q) & \cdots & x(q-p+2) \\ \vdots & \vdots & & \vdots \\ x(q+p-1) & x(q+p-2) & \cdots & x(q) \end{bmatrix} \begin{bmatrix} a_p(1) \\ a_p(2) \\ \vdots \\ a_p(p) \end{bmatrix} = - \begin{bmatrix} x(q+1) \\ x(q+2) \\ \vdots \\ x(q+p) \end{bmatrix} \tag{4.13}$$

Using matrix notation Eq. (4.13) may be expressed concisely as

$$\boxed{\mathbf{X}_q \bar{\mathbf{a}}_p = -\mathbf{x}_{q+1}} \tag{4.14}$$

where

$$\bar{\mathbf{a}}_p = \Big[a_p(1), a_p(2), \ldots, a_p(p)\Big]^T$$

is the vector of denominator coefficients,

$$\mathbf{x}_{q+1} = \big[x(q+1),\ x(q+2), \ldots, x(q+p)\big]^T$$

is a column vector of length p, and

$$\mathbf{X}_q = \begin{bmatrix} x(q) & x(q-1) & \cdots & x(q-p+1) \\ x(q+1) & x(q) & \cdots & x(q-p+2) \\ \vdots & \vdots & & \vdots \\ x(q+p-1) & x(q+p-2) & \cdots & x(q) \end{bmatrix}$$

is a $p \times p$ nonsymmetric Toeplitz matrix.[3]

With respect to solving Eq. (4.13), there are three cases of interest, depending upon whether or not the matrix $\mathbf{X}_q$ is invertible. Each case is considered separately below.

- **Case I.** $\mathbf{X}_q$ is nonsingular. If $\mathbf{X}_q$ is nonsingular then $\mathbf{X}_q^{-1}$ exists and the coefficients of $A_p(z)$ are unique and are given by

$$\bar{\mathbf{a}}_p = -\mathbf{X}_q^{-1}\mathbf{x}_{q+1}$$

  Since $\mathbf{X}_q$ is a nonsymmetric Toeplitz matrix, these equations may be solved efficiently using a recursion known as the Trench algorithm [18].

- **Case II.** $\mathbf{X}_q$ is singular and a solution to Eq. (4.14) exists. If there is a vector $\bar{\mathbf{a}}_p$ that solves Eq. (4.14) and if $\mathbf{X}_q$ is singular, then the solution is not unique. Specifically, when $\mathbf{X}_q$ is singular there are nonzero solutions to the homogeneous equation

$$\mathbf{X}_q \mathbf{z} = 0 \tag{4.15}$$

[3]The subscript q on the data matrix $\mathbf{X}_q$ indicates the index of $x(n)$ in the element in the upper-left corner. Similarly, the subscript $q+1$ on the data vector $\mathbf{x}_{q+1}$ indicates the index of $x(n)$ in the first element.

Therefore, for any solution $\bar{\mathbf{a}}_p$ to Eq. (4.14) and for any $\mathbf{z}$ that satisfies Eq. (4.15),

$$\widetilde{\mathbf{a}}_p = \bar{\mathbf{a}}_p + \mathbf{z}$$

is also a solution. In this case, the best choice for $\bar{\mathbf{a}}_p$ may be the one that has the fewest number of nonzero coefficients since this solution would provide the most efficient parametric representation for the signal.

- **Case III.** $\mathbf{X}_q$ is singular and no solution exists. In the formulation of Padé's method, it was assumed that the denominator polynomial in $H(z)$ could be written in the form

$$A_p(z) = 1 + \sum_{k=1}^{p} a_p(k) z^{-k}$$

In other words, it was assumed that the leading coefficient, $a_p(0)$, was nonzero and that $A_p(z)$ could be normalized so that $a_p(0) = 1$. However, if $\mathbf{X}_q$ is singular and no vector can be found that satisfies Eq. (4.14), then this nonzero assumption on $a_p(0)$ is incorrect. Therefore, if we set $a_p(0) = 0$, then the Padé equations that are derived from Eq. (4.12) become

$$\mathbf{X}_q \bar{\mathbf{a}}_p = 0 \tag{4.16}$$

Since $\mathbf{X}_q$ is singular, there is a nonzero solution to these homogeneous equations. Thus, as we will see in Example 4.3.3, the denominator coefficients may be found by solving Eq. (4.16).

Having solved Eq. (4.14) for $\bar{\mathbf{a}}_p$, the second step in the Padé approximation is to find the numerator coefficients $b_q(k)$. These coefficients are found from the first $(q+1)$ equations in Eq. (4.11),

$$\begin{bmatrix} x(0) & 0 & 0 & \cdots & 0 \\ x(1) & x(0) & 0 & \cdots & 0 \\ x(2) & x(1) & x(0) & \cdots & 0 \\ \vdots & \vdots & \vdots & & \vdots \\ x(q) & x(q-1) & x(q-2) & \cdots & x(q-p) \end{bmatrix} \begin{bmatrix} 1 \\ a_p(1) \\ a_p(2) \\ \vdots \\ a_p(p) \end{bmatrix} = \begin{bmatrix} b_q(0) \\ b_q(1) \\ b_q(2) \\ \vdots \\ b_q(q) \end{bmatrix} \tag{4.17}$$

With

$$\mathbf{a}_p = \Big[1,\ a_p(1),\ a_p(2),\ \cdots,\ a_p(p)\Big]^T$$

and

$$\mathbf{b}_q = \Big[b_q(0),\ b_q(1),\ b_q(2),\ \cdots,\ b_q(p)\Big]^T$$

these equations may be written in matrix form as follows

$$\boxed{\mathbf{X}_0 \mathbf{a}_p = \mathbf{b}_q} \tag{4.18}$$

Therefore, $\mathbf{b}_q$ may be found by simply multiplying $\mathbf{a}_p$ by the matrix $\mathbf{X}_0$. Equivalently, the coefficients $b_q(k)$ may be evaluated using Eq. (4.10) as follows

$$b_q(n) = x(n) + \sum_{k=1}^{p} a_p(k) x(n-k) \quad ; \quad n = 0, 1, \ldots, q \tag{4.19}$$

Table 4.1 ***The Padé Approximation for Modeling a Signal as the Unit Sample Response of Linear Shift-Invariant System Having p Poles and q Zeros***

Denominator coefficients

$$\begin{bmatrix} x(q) & x(q-1) & \cdots & x(q-p+1) \\ x(q+1) & x(q) & \cdots & x(q-p+2) \\ \vdots & \vdots & & \vdots \\ x(q+p-1) & x(q+p-2) & \cdots & x(q) \end{bmatrix} \begin{bmatrix} a_p(1) \\ a_p(2) \\ \vdots \\ a_p(p) \end{bmatrix} = - \begin{bmatrix} x(q+1) \\ x(q+2) \\ \vdots \\ x(q+p) \end{bmatrix}$$

Numerator coefficients

$$\begin{bmatrix} x(0) & 0 & 0 & \cdots & 0 \\ x(1) & x(0) & 0 & \cdots & 0 \\ x(2) & x(1) & x(0) & \cdots & 0 \\ \vdots & \vdots & \vdots & & \vdots \\ x(q) & x(q-1) & x(q-2) & \cdots & x(q-p) \end{bmatrix} \begin{bmatrix} 1 \\ a_p(1) \\ a_p(2) \\ \vdots \\ a_p(p) \end{bmatrix} = \begin{bmatrix} b_q(0) \\ b_q(1) \\ b_q(2) \\ \vdots \\ b_q(q) \end{bmatrix}$$

Thus, as summarized in Table 4.1, the Padé approximation method is a two-step approach for finding the model parameters. First, Eq. (4.14) is solved for the denominator coefficients $a_p(k)$. Then either Eq. (4.17) or Eq. (4.19) is used to find the numerator coefficients $b_q(k)$. A MATLAB program for the Padé approximation is given in Fig. 4.3.

Before looking at some examples, we consider the special case of an *all pole-model,*

$$H(z) = \frac{b(0)}{1 + \sum_{k=1}^{p} a_p(k) z^{-k}} \tag{4.20}$$

With $q = 0$, the Padé equations for the coefficients $a_p(k)$ become

$$\begin{bmatrix} x(0) & 0 & \cdots & 0 \\ x(1) & x(0) & \cdots & 0 \\ \vdots & \vdots & & \vdots \\ x(p-1) & x(p-2) & \cdots & x(0) \end{bmatrix} \begin{bmatrix} a_p(1) \\ a_p(2) \\ \vdots \\ a_p(p) \end{bmatrix} = - \begin{bmatrix} x(1) \\ x(2) \\ \vdots \\ x(p) \end{bmatrix} \tag{4.21}$$

The Padé Approximation

```
function [a,b] = pade(x,p,q)
%
  x    = x(:);
  if p+q>=length(x), error('Model order too large'), end
     X    = convm(x,p+1);
     Xq   = X(q+2:q+p+1,2:p+1);
     a    = [1;-Xq\X(q+2:q+p+1,1)];
     b    = X(1:q+1,1:p+1)*a;
     end;
```

Figure 4.3 *A* MATLAB *program for signal modeling using the Padé approximation. Note that this program calls* `convm.m` *(see Appendix).*

or,

$$\mathbf{X}_0 \bar{\mathbf{a}}_p = -\mathbf{x}_1$$

Since $\mathbf{X}_0$ is a lower triangular Toeplitz matrix the denominator coefficients may be found easily by back substitution using the recursion

$$a_p(k) = -\frac{1}{x(0)}\left[x(k) + \sum_{l=1}^{k-1} a_p(l)x(k-l)\right]$$

In order to gain some familiarity with the Padé approximation method and its properties, we now look at a few examples.

Example 4.3.2 *Padé Approximation*

Given a signal whose first six values are

$$\mathbf{x} = \begin{bmatrix}1, & 1.500, & 0.750, & 0.375, & 0.1875, & 0.0938\end{bmatrix}^T$$

let us use the Padé approximation to find a second-order all-pole model ($p = 2$ and $q = 0$), a second-order moving-average model ($p = 0$ and $q = 2$), and a model containing one pole and one zero ($p = q = 1$).

1. To find a second-order all-pole model, $p = 2$ and $q = 0$, the equations that must be solved are

$$\begin{bmatrix} x(0) & 0 & 0 \\ x(1) & x(0) & 0 \\ x(2) & x(1) & x(0) \end{bmatrix}\begin{bmatrix} 1 \\ a(1) \\ a(2) \end{bmatrix} = \begin{bmatrix} b(0) \\ 0 \\ 0 \end{bmatrix} \tag{4.22}$$

From the last two equations we have

$$\begin{bmatrix} x(0) & 0 \\ x(1) & x(0) \end{bmatrix}\begin{bmatrix} a(1) \\ a(2) \end{bmatrix} = -\begin{bmatrix} x(1) \\ x(2) \end{bmatrix} \tag{4.23}$$

Substituting the given values of $x(n)$ into Eq. (4.23)

$$\begin{bmatrix} 1 & 0 \\ 1.5 & 1 \end{bmatrix}\begin{bmatrix} a(1) \\ a(2) \end{bmatrix} = -\begin{bmatrix} 1.50 \\ 0.75 \end{bmatrix}$$

and solving for $a(1)$ and $a(2)$ gives

$$a(1) = -1.50 \qquad \text{and} \qquad a(2) = 1.50$$

Finally, using the first equation in Eq. (4.22) it follows that $b(0) = x(0) = 1$. Therefore, the model for $x(n)$ is

$$H(z) = \frac{1}{1 - 1.50z^{-1} + 1.50z^{-2}}$$

Note that this model is unstable, having a pair of complex poles that lie outside the unit circle. Evaluating the unit sample response we see that the model produces the following approximation for $x(n)$,

$$\hat{\mathbf{x}} = \begin{bmatrix}1, & 1.500, & 0.750, & -1.125, & -2.8125, & -2.5312\end{bmatrix}$$

Although $\hat{x}(n) = x(n)$ for $n = 0, 1, 2$, as it should, the model does not produce an accurate representation of $x(n)$ for $n = 3, 4, 5$.

2. For a second-order all-zero model, $p = 0$ and $q = 2$, the denominator polynomial is a constant, $A(z) = 1$, and the equations that need to be solved are trivial,

$$\begin{bmatrix} x(0) \\ x(1) \\ x(2) \end{bmatrix} = \begin{bmatrix} b(0) \\ b(1) \\ b(2) \end{bmatrix}$$

Therefore, the model is simply

$$H(z) = 1 + 1.50z^{-1} + 0.75z^{-2}$$

and $\hat{x}(n) = 0$ for $n > p + q = 2$.

3. Finally, let us consider a model having one pole and one zero, $p = q = 1$. In this case the model is of the form

$$H(z) = \frac{b(0) + b(1)z^{-1}}{1 + a(1)z^{-1}}$$

and the Padé equations are

$$\begin{bmatrix} x(0) & 0 \\ x(1) & x(0) \\ x(2) & x(1) \end{bmatrix} \begin{bmatrix} 1 \\ a(1) \end{bmatrix} = \begin{bmatrix} b(0) \\ b(1) \\ 0 \end{bmatrix} \tag{4.24}$$

The denominator coefficient $a(1)$ may be found from the last equation in Eq. (4.24),

$$x(2) + a(1)x(1) = 0$$

which gives

$$a(1) = -\frac{x(2)}{x(1)} = -0.5$$

Given $a(1)$, the coefficients $b(0)$ and $b(1)$ may be found from the first two equations in Eq. (4.24) as follows

$$\begin{bmatrix} b(0) \\ b(1) \end{bmatrix} = \begin{bmatrix} 1 & 0 \\ 1.5 & 1 \end{bmatrix} \begin{bmatrix} 1 \\ -0.5 \end{bmatrix} = \begin{bmatrix} 1 \\ 1 \end{bmatrix}$$

Therefore, the model is

$$H(z) = \frac{1 + z^{-1}}{1 - 0.5z^{-1}}$$

Since the unit sample response is

$$h(n) = (0.5)^n\, u(n) + (0.5)^{n-1}\, u(n-1) = \delta(n) + 1.5\,(0.5)^{n-1}\, u(n-1)$$

we see that this model matches all of the signal values from $n = 0$ to $n = 5$. This will not be true in general, however, unless the signal that is being modeled is the unit sample response of a linear shift-invariant filter and a sufficient number of poles and zeros are used in the Padé model.

If $\mathbf{X}_q$ is nonsingular, as in the previous example, then there are no difficulties in solving the Padé equations. The following example, however, illustrates what must be done when $\mathbf{X}_q$ is singular.

Example 4.3.3 *Singular Padé Approximation*

Let $x(n)$ be a signal that is to be approximated using a second-order model with $p = q = 2$, where the first five values of $x(n)$ are

$$\mathbf{x} = \left[1, 4, 2, 1, 3\right]^T$$

The Padé equations are

$$\begin{bmatrix} 1 & 0 & 0 \\ 4 & 1 & 0 \\ 2 & 4 & 1 \\ 1 & 2 & 4 \\ 3 & 1 & 2 \end{bmatrix} \begin{bmatrix} 1 \\ a(1) \\ a(2) \end{bmatrix} = \begin{bmatrix} b(0) \\ b(1) \\ b(2) \\ 0 \\ 0 \end{bmatrix} \tag{4.25}$$

From the last two equations in Eq. (4.25) we have

$$\begin{bmatrix} 2 & 4 \\ 1 & 2 \end{bmatrix} \begin{bmatrix} a(1) \\ a(2) \end{bmatrix} = -\begin{bmatrix} 1 \\ 3 \end{bmatrix}$$

which is a singular set of equations. In addition, since there are no values for $a(1)$ and $a(2)$ that satisfy these equations, it follows that the assumption that $a(0) = 1$ is incorrect. Therefore, setting $a(0) = 0$ gives

$$\begin{bmatrix} 2 & 4 \\ 1 & 2 \end{bmatrix} \begin{bmatrix} a(1) \\ a(2) \end{bmatrix} = \begin{bmatrix} 0 \\ 0 \end{bmatrix}$$

which has the nontrivial solution

$$a(1) = 2 \quad ; \quad a(2) = -1$$

Using the first three equations in Eq. (4.25) to determine the coefficients $b(k)$, we have

$$\begin{bmatrix} b(0) \\ b(1) \\ b(2) \end{bmatrix} = \begin{bmatrix} 1 & 0 & 0 \\ 4 & 1 & 0 \\ 2 & 4 & 1 \end{bmatrix} \begin{bmatrix} 0 \\ 2 \\ -1 \end{bmatrix} = \begin{bmatrix} 0 \\ 2 \\ 7 \end{bmatrix}$$

and the model becomes

$$H(z) = \frac{2z^{-1} + 7z^{-2}}{2z^{-1} - z^{-2}} = \frac{2 + 7z^{-1}}{2 - z^{-1}}$$

Note that a lower order model then originally proposed has been found, i.e., $p = 1$ and $q = 1$ instead of $p = q = 2$. Computing the inverse z-transform of $H(z)$ we find

$$h(n) = (\tfrac{1}{2})^n u(n) + \tfrac{7}{2}(\tfrac{1}{2})^{n-1} u(n-1) = \delta(n) + 8(\tfrac{1}{2})^n u(n-1)$$

Therefore, the first five values of $\hat{x}(n) = h(n)$ are

$$\hat{\mathbf{x}} = \left[1,\ 4,\ 2,\ 1,\ 0.5\right]^T$$

and the model produces a signal that only matches the data for $n = 0, 1, 2, 3$, i.e., $\hat{x}(4) \neq x(4)$. Thus, if $\mathbf{X}_q$ is singular and no solution to the Padé equations exists, then it is not always possible to match all of the signal values for $n = 0, 1, \ldots p + q$.

In the next example, we consider the application of the Padé approximation method to the design of digital filters [2]. The approach that is to be used is as follows. Given the unit sample response of an ideal digital filter, $i(n)$, we will find the system function of a realizable linear shift-invariant filter, $H(z) = B_q(z)/A_p(z)$, that has a unit sample response $h(n)$ that is equal to $i(n)$ for $n = 0, 1, \ldots, (p+q)$.[4] The hope is that if $h(n)$ "looks like" $i(n)$, then $H(e^{j\omega})$ will "look like" $I(e^{j\omega})$.

Example 4.3.4 *Filter Design Using the Padé Approximation*

Suppose that we would like to design a linear phase lowpass filter having a cutoff frequency of $\pi/2$. The frequency response of the ideal lowpass filter is

$$I(e^{j\omega}) = \begin{cases} e^{-jn_d\omega} & ; \quad |\omega| < \pi/2 \\ 0 & ; \quad \text{otherwise} \end{cases}$$

where n_d is the filter delay, which, for the moment, is left unspecified, and the unit sample response is

$$i(n) = \frac{\sin\left[(n - n_d)\pi/2\right]}{(n - n_d)\pi}$$

We will consider two designs, one with $p = 0$ and $q = 10$ and the other with $p = q = 5$. In both cases, $p + q + 1 = 11$ values of $i(n)$ may be matched exactly. Since $i(n)$ has its largest value at $n = n_d$ and decays as $1/n$ away from $n = n_d$, we will let $n_d = 5$ so that the model forces $h(n) = i(n)$ over the interval for which the energy in $i(n)$ is maximum. With $n_d = 5$, the first eleven values of $i(n)$ are

$$\mathbf{i} = \left[0.0637,\ 0,\ -0.1061,\ 0,\ 0.3183,\ 0.5,\ 0.3183,\ 0,\ -0.1061,\ 0,\ 0.0637\right]^T$$

With $p = 0$ and $q = 10$, the Padé approximation is

$$h(n) = \begin{cases} i(n) & ; \quad 0 \le n \le 10 \\ 0 & ; \quad \text{otherwise} \end{cases}$$

In other words, the filter is designed by multiplying the ideal unit sample response by a rectangular window. This is equivalent to the well-known and commonly used filter design technique known as windowing [8]. The frequency response of this filter is shown in Fig. 4.4 (solid line).

With $p = q = 5$, the coefficients $a(k)$ are found by solving Eq. (4.13). The solution is

$$\mathbf{a} = [1.0,\ -2.5256,\ 3.6744,\ -3.4853,\ 2.1307,\ -0.7034]^T$$

Next, the coefficients $b(k)$ are found from Eq. (4.17). The result is

$$\mathbf{b} = [0.0637,\ -0.1608,\ 0.1280,\ 0.0461,\ 0.0638,\ 0.0211]^T$$

The frequency response of this filter is shown in Fig. 4.4 (dotted line), and the error, $e'(n) = i(n) - h(n)$, is shown in Fig. 4.5. Note that this filter does not compare favorably with the filter designed by windowing. Also note that the extrapolation performed by the Padé approximation for $n > 11$ is not very accurate. It is this large difference between $i(n)$ and $h(n)$ that accounts for the poor frequency response characteristics.

---

[4]Note that in this application, $i(n)$ is used instead of $x(n)$ to represent the given data and $h(n)$ is used to denote the approximation to $i(n)$.

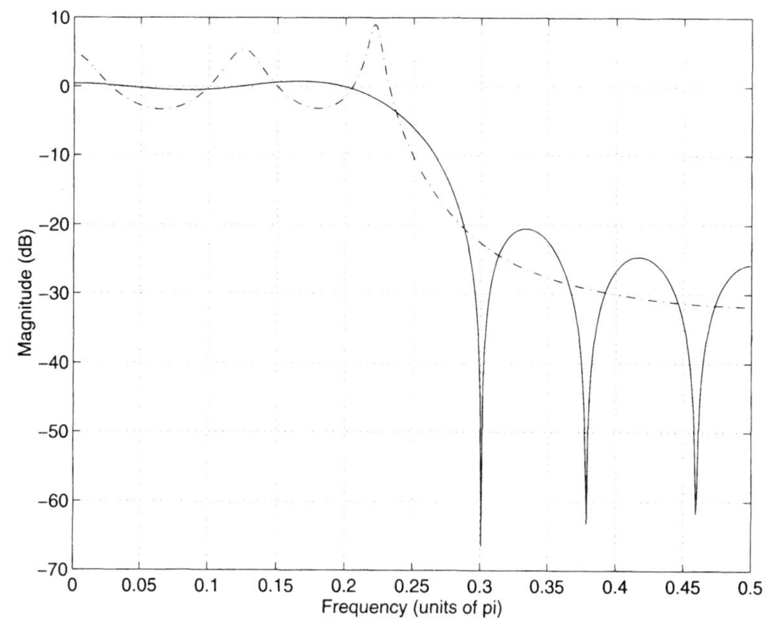

Figure 4.4 *Lowpass filter design using the Padé approximation. The frequency response of the filter having $p = 0$ poles and $q = 10$ zeros (solid line) and the frequency response of the filter having $p = 5$ poles and $q = 5$ zeros (dotted line).*

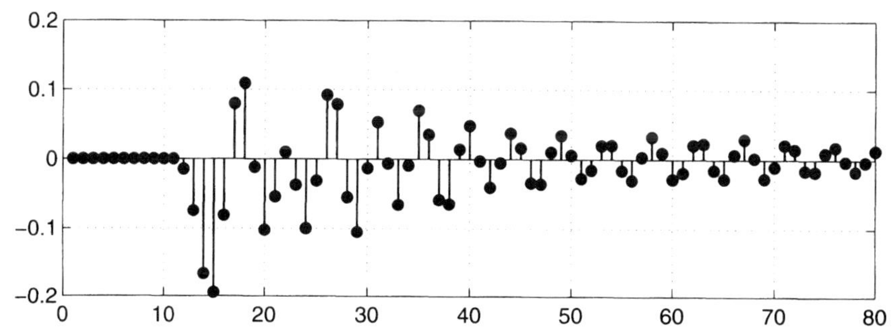

Figure 4.5 *The difference, $e'(n)$, between the unit sample response of the ideal lowpass filter, $i(n)$, and the unit sample response of the filter designed using the Padé approximation with $p = q = 5$.*

In conclusion, we may summarize the properties of the Padé approximation as follows. First, the model that is formed from the Padé approximation will always produce an exact fit to the data over the interval $[0,\ p+q]$, provided that $\mathbf{X}_q$ is nonsingular. However, since the data outside the interval $[0,\ p+q]$ is never considered in the modeling process, there is no guarantee on how accurate the model will be for $n > p+q$. In some cases, the match may be very good, while in others it may be poor. Second, for those cases in which $x(n)$ has a rational z-transform, the Padé approximation will give the correct model parameters, provided the model order is chosen to be large enough. Finally, since the Padé approximation forces the model to match the signal only over a limited range of values, the model that is generated is not guaranteed to be stable. This may be a serious problem in applications that rely on the model to synthesize or extrapolate the signal. For example, consider a telecommunications system in which a signal is segmented into subsequences of length N and the model parameters of each subsequence are transmitted across the channel, instead of the signal values themselves. The receiver then resynthesizes the signal by piecing together the unit sample responses of each of the transmitted filters. Although such a system is capable of realizing a significant reduction in the data transmission rate if $p + q \ll N$, if the model is unstable, then there is a risk of unacceptably large errors.

4.4 PRONY'S METHOD

The limitation with the Padé approximation is that it only uses values of the signal $x(n)$ over the interval $[0, p+q]$ to determine the model parameters and, over this interval, it models the signal without error. Since all of the degrees of freedom are used to force the error to zero for $n = 0, 1, \ldots, p+q$, there is no guarantee on how well the model will approximate the signal for $n > p+q$. An alternative is to relax the requirement that the model be exact over the interval $[0, p+q]$ in order to produce a better approximation to the signal for all n. In this section, we develop such an approach known as Prony's method. In Section 4.4.1 we consider the general problem of pole-zero modeling using Prony's method. As with the Padé approximation, Prony's method only requires solving a set of linear equations. In Section 4.4.2, we then introduce a modification to Prony's method known as Shanks' method, which differs from Prony's method in the approach that is used to find the zeros of the system function. Prony's method for the special case of all-pole signal modeling is then considered in Section 4.4.3. As we will see, all-pole models have a computational advantage over the more general pole-zero models. In Section 4.4.4 we look at the relationship between all-pole modeling and linear prediction and, finally, in Section 4.4.5 we look at the application of the all-pole Prony method to the design of FIR least squares inverse filters.

4.4.1 Pole-Zero Modeling

Suppose that a signal $x(n)$ is to be modeled as the unit sample response of a linear shift-invariant filter having a rational system function $H(z)$ with p poles and q zeros, i.e., $H(z) = B_q(z)/A_p(z)$, as in Eq. (4.3). For complete generality, let us assume that $x(n)$ is complex and allow the coefficients of the filter to be complex (the analysis in the case of real signals will be the same except that the complex conjugates will disappear). Thus, assuming that $x(n) = 0$ for $n < 0$ and that $x(n)$ is known for all values of $n \geq 0$, we would like to find the filter coefficients $a_p(k)$ and $b_q(k)$ that make $h(n)$, the unit sample response of the filter, as close as possible to $x(n)$ in the sense of minimizing the error

$$e'(n) = x(n) - h(n)$$

As we saw in Section 4.2, a least squares minimization of $e'(n)$ or, equivalently, a least squares minimization of

$$E'(z) = X(z) - \frac{B_q(z)}{A_p(z)} \tag{4.26}$$

results in a set of nonlinear equations for the filter coefficients. However, if we borrow the approach used in the Padé approximation and multiply both sides of Eq. (4.26) by $A_p(z)$, then we have a new error,

$$E(z) = A_p(z)E'(z) = A_p(z)X(z) - B_q(z)$$

that is *linear* in the filter coefficients. As illustrated in Fig. 4.6, in the time domain this error is the difference between $b_q(n)$ and the filtered signal $\hat{b}_q(n) = a_p(n) * x(n)$,

$$e(n) = a_p(n) * x(n) - b_q(n) = \hat{b}_q(n) - b_q(n) \tag{4.27}$$

Since $b_q(n) = 0$ for $n > q$, we may write this error explicitly for each value of n as follows

$$e(n) = \begin{cases} x(n) + \sum_{l=1}^{p} a_p(l)x(n-l) - b_q(n) & ; \quad n = 0, 1, \ldots, q \\ x(n) + \sum_{l=1}^{p} a_p(l)x(n-l) & ; \quad n > q \end{cases} \tag{4.28}$$

(recall that $A_p(z)$ is normalized so that $a_p(0) = 1$). Instead of setting $e(n) = 0$ for $n = 0, 1, \ldots, p + q$ as in the Padé approximation, Prony's method begins by finding the coefficients $a_p(k)$ that minimize the squared error[5]

$$\mathcal{E}_{p,q} = \sum_{n=q+1}^{\infty} |e(n)|^2 = \sum_{n=q+1}^{\infty} \left| x(n) + \sum_{l=1}^{p} a_p(l)x(n-l) \right|^2 \tag{4.29}$$

Note that $\mathcal{E}_{p,q}$ depends only on the coefficients, $a_p(k)$, and not on $b_q(k)$. Therefore, the coefficients that minimize this squared error may be found by setting the partial derivatives of $\mathcal{E}_{p,q}$ with respect to $a_p^*(k)$ equal to zero, as follows[6]

$$\frac{\partial \mathcal{E}_{p,q}}{\partial a_p^*(k)} = \sum_{n=q+1}^{\infty} \frac{\partial \left[e(n)e^*(n) \right]}{\partial a_p^*(k)} = \sum_{n=q+1}^{\infty} e(n) \frac{\partial e^*(n)}{\partial a_p^*(k)} = 0 \quad ; \quad k = 1, 2, \ldots, p \tag{4.30}$$

Since the partial derivative of $e^*(n)$ with respect to $a_p^*(k)$ is $x^*(n-k)$, Eq. (4.30) becomes

$$\boxed{\sum_{n=q+1}^{\infty} e(n)x^*(n-k) = 0 \quad ; \quad k = 1, 2, \ldots, p} \tag{4.31}$$

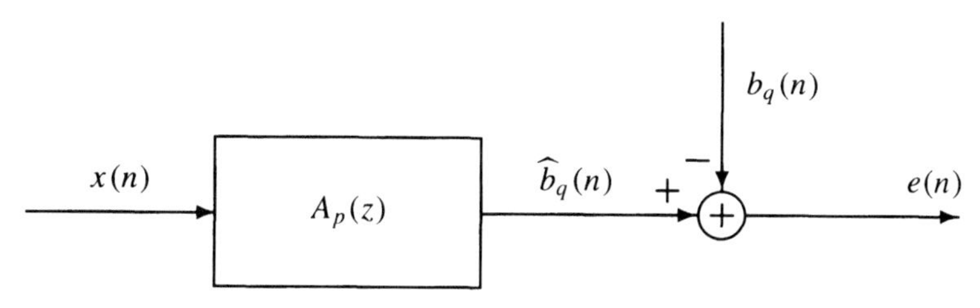

Figure 4.6 *System interpretation of Prony's method for signal modeling.*

[5]The subscripts on $\mathcal{E}_{p,q}$ are used to denote the order of the model, i.e., the number of poles and zeros. To be consistent with this notation, we should probably also denote the error sequence by $e_{p,q}(n)$. However, since this would significantly complicate the notation and add little in terms of clarity, we will continue writing the error simply as $e(n)$. In later chapters, however, most notably Chapters 5 and 6, we will use the notation $e_p(n)$ when discussing all-pole models ($q = 0$).

[6]See Section 2.3.10 for a discussion of how to find the minimum of a real-valued function of a complex vector and its conjugate.

which is known as the *orthogonality principle*.[7] Substituting Eq. (4.28) for $e(n)$ into Eq. (4.31) yields

$$\sum_{n=q+1}^{\infty}\left\{x(n)+\sum_{l=1}^{p}a_p(l)x(n-l)\right\}x^*(n-k)=0$$

or equivalently,

$$\sum_{l=1}^{p}a_p(l)\left[\sum_{n=q+1}^{\infty}x(n-l)x^*(n-k)\right]=-\sum_{n=q+1}^{\infty}x(n)x^*(n-k) \tag{4.32}$$

In order to simplify Eq. (4.32) notationally, let us define

$$\boxed{r_x(k,l)=\sum_{n=q+1}^{\infty}x(n-l)x^*(n-k)} \tag{4.33}$$

which is a conjugate symmetric function, $r_x(k,l)=r_x^*(l,k)$, that is similar to the sample autocorrelation sequence defined in Section 3.3.6. In terms of this (deterministic) autocorrelation sequence, Eq. (4.32) becomes

$$\boxed{\sum_{l=1}^{p}a_p(l)r_x(k,l)=-r_x(k,0) \quad ; \quad k=1,2,\ldots,p} \tag{4.34}$$

which is a set of p linear equations in the p unknowns $a_p(1),\ldots,a_p(p)$ referred to as the *Prony normal equations*. In matrix form, the normal equations are

$$\begin{bmatrix} r_x(1,1) & r_x(1,2) & r_x(1,3) & \cdots & r_x(1,p) \\ r_x(2,1) & r_x(2,2) & r_x(2,3) & \cdots & r_x(2,p) \\ r_x(3,1) & r_x(3,2) & r_x(3,3) & \cdots & r_x(3,p) \\ \vdots & \vdots & \vdots & & \vdots \\ r_x(p,1) & r_x(p,2) & r_x(p,3) & \cdots & r_x(p,p) \end{bmatrix}\begin{bmatrix} a_p(1) \\ a_p(2) \\ a_p(3) \\ \vdots \\ a_p(p) \end{bmatrix}=-\begin{bmatrix} r_x(1,0) \\ r_x(2,0) \\ r_x(3,0) \\ \vdots \\ r_x(p,0) \end{bmatrix} \tag{4.35}$$

which may be written concisely as follows

$$\boxed{\mathbf{R}_x\bar{\mathbf{a}}_p=-\mathbf{r}_x} \tag{4.36}$$

where $\mathbf{R}_x$ is a $p\times p$ Hermitian matrix of autocorrelations and $\mathbf{r}_x$ is the column vector

$$\mathbf{r}_x=\left[r_x(1,0),\ r_x(2,0),\ldots,r_x(p,0)\right]^T \tag{4.37}$$

[7] The origin of this terminology is as follows. Defining the (infinite-dimensional) vector $\mathbf{e}$ by

$$\mathbf{e}=\left[e(q+1),\ e(q+2),\ldots\right]^T$$

and the set of (infinite-dimensional) vectors $\mathbf{x}(k)$ by

$$\mathbf{x}(k)=\left[x(q+1-k),\ x(q+2-k),\ldots\right]^T$$

Eq. (4.31) says that the inner product of $\mathbf{e}$ and $\mathbf{x}(k)$ must be zero. Therefore, the orthogonality principle states that the error vector, $\mathbf{e}$, is *orthogonal* to the data vector, $\mathbf{x}(k)$.

We may express the Prony normal equations in a slightly different form as follows. With $\mathbf{X}_q$ the data matrix consisting of p infinite-dimensional column vectors,

$$\mathbf{X}_q = \begin{bmatrix} x(q) & x(q-1) & \cdots & x(q-p+1) \\ x(q+1) & x(q) & \cdots & x(q-p+2) \\ x(q+2) & x(q+1) & \cdots & x(q-p+3) \\ \vdots & \vdots & & \vdots \end{bmatrix} \tag{4.38}$$

note that the autocorrelation matrix $\mathbf{R}_x$ may be written in terms of $\mathbf{X}_q$ as follows

$$\mathbf{R}_x = \mathbf{X}_q^H \mathbf{X}_q \tag{4.39}$$

The vector of autocorrelations, $\mathbf{r}_x$, may also be expressed in terms of $\mathbf{X}_q$ as follows

$$\mathbf{r}_x = \mathbf{X}_q^H \mathbf{x}_{q+1}$$

where $\mathbf{x}_{q+1} = \big[x(q+1),\ x(q+2),\ x(q+3), \ldots\big]^T$. Thus, we have the following equivalent form for the Prony normal equations

$$\boxed{(\mathbf{X}_q^H \mathbf{X}_q)\bar{\mathbf{a}}_p = -\mathbf{X}_q^H \mathbf{x}_{q+1}} \tag{4.40}$$

If $\mathbf{R}_x$ is nonsingular[8] then the coefficients $a_p(k)$ that minimize $\mathcal{E}_{p,q}$ are

$$\bar{\mathbf{a}}_p = -\mathbf{R}_x^{-1}\mathbf{r}_x$$

or, equivalently,

$$\bar{\mathbf{a}}_p = -(\mathbf{X}_q^H \mathbf{X}_q)^{-1}\mathbf{X}_q^H \mathbf{x}_{q+1} \tag{4.41}$$

Let us now evaluate the minimum value for the modeling error $\mathcal{E}_{p,q}$. With

$$\begin{aligned} \mathcal{E}_{p,q} &= \sum_{n=q+1}^{\infty} |e(n)|^2 = \sum_{n=q+1}^{\infty} e(n)\left[x(n) + \sum_{k=1}^{p} a_p(k)x(n-k)\right]^* \\ &= \sum_{n=q+1}^{\infty} e(n)x^*(n) + \sum_{n=q+1}^{\infty} e(n)\left[\sum_{k=1}^{p} a_p(k)x(n-k)\right]^* \end{aligned} \tag{4.42}$$

it follows from the orthogonality principle that the second term in Eq. (4.42) is zero, i.e.,

$$\sum_{n=q+1}^{\infty} e(n)\left[\sum_{k=1}^{p} a_p(k)x(n-k)\right]^* = \sum_{k=1}^{p} a_p^*(k)\left[\sum_{n=q+1}^{\infty} e(n)x^*(n-k)\right] = 0$$

Thus, the minimum value of $\mathcal{E}_{p,q}$, which we will denote by $\epsilon_{p,q}$, is

$$\epsilon_{p,q} = \sum_{n=q+1}^{\infty} e(n)x^*(n) = \sum_{n=q+1}^{\infty}\left[x(n) + \sum_{k=1}^{p} a_p(k)x(n-k)\right]x^*(n) \tag{4.43}$$

[8]From the factorization of $\mathbf{R}_x$ given in Eq. (4.39), it follows that $\mathbf{R}_x$ is positive semidefinite. Although this does not guarantee that $\mathbf{R}_x$ is nonsingular, as we will see in Chapter 5, it does guarantee that the roots of the filter $A_p(z)$ will lie on or inside the unit circle (see, for example, Property 3 on p. 228).

which, in terms of the autocorrelation sequence $r_x(k, l)$, may be written as

$$\epsilon_{p,q} = r_x(0,0) + \sum_{k=1}^{p} a_p(k) r_x(0,k) \tag{4.44}$$

It is important to keep in mind that since the orthogonality principle is used in the derivation of Eq. (4.44), this expression is only valid for those coefficients $a_p(k)$ that satisfy the normal equations. To compute the modeling error for an arbitrary set of coefficients, it is necessary to use Eq. (4.42), which has no restrictions on $a_p(k)$.

Having derived an expression for the minimum error, the normal equations may be written in a slightly different form that will be useful later. If we bring the vector on the right-hand side of Eq. (4.35) to the left and absorb it into the autocorrelation matrix by adding a leading coefficient of 1 to the vector of coefficients $a_p(k)$ we have

$$\left[\begin{array}{c|cccc} r_x(1,0) & r_x(1,1) & r_x(1,2) & \cdots & r_x(1,p) \\ r_x(2,0) & r_x(2,1) & r_x(2,2) & \cdots & r_x(2,p) \\ r_x(3,0) & r_x(3,1) & r_x(3,2) & \cdots & r_x(3,p) \\ \vdots & \vdots & \vdots & & \vdots \\ r_x(p,0) & r_x(p,1) & r_x(p,2) & \cdots & r_x(p,p) \end{array}\right] \left[\begin{array}{c} 1 \\ \hline a_p(1) \\ a_p(2) \\ \vdots \\ a_p(p) \end{array}\right] = \left[\begin{array}{c} 0 \\ 0 \\ 0 \\ \vdots \\ 0 \end{array}\right] \tag{4.45}$$

Since the matrix on the left side of Eq. (4.45) has p rows and $p+1$ columns, we may make it square by augmenting these equations with the addition of Eq. (4.44) as follows:

$$\left[\begin{array}{ccccc} r_x(0,0) & r_x(0,1) & r_x(0,2) & \cdots & r_x(0,p) \\ \hline r_x(1,0) & r_x(1,1) & r_x(1,2) & \cdots & r_x(1,p) \\ r_x(2,0) & r_x(2,1) & r_x(2,2) & \cdots & r_x(2,p) \\ \vdots & \vdots & \vdots & & \vdots \\ r_x(p,0) & r_x(p,1) & r_x(p,2) & \cdots & r_x(p,p) \end{array}\right] \left[\begin{array}{c} 1 \\ a_p(1) \\ a_p(2) \\ \vdots \\ a_p(p) \end{array}\right] = \left[\begin{array}{c} \epsilon_{p,q} \\ \hline 0 \\ 0 \\ \vdots \\ 0 \end{array}\right] \tag{4.46}$$

Using matrix notation, Eq. (4.46) may be written as

$$\mathbf{R}_x \mathbf{a}_p = \epsilon_{p,q} \mathbf{u}_1 \tag{4.47}$$

where $\mathbf{R}_x$ is a Hermitian matrix having $p+1$ rows and $p+1$ columns, $\mathbf{u}_1$ is the unit vector

$$\mathbf{u}_1 = \left[1,\ 0, \ldots, 0\right]^T$$

and

$$\mathbf{a}_p = \left[1,\ a_p(1), \ldots, a_p(p)\right]^T$$

is the augmented vector of filter coefficients. Equation (4.47) is sometimes referred to as the *augmented normal equations*.

Once the coefficients $a_p(k)$ have been found, either by solving Eq. (4.36) or Eq. (4.47), the second step in Prony's method is to find the numerator coefficients. There are several different ways that this may be done. In what is referred to as Prony's method, these coefficients are found by setting the error

$$e(n) = a_p(n) * x(n) - b_q(n)$$

equal to zero for $n = 0, \ldots, q$. In other words, the coefficients are computed in exactly the

Table 4.2 ***Prony's Method of Modeling a Signal as the Unit Sample Response of a Linear Shift-Invariant System Having p Poles and q Zeros***

Normal equations

$$\sum_{l=1}^{p} a_p(l) r_x(k,l) = -r_x(k,0) \quad ; \quad k = 1, 2, \ldots, p$$

$$r_x(k,l) = \sum_{n=q+1}^{\infty} x(n-l)x^*(n-k) \quad ; \quad k, l \geq 0$$

Numerator

$$b_q(n) = x(n) + \sum_{k=1}^{p} a_p(k)x(n-k) \quad ; \quad n = 0, 1, \ldots, q$$

Minimum error

$$\epsilon_{p,q} = r_x(0,0) + \sum_{k=1}^{p} a_p(k) r_x(0,k)$$

same way as they are in the Padé approximation method,

$$b_q(n) = x(n) + \sum_{k=1}^{p} a_p(k)x(n-k) \tag{4.48}$$

Another approach, described in Section 4.4.2, is to use Shanks' method, which produces a set of coefficients $b_q(k)$ that minimize a least squares error. Yet another method is to formulate the MA parameter estimation problem as a pair of AR parameter estimation problems using Durbin's method, which is described in Section 4.7.3. Prony's method is summarized in Table 4.2.

Example 4.4.1 *Prony's Method*

Given the signal $x(n)$ consisting of a single pulse of length N,

$$x(n) = \begin{cases} 1 & ; \ n = 0, 1, \ldots, N-1 \\ 0 & ; \ \text{else} \end{cases}$$

let us use Prony's method to model $x(n)$ as the unit sample response of a linear shift-invariant filter having one pole and one zero,

$$H(z) = \frac{b(0) + b(1)z^{-1}}{1 + a(1)z^{-1}}$$

With $p = 1$, the normal equations are

$$a(1)r_x(1,1) = -r_x(1,0)$$

Therefore, with

$$r_x(1,1) = \sum_{n=2}^{\infty} x^2(n-1) = N - 1$$

$$r_x(1,0) = \sum_{n=2}^{\infty} x(n)x(n-1) = N - 2$$

then

$$a(1) = -\frac{r_x(1,0)}{r_x(1,1)} = -\frac{N-2}{N-1}$$

and the denominator of $H(z)$ becomes

$$A(z) = 1 - \frac{N-2}{N-1}z^{-1}$$

For the numerator coefficients, we have

$$\begin{aligned} b(0) &= x(0) = 1 \\ b(1) &= x(1) + a(1)x(0) = 1 - \frac{N-2}{N-1} = \frac{1}{N-1} \end{aligned}$$

Thus, the model for $x(n)$ is

$$H(z) = \frac{1 + \dfrac{1}{N-1}z^{-1}}{1 - \dfrac{N-2}{N-1}z^{-1}}$$

Finally, for the minimum squared error we have

$$\epsilon_{1,1} = r_x(0,0) + a(1)r_x(0,1)$$

Since

$$r_x(0,0) = \sum_{n=2}^{\infty} x^2(n) = N-2$$

then

$$\epsilon_{1,1} = (N-2) - \frac{N-2}{N-1}(N-2) = \frac{N-2}{N-1}$$

For example, if $N = 21$, then the model is

$$H(z) = \frac{1 + 0.05z^{-1}}{1 - 0.95z^{-1}}$$

and the unit sample response that is used to approximate $x(n)$ is

$$h(n) = \delta(n) + (0.95)^{n-1}u(n-1)$$

The error,

$$e(n) = a(n) * x(n) - b(n)$$

is shown in Fig. 4.7*a* and corresponds to a minimum squared error of

$$\epsilon_{1,1} = 0.95$$

By comparison, the error

$$e'(n) = x(n) - h(n)$$

which is the difference between the signal and the model, is shown in Fig. 4.7*b*. The sum of the squares of $e'(n)$ is

$$\sum_{n=0}^{\infty} [e'(n)]^2 \approx 4.5954$$

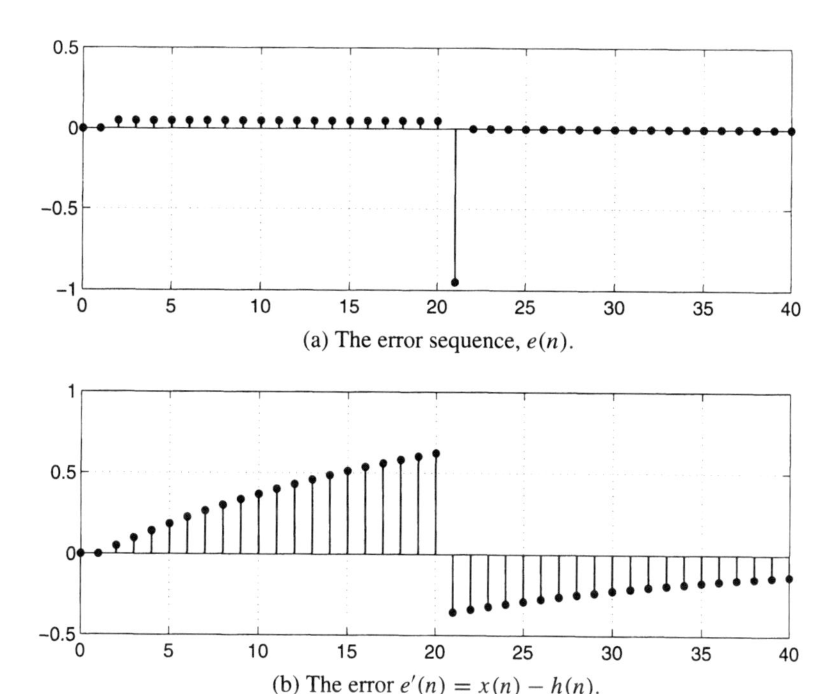

(a) The error sequence, $e(n)$.

(b) The error $e'(n) = x(n) - h(n)$.

Figure 4.7 *Prony's method to model a pulse, $x(n) = u(n) - u(n-21)$, as the unit sample response of a linear shift-invariant system having one pole and one zero ($p = q = 1$).*

Recall that it is this error that the direct method attempts to minimize. Finally, it is interesting to compare this model with that obtained using the Padé approximation. In Padé's method the model coefficients are found by solving Eq. (4.11), which, for our example, is

$$\begin{bmatrix} 1 & 0 \\ 1 & 1 \\ 1 & 1 \end{bmatrix} \begin{bmatrix} 1 \\ a(1) \end{bmatrix} = \begin{bmatrix} b(0) \\ b(1) \\ 0 \end{bmatrix}$$

Solving for $a(1)$, $b(0)$, and $b(1)$, we obtain the model

$$H(z) = \frac{1}{1 - z^{-1}}$$

which corresponds to a unit sample response of

$$h(n) = u(n)$$

Thus, the model error $e(n)$ is

$$e(n) = a(n) * x(n) - b(n) = x(n) - x(n-1) - \delta(n) = \begin{cases} 0 & ; \quad n < N \\ -1 & ; \quad n = N \\ 0 & ; \quad n > N \end{cases}$$

In the next example, we consider once again the design of a digital lowpass filter. This time, however, we will use Prony's method and compare it to the filter designed in the previous section using the Padé approximation.

Example 4.4.2 *Filter Design Using Prony's Method*

As in Example 4.3.4, let us design a linear phase lowpass filter having a cutoff frequency of $\pi/2$. The frequency response of the ideal lowpass filter is

$$I(e^{j\omega}) = \begin{cases} e^{-jn_d\omega} & ; \quad |\omega| < \pi/2 \\ 0 & ; \quad \text{otherwise} \end{cases}$$

which has a unit sample response

$$i(n) = \frac{\sin\left[(n - n_d)\pi/2\right]}{(n - n_d)\pi}$$

We will use Prony's method to approximate $i(n)$ using a fifth-order model containing $p = 5$ poles and $q = 5$ zeros. As in the previous example, the delay will be $n_d = 5$.

With $p = q = 5$, the coefficients $a(k)$ may be found by solving Eq. (4.36). The solution is[9]

$$\mathbf{a} = [1.0, \ -1.9167, \ 2.3923, \ -1.9769, \ 1.0537, \ -0.2970]^T$$

Next, the coefficients $b(k)$ are found from Eq. (4.48). The result is

$$\mathbf{b} = [0.0637, \ -0.1220, \ 0.0462, \ 0.0775, \ 0.1316, \ 0.0807]^T$$

The magnitude of the frequency response of this filter is shown in Fig. 4.8 (solid line). For comparison, also shown is the magnitude of the frequency response of a fifth-order digital elliptic filter that was designed to have a passband ripple of 0.1dB and a stopband attenuation of 40dB [8]. Although not as good as the elliptic filter, comparing this frequency response to that given in Fig. 4.4 we see that it is much better than the Padé approximation. Finally, shown in Fig. 4.9 is the difference, $e'(n) = i(n) - h(n)$ for $n \leq 80$, between the

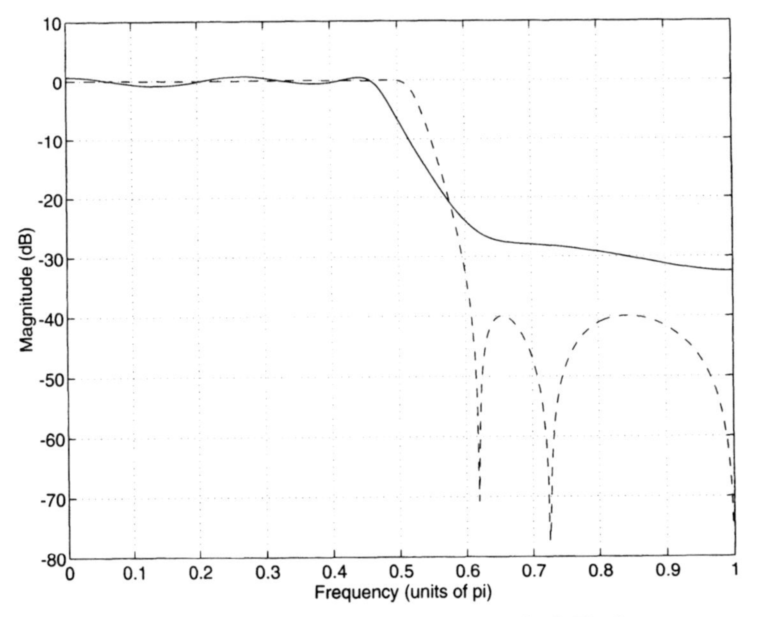

Figure 4.8 *Lowpass filter design using Prony's method. The frequency response of the filter having five poles and five zeros is shown (solid line) along with the frequency response of a fifth-order elliptic filter (dashed line).*

[9]This solution was obtained using the MATLAB program, `prony.m`, given in Fig. 4.10.

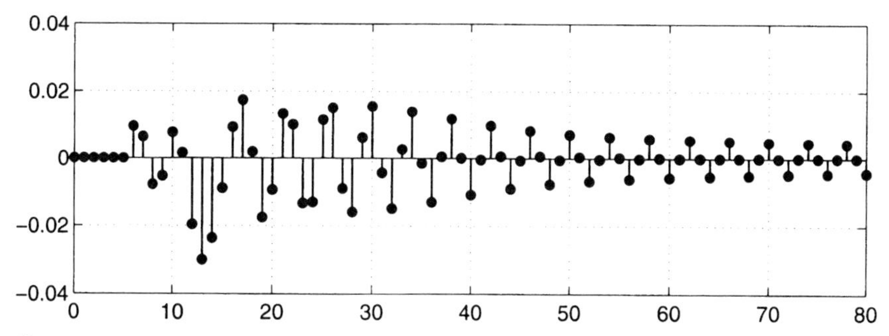

Figure 4.9 *The difference, $e'(n)$, between the unit sample response of the ideal lowpass filter, $i(n)$, and the unit sample response of the filter designed using Prony's method.*

ideal unit sample response and the unit sample response of the filter designed using Prony's method. The total squared error with this design is

$$\mathcal{E}_{LS} = \sum_n \left[e'(n)\right]^2 = 0.0838$$

compared to a squared error of 0.4426 for the Padé approximation. Comparing this design with Fig. 4.5 we see that the Prony design produces a much better approximation to the unit sample response of the ideal lowpass filter.

In the development of Prony's method given above, the normal equations were derived by differentiating the error $\mathcal{E}_{p,q}$ with respect to $a_p^*(k)$. We may also formulate Prony's method in terms of finding the least squares solution to a set of overdetermined linear equations. Specifically, note that in Prony's method we would like to find a set of coefficients $a_p(k)$ such that

$$x(n) + \sum_{k=1}^{p} a_p(k)x(n-k) = 0 \quad ; \quad n > q \tag{4.49}$$

i.e., we would like $e(n)$ to be equal to zero for $n > q$. In matrix form, these equations may be expressed as

$$\begin{bmatrix} x(q) & x(q-1) & \cdots & x(q-p+1) \\ x(q+1) & x(q) & \cdots & x(q-p+2) \\ x(q+2) & x(q+1) & \cdots & x(q-p+3) \\ \vdots & \vdots & & \vdots \end{bmatrix} \begin{bmatrix} a_p(1) \\ a_p(2) \\ \vdots \\ a_p(p) \end{bmatrix} = - \begin{bmatrix} x(q+1) \\ x(q+2) \\ x(q+3) \\ \vdots \end{bmatrix} \tag{4.50}$$

or, equivalently, as

$$\boxed{\mathbf{X}_q \bar{\mathbf{a}}_p = -\mathbf{x}_{q+1}} \tag{4.51}$$

where $\mathbf{X}_q$ is a convolution matrix, identical to the one that was defined in Eq. (4.38), and $\mathbf{x}_{q+1}$ is the data vector on the right side of Eq. (4.50). Since Eq. (4.51) represents an infinite set of linear equations in p unknowns, in general it is not possible to find values for $a_p(k)$ that solve these equations exactly. Therefore, suppose that we find the least squares solution to these equations. As discussed in Section 2.3.6 (p. 33), the least squares solution is found by solving the *normal equations*

$$(\mathbf{X}_q^H \mathbf{X}_q)\bar{\mathbf{a}}_p = -\mathbf{X}_q^H \mathbf{x}_{q+1} \tag{4.52}$$

However, these equations are identical to those given in Eq. (4.40). Therefore, finding the least squares solution to Eq. (4.51) and solving the normal equations in Eq. (4.36) are equivalent formulations of the Prony approximation problem. A MATLAB program for Prony's method that is based on finding the least squares solution to Eq. (4.51) with the numerator coefficients computed using Eq. (4.48) is given in Fig. 4.10.

4.4.2 Shanks' Method

In Prony's method, the moving average (numerator) coefficients are found by setting the error

$$e(n) = a_p(n) * x(n) - b_q(n)$$

equal to zero for $n = 0, 1, \ldots, q$. Although this forces the model to be exact over the interval $[0, q]$, it does not take into account the data for $n > q$. A better approach, suggested by Shanks [14], is to perform a least squares minimization of the model error

$$e'(n) = x(n) - \hat{x}(n)$$

over the entire length of the data record. To develop this idea, we begin by noting that the filter used to model $x(n)$ may be viewed as a cascade of two filters, $B_q(z)$ and $A_p(z)$,

$$H(z) = B_q(z)\left\{\frac{1}{A_p(z)}\right\}$$

As shown in Fig. 4.11, once $A_p(z)$ has been determined, we may compute $g(n)$, the unit sample response of the filter $1/A_p(z)$ using, for example, the recursion[10]

$$g(n) = \delta(n) - \sum_{k=1}^{p} a_p(k) g(n-k)$$

Prony's Method

```
function [a,b,err] = prony(x,p,q)
%
   x   = x(:);
   N   = length(x);
   if p+q>=length(x), error('Model order too large'), end
      X    = convm(x,p+1);
      Xq   = X(q+1:N+p-1,1:p);
      a    = [1;-Xq\X(q+2:N+p,1)];
      b    = X(1:q+1,1:p+1)*a;
      err  = x(q+2:N)'*X(q+2:N,1:p+1)*a;
      end;
```

Figure 4.10 *A* MATLAB *program for Prony's method of modeling a signal using p poles and q zeros. Note that this program calls* convm.m. *(See Appendix.)*

[10] Although we should probably use the notation $g_p(n)$, this extra subscript makes the notation in the development of Shanks' method unnecessarily complicated. Therefore, we will drop the subscript and simply write $g(n)$ for the input to the filter $B_q(z)$.

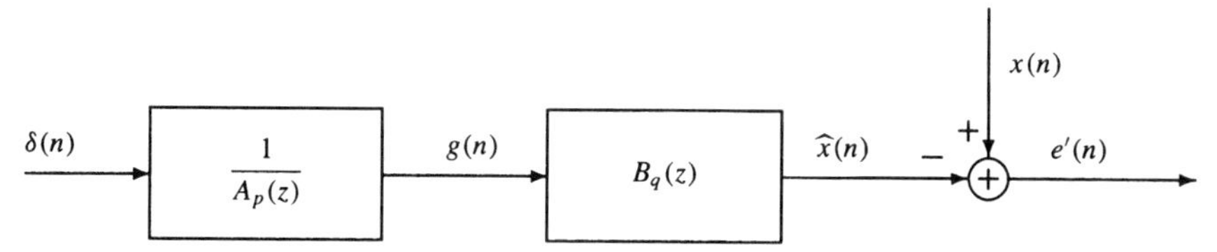

Figure 4.11 *Signal model for Shanks' method. The denominator $A_p(z)$ is found using Prony's method, and then the numerator $B_q(z)$ is found by minimizing the sum of the squares of the error $e'(n)$.*

with $g(n) = 0$ for $n < 0$. It then remains to find the coefficients of the filter $B_q(z)$ that produce the best approximation to $x(n)$ when the input to the filter is $g(n)$. Instead of forcing $e(n)$ to zero for the first $q+1$ values of n as in Prony's method, Shanks' method minimizes the squared error

$$\mathcal{E}_S = \sum_{n=0}^{\infty} |e'(n)|^2 \tag{4.53}$$

where

$$e'(n) = x(n) - \hat{x}(n) = x(n) - \sum_{l=0}^{q} b_q(l) g(n-l) \tag{4.54}$$

Note that although $\mathcal{E}_S$ is the same error used in the direct method, which led to a set of nonlinear equations, in Shanks' method the denominator coefficients are fixed and the minimization of $\mathcal{E}_S$ is carried out only with respect to the numerator coefficients $b_q(k)$. Since $e'(n)$ is *linear* in the coefficients $b_q(k)$, this problem is much easier to solve. To find the coefficients $b_q(k)$ that minimize $\mathcal{E}_S$, we set the partial derivative of $\mathcal{E}_S$ with respect to $b_q^*(k)$ equal to zero for $k = 0, 1, \ldots, q$ as follows

$$\frac{\partial \mathcal{E}_S}{\partial b_q^*(k)} = \sum_{n=0}^{\infty} e'(n) \frac{\partial [e'(n)]^*}{\partial b_q^*(k)} = -\sum_{n=0}^{\infty} e'(n) g^*(n-k) = 0 \quad ; \quad k = 0, 1, \ldots, q \tag{4.55}$$

Substituting Eq. (4.54) into Eq. (4.55) we have

$$-\sum_{n=0}^{\infty} \left\{ x(n) - \sum_{l=0}^{q} b_q(l) g(n-l) \right\} g^*(n-k) = 0$$

or

$$\sum_{l=0}^{q} b_q(l) \left[\sum_{n=0}^{\infty} g(n-l) g^*(n-k) \right] = \sum_{n=0}^{\infty} x(n) g^*(n-k) \quad ; \quad k = 0, 1, \ldots, q \tag{4.56}$$

As we did in Prony's method, if we let $r_g(k,l)$ denote the (deterministic) autocorrelation of $g(n)$,

$$r_g(k,l) = \sum_{n=0}^{\infty} g(n-l) g^*(n-k) \tag{4.57}$$

and if we define

$$r_{xg}(k) = \sum_{n=0}^{\infty} x(n) g^*(n-k) \tag{4.58}$$

to be the (deterministic) cross-correlation between $x(n)$ and $g(n)$, then Eq. (4.56) becomes

$$\sum_{l=0}^{q} b_q(l) r_g(k,l) = r_{xg}(k) \quad ; \quad k = 0, 1, \ldots, q \tag{4.59}$$

In matrix form, these equations are

$$\begin{bmatrix} r_g(0,0) & r_g(0,1) & r_g(0,2) & \cdots & r_g(0,q) \\ r_g(1,0) & r_g(1,1) & r_g(1,2) & \cdots & r_g(1,q) \\ r_g(2,0) & r_g(2,1) & r_g(2,2) & \cdots & r_g(2,q) \\ \vdots & \vdots & \vdots & & \vdots \\ r_g(q,0) & r_g(q,1) & r_g(q,2) & \cdots & r_g(q,q) \end{bmatrix} \begin{bmatrix} b_q(0) \\ b_q(1) \\ b_q(2) \\ \vdots \\ b_q(q) \end{bmatrix} = \begin{bmatrix} r_{xg}(0) \\ r_{xg}(1) \\ r_{xg}(2) \\ \vdots \\ r_{xg}(q) \end{bmatrix} \tag{4.60}$$

We may simplify these equations considerably, however, by observing that $r_g(k+1, l+1)$ may be expressed in terms of $r_g(k,l)$ as follows

$$\begin{aligned} r_g(k+1, l+1) &= \sum_{n=0}^{\infty} g(n - [l+1]) g^*(n - [k+1]) \\ &= \sum_{n=-1}^{\infty} g(n-l) g^*(n-k) = r_g(k,l) + g(-1-l) g^*(-1-k) \end{aligned} \tag{4.61}$$

Since $g(n) = 0$ for $n < 0$, the last term in Eq. (4.61) is zero when $k \geq 0$ or $l \geq 0$ and

$$r_g(k+1, l+1) = r_g(k,l)$$

Therefore, each term $r_g(k,l)$ in Eq. (4.60) depends only on the difference between the indices k and l. Consequently, with a slight abuse in notation, we will define

$$r_g(k-l) \equiv r_g(k,l) = \sum_{n=0}^{\infty} g(n-l) g^*(n-k) \tag{4.62}$$

Replacing $r_g(k,l)$ with $r_g(k-l)$ in Eq. (4.59), we have

$$\sum_{l=0}^{q} b_q(l) r_g(k-l) = r_{xg}(k) \quad ; \quad k = 0, \ldots, q \tag{4.63}$$

Since $r_x(k-l)$ is conjugate symmetric,

$$r_x(k-l) = r_x^*(l-k)$$

writing Eq. (4.63) in matrix form, we have

$$\begin{bmatrix} r_g(0) & r_g^*(1) & r_g^*(2) & \cdots & r_g^*(q) \\ r_g(1) & r_g(0) & r_g^*(1) & \cdots & r_g^*(q-1) \\ r_g(2) & r_g(1) & r_g(0) & \cdots & r_g^*(q-2) \\ \vdots & \vdots & \vdots & & \vdots \\ r_g(q) & r_g(q-1) & r_g(q-2) & \cdots & r_g(0) \end{bmatrix} \begin{bmatrix} b_q(0) \\ b_q(1) \\ b_q(2) \\ \vdots \\ b_q(q) \end{bmatrix} = \begin{bmatrix} r_{xg}(0) \\ r_{xg}(1) \\ r_{xg}(2) \\ \vdots \\ r_{xg}(q) \end{bmatrix}$$

or

$$\boxed{\mathbf{R}_g \mathbf{b}_q = \mathbf{r}_{xg}} \tag{4.64}$$

where $\mathbf{R}_g$ is a $(q+1) \times (q+1)$ Hermitian matrix.

To find the minimum squared error, we may proceed as in Prony's method and write

$$\begin{aligned}
\mathcal{E}_S = \sum_{n=0}^{\infty} |e'(n)|^2 &= \sum_{n=0}^{\infty} e'(n) \left[x(n) - \sum_{k=0}^{q} b_q(k) g(n-k) \right]^* \\
&= \sum_{n=0}^{\infty} e'(n) x^*(n) - \sum_{k=0}^{q} b_q^*(k) \left[\sum_{n=0}^{\infty} e'(n) g^*(n-k) \right]
\end{aligned} \tag{4.65}$$

Using the orthogonality of $e'(n)$ and $g(n)$ in Eq. (4.55), it follows that the second term in Eq. (4.65) is equal to zero. Therefore, the minimum error is

$$\begin{aligned}
\{\mathcal{E}_S\}_{\min} = \sum_{n=0}^{\infty} e'(n) x^*(n) &= \sum_{n=0}^{\infty} \left[x(n) - \sum_{k=0}^{q} b_q(k) g(n-k) \right] x^*(n) \\
&= \sum_{n=0}^{\infty} |x(n)|^2 - \sum_{k=0}^{q} b_q(k) \left[\sum_{n=0}^{\infty} g(n-k) x^*(n) \right]
\end{aligned} \tag{4.66}$$

In terms of $r_x(k)$ and $r_{xg}(k)$, the minimum error becomes

$$\{\mathcal{E}_S\}_{\min} = r_x(0) - \sum_{k=0}^{q} b_q(k) r_{gx}(-k)$$

or, since $r_{gx}(-k) = r_{xg}^*(k)$,

$$\boxed{\{\mathcal{E}_S\}_{\min} = r_x(0) - \sum_{k=0}^{q} b_q(k) r_{xg}^*(k)} \tag{4.67}$$

As we did with Prony's method, we may also formulate Shanks' method in terms of finding the least squares solution to a set of overdetermined linear equations. Specifically, setting $e'(n) = 0$ for $n \geq 0$ in Eq. (4.54) and writing these equations in matrix form we have

$$\begin{bmatrix} g(0) & 0 & 0 & \cdots & 0 \\ g(1) & g(0) & 0 & \cdots & 0 \\ g(2) & g(1) & g(0) & \cdots & 0 \\ \vdots & \vdots & \vdots & & \vdots \end{bmatrix} \begin{bmatrix} b_q(0) \\ b_q(1) \\ b_q(2) \\ \vdots \\ b_q(q) \end{bmatrix} = \begin{bmatrix} x(0) \\ x(1) \\ x(2) \\ \vdots \end{bmatrix} \tag{4.68}$$

or, equivalently, as

$$\boxed{\mathbf{G}_0 \mathbf{b}_q = \mathbf{x}_0} \tag{4.69}$$

The least squares solution is found by solving the linear equations

$$\boxed{(\mathbf{G}_0^H \mathbf{G}_0) \mathbf{b}_q = \mathbf{G}_0^H \mathbf{x}_0} \tag{4.70}$$

Table 4.3 Shanks' Method for Finding the Zeros in a Pole-Zero Model

Numerator

$$\sum_{l=0}^{q} b_q(l) r_g(k-l) = r_{xg}(k) \quad ; \quad k = 0, \ldots, q$$

Minimum error

$$\{\mathcal{E}_S\}_{\min} = r_x(0) - \sum_{k=0}^{q} b_q(k) r_{xg}^*(k)$$

Definitions

$$r_g(k-l) = \sum_{n=0}^{\infty} g(n-l) g^*(n-k)$$

$$r_{xg}(k) = \sum_{n=0}^{\infty} x(n) g^*(n-k)$$

$$g(n) = \delta(n) - \sum_{k=1}^{p} a_p(k) g(n-k)$$

From a computational point of view, Shanks' method is more involved than Prony's method. In addition to having to compute the sequence $g(n)$, it is necessary to evaluate the autocorrelation of $g(n)$ and the cross-correlation between $x(n)$ and $g(n)$. In spite of this extra computation, however, in some cases the reduction in the mean square error may be desirable, particularly in applications that do not require that the signal modeling be performed in real time.

Shanks' method is summarized in Table 4.3 and a MATLAB program is given in Fig. 4.12. This program is based on finding the least squares solution to the overdetermined set of linear equations given in Eq. (4.68).

Shanks' method

```
function [a,b,err] = shanks(x,p,q)
%
   x    = x(:);
   N    = length(x);
   if p+q>=length(x), error('Model order too large'), end
      a    = prony(x,p,q);
      u    = [1; zeros(N-1,1)];
      g    = filter(1,a,u);
      G    = convm(g,q+1);
      b    = G(1:N,:)\x;
      err = x'*x-x'*G(1:N,1:q+1)*b;
      end;
```

Figure 4.12 *A* MATLAB *program for Shanks' method of modeling a signal using p poles and q zeros. Note that this program calls* convm.m *and* prony.m *(see Appendix).*

Example 4.4.3 ***Shanks' Method***

As an example illustrating the use of Shanks' method, let us reconsider the problem in Example 4.4.1 of modeling a unit pulse of length N,

$$x(n) = \begin{cases} 1 & ; \ n = 0, 1, \ldots, N-1 \\ 0 & ; \ \text{else} \end{cases}$$

With $p = q = 1$, the solution to the Prony normal equations for $A(z)$ was found to be

$$A(z) = 1 - \frac{N-2}{N-1} z^{-1}$$

Using Shanks' method to find the numerator polynomial, $B(z) = b(0) + b(1)z^{-1}$, we begin by evaluating $g(n)$. Since

$$G(z) = \frac{1}{A(z)} = \frac{1}{1 - \dfrac{N-2}{N-1} z^{-1}}$$

it follows that

$$g(n) = \left(\frac{N-2}{N-1}\right)^n u(n)$$

For the autocorrelation of $g(n)$ we have

$$r_g(0) = \sum_{n=0}^{\infty} g^2(n) = \sum_{n=0}^{\infty} \left(\frac{N-2}{N-1}\right)^{2n} = \frac{1}{1 - \left(\dfrac{N-2}{N-1}\right)^2}$$

$$r_g(1) = \sum_{n=0}^{\infty} g(n)g(n-1) = \frac{N-2}{N-1} r_g(0)$$

and for the cross-correlation between $g(n)$ and $x(n)$ we have

$$r_{xg}(0) = \sum_{n=0}^{\infty} x(n)g(n) = \sum_{n=0}^{N-1} g(n) = \frac{1 - \left(\dfrac{N-2}{N-1}\right)^N}{1 - \left(\dfrac{N-2}{N-1}\right)} = (N-1)\left[1 - \left(\frac{N-2}{N-1}\right)^N\right]$$

$$r_{xg}(1) = \sum_{n=0}^{\infty} x(n)g(n-1) = \sum_{n=0}^{N-1} g(n-1) = (N-1)\left[1 - \left(\frac{N-2}{N-1}\right)^{N-1}\right]$$

Therefore,

$$\mathbf{R}_g = \frac{1}{1 - \left(\dfrac{N-2}{N-1}\right)^2} \begin{bmatrix} 1 & \dfrac{N-2}{N-1} \\ \dfrac{N-2}{N-1} & 1 \end{bmatrix}$$

and

$$\mathbf{r}_{xg} = (N-1)\begin{bmatrix} 1-\left(\dfrac{N-2}{N-1}\right)^{N} \\ 1-\left(\dfrac{N-2}{N-1}\right)^{N-1} \end{bmatrix}$$

Solving Eq. (4.60) for $b(0)$ and $b(1)$ we find

$$b(0) = 1$$

$$b(1) = (N-1)\left[\frac{1}{N-1} - \left(\frac{N-2}{N-1}\right)^{N-1} + \left(\frac{N-2}{N-1}\right)^{N+1}\right]$$

For example, with $N = 21$ the model becomes

$$H(z) = \frac{1 + 0.301z^{-1}}{1 - 0.950z^{-1}}$$

Finally, we have for the modeling error,

$$\{\mathcal{E}_S\}_{\min} = r_x(0) - b(0)r_{xg}(0) - b(1)r_{xg}(1) = 3.95$$

Recall that with Prony's method the model coefficients are $b(0) = 1$, $b(1) = 0.05$, and $a(1) = -0.95$ and the sum of the squares of $e'(n)$ is $\mathcal{E} = 4.5954$. Thus, Shanks' method results in a small improvement.

Although the coefficients $b_q(k)$ using Shanks' method are found by minimizing the error in Eq. (4.53), there may be instances in which other errors may be more appropriate. For example, it may be preferable to consider using the error

$$\mathcal{E} = \sum_{n=q+1}^{\infty} |e'(n)|^2$$

In this case, the interval over which the error is minimized, $[q+1, \infty)$, is the same as that used to derive the denominator coefficients $a_p(k)$. Alternatively, it may be preferable to consider minimizing the squared error over a finite interval,

$$\mathcal{E} = \sum_{n=0}^{N-1} |e'(n)|^2$$

This may be appropriate for signals that have a significant percentage of their energy over the interval $[0, N-1]$ or if the model is only going to be used to represent the signal over a finite interval. If this is the case, however, then one may also want to modify Prony's method to minimize the error $\mathcal{E}_{p,q}$ in Eq. (4.29) over the same interval.

4.4.3 All-Pole Modeling

In this section, we consider the special case of all-pole signal modeling using Prony's method. All-pole models are important for several reasons. First of all, in some applications the physical process by which a signal is generated will result in an all-pole (autoregressive) signal. In speech processing, for example, an acoustic tube model for speech production leads to an all-pole model for speech [11]. However, even in those applications for which it may not be possible to justify an all-pole model for the signal, one often finds an all-pole model being used. One reason for this is that all-pole models have been found to provide a sufficiently accurate representation for many different types of signals in many different applications. Another reason for the popularity of all-pole models is the special structure that is found in the all-pole Prony normal equations, which leads to fast and efficient algorithms for finding the all-pole parameters (see Chapter 5).

Let $x(n)$ be a signal, either real or complex, that is equal to zero for $n < 0$, and suppose that we would like to model $x(n)$ using an all-pole model of the form

$$H(z) = \frac{b(0)}{1 + \sum_{k=1}^{p} a_p(k) z^{-k}} \tag{4.71}$$

With Prony's method, the coefficients of the all-pole model, $a_p(k)$, are found by minimizing the error given in Eq. (4.29) with $q = 0$, i.e.,

$$\mathcal{E}_{p,0} = \sum_{n=1}^{\infty} \left| e(n) \right|^2 \tag{4.72}$$

where

$$e(n) = x(n) + \sum_{k=1}^{p} a_p(k) x(n-k)$$

Note that since $x(n) = 0$ for $n < 0$, then the error at time $n = 0$ is equal to $x(0)$, and, therefore, does not depend upon the coefficients $a_p(k)$. Thus, the coefficients that minimize $\mathcal{E}_{p,0}$ are the same as those that minimize the error[11]

$$\boxed{\mathcal{E}_p = \sum_{n=0}^{\infty} \left| e(n) \right|^2} \tag{4.73}$$

Since this error is more convenient to use than $\mathcal{E}_{p,0}$, we will formulate the all-pole modeling problem in terms of finding the coefficients $a_p(k)$ that minimize $\mathcal{E}_p$.

From our derivation of Prony's method, repeating the steps that led up to Eq. (4.34) we see that the all-pole normal equations for the coefficients $a_p(k)$ that minimize $\mathcal{E}_p$ are

$$\sum_{l=1}^{p} a_p(l) r_x(k,l) = -r_x(k,0) \quad ; \quad k = 1, 2, \ldots, p \tag{4.74}$$

where

$$r_x(k,l) = \sum_{n=0}^{\infty} x(n-l) x^*(n-k)$$

Note that in this definition for $r_x(k,l)$, the sum begins at $n = 0$ instead of $n = 1$ as we would have if we used the definition given in Eq. (4.33) with $q = 0$. This is due to the change that

[11] Since $e(n)$ is equal to zero for $n < 0$ this error may, in fact, be defined so that the sum extends from $n = -\infty$ to $n = \infty$.

was made in the limit on the sum in Eq. (4.73) for the error $\mathcal{E}_p$ that is being minimized. As in Shanks' method, the all-pole normal equations may be simplified considerably due to the underlying structure that is found in $r_x(k,l)$. Specifically, note that $r_x(k+1,l+1)$ may be expressed in terms of $r_x(k,l)$ as follows:

$$\begin{aligned} r_x(k+1,l+1) &= \sum_{n=0}^{\infty} x\big(n-[l+1]\big)x^*\big(n-[k+1]\big) \\ &= \sum_{n=-1}^{\infty} x(n-l)x^*(n-k) = r_x(k,l) + x(-1-l)x^*(-1-k) \end{aligned} \tag{4.75}$$

Since $x(n)$ is assumed to be equal to zero for $n < 0$, then the last term in Eq. (4.75) is equal to zero whenever $k \geq 0$ or $l \geq 0$. Therefore,

$$r_x(k+1,l+1) = r_x(k,l) \quad ; \quad k \geq 0 \text{ or } l \geq 0 \tag{4.76}$$

so that $r_x(k,l)$ depends only on the difference between the indices k and l. Consequently, with a slight abuse in notation we will define

$$\boxed{r_x(k-l) \equiv r_x(k,l) = \sum_{n=0}^{\infty} x(n-l)x^*(n-k)} \tag{4.77}$$

which is a conjugate symmetric function of k and l,

$$r_x(k-l) = r_x^*(l-k)$$

Replacing $r_x(k,l)$ with $r_x(k-l)$ in Eq. (4.74) we have

$$\boxed{\sum_{l=1}^{p} a_p(l) r_x(k-l) = -r_x(k) \quad ; \quad k = 1, \ldots, p} \tag{4.78}$$

which are the *all-pole normal equations*. Using the conjugate symmetry of $r_x(k)$, these equations may be written in matrix form as follows:

$$\begin{bmatrix} r_x(0) & r_x^*(1) & r_x^*(2) & \cdots & r_x^*(p-1) \\ r_x(1) & r_x(0) & r_x^*(1) & \cdots & r_x^*(p-2) \\ r_x(2) & r_x(1) & r_x(0) & \cdots & r_x^*(p-3) \\ \vdots & \vdots & \vdots & & \vdots \\ r_x(p-1) & r_x(p-2) & r_x(p-3) & \cdots & r_x(0) \end{bmatrix} \begin{bmatrix} a_p(1) \\ a_p(2) \\ a_p(3) \\ \vdots \\ a_p(p) \end{bmatrix} = - \begin{bmatrix} r_x(1) \\ r_x(2) \\ r_x(3) \\ \vdots \\ r_x(p) \end{bmatrix} \tag{4.79}$$

Note that the autocorrelation matrix multiplying the vector of coefficients $a_p(k)$ is a Hermitian Toeplitz matrix. As we will see in Chapter 5, this Toeplitz structure allows us to solve the all-pole normal equations efficiently using the Levinson-Durbin recursion. In addition to this, however, note that the Toeplitz structure implies that only $p+1$ values of the autocorrelation sequence need to be computed and stored, compared to $\frac{1}{2}p(p+1)$ values of $r_x(k,l)$ that are found on both sides of Eq. (4.35) when $q \neq 0$. Finally, the minimum modeling error for the all-pole model follows from Eq. (4.44) and is given by

$$\left\{\mathcal{E}_p\right\}_{\min} \equiv \epsilon_p = r_x(0) + \sum_{k=1}^{p} a_p(k) r_x^*(k) \tag{4.80}$$

From Eq. (4.50) note that the all-pole coefficients may also be determined by finding the least squares solution to the following set of overdetermined linear equations

$$\begin{bmatrix} x(0) & 0 & 0 & \cdots & 0 \\ x(1) & x(0) & 0 & \cdots & 0 \\ x(2) & x(1) & x(0) & \cdots & 0 \\ x(3) & x(2) & x(1) & \cdots & 0 \\ \vdots & \vdots & \vdots & & \vdots \end{bmatrix} \begin{bmatrix} a_p(1) \\ a_p(2) \\ a_p(3) \\ \vdots \\ a_p(p) \end{bmatrix} = - \begin{bmatrix} x(1) \\ x(2) \\ x(3) \\ x(4) \\ \vdots \end{bmatrix} \tag{4.81}$$

Having found the denominator coefficients, we need only determine the numerator coefficient $b(0)$ to complete the model. One possibility, as in the pole-zero Prony method, is to set

$$b(0) = x(0) \tag{4.82}$$

so that $\hat{x}(0) = x(0)$, i.e., the signal generated by the model at time $n = 0$ is equal to the value of the given data at time $n = 0$. Another approach, generally preferable to Eq. (4.82), is to choose $b(0)$ so that the energy in $x(n)$ is equal to the energy in $\hat{x}(n) = h(n)$, i.e.,

$$r_x(0) = r_h(0)$$

This prevents an anomalous or bad data value for $x(0)$ from affecting the scale factor $b(0)$. In order to satisfy this energy matching constraint, $b(0)$ must be chosen as follows

$$b(0) = \sqrt{\epsilon_p} \tag{4.83}$$

We will postpone the proof of this until Section 5.2.3 of Chapter 5 when we have a few more tools. At that time we will also show that this value for $b(0)$ ensures that $r_x(k) = r_h(k)$ for all $|k| \leq p$. This is referred to as the *autocorrelation matching property*. The equations for all-pole modeling using Prony's method are summarized in Table 4.4.

Table 4.4 ***All-Pole Modeling Using Prony's Method***

Normal equations

$$\sum_{l=1}^{p} a_p(l) r_x(k-l) = -r_x(k) \quad ; \quad k = 1, \ldots, p$$

$$r_x(k) = \sum_{n=0}^{\infty} x(n) x^*(n-k)$$

Numerator

$$b(0) = \sqrt{\epsilon_p}$$

Minimum error

$$\epsilon_p = r_x(0) + \sum_{k=1}^{p} a_p(k) r_x^*(k)$$

Example 4.4.4 *All-Pole Modeling Using Prony's Method*

Let us find a first-order all-pole model of the form

$$H(z) = \frac{b(0)}{1 + a(1)z^{-1}}$$

for the signal

$$x(n) = \delta(n) - \delta(n-1)$$

The autocorrelation sequence is

$$r_x(k) = \begin{cases} 2 & ; \quad k = 0 \\ -1 & ; \quad k = 1 \\ 0 & ; \quad \text{otherwise} \end{cases}$$

The all-pole normal equations for a first-order model are

$$r_x(0)a(1) = -r_x(1)$$

Therefore,

$$a(1) = -\frac{r_x(1)}{r_x(0)} = \frac{1}{2}$$

and the modeling error is

$$\epsilon_1 = r_x(0) + a(1)r_x(1) = 1.5$$

With $b(0)$ chosen to satisfy the energy matching constraint,

$$b(0) = \sqrt{1.5} = 1.2247$$

the model for $x(n)$ becomes

$$H(z) = \frac{1.2247}{1 + 0.5z^{-1}}$$

Let us now find the second-order all-pole model. In this case, the normal equations are

$$\begin{bmatrix} r_x(0) & r_x(1) \\ r_x(1) & r_x(0) \end{bmatrix} \begin{bmatrix} a(1) \\ a(2) \end{bmatrix} = -\begin{bmatrix} r_x(1) \\ r_x(2) \end{bmatrix}$$

Using the autocorrelation values computed above, the normal equations become

$$\begin{bmatrix} 2 & -1 \\ -1 & 2 \end{bmatrix} \begin{bmatrix} a(1) \\ a(2) \end{bmatrix} = \begin{bmatrix} 1 \\ 0 \end{bmatrix}$$

Solving for $a(0)$ and $a(1)$ we have

$$\begin{aligned} a(1) &= 2/3 \\ a(2) &= 1/3 \end{aligned}$$

For this second-order model, the modeling error is

$$\epsilon_2 = r_x(0) + a(1)r_x(1) + a(2)r_x(2) = 2 - 2/3 = 4/3$$

Again, with $b(0)$ chosen to satisfy the energy matching constraint,

$$b(0) = 2/\sqrt{3} = 1.1547$$

the model for $x(n)$ becomes

$$H(z) = \frac{1.1547}{1 + \frac{2}{3}z^{-1} + \frac{1}{3}z^{-2}}$$

Note that since $r_x(k) = 0$ for $k > 1$, for a pth-order all-pole model, the minimum modeling error is

$$\epsilon_p = r_x(0) + a_p(1)r_x(1)$$

In other words, ϵ_p depends only on the value of $a_p(1)$.

As we did in Section 4.4.1, the all-pole normal equations may be written in a different form that will be useful in later discussions. Specifically, bringing the vector on the right side of Eq. (4.79) to the left and absorbing it into the autocorrelation matrix by adding a coefficient of 1 to the vector of model parameters, the normal equations become

$$\left[\begin{array}{c:cccc} r_x(1) & r_x(0) & r_x^*(1) & \cdots & r_x^*(p-1) \\ r_x(2) & r_x(1) & r_x(0) & \cdots & r_x^*(p-2) \\ r_x(3) & r_x(2) & r_x(1) & \cdots & r_x^*(p-3) \\ \vdots & \vdots & \vdots & & \vdots \\ r_x(p) & r_x(p-1) & r_x(p-2) & \cdots & r_x(0) \end{array}\right] \left[\begin{array}{c} 1 \\ \hdashline a_p(1) \\ a_p(2) \\ \vdots \\ a_p(p) \end{array}\right] = \left[\begin{array}{c} 0 \\ 0 \\ 0 \\ \vdots \\ 0 \end{array}\right] \tag{4.84}$$

Since the matrix in Eq. (4.84) has p rows and $(p+1)$ columns, we will add an additional equation to make the matrix square while preserving the Toeplitz structure. The equation that accomplishes this is Eq. (4.80),

$$r_x(0) + \sum_{k=1}^{p} a_p(k)r_x^*(k) = \epsilon_p \tag{4.85}$$

With Eq. (4.85) incorporated into Eq. (4.84) the normal equations become

$$\left[\begin{array}{ccccc} r_x(0) & r_x^*(1) & r_x^*(2) & \cdots & r_x^*(p) \\ r_x(1) & r_x(0) & r_x^*(1) & \cdots & r_x^*(p-1) \\ r_x(2) & r_x(1) & r_x(0) & \cdots & r_x^*(p-2) \\ \vdots & \vdots & \vdots & & \vdots \\ r_x(p) & r_x(p-1) & r_x(p-2) & \cdots & r_x(0) \end{array}\right] \left[\begin{array}{c} 1 \\ a_p(1) \\ a_p(2) \\ \vdots \\ a_p(p) \end{array}\right] = \left[\begin{array}{c} \epsilon_p \\ 0 \\ 0 \\ \vdots \\ 0 \end{array}\right] \tag{4.86}$$

which, using vector notation, may be written as

$$\boxed{\mathbf{R}_x\mathbf{a}_p = \epsilon_p\mathbf{u}_1} \tag{4.87}$$

where, as before, $\mathbf{u}_1 = [1, 0, \ldots, 0]^T$ is a unit vector of length $p + 1$.

4.4.4 Linear Prediction

In this section, we will establish the equivalence between all-pole signal modeling and a problem known as *linear prediction*. Recall that Prony's method finds the set of all-pole parameters that minimizes the squared error

$$\mathcal{E}_p = \sum_{n=0}^{\infty} \left|e(n)\right|^2$$

where

$$e(n) = x(n) + \sum_{k=1}^{p} a_p(k)x(n-k)$$

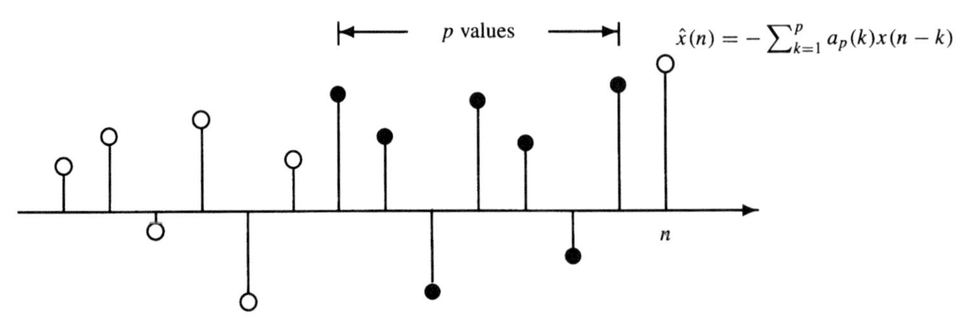

Figure 4.13 *Prony's method viewed as a problem of linear prediction in which a linear combination of the signal values $x(n-1), x(n-2), \ldots, x(n-p)$ are used to predict $x(n)$.*

If this error is expressed as follows

$$e(n) = x(n) - \hat{x}(n)$$

where

$$\hat{x}(n) = -\sum_{k=1}^{p} a_p(k)x(n-k)$$

then, as shown in Fig. 4.13, $\hat{x}(n)$ is a linear combination of the values of $x(n)$ over the interval $[n-p, n-1]$. Since $e(n)$ is the difference between $x(n)$ and $\hat{x}(n)$, minimizing the sum of the squares of $e(n)$ is equivalent to finding the coefficients $a_p(k)$ that make $\hat{x}(n)$ as close as possible to $x(n)$. Therefore, $\hat{x}(n)$ is an estimate, or prediction, of the signal $x(n)$ in terms of a linear combination of the p previous values of $x(n)$. As a result, the error $e(n)$ is referred to as the *linear prediction error* and the coefficients $a_p(k)$ are called the *linear prediction coefficients*. Furthermore, with

$$A_p(z) = 1 + \sum_{k=1}^{p} a_p(k)z^{-k}$$

since $x(n) * a_p(n) = e(n)$, then $A_p(z)$ is called the *Prediction Error Filter (PEF)*. In Chapter 6, we will look at the problem of linear prediction within the context of lattice filters, and in Chapter 7 will consider the problem of linear prediction within the context of Wiener filtering.

4.4.5 Application: FIR Least Squares Inverse Filters

Before continuing with our discussion of signal modeling techniques, we turn our attention briefly to another problem that is closely related to signal modeling: the design of an FIR least squares inverse filter. Given a filter $g(n)$, the inverse filter, which we will denote by $g^{-1}(n)$, is one for which[12]

$$g(n) * g^{-1}(n) = \delta(n)$$

or

$$G(z)G^{-1}(z) = 1$$

[12] Although $G^{-1}(z) = 1/G(z)$, the reader is cautioned not to interpret $g^{-1}(n)$ as $1/g(n)$. The superscript -1 is used simply as notation to indicate that when $g^{-1}(n)$ is convolved with $g(n)$ the result is $\delta(n)$.

The design of an inverse filter is of interest in a number of important applications. In a digital communication system, for example, a signal is to be transmitted across a nonideal channel. Assuming that the channel is linear and has a system function $G(z)$, to minimize the chance of making errors at the output of the receiver, we would like to design a *channel equalization filter* whose frequency response is the inverse, or approximate inverse, of the channel response $G(z)$. Thus, the goal is to find an equalizer $H(z)$ such that

$$G(z)H(z) \approx 1$$

A similar problem arises in seismic signal processing in which a recorded seismic signal is the convolution of a source signal $g(n)$ (wavelet) with the impulse response of a layered earth model [13]. Due to the nonimpulsive nature of the wavelet that is generated by the source, such as an airgun, the resolution of the signal will be limited. Therefore, in order to increase the amount of detail that is available in the recorded signal, a *spiking filter* $h(n)$ is designed so that

$$g(n) * h(n) \approx \delta(n)$$

i.e., $h(n)$ attempts to convert (shape) the wavelet $g(n)$ into an impulse [13].

In most applications, the inverse system $H(z) = 1/G(z)$ is not a practical solution. One of the difficulties with this solution is that, unless $G(z)$ is minimum phase, the inverse filter cannot be both causal and stable (see p. 18). Another limitation with the solution is that, in some applications, it may be necessary to constrain $H(z)$ to be an FIR filter. Since the inverse filter will be infinite in length unless $g(n)$ is an all-pole filter, constraining $h(n)$ to be FIR requires that we find the best approximation to the inverse filter.

We may formulate the FIR least squares inverse filtering problem as follows. If $g(n)$ is a causal filter that is to be equalized, the problem is to find an FIR filter $h_N(n)$ of length N such that

$$h_N(n) * g(n) \approx d(n)$$

where $d(n) = \delta(n)$. If we let

$$e(n) = d(n) - h_N(n) * g(n) = d(n) - \sum_{l=0}^{N-1} h_N(l)g(n-l) \tag{4.88}$$

be the difference between $d(n)$ and the equalized signal, $\hat{d}(n) = h_N(n) * g(n)$, as illustrated in Fig. 4.14, then we may use a least squares approach to minimize the sum of the squares of $e(n)$,

$$\mathcal{E}_N = \sum_{n=0}^{\infty} \left|e(n)\right|^2 = \sum_{n=0}^{\infty} \left|d(n) - \sum_{l=0}^{N-1} h_N(l)g(n-l)\right|^2 \tag{4.89}$$

Note that this error is the same as that which is minimized in Shanks' method, Eq. (4.54). In fact, comparing Figs. 4.11 and 4.14 we see that the two problems are identical. Therefore, from the solution that we derived for Shanks' method, Eq. (4.63), it follows that the solution for the optimum least squares inverse filter is

$$\boxed{\sum_{l=0}^{N-1} h_N(l) r_g(k-l) = r_{dg}(k) \quad ; \quad k = 0, 1, \ldots, N-1} \tag{4.90}$$

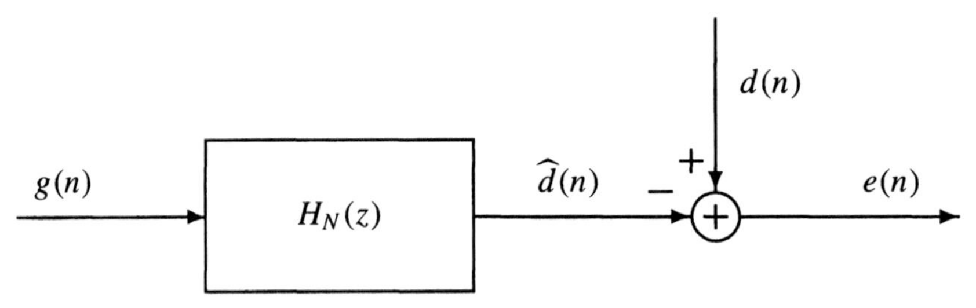

Figure 4.14 *A block diagram illustrating the design of an FIR least squares inverse filter for $g(n)$.*

where

$$r_g(k-l) = \sum_{n=0}^{\infty} g(n-l)g^*(n-k) = r_g^*(l-k)$$

and

$$r_{dg}(k) = \sum_{n=0}^{\infty} d(n)g^*(n-k) \tag{4.91}$$

Since $d(n) = \delta(n)$ and $g(n) = 0$ for $n < 0$, then $r_{dg}(k) = g^*(0)\delta(k)$ and Eq. (4.90), written in matrix form, becomes

$$\begin{bmatrix} r_g(0) & r_g^*(1) & r_g^*(2) & \cdots & r_g^*(N-1) \\ r_g(1) & r_g(0) & r_g^*(1) & \cdots & r_g^*(N-2) \\ r_g(2) & r_g(1) & r_g(0) & \cdots & r_g^*(N-3) \\ \vdots & \vdots & \vdots & & \vdots \\ r_g(N-1) & r_g(N-2) & r_g(N-3) & \cdots & r_g(0) \end{bmatrix} \begin{bmatrix} h_N(0) \\ h_N(1) \\ h_N(2) \\ \vdots \\ h_N(N-1) \end{bmatrix} = \begin{bmatrix} g^*(0) \\ 0 \\ 0 \\ \vdots \\ 0 \end{bmatrix} \tag{4.92}$$

or

$$\boxed{\mathbf{R}_g \mathbf{h}_N = g^*(0)\mathbf{u}_1} \tag{4.93}$$

which is a Toeplitz system of N linear equations in the N unknowns $h_N(n)$. From Eq. (4.67) it follows that the minimum squared error is

$$\{\mathcal{E}_N\}_{\min} = r_d(0) - \sum_{k=0}^{N-1} h_N(k) r_{dg}^*(k)$$

which, since $r_{dg}(k) = g^*(0)\delta(k)$, is equal to

$$\boxed{\{\mathcal{E}_N\}_{\min} = 1 - h_N(0)g(0)} \tag{4.94}$$

As we did in Shanks' method, we may also formulate the FIR least squares inverse filter in terms of finding the least squares solution to a set of overdetermined linear equations. Specifically, setting $e(n) = 0$ for $n \geq 0$ in Eq. (4.88) and writing these equations in matrix form we have

$$\begin{bmatrix} g(0) & 0 & 0 & \cdots & 0 \\ g(1) & g(0) & 0 & \cdots & 0 \\ g(2) & g(1) & g(0) & \cdots & 0 \\ g(3) & g(2) & g(1) & \cdots & 0 \\ \vdots & \vdots & \vdots & & \vdots \end{bmatrix} \begin{bmatrix} h_N(0) \\ h_N(1) \\ h_N(2) \\ \vdots \\ h_N(N-1) \end{bmatrix} = \begin{bmatrix} 1 \\ 0 \\ 0 \\ 0 \\ \vdots \end{bmatrix} \tag{4.95}$$

or

$$\boxed{\mathbf{G}_0 \mathbf{h}_N = \mathbf{u}_1} \tag{4.96}$$

where $\mathbf{G}_0$ is a matrix having N infinite-dimensional column vectors, and $\mathbf{u}_1$ is an infinite-dimensional column vector. The least squares solution is found by solving the linear equations

$$\boxed{(\mathbf{G}_0^H \mathbf{G}_0)\mathbf{h}_N = \mathbf{G}_0^H \mathbf{u}_1} \tag{4.97}$$

which is equivalent to Eq. (4.93).

Example 4.4.5 *FIR Least Squares Inverse*

Let us find the FIR least squares inverse for the system having a unit sample response

$$g(n) = \delta(n) - \alpha\delta(n-1)$$

where α is an arbitrary real number. Note that if $|\alpha| > 1$, then $G(z)$ has a zero that is outside the unit circle. In this case, $G(z)$ is not minimum phase and the inverse filter $1/G(z)$ cannot be both causal and stable. However, if $|\alpha| < 1$, then

$$G^{-1}(z) = \frac{1}{G(z)} = \frac{1}{1 - \alpha z^{-1}}$$

and the inverse filter is

$$g^{-1}(n) = \alpha^n u(n)$$

We begin by finding the least squares inverse of length $N = 2$. The autocorrelation of $g(n)$ is

$$r_g(k) = \begin{cases} 1+\alpha^2 & ; \quad k = 0 \\ -\alpha & ; \quad k = \pm 1 \\ 0 & ; \quad \text{else} \end{cases}$$

Therefore, Eq. (4.92) is

$$\begin{bmatrix} 1+\alpha^2 & -\alpha \\ -\alpha & 1+\alpha^2 \end{bmatrix} \begin{bmatrix} h(0) \\ h(1) \end{bmatrix} = \begin{bmatrix} 1 \\ 0 \end{bmatrix}$$

and the solution for $h(0)$ and $h(1)$ is

$$\begin{aligned} h(0) &= \frac{1+\alpha^2}{1+\alpha^2+\alpha^4} \\ h(1) &= \frac{\alpha}{1+\alpha^2+\alpha^4} \end{aligned}$$

with a squared error of

$$\{\mathcal{E}\}_{\min} = 1 - h(0) = \frac{\alpha^4}{1+\alpha^2+\alpha^4}$$

The system function of this least squares inverse filter is

$$H(z) = \frac{1+\alpha^2}{1+\alpha^2+\alpha^4} + \frac{\alpha}{1+\alpha^2+\alpha^4}z^{-1} = \frac{1+\alpha^2}{1+\alpha^2+\alpha^4}\left(1 + \frac{\alpha}{1+\alpha^2}z^{-1}\right)$$

which has a zero at

$$z_0 = -\frac{\alpha}{1+\alpha^2}$$

Note that since

$$|z_0| = \left|\frac{\alpha}{1+\alpha^2}\right| = \left|\frac{1}{\alpha+1/\alpha}\right| < 1$$

then the zero of $H(z)$ is inside the unit circle and $H(z)$ is minimum phase, regardless of whether the zero of $G(z)$ is inside or outside the unit circle.

Let us now look at the least squares inverse, $h_N(n)$, of length N. In this case Eq. (4.92) becomes

$$\begin{bmatrix} 1+\alpha^2 & -\alpha & 0 & \cdots & 0 \\ -\alpha & 1+\alpha^2 & -\alpha & \cdots & 0 \\ 0 & -\alpha & 1+\alpha^2 & \cdots & 0 \\ \vdots & \vdots & \vdots & & \vdots \\ 0 & 0 & 0 & \cdots & 1+\alpha^2 \end{bmatrix} \begin{bmatrix} h_N(0) \\ h_N(1) \\ h_N(2) \\ \vdots \\ h_N(N-1) \end{bmatrix} = \begin{bmatrix} 1 \\ 0 \\ 0 \\ \vdots \\ 0 \end{bmatrix} \tag{4.98}$$

Solving these equations for arbitrary α and N may be accomplished as follows [12]. For $n = 1, 2, \ldots, N-2$ these equations may be represented by the homogeneous difference equation,

$$-\alpha h_N(n-1) + (1+\alpha^2)h_N(n) - \alpha h_N(n+1) = 0$$

The general solution to this equation is

$$h_N(n) = c_1\alpha^n + c_2\alpha^{-n} \tag{4.99}$$

where c_1 and c_2 are constants that are determined by the boundary conditions at $n = 0$ and $n = N-1$, i.e., the first and last equation in Eq. (4.98),

$$(1+\alpha^2)h_N(0) - \alpha h_N(1) = 1$$

and

$$-\alpha h_N(N-2) + (1+\alpha^2)h_N(N-1) = 0 \tag{4.100}$$

Substituting Eq. (4.99) into Eq. (4.100) we have

$$(1+\alpha^2)\left[c_1 + c_2\right] - \alpha\left[c_1\alpha + c_2\alpha^{-1}\right] = 1$$

and

$$-\alpha\left[c_1\alpha^{N-2} + c_2\alpha^{-(N-2)}\right] + (1+\alpha^2)\left[c_1\alpha^{N-1} + c_2\alpha^{-(N-1)}\right] = 0$$

which, after canceling common terms, may be simplified to

$$c_1 + \alpha^2 c_2 = 1$$

and

$$\alpha^{N+1}c_1 + \alpha^{-(N-1)}c_2 = 0$$

or,

$$\begin{bmatrix} 1 & \alpha^2 \\ \alpha^{N+1} & \alpha^{-(N-1)} \end{bmatrix} \begin{bmatrix} c_1 \\ c_2 \end{bmatrix} = \begin{bmatrix} 1 \\ 0 \end{bmatrix}$$

The solution for c_1 and c_2 is

$$\begin{bmatrix} c_1 \\ c_2 \end{bmatrix} = \frac{1}{\alpha^{-(N-1)} - \alpha^{N+3}} \begin{bmatrix} \alpha^{-(N-1)} \\ -\alpha^{N+1} \end{bmatrix}$$

Therefore, $h_N(n)$ is

$$h_N(n) = \begin{cases} \dfrac{\alpha^{n-N} - \alpha^{N-n}}{\alpha^{-N} - \alpha^{N+2}} & ; \quad 0 \le n \le N-1 \\ 0 & ; \quad \text{else} \end{cases}$$

Finally, with

$$h_N(0) = \frac{\alpha^{-N} - \alpha^{N}}{\alpha^{-N} - \alpha^{N+2}}$$

it follows that the squared error is

$$\{\mathcal{E}_N\}_{\min} = 1 - h_N(0)g^*(0) = 1 - \frac{\alpha^{-N} - \alpha^{N}}{\alpha^{-N} - \alpha^{N+2}} = \frac{\alpha^{N} - \alpha^{N+2}}{\alpha^{-N} - \alpha^{N+2}}$$

Let us now look at what happens asymptotically as $N \to \infty$. If $|\alpha| < 1$, then

$$\lim_{N\to\infty} h_N(n) = \frac{\alpha^{n-N}}{\alpha^{-N}} = \alpha^n \quad ; \quad n \ge 0$$

which is the inverse filter, i.e.,

$$\lim_{N\to\infty} h_N(n) = \alpha^n u(n) = g^{-1}(n)$$

and

$$\lim_{N\to\infty} H_N(z) = \frac{1}{1 - \alpha z^{-1}}$$

In addition,

$$\lim_{N\to\infty} \{\mathcal{E}_N\} = 0$$

However, if $|\alpha| > 1$, then

$$\lim_{N\to\infty} h_N(n) = \frac{\alpha^{N-n}}{\alpha^{N+2}} = \alpha^{-n-2} \quad ; \quad n \ge 0$$

and

$$\lim_{N\to\infty} H_N(z) = \frac{\alpha^{-2}}{1 - \alpha^{-1}z^{-1}}$$

which is *not* the inverse filter. The squared error in this case is

$$\lim_{N\to\infty} \{\mathcal{E}_N\} = 1 - \frac{1}{\alpha^2}$$

Note that although $\hat{d}(n) = h_N(n) * g(n)$ does not converge to $\delta(n)$ as $N \to \infty$, taking the limit of $\hat{D}_N(z)$ as $N \to \infty$ we have

$$\lim_{N\to\infty} \hat{D}_N(z) = \lim_{N\to\infty} \hat{H}_N(z)G(z) = \frac{1}{\alpha}\left(\frac{1 - \alpha z^{-1}}{\alpha - z^{-1}}\right)$$

which is an allpass filter, i.e.,

$$|\hat{D}_N(e^{j\omega})| = \frac{1}{\alpha}$$

In many cases, constraining the least squares inverse filter to minimize the difference between $h_N(n) * g(n)$ and $\delta(n)$ is overly restrictive. For example, if a delay may be tolerated, then we may consider minimizing the error

$$e(n) = \delta(n - n_0) - h_N(n) * g(n)$$

i.e., $d(n) = \delta(n - n_0)$ in Eq. (4.88). In most cases, a nonzero delay will produce a better approximate inverse filter and, in many cases, the improvement will be substantial. The least squares inverse filter with delay is found by solving Eq. (4.90) with $d(n) = \delta(n - n_0)$. In this case, since

$$r_{dg}(k) = \sum_{n=0}^{\infty} d(n)g^*(n-k) = \sum_{n=0}^{\infty} \delta(n - n_0)g^*(n-k) = g^*(n_0 - k)$$

then the equations that define the coefficients of the FIR least squares inverse filter are

$$\mathbf{R}_g \mathbf{h}_N = \begin{bmatrix} g^*(n_0) \\ g^*(n_0 - 1) \\ \vdots \\ g^*(0) \\ 0 \\ \vdots \\ 0 \end{bmatrix} \tag{4.101}$$

and the minimum squared error is

$$\{\mathcal{E}_N\}_{\min} = 1 - \sum_{k=0}^{n_0} h_N(k) g^*(n_0 - k)$$

Generalizing Eq. (4.96) to the case of a least squares inverse with a delay, we have

$$\boxed{\mathbf{G}_0 \mathbf{h}_N = \mathbf{u}_{n_0+1}} \tag{4.102}$$

where $\mathbf{u}_{n_0+1}$ is a unit vector with a one in the $(n_0 + 1)$st position. A MATLAB program for the design of an FIR least squares inverse filter with delay is given in Fig. 4.15. This program is based on finding the least squares solution to Eq. (4.102).

FIR Least Squares Inverse

```
function [h,err] = spike(g,n0,n)
%
   g   = g(:);
   m   = length(g);
   if m+n-1<=n0, error('Delay too large'), end
      G    = convm(g,n);
      d    = zeros(m+n-1,1);
      d(n0+1) = 1;
      h    = G\d;
      err  = 1 - G(n0+1,:)*h;
      end;
```

Figure 4.15 *A* MATLAB *program for finding the FIR least squares inverse filter for approximating a unit sample at time $n = n_0$, i.e., $\delta(n - n_0)$. Note that this program calls* `convm.m` *(see Appendix).*

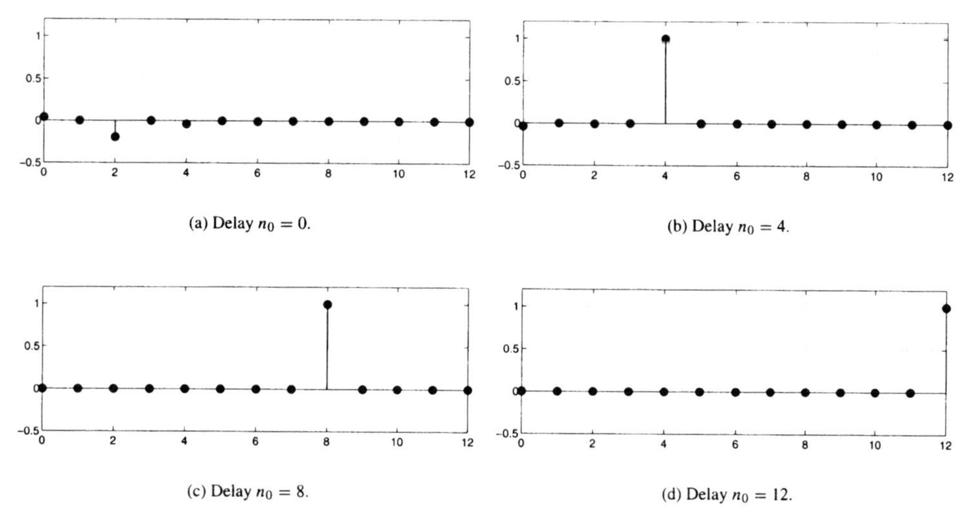

(a) Delay $n_0 = 0$.

(b) Delay $n_0 = 4$.

(c) Delay $n_0 = 8$.

(d) Delay $n_0 = 12$.

Figure 4.16 *The FIR least squares inverse for $g(n) = -0.2\delta(n) + \delta(n-2)$. Shown in each figure is the equalized signal, $\hat{d}(n) = g(n) * h(n)$, for different values of the delay n_0.*

Example 4.4.6 ***Least Squares Inverse with Delay***

Let us find the FIR least squares inverse filter of length $N = 11$ for the system

$$G(z) = -0.20 + z^{-2}$$

which has a unit sample response

$$g(n) = -0.20\delta(n) + \delta(n-2)$$

Note that $G(z)$ has two zeros that are outside the unit circle and, therefore, is not minimum phase. Since $h(n)$ is of length 11 and $g(n)$ is of length 3, then

$$\hat{d}(n) = h(n) * g(n)$$

is of length 13. Therefore, the delay n_0 is restricted to the values of $n_0 = 0, 1, \ldots, 12$. Using the MATLAB program in Fig. 4.15 we find that, without a delay ($n_0 = 0$), the squared error is $\mathcal{E} = 0.960$ and that the FIR least squares inverse filter is poor. Specifically, shown in Fig. 4.16*a* is the equalized signal, $\hat{d}(n) = g(n)*h(n)$, which ideally, should be equal to a unit sample, $\delta(n)$. Shown in Fig. 4.16*b–d*, on the other hand, are the equalized signals for delays of $n_0 = 4$, 8, and 12. The corresponding squared errors are $\mathcal{E} = 1.5 \times 10^{-3}$, 2.457×10^{-6}, and 3.9322×10^{-9}, respectively. For this particular example, a delay of $n_0 = 12$ results in the best least squares inverse filter in the sense of producing the smallest squared error.

So far we have only considered the problem of finding the least squares inverse filter $h_N(n)$ so that, for a given $g(n)$,

$$h_N(n) * g(n) \approx \delta(n - n_0)$$

However, this problem may be easily generalized to one of finding the least squares inverse filter $h_N(n)$ such that

$$h_N(n) * g(n) \approx d(n)$$

where $d(n)$ is an arbitrary sequence. In fact, note that since no assumptions were made about the form of $d(n)$ in the derivation leading up to Eq. (4.90), the solution to this generalized problem is

$$\mathbf{R}_g \mathbf{h}_N = \mathbf{r}_{dg}$$

where $\mathbf{r}_{dg}$ is a vector containing the cross-correlation $r_{dg}(k)$ as defined in Eq. (4.91). As in Eq. (4.102) for the special case of $d(n) = \delta(n - n_0)$, the FIR least squares inverse filter is the least squares solution to the set of overdetermined linear equations

$$\boxed{\mathbf{G}_0 \mathbf{h}_N = \mathbf{d}} \tag{4.103}$$

where $\mathbf{d}$ is a vector containing the signal values $d(n)$. In Chapter 7 where we consider the problem of Wiener filtering, we will be looking at the stochastic version of this problem.

4.5 ITERATIVE PREFILTERING*

In Section 4.2 we saw how the direct method of performing a least squares minimization of the error

$$E'(z) = X(z) - \frac{B_q(z)}{A_p(z)} \tag{4.104}$$

led to a set of nonlinear equations in the coefficients $a_p(k)$ and $b_q(k)$. In this section we consider an iterative approach to minimizing Eq. (4.104) known as *iterative prefiltering*. Originally developed by Steiglitz and McBride for system identification and later applied to pole-zero modeling of speech [15, 16], this method is sometimes referred to as the Steiglitz-McBride method.

The basic idea of iterative prefiltering may be developed as an extension of Shanks' method as follows. Recall that in Shanks' method, once the coefficients of $A_p(z)$ have been found using Prony's method, the coefficients of $B_q(z)$ are determined by performing a least squares minimization of the error

$$E'(z) = X(z) - B_q(z)G(z) \tag{4.105}$$

where

$$G(z) = \frac{1}{A_p(z)}$$

Since $E'(z)$ is *linear* in the coefficients $b_q(k)$, this minimization is straightforward. Note, however, that if we rewrite Eq. (4.105) as

$$E'(z) = \Big[X(z)A_p(z) - B_q(z)\Big]G(z) \tag{4.106}$$

then, once $G(z)$ fixed, $E'(z)$ becomes *linear* in $A_p(z)$ as well as $B_q(z)$. Therefore, we may easily perform a least squares minimization of $E'(z)$ with respect to both $a_p(k)$ and $b_q(k)$, thereby refining the estimate of $a_p(k)$. Having found a new set of coefficients for $A_p(z)$, we may then update $G(z)$ and repeat the process, thus forming an iteration. Therefore, the iteration proceeds as follows. Let $B_q^{(i)}(z)$ and $A_p^{(i)}(z)$ be the estimates of the numerator and denominator of $H(z)$, respectively, at the ith iteration. With $A_p^{(i)}(z)$ fixed, new estimates $B_q^{(i+1)}(z)$ and $A_p^{(i+1)}(z)$ are found by performing a least squares minimization of

$$E^{(i+1)}(z) = \frac{A_p^{(i+1)}(z)X(z) - B_q^{(i+1)}(z)}{A_p^{(i)}(z)} \tag{4.107}$$

Given the new estimate $A_p^{(i+1)}(z)$, the denominator of $E^{(i+1)}(z)$ is then updated and the process repeated.

In order to formalize this iteration, we begin by rewriting Eq. (4.107) as follows:

$$E^{(i+1)}(z) = \left[A_p^{(i+1)}(z)X(z) - B_q^{(i+1)}(z)\right]G^{(i)}(z) \tag{4.108}$$

where $G^{(i)}(z) = 1/A_p^{(i)}(z)$. If we define

$$F^{(i)}(z) = X(z)G^{(i)}(z)$$

which is the signal that is formed by filtering $x(n)$ with $g^{(i)}(n)$, then $E^{(i+1)}(z)$ becomes

$$E^{(i+1)}(z) = A_p^{(i+1)}(z)F^{(i)}(z) - B_q^{(i+1)}(z)G^{(i)}(z) \tag{4.109}$$

which, in the time domain, is

$$e^{(i+1)}(n) = a_p^{(i+1)}(n) * f^{(i)}(n) - b_q^{(i+1)}(n) * g^{(i)}(n) \tag{4.110}$$

Note that both sequences, $g^{(i)}(n)$ and $f^{(i)}(n)$, may be computed recursively as follows,

$$g^{(i)}(n) = \delta(n) - \sum_{k=1}^{p} a_p^{(i)}(k)g^{(i)}(n-k) \quad ; \quad n \geq 0$$

$$f^{(i)}(n) = x(n) - \sum_{k=1}^{p} a_p^{(i)}(k)f^{(i)}(n-k) \quad ; \quad n \geq 0$$

where $g^{(i)}(n) = f^{(i)}(n) = 0$ for $n < 0$. With $f^{(i)}(n)$ and $g^{(i)}(n)$ fixed, the coefficients $a_p^{(i+1)}(k)$ and $b_q^{(i+1)}(k)$ are found by minimizing the error

$$\mathcal{E}^{(i+1)} = \sum_{n=0}^{\infty} \left|e^{(i+1)}(n)\right|^2$$

As we have done before, a set of linear equations for the coefficients that minimize $\mathcal{E}^{(i+1)}$ may be derived by differentiating this error with respect to $[a_p^{(i+1)}(k)]^*$ and $[b_q^{(i+1)}(k)]^*$. Differentiating and setting the derivatives equal to zero results in the following *orthogonality conditions*:

$$\sum_{n=0}^{\infty} e^{(i+1)}(n)\left[f^{(i)}(n-k)\right]^* = 0 \quad ; \quad k = 1, 2, \ldots p$$

$$\sum_{n=0}^{\infty} e^{(i+1)}(n)\left[g^{(i)}(n-k)\right]^* = 0 \quad ; \quad k = 0, 1 \ldots q \tag{4.111}$$

Substituting Eq. (4.110) for $e^{(i+1)}(n)$ into Eq. (4.111) we obtain a set of linear equations that may be solved for $a_p^{(i+1)}(k)$ and $b_q^{(i+1)}(k)$. Instead of pursuing this approach, we will formulate the minimization of $\mathcal{E}^{(i+1)}$ as a problem of finding the least squares solution to a set of overdetermined linear equations. This may be done as follows. Note that in the iterative prefiltering method, we would like to find a set of coefficients, $a_p^{(i+1)}(k)$ and

$b_q^{(i+1)}(k)$, that force the error $e^{(i+1)}(n)$ to zero for $n \geq p$. From Eq. (4.110) this implies that

$$f^{(i)}(n) + \sum_{k=1}^{p} a_p^{(i+1)}(k) f^{(i)}(n-k) - \sum_{k=0}^{q} b_q^{(i+1)}(k) g^{(i)}(n-k) = 0 \quad ; \quad n \geq p \tag{4.112}$$

Writing these equations in matrix form we have

$$\begin{bmatrix} \mathbf{F}^{(i)} & \vdots & \mathbf{G}^{(i)} \end{bmatrix} \begin{bmatrix} \bar{\mathbf{a}}_p^{(i+1)} \\ \hline -\mathbf{b}_q^{(i+1)} \end{bmatrix} = -\mathbf{f}^{(i)} \tag{4.113}$$

where

$$\mathbf{F}^{(i)} = \begin{bmatrix} f^{(i)}(p-1) & f^{(i)}(p-2) & \cdots & f^{(i)}(0) \\ f^{(i)}(p) & f^{(i)}(p-1) & \cdots & f^{(i)}(1) \\ f^{(i)}(p+1) & f^{(i)}(p) & \cdots & f^{(i)}(2) \\ \vdots & \vdots & & \vdots \end{bmatrix} \tag{4.114}$$

$$\mathbf{G}^{(i)} = \begin{bmatrix} g^{(i)}(p) & g^{(i)}(p-1) & \cdots & g^{(i)}(p-q) \\ g^{(i)}(p+1) & g^{(i)}(p) & \cdots & g^{(i)}(p-q+1) \\ g^{(i)}(p+2) & g^{(i)}(p+1) & \cdots & g^{(i)}(p-q+2) \\ \vdots & \vdots & & \vdots \end{bmatrix} \tag{4.115}$$

and

$$\mathbf{f}^{(i)} = \left[f^{(i)}(p),\ f^{(i)}(p+1),\ f^{(i)}(p+2) \cdots \right]^T \tag{4.116}$$

Since Eq. (4.113) is a set of overdetermined linear equations, we seek a least squares solution, which, as discussed in Section 2.3.6, is found by solving

$$\begin{bmatrix} \mathbf{F}^{(i)} & \vdots & \mathbf{G}^{(i)} \end{bmatrix}^H \begin{bmatrix} \mathbf{F}^{(i)} & \vdots & \mathbf{G}^{(i)} \end{bmatrix} \begin{bmatrix} \bar{\mathbf{a}}_p^{(i+1)} \\ \hline -\mathbf{b}_q^{(i+1)} \end{bmatrix} = -\begin{bmatrix} \mathbf{F}^{(i)} & \vdots & \mathbf{G}^{(i)} \end{bmatrix}^H \mathbf{f}^{(i)} \tag{4.117}$$

Multiplying out the matrix product leads to

$$\begin{bmatrix} (\mathbf{F}^{(i)})^H \mathbf{F}^{(i)} & (\mathbf{F}^{(i)})^H \mathbf{G}^{(i)} \\ (\mathbf{G}^{(i)})^H \mathbf{F}^{(i)} & (\mathbf{G}^{(i)})^H \mathbf{G}^{(i)} \end{bmatrix} \begin{bmatrix} \bar{\mathbf{a}}_p^{(i+1)} \\ \hline -\mathbf{b}_q^{(i+1)} \end{bmatrix} = -\begin{bmatrix} \mathbf{F}^{(i)} & \vdots & \mathbf{G}^{(i)} \end{bmatrix}^H \mathbf{f}^{(i)} \tag{4.118}$$

which is equivalent to the set of linear equations that would be derived by differentiation as discussed above. In fact, these equations have the form

$$\begin{bmatrix} \mathbf{R}_f^{(i)} & \mathbf{R}_{gf}^{(i)} \\ \mathbf{R}_{fg}^{(i)} & \mathbf{R}_g^{(i)} \end{bmatrix} \begin{bmatrix} \bar{\mathbf{a}}_p^{(i+1)} \\ \hline -\mathbf{b}_q^{(i+1)} \end{bmatrix} = -\begin{bmatrix} \mathbf{r}_f^{(i)} \\ \hline \mathbf{r}_{fg}^{(i)} \end{bmatrix} \tag{4.119}$$

where $\mathbf{R}_f^{(i)}$ and $\mathbf{R}_g^{(i)}$ contain the autocorrelations of $g^{(i)}(n)$ and $f^{(i)}(n)$, respectively, and $\mathbf{R}_{fg}^{(i)}$ is a vector of cross-correlations between $f^{(i)}(n)$ and $g^{(i)}(n)$. A MATLAB program for the iterative prefiltering method that is based on finding the least squares solution to Eq. (4.113) is given in Fig. 4.17.

Iterative Prefiltering

```
function [a,b,err] = ipf(x,p,q,n,a)
%
   x   = x(:);
   N   = length(x);
   if p+q>=length(x), error('Model order too large'), end
   if nargin < 5
      a   = prony(x,p,q);  end;
   delta = [1; zeros(N-1,1)];
   for i=1:n
      f   = filter(1,a,x);
      g   = filter(1,a,delta);
      u   = convm(f,p+1);
      v   = convm(g,q+1);
      ab  = -[u(1:N,2:p+1) -v(1:N,:) ]\u(1:N,1);
      a   = [1; ab(1:p)];
      b   = ab(p+1:p+q+1);
      err = norm( u(1:N,1) + [u(1:N,2:p+1) -v(1:N,:)]*ab);
      end;
```

Figure 4.17 *A* MATLAB *program for the Iterative Prefiltering (Steiglitz-McBride) method. Note that this program calls* `convm.m` *and* `prony.m` *(see Appendix).*

As with any iterative algorithm, an important issue concerns the conditions for which this iteration converges. Although no general conditions have been given to guarantee convergence, this technique has been observed to either converge or to reach an acceptable solution within five to ten iterations [6]. Since the rate of convergence will depend, in general, on how close the initial guess is to the optimum solution, it is important to initialize the iteration with a set of coefficients $a_p^{(0)}(k)$ that is as accurate as possible using, for example, Prony's method.

4.6 FINITE DATA RECORDS

In Section 4.4, we considered the problem of modeling a signal $x(n)$ using Prony's method. Since Prony's method assumes that $x(n)$ is known for all n, the coefficients $a_p(k)$ are found by minimizing the sum of the squares of $e(n)$ for all $n > q$. What happens, however, when $x(n)$ is only known or measured for values of n over a finite interval such as $[0, N]$? In this case, since $e(n)$ cannot be evaluated for $n > N$, it is not possible to minimize the sum of the squares of $e(n)$ over any interval that includes values of $n > N$, unless some assumptions are made about the values of $x(n)$ outside the interval $[0, N]$. A similar problem arises in applications for which we are only interested in modeling a signal over a finite interval. Suppose, for example, that the properties of the signal that we wish to model are slowly varying in time. It may be appropriate, in this case, to consider segmenting the signal into nonoverlapping subsequences, $x_i(n)$, and to find a model for each subsequence. Again, finding the denominator coefficients by minimizing the sum of the squares of the error for all $n > q$ would not be appropriate. In this section we look at the problem of signal modeling when only a finite data record is available. We consider two approaches: the *autocorrelation method* and the *covariance method*. In the autocorrelation method, the data outside the interval $[0, N]$ is set to zero by applying a data window to $x(n)$ and then Prony's method is used to find a model for the windowed signal. With the covariance method, on the other hand, instead of using a data window, the denominator coefficients are found by minimizing an error that does not depend upon the values of $x(n)$ outside the interval $[0, N]$. As we will soon see, there are advantages and disadvantages with each method. Since the autocorrelation and covariance methods are typically used for all-pole signal modeling, our development will be in terms of all-pole models. At the end of the section we will indicate how these methods may be used in the context of a more general signal model.

4.6.1 The Autocorrelation Method

Suppose that a signal $x(n)$, possibly complex, is only known over the finite interval $[0, N]$ and that $x(n)$ is to be approximated using an all-pole model of the form given in Eq. (4.71). Using Prony's method to find an all-pole model for $x(n)$, the coefficients $a_p(k)$ are found that minimize the error

$$\mathcal{E}_p = \sum_{n=0}^{\infty} |e(n)|^2$$

where

$$e(n) = x(n) + \sum_{k=1}^{p} a_p(k) x(n-k) \tag{4.120}$$

However, if $x(n)$ is unknown or unspecified outside the interval $[0, N]$, then $e(n)$ cannot be evaluated for $n < p$ or for $n > N$. Therefore, it is not possible to minimize $\mathcal{E}_p$ without modifying Prony's method or making some assumptions about the values of $x(n)$ outside the interval $[0, N]$. One way around this difficulty is to assume that $x(n) = 0$ for $n < 0$ and for $n > N$ and then use Prony's method to find the coefficients $a_p(k)$. This approach, known as the *autocorrelation method*, effectively forms a new signal, $\tilde{x}(n)$, by applying a rectangular window to $x(n)$, i.e.,

$$\tilde{x}(n) = x(n) w_R(n)$$

where

$$w_R(n) = \begin{cases} 1 & ; \quad n = 0, 1, \ldots, N \\ 0 & ; \quad \text{otherwise} \end{cases}$$

and then uses Prony's method to find an all-pole model for $\tilde{x}(n)$. The *normal equations* for the coefficients $a_p(k)$ that minimize $\mathcal{E}_p$ are the same as those given in Eq. (4.79) for Prony's method except that $r_x(k)$ is computed using $\tilde{x}(n)$ instead of $x(n)$,

$$r_x(k) = \sum_{n=0}^{\infty} \tilde{x}(n) \tilde{x}^*(n-k) = \sum_{n=k}^{N} x(n) x^*(n-k) \quad ; \quad k = 0, 1, \ldots, p \tag{4.121}$$

Since the autocorrelation method simply involves the application of Prony's method to a windowed signal, the Toeplitz structure of the normal equations is preserved. Therefore, the Levinson-Durbin recursion presented in Chapter 5 may be used to solve the *autocorrelation normal equations*. The equations for all-pole signal modeling using the autocorrelation method are summarized in Table 4.5.

Table 4.5 ***All-Pole Modeling Using the Autocorrelation Method***

Normal equations	$$\sum_{l=1}^{p} a_p(l) r_x(k-l) = -r_x(k) \quad ; \quad k = 1, 2, \ldots, p$$ $$r_x(k) = \sum_{n=k}^{N} x(n) x^*(n-k) \quad ; \quad k \geq 0$$
Minimum error	$$\epsilon_p = r_x(0) + \sum_{k=1}^{p} a_p(k) r_x^*(k)$$

Example 4.6.1 *The Autocorrelation Method*

Consider the signal $x(n)$ whose first $N+1$ values are

$$\mathbf{x} = \left[1,\ \beta,\ \beta^2, \ldots, \beta^N\right]^T$$

where β is an arbitrary complex number. Assuming that $x(n)$ is either unknown or unspecified outside the interval $[0, N]$, let us use the autocorrelation method to find a first-order all-pole model of the form

$$H(z) = \frac{b(0)}{1 + a(1) z^{-1}}$$

With $p = 1$, it follows from the all-pole normal equations, Eq. (4.79), that

$$a(1) = -r_x(1)/r_x(0)$$

Since

$$r_x(k) = \sum_{n=k}^{N} x(n) x^*(n-k) = \sum_{n=k}^{N} \beta^n (\beta^*)^{n-k} = \beta^k \sum_{n=0}^{N-k} |\beta|^{2n} = \beta^k \frac{1 - |\beta|^{2(N-k+1)}}{1 - |\beta|^2}$$

and $r_x(k) = 0$ for $k > N$, then

$$r_x(1) = \beta \frac{1 - |\beta|^{2N}}{1 - |\beta|^2} \qquad \text{and} \qquad r_x(0) = \frac{1 - |\beta|^{2N+2}}{1 - |\beta|^2}$$

Therefore, for $a(1)$ we have

$$a(1) = -\beta \frac{1 - |\beta|^{2N}}{1 - |\beta|^{2N+2}}$$

Note that if $|\beta| < 1$, then $|a(1)| < 1$ and the model is stable. However, if β is replaced with $1/\beta^*$, then we have exactly the same model. Therefore, it follows that $|a(1)| < 1$ and the model is stable for any value of β. Furthermore, note that if $|\beta| < 1$, then

$$\lim_{N \to \infty} a(1) = -\beta$$

If, on the other hand, $|\beta| > 1$, then the model approaches the minimum phase solution,

$$\lim_{N\to\infty} a(1) = -1/\beta$$

Finally, after a bit of algebra, it follows that the model error is

$$\epsilon_1 = r_x(0) + a(1)r_x^*(1) = \frac{1-|\beta|^{4N+2}}{1-|\beta|^{2N+2}}$$

Note that if $|\beta| < 1$, then the error approaches one as $N \to \infty$. This is due to the term $e(0)$ that is included in the error ϵ_1. Thus, although $e(n) = 0$ for $n > 0$, since $e(0) = x(0)$, then the minimum value for the squared error is $\epsilon_1 = 1$.

Applying a window to $x(n)$ in the autocorrelation method forces $x(n)$ to zero for $n < 0$ and $n > N$, even if $x(n)$ is nonzero outside the interval $[0, N]$. As a result, the accuracy of the model is compromised. For example, although the signal $x(n) = \beta^n u(n)$ is the unit sample response of the all-pole filter

$$H(z) = \frac{1}{1-\beta z^{-1}}$$

as we saw in the previous example, setting $x(n) = 0$ for $n > N$ modifies the signal and biases the solution found using the autocorrelation method. On the other hand, it may be shown that windowing the data has the advantage of ensuring that the model will be stable, i.e., the poles of $H(z)$ will be inside the unit circle. This property may be of importance in applications in which the model is used to synthesize or extrapolate a signal over long periods of time since an unstable model will result in the signal becoming unbounded. Although many different proofs of this property have been given, we will postpone the proof of this very important result until Chapter 5, after we have developed the Levinson-Durbin recursion and a few of the properties of this recursion.

As we did with Prony's method, the autocorrelation method may also be expressed in terms of finding the least squares solution to a set of overdetermined linear equations. Specifically, note that in the autocorrelation method we would like to find a set of coefficients $a_p(k)$ so that the error, $e(n)$, is equal to zero for $n > 0$, i.e.,

$$e(n) = x(n) + \sum_{k=1}^{p} a_p(k)x(n-k) = 0 \quad ; \quad n > 0 \tag{4.122}$$

In matrix form, Eq. (4.122) may be expressed as

$$\boxed{\mathbf{X}_p\bar{\mathbf{a}}_p = -\mathbf{x}_1} \tag{4.123}$$

where

$$\mathbf{X}_p = \left[\begin{array}{ccccc} x(0) & 0 & 0 & \cdots & 0 \\ x(1) & x(0) & 0 & \cdots & 0 \\ x(2) & x(1) & x(0) & \cdots & 0 \\ \vdots & \vdots & \vdots & & \vdots \\ \hline x(p-1) & x(p-2) & x(p-3) & \cdots & x(0) \\ x(p) & x(p-1) & x(p-2) & \cdots & x(1) \\ \vdots & \vdots & \vdots & & \vdots \\ x(N-2) & x(N-3) & x(N-4) & \cdots & x(N-p-1) \\ x(N-1) & x(N-2) & x(N-3) & \cdots & x(N-p) \\ \hline x(N) & x(N-1) & x(N-2) & \cdots & x(N-p+1) \\ 0 & x(N) & x(N-1) & \cdots & x(N-p+2) \\ \vdots & \vdots & \vdots & & \vdots \\ 0 & 0 & 0 & \cdots & x(N) \end{array}\right] \tag{4.124}$$

is an $(N+p) \times p$ matrix and

$$\mathbf{x}_1 = \left[x(1),\ x(2), \ldots, x(N),\ 0, \ldots, 0\right]^T \tag{4.125}$$

is a vector of length $(N+p)$. The least squares solution is thus determined by solving the linear equations

$$(\mathbf{X}_p^H \mathbf{X}_p)\bar{\mathbf{a}}_p = \mathbf{X}_p^H \mathbf{x}_1$$

A MATLAB program for all-pole signal modeling using the autocorrelation method is given in Fig. 4.18.

Although typically associated with all-pole modeling, the approach used in the autocorrelation method may also be used for pole-zero modeling. The procedure simply requires solving the Prony normal equations, Eq. (4.34), with the autocorrelation $r_x(k, l)$ computed as in Eq. (4.33) using windowed data. Unlike the all-pole case, however, if $q > 0$, then the normal equations will no longer be Toeplitz.

The Autocorrelation Method

```
function [a,err] = acm(x,p)
%
   x   = x(:);
   N   = length(x);
   if p>=length(x), error('Model order too large'), end
      X   = convm(x,p+1)
      Xq  = X(1:N+p-1,1:p)
      a   = [1;-Xq\X(2:N+p,1)];
      err = abs(X(1:N+p,1)'*X*a);
      end;
```

Figure 4.18 *A* MATLAB *program for finding the coefficients of an all-pole model for a signal $x(n)$ using the autocorrelation method. Note that this program calls* convm.m *(see Appendix).*

Finally, it should be pointed out that windows other than a rectangular window may be applied to $x(n)$. The reason for considering a different window would be to minimize the edge effects in the autocorrelation method. Specifically, note that for $n = 0, 1, \ldots, p-1$ the prediction of $x(n)$ is based on fewer than p values of $x(n)$, i.e.,

$$e(n) = x(n) - \hat{x}(n)$$

where

$$\hat{x}(n) = -\sum_{k=1}^{n} a_p(k)x(n-k) \quad ; \quad n < p$$

Therefore, $e(n)$ will typically be larger for $n < p$ than it would be if the prediction were based on p values. The same will be true for $n > N - p$ since again the prediction is based on fewer than p values. As a result, $\mathcal{E}_p$ may be disproportionately influenced by the large errors at the edge of the window. To reduce the contribution of these error to $\mathcal{E}_p$, a window with a taper is often used. This is particularly true and important, for example, in all-pole modeling of speech [4, 11].

4.6.2 The Covariance Method

As we saw in the previous section, all-pole modeling with the autocorrelation method forces the signal to zero outside the interval $[0, N]$ by applying a window to $x(n)$. This may not always be the best approach to take, however, particularly in light of the fact that an all-pole model implies that the signal is infinite in length. In this section we consider another method for all-pole signal modeling, known as the *covariance method*, that does not make any assumptions about the data outside the given observation interval and does not apply a window to the data. As a result, the model obtained with the covariance method is typically more accurate than the autocorrelation method. With the covariance method, however, we lose the Toeplitz structure of the normal equations along with the guaranteed stability of the all-pole model.

To develop the covariance method, recall that all-pole modeling using Prony's method requires the minimization of $\mathcal{E}_p$ in Eq. (4.73), which is a sum of the squares of the error $e(n)$. However, as we see from the definition of $e(n)$ given in Eq. (4.120), in order to evaluate $e(n)$ at time n, it is necessary that the values of $x(n)$ be known at times $n, n-1, \ldots, n-p$. If we assume that $x(n)$ is only specified for $n = 0, 1, \ldots, N$ then, as illustrated in Fig. 4.19, the error may only be evaluated for values of n in the interval $[p, N]$. Therefore, suppose that we modify the error $\mathcal{E}_p$ so that the sum includes only those values of $e(n)$ that may be computed from the given data, i.e.,

$$\boxed{\mathcal{E}_p^C = \sum_{n=p}^{N} |e(n)|^2} \tag{4.126}$$

Minimizing $\mathcal{E}_p^C$ avoids having to make any assumptions about the data outside the interval $[0, N]$ and does not require that the data be windowed. Finding the all-pole model that minimizes $\mathcal{E}_p^C$ is known as the *covariance method*. Since the only difference between the covariance method and Prony's method is in the limits on the sum for the error, the *covariance normal equations* are identical to those in Prony's method,

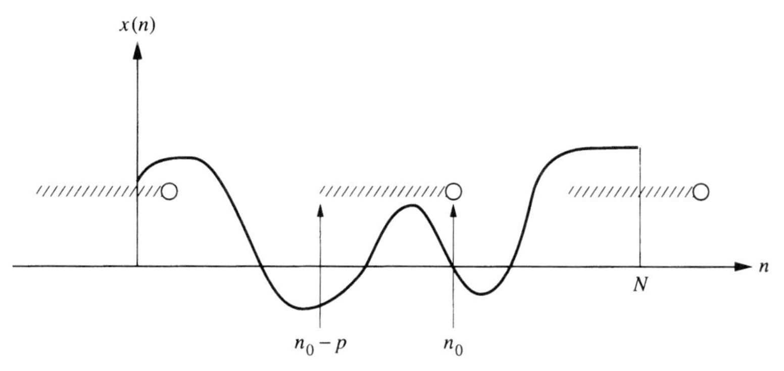

Figure 4.19 *Evaluation of the error,* $e(n) = x(n) + \sum_{k=1}^{p} a_p(k)x(n-k)$. *If* $x(n)$ *is only known over the interval* $[0, N]$, *then* $e(n)$ *may only be evaluated from* $n = p$ *to* $n = N$. *To evaluate the error for* $n < p$ *it is necessary that* $x(n)$ *be known for* $n < 0$, *whereas for* $n > N$ *it is necessary that* $x(n)$ *be known for* $n > N$.

$$\begin{bmatrix} r_x(1,1) & r_x(1,2) & r_x(1,3) & \cdots & r_x(1,p) \\ r_x(2,1) & r_x(2,2) & r_x(2,3) & \cdots & r_x(2,p) \\ r_x(3,1) & r_x(3,2) & r_x(3,3) & \cdots & r_x(3,p) \\ \vdots & \vdots & \vdots & & \vdots \\ r_x(p,1) & r_x(p,2) & r_x(p,3) & \cdots & r_x(p,p) \end{bmatrix} \begin{bmatrix} a_p(1) \\ a_p(2) \\ a_p(3) \\ \vdots \\ a_p(p) \end{bmatrix} = - \begin{bmatrix} r_x(1,0) \\ r_x(2,0) \\ r_x(3,0) \\ \vdots \\ r_x(p,0) \end{bmatrix} \tag{4.127}$$

except that the computation of the autocorrelation sequence $r_x(k, l)$ is modified as follows:

$$\boxed{r_x(k,l) = \sum_{n=p}^{N} x(n-l)x^*(n-k)} \tag{4.128}$$

Unlike the autocorrelation normal equations, the covariance normal equations are not Toeplitz. This follows by noting that

$$\begin{aligned} r_x(k+1, l+1) &= \sum_{n=p}^{N} x(n-[l+1])x^*(n-[k+1]) = \sum_{m=p-1}^{N-1} x(m-l)x^*(m-k) \\ &= r_x(k,l) - x(N-l)x^*(N-k) + x(p-1-l)x^*(p-1-k) \end{aligned} \tag{4.129}$$

Therefore, the Levinson-Durbin recursion cannot be used to solve Eq. (4.127). Although there is a fast covariance algorithm that is more efficient than Gaussian elimination, it is more complicated and computationally less efficient than the Levinson-Durbin recursion [6, 7]. Finally, the covariance modeling error is

$$\left\{\mathcal{E}_p^C\right\}_{\min} = r_x(0,0) + \sum_{k=1}^{p} a_p(k) r_x(0,k)$$

The covariance method is summarized in Table 4.6.

As with the autocorrelation method, the covariance method may be formulated as a least squares approximation problem. Specifically, note that if we set the error in Eq. (4.122) equal

Table 4.6 *All-Pole Modeling Using the Covariance Method*

Normal equations	$\sum_{l=1}^{p} a_p(l) r_x(k,l) = -r_x(k,0) \quad ; \quad k = 1, 2, \ldots, p$ $r_x(k,l) = \sum_{n=p}^{N} x(n-l)x^*(n-k) \quad ; \quad k, l > 0$
Minimum error	$\left\{\mathcal{E}_p^C\right\}_{\min} = r_x(0,0) + \sum_{k=1}^{p} a_p(k) r_x(0,k)$

The Covariance Method

```
function [a,err] = covm(x,p)
%
  x   = x(:);
  N   = length(x);
  if p>=length(x), error('Model order too large'), end
     X   = convm(x,p+1);
     Xq  = X(p:N-1,1:p);
     a   = [1;-Xq\X(p+1:N,1)];
     err = abs(X(p+1:N,1)'*X(p+1:N,:)*a);
     end;
```

Figure 4.20 *A* MATLAB *program for finding the coefficients of an all-pole model for a signal $x(n)$ using the covariance method. Note that this program calls* `convm.m` *(see Appendix).*

to zero for n in the interval $[p, N]$ we have a set of $(N-p+1)$ linear equations of the form

$$\begin{bmatrix} x(p-1) & x(p-2) & \cdots & x(0) \\ x(p) & x(p-1) & \cdots & x(1) \\ \vdots & \vdots & & \vdots \\ x(N-2) & x(N-3) & \cdots & x(N-p-1) \\ x(N-1) & x(N-2) & \cdots & x(N-p) \end{bmatrix} \begin{bmatrix} a_p(1) \\ a_p(2) \\ \vdots \\ a_p(p) \end{bmatrix} = - \begin{bmatrix} x(p) \\ x(p+1) \\ \vdots \\ x(N-1) \\ x(N) \end{bmatrix} \tag{4.130}$$

Note that these equations correspond to the subset of $N-p+1$ equations in Eq. (4.123) that do not require information about $x(n)$ outside the range $[0, N]$. Since these equations are overdetermined, we may use the same approach described for the autocorrelation method and find the least squares best approximation. A MATLAB program for finding this least squares solution is given in Fig. 4.20.

Example 4.6.2 *The Covariance Method*

Let us again consider the problem of finding a first-order all-pole model for the signal $x(n)$ whose first $N+1$ values are

$$\mathbf{x} = \left[1, \ \beta, \ \beta^2, \ldots, \beta^N\right]^T$$

where β is an arbitrary complex number. With

$$H(z) = \frac{b(0)}{1 + a(1)z^{-1}}$$

the value for $a(1)$ using the covariance method is

$$a(1) = -r_x(1,0)/r_x(1,1)$$

where

$$r_x(k,l) = \sum_{n=1}^{N} x(n-l)x^*(n-k)$$

Since

$$r_x(1,1) = \sum_{n=1}^{N} x(n-1)x^*(n-1) = \sum_{n=0}^{N-1} |x(n)|^2 = \frac{1-|\beta|^{2N}}{1-|\beta|^2}$$

$$r_x(1,0) = \sum_{n=1}^{N} x(n)x^*(n-1) = \beta\frac{1-|\beta|^{2N}}{1-|\beta|^2}$$

then $a(1) = -\beta$. With $b(0) = 1$ so that $x(0) = \hat{x}(0)$ the model becomes

$$H(z) = \frac{1}{1-\beta z^{-1}}$$

Note that the model will be stable if and only if $|\beta| < 1$.

In the previous example, the covariance method was applied to the first $N+1$ values of the signal $x(n) = \beta^n u(n)$, which is the unit sample response of the all-pole filter

$$H(z) = \frac{1}{1-\beta z^{-1}}$$

Unlike the autocorrelation method, the covariance method was able to find the pole location exactly. This will be true, in general, when modeling a signal having an all-pole z-transform

$$X(z) = \frac{b(0)}{1 + \sum_{k=1}^{p} a(k)z^{-k}}$$

provided a sufficient number of values of $x(n)$ are given. In the next example, we look at the models that are produced when the autocorrelation and covariance methods are applied to the signal $x(n) = (-1)^n$ for $n = 0, 1, \ldots, N$.

Example 4.6.3 *Comparison of the Autocorrelation and Covariance Methods*

Suppose that a signal $x(n)$ is to be modeled as the unit sample response of a second-order all-pole filter,

$$H(z) = \frac{b(0)}{1 + a(1)z^{-1} + a(2)z^{-2}}$$

The first 20 values of this signal are measured and found to be

$$\mathbf{x} = \left[1,\ -1,\ 1,\ -1, \ldots, 1,\ -1\right]$$

To solve for the coefficients $a(k)$ using the autocorrelation method, Prony's method is applied to the windowed sequence

$$\tilde{x}(n) = \begin{cases} x(n) & ; \quad n = 0, 1, \dots, N \\ 0 & ; \quad \text{otherwise} \end{cases}$$

The normal equations are

$$\begin{bmatrix} r_x(0) & r_x(1) \\ r_x(1) & r_x(0) \end{bmatrix} \begin{bmatrix} a(1) \\ a(2) \end{bmatrix} = - \begin{bmatrix} r_x(1) \\ r_x(2) \end{bmatrix}$$

where

$$r_x(k) = \sum_{n=0}^{\infty} \tilde{x}(n)\tilde{x}(n-k) = \sum_{n=k}^{N} x(n)x(n-k)$$

Evaluating this sum for $k = 0, 1$, and 2, we find

$$r_x(0) = 20 \qquad r_x(1) = -19 \qquad r_x(2) = 18$$

Substituting these values into the normal equations, we have

$$\begin{bmatrix} 20 & -19 \\ -19 & 20 \end{bmatrix} \begin{bmatrix} a(1) \\ a(2) \end{bmatrix} = - \begin{bmatrix} -19 \\ 18 \end{bmatrix}$$

Solving for $a(1)$ and $a(2)$, we find

$$a(1) = 0.9744 \quad ; \quad a(2) = 0.0256$$

Therefore, the denominator polynomial is

$$A(z) = 1 + 0.9744z^{-1} + 0.0256z^{-2}$$

From Eq. (4.80) we see that the modeling error is

$$\epsilon_p = r_x(0) + a(1)r_x(1) + a(2)r_x(2) = 1.9487$$

Setting $b(0) = \sqrt{\epsilon_1}$ to satisfy the energy matching constraint the model becomes

$$H(z) = \frac{1.3960}{1 + 0.9744z^{-1} + 0.0256z^{-2}}$$

Suppose, now, that the covariance method is used to find the all-pole model parameters. In this case, the normal equations take the form

$$\begin{bmatrix} r_x(1,1) & r_x(1,2) \\ r_x(2,1) & r_x(2,2) \end{bmatrix} \begin{bmatrix} a(1) \\ a(2) \end{bmatrix} = - \begin{bmatrix} r_x(1,0) \\ r_x(2,0) \end{bmatrix} \tag{4.131}$$

where the autocorrelations are

$$r_x(k,l) = \sum_{n=p}^{N} x(n-l)x(n-k) = \sum_{n=2}^{19} x(n-l)x(n-k)$$

Evaluating the sum, we find

$$r_x(1,1) = r_x(2,2) = r_x(2,0) = 18 \qquad r_x(1,2) = r_x(2,1) = r_x(1,0) = -18$$

which, when substituted into the normal equations, gives

$$\begin{bmatrix} 18 & -18 \\ -18 & 18 \end{bmatrix} \begin{bmatrix} a(1) \\ a(2) \end{bmatrix} = - \begin{bmatrix} -18 \\ 18 \end{bmatrix}$$

Note that these equations are singular, indicating that a lower model order is possible. Therefore, setting $a(2) = 0$ and solving for $a(1)$ we find

$$a(1) = -r_x(1,0)/r_x(1,1) = 1$$

Thus,

$$A(z) = 1 + z^{-1}$$

and, with $b(0) = 1$, the model is

$$H(z) = \frac{1}{1 + z^{-1}}$$

As a result, the model produces an exact fit to the data for $n = 0, 1, \ldots, N$.

Before concluding, we mention some modifications and generalizations to the covariance method that may be appropriate in certain applications. Recall that in the development of the covariance method it was assumed that $x(n)$ is only known or specified over the finite interval $[0, N-1]$. As a result, Prony's error was modified so that it included only those values of $e(n)$ that could be evaluated from the given data. Specifically, the covariance method minimizes the error

$$\mathcal{E}_p^C = \sum_{n=p}^{N} |e(n)|^2 \tag{4.132}$$

However, if $x(n)$ begins at time $n = 0$, i.e., $x(n) = 0$ for $n < 0$, then $e(n)$ may be evaluated for all $n \leq N$. Therefore, the covariance method may be modified to minimize the error

$$\mathcal{E}_p^{C^-} = \sum_{n=0}^{N} |e(n)|^2$$

In this case, the all-pole coefficients are found by solving the covariance normal equations in Eq. (4.127) with the autocorrelations $r_x(k,l)$ computed as follows

$$r_x(k,l) = \sum_{n=0}^{N} x(n-l)x^*(n-k) \tag{4.133}$$

This method is sometimes referred to as the *prewindowed covariance method*. In a similar vein, if $x(n)$ is known to be equal to zero for $n > N$, then we may minimize the error

$$\mathcal{E}_p^{C^+} = \sum_{n=p}^{\infty} |e(n)|^2$$

which leads to the *post-windowed covariance method*. Again, the all-pole coefficients are found by solving Eq. (4.127) with the autocorrelation sequence computed as in Eq. (4.133) with the sum extending from $n = p$ to $n = \infty$.

Each form of covariance method may also be modified to accommodate the case in which $x(n)$ is given over an arbitrary interval, say $[L, U]$. For example, with the covariance method the error to be minimized would become

$$\mathcal{E}_p^C = \sum_{n=L+p}^{U} |e(n)|^2$$

and the normal equations that must be solved are the same as Eq. (4.127) with

$$r_x(k,l) = \sum_{n=L+p}^{U} x(n-l)x^*(n-k)$$

Finally, although derived only for the case of all-pole modeling, the covariance method may also be used for pole-zero modeling. In this case, Eq. (4.132) would be modified as follows

$$\mathcal{E}_{p,q}^C = \sum_{\max(q+1,p)}^{N} \left|e(n)\right|^2$$

and the coefficients are found by solving the normal equations Eq. (4.35) with

$$r_x(k,l) = \sum_{\max(q+1,p)}^{N} x(n-l)x^*(n-k)$$

4.7 STOCHASTIC MODELS

In the preceding sections, we looked at different approaches for modeling deterministic signals. In each case, the values of $x(n)$ were known, either for all n or for values of n over some fixed finite interval. In some applications, however, it is necessary to develop models for random processes—signals whose values are unknown and may only be described probabilistically. Examples include signals whose time evolution is affected or driven by random or unknown factors, as is the case for electrocardiograms, unvoiced speech, population statistics, sunspot numbers, economic data, seismograms, and sonar data. Models for random processes differ from those for deterministic signals in two ways. First, since a random process may only be characterized statistically and the values of $x(n)$ are only known in a probabilistic sense, the errors that are minimized for deterministic signals are no longer appropriate. Recall, for example, that with Prony's method the coefficients $a_p(k)$ are found by minimizing the deterministic squared error

$$\mathcal{E}_p = \sum_{n=q+1}^{\infty} \left|e(n)\right|^2 = \sum_{n=q+1}^{\infty} \left|x(n) + \sum_{k=1}^{p} a_p(k)x(n-k)\right|^2 \tag{4.134}$$

Therefore, if $x(n)$ is only known probabilistically, then it is not feasible to consider minimizing $\mathcal{E}_p$. The second difference is in the characteristics of the signal that is used as the input to the system that is used to model $x(n)$. Whereas for deterministic signals the input signal was chosen to be a unit sample, for a random process the input signal must be a random process. Typically, this input will be taken to be unit variance white noise.

With these two differences in mind, in this section we consider the problem of modeling a wide-sense stationary random process. As we will see, the Yule-Walker equations will play an important role in the development of stochastic modeling algorithms. We begin, in Section 4.7.1, with the problem of modeling an autoregressive moving average process. We will develop two approaches, the Modified Yule-Walker Equation (MYWE) method and the least squares MYWE method. Both of these approaches are based on solving the Yule-Walker equations. Then, in Section 4.7.2, we consider the special case of autoregressive (all-pole) processes. Here we will see that all-pole modeling of a random process is similar to all-pole modeling of a deterministic signal. The primary difference is in how the autocorrelations are defined in the stochastic all-pole normal equations. Finally, in Section 4.7.3, we consider the problem of modeling a moving average (all-zero) process and develop two new approaches. The first, *spectral factorization*, requires factoring a polynomial of order $2q$ for a qth-order moving average process. The second, *Durbin's method*, avoids polynomial rooting by modeling the moving average process with a high-order all-pole model and then forming an estimate of the moving average coefficients from the all-pole coefficients.

4.7.1 Autoregressive Moving Average Models

As we saw in Section 3.6.1, a wide-sense stationary ARMA(p, q) process may be generated by filtering unit variance white noise $v(n)$ with a causal linear shift-invariant filter having p poles and q zeros,

$$H(z) = \frac{B_q(z)}{A_p(z)} = \frac{\displaystyle\sum_{k=0}^{q} b_q(k)z^{-k}}{1 + \displaystyle\sum_{k=1}^{p} a_p(k)z^{-k}}$$

Therefore, a random process $x(n)$ may be modeled as an ARMA(p, q) process using the model shown in Fig. 4.21 where $v(n)$ is unit variance white noise. To find the filter coefficients that produce the best approximation $\hat{x}(n)$ to $x(n)$ we could take the approach used in Section 4.2, replacing the least squares error $\mathcal{E}_{LS}$ with a mean square error

$$\xi_{MS} = E\left\{\left|x(n) - \hat{x}(n)\right|^2\right\}$$

However, this would lead to a set of nonlinear equations just as it did in the least squares method of Section 4.2. Therefore, we will consider another approach.

In Section 3.6.1 (p. 110), we saw that the autocorrelation sequence of an ARMA(p, q) process satisfies the Yule-Walker equations[13]

$$\boxed{r_x(k) + \sum_{l=1}^{p} a_p(l) r_x(k-l) = c_q(k)} \tag{4.135}$$

where the sequence $c_q(k)$ is the convolution of $b_q(k)$ and $h^*(-k)$,

$$c_q(k) = b_q(k) * h^*(-k) = \sum_{l=0}^{q-k} b_q(l+k) h^*(l) \tag{4.136}$$

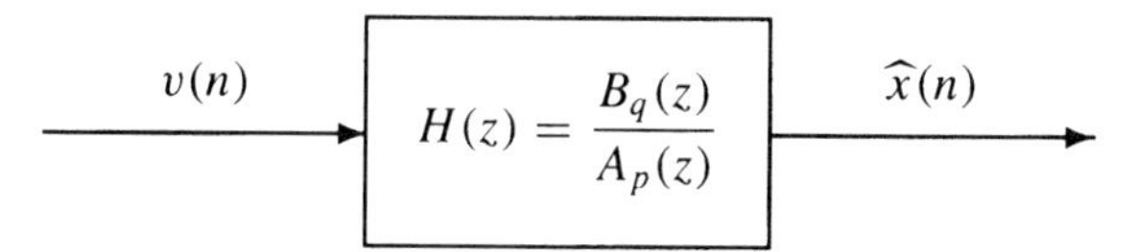

Figure 4.21 *Modeling a random process $x(n)$ as the response of a linear shift-invariant filter to unit variance white noise.*

[13]Note that σ_v^2 in Eq. (3.115) has been set equal to one since here we are assuming that $v(n)$ has unit variance.

and $r_x(k)$ is a statistical autocorrelation,

$$r_x(k) = E\{x(n)x^*(n-k)\}$$

Since $h(n)$ is assumed to be causal, then $c_q(k) = 0$ for $k > q$ and the Yule-Walker equations for $k > q$ are a function only of the coefficients $a_p(k)$,

$$\boxed{r_x(k) + \sum_{l=1}^{p} a_p(l) r_x(k-l) = 0 \quad ; \quad k > q} \tag{4.137}$$

Expressing Eq. (4.137) in matrix form for $k = q+1, q+2, \ldots, q+p$ we have

$$\begin{bmatrix} r_x(q) & r_x(q-1) & \cdots & r_x(q-p+1) \\ r_x(q+1) & r_x(q) & \cdots & r_x(q-p+2) \\ \vdots & \vdots & & \vdots \\ r_x(q+p-1) & r_x(q+p-2) & \cdots & r_x(q) \end{bmatrix} \begin{bmatrix} a_p(1) \\ a_p(2) \\ \vdots \\ a_p(p) \end{bmatrix} = - \begin{bmatrix} r_x(q+1) \\ r_x(q+2) \\ \vdots \\ r_x(q+p) \end{bmatrix} \tag{4.138}$$

which is a set of p linear equations in the p unknowns, $a_p(k)$. These equations, referred to as the *Modified Yule-Walker equations*, allow us to find the coefficients $a_p(k)$ in the filter $H(z)$ from the autocorrelation sequence $r_x(k)$ for $k = q, q+1, \ldots, q+p$. This approach, therefore, is called the *Modified Yule-Walker Equation* (MYWE) method. If the autocorrelations are unknown, then they may be replaced with estimated autocorrelations $\hat{r}_x(k)$ using a sample realization of the process.

Comparing the modified Yule-Walker equations to the Padé equations for the denominator coefficients given in Eq. (4.13), we see that the two sets of equations have exactly the same form. In fact, the only difference between them is in the definition of the "data sequence." In the modified Yule-Walker equations, the "data" consists of the sequence of autocorrelations $r_x(k)$, whereas in the Padé approximation, the data are the values of $x(n)$. As in the Padé approximation method, since the matrix in Eq. (4.138) is a nonsymmetric Toeplitz matrix, these equations may be solved efficiently for the coefficients $a_p(k)$ using the Trench algorithm [18].

Once the coefficients $a_p(k)$ have been determined, the next step is to find the MA coefficients, $b_q(k)$. There are several approaches that may be used to accomplish this. If $x(n)$ is an ARMA(p, q) process with power spectrum

$$P_x(z) = \frac{B_q(z)B_q^*(1/z^*)}{A_p(z)A_p^*(1/z^*)}$$

then filtering $x(n)$ with a linear shift-invariant filter having a system function $A_p(z)$ produces an MA(q) process, $y(n)$, that has a power spectrum

$$P_y(z) = B_q(z)B_q^*(1/z^*)$$

Therefore, the moving average parameters $b_q(k)$ may be estimated from $y(n)$ using one of the moving average modeling techniques described in Section 4.7.3. Alternatively, we may avoid filtering $x(n)$ explicitly and solve for $b_q(k)$ as follows. Given $a_p(k)$, the Yule-Walker equations may be used to find the values of the sequence $c_q(k)$ for $k = 0, 1, \ldots, q$,

$$\begin{bmatrix} r_x(0) & r_x^*(1) & \cdots & r_x^*(p) \\ r_x(1) & r_x(0) & \cdots & r_x^*(p-1) \\ \vdots & \vdots & & \vdots \\ r_x(q) & r_x(q+1) & \cdots & r_x(0) \end{bmatrix} \begin{bmatrix} 1 \\ a_p(1) \\ \vdots \\ a_p(p) \end{bmatrix} = \begin{bmatrix} c_q(0) \\ c_q(1) \\ \vdots \\ c_q(q) \end{bmatrix}$$

which may be written as

$$\boxed{\mathbf{R}_x \mathbf{a}_p = \mathbf{c}_q} \qquad (4.139)$$

Since $c_q(k) = 0$ for $k > q$ the sequence $c_q(k)$ is then known for all $k \geq 0$. We will denote the z-transform of this *causal* or *positive-time* part of $c_q(k)$ by $\left[C_q(z)\right]_+$,

$$\left[C_q(z)\right]_+ = \sum_{k=0}^{\infty} c_q(k) z^{-k}$$

Similarly, we will denote the *anticausal* or *negative-time* part by $\left[C_q(z)\right]_-$,

$$\left[C_q(z)\right]_- = \sum_{k=-\infty}^{-1} c_q(k) z^{-k} = \sum_{k=1}^{\infty} c_q(-k) z^{k}$$

Recall that $c_q(k)$ is the convolution of $b_q(k)$ with $h^*(-k)$. Therefore,

$$C_q(z) = B_q(z) H^*(1/z^*) = B_q(z) \frac{B_q^*(1/z^*)}{A_p^*(1/z^*)}$$

Multiplying $C_q(z)$ by $A_p^*(1/z^*)$ we have

$$\boxed{P_y(z) \equiv C_q(z) A_p^*(1/z^*) = B_q(z) B_q^*(1/z^*)} \qquad (4.140)$$

which is the power spectrum of an MA(q) process. Since $a_p(k)$ is zero for $k < 0$, then $A_p^*(1/z^*)$ contains only positive powers of z. Therefore, with

$$P_y(z) = C_q(z) A_p^*(1/z^*) = \left[C_q(z)\right]_+ A_p^*(1/z^*) + \left[C_q(z)\right]_- A_p^*(1/z^*) \qquad (4.141)$$

since both $\left[C_q(z)\right]_-$ and $A_p^*(1/z^*)$ are polynomials that contain only positive powers of z, then the causal part of $P_y(z)$ is equal to

$$\boxed{\left[P_y(z)\right]_+ = \left[\left[C_q(z)\right]_+ A_p^*(1/z^*)\right]_+} \qquad (4.142)$$

Thus, although $c_q(k)$ is unknown for $k < 0$, the causal part of $P_y(z)$ may be computed from the causal part of $c_q(k)$ and the AR coefficients $a_p(k)$. Using the conjugate symmetry of $P_y(z)$ we may then determine $P_y(z)$ for all z. Finally, performing a spectral factorization of $P_y(z)$,

$$P_y(z) = B_q(z) B_q^*(1/z^*) \tag{4.143}$$

produces the polynomial $B_q(z)$. The procedure is illustrated in the following example.

Example 4.7.1 *The MYWE Method for Modeling an ARMA(1,1) Process*

Suppose that we would like to find an ARMA(1,1) model for a real-valued random process $x(n)$ having autocorrelation values

$$r_x(0) = 26 \quad ; \quad r_x(1) = 7 \quad ; \quad r_x(2) = 7/2$$

The Yule-Walker equations are

$$\begin{bmatrix} r_x(0) & r_x(1) \\ r_x(1) & r_x(0) \\ r_x(2) & r_x(1) \end{bmatrix} \begin{bmatrix} 1 \\ a_1(1) \end{bmatrix} = \begin{bmatrix} c_1(0) \\ c_1(1) \\ 0 \end{bmatrix} \tag{4.144}$$

Thus, the modified Yule-Walker equations are

$$r_x(1) a_1(1) = -r_x(2)$$

which gives $a_1(1) = -r_x(2)/r_x(1) = -1/2$.

For the moving average coefficients we begin by computing $c_1(0)$ and $c_1(1)$ using the Yule-Walker equations as follows

$$\begin{bmatrix} r_x(0) & r_x(1) \\ r_x(1) & r_x(0) \end{bmatrix} \begin{bmatrix} 1 \\ a_1(1) \end{bmatrix} = \begin{bmatrix} c_1(0) \\ c_1(1) \end{bmatrix}$$

With the given values for $r_x(k)$, using $a_1(1) = -1/2$, we find

$$\begin{bmatrix} c_1(0) \\ c_1(1) \end{bmatrix} = \begin{bmatrix} 26 & 7 \\ 7 & 26 \end{bmatrix} \begin{bmatrix} 1 \\ -1/2 \end{bmatrix} = \begin{bmatrix} 45/2 \\ -6 \end{bmatrix}$$

and

$$\big[C_1(z)\big]_+ = \tfrac{45}{2} - 6z^{-1}$$

Multiplying by $A_1^*(1/z^*) = 1 - 0.5z$ we have

$$\big[C_1(z)\big]_+ A_1^*(1/z^*) = \left(\tfrac{45}{2} - 6z^{-1}\right)(1 - 0.5z) = -\tfrac{45}{4}z + \tfrac{51}{2} - 6z^{-1}$$

Therefore, the causal part of $P_y(z)$ is

$$\big[P_y(z)\big]_+ = \Big[\big[C_1(z)\big]_+ A_1^*(1/z^*)\Big]_+ = \tfrac{51}{2} - 6z^{-1}$$

Using the symmetry of $P_y(z)$, we have

$$C_1(z) A_1^*(1/z^*) = B_1(z) B_1^*(1/z^*) = -6z + \tfrac{51}{2} - 6z^{-1}$$

Performing a spectral factorization gives

$$B_1(z) B_1^*(1/z^*) = 24(1 - \tfrac{1}{4}z^{-1})(1 - \tfrac{1}{4}z)$$

so the ARMA(1,1) model is

$$H(z) = 2\sqrt{6}\,\frac{1 - 0.25z^{-1}}{1 - 0.5z^{-1}}$$

Just as with the Padé approximation, the MYWE method only uses the values of $r_x(k)$ for $k = q, q+1, \ldots, q+p$ to estimate the coefficients $a_p(k)$. If $r_x(k)$ is unknown and must be estimated from the data, then the accuracy of the model will depend on how accurately $r_x(k)$ may be estimated. However, suppose that we may estimate $r_x(k)$ for $k > p+q$. This would certainly be possible, for example, if $N \gg p+q$ where N is the length of $x(n)$. How may these estimates be used to improve the accuracy of the coefficients $a_p(k)$? As we did in extending the Padé approximation method to Prony's method, we may consider the problem of finding the set of coefficients $a_p(k)$ that produces the "best fit," in a least squares sense, to a given set of autocorrelation values. For example, given the autocorrelation sequence $r_x(k)$ for $k = 0, 1, \ldots, L$, we may form the set of *extended Yule-Walker equations* by evaluating Eq. (4.135) for $k = 0, 1, \ldots, L$ as follows:

$$\begin{bmatrix} r_x(0) & r_x^*(1) & r_x^*(2) & \cdots & r_x^*(p) \\ r_x(1) & r_x(0) & r_x^*(1) & \cdots & r_x^*(p-1) \\ \vdots & \vdots & \vdots & & \vdots \\ r_x(q) & r_x(q-1) & r_x(q-2) & \cdots & r_x(q-p) \\ r_x(q+1) & r_x(q) & r_x(q-1) & \cdots & r_x(q-p+1) \\ \vdots & \vdots & \vdots & & \vdots \\ r_x(L) & r_x(L-1) & r_x(L-2) & \cdots & r_x(L-p) \end{bmatrix} \begin{bmatrix} 1 \\ a_p(1) \\ a_p(2) \\ \vdots \\ a_p(p) \end{bmatrix} = \begin{bmatrix} c_q(0) \\ c_q(1) \\ \vdots \\ c_q(q) \\ 0 \\ \vdots \\ 0 \end{bmatrix} \tag{4.145}$$

From the last $L - q$ equations we have

$$\begin{bmatrix} r_x(q) & r_x(q-1) & \cdots & r_x(q-p+1) \\ r_x(q+1) & r_x(q) & \cdots & r_x(q-p+2) \\ \vdots & \vdots & & \vdots \\ r_x(L-1) & r_x(L-2) & \cdots & r_x(L-p) \end{bmatrix} \begin{bmatrix} a_p(1) \\ a_p(2) \\ \vdots \\ a_p(p) \end{bmatrix} = - \begin{bmatrix} r_x(q+1) \\ r_x(q+2) \\ \vdots \\ r_x(L) \end{bmatrix} \tag{4.146}$$

or

$$\boxed{\mathbf{R}_q \bar{\mathbf{a}}_p = -\mathbf{r}_{q+1}} \tag{4.147}$$

which is an overdetermined set of linear equations in the unknowns $a_p(k)$. As we did in Prony's method (p. 154), we may find the least squares solution, which is found by solving the equations

$$\boxed{(\mathbf{R}_q^H \mathbf{R}_q)\bar{\mathbf{a}}_p = -\mathbf{R}_q^H \mathbf{r}_{q+1}} \tag{4.148}$$

where $\left(\mathbf{R}_q^H \mathbf{R}_q\right)$ is a $p \times p$ Hermitian Toeplitz matrix whose elements are the autocorrelations of the sequence $r_x(k)$.

4.7.2 Autoregressive Models

A wide-sense stationary autoregressive process of order p is a special case of an ARMA(p, q) process in which $q = 0$. An AR(p) process may be generated by filtering unit variance white noise, $v(n)$, with an all-pole filter of the form

$$H(z) = \frac{b(0)}{1 + \sum_{k=1}^{p} a_p(k) z^{-k}} \tag{4.149}$$

Just as with an ARMA process, the autocorrelation sequence of an AR process satisfies the Yule-Walker equations, which are

$$\boxed{r_x(k) + \sum_{l=1}^{p} a_p(l) r_x(k-l) = |b(0)|^2 \delta(k) \quad ; \quad k \geq 0} \tag{4.150}$$

Writing these equations in matrix form for $k = 1, 2, \ldots p$, using the conjugate symmetry of $r_x(k)$, we have

$$\begin{bmatrix} r_x(0) & r_x^*(1) & r_x^*(2) & \cdots & r_x^*(p-1) \\ r_x(1) & r_x(0) & r_x^*(1) & \cdots & r_x^*(p-2) \\ r_x(2) & r_x(1) & r_x(0) & \cdots & r_x^*(p-3) \\ \vdots & \vdots & \vdots & & \vdots \\ r_x(p-1) & r_x(p-2) & r_x(p-3) & \cdots & r_x(0) \end{bmatrix} \begin{bmatrix} a_p(1) \\ a_p(2) \\ a_p(3) \\ \vdots \\ a_p(p) \end{bmatrix} = - \begin{bmatrix} r_x(1) \\ r_x(2) \\ r_x(3) \\ \vdots \\ r_x(p) \end{bmatrix} \tag{4.151}$$

Therefore, given the autocorrelations $r_x(k)$ for $k = 0, 1 \ldots, p$ we may solve Eq. (4.151) for the AR coefficients $a_p(k)$. This approach is referred to as the *Yule-Walker method.* If we compare Eq. (4.151) to the normal equations for all-pole modeling of a deterministic signal using Prony's method, Eq. (4.79), we see that the two sets of equations are identical. The only difference between them is in how the autocorrelation $r_x(k)$ is defined. In the all-pole Prony method, $r_x(k)$ is a deterministic autocorrelation, whereas in the Yule-Walker method $r_x(k)$ is a statistical autocorrelation. Finally, to determine the coefficient $b(0)$ in the AR model we may use the Yule-Walker equation for $k = 0$ as follows

$$|b(0)|^2 = r_x(0) + \sum_{k=1}^{p} a_p(k) r_x(k) \tag{4.152}$$

In most applications, the statistical autocorrelation $r_x(k)$ is unknown and must be estimated from a sample realization of the process. For example, given $x(n)$ for $0 \leq n < N$, we may estimate $r_x(k)$ using the sample autocorrelation

$$\hat{r}_x(k) = \frac{1}{N} \sum_{n=0}^{N-1} x(n) x(n-k) \tag{4.153}$$

However, note that once we replace the statistical autocorrelation $r_x(k)$ with this estimate, we have come full circle back to the autocorrelation method. Therefore, in spite of the important philosophical differences between deterministic and stochastic all-pole signal modeling, the two approaches become equivalent when the autocorrelation sequence must be estimated.

4.7.3 Moving Average Models

A moving average process is another special case of ARMA(p, q) process. An MA(q) process may be generated by filtering unit variance white noise $v(n)$ with an FIR filter of order q as follows:

$$x(n) = \sum_{k=0}^{q} b_q(k)v(n-k)$$

The Yule-Walker equations relating the autocorrelation sequence to the filter coefficients $b_q(k)$ are

$$r_x(k) = b_q(k) * b_q^*(-k) = \sum_{l=0}^{q-|k|} b_q(l+|k|)b_q^*(l) \qquad (4.154)$$

Note that, unlike the case for an autoregressive process, these equations are nonlinear in the model coefficients, $b_q(k)$. Therefore, even if the autocorrelation sequence were known exactly, finding the coefficients $b_q(k)$ may be difficult. Instead of attempting to solve the Yule-Walker equations directly, another approach is to perform a spectral factorization of the power spectrum $P_x(z)$. Specifically, since the autocorrelation of an MA(q) process is equal to zero for $|k| > q$, the power spectrum is a polynomial of the form

$$\begin{aligned} P_x(z) &= \sum_{k=-q}^{q} r_x(k)z^{-k} = B_q(z)B_q^*(1/z^*) \\ &= |b_q(0)|^2 \prod_{k=1}^{q}(1-\beta_k z^{-1}) \prod_{k=1}^{q}(1-\beta_k^* z) \end{aligned} \qquad (4.155)$$

where

$$B_q(z) = \sum_{k=0}^{q} b_q(k)z^{-k}$$

Using the spectral factorization given in Eq. (3.102), $P_x(z)$ may also be factored as follows

$$P_x(z) = \sigma_0^2\, Q(z)\, Q^*(1/z^*) = \sigma_0^2 \prod_{k=1}^{q}(1-\alpha_k z^{-1}) \prod_{k=1}^{q}(1-\alpha_k^* z) \qquad (4.156)$$

where $Q(z)$ is a minimum phase monic polynomial of degree q, i.e., $|\alpha_k| \le 1$.[14] Comparing Eqs. (4.155) and (4.156) it follows that $b_q(0) = \sigma_0$ and that $Q(z)$ is the minimum phase version of $B_q(z)$ that is formed by replacing each zero of $B_q(z)$ that lies outside the unit circle with one that lies inside the unit circle at the conjugate reciprocal location. Thus, given the autocorrelation sequence of an MA(q) process, we may find a model for $x(n)$ as follows. From the autocorrelation sequence $r_x(k)$ we form the polynomial $P_x(z)$ and factor it into a product of a minimum phase polynomial, $Q(z)$, and a maximum phase polynomial $Q^*(1/z^*)$ as in Eq. (4.156). The process $x(n)$ may then be modeled as the output of the minimum phase FIR filter

$$H(z) = \sigma_0 Q(z) = \sigma_0 \sum_{k=0}^{q} q(k)z^{-k}$$

[14] Since there is nothing that prohibits $B_q(z)$ and thus $Q(z)$ from having zeros on the unit circle, here we are allowing the factors of $Q(z)$ to have roots on the unit circle.

driven by unit variance white noise. It should be pointed out, however, that this model is not unique. For example, we may replace any one or more factors $(1-\alpha_k z^{-1})$ of $Q(z)$ with factors of the form $(1-\alpha_k^* z)$. The following example illustrates how spectral factorization may be used to find a moving average model for a simple process.

Example 4.7.2 *Moving Average Model Using Spectral Factorization*

Consider the MA(1) process that has an autocorrelation sequence given by

$$r_x(k) = 17\delta(k) + 4\big[\delta(k-1) + \delta(k+1)\big]$$

The power spectrum is the second-order polynomial

$$P_x(z) = 17 + 4z^{-1} + 4z = z\big[4 + 17z^{-1} + 4z^{-2}\big]$$

Performing a spectral factorization of $P_x(z)$ we find

$$P_x(z) = z\big(4 + z^{-1}\big)\big(1 + 4z^{-1}\big) = \big(4 + z^{-1}\big)\big(4 + z\big)$$

Therefore, $x(n)$ may be modeled as the output of an FIR filter having a system function equal to either

$$H(z) = 4 + z^{-1}$$

or

$$H(z) = 1 + 4z^{-1}$$

As an alternative to spectral factorization, a moving average model for a process $x(n)$ may also be developed using Durbin's method [5]. This approach begins by finding a high-order all-pole model $A_p(z)$ for the moving average process. Then, by considering the coefficients of the all-pole model $a_p(k)$ to be a new "data set," the coefficients of a qth-order moving average model are determined by finding a qth-order all-pole model for the sequence $a_p(k)$. More specifically, let $x(n)$ be a moving average process of order q with

$$B_q(z) = \sum_{k=0}^{q} b_q(k) z^{-k}$$

so that

$$x(n) = \sum_{k=0}^{q} b_q(k) w(n-k)$$

where $w(n)$ is white noise. Suppose that we find a pth-order all-pole model for $x(n)$ and that p is large enough so that[15]

$$B_q(z) \approx \frac{1}{A_p(z)} = \frac{1}{a_p(0) + \sum_{k=1}^{p} a_p(k) z^{-k}} \tag{4.157}$$

For example, if

$$B_1(z) = b(0) - b(1)z^{-1}$$

[15]Note that the coefficient that normally appears in the numerator of the all-pole model has been absorbed into the denominator, thereby making the coefficient $a_p(0)$ some value other than one, in general.

and $|b(0)| > |b(1)|$ then $1/B_1(z)$ may be expanded in a power series as follows

$$\frac{1}{B_1(z)} = \frac{1}{b(0) - b(1)z^{-1}} = \frac{1}{b(0)} \sum_{k=0}^{\infty} \beta^k z^{-k}$$

where

$$\beta = \frac{b(1)}{b(0)}$$

Therefore, if p is sufficiently large so that $\beta^p \approx 0$ then $B_1(z)$ may be approximated by the expansion

$$B_1(z) \approx \frac{b(0)}{1 + \beta z^{-1} + \cdots + \beta^p z^{-p}} \tag{4.158}$$

Once a high-order all-pole model for $x(n)$ has been found, it is then necessary to estimate the MA coefficients $b_q(k)$ from the all-pole coefficients $a_p(k)$. From Eq. (4.157) we see that since

$$A_p(z) \approx \frac{1}{B_q(z)} = \frac{1}{b_q(0) + \sum_{k=1}^{q} b_q(k) z^{-k}}$$

then $1/B_q(z)$ represents a qth-order all-pole model for the "data" $a_p(k)$. The coefficients of the all-pole model for $a_p(k)$ are then taken as the coefficients of the moving average model. Thus, Durbin's method may be summarized as follows.

1. Given $x(n)$ for $n = 0, 1, \ldots, N-1$ or $r_x(k)$ for $k = 0, 1, \ldots, N-1$, a pth-order all-pole model is found and the coefficients are normalized, as in Eq. (4.157), by the gain (numerator coefficient) of the all-pole model. Typically, the model order p is chosen so that it is at least four times the order q of the moving average process [17].
2. Using the AR coefficients derived in Step 1 as data, a qth-order all-pole model for the sequence $a_p(k)$ is then found. The resulting coefficients, after dividing by the gain term, are the coefficients of the moving average model.

A MATLAB program for Durbin's method that uses the autocorrelation method for both all-pole modeling problems is given in Fig. 4.22.

Durbin's Method

```
function b = durbin(x,p,q)
%
   x    = x(:);
   if p>=length(x), error('Model order too large'), end
   [a,epsilon] = acm(x,p);
   [b,epsilon] = acm(a/sqrt(epsilon),q);
   b = b/sqrt(epsilon)
   end;
```

Figure 4.22 *A* MATLAB *program for finding the moving average coefficients of an all-zero model for a signal $x(n)$ using Durbin's method. Note that this program calls* `acm.m` *(see Appendix).*

Example 4.7.3 *The Durbin Algorithm*

To illustrate how the Durbin algorithm is used to find a moving average model, consider the signal $x(n)$ that has a z-transform

$$X(z) = 1 - \beta z^{-1}$$

where β is a real number with $|\beta| < 1$. The first step in Durbin's method requires that we find a high-order all-pole model for $x(n)$. Since

$$\frac{1}{X(z)} = \frac{1}{1 - \beta z^{-1}} = \sum_{k=0}^{\infty} \beta^k z^{-k}$$

then the all-pole model for $x(n)$ using the autocorrelation method with $p \gg 1$ is approximately

$$H(z) = \frac{1}{1 + \beta z^{-1} + \cdots + \beta^p z^{-p}}$$

The next step is to fit a qth-order all-pole model to the sequence $a_p(k) = \beta^k$. As we saw in Example 4.6.1 (p. 179) a first-order all-pole model for this sequence is

$$H(z) = \frac{\sqrt{\epsilon_1}}{1 + a(1)z^{-1}}$$

where

$$a(1) = -\beta \frac{1 - \beta^{2p}}{1 - \beta^{2(p+1)}}$$

and

$$\epsilon_1 = \frac{1 - \beta^{4p+2}}{1 - \beta^{2p+2}}$$

Therefore, assuming that $\beta^{2p} \ll 1$, so that $\epsilon_1 \approx 1$, the first-order moving average model is

$$B_1(z) = 1 - \beta \frac{1 - \beta^{2p}}{1 - \beta^{2(p+1)}} z^{-1}$$

Note that if, instead of the autocorrelation method in the second step, we were to use the covariance method, then we would have, as we saw in Example 4.6.2 (p. 184),

$$a(1) = -\beta$$

and the first-order moving average model would be

$$B_1(z) = 1 - \beta z^{-1}$$

4.7.4 Application: Power Spectrum Estimation

Spectrum estimation is an important application of stochastic signal modeling. As we saw in Section 3.3.8, the power spectrum of a wide-sense stationary random process $x(n)$ is

$$P_x(e^{j\omega}) = \sum_{k=-\infty}^{\infty} r_x(k) e^{-jk\omega} \tag{4.159}$$

where

$$r_x(k) = E\{x(n)x^*(n-k)\} \tag{4.160}$$

is the autocorrelation of $x(n)$. Since $r_x(k)$ is generally unknown, spectrum estimation is concerned with the estimation of $P_x(e^{j\omega})$ from a sample realization of $x(n)$ for $n = 0, 1, \ldots, N$.

One approach that may be used to estimate $P_x(e^{j\omega})$ is to estimate $r_x(k)$ from $x(n)$ and then use this estimate in Eq. (4.159). However, with only $N+1$ values of $x(n)$ the autocorrelation may only be estimated for lags $|k| \le N$, and the power spectrum estimate would have the form

$$\hat{P}_x(e^{j\omega}) = \sum_{k=-N}^{N} \hat{r}_x(k)e^{-jk\omega} \tag{4.161}$$

This estimate is limited by two factors. First, since estimated autocorrelations are used instead of the true values, the accuracy of $\hat{P}_x(e^{j\omega})$ will be limited by the accuracy of the estimates $\hat{r}_x(k)$. Second, since $\hat{P}_x(e^{j\omega})$ does not include any estimates of $r_x(k)$ for $|k| > N$, the power spectrum estimate will be limited in resolution unless $r_x(k) \approx 0$ for $|k| > N$.

The estimation of $P_x(e^{j\omega})$ may be facilitated if something is known about the process $x(n)$ in addition to the signal values. For example, suppose that $x(n)$ is known to be an autoregressive process of order p. Since the power spectrum of an AR(p) process is

$$P_x(e^{j\omega}) = \frac{|b(0)|^2}{\left|1 + \sum_{k=1}^{p} a_p(k)e^{-jk\omega}\right|^2} \tag{4.162}$$

then we may use the Yule-Walker method with estimated autocorrelations to estimate the coefficients $a_p(k)$ and $b(0)$, and then use these estimates in Eq. (4.162) as follows:

$$\hat{P}_x(e^{j\omega}) = \frac{|\hat{b}(0)|^2}{\left|1 + \sum_{k=1}^{p} \hat{a}_p(k)e^{-jk\omega}\right|^2} \tag{4.163}$$

Assuming that the estimates of the model coefficients are sufficiently accurate, this approach may result in a significantly improved spectrum estimate.

Example 4.7.4 *AR Spectrum Estimation*

Shown in Fig. 4.23 are $N = 64$ samples of an AR(4) process that is generated by filtering unit variance white noise with the fourth-order all-pole filter

$$H(z) = \frac{b(0)}{1 + \sum_{k=1}^{4} a(k)z^{-k}}$$

where $b(0) = 1$ and

$$a(1) = 0.7348 \quad ; \quad a(2) = 1.8820 \quad ; \quad a(3) = 0.7057 \quad ; \quad a(4) = 0.8851$$

which corresponds to a filter having a pair of poles at $z = 0.98e^{\pm j\pi/2}$ and a pair of poles at $z = 0.96e^{\pm j5\pi/8}$. Estimating the autocorrelation sequence for $|k| < N$ using

$$\hat{r}_x(k) = \frac{1}{N}\sum_{n=0}^{N-1} x(n)x(n-k)$$

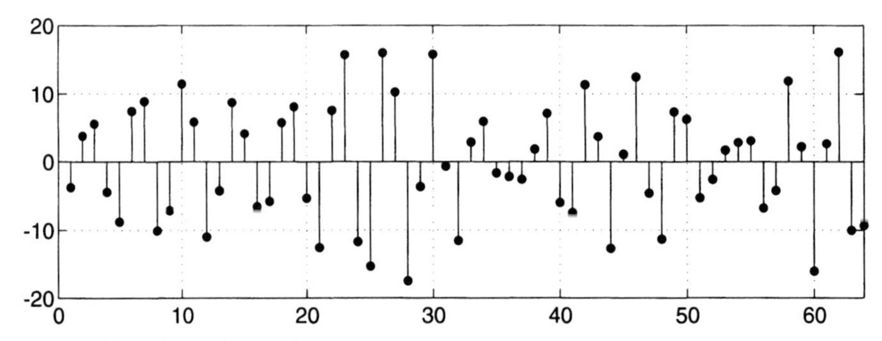

Figure 4.23 *An AR(4) random process.*

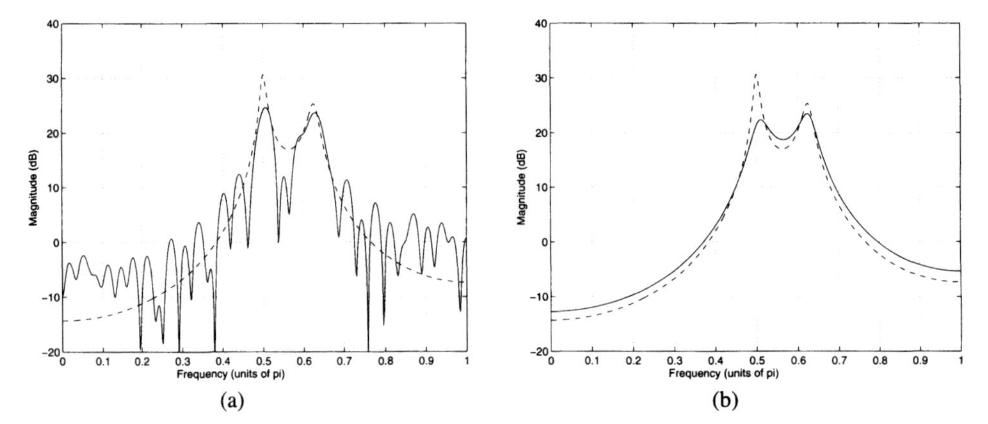

Figure 4.24 *Power spectrum estimation. (a) The estimate of the power spectrum by using estimated autocorrelations. (b) The estimate of the power spectrum using the Yule-Walker method. In both figures the true power spectrum is shown by the dashed line.*

and substituting these estimates into Eq. (4.161) we obtain the estimate of the spectrum shown in Fig. 4.24*a*.

On the other hand, using these estimates of $r_x(k)$ in the Yule-Walker equations, Eq. (4.151), we find the following estimates for $b(0)$ and $a(k)$,

$$b(0) = 1.4844 \ ; \ a(1) = 0.7100 \ ; \ a(2) = 1.6198 \ ; \ a(3) = 0.6193 \ ; \ a(4) = 0.6908$$

Incorporating these estimates into Eq. (4.162) gives the spectrum estimate shown in Fig. 4.24*b*.

As we will see in Chapter 8, there are two limitations with this model-based approach to spectrum estimation. First of all, it assumes that we know how $x(n)$ should be modeled. In Example 4.7.4, for example, it was assumed that we knew that $x(n)$ was an AR(p) process. If $x(n)$ is not an autoregressive process and cannot be sufficiently approximated as an AR process, then this approach will lead to an erroneous spectrum estimate. The second problem lies in the requirement that the order of the process be known. If p is unknown, then some criterion must be used to select an appropriate value for the model order. If this criterion does not produce a precise estimate for p, then $x(n)$ will either be overmodeled

(p too large) or undermodeled (p too small). Both cases will generally result in inaccurate spectrum estimates.

4.8 SUMMARY

In this chapter, we presented a number of different approaches to signal modeling. The first part of the chapter was concerned with modeling deterministic signals. Beginning with the least squares (direct) method, we found that this approach was only of limited use since it required finding the solution to a set of nonlinear equations. However, in those instances where it is necessary to find the true least squares solution, we saw that the method of iterative prefiltering could be used. We then looked at the Padé approximation, which only requires finding the solution to a set of linear equations and produces a model that exactly matches the data for the first $p + q + 1$ values, where p is the number of poles and q is the number of zeros in the model. In most applications, however, this property of perfect matching over a finite interval is overly restrictive and compromises the accuracy of the model for values of n outside this interval. We then developed Prony's method, which relaxes the perfect matching property of the Padé approximation and finds a model that is more accurate, on the average, for all values of n. As with the Padé approximation, Prony's method only requires finding the solution to a set of linear equations. After discussing Shanks' method, which modifies Prony's method in the way that the numerator coefficients are determined, we looked at the special case of all-pole signal modeling using Prony's method. The advantage of an all-pole model is that the normal equations are Toeplitz and may therefore be easily solved using the Levinson-Durbin recursion (Chapter 5). The Toeplitz structure also provides an advantage in terms of storage and reduces the amount of computations required to evaluate the autocorrelations in the normal equations. Finally, we considered the autocorrelation and covariance methods, which address the issue of how to determine the model when only a finite length data record is available. Here we saw that there is a tradeoff between computational efficiency and model accuracy. Specifically, whereas the autocorrelation method preserves the Toeplitz structure of the all-pole normal equations and guarantees a stable model, it requires that the data be windowed, which results in a decrease in the accuracy of the model. The covariance method, on the other hand, does not apply a window to the data and is therefore more accurate in its determination of the model coefficients, but the Toeplitz structure of the normal equations is lost along with the guaranteed stability of the model.

In the second part of the chapter, we considered briefly the problem of modeling a random process. Beginning with an ARMA process, we saw that the Yule-Walker equations could be used to find the coefficients of the model. In the case of an autoregressive process, the Yule-Walker equations are identical to the Prony all-pole normal equations. Therefore, with estimated autocorrelations, stochastic all-pole modeling is essentially equivalent to deterministic all-pole signal modeling. Finally, for a moving average process, since the Yule-Walker equations are nonlinear in the filter coefficients we looked at two methods for deriving the model parameters. The first involved a spectral factorization of the power spectrum of the process, whereas the second derived the model through a succession of all-pole models.

Before concluding this chapter it should be pointed out that an important problem in signal modeling has been, to this point, overlooked. This is the problem of model order estimation. In each case that we have considered thus far, we have assumed that a model of a given order was to be found. In the absence of any information about the correct model order, it becomes necessary to estimate what an appropriate model order should be. As one might expect, misleading information and inaccurate models may result if an inappropriate model order is used. At this point, however, we will simply note that although a number of different model order estimation techniques are available, all of them are somewhat limited in terms of their robustness and accuracy (A more detailed discussion of model order estimation may be found in Chapter 8).

References

1. B. S. Atal and J. R. Remda, "A new model of LPC excitation for producing natural-sounding speech at low bit rates," *Proc. IEEE Int. Conf. on Acoust., Speech, Sig. Proc.*, Paris, pp. 614–617, May 1982.
2. F. Brophy and A. C. Salazar, "Considerations of the Padé approximant technique in the synthesis of recursive digital filters," *IEEE Trans. Audio and Electroacoust*, pp. 500–505, December 1973.
3. R. V. Churchill and J. W. Brown, *Introduction to Complex Variables and Applications*, 4th ed., McGraw-Hill, New York, 1984.
4. J. R. Deller, J. G. Proakis, and J. H. L. Hansen, *Discrete-time Processing of Speech Signals*, MacMillan, New York, 1993.
5. J. Durbin, "Efficient estimation of parameters in moving-average models," *Biometrica*, vol. 46, pp. 306–316, 1959.
6. J. H. McClellan, "Parametric Signal Modeling," Chapter 1 in *Advanced Topics in Signal Processing*, Prentice-Hall, Englewood Cliffs, NJ, 1988.
7. M. Morf, B. Dickenson, T. Kailath, and A. Vieira, "Efficient solution of covariance equations for linear prediction," *IEEE Trans. Acoust., Speech, Sig. Proc.*, vol. 25, pp. 429–433, October 1977.
8. A. V. Oppenheim and R. W. Schafer, *Discrete-Time Signal Processing*, Prentice-Hall, Englewood Cliffs, NJ, 1989.
9. H. E. Padé, "Sur la représentation approchée d'une fonction par des fractions rationelles," *Annales Scientifique de l'Ecole Normale Supérieure*, vol. 9, no. 3 (supplement), pp. 1–93, 1992.
10. G. R. B. Prony, "Essai expémental et analytique sur les lois de la dilatabilité de fluides elastiques et sur celles de la force expansion de la vapeur de l'alcool, à différentes températures," *Journal de l'Ecole Polytechnique* (Paris), vol. 1, no. 2, pp. 24–76, 1795.
11. L. R. Rabiner and R. W. Schafer, *Digital Processing of Speech Signals*, Prentice-Hall, Englewood Cliffs, NJ, 1978.
12. R. A. Roberts and C. T. Mullis, *Digital Signal Processing*, Addison Wesley, Reading, MA. 1987.
13. E. A. Robinson and S. Treitel, *Geophysical Signal Analysis*, Prentice-Hall, Englewood Cliffs, NJ, 1980.
14. J. L. Shanks, "Recursion filters for digital processing," *Geophysics*, vol. 32, pp. 33–51, February 1967.
15. K. Steiglitz and L. E. McBride, "A technique for the identification of linear systems," *IEEE Trans. on Automatic Control*, vol. AC-10, pp. 461–464, October 1965.
16. K. Steiglitz, "On the simultaneous estimation of poles and zeros in speech analysis," *IEEE Trans. Acoust., Speech, Sig. Proc.*, vol. 25, pp. 229–234, June 1977.
17. C. W. Therrien, *Discrete Random Signals and Statistical Signal Processing*, Prentice-Hall, Englewood Cliffs, NJ, 1992.
18. W. F. Trench, "An algorithm for the inversion of finite Toeplitz matrices," *J. SIAM*, vol. 12, no. 3, pp. 512–522, 1964.

4.9 PROBLEMS

4.1. Find the Padé approximation of second-order to a signal $x(n)$ that is given by

$$\mathbf{x} = \left[2,\ 1,\ 0,\ -1,\ 0,\ 1,\ 0,\ -1,\ 0,\ 1,\ \ldots\right]^T$$

i.e., $x(0) = 2$, $x(1) = 1$, $x(2) = 0$, and so on. In other words, using an approximation of the form

$$H(z) = \frac{b(0) + b(1)z^{-1} + b(2)z^{-2}}{1 + a(1)z^{-1} + a(2)z^{-2}}$$

find the coefficients $b(0)$, $b(1)$, $b(2)$, $a(1)$, and $a(2)$.

4.2. A third-order all-pole Padé approximation to a signal $x(n)$ has been found to be

$$H(z) = \frac{1}{1 + 2z^{-1} + z^{-2} + 3z^{-3}}$$

What information about $x(n)$ can be determined from this model?

4.3. Suppose that a signal $x(n)$ is known to be of the form

$$x(n) = \sum_{k=1}^{L} c_k(\lambda_k)^n u(n)$$

where the λ_k's are distinct complex numbers.

(a) Show that the Padé approximation method can be used to determine the parameters c_k and λ_k for $k = 1, 2, \ldots, L$. Is the answer unique?

(b) The first eight values of a signal $x(n)$, which is known to be of the form given above with $L = 3$, are

$$\mathbf{x} = \left[32,\ 16,\ 8,\ 12,\ 18,\ 33,\ 64.5,\ 128.25\right]^T$$

Determine c_k and λ_k for $k = 1, 2, 3$.

4.4. A consumer electronics device includes a DSP chip that contains a linear shift-invariant digital filter that is implemented in ROM. In order to perform some reverse engineering on the product, it is necessary to determine the system function of the filter. Therefore, the unit sample response is measured and it is determined that the first eight values of $h(n)$ are as listed in the following table.

Unit sample response								
n	0	1	2	3	4	5	6	7
$h(n)$	−1	2	3	2	1	2	0	1

Having no knowledge of the order of the filter, it is assumed that $H(z)$ contains two poles and two zeros.

(a) Based on this assumption, determine a candidate system function, $H(z)$, for the filter.

(b) Based on the solution found in (a) and the given values for $h(n)$, is it possible to determine whether or not the hypothesis about the order of the system is correct? Explain.

4.5. The Padé approximation models a signal as the response of a filter to a unit sample input, $\delta(n)$. Suppose, however, that we would like to model a signal $x(n)$ as the *step response* of a filter as shown in the following figure

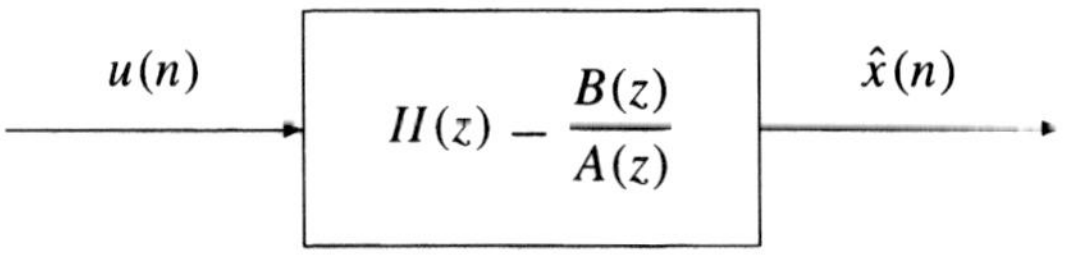

$$u(n) \longrightarrow \boxed{H(z) = \frac{B(z)}{A(z)}} \longrightarrow \hat{x}(n)$$

In the following, assume that $H(z)$ is a second-order filter having a system function of the form

$$H(z) = \frac{b(0) + b(1)z^{-1} + b(2)z^{-2}}{1 + a(1)z^{-1} + a(2)z^{-2}}$$

(a) Using the Padé approximation method with a *unit step* input, derive the set of equations that must be solved so that

$$\hat{x}(n) = x(n) \quad \text{for} \quad n = 0, 1, \ldots, 4$$

(b) If the first eight values of $x(n)$ are

$$\mathbf{x} = \begin{bmatrix} 1, & 0, & 2, & -1, & 2, & 0, & 1, & 2 \end{bmatrix}^T$$

find $b(0)$, $b(1)$, $b(2)$, $a(1)$, and $a(2)$.

4.6. With a real-valued signal $x(n)$ known only for $n = 0, 1, \ldots, N$, the *backwards covariance method* finds the coefficients of the all-pole model that minimize the *backward prediction error*

$$\mathcal{E}_p^- = \sum_{n=p}^{N} [e_p^-(n)]^2$$

where

$$e_p^-(n) = x(n-p) + \sum_{k=1}^{p} a_p(k)x(n+k-p)$$

(a) Show that the coefficients $a_p(k)$ that minimize $\mathcal{E}_p^-$ satisfy a set of normal equations of the form

$$\mathbf{R}_x \bar{\mathbf{a}}_p = -\mathbf{r}_x$$

where

$$\bar{\mathbf{a}}_p = \begin{bmatrix} a_p(1), & a_p(2), & \ldots, & a_p(p) \end{bmatrix}^T$$

and find explicit expressions for the entries in $\mathbf{R}_x$ and $\mathbf{r}_x$.

(b) Is the solution to the backwards covariance method the same as the solution to the covariance method? Why or why not?

(c) Consider a new error that is the sum of the forward and backward prediction errors,

$$\mathcal{E}_p^B = \sum_{n=p}^{N} \left\{ [e_p^+(n)]^2 + [e_p^-(n)]^2 \right\}$$

where $e_p^-(n)$ is the backwards prediction error defined above, and $e_p^+(n)$ is the forward prediction error used in the covariance method,

$$e_p^+(n) = x(n) + \sum_{k=1}^{p} a_p(k)x(n-k)$$

Derive the normal equations for the coefficients that minimize $\mathcal{E}_p^B$. (This approach is known as the *Modified Covariance Method.*)

(d) Consider the signal

$$x(n) = \beta^n \quad ; \quad n = 0, 1, \ldots, N$$

With $p = 1$ find the first-order all-pole model that minimizes $\mathcal{E}_p^B$ and determine the value of $\mathcal{E}_p^B$. For what values of β is the model stable? What happens to the model and the modeling error as $N \to \infty$?

4.7. Suppose that we would like to derive a rational model for an unknown system S using the approach shown in the following figure,

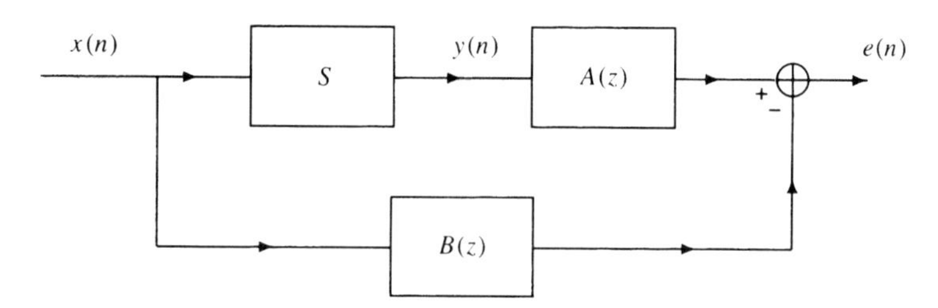

For a given input $x(n)$ the output of the system, $y(n)$, is observed. The coefficients of the two FIR filters $A(z)$ and $B(z)$ that minimize the sum of the squares of the error signal $e(n)$ are then to be determined. Assume that the sum is for all $n \geq 0$ as in Eq. (4.73).

(a) Derive the normal equations that define the optimal solution for the coefficients of $A(z)$ and $B(z)$.

(b) The philosophy of this method is that if the error is small then $B(z)/A(z)$ is a reasonable model for S. Suppose that S is a linear shift-invariant system with a rational system function. Show that this method will identify the parameters of S exactly assuming that the orders of the filters $A(z)$ and $B(z)$ are chosen appropriately [15].

4.8. Consider a signal, $x(n)$, which is the unit sample response of a causal all-pole filter with system function

$$H(z) = \frac{1}{(1 + 0.5z^{-1})(1 + 0.75z^{-1})(1 + 2z^{-1})}$$

We observe $x(n)$ over the interval $0 \leq n \leq N$ where $N \gg 1$.

(a) Using the covariance method, we determine a third-order all-pole model for $x(n)$. What, if anything, can you say about the location of the poles in the model? Do the pole locations depend on N? If so, where do the poles go as $N \to \infty$?

(b) Repeat part (a) for the case in which you use the autocorrelation method.

4.9. Equation (4.129) may be used to reduce the amount of computation required to set-up the covariance normal equations.

(a) Show that the elements along the main diagonal may be computed recursively beginning with $r_x(1, 1)$.

(b) Show how the elements along the lower diagonals may be computed recursively beginning with $r_x(k, 1)$. How may the terms along the upper diagonals be obtained?

(c) Determine how many multiplies and adds are necessary to set-up the covariance normal equations (do not forget the evaluation of the vector on the right-hand side).

4.10. We want to model a signal $x(n)$ using an all-pole model of the form

$$H(z) = \frac{b(0)}{1 + z^{-N}\left[\sum_{k=1}^{p} a_p(k)z^{-k}\right]}$$

For example, with $p = 2$ the model is

$$H(z) = \frac{b(0)}{1 + a(1)z^{-N-1} + a(2)z^{-N-2}}$$

Derive the normal equations that define the coefficients $a_p(k)$ that minimize the Prony error

$$\mathcal{E}_p = \sum_{n=0}^{\infty} |e(n)|^2$$

where

$$e(n) = x(n) + \sum_{l=1}^{p} a_p(l)x(n - l - N)$$

and derive an expression for the minimum error.

4.11. Suppose that we would like to model a signal $x(n)$ as shown in the following figure.

$$\delta(n) \longrightarrow \boxed{H(z)} \longrightarrow \hat{x}(n)$$

where $h(n)$ is an all-pole filter that has a system function of the form

$$H(z) = \frac{b(0)}{1 + \sum_{k=1}^{p} a_p(k)z^{-2k}}$$

Modify the Prony normal equations so that one can determine the coefficients $a_p(k)$ in $H(z)$ from a sequence of signal values, $x(n)$.

4.12. Suppose that we would like to model a signal $x(n)$ that we believe to be quasiperiodic. Based on our observations of $x(n)$ we estimate the autocorrelations through lag $k = 10$ to be

$$r_x(k) = \left[1.0,\ 0.4,\ 0.4,\ 0.3,\ 0.2,\ 0.9,\ 0.4,\ 0.4,\ 0.2,\ 0.1,\ 0.7\right]^T$$

(a) In formulating an all-pole model to take into account the suspected periodicity let us consider a two-coefficient model of the form

$$H(z) = \frac{b(0)}{1 + a(5)z^{-5} + a(10)z^{-10}}$$

Find the values for the coefficients $a(5)$ and $a(10)$ that minimize the all-pole modeling error.

(b) Compare the error obtained with the model found in (a) to the error that is obtained with a model of the form

$$H(z) = \frac{b(0)}{1 + a(1)z^{-1} + a(2)z^{-2}}$$

(c) Now consider an all-pole model of the form

$$H(z) = \frac{b(0)}{1 + a(N)z^{-N}}$$

where both $a(N)$ and N are considered to be model parameters. Find the value for $a(N)$ and N that minimize the all-pole modeling error and evaluate the modeling error.

4.13. We would like to build a predictor of digital waveforms. Such a system would form an estimate of a later sample (say n_0 samples later) by observing p consecutive data samples. Thus we would set

$$\hat{x}(n + n_0) = \sum_{k=1}^{p} a_p(k)x(n - k)$$

The predictor coefficients $a_p(k)$ are to be chosen to minimize

$$\mathcal{E}_p = \sum_{n=0}^{\infty} [x(n + n_0) - \hat{x}(n + n_0)]^2$$

(a) Derive the equations that define the optimum set of coefficients $a_p(k)$.

(b) If $n_0 = 0$, how is your formulation of this problem different from Prony's method?

4.14. You are told that it is always possible to determine whether or not a causal all-pole filter is stable from a finite number of values of its unit sample response. For example, if $H(z)$ is a pth-order all-pole filter, given $h(n)$ for $n = 0, 1, \ldots, N$, then the stability of $H(z)$ may be determined. If this is true, explain the procedure and list any conditions that must be placed on p or N. If false, explain why it cannot be done.

4.15. Let $H(z)$ be a first-order model for a real-valued signal $x(n) = S(n) + S(n - 1)$,

$$H(z) = \frac{b(0)}{1 - a(1)z^{-1}}$$

and let

$$\mathcal{E}_{LS} = \sum_{n=0}^{N-1} [x(n) - h(n)]^2$$

be the error that is to be minimized. By setting the derivatives of $\mathcal{E}_{LS}$ with respect to $b(0)$ and $a(1)$ equal to zero, try to find an analytic solution for the values of $b(0)$ and $a(1)$ that minimize $\mathcal{E}_{LS}$. (This problem illustrates how difficult the direct method of signal modeling may be, even for a first-order model.)

4.16. We have a signal $x(n)$ for which we would like to obtain an all-pole model of the form

$$H(z) = \frac{b(0)}{1 + a(1)z^{-1} + a(2)z^{-2}}$$

Using the autocorrelation method, find explicit formulas for $b(0)$, $a(1)$, and $a(2)$ in terms of $r_x(0)$, $r_x(1)$, and $r_x(2)$.

4.17. If one is modeling a signal $x(n)$ whose transform, $X(z)$, contains zeros, then an all-pole model may be used to effectively model a zero with an infinite number of poles. In this problem we look at how a zero is modeled with the autocorrelation method. Let $x(n) = \delta(n) - \alpha\delta(n-1)$ where $|\alpha| < 1$ and α is real.

(a) Determine the *pth*-order all-pole model $A_p(z)$ for $x(n)$ where p is an arbitrary positive integer, and find the value of the squared error $\mathcal{E}_p$.

(b) For the all-pole model determined in part (a), what is the limit of $A_p(z)$ as $p \to \infty$? What does $\mathcal{E}_p$ converge to as $p \to \infty$? Justify your answers.

(c) Repeat parts (a) and (b) for $|\alpha| > 1$.

4.18. Find a closed-form expression for the FIR least squares inverse filter of length N for each of the following systems.

1. $G(z) = \dfrac{1}{1 - \alpha z^{-1}} \quad ; \quad |\alpha| < 1.$
2. $G(z) = 1 - z^{-1}.$
3. $G(z) = \dfrac{\alpha - z^{-1}}{1 - \alpha z^{-1}} \quad ; \quad |\alpha| < 1.$

4.19. An important application of least squares inverse filtering is deconvolution, which is concerned with the recovery of a signal $d(n)$ that has been convolved with a filter $g(n)$

$$x(n) = d(n) * g(n)$$

The problem is to design a filter $h_N(n)$ that may be used to produce an estimate of $d(n)$ from $x(n)$,

$$\hat{d}(n) = x(n) * h_N(n)$$

One of the difficulties, however, is that noise in the observed signal may be amplified by the filter. For example, if we observe

$$y(n) = d(n) * g(n) + v(n)$$

then the filtered observations become

$$y(n) * h_N(n) = \hat{d}(n) + v(n) * h_N(n) = \hat{d}(n) + u(n)$$

where

$$u(n) = v(n) * h_N(n)$$

is the filtered noise. One way to reduce this noise is to design a least squares inverse filter that minimizes

$$\mathcal{E} = \sum_{n=0}^{\infty} |e(n)|^2 + \lambda E\{|u(n)|^2\}$$

where

$$e(n) = \delta(n - n_0) - h_N(n) * g(n)$$

and $\lambda > 0$ is a parameter that is to be selected. Note that for large values of λ, minimizing $\mathcal{E}$ will force a large reduction in the filtered noise at the expense of a decrease in resolution, i.e., larger $e(n)$, whereas smaller values of λ lead to higher resolution and larger noise.

(a) Assume that $v(n)$ is zero-mean white noise with a variance σ_v^2. Show that

$$E\{|u(n)|^2\} = \sigma_v^2 \mathbf{h}_N^H \mathbf{h}_N$$

where $\mathbf{h}_N$ is a vector containing the coefficients of the filter $h_N(n)$.

(b) Derive the normal equations that result from minimizing the error

$$\mathcal{E} = \mathbf{e}^H \mathbf{e} + \lambda \sigma_v^2 \mathbf{h}_N^H \mathbf{h}_N$$

where $\mathbf{e} = \left[e(0),\ e(1),\ \ldots\right]^T$, and show that they may be written in the form

$$\left(\mathbf{R}_g + \alpha \mathbf{I}\right)\mathbf{h}_N = \mathbf{g}_{n_0}^*$$

where $\alpha > 0$ is a *prewhitening parameter* that depends upon the values of λ, and $\mathbf{g}_{n_0}^*$ is the vector on the right-side of Eq. (4.101).

4.20. We are given a signal, $x(n)$, that we want to model as the unit sample response of an all-pole filter. We have reason to believe that the signal is periodic and, consequently, the poles of the model should lie on the unit circle. Thus, assuming a second-order model for the signal, the system function is constrained to have the form

$$H(z) = \frac{b(0)}{1 + a(1)z^{-1} + z^{-2}}$$

With $|a(1)| < 2$ this model produces a pair of poles on the unit circle at an angle θ defined by

$$2\cos\theta = -a(1)$$

(a) Using the autocorrelation method, derive the normal equations that define the value of $a(1)$ that minimizes the error

$$\mathcal{E}_p = \sum_{n=0}^{\infty} e^2(n)$$

(b) Find an expression for the minimum error, $\{\mathcal{E}_p\}_{\min}$.

4.21. Voiced speech may be modeled as the output of an all-pole filter driven by an impulse train

$$p_{n_0}(n) = \sum_{k=1}^{K} \delta(n - kn_0)$$

where the time between pulses, n_0, is known as the *pitch period*. Suppose that we have a segment of voiced speech and suppose that we know the pitch period, n_0. We extract a subsequence, $x(n)$, of length $N = 2n_0$ and model this signal as shown in the following figure

$$p_{n_0}(n) \longrightarrow \boxed{\frac{b(0)}{1+\sum_{k=1}^{p} a_p(k)z^{-k}}} \longrightarrow \hat{x}(n)$$

where the input, $p_{n_0}(n)$, consists of two pulses,

$$p_{n_0}(n) = \delta(n) + \delta(n - n_0)$$

Find the normal equations that define the coefficients $a_p(k)$ that minimize the error

$$\mathcal{E}_p = \sum_{n=0}^{N} e^2(n)$$

where

$$e(n) = a_p(n) * x(n) - b(n) * p_{n_0}(n)$$

and $b(n) = b(0)\delta(n)$.

4.22. You would like to design a linear predictor of a signal $x(n)$ but, due to hardware constraints, are limited to a two-tap predictor. However, since delays can be tolerated in the predictor, you decide to design a predictor of the form

$$\hat{x}(n) = a(1)x(n - N_1) + a(2)x(n - N_2)$$

where N_1 and N_2 are positive integers. The goal is to find the values of $a(1)$, $a(2)$, N_1, and N_2 that minimize the mean-square error $E\{e^2(n)\}$ where

$$e(n) = x(n) - \hat{x}(n)$$

Assuming that $x(n)$ is a zero mean wide-sense stationary process, you estimate the autocorrelation of $x(n)$ and find that the values of $r_x(k)$ for $k = 0, 1, \ldots, 7$ are as given in the following table. For $k > 7$, it is determined that the autocorrelation is approximately zero.

Autocorrelation values								
k	0	1	2	3	4	5	6	7
$r_x(k)$	1.0	−0.1	0.0	−0.5	−0.2	0.6	0.2	0.2

(a) If you were to design an optimum predictor of the form $\hat{x}(n) = a(1)x(n-1)$, what would be the mean-square error in your prediction of $x(n)$? What about for the predictor $\hat{x}(n) = a(1)x(n-3)$?

(b) Derive a general expression for the minimum mean-square error for a predictor of the form $\hat{x}(n) = a(1)x(n - N_1)$ with your expression given only in terms of $r_x(k)$. What value of N_1 minimizes the mean-square error?

(c) Find the values of $a(1)$, $a(2)$, N_1, and N_2 in the two-coefficient linear predictor defined above that minimize the mean-square error $E\{e^2(n)\}$.

(d) Find the value for the mean-square prediction error for your predictor designed in part (c).

4.23. If $r_x(0) = 1$, $r_x(1) = 0.5$, and $r_x(2) = 0.75$, find the values of $a(1)$, $a(2)$, and $b(0)$ in the following AR(2) model for $x(n)$,

$$x(n) + a(1)x(n-1) + a(2)x(n-2) = b(0)w(n)$$

where $w(n)$ is unit variance white noise.

4.24. Use the method of spectral factorization to find a moving average model of order 2 for a process whose autocorrelation sequence is

$$\mathbf{r}_x = \begin{bmatrix}3, & 1.5, & 1\end{bmatrix}^T$$

4.25. Suppose that the first five values in the autocorrelation sequence for the process $x(n)$ are

$$\mathbf{r}_x = \begin{bmatrix}3, & 9/4, & 9/8, & 9/16, & 9/32 & \dots\end{bmatrix}^T$$

(a) Use the modified Yule-Walker equation method to find an ARMA(1,1) model for $x(n)$.

(b) Are the given values in the autocorrelation sequence consistent with the model that you found in part (a)?

Computer Exercises

C4.1. In this problem, you are to consider the design of lowpass filters using the Padé and Prony methods.

(a) Suppose you would like to design a lowpass filter with a cutoff frequency $\omega_c = \pi/2$. Using the Padé approximation with $p + q + 1 = 20$, compare the designs that result when $p = 0, 2, 4, \dots, 20$. Which design is the best? (Note: one of the parameters that you will need to experiment with in your design is the delay, n_0, of the ideal unit sample response.)

(b) Repeat part (a) for a narrowband lowpass filter that has a cutoff frequency $\omega_c = \pi/16$. Which design is the best and how do these filters compare to those designed in part (a)?

(c) Repeat part (a) using Prony's method and compute the minimum error $\epsilon_{p,q}$ for each filter. Which filter is the best? Compare your designs to those obtained using Padé approximation.

(d) Using the m-file `ellip.m` in the Signal Processing Toolbox, design a tenth-order elliptic lowpass filter with a cutoff frequency $\omega_c = \pi/2$, and evaluate the Prony error. Compare this filter to the best Prony filter found in part (c) and explain what you observe.

C4.2. We have seen that the direct method of signal modeling leads to a set of nonlinear equations that need to be solved for the model coefficients. Iterative prefiltering, however, is an approach that may be used to avoid having to solve these nonlinear equations. In this exercise we look at the method of iterative prefiltering and compare it to Prony's method.

(a) Let

$$H(z) = \frac{1 - 0.9z^{-1} + 0.81z^{-2}}{1 - 1.978z^{-1} + 2.853z^{-2} - 1.877z^{-3} + 0.9036z^{-4}}$$

be the system function of a linear shift-invariant system. Generate the first 100 samples of the unit sample response $h(n)$ of this filter.

(b) Using the method of iterative prefiltering, find a two-zero, four-pole model for $h(n)$. How many iterations are required for the coefficients to converge? What happens if $h(n)$ is over-modeled using $p = q = 4$? What about $p = q = 5$?

(c) The model found in part (b) assumes exact measurements of the unit sample response $h(n)$. Suppose that the measurements of $h(n)$ are noisy and you observe

$$y(n) = h(n) + v(n)$$

where $v(n)$ is white Gaussian noise with a variance σ_v^2. Repeat part (b) using these noisy measurements with $\sigma_v^2 = 0.0001, 0.001, 0.01$. Comment on the accuracy of your models and the sensitivity of the coefficients to the noise variance.

(d) Repeat parts (b) and (c) using Prony's method and compare your results with those obtained using iterative prefiltering. Which method works the best?

C4.3. In this problem, we look briefly at the problem of deconvolution using FIR least squares inverse filters. Suppose that we have recorded a signal, $y(n)$, that is known to have been blurred by a filter having a unit sample response

$$g(n) = \begin{cases} \cos\big(0.2[n-25]\big)\exp\big\{-0.01[n-25]^2\big\} & ; \quad 0 \le n \le 50 \\ 0 & ; \quad \text{otherwise} \end{cases}$$

The signal that is to be recovered from $y(n)$ is a sequence of impulses,

$$x(n) = \sum_{k=1}^{10} x(k)\delta(n - n_k)$$

where the values of $x(k)$ and n_k are as listed in the following table.

n_k	25	40	55	65	85	95	110	130	140	155
$x(k)$	1	0.8	0.7	0.5	0.7	0.2	0.9	0.5	0.6	0.3

(a) Make a plot the observed signal $y(n) = g(n) * x(n)$ and determine how accurately the amplitudes and locations of the impulses in $x(n)$ may be estimated by simply looking at the peaks of $y(n)$.

(b) Using the m-file `spike.m`, design the least squares inverse filter $h_N(n)$ of length $N = 50$ that has the optimum spiking delay.

(c) Filter $y(n)$ with your optimum spiking filter and plot the output of the filter $\hat{x}(n) = h_N(n) * y(n)$. What are your estimated values for the amplitudes and locations of the impulses in $x(n)$?

(d) Your results in part (c) assume noise-free observations of $y(n) = g(n) * x(n)$. Suppose these measurements are noisy,

$$\tilde{y}(n) = g(n) * x(n) + v(n)$$

where $v(n)$ is white Gaussian noise with variance σ_v^2. Repeat part (c) using $\tilde{y}(n)$ with $\sigma_v^2 = .0001$ and $\sigma_v^2 = .001$ and comment on the accuracy of your estimates of $x(k)$ and n_k.

(e) As discussed in Problem 4.19, the effect of measurement noise may be reduced by incorporating a *prewhitening parameter* α in the design of the least squares inverse filter. Write a MATLAB m-file or modify `spike.m` to allow for noise reduction in the least squares inverse filter design. Using this m-file, repeat your experiments in part (d) using different values for the prewhitening parameter. Comment of the effectiveness of α in reducing the noise. What values for α seem to work the best?

(f) Your results in parts (b) and (c) assume perfect knowledge of $g(n)$. Repeat the design of your least squares inverse filter assuming that $g(n)$ has been measured in the presence of noise, i.e., you are given

$$\hat{g}(n) = g(n) + w(n)$$

where $w(n)$ is white noise that is uniformly distributed between $[-.005, .005]$. Filter $y(n)$ with your optimum spiking filter and plot the output of the filter $\hat{x}(n) = h_N(n) * y(n)$. How accurate are your estimates of the amplitudes and locations of the impulses in $x(n)$?

C4.4. In this exercise, we consider the problem of finding a moving average model for a signal $x(n)$.

(a) In Fig. 4.22 is an m-file to find the moving average coefficients $b_q(k)$ of a process $x(n)$ using Durbin's method. Write an m-file to find these coefficients using the method of spectral factorization.

(b) Generate $N = 256$ samples of the process

$$x(n) = w(n) + 0.9w(n-2)$$

where $w(n)$ is unit variance white Gaussian noise. Using the method of spectral factorization, find a second-order moving average model for $x(n)$.

(c) Repeat part (b) using Durbin's method. Compare your results with the method of spectral factorization and discuss how the accuracy of your model is affected by the order of the all-pole model used in the first step of Durbin's method.

(d) Modify the m-file `durbin.m` by replacing the autocorrelation method in the second step of Durbin's method with the covariance method. Find a second-order moving average model for $x(n)$ and compare your results with those obtained in part (b). Repeat for other moving average processes and discuss your findings.

(e) Replace the autocorrelation method in both steps of Durbin's method with the covariance method and repeat part (d).

C4.5. In this exercise, we consider the problem of finding an autoregressive moving average model for a process $x(n)$.

(a) Write an m-file to find an ARMA(p, q) model for a process $x(n)$ using the modified Yule-Walker equation method, given the autocorrelation sequence $r_x(k)$.

(b) Generate 100 samples of an ARMA(4,2) process $x(n)$ by filtering unit-variance white Gaussian noise with

$$H(z) = \frac{1 - .9z^{-1} + 0.81z^{-2}}{1 - 1.978z^{-1} + 2.853z^{-2} - 1.877z^{-3} + 0.904z^{-4}}$$

and make a plot of the power spectrum of $x(n)$.

(c) Using your m-file for the modified Yule-Walker equation method in part (a), find an ARMA(4,2) model for $x(n)$. Compare your model with the coefficients of $H(z)$. Repeat for ten different realizations of the process $x(n)$ and examine the consistency of your model coefficients. Are your estimates of $a_p(k)$ and $b_q(k)$ close to the correct values, on the average? How much variation is there from one realization to the next?

THE LEVINSON RECURSION

5

5.1 INTRODUCTION

In Chapter 4, we saw that several different signal modeling problems require finding the solution to a set of linear equations of the form

$$\mathbf{R}_x \mathbf{a}_p = \mathbf{b} \tag{5.1}$$

where $\mathbf{R}_x$ is a Toeplitz matrix. In the Padé approximation method, for example, the denominator coefficients $a_p(k)$ which are represented by the vector $\mathbf{a}_p$ are found by solving a set of Toeplitz equations of the form (5.1) where $\mathbf{R}_x$ is a *non-symmetric* Toeplitz matrix containing the signal values $x(q), x(q+1), \ldots, x(q+p-1)$ in the first column and the signal values $x(q), x(q-1), \ldots, x(q-p+1)$ in the first row. In addition, the vector $\mathbf{b}$ contains the signal values $x(q+1), x(q+2), \ldots, x(q+p)$ and is therefore tightly constrained by the values in the matrix $\mathbf{R}_x$. A similar set of linear equations is also found in the modified Yule-Walker equations used in modeling an ARMA process. We saw that Toeplitz equations also arise in all-pole modeling of deterministic signals using either Prony's method or the autocorrelation method and in all-pole modeling of stochastic processes using the Yule-Walker method. Unlike the Padé approximation, however, in these cases $\mathbf{R}_x$ is a *Hermitian* Toeplitz matrix of autocorrelation values $r_x(0), r_x(1), \ldots, r_x(p-1)$. In addition, since $\mathbf{b} = -\big[r_x(1), \ldots, r_x(p)\big]^T$, the vector on the right side of Eq. (5.1) is again tightly constrained by the values in the Toeplitz matrix $\mathbf{R}_x$. In Shanks' method for finding the numerator coefficients, we again find a set of Hermitian Toeplitz equations. In this case, however, unlike the previous examples, the vector $\mathbf{b}$ is not constrained by the values in the matrix $\mathbf{R}_x$. In Chapter 7, Toeplitz equations will again be encountered when we consider the design of FIR Wiener filters. As in Shanks' method, $\mathbf{R}_x$ will be a Hermitian Toeplitz matrix but the vector $\mathbf{b}$ will be, in general, independent of the values in $\mathbf{R}_x$.

Due to the importance of solving Toeplitz equations in a variety of different problems, in this chapter, we look at efficient algorithms for solving these equations. In the process of deriving these algorithms, we will also discover a number of interesting properties of the solutions to these equations and will gain some insight into how other approaches to signal modeling may be developed. We begin, in Section 5.2, with the derivation of the Levinson-Durbin recursion. This recursion may be used to solve the Prony all-pole normal equations and the autocorrelation normal equations. The Levinson-Durbin recursion will also lead us to several interesting results including the lattice filter structure, the Schur-Cohn stability test for digital filters, the Cholesky decomposition of a Toeplitz matrix, and a procedure for recursively computing the inverse of a Toeplitz matrix. In Section 5.3, we develop the Levinson recursion for solving a general set of Hermitian Toeplitz equations in which the vector **b** is unconstrained. The Levinson recursion may be used in Shanks' method, and it may be used to solve the general FIR Wiener filtering problem developed in Chapter 7. Finally, in Section 5.4 we derive the split Levinson recursion. This recursion is slightly more efficient than the Levinson-Durbin recursion and introduces the idea of *singular predictor polynomials* and *line spectral pairs* that are of interest in speech processing applications.

5.2 THE LEVINSON-DURBIN RECURSION

In 1947, N. Levinson presented a recursive algorithm for solving a general set of linear symmetric Toeplitz equations $\mathbf{R}_x\mathbf{a} = \mathbf{b}$. Appearing in an expository paper on the Wiener linear prediction problem, Levinson referred to the algorithm as a "mathematically trivial procedure" [16]. Nevertheless, this recursion has led to a number of important discoveries including the lattice filter structure, which has found widespread application in speech processing, spectrum estimation, and digital filter implementations. Later, in 1961, Durbin improved the Levinson recursion for the special case in which the right-hand side of the Toeplitz equations is a unit vector [7]. In this section we develop this algorithm, known as the *Levinson-Durbin recursion*. In addition, we will explore some of the properties of the recursion, show how it leads to a lattice filter structure for digital filtering, and prove that that the all-pole model derived from the autocorrelation method is stable.

5.2.1 Development of the Recursion

All-pole modeling using Prony's method or the autocorrelation method requires that we solve the normal equations which, for a pth-order model, are

$$r_x(k) + \sum_{l=1}^{p} a_p(l) r_x(k-l) = 0 \quad ; \quad k = 1, 2, \ldots, p \tag{5.2}$$

where the modeling error is

$$\epsilon_p = r_x(0) + \sum_{l=1}^{p} a_p(l) r_x(l) \tag{5.3}$$

Combining Eqs. (5.2) and (5.3) into matrix form we have

$$\begin{bmatrix} r_x(0) & r_x^*(1) & r_x^*(2) & \cdots & r_x^*(p) \\ r_x(1) & r_x(0) & r_x^*(1) & \cdots & r_x^*(p-1) \\ r_x(2) & r_x(1) & r_x(0) & \cdots & r_x^*(p-2) \\ \vdots & \vdots & \vdots & & \vdots \\ r_x(p) & r_x(p-1) & r_x(p-2) & \cdots & r_x(0) \end{bmatrix} \begin{bmatrix} 1 \\ a_p(1) \\ a_p(2) \\ \vdots \\ a_p(p) \end{bmatrix} = \epsilon_p \begin{bmatrix} 1 \\ 0 \\ 0 \\ \vdots \\ 0 \end{bmatrix} \tag{5.4}$$

which is a set of $p+1$ linear equations in the $p+1$ unknowns $a_p(1), a_p(2), \ldots, a_p(p)$ and ϵ_p. Equivalently, Eq. (5.4) may be written as

$$\boxed{\mathbf{R}_p \mathbf{a}_p = \epsilon_p \mathbf{u}_1} \tag{5.5}$$

where $\mathbf{R}_p$ is a $(p+1) \times (p+1)$ Hermitian Toeplitz matrix and $\mathbf{u}_1 = [1, 0, \ldots, 0]^T$ is a unit vector with 1 in the first position. In the special case of real data, the $\mathbf{R}_x$ is a symmetric Toeplitz matrix.

The Levinson-Durbin recursion for solving Eq. (5.5) is an algorithm that is recursive in the model order. In other words, the coefficients of the $(j+1)$st-order all-pole model, $\mathbf{a}_{j+1}$, are found from the coefficients of the j-pole model, $\mathbf{a}_j$. We begin, therefore, by showing how the solution to the jth-order normal equations may be used to derive the solution to the $(j+1)$st-order equations. Let $a_j(i)$ be the solution to the jth-order normal equations

$$\begin{bmatrix} r_x(0) & r_x^*(1) & r_x^*(2) & \cdots & r_x^*(j) \\ r_x(1) & r_x(0) & r_x^*(1) & \cdots & r_x^*(j-1) \\ r_x(2) & r_x(1) & r_x(0) & \cdots & r_x^*(j-2) \\ \vdots & \vdots & \vdots & & \vdots \\ r_x(j) & r_x(j-1) & r_x(j-2) & \cdots & r_x(0) \end{bmatrix} \begin{bmatrix} 1 \\ a_j(1) \\ a_j(2) \\ \vdots \\ a_j(j) \end{bmatrix} = \begin{bmatrix} \epsilon_j \\ 0 \\ 0 \\ \vdots \\ 0 \end{bmatrix} \tag{5.6}$$

which, in matrix notation is

$$\mathbf{R}_j \mathbf{a}_j = \epsilon_j \mathbf{u}_1 \tag{5.7}$$

Given $\mathbf{a}_j$, we want to derive the solution to the $(j+1)$st-order normal equations,

$$\mathbf{R}_{j+1} \mathbf{a}_{j+1} = \epsilon_{j+1} \mathbf{u}_1 \tag{5.8}$$

The procedure for doing this is as follows. Suppose that we append a zero to the vector $\mathbf{a}_j$ and multiply the resulting vector by $\mathbf{R}_{j+1}$. The result is

$$\begin{bmatrix} r_x(0) & r_x^*(1) & r_x^*(2) & \cdots & r_x^*(j) & r_x^*(j+1) \\ r_x(1) & r_x(0) & r_x^*(1) & \cdots & r_x^*(j-1) & r_x^*(j) \\ r_x(2) & r_x(1) & r_x(0) & \cdots & r_x^*(j-2) & r_x^*(j-1) \\ \vdots & \vdots & \vdots & & \vdots & \vdots \\ r_x(j) & r_x(j-1) & r_x(j-2) & \cdots & r_x(0) & r_x^*(1) \\ r_x(j+1) & r_x(j) & r_x(j-1) & \cdots & r_x(1) & r_x(0) \end{bmatrix} \begin{bmatrix} 1 \\ a_j(1) \\ a_j(2) \\ \vdots \\ a_j(j) \\ 0 \end{bmatrix} = \begin{bmatrix} \epsilon_j \\ 0 \\ 0 \\ \vdots \\ 0 \\ \gamma_j \end{bmatrix} \tag{5.9}$$

where the parameter γ_j is

$$\boxed{\gamma_j = r_x(j+1) + \sum_{i=1}^{j} a_j(i) r_x(j+1-i)} \tag{5.10}$$

Note that if $\gamma_j = 0$, then the right side of Eq. (5.9) is a scaled unit vector and $\mathbf{a}_{j+1} = [1, a_j(1), \ldots, a_j(j), 0]^T$ is the solution to the $(j+1)$st-order normal equations (5.8). In general, however, $\gamma_j \neq 0$ and $[1, a_j(1), \ldots, a_j(j), 0]^T$ is not the solution to Eq. (5.8).

The key step in the derivation of the Levinson-Durbin recursion is to note that the Hermitian Toeplitz property of $\mathbf{R}_{j+1}$ allows us to rewrite Eq. (5.9) in the equivalent form

$$\begin{bmatrix} r_x(0) & r_x(1) & r_x(2) & \cdots & r_x(j) & r_x(j+1) \\ r_x^*(1) & r_x(0) & r_x(1) & \cdots & r_x(j-1) & r_x(j) \\ r_x^*(2) & r_x^*(1) & r_x(0) & \cdots & r_x(j-2) & r_x(j-1) \\ \vdots & \vdots & \vdots & & \vdots & \vdots \\ r_x^*(j) & r_x^*(j-1) & r_x^*(j-2) & \cdots & r_x(0) & r_x(1) \\ r_x^*(j+1) & r_x^*(j) & r_x^*(j-1) & \cdots & r_x^*(1) & r_x(0) \end{bmatrix} \begin{bmatrix} 0 \\ a_j(j) \\ a_j(j-1) \\ \vdots \\ a_j(1) \\ 1 \end{bmatrix} = \begin{bmatrix} \gamma_j \\ 0 \\ 0 \\ \vdots \\ 0 \\ \epsilon_j \end{bmatrix} \tag{5.11}$$

Taking the complex conjugate of Eq. (5.11) and combining the resulting equation with Eq. (5.9), it follows that, for any (complex) constant Γ_{j+1},

$$\mathbf{R}_{j+1}\left\{\begin{bmatrix} 1 \\ a_j(1) \\ a_j(2) \\ \vdots \\ a_j(j) \\ 0 \end{bmatrix} + \Gamma_{j+1}\begin{bmatrix} 0 \\ a_j^*(j) \\ a_j^*(j-1) \\ \vdots \\ a_j^*(1) \\ 1 \end{bmatrix}\right\} = \begin{bmatrix} \epsilon_j \\ 0 \\ 0 \\ \vdots \\ 0 \\ \gamma_j \end{bmatrix} + \Gamma_{j+1}\begin{bmatrix} \gamma_j^* \\ 0 \\ 0 \\ \vdots \\ 0 \\ \epsilon_j^* \end{bmatrix} \tag{5.12}$$

Since we want to find the vector $\mathbf{a}_{j+1}$ which, when multiplied by $\mathbf{R}_{j+1}$, yields a scaled unit vector, note that if we set

$$\boxed{\Gamma_{j+1} = -\frac{\gamma_j}{\epsilon_j^*}} \tag{5.13}$$

then Eq. (5.12) becomes

$$\mathbf{R}_{j+1}\mathbf{a}_{j+1} = \epsilon_{j+1}\mathbf{u}_1$$

where

$$\mathbf{a}_{j+1} = \begin{bmatrix} 1 \\ a_j(1) \\ a_j(2) \\ \vdots \\ a_j(j) \\ 0 \end{bmatrix} + \Gamma_{j+1}\begin{bmatrix} 0 \\ a_j^*(j) \\ a_j^*(j-1) \\ \vdots \\ a_j^*(1) \\ 1 \end{bmatrix} \tag{5.14}$$

which is the solution to the $(j+1)$st-order normal equations. Furthermore,

$$\boxed{\epsilon_{j+1} = \epsilon_j + \Gamma_{j+1}\gamma_j^* = \epsilon_j\left[1 - |\Gamma_{j+1}|^2\right]} \tag{5.15}$$

is the $(j+1)$st-order modeling error.[1] If we define $a_j(0) = 1$ and $a_j(j+1) = 0$ then Eq. (5.14), referred to as the *Levinson order-update equation*, may be expressed as

$$\boxed{a_{j+1}(i) = a_j(i) + \Gamma_{j+1}a_j^*(j-i+1) \quad ; \quad i = 0, 1, \ldots j+1} \tag{5.16}$$

All that is required to complete the recursion is to define the conditions necessary to initialize the recursion. These conditions are given by the solution for the model of order $j = 0$,

$$\begin{aligned} a_0(0) &= 1 \\ \epsilon_0 &= r_x(0) \end{aligned} \tag{5.17}$$

In summary, the steps of the Levinson-Durbin recursion are as follows. The recursion is first initialized with the zeroth-order solution, Eq. (5.17). Then, for $j = 0, 1, \ldots, p-1$, the $(j+1)$st-order model is found from the jth-order model in three steps. The first step is to use Eqs. (5.10) and (5.13) to determine the value of Γ_{j+1}, which is referred to as

[1] Since ϵ_j is real, then the complex conjugate in Eq. (5.13) may be dropped.

Table 5.1 ***The Levinson-Durbin Recursion***

1. Initialize the recursion
   (a) $a_0(0) = 1$
   (b) $\epsilon_0 = r_x(0)$
2. For $j = 0, 1, \ldots, p-1$
   (a) $\gamma_j = r_x(j+1) + \sum_{i=1}^{j} a_j(i) r_x(j-i+1)$
   (b) $\Gamma_{j+1} = -\gamma_j / \epsilon_j$
   (c) For $i = 1, 2, \ldots, j$
   $$a_{j+1}(i) = a_j(i) + \Gamma_{j+1} a_j^*(j-i+1)$$
   (d) $a_{j+1}(j+1) = \Gamma_{j+1}$
   (e) $\epsilon_{j+1} = \epsilon_j \big[1 - |\Gamma_{j+1}|^2\big]$
3. $b(0) = \sqrt{\epsilon_p}$

the $(j+1)$st *reflection coefficient*. The next step of the recursion is to use the Levinson order-update equation to compute the coefficients $a_{j+1}(i)$ from $a_j(i)$. The final step of the recursion is to update the error, ϵ_{j+1}, using Eq. (5.15). This error may also be written in two equivalent forms that will be useful in later discussions. The first is

$$\epsilon_{j+1} = \epsilon_j \big[1 - |\Gamma_{j+1}|^2\big] = r_x(0) \prod_{i=1}^{j+1} \big[1 - |\Gamma_i|^2\big] \tag{5.18}$$

and the second, which follows from Eq. (5.3), is

$$\epsilon_{j+1} = r_x(0) + \sum_{i=1}^{j+1} a_{j+1}(i) r_x(i) \tag{5.19}$$

The complete recursion is listed in Table 5.1 and a MATLAB program is given in Fig. 5.1.[2]

Example 5.2.1 *Solving the Autocorrelation Normal Equations*

Let us use the Levinson-Durbin recursion to solve the autocorrelation normal equations and find a third-order all-pole model for a signal having autocorrelation values

$$r_x(0) = 1, \quad r_x(1) = 0.5, \quad r_x(2) = 0.5, \quad r_x(3) = 0.25$$

The normal equations for the third-order all-pole model are

$$\begin{bmatrix} 1 & 0.5 & 0.5 \\ 0.5 & 1 & 0.5 \\ 0.5 & 0.5 & 1 \end{bmatrix} \begin{bmatrix} a_3(1) \\ a_3(2) \\ a_3(3) \end{bmatrix} = - \begin{bmatrix} 0.5 \\ 0.5 \\ 0.25 \end{bmatrix}$$

Using the Levinson-Durbin recursion we have

[2] The convention used in the Levinson-Durbin recursion varies a bit from one author to another. Some authors, for example, use $-\Gamma_{j+1}$ in Eq. (5.12), while others use Γ_{j+1}^*. This results, of course, in different sign conventions for Γ_{j+1}.

The Levinson-Durbin Recursion

```
function [a,epsilon]=rtoa(r)
%
   r=r(:);
   p=length(r)-1;
   a=1;
   epsilon=r(1);
   for j=2:p+1;
       gamma=-r(2:j)'*flipud(a)/epsilon;
       a=[a;0] + gamma*[0;conj(flipud(a))];
       epsilon=epsilon*(1 - abs(gamma)^2);
       end
```

Figure 5.1 *A* MATLAB *program for solving a set of Toeplitz equations using the Levinson-Durbin Recursion. This m-file provides a mapping from the autocorrelation sequence $r_x(k)$ to the filter coefficients $a_p(k)$, hence the name* `rtoa`.

1. First-order model:

$$\gamma_0 = r_x(1)$$

$$\Gamma_1 = -\frac{\gamma_0}{\epsilon_0} = -\frac{r_x(1)}{r_x(0)} = -\tfrac{1}{2} \quad ; \quad \epsilon_1 = r_x(0)\big[1 - \Gamma_1^2\big] = \tfrac{3}{4}$$

and

$$\mathbf{a}_1 = \begin{bmatrix} 1 \\ \Gamma_1 \end{bmatrix} = \begin{bmatrix} 1 \\ -1/2 \end{bmatrix}$$

2. Second-order model:

$$\gamma_1 = r_x(2) + a_1(1) r_x(1) = \tfrac{1}{4}$$

$$\Gamma_2 = -\gamma_1/\epsilon_1 = -\tfrac{1}{3} \quad ; \quad \epsilon_2 = \epsilon_1\big[1 - \Gamma_2^2\big] = \tfrac{2}{3}$$

and

$$\mathbf{a}_2 = \begin{bmatrix} 1 \\ -1/2 \\ 0 \end{bmatrix} - 1/3 \begin{bmatrix} 0 \\ -1/2 \\ 1 \end{bmatrix} = \begin{bmatrix} 1 \\ -1/3 \\ -1/3 \end{bmatrix}$$

3. Third-order model:

$$\gamma_2 = r_x(3) + a_2(1) r_x(2) + a_2(2) r_x(1) = -\tfrac{1}{12}$$

$$\Gamma_3 = -\gamma_2/\epsilon_2 = \tfrac{1}{8} \quad ; \quad \epsilon_3 = \epsilon_2\big[1 - \Gamma_3^2\big] = \tfrac{21}{32}$$

and

$$\mathbf{a}_3 = \begin{bmatrix} 1 \\ -1/3 \\ -1/3 \\ 0 \end{bmatrix} + \frac{1}{8} \begin{bmatrix} 0 \\ -1/3 \\ -1/3 \\ 1 \end{bmatrix} = \begin{bmatrix} 1 \\ -3/8 \\ -3/8 \\ 1/8 \end{bmatrix}$$

Finally, with

$$b(0) = \sqrt{\epsilon_3} = \frac{1}{8}\sqrt{42}$$

the all-pole model becomes

$$H_3(z) = \frac{\sqrt{42}/8}{1 - \frac{3}{8}z^{-1} - \frac{3}{8}z^{-2} + \frac{1}{8}z^{-3}}$$

Having found the third-order all-pole model for $x(n)$, let us determine what the next value in the autocorrelation sequence would be if $H_3(z)$ were the correct model for $x(n)$, i.e., if $x(n)$ is the inverse z-transform of $H_3(z)$. In this case, the model will remain unchanged for $k > 3$, i.e., $\Gamma_k = 0$ and, therefore, $\gamma_k = 0$ for $k > 3$. Thus, it follows from Eq. (5.10) by setting $j = 3$ that

$$r_x(4) + \sum_{i=1}^{3} a_3(i) r_x(4-i) = 0$$

which may be solved for $r_x(4)$ as follows

$$r_x(4) = -\sum_{i=1}^{3} a_3(i) r_x(4-i) = 7/32$$

Successive values of $r_x(k)$ may be determined in a similar fashion. For example, if $k \geq 4$, then

$$r_x(k) = -\sum_{i=1}^{3} a_3(i) r_x(k-i)$$

which are the Yule-Walker equations for an AR(3) process.

It follows from our discussions in Section 3.6.2 of Chapter 3 that if white noise with a variance of σ_w^2 is filtered with a first-order all-pole filter

$$H(z) = \frac{1}{1 - \alpha z^{-1}}$$

then the output will be an AR(1) process with an autocorrelation sequence of the form

$$r_x(k) = \frac{\sigma_w^2}{1-\alpha^2} \alpha^{|k|}$$

In the following example, the Levinson-Durbin recursion is used to find the sequence of reflection coefficients Γ_k and model errors ϵ_k that are associated with this process.

Example 5.2.2 ***The Reflection Coefficients for an AR(1) Process***

Let $x(n)$ be an AR(1) process with an autocorrelation

$$\mathbf{r}_x = \frac{\sigma_w^2}{1-\alpha^2} [1, \alpha, \alpha^2, \ldots, \alpha^p]^T$$

To find the reflection coefficients associated with this vector of autocorrelations we may use the Levinson-Durbin recursion as follows. Initializing the recursion with

$$\mathbf{a}_0 = 1 \quad ; \quad \epsilon_0 = r_x(0) = \frac{\sigma_w^2}{1-\alpha^2}$$

it follows that the first reflection coefficient is

$$\Gamma_1 = -\frac{r_x(1)}{r_x(0)} = -\alpha$$

and that

$$\epsilon_1 = \epsilon_0\big[1 - \Gamma_1^2\big] = \sigma_w^2$$

Thus, for the first-order model

$$\mathbf{a}_1 = \begin{bmatrix} 1 \\ -\alpha \end{bmatrix}$$

For the second-order model

$$\gamma_1 = r_x(2) + a_1(1)r_x(1) = 0$$

and

$$\Gamma_2 = 0 \quad ; \quad \epsilon_2 = \epsilon_1 = \sigma_w^2$$

so the coefficients of the model are

$$\mathbf{a}_2 = \begin{bmatrix} 1 \\ -\alpha \\ 0 \end{bmatrix}$$

Continuing, we may show by induction that $\Gamma_j = 0$ and $\epsilon_j = \sigma_w^2$ for all $j > 1$. Specifically, suppose that at the jth step of the recursion $\Gamma_j = 0$ and

$$\mathbf{a}_j = [1, \ -\alpha, \ 0, \ldots, \ 0]^T$$

Since

$$\gamma_j = r_x(j+1) + \sum_{i=1}^{j} a_j(i) r_x(j-i+1) = r_x(j+1) - \alpha r_x(j)$$

we see from the form of the autocorrelation sequence that $\gamma_j = 0$. Therefore, $\Gamma_{j+1} = 0$ and $\mathbf{a}_{j+1}$ is formed by appending a zero to $\mathbf{a}_j$ and the result follows. In summary, for an AR(1) process,

$$\begin{aligned}
\mathbf{\Gamma}_p &= \big[-\alpha, \ 0, \ 0, \ \ldots, \ 0\big]^T \\
\mathbf{a}_p &= \big[1, \ -\alpha, \ 0, \ \ldots, \ 0\big]^T \\
\boldsymbol{\epsilon}_p &= \left[\frac{\sigma_w^2}{1-\alpha^2}, \ \sigma_w^2, \ \sigma_w^2, \ \ldots, \ \sigma_w^2\right]^T
\end{aligned}$$

We now look at the computational complexity of the Levinson-Durbin recursion and compare it to Gaussian elimination for solving the pth-order autocorrelation normal equations. Using Gaussian elimination to solve a set of p linear equations in p unknowns, approximately $\frac{1}{3}p^3$ multiplications and divisions are required. With the Levinson-Durbin recursion, on the other hand, at the jth step of the recursion $2j + 2$ multiplications, 1 division, and $2j + 1$ additions are necessary. Since there are p steps in the recursion, the total number of multiplications and divisions is[3]

$$\sum_{j=0}^{p-1}(2j+3) = p^2 + 2p$$

and the number of additions is

$$\sum_{j=0}^{p-1}(2j+1) = p^2$$

Therefore, the number of multiplications and divisions is proportional to p^2 for the Levinson-Durbin recursion compared with p^3 for Gaussian elimination. Another advantage of the Levinson-Durbin recursion over Gaussian elimination is that it requires less memory for data storage. Specifically, whereas Gaussian elimination requires p^2 memory locations, the Levinson-Durbin recursion requires only $2(p+1)$ locations: $p+1$ for the autocorrelation sequence $r_x(0), \ldots, r_x(p)$, and p for the model parameters $a_p(1), \ldots, a_p(p)$, and one for the error ϵ_p.

In spite of the increased efficiency of the Levinson-Durbin recursion over Gaussian elimination, it should be pointed out that solving the normal equations may only be a small fraction of the total computational cost in the modeling process. For example, note that for a signal of length N, computing the autocorrelation values $r_x(0), \ldots, r_x(p)$ requires approximately $N(p+1)$ multiplications and additions. Therefore, if $N \gg p$, then the cost associated with finding the autocorrelation sequence will dominate the computational requirements of the modeling algorithm.

5.2.2 The Lattice Filter

One of the by-products of the Levinson-Durbin recursion is the lattice structure for digital filters. Lattice filters are used routinely in digital filter implementations as a result of a number of interesting and important properties that they possess. These properties include a modular structure, low sensitivity to parameter quantization effects, and a simple method to ensure filter stability. In this section, we show how the Levinson order-update equation may be used to derive the lattice filter structure for FIR digital filters. In Chapter 6, other lattice filters structures will be developed, including the all-pole lattice and the pole-zero lattice filters, and it will be shown how these filters may be used for signal modeling.

The derivation of the lattice filter begins with the Levinson order-update equation given in Eq. (5.16). First, however, it will be convenient to define the *reciprocal vector*, denoted by $\mathbf{a}_j^R$, which is the vector that is formed by reversing the order of the elements in $\mathbf{a}_j$ and taking the complex conjugate,

$$\mathbf{a}_j = \begin{bmatrix} 1 \\ a_j(1) \\ a_j(2) \\ \vdots \\ a_j(j-1) \\ a_j(j) \end{bmatrix} \Longrightarrow \begin{bmatrix} a_j^*(j) \\ a_j^*(j-1) \\ a_j^*(j-2) \\ \vdots \\ a_j^*(1) \\ 1 \end{bmatrix} = \mathbf{a}_j^R$$

[3]Recall the summations given in Table 2.3, p. 16.

or,

$$a_j^R(i) = a_j^*(j-i) \tag{5.20}$$

for $i = 0, 1, \ldots, j$. Using the reciprocal vector in the Levinson order-update equation, Eq. (5.16) becomes

$$a_{j+1}(i) = a_j(i) + \Gamma_{j+1} a_j^R(i-1) \tag{5.21}$$

With $A_j(z)$ the z-transform of $a_j(i)$ and $A_j^R(z)$ the z-transform of the reciprocal sequence $a_j^R(i)$, it follows from Eq. (5.20) that $A_j(z)$ is related to $A_j^R(z)$ by

$$A_j^R(z) = z^{-j} A_j^*(1/z^*) \tag{5.22}$$

Rewriting Eq. (5.21) in terms of $A_j(z)$ and $A_j^R(z)$ gives

$$A_{j+1}(z) = A_j(z) + \Gamma_{j+1} z^{-1} A_j^R(z) \tag{5.23}$$

which is an order-update equation for $A_j(z)$.

The next step is to derive an order-update equation for $a_{j+1}^R(i)$ and $A_{j+1}^R(z)$. Beginning with Eq. (5.21), note that if we take the complex conjugate of both sides of the equation and replace i with $j - i + 1$, we have

$$a_{j+1}^*(j-i+1) = a_j^*(j-i+1) + \Gamma_{j+1}^* a_j(i) \tag{5.24}$$

Incorporating the definition of the reversed vector, Eq. (5.20), into Eq. (5.24) gives the desired update equation for $a_{j+1}^R(i)$,

$$a_{j+1}^R(i) = a_j^R(i-1) + \Gamma_{j+1}^* a_j(i) \tag{5.25}$$

Expressed in the z-domain, Eq. (5.25) gives the desired update equation for $A_{j+1}^R(z)$,

$$A_{j+1}^R(z) = z^{-1} A_j^R(z) + \Gamma_{j+1}^* A_j(z) \tag{5.26}$$

In summary, we have a pair of coupled difference equations,

$$\begin{aligned} a_{j+1}(n) &= a_j(n) + \Gamma_{j+1} a_j^R(n-1) \\ a_{j+1}^R(n) &= a_j^R(n-1) + \Gamma_{j+1}^* a_j(n) \end{aligned} \tag{5.27}$$

for updating $a_j(n)$ and $a_j^R(n)$, along with a pair of coupled equations for updating the system functions $A_j(z)$ and $A_j^R(z)$, which may be written in matrix form as

$$\begin{bmatrix} A_{j+1}(z) \\ A_{j+1}^R(z) \end{bmatrix} = \begin{bmatrix} 1 & \Gamma_{j+1} z^{-1} \\ \Gamma_{j+1}^* & z^{-1} \end{bmatrix} \begin{bmatrix} A_j(z) \\ A_j^R(z) \end{bmatrix} \tag{5.28}$$

Both of these representations describe to the two-port network shown in Fig. 5.2. This two-port represents the basic module used to implement an FIR lattice filter. Cascading p such lattice filter modules with reflection coefficients $\Gamma_1, \Gamma_2, \ldots, \Gamma_p$, forms the pth-order lattice filter shown in Fig. 5.3. Note that the system function between the input $\delta(n)$ and the output $a_p(n)$ is $A_p(z)$, whereas the system function relating the input to the output $a_p^R(n)$ is $A_p^R(z)$.

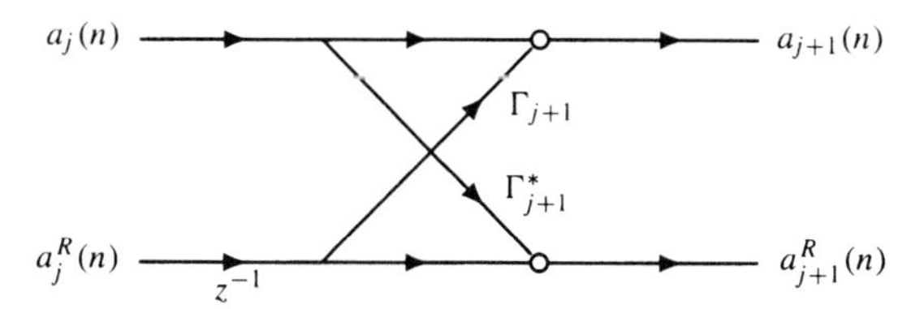

Figure 5.2 *A two-port network relating $a_j(n)$ and $a^R_j(n)$ to $a_{j+1}(n)$ and $a^R_{j+1}(n)$. This network is a single stage of an FIR lattice filter.*

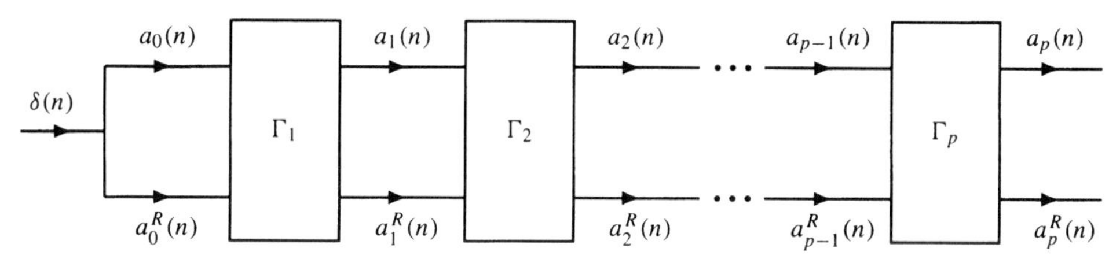

Figure 5.3 *A pth-order FIR lattice filter built up as a cascade of p two-port networks.*

5.2.3 Properties

In this section, we will establish some important properties of the reflection coefficient sequence that is generated by the Levinson-Durbin recursion and we will establish an important property of the solution to the autocorrelation normal equations. Specifically, we will show that the roots of $A_p(z)$ will be inside the unit circle if and only if the reflection coefficients are bounded by one in magnitude or, equivalently, if and only if $\mathbf{R}_p$ is positive definite. We will also show that the all-pole model obtained from the autocorrelation method is guaranteed to be stable. Finally, we will establish the *autocorrelation matching property*, which states that if we set $b(0) = \sqrt{\epsilon_p}$ in the autocorrelation method and if $h(n)$ is the inverse z-transform of

$$H(z) = \frac{b(0)}{A_p(z)}$$

then the autocorrelation of $h(n)$ is equal to the autocorrelation of $x(n)$.

The first property of the reflection coefficients is the following.

> ***Property 1.*** The reflection coefficients that are generated by the Levinson-Durbin recursion to solve the autocorrelation normal equations are bounded by one in magnitude, $|\Gamma_j| \leq 1$.

This property follows immediately from the fact that since ϵ_j is the minimum squared error,

$$\epsilon_j = \{\mathcal{E}_j\}_{\min} \tag{5.29}$$

with

$$\mathcal{E}_j = \sum_{n=0}^{\infty} |e(n)|^2$$

where $e(n)$ is the jth-order modeling error, then $\epsilon_j \geq 0$. Therefore, since

$$\epsilon_j = \epsilon_{j-1}\left[1 - |\Gamma_j|^2\right]$$

it follows that if $\epsilon_j \geq 0$ and $\epsilon_{j-1} \geq 0$, then $|\Gamma_j| \leq 1$. ■

It is important to keep in mind that the validity of Property 1 relies on the nonnegativity of ϵ_j. Implicit in this nonnegativity is the assumption that the autocorrelation values $r_x(k)$ in $\mathbf{R}_p$ are computed according to Eq. (4.121). It is not necessarily true, for example, that $\epsilon_j \geq 0$ and $|\Gamma_j| \leq 1$ if an arbitrary sequence of numbers $r_x(0), r_x(1), \ldots, r_x(p-1)$ are used in the matrix $\mathbf{R}_p$. To illustrate this point, consider what happens if we set $r_x(0) = 1$ and $r_x(1) = 2$ in the normal equations and solve for the first-order model. In this case,

$$\Gamma_1 = -r_x(1)/r_x(0) = -2$$

which has a magnitude that is greater than one. This example is not in conflict with Property 1, however, since for any valid autocorrelation sequence, $|r_x(0)| \geq |r_x(1)|$. As we will soon discover (see Property 7 on p. 253), the unit magnitude constraint on the reflection coefficients Γ_j and the nonnegativity of the sequence ϵ_j are tied to the positive definiteness of the matrix $\mathbf{R}_p$.

The next property establishes a relationship between the locations of the roots of the polynomial $A_p(z)$ and the magnitudes of the reflection coefficients Γ_j.

Property 2. If $a_p(k)$ is a set of model parameters and Γ_j is the corresponding set of reflection coefficients, then

$$A_p(z) = 1 + \sum_{k=1}^{p} a_p(k)z^{-k}$$

will be a *minimum phase polynomial* (all of the roots of $A_p(z)$ will be *inside* the unit circle) if and only if $|\Gamma_j| < 1$ for all j. Furthermore, if $|\Gamma_j| \leq 1$ for all j then the roots of $A_p(z)$ must lie either inside or on the unit circle.

There are many different ways to establish this property [14, 20, 24]. One way, as demonstrated below, is to use the *encirclement principle* (*principle of the argument*) from complex analysis [3].

Encirclement Principle. Given a rational function of z

$$P(z) = \frac{B(z)}{A(z)}$$

let C be a simple closed curve in the z plane as shown in Fig. 5.4. As the path C is traversed in a counterclockwise direction, a closed curve is generated in the $P(z)$ plane that encircles the origin $(N_z - N_p)$ times in a counterclockwise direction where N_z is the number of *zeros inside* C and N_p is the number of *poles inside* C.[4]

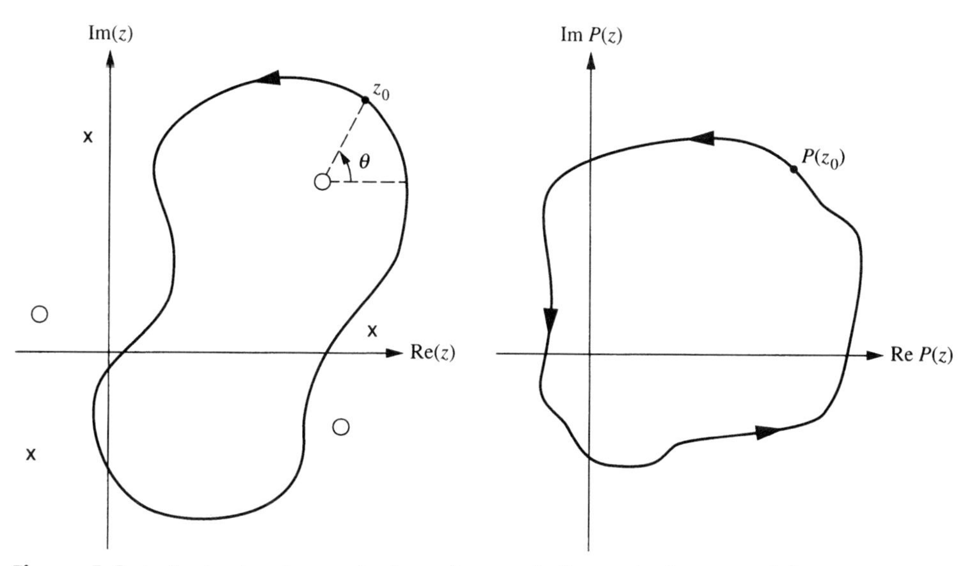

Figure 5.4 *A simple closed curve in the z-plane encircling a simple zero and the corresponding contour in the $P(z)$ plane.*

The encirclement principle may be interpreted geometrically as follows. Since C is a simple closed curve, beginning and ending at z_0, then $P(z)$ will also be a closed curve, beginning and ending at the same point, $P(z_0)$. If there is a zero contained within the contour then the angle θ from the zero to a point on the contour will increase by 2π as the contour is traversed. In the $P(z)$ plane, therefore, the curve that is generated will encircle the origin in a counterclockwise direction. Similarly, if there is a pole inside the contour, then the angle will *decrease* by 2π as C is traversed, which produces a curve in the $P(z)$ plane that encircles the origin in the clockwise direction. With N_z zeros and N_p poles inside C, the change in angle is $2\pi(N_z - N_p)$ and the encirclement principle follows. Although not stated in the encirclement principle, it should be clear that if there is a zero of $P(z)$ on the contour C, then the curve in the $P(z)$ plane will pass *through* the origin.

With the encirclement principle, we may use induction to establish Property 2. Specifically, for a first-order model,

$$A_1(z) = 1 + \Gamma_1 z^{-1}$$

and it follows that $A_1(z)$ will be minimum phase if and only if $|\Gamma_1| < 1$. Let us now assume that $A_j(z)$ is minimum phase and show that $A_{j+1}(z)$ will be minimum phase if and only if $|\Gamma_{j+1}| < 1$. To accomplish this, we will use the z-domain representation of the Levinson order-update equation given in Eq. (5.23), which is repeated below for convenience

$$A_{j+1}(z) = A_j(z) + \Gamma_{j+1} z^{-1} A_j^R(z)$$

Dividing both sides of this update equation by $A_j(z)$ we have

$$P(z) = \frac{A_{j+1}(z)}{A_j(z)} = 1 + \Gamma_{j+1} z^{-1} \frac{A_j^R(z)}{A_j(z)} \tag{5.30}$$

[4]If $N_z - N_p$ is negative then the curve $P(z)$ encircles the origin in a clockwise direction $|N_z - N_p|$ times.

We will now show that if C is a closed contour that traverses the unit circle then the number of times that $P(z)$ encircles the origin in a clockwise direction is equal to the number of zeros of $A_{j+1}(z)$ that are *outside* the unit circle. Since $A_j(z)$ is assumed to be minimum phase, $A_j(z)$ has j zeros *inside* the unit circle and j poles at $z = 0$. Now let us assume that $A_{j+1}(z)$ has l zeros *outside* the unit circle and $j + 1 - l$ zeros *inside*. Since $A_{j+1}(z)$ will have $j + 1$ poles at $z = 0$, it follows that the number of times that $P(z)$ encircles the origin in a counter clockwise direction is

$$N_z - N_p = (j + 1 - l) - (j + 1) = -l$$

or, l times in a clockwise direction. Now, note that since the contour C is the unit circle, then $z = e^{j\omega}$ and

$$\left|\Gamma_{j+1} z^{-1} \frac{A_j^R(z)}{A_j(z)}\right|_C = |\Gamma_{j+1}| \left|\frac{A_j^R(e^{j\omega})}{A_j(e^{j\omega})}\right|_C = |\Gamma_{j+1}|$$

Therefore, it follows from Eq. (5.30) that $P(z)$ traces out a curve which is a circle of radius $|\Gamma_{j+1}|$ that is centered at $z = 1$ (see Fig. 5.5). Thus, if $|\Gamma_{j+1}| < 1$, then $P(z)$ does not encircle or pass through the origin. Conversely, if $|\Gamma_{j+1}| > 1$ then $P(z)$ encircles the origin and $A_{j+1}(z)$ will not have all of its zeros inside the unit circle. Consequently, if $A_j(z)$ is minimum phase, then $A_{j+1}(z)$ will be minimum phase if and only if $|\Gamma_{j+1}| < 1$. ∎

There is an intimate connection between the positive definiteness of the Toeplitz matrix, $\mathbf{R}_p$, and the minimum phase property of the solution to the normal equations. In particular, the following property states that $A_p(z)$ will be minimum phase if and only if $\mathbf{R}_p$ is positive definite (see also Section 5.2.7).

Property 3. If $\mathbf{a}_p$ is the solution to the Toeplitz normal equations $\mathbf{R}_p\mathbf{a}_p = \epsilon_p\mathbf{u}_1$, then $A_p(z)$ will be minimum phase if and only if $\mathbf{R}_p$ is positive definite, $\mathbf{R}_p > 0$.

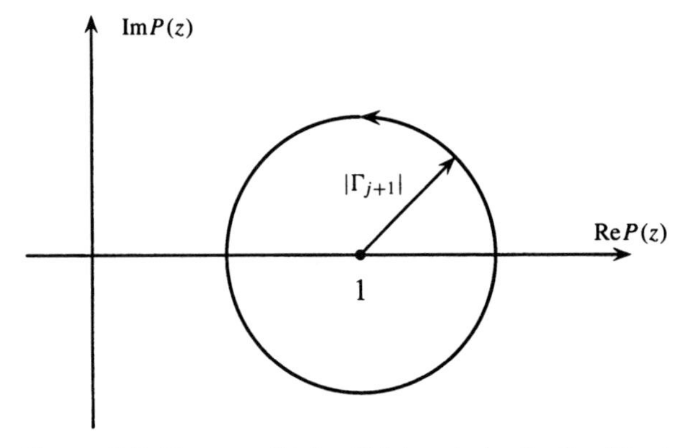

Figure 5.5 *The curve $P(z)$, which is a circle of radius $|\Gamma_{j+1}|$ that is centered at $z = 1$.*

To establish this property, let α be a root of the polynomial $A_p(z)$. In general, α will be complex and we may factor $A_p(z)$ as follows

$$A_p(z) = 1 + \sum_{k=1}^{p} a_p(k) z^{-k} = (1 - \alpha z^{-1})(1 + b_1 z^{-1} + b_2 z^{-2} + \cdots + b_{p-1} z^{-(p-1)})$$

We will show that if $\mathbf{R}_p > 0$, then $|\alpha| < 1$. Writing the vector $\mathbf{a}_p$ in factored form as

$$\mathbf{a}_p = \begin{bmatrix} 1 \\ a_p(1) \\ a_p(2) \\ \vdots \\ a_p(p) \end{bmatrix} = \begin{bmatrix} 1 & 0 \\ b_1 & 1 \\ b_2 & b_1 \\ \vdots & \vdots \\ 0 & b_{p-1} \end{bmatrix} \begin{bmatrix} 1 \\ -\alpha \end{bmatrix} = \mathbf{B} \begin{bmatrix} 1 \\ -\alpha \end{bmatrix} \tag{5.31}$$

it follows that

$$\mathbf{R}_p \mathbf{a}_p = \mathbf{R}_p \mathbf{B} \begin{bmatrix} 1 \\ -\alpha \end{bmatrix} = \epsilon_p \mathbf{u}_1 \tag{5.32}$$

Multiplying Eq. (5.32) on the left by $\mathbf{B}^H$ we have

$$\mathbf{B}^H \mathbf{R}_p \mathbf{B} \begin{bmatrix} 1 \\ -\alpha \end{bmatrix} = \epsilon_p \begin{bmatrix} 1 \\ 0 \end{bmatrix}$$

Since $\mathbf{B}$ has full rank, if $\mathbf{R}_p$ is positive definite, then $\mathbf{B}^H \mathbf{R}_p \mathbf{B}$ will be positive definite (see p. 40). Therefore,

$$\mathbf{B}^H \mathbf{R}_p \mathbf{B} = \begin{bmatrix} s_0 & s_1 \\ s_1^* & s_0 \end{bmatrix} > 0$$

which implies that

$$|s_0|^2 > |s_1|^2 \tag{5.33}$$

Since

$$\mathbf{B}^H \mathbf{R}_p \mathbf{B} \begin{bmatrix} 1 \\ -\alpha \end{bmatrix} = \begin{bmatrix} s_0 - s_1 \alpha \\ s_1^* - s_0 \alpha \end{bmatrix} = \epsilon_p \begin{bmatrix} 1 \\ 0 \end{bmatrix}$$

then $s_0 \alpha = s_1^*$ and

$$\alpha = \frac{s_1^*}{s_0} \tag{5.34}$$

Thus, Eq. (5.33) implies that $|\alpha| < 1$ and $A_p(z)$ is minimum phase. Conversely, if $A_p(z)$ is not minimum phase, then $A_p(z)$ will have a root α with $|\alpha| > 1$. Therefore, from Eq. (5.34) it follows that $|s_1| > |s_0|$ and $\mathbf{R}_p$ is not positive definite. ∎

Property 1 states that the reflection coefficients that are generated in the course of solving the autocorrelation normal equations are bounded by one in magnitude. Property 2 states that a polynomial $A_p(z)$ will have all of its roots inside the unit circle if its reflection coefficients are less than one in magnitude. Combining Properties 1 and 2 leads to the following property.

Property 4. The autocorrelation method produces a stable all-pole model.

In the next property, we consider what happens to the roots of the polynomial $A_p(z)$ when $|\Gamma_p| = 1$ with the remaining terms in the reflection coefficient sequence being less than one in magnitude.

Property 5. Let $a_p(k)$ be a set of filter coefficients and Γ_j the corresponding set of reflection coefficients. If $|\Gamma_j| < 1$ for $j = 1, \ldots, p-1$ and $|\Gamma_p| = 1$, then the polynomial

$$A_p(z) = 1 + \sum_{k=1}^{p} a_p(k) z^{-k}$$

has all of its roots *on* the unit circle.

This property may be established easily as follows. Let Γ_j be a sequence of reflection coefficients with $|\Gamma_j| < 1$ for $j < p$ and $|\Gamma_p| = 1$. Since $|\Gamma_j| \leq 1$ for all j, it follows from Property 2 that all of the zeros of $A_p(z)$ are on or inside the unit circle. If we denote the zeros of $A_p(z)$ by z_k, then

$$A_p(z) = \prod_{k=1}^{p} (1 - z_k z^{-1})$$

By equating the coefficients of z^{-p} on both sides of the equation, we have

$$a_p(p) = \prod_{k=1}^{p} z_k \tag{5.35}$$

However, since $a_p(p) = \Gamma_p$, if $|\Gamma_p| = 1$, then

$$\prod_{k=1}^{p} |z_k| = 1 \tag{5.36}$$

Therefore, if there is a zero of $A_p(z)$ that has a magnitude that is less than one, then there must be *at least one* zero with a magnitude greater than one, otherwise the product of the roots would not equal one. This, however, contradicts the requirement that all of the roots must lie on or inside the unit circle and Property 5 is established. ■

This property may be used to develop a constrained lattice filter for detecting sinusoids in noise (see computer exercise C6.1 in Chapter 6). It also forms the basis for an efficient algorithm in spectrum estimation for finding the Pisarenko harmonic decomposition (Chapter 8) of a stationary random process [11].

The next property relates the autocorrelation sequence of $x(n)$ to the autocorrelation sequence of $h(n)$, the unit sample response of the all-pole filter that is used to model $x(n)$. Specifically, as we demonstrate below, if $b(0)$, the numerator coefficient in $H(z)$, is selected so that $x(n)$ satisfies the energy matching constraint, $r_x(0) = r_h(0)$, then $b(0) = \sqrt{\epsilon_p}$ and the autocorrelation sequences are equal for $|k| \leq p$. This is the so-called *autocorrelation matching property*.

> ***Property 6—Autocorrelation matching property.*** If $b(0)$ is chosen to satisfy the energy matching constraint, then $b(0) = \sqrt{\epsilon_p}$ and the autocorrelation sequences of $x(n)$ and $h(n)$ are equal for $|k| \leq p$.

In order to establish this property, we will follow the approach given in [18]. We begin by noting that since $h(n)$ is the unit sample response of the all-pole model for $x(n)$ with

$$H(z) = \frac{b(0)}{1 + \sum_{k=1}^{p} a_p(k) z^{-k}}$$

where the coefficients $a_p(k)$ are solutions to the normal equations

$$\mathbf{R}_x \mathbf{a}_p = \epsilon_p \mathbf{u}_1 \tag{5.37}$$

then $h(n)$ satisfies the difference equation

$$h(n) + \sum_{k=1}^{p} a_p(k) h(n-k) = b(0)\delta(n) \tag{5.38}$$

Multiplying both sides of Eq. (5.38) by $h^*(n-l)$ and summing over n yields

$$\begin{aligned} \sum_{n=0}^{\infty} h(n)h^*(n-l) + \sum_{k=1}^{p} a_p(k) \sum_{n=0}^{\infty} h(n-k)h^*(n-l) &= b(0) \sum_{n=0}^{\infty} \delta(n) h^*(n-l) \\ &= b(0) h^*(-l) \end{aligned} \tag{5.39}$$

With

$$r_h(l) = \sum_{n=0}^{\infty} h(n) h^*(n-l) = r_h^*(-l)$$

we may rewrite Eq. (5.39) in terms of $r_h(l)$ as follows

$$r_h(l) + \sum_{k=1}^{p} a_p(k) r_h(l-k) = b(0) h^*(-l)$$

Since $h(n)$ is causal, then $h(n) = 0$ for $n < 0$ and $h(0) = b(0)$. Therefore, for $l \geq 0$, it follows that

$$r_h(l) + \sum_{k=1}^{p} a_p(k) r_h(l-k) = |b(0)|^2 \delta(l) \quad ; \quad l \geq 0 \tag{5.40}$$

Using the conjugate symmetry of $r_h(l)$ we may write Eq. (5.40) in matrix form as

$$\begin{bmatrix} r_h(0) & r_h^*(1) & r_h^*(2) & \cdots & r_h^*(p) \\ r_h(1) & r_h(0) & r_h^*(1) & \cdots & r_h^*(p-1) \\ r_h(2) & r_h(1) & r_h(0) & \cdots & r_h^*(p-2) \\ \vdots & \vdots & \vdots & & \vdots \\ r_h(p) & r_h(p-1) & r_h(p-2) & \cdots & r_h(0) \end{bmatrix} \begin{bmatrix} 1 \\ a_p(1) \\ a_p(2) \\ \vdots \\ a_p(p) \end{bmatrix} = |b(0)|^2 \begin{bmatrix} 1 \\ 0 \\ 0 \\ \vdots \\ 0 \end{bmatrix} \tag{5.41}$$

or

$$\mathbf{R}_h \mathbf{a}_p = |b(0)|^2 \mathbf{u}_1 \tag{5.42}$$

Therefore, $\mathbf{a}_p$ is the solution to each of the following sets of Toeplitz equations,

$$\begin{aligned} \mathbf{R}_h \mathbf{a}_p &= |b_0|^2 \mathbf{u}_1 \\ \mathbf{R}_x \mathbf{a}_p &= \epsilon_p \mathbf{u}_1 \end{aligned} \tag{5.43}$$

Now, suppose that $b(0)$ is chosen to satisfy the energy matching constraint,

$$r_x(0) = \sum_{n=0}^{\infty} |x(n)|^2 = \sum_{n=0}^{\infty} |h(n)|^2 = r_h(0) \tag{5.44}$$

By induction we may then show that $r_x(k) = r_h(k)$ for $k > 0$ as follows. Using the Levinson-Durbin recursion to solve Eq. (5.43) we have

$$a_1(1) = -\frac{r_x(1)}{r_x(0)} = -\frac{r_h(1)}{r_h(0)}$$

Since $r_x(0) = r_h(0)$, it follows that $r_x(1) = r_h(1)$. Now, assume that $r_x(k) = r_h(k)$ for $k = 1, 2, \ldots, j$. Using Eqs. (5.10) and (5.13) we have

$$\begin{aligned} r_x(j+1) &= -\Gamma_{j+1}\epsilon_j - \sum_{i=1}^{j} a_j(i) r_x(j+1-i) \\ r_h(j+1) &= -\Gamma_{j+1}\epsilon_j - \sum_{i=1}^{j} a_j(i) r_h(j+1-i) \end{aligned} \tag{5.45}$$

Therefore, since $r_x(i) = r_h(i)$ for $i = 1, 2, \ldots, j$, then $r_x(j+1) = r_h(j+1)$. Finally, from Eq. (5.41) note that

$$|b(0)|^2 = r_h(0) + \sum_{k=1}^{p} a_p(k) r_h(k)$$

With $r_h(k) = r_x(k)$ it follows that

$$|b(0)|^2 = r_x(0) + \sum_{k=1}^{p} a_p(k) r_x(k) = \epsilon_p \tag{5.46}$$

Therefore, $b(0) = \sqrt{\epsilon_p}$ and the autocorrelation matching property is established. ■

5.2.4 The Step-Up and Step-Down Recursions

The Levinson-Durbin recursion solves the autocorrelation normal equations and produces an all-pole model for a signal from the autocorrelation sequence $r_x(k)$. In addition to the model parameters, the recursion generates a set of reflection coefficients, $\Gamma_1, \Gamma_2, \ldots, \Gamma_p$, along with the final modeling error, ϵ_p. Thus, the recursion may be viewed as a mapping

$$\{r_x(0), r_x(1), \ldots, r_x(p)\} \xrightarrow{LEV} \left\{ \begin{array}{c} a_p(1), a_p(2), \ldots, a_p(p), b(0) \\ \\ \Gamma_1, \Gamma_2, \ldots, \Gamma_p, \epsilon_p \end{array} \right.$$

from an autocorrelation sequence to a set of model parameters and to a set of reflection coefficients. There are many instances in which it would be convenient to be able to derive the reflection coefficients, Γ_j, from the filter coefficients $a_p(k)$, and vice versa. In this section, we show how one set of parameters may be derived from the other in a recursive fashion. The recursions for performing these transformations are called the *step-up* and *step-down* recursions.

The Levinson order-update equation given in Eq. (5.16) is a recursion for deriving the filter coefficients $a_p(k)$ from the reflection coefficients, Γ_j. Specifically, since

$$a_{j+1}(i) = a_j(i) + \Gamma_{j+1} a_j^*(j - i + 1) \tag{5.47}$$

then the filter coefficients $a_{j+1}(i)$ may be easily found from $a_j(i)$ and Γ_{j+1}. The recursion is initialized by setting $a_0(0) = 1$ and, after the coefficients $a_p(k)$ have been determined, the recursion is completed by setting $b(0) = \sqrt{\epsilon_p}$. Since Eq. (5.47) defines how the model parameters for a jth-order filter may be updated (stepped-up) to a $(j+1)$st-order filter given Γ_{j+1}, Eq. (5.47) is referred to as the *step-up recursion*. The recursion, represented pictorially as

$$\{\Gamma_1, \Gamma_2, \ldots, \Gamma_p, \epsilon_p\} \xrightarrow{Step-up} \{a_p(1), a_p(2), \ldots, a_p(p), b(0)\}$$

is summarized in Table 5.2 and a MATLAB program is given in Fig 5.6.

Table 5.2 ***The Step-up Recursion***

1. Initialize the recursion: $a_0(0) = 1$
2. For $j = 0, 1 \ldots, p - 1$
   (a) For $i = 1, 2, , \ldots, j$
   $$a_{j+1}(i) = a_j(i) + \Gamma_{j+1} a_j^*(j - i + 1)$$
   (b) $a_{j+1}(j + 1) = \Gamma_{j+1}$
3. $b(0) = \sqrt{\epsilon_p}$

The Step-Up Recursion

```
function a=gtoa(gamma)
%
   a=1;
   gamma=gamma(:);
   p=length(gamma);
   for j=2:p+1;
      a=[a;0] + gamma(j-1)*[0;conj(flipud(a))];
      end
```

Figure 5.6 *A MATLAB program for the step-up recursion to find the filter coefficients $a_p(k)$ from the reflection coefficients.*

Example 5.2.3 *The Step-Up Recursion*

Given the reflection coefficient sequence

$$\mathbf{\Gamma}_2 = \left[\Gamma_1, \Gamma_2\right]^T$$

the first- and second-order all-pole models may be found from the step-up recursion as follows. From Eq. (5.14), the first-order model is

$$\mathbf{a}_1 = \begin{bmatrix} 1 \\ a_1(1) \end{bmatrix} = \begin{bmatrix} 1 \\ 0 \end{bmatrix} + \Gamma_1 \begin{bmatrix} 0 \\ 1 \end{bmatrix} = \begin{bmatrix} 1 \\ \Gamma_1 \end{bmatrix}$$

Again using Eq. (5.14) we have for $\mathbf{a}_2$,

$$\mathbf{a}_2 = \begin{bmatrix} 1 \\ a_2(1) \\ a_2(2) \end{bmatrix} = \begin{bmatrix} 1 \\ a_1(1) \\ 0 \end{bmatrix} + \Gamma_2 \begin{bmatrix} 0 \\ a_1^*(1) \\ 1 \end{bmatrix}$$

$$= \begin{bmatrix} 1 \\ \Gamma_1 \\ 0 \end{bmatrix} + \Gamma_2 \begin{bmatrix} 0 \\ \Gamma_1^* \\ 1 \end{bmatrix} = \begin{bmatrix} 1 \\ \Gamma_1 + \Gamma_1^*\Gamma_2 \\ \Gamma_2 \end{bmatrix}$$

Therefore, in terms of the reflection coefficients Γ_1 and Γ_2, the general form for the second-order all-pole model is

$$A_2(z) = 1 + (\Gamma_1 + \Gamma_1^*\Gamma_2)z^{-1} + \Gamma_2 z^{-2}$$

Given a second-order model, $\mathbf{a}_2$, note that if a third reflection coefficient, Γ_3, is appended to the sequence, then $\mathbf{a}_3$ may be easily found from $\mathbf{a}_2$ as follows

$$\mathbf{a}_3 = \begin{bmatrix} 1 \\ a_3(1) \\ a_3(2) \\ a_3(3) \end{bmatrix} = \begin{bmatrix} 1 \\ a_2(1) \\ a_2(2) \\ 0 \end{bmatrix} + \Gamma_3 \begin{bmatrix} 0 \\ a_2^*(2) \\ a_2^*(1) \\ 1 \end{bmatrix}$$

$$= \begin{bmatrix} 1 \\ a_2(1) + \Gamma_3 a_2^*(2) \\ a_2(2) + \Gamma_3 a_2^*(1) \\ \Gamma_3 \end{bmatrix} = \begin{bmatrix} 1 \\ (\Gamma_1 + \Gamma_1^*\Gamma_2) + \Gamma_3\Gamma_2^* \\ \Gamma_2 + \Gamma_3(\Gamma_1^* + \Gamma_1\Gamma_2^*) \\ \Gamma_3 \end{bmatrix}$$

Thus, it is always possible to increase the order of the filter without having to solve again for the lower-order filters.

The step-up recursion is a mapping, $\mathcal{L} : \mathbf{\Gamma}_p \longrightarrow \mathbf{a}_p$, from a reflection coefficient sequence $\mathbf{\Gamma}_p$ to a filter coefficient sequence $\mathbf{a}_p$. In most cases, this mapping may be inverted, $\mathcal{L}^{-1} : \mathbf{a}_p \longrightarrow \mathbf{\Gamma}_p$, and the reflection coefficients determined from the model parameters, $\mathbf{a}_p$. The mapping that accomplishes this is called the *step-down* or *backward Levinson* recursion. The procedure for finding the reflection coefficients is based on the fact that since

$$\Gamma_j = a_j(j) \tag{5.48}$$

then the reflection coefficients may be computed by running the Levinson-Durbin recursion *backwards*. Specifically, beginning with $\mathbf{a}_p$ we set $\Gamma_p = a_p(p)$. Then, we recursively find each of the lower-order models, $\mathbf{a}_j$, for $j = p-1, p-2, \ldots, 1$ and set $\Gamma_j = a_j(j)$ as illustrated below

$$\begin{array}{lllllll}
\mathbf{a}_p & = & a_p(1) & a_p(2) & \cdots & a_p(p-2) & a_p(p-1) \quad \boxed{\Gamma_p} \\
\mathbf{a}_{p-1} & = & a_{p-1}(1) & a_{p-1}(2) & \cdots & a_{p-1}(p-2) & \boxed{\Gamma_{p-1}} \\
\mathbf{a}_{p-2} & = & a_{p-2}(1) & a_{p-2}(2) & \cdots & \boxed{\Gamma_{p-2}} & \\
\vdots & & \vdots & \vdots & & & \\
\mathbf{a}_2 & = & a_2(1) & \boxed{\Gamma_2} & & & \\
\mathbf{a}_1 & = & \boxed{\Gamma_1} & & & &
\end{array}$$

To see how the jth-order model may be derived from the $(j+1)$st-order model, let us assume that the coefficients $a_{j+1}(i)$ are known. The Levinson order-update equation, Eq. (5.16), expresses $a_{j+1}(i)$ in terms of $a_j(i)$ and $a_j(j-i+1)$ as follows

$$a_{j+1}(i) = a_j(i) + \Gamma_{j+1} a_j^*(j-i+1) \tag{5.49}$$

which provides us with one equation in the two unknowns, $a_j(i)$ and $a_j^*(j-i+1)$. However, if we again use the Levinson order-update equation to express $a_{j+1}(j-i+1)$ in terms of $a_j(j-i+1)$ and $a_j^*(i)$ we have

$$a_{j+1}(j-i+1) = a_j(j-i+1) + \Gamma_{j+1} a_j^*(i)$$

which, after taking complex conjugates,

$$a_{j+1}^*(j-i+1) = a_j^*(j-i+1) + \Gamma_{j+1}^* a_j(i) \tag{5.50}$$

gives us a second linear equation in the same two unknowns. Writing Eqs. (5.49) and (5.50) in matrix form we have

$$\begin{bmatrix} a_{j+1}(i) \\ a_{j+1}^*(j-i+1) \end{bmatrix} = \begin{bmatrix} 1 & \Gamma_{j+1} \\ \Gamma_{j+1}^* & 1 \end{bmatrix} \begin{bmatrix} a_j(i) \\ a_j^*(j-i+1) \end{bmatrix} \tag{5.51}$$

If $|\Gamma_{j+1}| \neq 1$, then the matrix in Eq. (5.51) is invertible and we may solve uniquely for $a_j(i)$.[5] The solution is

$$\boxed{a_j(i) = \frac{1}{1-|\Gamma_{j+1}|^2}\Big[a_{j+1}(i) - \Gamma_{j+1} a_{j+1}^*(j-i+1)\Big]} \tag{5.52}$$

which is the *step-down recursion*. This recursion may also be written in vector form as follows

$$\begin{bmatrix} a_j(1) \\ a_j(2) \\ \vdots \\ a_j(j) \end{bmatrix} = \frac{1}{1-|\Gamma_{j+1}|^2} \left\{ \begin{bmatrix} a_{j+1}(1) \\ a_{j+1}(2) \\ \vdots \\ a_{j+1}(j) \end{bmatrix} - \Gamma_{j+1} \begin{bmatrix} a_{j+1}^*(j) \\ a_{j+1}^*(j-1) \\ \vdots \\ a_{j+1}^*(1) \end{bmatrix} \right\} \tag{5.53}$$

Once the sequence $a_j(i)$ has been found, we set $\Gamma_j = a_j(j)$ and the recursion continues by finding the next lower-order polynomial. The step-down recursion, represented pictorially

[5] If $|\Gamma_{j+1}| = 1$, then Eq. (5.51) is a singular set of equations and the mapping is not uniquely invertible.

as

$$\{a_p(1), a_p(2), \ldots, a_p(p), b(0)\} \overset{Step-down}{\longrightarrow} \{\Gamma_1, \Gamma_2, \ldots, \Gamma_p, \epsilon_p\}$$

is summarized in Table 5.3.

Another and perhaps more elegant way to derive the step-down recursion is to use Eq. (5.28) to solve for the jth-order polynomial $A_j(z)$ in terms of $A_{j+1}(z)$. The solution is easily seen to be

$$\begin{bmatrix} A_j(z) \\ A_j^R(z) \end{bmatrix} = \frac{1}{z^{-1}\left(1 - |\Gamma_{j+1}|^2\right)} \begin{bmatrix} z^{-1} & -\Gamma_{j+1} z^{-1} \\ -\Gamma_{j+1}^* & 1 \end{bmatrix} \begin{bmatrix} A_{j+1}(z) \\ A_{j+1}^R(z) \end{bmatrix} \tag{5.54}$$

Therefore,

$$\boxed{A_j(z) = \frac{1}{1 - |\Gamma_{j+1}|^2}\Big[A_{j+1}(z) - \Gamma_{j+1} A_{j+1}^R(z)\Big]} \tag{5.55}$$

which is the z-domain formulation of Eq. (5.52). A MATLAB program for the step-down recursion is given in Fig. 5.7.

Table 5.3 ***The Step-down Recursion***

1. Set $\Gamma_p = a_p(p)$
2. For $j = p-1, p-2, \ldots, 1$
   (a) For $i = 1, 2, \ldots, j$
   $$a_j(i) = \frac{1}{1 - |\Gamma_{j+1}|^2}\Big[a_{j+1}(i) - \Gamma_{j+1} a_{j+1}^*(j-i+1)\Big]$$
   (b) Set $\Gamma_j = a_j(j)$
   (c) If $|\Gamma_j| = 1$, Quit.
3. $\epsilon_p = b^2(0)$

The Step-Down Recursion

```
function gamma=atog(a)
%
   a=a(:);
   p=length(a);
   a=a(2:p)/a(1);
   gamma(p-1)=a(p-1);
   for j=p-1:-1:2;
      a=(a(1:j-1) - gamma(j)*flipud(conj(a(1:j-1))))./ ...
        (1 - abs(gamma(j))^2);
      gamma(j-1)=a(j-1);
      end
```

Figure 5.7 *A MATLAB program for the step-down recursion to find the reflection coefficients from a set of filter coefficients $a_p(k)$.*

Example 5.2.4 *The Step-Down Recursion*

Suppose that we would like to implement the third-order FIR filter

$$H(z) = 1 + 0.5z^{-1} - 0.1z^{-2} - 0.5z^{-3}$$

using a lattice filter structure. Using the step-down recursion, the reflection coefficients may be found from the vector

$$\mathbf{a}_3 = \begin{bmatrix} 1, & 0.5, & -0.1, & -0.5 \end{bmatrix}^T$$

and then implemented using the structure shown in Fig. 5.3. The step-down recursion begins by setting $\Gamma_3 = a_3(3) = -0.5$. Then, the coefficients of the second-order polynomial $\mathbf{a}_2 = [1, a_2(1), a_2(2)]^T$ are found from $a_3(i)$ using Eq. (5.53) with $j = 2$ as follows

$$\begin{bmatrix} a_2(1) \\ a_2(2) \end{bmatrix} = \frac{1}{1 - \Gamma_3^2} \left\{ \begin{bmatrix} a_3(1) \\ a_3(2) \end{bmatrix} - \Gamma_3 \begin{bmatrix} a_3(2) \\ a_3(1) \end{bmatrix} \right\}$$

From the given values for $\mathbf{a}_3$ and Γ_{j+1} we find

$$\begin{bmatrix} a_2(1) \\ a_2(2) \end{bmatrix} = \frac{1}{1 - 0.25} \left\{ \begin{bmatrix} 0.5 \\ -0.1 \end{bmatrix} + 0.5 \begin{bmatrix} -0.1 \\ 0.5 \end{bmatrix} \right\} = \begin{bmatrix} 0.6 \\ 0.2 \end{bmatrix}$$

and $\Gamma_2 = a_2(2) = 0.2$. Having found $\mathbf{a}_2$, Eq. (5.52) may again be used to find $\mathbf{a}_1$ as follows

$$a_1(1) = \frac{1}{1 - \Gamma_2^2} \left[a_2(1) - \Gamma_2 a_2(1) \right] = 0.5$$

Thus, $\Gamma_1 = a_1(1) = 0.5$ and the reflection coefficient sequence is

$$\mathbf{\Gamma} = \begin{bmatrix} 0.5, & 0.2, & -0.5 \end{bmatrix}^T$$

Therefore, a lattice filter having the given system function is as shown in Fig. 5.8.

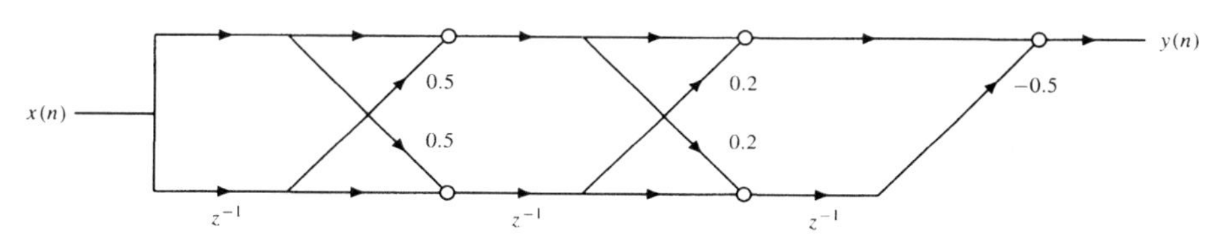

Figure 5.8 *A lattice filter implementation of the filter* $H(z) = 1 + 0.5z^{-1} - 0.1z^{-2} - 0.5z^{-3}$.

An interesting and useful application of the step-down recursion is the *Schur-Cohn stability test* for digital filters [27]. This test is based on Property 2 (p. 226), which states that the roots of a polynomial will lie inside the unit circle if and only if the magnitudes of the reflection coefficients are less than one. Therefore, given a causal, linear shift-invariant filter with a rational system function,

$$H(z) = \frac{B(z)}{A(z)}$$

the filter may be tested for stability as follows. First, the step-down recursion is applied to the coefficients of the denominator polynomial $A(z)$ to generate a reflection coefficient sequence, Γ_j. The filter will then be stable if and only if all of the reflection coefficients are less than one in magnitude. The following example illustrates the procedure.

Example 5.2.5 *The Schur-Cohn Stability Test*

Let us use the Schur-Cohn stability test to check the stability of the filter

$$H(z) = \frac{1}{2 + 4z^{-1} - 3z^{-2} + z^{-3}}$$

First note that the leading coefficient in the denominator polynomial is not equal to one. Therefore, we begin by rewriting $H(z)$ as follows

$$H(z) = \frac{0.5}{1 + 2z^{-1} - 1.5z^{-2} + 0.5z^{-3}}$$

With $\Gamma_3 = a_3(3) = 0.5$ we then solve for the second-order polynomial $A_2(z) = 1 + a_2(1)z^{-1} + a_2(2)z^{-2}$ using Eq. (5.53) as follows

$$\begin{aligned}
\begin{bmatrix} a_2(1) \\ a_2(2) \end{bmatrix} &= \frac{1}{1-\Gamma_3^2}\left\{\begin{bmatrix} a_3(1) \\ a_3(2) \end{bmatrix} - \Gamma_3 \begin{bmatrix} a_3(2) \\ a_3(1) \end{bmatrix}\right\} \\
&= \frac{4}{3}\left\{\begin{bmatrix} 2 \\ -1.5 \end{bmatrix} - 0.5 \begin{bmatrix} -1.5 \\ 2 \end{bmatrix}\right\} \\
&= \begin{bmatrix} 11/3 \\ -10/3 \end{bmatrix}
\end{aligned}$$

Therefore, since $\Gamma_2 = a_2(2) = -10/3$, then $|\Gamma_2| > 1$ and the filter is *unstable*.

5.2.5 The Inverse Levinson-Durbin Recursion

We have seen how to derive the reflection coefficients and the model parameters from an autocorrelation sequence. We have also derived recursions that map a sequence of reflection coefficients into a sequence of model parameters and vice versa. It is also possible to recursively compute the autocorrelation sequence from the reflection coefficients $\{\Gamma_1, \Gamma_2, \ldots, \Gamma_p\}$ and ϵ_p or from the coefficients of the pth-order model $a_p(k)$ and $b(0)$. The recursion for doing this is referred to as the *inverse Levinson recursion.* To see how the recursion works, assume that we are given the reflection coefficient sequence Γ_j for $j = 1, \ldots, p$ and the pth-order error ϵ_p. Since

$$\epsilon_p = r_x(0) \prod_{i=1}^{p} (1 - |\Gamma_i|^2) \tag{5.56}$$

then the first term in the autocorrelation sequence is

$$\boxed{r_x(0) = \frac{\epsilon_p}{\prod_{i=1}^{p} (1 - |\Gamma_i|^2)}} \tag{5.57}$$

Table 5.4 ***The Inverse Levinson-Durbin Recursion***

1. Initialize the recursion
   (a) $r_x(0) = \epsilon_p / \prod_{i=1}^{p}(1 - |\Gamma_i|^2)$
   (b) $a_0(0) = 1$
2. For $j = 0, 1, \ldots, p-1$
   (a) For $i = 1, 2, \ldots, j$
       $a_{j+1}(i) = a_j(i) + \Gamma_{j+1} a_j^*(j-i+1)$
   (b) $a_{j+1}(j+1) = \Gamma_{j+1}$
   (c) $r_x(j+1) = -\sum_{i=1}^{j+1} a_{j+1}(i) r_x(j+1-i)$
3. Done

This, along with the zeroth-order model

$$\boxed{\mathbf{a}_0 = 1} \qquad (5.58)$$

initializes the recursion.

Now suppose that the first j terms of the autocorrelation sequence are known, along with the jth-order filter coefficients $a_j(i)$. We will now show how to find the next term in the sequence, $r_x(j+1)$. First we use the step-up recursion to find the coefficients $a_{j+1}(i)$ from Γ_{j+1} and $a_j(i)$. Then, setting $p = j+1$ and $k = j+1$ in Eq. (5.2) we have the following expression for $r_x(j+1)$:

$$\boxed{r_x(j+1) = -\sum_{i=1}^{j+1} a_{j+1}(i) r_x(j+1-i)} \qquad (5.59)$$

Since the sum on the right only involves known autocorrelation values, this expression may be used to determine $r_x(j+1)$, which completes the recursion. The inverse Levinson-Durbin recursion, which may be represented pictorially as

$$\{\Gamma_1, \Gamma_2, \ldots, \Gamma_p, \epsilon_p\} \overset{(LEV)^{-1}}{\longrightarrow} \{r_x(0), r_x(1), \ldots, r_x(p)\}$$

is summarized in Table 5.4.

Example 5.2.6 ***The Inverse Levinson-Durbin Recursion***

Given the sequence of reflection coefficients, $\Gamma_1 = \Gamma_2 = \Gamma_3 = \frac{1}{2}$ and a model error of $\epsilon_3 = 2(\frac{3}{4})^3$, let us find the autocorrelation sequence $\mathbf{r}_x = [r_x(0), r_x(1), r_x(2), r_x(3)]^T$. Initializing the recursion with

$$r_x(0) = \frac{\epsilon_3}{\prod_{i=1}^{3}(1 - \Gamma_i^2)} = 2$$

we begin by finding the first-order model, which is given by

$$\mathbf{a}_1 = \begin{bmatrix} 1 \\ \Gamma_1 \end{bmatrix} = \begin{bmatrix} 1 \\ 0.5 \end{bmatrix}$$

Therefore,

$$r_x(1) = -r_x(0)\Gamma_1 = -1.0$$

Updating the model parameters using the step-up recursion we have

$$\mathbf{a}_2 = \begin{bmatrix} 1 \\ a_1(1) \\ 0 \end{bmatrix} + \Gamma_2 \begin{bmatrix} 0 \\ a_1(1) \\ 1 \end{bmatrix} = \begin{bmatrix} 1 \\ 1/2 \\ 0 \end{bmatrix} + \frac{1}{2}\begin{bmatrix} 0 \\ 1/2 \\ 1 \end{bmatrix} = \begin{bmatrix} 1 \\ 3/4 \\ 1/2 \end{bmatrix}$$

Thus,

$$r_x(2) = -a_2(1)r_x(1) - a_2(2)r_x(0) = 3/4 - 1 = -1/4$$

Applying the step-up recursion to $\mathbf{a}_2$ we have

$$\mathbf{a}_3 = \begin{bmatrix} 1 \\ a_2(1) \\ a_2(2) \\ 0 \end{bmatrix} + \Gamma_3 \begin{bmatrix} 0 \\ a_2(2) \\ a_2(1) \\ 1 \end{bmatrix} = \begin{bmatrix} 1 \\ 3/4 \\ 1/2 \\ 0 \end{bmatrix} + \frac{1}{2}\begin{bmatrix} 0 \\ 1/2 \\ 3/4 \\ 1 \end{bmatrix} = \begin{bmatrix} 1 \\ 1 \\ 7/8 \\ 1/2 \end{bmatrix}$$

Finally, for $r_x(3)$ we have

$$r_x(3) = -a_3(1)r_x(2) - a_3(2)r_x(1) - a_3(3)r_x(0) = 1/4 + 7/8 - 1 = 1/8$$

and the autocorrelation sequence is

$$\mathbf{r}_x = \left[2, \ -1, \ -1/4, \ 1/8\right]^T$$

The inverse Levinson recursion described above provides a procedure for finding the autocorrelation sequence from the reflection coefficients and ϵ_p. If, instead of the reflection coefficients, we were given the filter coefficients $a_p(k)$, then the autocorrelation sequence may still be determined by using the step-down recursion to find the sequence of reflection coefficients and then using the inverse Levinson recursion. MATLAB programs for the inverse Levinson recursion to find $r_x(k)$ from either the reflection coefficients Γ_j or the filter coefficients $a_p(k)$ are given in Fig. 5.9.

In summary, combining the results of Sections 5.2.4 and 5.2.5 we see that there is an equivalence between three sets of parameters: the autocorrelation sequence $r_x(k)$, the all-pole model $a_p(k)$ and $b(0)$, and the reflection coefficients Γ_j along with ϵ_p. This equivalence is illustrated in Fig. 5.10, which shows how one set of parameters may be derived from another.

*5.2.6 The Schur Recursion**

The Levinson-Durbin recursion provides an efficient solution to the autocorrelation normal equations. For a pth-order model, this recursion requires on the order of p^2 arithmetic operations compared to an order of p^3 operations for Gaussian elimination. Therefore, with a single processor that takes one unit of time for an arithmetic operation, the Levinson-Durbin recursion would require on the order of p^2 units of time to solve the normal equations. With the increasing use of VLSI technology and parallel processing architectures

The Inverse Levinson-Durbin Recursions

```
function r=gtor(gamma,epsilon)
%
   p=length(gamma);
   aa=gamma(1);
   r=[1 -gamma(1)];
   for j=2:p;
      aa=[aa;0]+gamma(j)*[conj(flipud(aa));1];
      r=[r -fliplr(r)*aa];
      end;
   if nargin == 2,
   r = r*epsilon/prod(1-abs(gamma).^2);
   end;

function r=ator(a,b)
%
   p=length(a)-1;
   gamma=atog(a);
   r=gtor(gamma);
   if nargin == 2,
   r = r*sqrt(b)/prod(1-abs(gamma).^2);
   end;
```

Figure 5.9 MATLAB *programs to find the autocorrelation sequence $r_x(k)$ from either the reflection coefficients Γ_j and the modeling error ϵ_p, or the filter coefficients $a_p(k)$ and $b(0)$.*

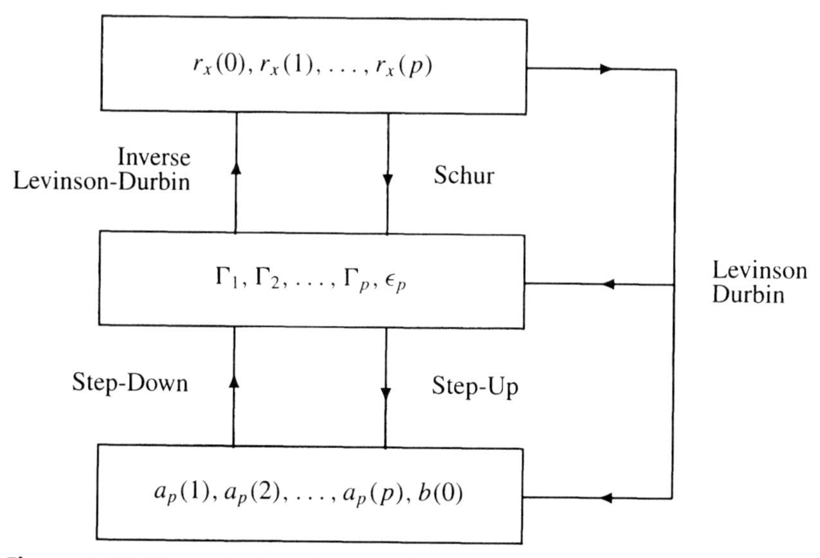

Figure 5.10 *The equivalence between the autocorrelation sequence, the all-pole model, and the reflection coefficients.*

in signal processing applications, an interesting question to ask is how much time would be necessary to solve the normal equations if we were to use p processors in parallel. Although one would hope that the computation time would be proportional to p rather than p^2, this is not the case. In fact, it is not difficult to see that with p processors the

Levinson-Durbin recursion would solve the normal equations in a time that is proportional to $p \log_2 p$ instead of p. The reason for this is the requirement to compute the inner product

$$\gamma_j = r_x(j+1) + \sum_{i=1}^{j} a_j(i) r_x(j+1-i) = \left[r_x(j+1),\ r_x(j),\ \ldots,\ r_x(1) \right] \begin{bmatrix} 1 \\ a_j(1) \\ \vdots \\ a_j(j) \end{bmatrix} \tag{5.60}$$

which is the primary computational bottleneck in the Levinson-Durbin recursion. Although the j multiplications may be performed in one unit of time using j processors, the j additions require $\log_2 j$ units of time. Therefore, with p processors, finding the solution to the pth-order normal equations requires on the order of $p \log_2 p$ units of time.

In this section, we derive another algorithm for solving the normal equations known as the *Schur recursion*. This recursion, originally presented in 1917 as a procedure for testing a polynomial to see if it is analytic and bounded in the unit disk [9, 22], avoids the inner product in the calculation of γ_j and is therefore better suited to parallel processing. Unlike the Levinson-Durbin recursion, which finds the filter coefficients $a_p(k)$ in addition to the reflection coefficients Γ_j, the Schur recursion only solves for the reflection coefficients. Thus, the Schur recursion is a mapping from an autocorrelation sequence $r_x(k)$ to a reflection coefficient sequence

$$\left\{ r_x(0),\ r_x(1),\ \ldots, r_x(p) \right\} \stackrel{Schur}{\longrightarrow} \left\{ \Gamma_1,\ \Gamma_2,\ \ldots,\ \Gamma_p,\ \epsilon_p \right\} \tag{5.61}$$

By avoiding the computation of the filter coefficients, $a_p(k)$, the Schur recursion is slightly more efficient than the Levinson-Durbin recursion in terms of the number of multiplications, even in a single processor implementation of the algorithm.

The development of the Schur recursion begins with the autocorrelation normal equations for a jth-order model

$$r_x(k) + \sum_{l=1}^{j} a_j(l) r_x(k-l) = 0 \quad ; \quad k = 1, 2, \ldots, j \tag{5.62}$$

If we set $a_j(0) = 1$ then the jth-order normal equations may be written as

$$\sum_{l=0}^{j} a_j(l) r_x(k-l) = 0 \quad ; \quad k = 1, 2, \ldots, j \tag{5.63}$$

which we will refer to as the *orthogonality condition*. Note that the left side of Eq. (5.63) is the convolution of the finite length sequence $a_j(k)$ with the autocorrelation sequence $r_x(k)$. Therefore, let us define $g_j(k)$ to be the sequence that is formed by convolving $a_j(k)$ with $r_x(k)$,

$$\boxed{g_j(k) = \sum_{l=0}^{j} a_j(l) r_x(k-l) = a_j(k) * r_x(k)} \tag{5.64}$$

Thus, as shown in Fig. 5.11, $g_j(k)$ is the output of the jth-order prediction error filter $A_j(z)$ when the input is the autocorrelation sequence $r_x(k)$ and the orthogonality condition states

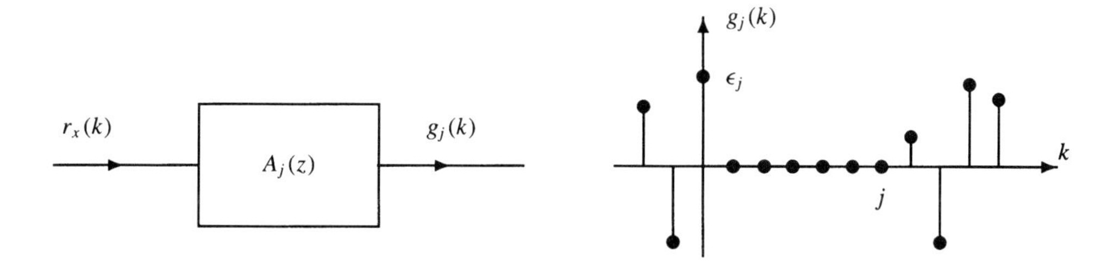

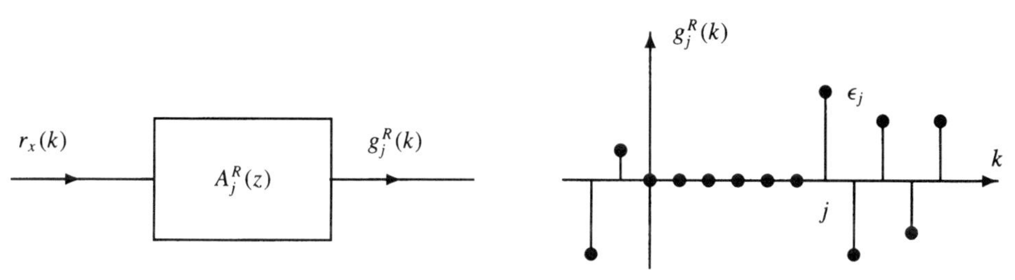

Figure 5.11 *The sequences $g_j(k)$ and $g_j^R(k)$ expressed as the output of the filter $A_j(z)$ and $A_j^R(z)$, respectively, when the input is the autocorrelation sequence $r_x(k)$. Due to the orthogonality condition, $g_j(k) = 0$ for $k = 1, \ldots, j$ and $g_j^R(k) = 0$ for $k = 0, \ldots, j-1$.*

that $g_j(k)$ is equal to zero for $k = 1, \ldots, j$,

$$g_j(k) = 0 \quad ; \quad k = 1, 2, \ldots, j \tag{5.65}$$

In addition, since

$$g_j(0) = \sum_{l=0}^{j} a_j(l) r_x(l) = \epsilon_j \tag{5.66}$$

where the last equality follows from Eq. (5.3), then $g_j(0)$ is equal to the jth-order modeling (prediction) error.

Now consider what happens when $a_j(k)$ is replaced with $a_j^R(k)$ and $g_j^R(k)$ is the convolution of $a_j^R(k)$ with $r_x(k)$,

$$\boxed{g_j^R(k) = \sum_{l=0}^{j} a_j^R(l) r_x(k-l) = a_j^R(k) * r_x(k)} \tag{5.67}$$

Thus, $g_j^R(k)$ is the response of the filter $A_j^R(z)$ to the input $r_x(k)$. Since $a_j^R(k) = a_j^*(j-k)$, then

$$g_j^R(k) = \sum_{l=0}^{j} a_j^*(j-l) r_x(k-l) = \sum_{l=0}^{j} a_j^*(l) r_x(k-j+l) \tag{5.68}$$

Using the conjugate symmetry of the autocorrelation sequence, Eq. (5.68) becomes

$$g_j^R(k) = \sum_{l=0}^{j} a_j^*(l) r_x^*([j-k]-l) = g_j^*(j-k) \tag{5.69}$$

Therefore, it follows from Eqs. (5.65) and (5.66) that

$$g_j^R(k) = 0 \quad ; \quad k = 0, 1, \ldots, j-1 \tag{5.70}$$

and

$$g_j^R(j) = \epsilon_j \tag{5.71}$$

These constraints on $g_j^R(k)$ are illustrated in Fig. 5.11.

The next step is to use the Levinson Durbin recursion to show how the sequences $g_j(k)$ and $g_j^R(k)$ may be updated to form $g_{j+1}(k)$ and $g_{j+1}^R(k)$. Since $g_{j+1}(k) = a_{j+1}(k) * r_x(k)$, using the Levinson order-update equation for $a_{j+1}(k)$ we have

$$\begin{aligned} g_{j+1}(k) &= a_{j+1}(k) * r_x(k) = \left[a_j(k) + \Gamma_{j+1} a_j^R(k-1)\right] * r_x(k) \\ &= g_j(k) + \Gamma_{j+1} g_j^R(k-1) \end{aligned} \tag{5.72}$$

In a similar fashion, since $g_{j+1}^R(k) = a_{j+1}^R(k) * r_x(k)$, application of the Levinson order-update equation gives

$$\begin{aligned} g_{j+1}^R(k) &= a_{j+1}^R(k) * r_x(k) = \left[a_j^R(k-1) + \Gamma_{j+1}^* a_j(k)\right] * r_x(k) \\ &= g_j^R(k-1) + \Gamma_{j+1}^* g_j(k) \end{aligned} \tag{5.73}$$

Equations (5.72) and (5.73) are recursive update equations for $g_j(k)$ and $g_j^R(k)$ and are identical in form to the update equations given in Eq. (5.27) for $a_j(k)$ and $a_j^R(k)$. The only difference between the two recursions is in the initial condition. Specifically, for $a_j(k)$ we have $a_0(k) = a_0^R(k) = \delta(k)$, whereas for $g_j(k)$ we have $g_0(k) = g_0^R(k) = r_x(k)$. Taking the z-transform of Eqs. (5.72) and (5.73) and putting them in matrix form gives

$$\begin{bmatrix} G_{j+1}(z) \\ G_{j+1}^R(z) \end{bmatrix} = \begin{bmatrix} 1 & \Gamma_{j+1} z^{-1} \\ \Gamma_{j+1}^* & z^{-1} \end{bmatrix} \begin{bmatrix} G_j(z) \\ G_j^R(z) \end{bmatrix} \tag{5.74}$$

A lattice filter interpretation of these update equations is shown in Fig. 5.12.

Recall that our goal is to derive a recursion that will take a sequence of autocorrelation values and generate the corresponding sequence of reflection coefficients. Given the first j reflection coefficients, the lattice filter shown in Fig. 5.12 allows one to determine the sequences $g_j(k)$ and $g_j^R(k)$ from the autocorrelation sequence. All that remains is to derive a method to find the reflection coefficient, Γ_{j+1}, from $g_j(k)$ and $g_j^R(k)$. This may be done as follows. Since $g_{j+1}(j+1) = 0$, evaluating Eq. (5.72) for $k = j+1$ we have

$$g_{j+1}(j+1) = g_j(j+1) + \Gamma_{j+1} g_j^R(j) = 0$$

Therefore,

$$\boxed{\Gamma_{j+1} = -\frac{g_j(j+1)}{g_j^R(j)}} \tag{5.75}$$

and the recursion is complete. In summary, the Schur recursion begins by initializing $g_0(k)$ and $g_0^R(k)$ for $k = 1, 2, \ldots, p$ with $r_x(k)$,

$$g_0(k) = g_0^R(k) = r_x(k)$$

The first reflection coefficient is then computed from the ratio

$$\Gamma_1 = -\frac{g_0(1)}{g_0^R(0)} = -\frac{r_x(1)}{r_x(0)}$$

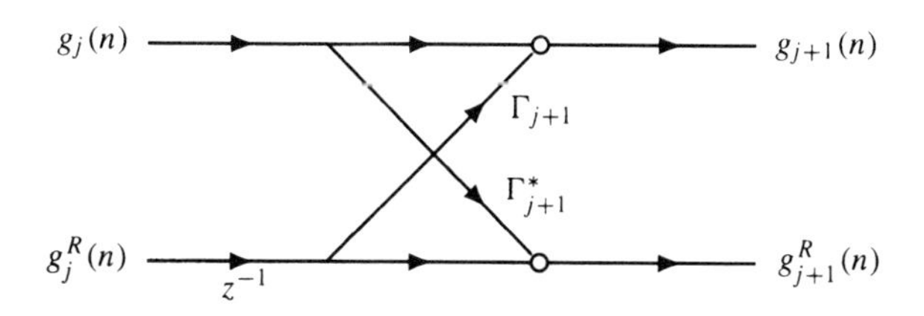

(a) Lattice filter module for updating $g_j(k)$ and $g_j^R(k)$.

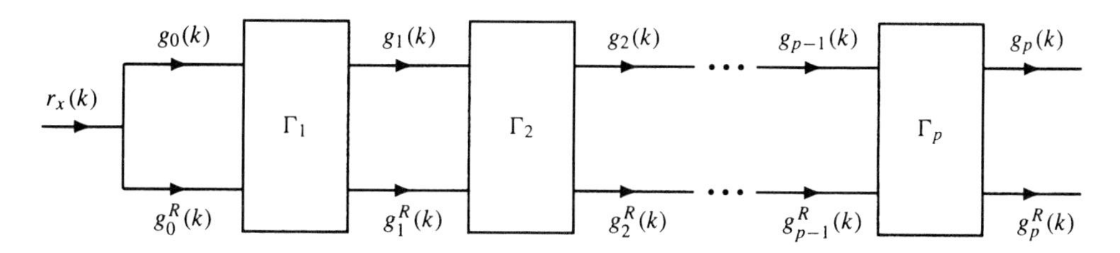

(b) A pth-order lattice filter for generating $g_p(k)$ and $g_p^R(k)$ from $r_x(k)$.

Figure 5.12 *The lattice filter used in the Schur recursion to generate the sequences $g_p(k)$ and $g_p^R(k)$ from the autocorrelation sequence.*

Given Γ_1, a lattice filter is then used to generate the sequences $g_1(k)$ and $g_1^R(k)$. The second reflection coefficient is then computed from the ratio

$$\Gamma_2 = -\frac{g_1(2)}{g_1^R(1)}$$

which, in turn, is used in the lattice filter to generate the sequences $g_2(k)$ and $g_2^R(k)$ and so on. In updating the sequences $g_{j+1}(k)$ and $g_{j+1}^R(k)$, note that since $g_{j+1}(k) = 0$ for $k = 1, 2, \ldots, j+1$ and $g_{j+1}^R(k) = 0$ for $k = 0, 1, \ldots, j$, not all p values need to be updated. Specifically, all that is required is to evaluate $g_{j+1}(k)$ for $k > j+1$ and $g_{j+1}^R(k)$ for $k > j$. The complete recursion is given in Table 5.5.

Additional insight into the operation of the Schur recursion may be gained if it is formulated using vector notation. Let $\mathbf{g}_j$ and $\mathbf{g}_j^R$ be the vectors that contain the sequences $g_j(k)$ and $g_j^R(k)$, respectively, and consider the $2 \times (p+1)$ matrix that has $\mathbf{g}_j$ in the first

Table 5.5 ***The Schur Recursion***

1. Set $g_0(k) = g_0^R(k) = r_x(k)$ for $k = 0, 1, \ldots, p$.
2. For $j = 0, 1, \ldots, p-1$
   (a) Set $\Gamma_{j+1} = -g_j(j+1)/g_j^R(j)$
   (b) For $k = j+2, \ldots, p$
   $$g_{j+1}(k) = g_j(k) + \Gamma_{j+1} g_j^R(k-1)$$
   (c) For $k = j+1, \ldots, p$
   $$g_{j+1}^R(k) = g_j^R(k-1) + \Gamma_{j+1}^* g_j(k)$$
3. $\epsilon_p = g_p^R(p)$

row and $\mathbf{g}_j^R$ in the second row

$$\begin{bmatrix} \mathbf{g}_j^T \\ (\mathbf{g}_j^R)^T \end{bmatrix} = \begin{bmatrix} g_j(0) & g_j(1) & \cdots & g_j(p) \\ g_j^R(0) & g_j^R(1) & \cdots & g_j^R(p) \end{bmatrix}$$

From Eqs. (5.65), (5.66), and (5.70), we see that this matrix has the form[6]

$$\begin{bmatrix} \mathbf{g}_j^T \\ (\mathbf{g}_j^R)^T \end{bmatrix} = \begin{bmatrix} \epsilon_j & \cdots & 0 & 0 & g_j(j+1) & \cdots & g_j(p) \\ 0 & \cdots & 0 & g_j^R(j) & g_j^R(j+1) & \cdots & g_j^R(p) \end{bmatrix} \tag{5.76}$$

We are now going to perform a sequence of operations on this matrix that correspond to those expressed in the z-domain by Eq. (5.74). These operations will comprise one iteration of the Schur recursion. The first operation is to shift the second row of the matrix to the right by one with $g_j^R(-1)$ entering as the first element of the second row (this corresponds to a delay of $g_j^R(k)$ by one or a multiplication of $G_j^R(z)$ by z^{-1}). Note that after this shift, the ratio of the two terms in column number $(j+2)$ is equal to $-\Gamma_{j+1}$. Therefore, evaluating the reflection coefficient we may then form the matrix

$$\boldsymbol{\Theta}_{j+1} = \begin{bmatrix} 1 & \Gamma_{j+1} \\ \Gamma_{j+1}^* & 1 \end{bmatrix} \tag{5.77}$$

and multiply the shifted matrix by $\boldsymbol{\Theta}_{j+1}$ to generate the updated sequences $\mathbf{g}_{j+1}$ and $\mathbf{g}_{j+1}^R$ as follows

$$\begin{aligned} \begin{bmatrix} \mathbf{g}_{j+1}^T \\ (\mathbf{g}_{j+1}^R)^T \end{bmatrix} &= \begin{bmatrix} 1 & \Gamma_{j+1} \\ \Gamma_{j+1}^* & 1 \end{bmatrix} \begin{bmatrix} \epsilon_j & \cdots & 0 & g_j(j+1) & \cdots & g_j(p) \\ g_j^R(-1) & \cdots & 0 & g_j^R(j) & \cdots & g_j^R(p-1) \end{bmatrix} \\ &= \begin{bmatrix} \epsilon_{j+1} & \cdots & 0 & 0 & g_{j+1}(j+2) & \cdots & g_{j+1}(p) \\ 0 & \cdots & 0 & g_{j+1}^R(j+1) & g_{j+1}^R(j+2) & \cdots & g_{j+1}^R(p) \end{bmatrix} \end{aligned} \tag{5.78}$$

This completes one step of the recursion. We may, however, simplify these operations slightly by noting that the first column of the matrix in Eq. (5.76) never enters into the calculations. Therefore, we may suppress the evaluation of these entries by initializing the first element in the first column to zero and by bringing in a zero into the second row each time that it is shifted to the right. Note that since $\epsilon_j = g_j^R(j)$, then the error is not discarded with this simplification. In summary, the steps described above lead to the following matrix formulation of the Schur recursion. Beginning with

$$\mathbf{G}_0 = \begin{bmatrix} 0 & r_x(1) & r_x(2) & \cdots & r_x(p) \\ r_x(0) & r_x(1) & r_x(2) & \cdots & r_x(p) \end{bmatrix} \tag{5.79}$$

which is referred to as the *generator matrix*, a new matrix, $\widetilde{\mathbf{G}}_0$, is formed by shifting the second row of $\mathbf{G}_0$ to the right by one

$$\widetilde{\mathbf{G}}_0 = \begin{bmatrix} 0 & r_x(1) & r_x(2) & \cdots & r_x(p) \\ 0 & r_x(0) & r_x(1) & \cdots & r_x(p-1) \end{bmatrix}$$

Setting Γ_1 equal to the negative of the ratio of the two terms in the second column of $\widetilde{\mathbf{G}}_0$,

[6]We could also use Eq. (5.71) to set $g_j^R(j) = \epsilon_j$ in this matrix, but this would make the next step of computing the reflection coefficient Γ_{j+1} using Eq. (5.75) a little more obscure.

we then form the matrix

$$\mathbf{\Theta}_1 = \begin{bmatrix} 1 & \Gamma_1 \\ \Gamma_1^* & 1 \end{bmatrix}$$

and evaluate $\mathbf{G}_1$ as follows

$$\mathbf{G}_1 = \mathbf{\Theta}_1 \widetilde{\mathbf{G}}_0 = \begin{bmatrix} 0 & 0 & g_1(2) & \cdots & g_1(p) \\ 0 & g_1^R(1) & g_1^R(2) & \cdots & g_1^R(p) \end{bmatrix} \tag{5.80}$$

The recursion then repeats these three steps where, in general, at the jth step we

1. Shift the second row of $\mathbf{G}_j$ to the right by one,
2. Compute Γ_{j+1} as the negative of the ratio of the two terms in the $(j+2)$nd column,
3. Multiply $\widetilde{\mathbf{G}}_j$ by $\mathbf{\Theta}_{j+1}$ to form $\mathbf{G}_{j+1}$.

The following example illustrates the procedure.

Example 5.2.7 ***The Schur Recursion***

Given the autocorrelation sequence $\mathbf{r}_x = [2, \ -1, \ -1/4, \ 1/8]^T$, let us use the Schur recursion to find the reflection coefficients Γ_1, Γ_2, and Γ_3.

1. **Step 1:** We begin the Schur recursion by initializing the generator matrix $\mathbf{G}_0$ to

$$\mathbf{G}_0 = \begin{bmatrix} 0 & r_x(1) & r_x(2) & r_x(3) \\ r_x(0) & r_x(1) & r_x(2) & r_x(3) \end{bmatrix} = \begin{bmatrix} 0 & -1 & -1/4 & 1/8 \\ 2 & -1 & -1/4 & 1/8 \end{bmatrix}$$

2. **Step 2:** Forming the shifted matrix

$$\widetilde{\mathbf{G}}_0 = \begin{bmatrix} 0 & -1 & -1/4 & 1/8 \\ 0 & 2 & -1 & -1/4 \end{bmatrix}$$

it follows that $\Gamma_1 = 0.5$ and

$$\mathbf{\Theta}_1 = \begin{bmatrix} 1 & 0.5 \\ 0.5 & 1 \end{bmatrix}$$

This step only requires one division.

3. **Step 3:** Forming the product $\mathbf{\Theta}_1 \widetilde{\mathbf{G}}_0$ we find

$$\mathbf{G}_1 = \mathbf{\Theta}_1 \widetilde{\mathbf{G}}_0 = \begin{bmatrix} 1 & 0.5 \\ 0.5 & 1 \end{bmatrix} \begin{bmatrix} 0 & -1 & -1/4 & 1/8 \\ 0 & 2 & -1 & -1/4 \end{bmatrix}$$

$$= \begin{bmatrix} 0 & 0 & -3/4 & 0 \\ 0 & 3/2 & -9/8 & -3/16 \end{bmatrix}$$

Since we know that the second element in the first row of $\mathbf{G}_1$ will be equal to zero after multiplying $\widetilde{\mathbf{G}}_0$ by $\mathbf{\Theta}_1$, this step of the recursion only requires five multiplications.

4. **Step 4:** From the shifted matrix

$$\widetilde{\mathbf{G}}_1 = \begin{bmatrix} 0 & 0 & -3/4 & 0 \\ 0 & 0 & 3/2 & -9/8 \end{bmatrix}$$

we see that $\Gamma_2 = 0.5$ and

$$\mathbf{\Theta}_2 = \begin{bmatrix} 1 & 0.5 \\ 0.5 & 1 \end{bmatrix}$$

Again, in this step, we have one division.

5. **Step 5:** Forming the product $\Theta_2\widetilde{\mathbf{G}}_1$ we have

$$\mathbf{G}_2 = \Theta_2\widetilde{\mathbf{G}}_1 = \begin{bmatrix} 1 & 0.5 \\ 0.5 & 1 \end{bmatrix}\begin{bmatrix} 0 & 0 & -3/4 & 0 \\ 0 & 0 & 3/2 & -9/8 \end{bmatrix} = \begin{bmatrix} 0 & 0 & 0 & -9/16 \\ 0 & 0 & 9/8 & -9/8 \end{bmatrix}$$

   For this step we have three multiplications.

6. **Step 6:** Finally, forming the shifted matrix $\widetilde{\mathbf{G}}_2$,

$$\widetilde{\mathbf{G}}_2 = \begin{bmatrix} 0 & 0 & 0 & -9/16 \\ 0 & 0 & 0 & 9/8 \end{bmatrix}$$

   we find that $\Gamma_3 = 0.5$,

$$\Theta_3 = \begin{bmatrix} 1 & 0.5 \\ 0.5 & 1 \end{bmatrix}$$

   and

$$\mathbf{G}_3 = \Theta_3\widetilde{\mathbf{G}}_2 = \begin{bmatrix} 1 & 0.5 \\ 0.5 & 1 \end{bmatrix}\begin{bmatrix} 0 & 0 & 0 & -9/16 \\ 0 & 0 & 0 & 9/8 \end{bmatrix} = \begin{bmatrix} 0 & 0 & 0 & 0 \\ 0 & 0 & 0 & 27/32 \end{bmatrix}$$

   Therefore, the error is

$$\epsilon_3 = 27/32$$

   In this last step, we have one division and one multiplication.

Counting the number of arithmetic operations we find that this recursion required 9 multiplications and 3 divisions.

Unlike the Levinson-Durbin recursion, the Schur recursion is well suited to parallel implementation. In particular, Kung and Hu [13] developed the pipelined structure shown in Fig. 5.13 for implementing the Schur recursion. This structure consists of a cascade of p modular cells, each containing an upper processing unit that computes the sequence values $g_j(k)$ and a lower processing unit that computes the values $g_j^R(k)$. The upper processing unit in cell 1 differs from the others in that it is a *divider cell* for computing the reflection coefficients Γ_j. The operation of this pipelined architecture is as follows.

1. The upper processing units are initialized with the autocorrelation values $r_x(1), \ldots, r_x(p)$ and the lower units with $r_x(0), \ldots, r_x(p-1)$ as shown in Fig. 5.13*a*. This step loads the cells with the values of the matrix $\widetilde{\mathbf{G}}_0$.
2. The divider cell computes the first reflection coefficient by dividing the contents in the upper unit of cell 1 with the contents of the lower unit,

$$\Gamma_1 = -r_x(1)/r_x(0)$$

   This reflection coefficient is then propagated simultaneously to each of the other cells.

3. The upper units of each cell are updated as follows

$$g_1(k) = r_x(k) + \Gamma_1 r_x(k-1) \quad ; \quad k = 1, \ldots, p$$

   and the lower units updated as

$$g_1^R(k) = r_x(k-1) + \Gamma_1^* r_x(k) \quad ; \quad k = 1, \ldots, p$$

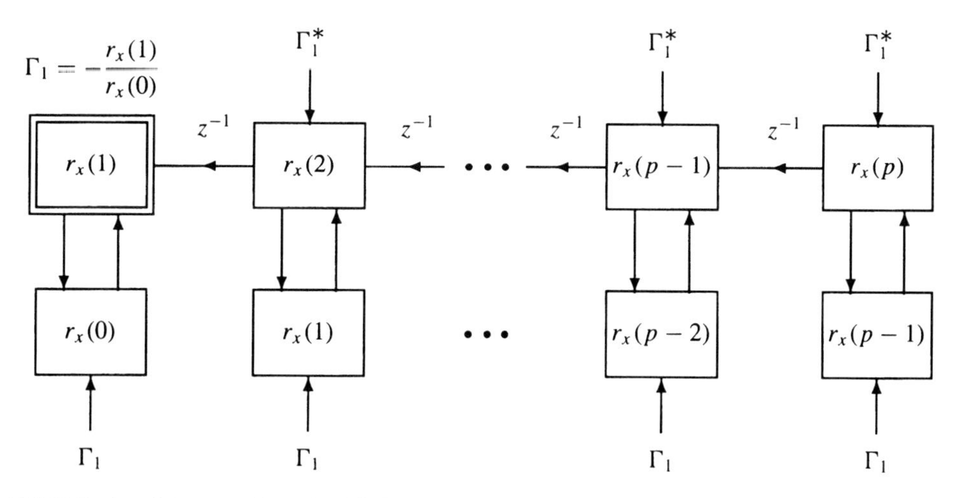

(a) Initialization of the processing units with the autocorrelation sequence $r_x(k)$. The divider cell computes the first reflection coefficient Γ_1 which is then simultaneously propagated to each of the other units.

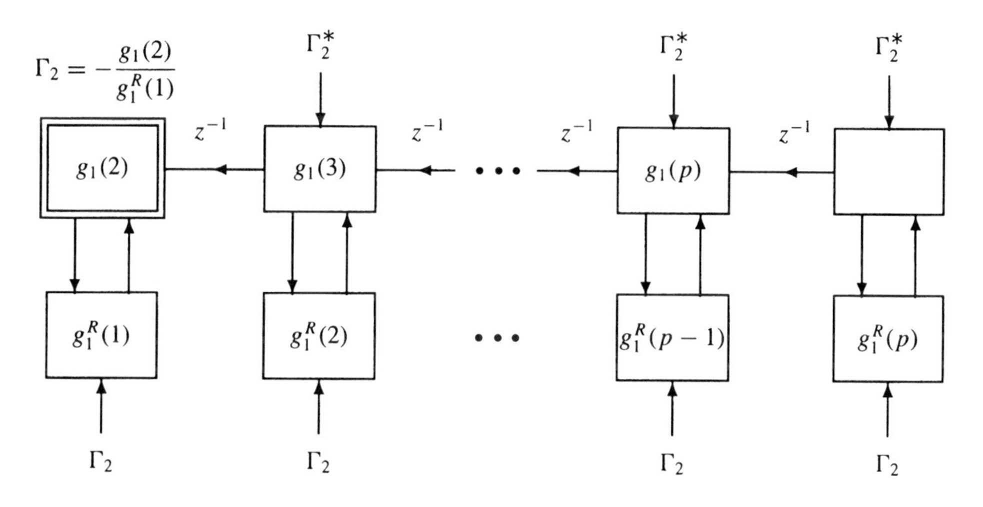

(b) State of the Schur pipeline after the first update.

Figure 5.13 *A pipelined structure for implementing the Schur recursion.*

This operation corresponds to the updating of $\widetilde{\mathbf{G}}_0$ by multiplying by $\mathbf{\Theta}_1$ to form the new generator matrix $\mathbf{G}_1$.

4. The contents of the upper units in each cell are then shifted to the left by one cell. Note that the last cell in the top row is left vacated.
5. The next reflection coefficient is then computed in the divider cell by dividing the contents in the upper unit of cell 1 with the contents of the lower unit. This reflection coefficient is then propagated simultaneously to each of the other cells. At this point, the pipeline structure is in the state shown in Fig. 5.13*b*.
6. The upper units of each cell are again updated using

$$g_{j+1}(k) = g_j(k) + \Gamma_{j+1} g_j^R(k-1)$$

and the lower units using

$$g_{j+1}^R(k) = g_j^R(k-1) + \Gamma_{j+1}^* g_j(k)$$

Steps 4, 5 and 6 are repeated until all of the reflection coefficients have been computed.

We conclude this discussion of the Schur recursion with an evaluation of the computational requirements of the algorithm. Note that at the jth stage of the recursion, one division is required to determine Γ_{j+1} and $p-j-1$ multiplications along with $p-j-1$ additions are necessary to evaluate the terms in the sequence $g_{j+1}(k)$ (step 2b in Table 5.5). In addition, $(p-j)$ multiplications and $(p-j)$ additions are necessary in order to evaluate the terms in the sequence $g_{j+1}^R(k)$ (step 2c in Table 5.5). With p steps in the recursion, the total number of multiplications and divisions is

$$p + 2\sum_{j=0}^{p-1}(p-j) - p = p^2 + p$$

along with p^2 additions. Thus, in comparing the Schur recursion to the Levinson-Durbin recursion (p. 223), we see that although the same number of additions are required with the Schur recursion, there are p fewer multiplications.

5.2.7 The Cholesky Decomposition

We have seen that the Levinson-Durbin recursion is an efficient algorithm for solving the autocorrelation normal equations. It may also be used to perform a Cholesky decomposition of the Hermitian Toeplitz autocorrelation matrix $\mathbf{R}_p$. The Cholesky (LDU) decomposition of a Hermitian matrix $\mathbf{C}$ is a factorization of the form

$$\mathbf{C} = \mathbf{L}\mathbf{D}\mathbf{L}^H \tag{5.81}$$

where $\mathbf{L}$ is a *lower triangular* matrix with ones along the diagonal and $\mathbf{D}$ is a *diagonal* matrix. If the diagonal terms of the matrix $\mathbf{D}$ are nonnegative ($\mathbf{C}$ is positive semidefinite), then $\mathbf{D}$ may be split into a product of two matrices by taking square roots

$$\mathbf{D} = \mathbf{D}^{1/2}\mathbf{D}^{1/2} \tag{5.82}$$

and the Cholesky decomposition of $\mathbf{C}$ may be expressed as the product of an upper triangular and a lower triangular matrix as follows

$$\mathbf{C} = \left(\mathbf{L}\mathbf{D}^{1/2}\right)\left(\mathbf{D}^{1/2}\mathbf{L}^H\right) \tag{5.83}$$

With a Cholesky decomposition of the autocorrelation matrix we will easily be able to establish the equivalence between the positive definiteness of $\mathbf{R}_p$, the positivity of the error sequence ϵ_j, and the unit magnitude constraint on the reflection coefficients Γ_j. In addition, we will be able to derive a closed-form expression for the inverse of the autocorrelation matrix as well as a recursive algorithm for inverting a Toeplitz matrix.

To derive the Cholesky decomposition of $\mathbf{R}_p$, consider the $(p+1) \times (p+1)$ *upper triangular* matrix

$$\mathbf{A}_p = \begin{bmatrix} 1 & a_1^*(1) & a_2^*(2) & \cdots & a_p^*(p) \\ 0 & 1 & a_2^*(1) & \cdots & a_p^*(p-1) \\ 0 & 0 & 1 & \cdots & a_p^*(p-2) \\ \vdots & \vdots & \vdots & & \vdots \\ 0 & 0 & 0 & \cdots & 1 \end{bmatrix} \tag{5.84}$$

This matrix is formed from the vectors $\mathbf{a}_0, \mathbf{a}_1, \ldots, \mathbf{a}_p$ that are produced when the Levinson-Durbin recursion is applied to the autocorrelation sequence $r_x(0), \ldots, r_x(p)$. Note that the jth column of $\mathbf{A}_p$ contains the filter coefficients, $\mathbf{a}_{j-1}^R$, padded with zeros. Since

$$\mathbf{R}_j \mathbf{a}_j^R = \epsilon_j \mathbf{u}_j \tag{5.85}$$

where $\mathbf{u}_j = [0,\ 0,\ \ldots,\ 1]^T$ is a unit vector of length $j+1$ with a one in the final position, then

$$\mathbf{R}_p \mathbf{A}_p = \begin{bmatrix} \epsilon_0 & 0 & 0 & \cdots & 0 \\ * & \epsilon_1 & 0 & \cdots & 0 \\ * & * & \epsilon_2 & \cdots & 0 \\ \vdots & \vdots & \vdots & & \vdots \\ * & * & * & \cdots & \epsilon_p \end{bmatrix} \tag{5.86}$$

which is a lower triangular matrix with the prediction errors ϵ_j along the diagonal (an asterisk is used to indicate those elements that are, in general, nonzero). Although we have not yet established Eq. (5.85), it may be inferred from the Hermitian Toeplitz property of $\mathbf{R}_p$. For example, since $\mathbf{J}\mathbf{R}_p^*\mathbf{J} = \mathbf{R}_p$ then

$$\mathbf{R}_j \mathbf{a}_j = (\mathbf{J}\mathbf{R}_j^*\mathbf{J})\mathbf{a}_j = \epsilon_j \mathbf{u}_1 \tag{5.87}$$

Multiplying on the left by $\mathbf{J}$ and using the fact that $\mathbf{J}^2 = \mathbf{I}$ and $\mathbf{J}\mathbf{a}_j = (\mathbf{a}_j^R)^*$, Eq. (5.85) follows from Eq. (5.87) after taking complex conjugates.

Since the product of two lower triangular matrices is a lower triangular matrix, if we multiply $\mathbf{R}_p\mathbf{A}_p$ on the left by the lower triangular matrix $\mathbf{A}_p^H$, then we obtain another lower triangular matrix, $\mathbf{A}_p^H\mathbf{R}_p\mathbf{A}_p$. Note that since the terms along the diagonal of $\mathbf{A}_p$ are equal to one, the diagonal of $\mathbf{A}_p^H\mathbf{R}_p\mathbf{A}_p$ will be the same as that of $\mathbf{R}_p\mathbf{A}_p$, and $\mathbf{A}_p^H\mathbf{R}_p\mathbf{A}_p$ will also have the form

$$\mathbf{A}_p^H \mathbf{R}_p \mathbf{A}_p = \begin{bmatrix} \epsilon_0 & 0 & 0 & \cdots & 0 \\ * & \epsilon_1 & 0 & \cdots & 0 \\ * & * & \epsilon_2 & \cdots & 0 \\ \vdots & \vdots & \vdots & & \vdots \\ * & * & * & \cdots & \epsilon_p \end{bmatrix} \tag{5.88}$$

Even more importantly, however, is the observation that since $\mathbf{A}_p^H\mathbf{R}_p\mathbf{A}_p$ is Hermitian, then the matrix on the right side of Eq. (5.88) must also be Hermitian. Therefore, the terms below the diagonal are zero and

$$\boxed{\mathbf{A}_p^H \mathbf{R}_p \mathbf{A}_p = \mathbf{D}_p} \tag{5.89}$$

where $\mathbf{D}_p$ is a diagonal matrix

$$\mathbf{D}_p = \text{diag}\{\epsilon_0, \epsilon_1, \ldots, \epsilon_p\}$$

Since $\mathbf{A}_p^H$ is a lower triangular matrix with ones along the diagonal, then $\mathbf{A}_p^H$ is nonsingular with $\det(\mathbf{A}_p) = 1$ and the inverse of $\mathbf{A}_p^H$ is also a lower triangular matrix. Denoting the inverse of $\mathbf{A}_p^H$ by $\mathbf{L}_p$, if we multiply Eq. (5.89) on the left by $\mathbf{L}_p$ and on the right by $\mathbf{L}_p^H$

we have

$$\boxed{\mathbf{R}_p = \mathbf{L}_p \mathbf{D}_p \mathbf{L}_p^H} \tag{5.90}$$

which is the desired factorization.

An interesting and important property that may be established from Eq. (5.89) is that the determinant of the autocorrelation matrix $\mathbf{R}_p$ is equal to the product of the modeling errors,

$$\boxed{\det \mathbf{R}_p = \prod_{k=0}^{p} \epsilon_k} \tag{5.91}$$

To see this, note that if we take the determinant of both sides of Eq. (5.89) and use the fact that $\det(\mathbf{A}_p) = 1$, then

$$\begin{aligned} \det \mathbf{D}_p &= \det\left(\mathbf{A}_p^H \mathbf{R}_p \mathbf{A}_p\right) \\ &= \left(\det \mathbf{A}_p^H\right)\left(\det \mathbf{R}_p\right)\left(\det \mathbf{A}_p\right) \\ &= \det \mathbf{R}_p \end{aligned} \tag{5.92}$$

and Eq. (5.91) follows since

$$\det \mathbf{D}_p = \prod_{k=0}^{p} \epsilon_k$$

Using Eq. (5.91) we may also show that the positivity of the autocorrelation matrix $\mathbf{R}_p$ is tied to the positivity of the error sequence ϵ_k. In particular, note that $\mathbf{R}_p$ will be positive definite if and only if

$$\det \mathbf{R}_j > 0 \quad ; \quad j = 0, 1, \ldots, p \tag{5.93}$$

Therefore, since

$$\det \mathbf{R}_j = \prod_{k=0}^{j} \epsilon_k \tag{5.94}$$

it follows that $\mathbf{R}_p$ will be positive definite if and only if $\epsilon_k > 0$ for $k = 0, 1, \ldots, p$. Furthermore, $\mathbf{R}_p$ will be singular, $\det(\mathbf{R}_p) = 0$, if $\epsilon_k = 0$ for some k. This relationship between the positive definiteness of $\mathbf{R}_p$ and the positivity of the error sequence ϵ_k may also be established using Sylvester's *law of inertia* [21], which states that if $\mathbf{A}_p$ is a nonsingular matrix, then $\mathbf{A}_p^H \mathbf{R}_p \mathbf{A}_p$ has the same number of positive eigenvalues as $\mathbf{R}_p$, the same number of negative eigenvalues, and the same number of zero eigenvalues. Since $\mathbf{A}_p^H \mathbf{R}_p \mathbf{A}_p = \mathbf{D}_p$ with $\mathbf{A}_p$ being nonsingular, then $\mathbf{R}_p$ and $\mathbf{D}_p$ have the same number of positive, negative, and zero eigenvalues. Since the eigenvalues of $\mathbf{D}_p$ are equal to ϵ_k it follows that $\mathbf{R}_p > 0$ if and only if $\epsilon_k > 0$ for all k. In summary, we have established the following fundamental property of Hermitian Toeplitz matrices.

Property 7. For any $(p+1) \times (p+1)$ Hermitian Toeplitz matrix $\mathbf{R}_p$, the following are equivalent:

1. $\mathbf{R}_p > 0$
2. $\epsilon_j > 0, \ j = 1, \ldots, p$
3. $|\Gamma_j| < 1, \ j = 1, \ldots, p$

Furthermore, this equivalence remains valid if the strict inequalities are replaced with less than or equal to inequalities.

In the following example we use this property to show how a Toeplitz matrix may be tested to see whether or not it is positive definite.

Example 5.2.8 ***Testing for Positive Definiteness***

Consider the 3×3 symmetric Toeplitz matrix

$$\mathbf{R} = \begin{bmatrix} 1 & \alpha & \beta \\ \alpha & 1 & \alpha \\ \beta & \alpha & 1 \end{bmatrix}$$

Let us find the values of α and β for which $\mathbf{R}$ is positive definite.

The first row of $\mathbf{R}$ may be considered to be the first three terms of an autocorrelation sequence, $\mathbf{r} = \left[1, \alpha, \beta\right]^T$. Using the Levinson-Durbin recursion we may express the values of the reflection coefficients, Γ_1 and Γ_2, in terms of α and β. Since

$$\Gamma_1 = -r(1)/r(0) = -\alpha$$

then

$$\mathbf{a}_1 = [1, -\alpha]^T \quad \text{and} \quad \epsilon_1 = 1 - \alpha^2$$

Therefore,

$$\gamma_1 = r_x(2) + a_1(1) r_x(1) = \beta - \alpha^2$$

and

$$\Gamma_2 = -\frac{\gamma_1}{\epsilon_1} = \frac{\alpha^2 - \beta}{1 - \alpha^2}$$

Now, in order for $\mathbf{R}$ to be positive definite it is necessary that $|\Gamma_1| < 1$ and $|\Gamma_2| < 1$, which implies that

$$\text{(i)} \qquad |\alpha| < 1$$

$$\text{(ii)} \qquad \left| \frac{\alpha^2 - \beta}{1 - \alpha^2} \right| < 1$$

From the second condition, it follows that

$$-1 < \frac{\alpha^2 - \beta}{1 - \alpha^2} < 1$$

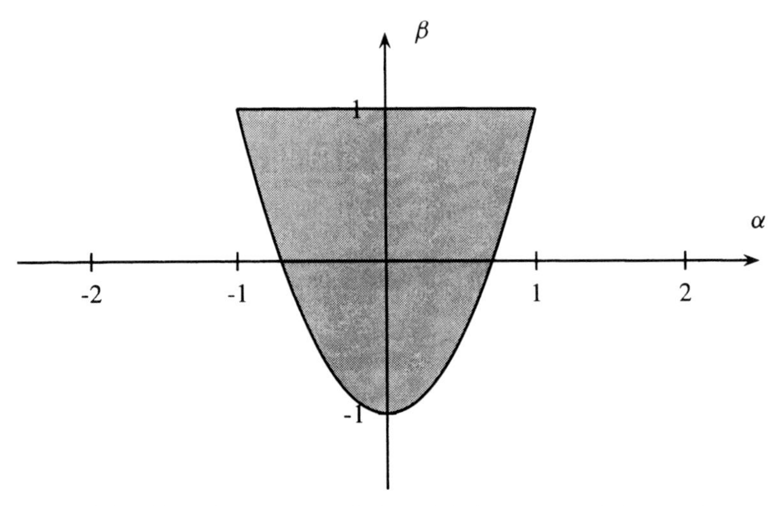

Figure 5.14 *The values for α and β for which the autocorrelation matrix associated with the autocorrelation sequence $\mathbf{r} = [1, \alpha, \beta]^T$ is positive definite.*

which, since $|\alpha| < 1$, becomes

$$\alpha^2 - 1 < \alpha^2 - \beta < 1 - \alpha^2$$

The left-hand inequality gives

$$\beta < 1$$

and the right-hand inequality is equivalent to

$$\beta > 2\alpha^2 - 1$$

Therefore, the allowable values for α and β lie within the shaded region shown in Fig. 5.14.

5.2.8 The Autocorrelation Extension Problem

In this section, we consider the *autocorrelation extension problem*, which addresses the following question: Given the first $(p+1)$ values of an autocorrelation sequence, $r_x(k)$ for $k = 0, 1, \ldots, p$, how may we extend (extrapolate) this *partial autocorrelation sequence* for $k > p$ in such a way that the extrapolated sequence is a valid autocorrelation sequence? In Section 3.3.5 of Chapter 3 we saw that in order for a finite sequence of numbers to be a valid partial autocorrelation sequence, the autocorrelation matrix formed from this sequence must be nonnegative definite, $\mathbf{R}_p \geq 0$. Therefore, any extension must preserve this nonnegative definite property, i.e., $\mathbf{R}_{p+1} \geq 0$, and $\mathbf{R}_{p+2} \geq 0$, and so on. In Section 3.6.1 of Chapter 3 we discovered one possible way to extrapolate a partial autocorrelation sequence. Specifically, given $r_x(k)$ for $k = 0, 1, \ldots, p$, if $r_x(k)$ is extrapolated for $k > p$ according to the recursion

$$r_x(k) = -\sum_{l=1}^{p} a_p(l) r_x(k-l) \tag{5.95}$$

where the coefficients $a_p(l)$ are the solution to the Yule-Walker (normal) equations

$$\mathbf{R}_p \mathbf{a}_p = \epsilon_p \mathbf{u}_1 \tag{5.96}$$

then the resulting sequence will be a valid autocorrelation sequence. This follows from the fact that this extrapolation generates the autocorrelation sequence of the AR(p) process that is consistent with the given autocorrelation values, i.e., $r_x(k)$ for $|k| \leq p$. Thus, a finite sequence of numbers is *extendible* if and only if $\mathbf{R}_p \geq 0$ and, if extendible, one possible extension is given by Eq. (5.95). We have not yet addressed, however, the question of whether or not other extensions are possible. Equivalently, given a partial autocorrelation sequence $r_x(k)$ for $k = 0, 1, \ldots, p$, what values of $r_x(p+1)$ will produce a valid partial autocorrelation sequence with $\mathbf{R}_{p+1} \geq 0$? The answers to these questions may be deduced from Property 7 (p. 253), which states that if $\mathbf{R}_p$ is nonnegative definite, then $\mathbf{R}_{p+1}$ will be nonnegative definite if and only if $|\Gamma_{p+1}| \leq 1$. Therefore, by expressing $r_x(p+1)$ in terms of the reflection coefficient Γ_{p+1} we may place a bound on the allowable values for $r_x(p+1)$. The desired relationship follows from Eqs. (5.10) and (5.13), which gives, for $r_x(p+1)$,

$$r_x(p+1) = -\Gamma_{p+1}\epsilon_p - \sum_{k=1}^{p} a_p(k) r_x(p-k+1) \tag{5.97}$$

With Γ_{p+1} a complex number that is bounded by one in magnitude, $|\Gamma_{p+1}| \leq 1$, Eq. (5.97) implies that $r_x(p+1)$ is constrained to lie within or on a circle of radius ϵ_p that is centered at $C = -\sum_{k=1}^{p} a_p(k) r_x(p-k+1)$, as shown in Fig. 5.15. In the real case, the range of allowable values is

$$-\epsilon_p - \sum_{k=1}^{p} a_p(k) r_x(p-k+1) \leq r_x(p+1) \leq \epsilon_p - \sum_{k=1}^{p} a_p(k) r_x(p-k+1) \tag{5.98}$$

There are two special cases of autocorrelation sequence extension that are worth some discussion. The first occurs when we set $\Gamma_{p+1} = 0$. In this case, the extrapolation is performed according to Eq. (5.95) with

$$r_x(p+1) = -\sum_{l=1}^{p} a_p(l) r_x(p+1-l)$$

Furthermore, if the extrapolation of $r_x(k)$ is continued in this manner for all $k > p$, then the resulting autocorrelation sequence will correspond to an AR(p) process that has a power

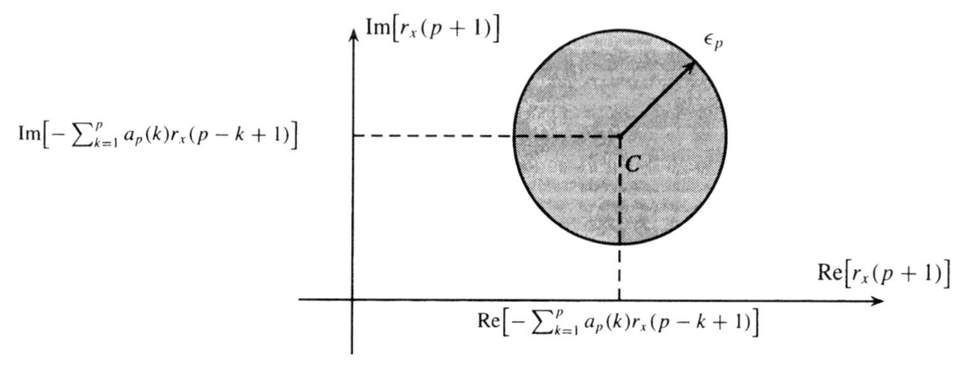

Figure 5.15 *The allowable values for the autocorrelation coefficient $r_x(p+1)$ given the partial autocorrelation sequence $r_x(k)$ for $k = 0, 1, \ldots, p$ are shown to lie within the circle of radius ϵ_p centered at C.*

spectrum equal to

$$P_x(e^{j\omega}) = \frac{|b(0)|^2}{\left|1 + \sum_{k=1}^{p} a_p(k)e^{-jk\omega}\right|^2} \tag{5.99}$$

The second case occurs when $|\Gamma_{p+1}| = 1$. When the magnitude of Γ_{p+1} is equal to one, $r_x(p+1)$ lies *on* the circle shown in Fig. 5.15, the $(p+1)$st-order model error is zero,

$$\epsilon_{p+1} = \epsilon_p\left[1 - |\Gamma_{p+1}|^2\right] = 0 \tag{5.100}$$

the matrix $\mathbf{R}_{p+1}$ is singular, and the predictor polynomial $A_{p+1}(z)$ has all of its roots on the unit circle (see Property 5, p. 230). As we will see in Section 8.6 of Chapter 8, this case will be important for frequency estimation and for modeling signals as a sum of complex exponentials.

Example 5.2.9 *Autocorrelation Extension*

Given the partial autocorrelation sequence $r_x(0) = 1$ and $r_x(1) = 0.5$, let us find the set of allowable values for $r_x(2)$, assuming that $r_x(2)$ is real. For the first-order model we have

$$\Gamma_1 = -\frac{r_x(1)}{r_x(0)} = -0.5$$

and $a_1(1) = \Gamma_1 = -0.5$. Thus, with a first-order modeling error of

$$\epsilon_1 = r_x(0)\left[1 - |\Gamma_1|^2\right] = 0.75$$

we have

$$r_x(2) = -\Gamma_2\epsilon_1 - a_1(1)r_x(1) = -0.75\Gamma_2 + 0.25$$

Therefore, with $|\Gamma_2| \leq 1$ it follows that

$$-0.5 \leq r_x(2) \leq 1$$

In the special case of $\Gamma_2 = 0$, $r_x(2) = 0.25$ and, in the extreme cases of $\Gamma_2 = \pm 1$, the autocorrelation values are $r_x(2) = -0.5$ and $r_x(2) = 1$.

5.2.9 Inverting a Toeplitz Matrix*

In this section, we will derive a recursive algorithm for inverting a Toeplitz matrix $\mathbf{R}_p$ and show how this recursion may be used to derive the Levinson recursion for solving a general set of Toeplitz equations

$$\mathbf{R}_p\mathbf{x} = \mathbf{b}$$

We begin by showing how the decomposition derived in Section 5.2.7 may be used to express the inverse of a Toeplitz matrix in terms of the vectors $\mathbf{a}_j$ and the errors ϵ_j. The Levinson-Durbin recursion will then be applied to this expression for $\mathbf{R}_p^{-1}$ to derive the *Toeplitz matrix inversion recursion.*

Let $\mathbf{R}_p$ be a nonsingular Hermitian Toeplitz matrix. Using the decomposition given in Eq. (5.89), taking the inverse of both sides, we have

$$\left(\mathbf{A}_p^H\mathbf{R}_p\mathbf{A}_p\right)^{-1} = \mathbf{A}_p^{-1}\mathbf{R}_p^{-1}\mathbf{A}_p^{-H} = \mathbf{D}_p^{-1} \tag{5.101}$$

Multiplying both sides of this equation by $\mathbf{A}_p$ on the left and by $\mathbf{A}_p^H$ on the right gives the desired expression for $\mathbf{R}_p^{-1}$,

$$\mathbf{R}_p^{-1} = \mathbf{A}_p \mathbf{D}_p^{-1} \mathbf{A}_p^H \tag{5.102}$$

Since $\mathbf{D}_p$ is a diagonal matrix, $\mathbf{D}_p^{-1}$ is easily computed. Therefore, finding the inverse of $\mathbf{R}_p$ simply involves applying the Levinson-Durbin recursion to the sequence $r_x(0), \ldots, r_x(p)$, forming the matrix $\mathbf{A}_p$, and performing the matrix product in Eq. (5.102).

Example 5.2.10 *Inverting a Toeplitz Matrix*

Let us find the inverse of the 3×3 autocorrelation matrix

$$\mathbf{R}_2 = \begin{bmatrix} 2 & 0 & 1 \\ 0 & 2 & 0 \\ 1 & 0 & 2 \end{bmatrix}$$

We begin by using the Levinson-Durbin recursion to find the vectors $\mathbf{a}_j$ and the errors ϵ_j for $j = 0, 1, 2$. For $j = 0$, we have $\mathbf{a}_0 = 1$ and $\epsilon_0 = r_x(0) = 2$. With

$$\Gamma_1 = -r_x(1)/r_x(0) = 0$$

the first-order model is

$$\mathbf{a}_1 = \begin{bmatrix} 1 \\ 0 \end{bmatrix}$$

with an error

$$\epsilon_1 = \epsilon_0\left[1 - \Gamma_1^2\right] = 2$$

For the second-order model,

$$\gamma_1 = r_x(2) + a_1(1)r_x(1) = 1$$

and

$$\Gamma_2 = -\frac{\gamma_1}{\epsilon_1} = -1/2$$

Thus,

$$\mathbf{a}_2 = \begin{bmatrix} 1 \\ 0 \\ -1/2 \end{bmatrix}$$

and

$$\epsilon_2 = \epsilon_1\left[1 - \Gamma_2^2\right] = 3/2$$

Finally, using Eq. (5.102) we have for the inverse of $\mathbf{R}_2$,

$$\mathbf{R}_2^{-1} = \mathbf{A}_2 \mathbf{D}_2^{-1} \mathbf{A}_2^T = \begin{bmatrix} 1 & 0 & -1/2 \\ 0 & 1 & 0 \\ 0 & 0 & 1 \end{bmatrix} \begin{bmatrix} 1/2 & 0 & 0 \\ 0 & 1/2 & 0 \\ 0 & 0 & 2/3 \end{bmatrix} \begin{bmatrix} 1 & 0 & 0 \\ 0 & 1 & 0 \\ -1/2 & 0 & 1 \end{bmatrix}$$

$$= \begin{bmatrix} 1/2 & 0 & -1/3 \\ 0 & 1/2 & 0 \\ 0 & 0 & 2/3 \end{bmatrix} \begin{bmatrix} 1 & 0 & 0 \\ 0 & 1 & 0 \\ -1/2 & 0 & 1 \end{bmatrix}$$

$$= \begin{bmatrix} 2/3 & 0 & -1/3 \\ 0 & 1/2 & 0 \\ -1/3 & 0 & 2/3 \end{bmatrix}$$

Note that the inverse of a Toeplitz matrix is not, in general, Toeplitz. However, as discussed in Section 2.3.7 (p. 38) and as illustrated in this example, the inverse of a symmetric Toeplitz matrix is centrosymmetric [17].

In the next example, Eq. (5.102) is used to find the inverse autocorrelation matrix of a first-order autoregressive process.

Example 5.2.11 ***The Inverse Autocorrelation Matrix for an AR(1) Process***

Let $\mathbf{R}_p$ be the autocorrelation matrix of an AR(1) process that is generated by the difference equation

$$x(n) = \alpha x(n-1) + w(n)$$

where $|\alpha| < 1$ and $w(n)$ is zero mean white noise with a variance σ_w^2. Thus, the autocorrelation sequence of $x(n)$ is

$$r_x(k) = \frac{\sigma_w^2}{1-\alpha^2}\alpha^{|k|}$$

and the autocorrelation matrix is

$$\mathbf{R}_p = \frac{\sigma_w^2}{1-\alpha^2}\begin{bmatrix} 1 & \alpha & \alpha^2 & \cdots & \alpha^p \\ \alpha & 1 & \alpha & \cdots & \alpha^{p-1} \\ \alpha^2 & \alpha & 1 & \cdots & \alpha^{p-2} \\ \vdots & \vdots & \vdots & & \vdots \\ \alpha^p & \alpha^{p-1} & \alpha^{p-2} & \cdots & 1 \end{bmatrix}$$

From Example 5.2.2 (p. 221) we know that the jth-order model for $x(n)$ is

$$\mathbf{a}_j = \left[1, -\alpha, 0, \ldots, 0\right]^T$$

and that the sequence of modeling errors is

$$\epsilon_0 = \frac{\sigma_w^2}{1-\alpha^2}$$

and

$$\epsilon_k = \sigma_w^2 \quad ; \quad k \geq 1$$

Therefore,

$$\mathbf{A}_p = \begin{bmatrix} 1 & -\alpha & 0 & \cdots & 0 & 0 \\ 0 & 1 & -\alpha & \cdots & 0 & 0 \\ 0 & 0 & 1 & \cdots & 0 & 0 \\ \vdots & \vdots & \vdots & & \vdots & \vdots \\ 0 & 0 & 0 & \cdots & 1 & -\alpha \\ 0 & 0 & 0 & \cdots & 0 & 1 \end{bmatrix}$$

and

$$\mathbf{D}_p^{-1} = \text{diag}\left\{\epsilon_0^{-1}, \epsilon_1^{-1}, \ldots, \epsilon_{p-1}^{-1}, \epsilon_p^{-1}\right\} = \frac{1}{\sigma_w^2}\text{diag}\left\{(1-\alpha^2),\ 1,\ 1, \ldots,\ 1\right\}$$

From Eq. (5.102) it follows that

$$\mathbf{R}_p^{-1} = \frac{1}{\sigma_w^2}\begin{bmatrix} 1 & -\alpha & 0 & \cdots & 0 & 0 \\ 0 & 1 & -\alpha & \cdots & 0 & 0 \\ 0 & 0 & 1 & \cdots & 0 & 0 \\ \vdots & \vdots & \vdots & & \vdots & \vdots \\ 0 & 0 & 0 & \cdots & 1 & -\alpha \\ 0 & 0 & 0 & \cdots & 0 & 1 \end{bmatrix}\begin{bmatrix} 1-\alpha^2 & 0 & \cdots & 0 & 0 \\ 0 & 1 & \cdots & 0 & 0 \\ 0 & 0 & \cdots & 0 & 0 \\ \vdots & \vdots & & \vdots & \vdots \\ 0 & 0 & \cdots & 1 & 0 \\ 0 & 0 & \cdots & 0 & 1 \end{bmatrix}$$

$$\times \begin{bmatrix} 1 & 0 & \cdots & 0 & 0 \\ -\alpha & 1 & \cdots & 0 & 0 \\ 0 & -\alpha & \cdots & 0 & 0 \\ \vdots & \vdots & & \vdots & \vdots \\ 0 & 0 & \cdots & 1 & 0 \\ 0 & 0 & \cdots & -\alpha & 1 \end{bmatrix}$$

$$= \frac{1}{\sigma_w^2}\begin{bmatrix} 1-\alpha^2 & -\alpha & \cdots & 0 & 0 \\ 0 & 1 & \cdots & 0 & 0 \\ 0 & 0 & \cdots & 0 & 0 \\ \vdots & \vdots & & \vdots & \vdots \\ 0 & 0 & \cdots & 1 & -\alpha \\ 0 & 0 & \cdots & 0 & 1 \end{bmatrix}\begin{bmatrix} 1 & 0 & \cdots & 0 & 0 \\ -\alpha & 1 & \cdots & 0 & 0 \\ 0 & -\alpha & \cdots & 0 & 0 \\ \vdots & \vdots & & \vdots & \vdots \\ 0 & 0 & \cdots & 1 & 0 \\ 0 & 0 & \cdots & -\alpha & 1 \end{bmatrix}$$

$$= \frac{1}{\sigma_w^2}\begin{bmatrix} 1 & -\alpha & \cdots & 0 & 0 \\ -\alpha & 1+\alpha^2 & \cdots & 0 & 0 \\ 0 & -\alpha & \cdots & 0 & 0 \\ \vdots & \vdots & & \vdots & \vdots \\ 0 & 0 & \cdots & 1+\alpha^2 & -\alpha \\ 0 & 0 & \cdots & -\alpha & 1 \end{bmatrix}$$

Thus, we see that $\mathbf{R}_p^{-1}$ is centrosymmetric and *tri-diagonal.*

Equation (5.102) provides a closed form expression for the inverse of a Hermitian Toeplitz matrix in terms of the matrices $\mathbf{A}_p$ and $\mathbf{D}_p$. However, since these matrices may be built up recursively using the Levinson-Durbin recursion, we may fold this recursion into the evaluation of Eq. (5.102) and derive a method for recursively computing the inverse of $\mathbf{R}_p$. In particular, given

$$\mathbf{R}_j^{-1} = \mathbf{A}_j \mathbf{D}_j^{-1} \mathbf{A}_j^H \tag{5.103}$$

we may find

$$\mathbf{R}_{j+1}^{-1} = \mathbf{A}_{j+1} \mathbf{D}_{j+1}^{-1} \mathbf{A}_{j+1}^H \tag{5.104}$$

To see how this may be accomplished, let us partition $\mathbf{A}_{j+1}$ and $\mathbf{D}_{j+1}$ as follows[7]

$$\mathbf{A}_{j+1} = \left[\begin{array}{ccc|c} & & & a_{j+1}^*(j+1) \\ & \mathbf{A}_j & & \vdots \\ & & & a_{j+1}^*(1) \\ \hline 0 & \cdots & 0 & 1 \end{array}\right] = \left[\begin{array}{c|c} \mathbf{A}_j & \bar{\mathbf{a}}_{j+1}^R \\ \hline \mathbf{0}^T & 1 \end{array}\right] \tag{5.105}$$

and

$$\mathbf{D}_{j+1}^{-1} = \left[\begin{array}{ccc|c} & & & 0 \\ & \mathbf{D}_j^{-1} & & \vdots \\ & & & 0 \\ \hline 0 & \cdots & 0 & \epsilon_{j+1}^{-1} \end{array}\right] \tag{5.106}$$

Incorporating Eqs. (5.105) and (5.106) into Eq. (5.104) we have

$$\begin{aligned}
\mathbf{R}_{j+1}^{-1} &= \left[\begin{array}{c|c} \mathbf{A}_j & \bar{\mathbf{a}}_{j+1}^R \\ \hline \mathbf{0}^T & 1 \end{array}\right] \left[\begin{array}{c|c} \mathbf{D}_j^{-1} & \mathbf{0} \\ \hline \mathbf{0}^T & \epsilon_{j+1}^{-1} \end{array}\right] \left[\begin{array}{c|c} \mathbf{A}_j^H & \mathbf{0} \\ \hline (\bar{\mathbf{a}}_{j+1}^R)^H & 1 \end{array}\right] \\
&= \left[\begin{array}{c|c} \mathbf{A}_j & \bar{\mathbf{a}}_{j+1}^R \\ \hline \mathbf{0}^T & 1 \end{array}\right] \left[\begin{array}{c|c} \mathbf{D}_j^{-1}\mathbf{A}_j^H & \mathbf{0} \\ \hline \epsilon_{j+1}^{-1}(\bar{\mathbf{a}}_{j+1}^R)^H & \epsilon_{j+1}^{-1} \end{array}\right] \\
&= \left[\begin{array}{c|c} \mathbf{A}_j\mathbf{D}_j^{-1}\mathbf{A}_j^H + \epsilon_{j+1}^{-1}\bar{\mathbf{a}}_{j+1}^R(\bar{\mathbf{a}}_{j+1}^R)^H & \epsilon_{j+1}^{-1}\bar{\mathbf{a}}_{j+1}^R \\ \hline \epsilon_{j+1}^{-1}(\bar{\mathbf{a}}_{j+1}^R)^H & \epsilon_{j+1}^{-1} \end{array}\right]
\end{aligned} \tag{5.107}$$

Splitting this expression for $\mathbf{R}_{j+1}^{-1}$ into a sum as follows

$$\mathbf{R}_{j+1}^{-1} = \left[\begin{array}{c|c} \mathbf{A}_j\mathbf{D}_j^{-1}\mathbf{A}_j^H & \mathbf{0} \\ \hline \mathbf{0}^T & 0 \end{array}\right] + \epsilon_{j+1}^{-1} \left[\begin{array}{c|c} \bar{\mathbf{a}}_{j+1}^R(\bar{\mathbf{a}}_{j+1}^R)^H & \bar{\mathbf{a}}_{j+1}^R \\ \hline (\bar{\mathbf{a}}_{j+1}^R)^H & 1 \end{array}\right] \tag{5.108}$$

we see that $\mathbf{R}_j^{-1}$ appears in the upper-left corner of the first term and that the second term may be written as an outer product

$$\begin{aligned}
\epsilon_{j+1}^{-1} \left[\begin{array}{c|c} \bar{\mathbf{a}}_{j+1}^R(\bar{\mathbf{a}}_{j+1}^R)^H & \bar{\mathbf{a}}_{j+1}^R \\ \hline (\bar{\mathbf{a}}_{j+1}^R)^H & 1 \end{array}\right] &= \epsilon_{j+1}^{-1} \left[\begin{array}{c} \bar{\mathbf{a}}_{j+1}^R \\ \hline 1 \end{array}\right] \left[\begin{array}{c|c} (\bar{\mathbf{a}}_{j+1}^R)^H & 1 \end{array}\right] \\
&= \frac{1}{\epsilon_{j+1}} \mathbf{a}_{j+1}^R (\mathbf{a}_{j+1}^R)^H
\end{aligned} \tag{5.109}$$

[7]**Note to the reader:** Although there is nothing difficult about the following derivation from a mathematical point of view, from a notational point of view it is a bit complex. Since there is not much to be gained in going through this derivation, other than a feeling of accomplishment, the reader may wish to jump to the main results which are contained in Eqs. (5.110) and (5.114). Although the discussion following Eq. (5.114) uses these results to derive the general Levinson recursion, this recursion is re-derived in the following section, without using these results.

Therefore, we have

$$\mathbf{R}_{j+1}^{-1} = \left[\begin{array}{c:c} \mathbf{R}_j^{-1} & \mathbf{0} \\ \hdashline \mathbf{0}^T & 0 \end{array}\right] + \frac{1}{\epsilon_{j+1}} \mathbf{a}_{j+1}^R (\mathbf{a}_{j+1}^R)^H \tag{5.110}$$

which is the desired update equation for $\mathbf{R}_j^{-1}$. The recursion is initialized by setting $\mathbf{R}_0^{-1} = 1$, $\epsilon_0 = r_x(0)$, and $\mathbf{a}_0 = 1$. Then, from $\mathbf{R}_j^{-1}$, ϵ_j, and $\mathbf{a}_j$, the Levinson-Durbin recursion is used to determine $\mathbf{a}_{j+1}$ and ϵ_{j+1} which, in turn, are then used in Eq. (5.110) to perform the update of $\mathbf{R}_j^{-1}$. The following example illustrates the procedure.

Example 5.2.12 ***Recursive Evaluation of an Inverse Autocorrelation Matrix***

In this example we show how to recursively compute the inverse of the 3×3 autocorrelation matrix

$$\mathbf{R}_2 = \begin{bmatrix} 1 & 1/2 & 1/2 \\ 1/2 & 1 & 1/2 \\ 1/2 & 1/2 & 1 \end{bmatrix}$$

Initializing the recursion we have

$$\mathbf{a}_0 = 1 \quad ; \quad \epsilon_0 = r_x(0) = 1 \quad ; \quad \mathbf{R}_0^{-1} = 1$$

Then, using the Levinson-Durbin recursion to calculate the first-order model

$$\Gamma_1 = -r_x(1)/r_x(0) = -1/2$$

and

$$\mathbf{a}_1 = \begin{bmatrix} 1 \\ -1/2 \end{bmatrix}$$

Computing the first-order error

$$\epsilon_1 = \epsilon_0 \left[1 - \Gamma_1^2\right] = 3/4$$

we find the inverse of $\mathbf{R}_1$ to be

$$\begin{aligned} \mathbf{R}_1^{-1} &= \begin{bmatrix} \mathbf{R}_0^{-1} & 0 \\ 0 & 0 \end{bmatrix} + \frac{1}{\epsilon_1} \mathbf{a}_1^R (\mathbf{a}_1^R)^T \\ &= \begin{bmatrix} 1 & 0 \\ 0 & 0 \end{bmatrix} + \frac{4}{3} \begin{bmatrix} -1/2 \\ 1 \end{bmatrix} \begin{bmatrix} -1/2 & 1 \end{bmatrix} \\ &= \begin{bmatrix} 4/3 & -2/3 \\ -2/3 & 4/3 \end{bmatrix} \end{aligned} \tag{5.111}$$

Using the Levinson-Durbin we next generate the second-order model. With

$$\begin{aligned}\gamma_1 &= r_x(2) + a_1(1)r_x(1) = 1/4\\ \Gamma_2 &= -\gamma_1/\epsilon_1 = -1/3\\ \epsilon_2 &= \epsilon_1\big[1-\Gamma_2^2\big] = 2/3\end{aligned}$$

we find, for $\mathbf{a}_2$,

$$\mathbf{a}_2 = \begin{bmatrix} 1 \\ -1/2 \\ 0 \end{bmatrix} - \frac{1}{3}\begin{bmatrix} 0 \\ -1/2 \\ 1 \end{bmatrix} = \begin{bmatrix} 1 \\ -1/3 \\ -1/3 \end{bmatrix}$$

Therefore the inverse of $\mathbf{R}_2$ is

$$\mathbf{R}_2^{-1} = \begin{bmatrix} \mathbf{R}_1^{-1} & 0 \\ 0 & 0 \end{bmatrix} + \frac{1}{\epsilon_2}\mathbf{a}_2^R(\mathbf{a}_2^R)^T$$

$$= \begin{bmatrix} 4/3 & -2/3 & 0 \\ -2/3 & 4/3 & 0 \\ 0 & 0 & 0 \end{bmatrix} + \frac{3}{2}\begin{bmatrix} -1/3 \\ -1/3 \\ 1 \end{bmatrix}\begin{bmatrix} -1/3 & -1/3 & 1 \end{bmatrix}$$

$$= \begin{bmatrix} 3/2 & -1/2 & -1/2 \\ -1/2 & 3/2 & -1/2 \\ -1/2 & -1/2 & 3/2 \end{bmatrix} \tag{5.112}$$

which is the desired result.

A slight variation of the recursion given in Eq. (5.110) may be derived by exploiting the Toeplitz structure of $\mathbf{R}_p$. Specifically, since

$$\mathbf{J}\mathbf{R}_p\mathbf{J} = \mathbf{R}_p^*$$

where $\mathbf{J}$ is the exchange matrix, taking the inverse of both sides of this equation, using the fact that $\mathbf{J}^{-1} = \mathbf{J}$, we find

$$\mathbf{R}_p^{-1} = \mathbf{J}(\mathbf{R}_p^{-1})^*\mathbf{J}$$

Applying this transformation to the expression for $\mathbf{R}_{j+1}^{-1}$ given in Eq. (5.110) we have

$$\mathbf{R}_{j+1}^{-1} = \mathbf{J}(\mathbf{R}_{j+1}^{-1})^*\mathbf{J} = \mathbf{J}\left[\begin{array}{c|c} (\mathbf{R}_j^{-1})^* & \mathbf{0} \\ \hline \mathbf{0}^T & 0 \end{array}\right]\mathbf{J} + \frac{1}{\epsilon_{j+1}}\mathbf{J}(\mathbf{a}_{j+1}^R)^*(\mathbf{a}_{j+1}^R)^T\mathbf{J} \tag{5.113}$$

Simplifying the expression on the right by multiplying by the exchange matrices, we find

$$\mathbf{R}_{j+1}^{-1} = \left[\begin{array}{c|c} 0 & \mathbf{0}^T \\ \hline \mathbf{0} & \mathbf{R}_j^{-1} \end{array}\right] + \frac{1}{\epsilon_{j+1}} \mathbf{a}_{j+1}\mathbf{a}_{j+1}^H \tag{5.114}$$

which is the desired recursion.

In the next section, we derive the Levinson recursion which may be used to solve a general set of Toeplitz equations of the form

$$\mathbf{R}_p \mathbf{x}_p = \mathbf{b}_p \tag{5.115}$$

where $\mathbf{b}_p = \left[b(0),\ b(1),\ \ldots,\ b(p)\right]^T$ is an arbitrary vector. Here, we will derive an equivalent form of the recursion that is based on the recursion given in Eq. (5.110). The derivation proceeds as follows. Suppose that we are given the solution to the Toeplitz equations $\mathbf{R}_j\mathbf{x}_j = \mathbf{b}_j$ and that we want to derive the solution to

$$\mathbf{R}_{j+1}\mathbf{x}_{j+1} = \mathbf{b}_{j+1} \tag{5.116}$$

where

$$\mathbf{b}_{j+1} = \left[\begin{array}{c} \mathbf{b}_j \\ \hline b_{j+1} \end{array}\right]$$

Applying Eq. (5.110) to Eq. (5.116) we have

$$\begin{aligned} \mathbf{x}_{j+1} &= \left[\begin{array}{c|c} \mathbf{R}_j^{-1} & \mathbf{0} \\ \hline \mathbf{0}^T & 0 \end{array}\right] \mathbf{b}_{j+1} + \frac{1}{\epsilon_{j+1}} \mathbf{a}_{j+1}^R \left(\mathbf{a}_{j+1}^R\right)^H \mathbf{b}_{j+1} \\ &= \left[\begin{array}{c} \mathbf{x}_j \\ \hline 0 \end{array}\right] + \frac{1}{\epsilon_{j+1}} \left[\left(\mathbf{a}_{j+1}^R\right)^H \mathbf{b}_{j+1}\right] \mathbf{a}_{j+1}^R \end{aligned} \tag{5.117}$$

or

$$\mathbf{x}_{j+1} = \left[\begin{array}{c} \mathbf{x}_j \\ \hline 0 \end{array}\right] + \frac{\alpha_{j+1}}{\epsilon_{j+1}} \mathbf{a}_{j+1}^R \tag{5.118}$$

where

$$\alpha_{j+1} = \left(\mathbf{a}_{j+1}^R\right)^H \mathbf{b}_{j+1} = \sum_{k=0}^{j+1} b_{j+1}(k) a_{j+1}(j+1-k) \tag{5.119}$$

Equations (5.118) and (5.119) along with the Levinson-Durbin recursion for performing the updates of $\mathbf{a}_j$ constitute the Levinson recursion.

5.3 THE LEVINSON RECURSION

We have seen how the Toeplitz structure of the all-pole normal equations is exploited in the Levinson-Durbin recursion to efficiently solve a set of equations of the form

$$\mathbf{R}_p \mathbf{a}_p = \epsilon_p \mathbf{u}_1 \tag{5.120}$$

where $\mathbf{R}_p$ is a $(p+1) \times (p+1)$ Hermitian Toeplitz matrix and $\mathbf{u}_1$ is a unit vector. There are many applications, however, that require solving a more general set of Toeplitz equations

$$\mathbf{R}_p \mathbf{x}_p = \mathbf{b} \tag{5.121}$$

where the vector on the right-hand side is arbitrary. For example, as we saw in Chapter 4, Shanks' method requires solving a set of linear equations of this form. In addition, we will see in Chapter 7 that the design of an FIR digital Wiener filter requires finding the solution to the Wiener-Hopf equation

$$\mathbf{R}_p \mathbf{a}_p = \mathbf{r}_{dx}$$

where $\mathbf{R}_p$ is a Toeplitz matrix and $\mathbf{r}_{dx}$ is a vector of cross-correlations that is, in general, unrelated to the elements in $\mathbf{R}_p$. Therefore, in this section, we derive the original recursion developed by N. Levinson in 1947 for solving a general set of Hermitian Toeplitz equations of the form given in Eq. (5.121) where $\mathbf{b}$ is an arbitrary vector. This recursion, known as the Levinson recursion, simultaneously solves the following two sets of linear equations

$$\begin{bmatrix} r_x(0) & r_x^*(1) & r_x^*(2) & \cdots & r_x^*(j) \\ r_x(1) & r_x(0) & r_x^*(1) & \cdots & r_x^*(j-1) \\ r_x(2) & r_x(1) & r_x(0) & \cdots & r_x^*(j-2) \\ \vdots & \vdots & \vdots & & \vdots \\ r_x(j) & r_x(j-1) & r_x(j-2) & \cdots & r_x(0) \end{bmatrix} \begin{bmatrix} 1 \\ a_j(1) \\ a_j(2) \\ \vdots \\ a_j(j) \end{bmatrix} = \begin{bmatrix} \epsilon_j \\ 0 \\ 0 \\ \vdots \\ 0 \end{bmatrix} \tag{5.122}$$

and

$$\begin{bmatrix} r_x(0) & r_x^*(1) & r_x^*(2) & \cdots & r_x^*(j) \\ r_x(1) & r_x(0) & r_x^*(1) & \cdots & r_x^*(j-1) \\ r_x(2) & r_x(1) & r_x(0) & \cdots & r_x^*(j-2) \\ \vdots & \vdots & \vdots & & \vdots \\ r_x(j) & r_x(j-1) & r_x(j-2) & \cdots & r_x(0) \end{bmatrix} \begin{bmatrix} x_j(0) \\ x_j(1) \\ x_j(2) \\ \vdots \\ x_j(j) \end{bmatrix} = \begin{bmatrix} b(0) \\ b(1) \\ b(2) \\ \vdots \\ b(j) \end{bmatrix} \tag{5.123}$$

for $j = 0, 1, \ldots, p$. Since Eq. (5.122) corresponds to the all-pole normal equations, we will find that, embedded within the Levinson recursion, is the Levinson-Durbin recursion. In Section 5.2.9 we used the Toeplitz matrix inverse recursion to derive one form of the Levinson recursion. Here, we will rederive the recursion without using this matrix inverse recursion. The derivation, in fact, will closely resemble that used for the Levinson-Durbin recursion in Section 5.2.

The Levinson recursion begins by finding the solution to Eqs. (5.122) and (5.123) for $j = 0$. The result is clearly seen to be

$$\epsilon_0 = r_x(0) \qquad \text{and} \qquad x_0(0) = b(0)/r_x(0)$$

Now let us assume that we know the solution to the jth-order equations and derive the solution to the $(j+1)$st-order equations. Given the solution $\mathbf{a}_j$ to Eq. (5.122) the Levinson-

Durbin update equation (5.14) gives us the solution $\mathbf{a}_{j+1}$ to the $(j+1)$st-order equations

$$\begin{bmatrix} r_x(0) & r_x^*(1) & \cdots & r_x^*(j) & r_x^*(j+1) \\ r_x(1) & r_x(0) & \cdots & r_x^*(j-1) & r_x^*(j) \\ \vdots & \vdots & & \vdots & \vdots \\ r_x(j) & r_x(j-1) & \cdots & r_x(0) & r_x^*(1) \\ r_x(j+1) & r_x(j) & \cdots & r_x(1) & r_x(0) \end{bmatrix} \begin{bmatrix} 1 \\ a_{j+1}(1) \\ \vdots \\ a_{j+1}(j) \\ a_{j+1}(j+1) \end{bmatrix} = \begin{bmatrix} \epsilon_{j+1} \\ 0 \\ \vdots \\ 0 \\ 0 \end{bmatrix} \tag{5.124}$$

Exploiting the Hermitian Toeplitz property of $\mathbf{R}_{j+1}$, if we take the complex conjugate of Eq. (5.124) and reverse the order of the rows and columns of $\mathbf{R}_{j+1}$, Eq. (5.124) may be written in the equivalent form

$$\begin{bmatrix} r_x(0) & r_x^*(1) & \cdots & r_x^*(j) & r_x^*(j+1) \\ r_x(1) & r_x(0) & \cdots & r_x^*(j-1) & r_x^*(j) \\ \vdots & \vdots & & \vdots & \vdots \\ r_x(j) & r_x(j-1) & \cdots & r_x(0) & r_x^*(1) \\ r_x(j+1) & r_x(j) & \cdots & r_x(1) & r_x(0) \end{bmatrix} \begin{bmatrix} a_{j+1}^*(j+1) \\ a_{j+1}^*(j) \\ \vdots \\ a_{j+1}^*(1) \\ 1 \end{bmatrix} = \begin{bmatrix} 0 \\ 0 \\ \vdots \\ 0 \\ \epsilon_{j+1}^* \end{bmatrix} \tag{5.125}$$

Since it is assumed that the solution $\mathbf{x}_j$ to Eq. (5.123) is known, suppose that we append a zero to $\mathbf{x}_j$ and then multiply this extended vector by $\mathbf{R}_{j+1}$, as we did in the Levinson-Durbin recursion. This leads to the following set of linear equations

$$\begin{bmatrix} r_x(0) & r_x^*(1) & \cdots & r_x^*(j) & r_x^*(j+1) \\ r_x(1) & r_x(0) & \cdots & r_x^*(j-1) & r_x^*(j) \\ \vdots & \vdots & & \vdots & \vdots \\ r_x(j) & r_x(j-1) & \cdots & r_x(0) & r_x^*(1) \\ r_x(j+1) & r_x(j) & \cdots & r_x(1) & r_x(0) \end{bmatrix} \begin{bmatrix} x_j(0) \\ x_j(1) \\ \vdots \\ x_j(j) \\ 0 \end{bmatrix} = \begin{bmatrix} b(0) \\ b(1) \\ \vdots \\ b(j) \\ \delta_j \end{bmatrix} \tag{5.126}$$

where δ_j is given by

$$\delta_j = \sum_{i=0}^{j} r_x(j+1-i)x_j(i)$$

Since δ_j is not, in general, equal to $b(j+1)$, the extended vector $[\mathbf{x}_j^T, 0]$ is not the solution to the $(j+1)$st-order equations. Consider what happens, however, if we form the sum

$$\mathbf{R}_{j+1} \left\{ \begin{bmatrix} x_j(0) \\ x_j(1) \\ \vdots \\ x_j(j) \\ 0 \end{bmatrix} + q_{j+1} \begin{bmatrix} a_{j+1}^*(j+1) \\ a_{j+1}^*(j) \\ \vdots \\ a_{j+1}^*(1) \\ 1 \end{bmatrix} \right\} = \begin{bmatrix} b(0) \\ b(1) \\ \vdots \\ b(j) \\ \delta_j + q_{j+1}\epsilon_{j+1}^* \end{bmatrix} \tag{5.127}$$

where q_{j+1} is an arbitrary complex constant. If we set

$$q_{j+1} = \frac{b(j+1) - \delta_j}{\epsilon^*_{j+1}} \tag{5.128}$$

then

$$\delta_j + q_{j+1}\epsilon^*_{j+1} = b(j+1)$$

and the vector

$$\mathbf{x}_{j+1} = \begin{bmatrix} x_j(0) \\ x_j(1) \\ \vdots \\ x_j(j) \\ 0 \end{bmatrix} + q_{j+1} \begin{bmatrix} a^*_{j+1}(j+1) \\ a^*_{j+1}(j) \\ \vdots \\ a^*_{j+1}(1) \\ 1 \end{bmatrix} \tag{5.129}$$

is the desired solution. Note that since ϵ_j is real, the complex conjugate in Eq. (5.128) may be neglected. Thus, the Levinson recursion, as summarized in Table 5.6, is defined by Eqs. (5.128) and (5.129) along with the Levinson-Durbin recursion.

Example 5.3.1 ***The Levinson Recursion***

Let us use the Levinson recursion to solve the following set of Toeplitz equations,

$$\begin{bmatrix} 4 & 2 & 1 \\ 2 & 4 & 2 \\ 1 & 2 & 4 \end{bmatrix} \begin{bmatrix} x(0) \\ x(1) \\ x(2) \end{bmatrix} = \begin{bmatrix} 9 \\ 6 \\ 12 \end{bmatrix}$$

The solution is as follows.

1. Initialization of the recursion.

$$\begin{aligned} \epsilon_0 &= r(0) = 4 \\ x_0(0) &= b(0)/r(0) = 9/4 \end{aligned}$$

2. For $j = 0$,

$$\begin{aligned} \gamma_0 &= r(1) = 2 \\ \Gamma_1 &= -\gamma_0/\epsilon_0 = -1/2 \end{aligned}$$

Therefore we have, for the first-order model,

$$\mathbf{a}_1 = \begin{bmatrix} 1 \\ \Gamma_1 \end{bmatrix} = \begin{bmatrix} 1 \\ -1/2 \end{bmatrix}$$

Continuing we have,

$$\begin{aligned} \epsilon_1 &= \epsilon_0\Big[1 - |\Gamma_1|^2\Big] = 4(1 - 1/4) = 3 \\ \delta_0 &= x_0(0)r(1) = 2(9/4) = 9/2 \\ q_1 &= \Big[b(1) - \delta_0\Big]/\epsilon_1 = (6 - 9/2)/3 = 1/2 \end{aligned}$$

Table 5.6 ***The Levinson Recursion***

1. Initialize the recursion
   (a) $a_0(0) = 1$
   (b) $x_0(0) = b(0)/r_x(0)$
   (c) $\epsilon_0 = r_x(0)$
2. For $j = 0, 1, \ldots, p-1$
   (a) $\gamma_j = r_x(j+1) + \sum_{i=1}^{j} a_j(i) r_x(j-i+1)$
   (b) $\Gamma_{j+1} = -\gamma_j/\epsilon_j$
   (c) For $i = 1, 2, \ldots, j$
   $$a_{j+1}(i) = a_j(i) + \Gamma_{j+1} a_j^*(j-i+1)$$
   (d) $a_{j+1}(j+1) = \Gamma_{j+1}$
   (e) $\epsilon_{j+1} = \epsilon_j[1 - |\Gamma_{j+1}|^2]$
   (f) $\delta_j = \sum_{i=0}^{j} x_j(i) r_x(j-i+1)$
   (g) $q_{j+1} = [b(j+1) - \delta_j]/\epsilon_{j+1}$
   (h) For $i = 0, 1, \ldots, j$
   $$x_{j+1}(i) = x_j(i) + q_{j+1} a_{j+1}^*(j-i+1)$$
   (i) $x_{j+1}(j+1) = q_{j+1}$

which leads to the first-order solution

$$\mathbf{x}_1 = \begin{bmatrix} x_0(0) \\ 0 \end{bmatrix} + q_1 \begin{bmatrix} a_1(1) \\ 1 \end{bmatrix} = \begin{bmatrix} 9/4 \\ 0 \end{bmatrix} + 1/2 \begin{bmatrix} -1/2 \\ 1 \end{bmatrix} = \begin{bmatrix} 2 \\ 1/2 \end{bmatrix}$$

3. For $j = 1$ we find that

$$\gamma_1 = r(2) + a_1(1) r(1) = 1 + 2(-1/2) = 0$$

Therefore, $\Gamma_2 = 0$ and the second-order model is simply

$$\mathbf{a}_2 = \begin{bmatrix} \mathbf{a}_1 \\ 0 \end{bmatrix}$$

In addition, $\epsilon_2 = \epsilon_1$ and

$$\delta_1 = \Big[x_1(0),\ x_1(1)\Big] \begin{bmatrix} r(2) \\ r(1) \end{bmatrix} = 2 + 2(1/2) = 3$$

$$q_2 = \Big[b(2) - \delta_1\Big]/\epsilon_2 = (12-3)/3 = 3$$

Thus, the desired solution is

$$\mathbf{x}_2 = \begin{bmatrix} x_1(0) \\ x_1(1) \\ 0 \end{bmatrix} + q_2 \begin{bmatrix} a_2(2) \\ a_2(1) \\ 1 \end{bmatrix} = \begin{bmatrix} 2 \\ 1/2 \\ 0 \end{bmatrix} + 3 \begin{bmatrix} 0 \\ -1/2 \\ 1 \end{bmatrix} = \begin{bmatrix} 2 \\ -1 \\ 3 \end{bmatrix}$$

The General Levinson Recursion

```
function x=glev(r,b)
%
   r=r(:);
   p=length(b);
   a-1;
   x=b(1)/r(1);
   epsilon=r(1);
   for j=2:p;
       g=r(2:j)'*flipud(a);
       gamma=-g/epsilon;
       a=[a;0] + gamma*[0;conj(flipud(a))];
       epsilon=epsilon*(1 - abs(gamma)^2);
       delta=r(2:j)'*flipud(x);
       q=(b(j)-delta)/epsilon;
       x=[x;0] + q*[conj(flipud(a))];
       end
```

Figure 5.16 *A* MATLAB *program for solving a set of Toeplitz equations of the form* $\mathbf{R}_p\mathbf{x}_p = \mathbf{b}$ *using the general Levinson recursion.*

We conclude this section by looking at the computational complexity of the Levinson recursion. From Table 5.6 we see that at jth step of the recursion there are two divisions, $4(j+1)$ multiplications, and $4j+3$ additions. Since there are p steps in the recursion, the total number of multiplications and divisions that are required is

$$\sum_{j=0}^{p-1}(4j+6) = 2p^2 + 4p$$

and the number of additions is

$$\sum_{j=0}^{p-1}(4j+3) = 2p^2 + p$$

Therefore, the number of multiplications and divisions is proportional to p^2 for the Levinson recursion compared with p^3 for Gaussian elimination, and twice that required for the Levinson-Durbin recursion. A MATLAB program for the general Levinson recursion is provided in Fig. 5.16.

5.4 THE SPLIT LEVINSON RECURSION*

The Levinson-Durbin recursion provides an efficient algorithm for solving the pth-order autocorrelation normal equations given in Eq. (5.5) using $p^2 + 2p$ multiplications and p^2 additions. It is possible, however, to reduce the number of multiplications by approximately a factor of 2 while maintaining approximately the same number of additions using what is known as the *Split Levinson Recursion* [5]. The key to achieving this reduction is to replace the predictor polynomials $\mathbf{a}_j$ with *symmetric* vectors, $\mathbf{s}_j$, and use these instead of $\mathbf{a}_j$ in a "Levinson-like" recursion. Due to the symmetry of $\mathbf{s}_j$, only half of the coefficients need to be computed. What needs to be done to make this procedure work is to develop a recursion for the symmetric vectors along with a procedure to find $\mathbf{a}_p$ from the symmetric vectors. The split Levinson algorithm was originally developed by Delsarte and Genin for the case of real data [5]. Later they extended this work by developing a split Schur recursion along with a split lattice filter structure [6]. Although split algorithms have also been developed for the complex case [1, 2, 12], these algorithms are a bit more complicated. Therefore, in this section we will restrict our attention to the case of real data and follow the development originally presented in [5]. In the following chapter, we derive the split lattice filter structure.

From our development of the Levinson-Durbin recursion in Section 5.2.1, we have seen that the solution to the pth-order normal equations is built up recursively from the sequence of reflection coefficients Γ_j for $j = 1, \ldots, p$ and the prediction error filters $A_j(z)$ as follows

$$A_j(z) = A_{j-1}(z) + \Gamma_j z^{-1} A_{j-1}^R(z) \quad ; \quad A_0(z) = 1 \tag{5.130}$$

where $A_j^R(z) = z^{-j} A_j(z^{-1})$ is the *reciprocal polynomial*.[8] If we replace z with z^{-1} in Eq. (5.130) and multiply both sides by z^{-j} we find, as we did in Section 5.2.1, that $A_j^R(z)$ satisfies a similar recursion given by

$$A_j^R(z) = z^{-1} A_{j-1}^R(z) + \Gamma_j A_{j-1}(z) \quad ; \quad A_0^R(z) = 1 \tag{5.131}$$

Let us now define a second set of polynomials, $S_j(z)$ and $S_j^\star(z)$, that are formed by setting the reflection coefficient in Eq. (5.130) equal to $\Gamma_j = 1$ and $\Gamma_j = -1$, respectively,

$$S_j(z) = A_{j-1}(z) + z^{-1} A_{j-1}^R(z) \tag{5.132}$$

$$S_j^\star(z) = A_{j-1}(z) - z^{-1} A_{j-1}^R(z) \tag{5.133}$$

These polynomials are referred to as the *singular predictor polynomials*.[9] By adding together Eqs. (5.130) and (5.131) we may relate $S_j(z)$ to $A_j(z)$ and $A_j^R(z)$ as follows,

$$(1 + \Gamma_j) S_j(z) = A_j(z) + A_j^R(z) \tag{5.134}$$

Similarly, if we subtract Eq. (5.131) from Eq. (5.130) we find that the singular polynomial $S_j^\star(z)$ is related to $A_j(z)$ and $A_j^R(z)$ by

$$(1 - \Gamma_j) S_j^\star(z) = A_j(z) - A_j^R(z) \tag{5.135}$$

If we denote the coefficients of the singular predictor polynomials $S_j(z)$ and $S_j^\star(z)$ by $s_j(i)$ and $s_j^\star(i)$, respectively,

$$S_j(z) = \sum_{i=0}^{j} s_j(i) z^{-i}$$

and

$$S_j^\star(z) = \sum_{i=0}^{j} s_j^\star(i) z^{-i}$$

[8] Recall that we are assuming that the coefficients $a_j(k)$ are real. Therefore, $A_j(z)$ is conjugate symmetric, $A_j(z) = A_j^*(z^*)$ and, as a result, the reciprocal polynomial has the form given here [compare this to the definition of $A_j^R(z)$ given in Eq. (5.22)].

[9] It is interesting to note that the singular predictor polynomials are identical to the *line spectral pairs* that are of interest in speech processing [4,23].

and let $\mathbf{s}_j = [s_j(0), \ldots, s_j(j)]^T$ and $\mathbf{s}_j^\star = [s_j^\star(0), \ldots, s_j^\star(j)]^T$ be the singular predictor vectors, then it follows from Eqs. (5.134) and (5.135) that

$$(1 + \Gamma_j)\mathbf{s}_j = \begin{bmatrix} 1 \\ a_j(1) \\ \vdots \\ a_j(j-1) \\ a_j(j) \end{bmatrix} + \begin{bmatrix} a_j(j) \\ a_j(j-1) \\ \vdots \\ a_j(1) \\ 1 \end{bmatrix} \tag{5.136}$$

and

$$(1 - \Gamma_j)\mathbf{s}_j^\star = \begin{bmatrix} 1 \\ a_j(1) \\ \vdots \\ a_j(j-1) \\ a_j(j) \end{bmatrix} - \begin{bmatrix} a_j(j) \\ a_j(j-1) \\ \vdots \\ a_j(1) \\ 1 \end{bmatrix} \tag{5.137}$$

Therefore, $\mathbf{s}_j$ is a *symmetric* vector that is formed by taking the symmetric (even) part of the vector $\mathbf{a}_j$, and $\mathbf{s}_j^\star$ is an *antisymmetric* vector corresponding to the antisymmetric (odd) part of $\mathbf{a}_j$.[10] Thus, $S_j(z)$ is a *symmetric* polynomial

$$S_j(z) = z^{-j} S_j(z^{-1})$$

and $S_j^\star(z)$ is an *antisymmetric* polynomial,

$$S_j^\star(z) = -z^{-j} S_j^\star(z^{-1})$$

and as a result, both have linear phase.[11]

We now want to show that $\mathbf{s}_j$ and $\mathbf{s}_j^\star$ satisfy a set of Toeplitz equations. Clearly from Eq. (5.136) it follows that

$$(1 + \Gamma_j)\mathbf{R}_j\mathbf{s}_j = \mathbf{R}_j\mathbf{a}_j + \mathbf{R}_j\mathbf{a}_j^R$$

Therefore, since

$$\mathbf{R}_j\mathbf{a}_j = \epsilon_j[1, 0, \ldots, 0]^T$$

and

$$\mathbf{R}_j\mathbf{a}_j^R = \epsilon_j[0, \ldots, 0, 1]^T$$

then

$$\boxed{\mathbf{R}_j\mathbf{s}_j = \tau_j\Big[1, 0, \ldots, 0, 1\Big]^T} \tag{5.138}$$

[10]It is for this reason that the algorithm we are developing is called a "split" Levinson recursion, i.e., it *splits* the predictor polynomial $\mathbf{a}_j$ into its *symmetric* and *antisymmetric* parts.

[11]We may also see this from the definitions of $S_j(z)$ and $S_j^\star(z)$ given in Eqs. (5.132) and (5.133).

where we have defined

$$\tau_j = \frac{\epsilon_j}{1 + \Gamma_j} \tag{5.139}$$

Note that τ_j may be evaluated from $r_x(i)$ and the coefficients of the singular predictor vector $\mathbf{s}_j$ as follows

$$\boxed{\tau_j = \sum_{i=0}^{j} r_x(i) s_j(i)} \tag{5.140}$$

Similarly for $\mathbf{s}_j^\star$ we find that $\mathbf{s}_j^\star$ satisfies the Toeplitz equations

$$\boxed{\mathbf{R}_j \mathbf{s}_j^\star = \tau_j^\star \Big[1, 0, \ldots, 0, -1\Big]^T} \tag{5.141}$$

where

$$\tau_j^\star = \frac{\epsilon_j}{1 - \Gamma_j} \tag{5.142}$$

Given that $\mathbf{s}_j$ and $\mathbf{s}_j^\star$ satisfy a set of Toeplitz equations, we may now derive a recursion for the singular predictor polynomials $S_j(z)$, similar to what we did for $A_j(z)$ and $A_j^R(z)$ in the Levinson-Durbin recursion. To do this, we must first show how to express $A_j(z)$ in terms of $S_j(z)$ and $S_{j+1}(z)$. Beginning with Eq. (5.134) we have

$$A_j(z) = (1 + \Gamma_j) S_j(z) - A_j^R(z) \tag{5.143}$$

Replacing j by $(j+1)$ in Eq. (5.132) and solving for $A_j^R(z)$ gives

$$A_j^R(z) = z S_{j+1}(z) - z A_j(z) \tag{5.144}$$

Substituting Eq. (5.144) into Eq. (5.143) we find

$$A_j(z) = (1 + \Gamma_j) S_j(z) - z S_{j+1}(z) + z A_j(z)$$

Therefore, solving for $A_j(z)$,

$$\boxed{(1 - z) A_j(z) = (1 + \Gamma_j) S_j(z) - z S_{j+1}(z)} \tag{5.145}$$

which is the desired relation. Using Eq. (5.145) we may now derive a recursion for $S_j(z)$. We begin with the Levinson-Durbin recursion

$$A_j(z) = A_{j-1}(z) + \Gamma_j z^{-1} A_{j-1}^R(z) \tag{5.146}$$

and note from Eq. (5.132) that $A_{j-1}^R(z)$ may be expressed in terms of $A_{j-1}(z)$ and $S_j(z)$ as follows,

$$A_{j-1}^R(z) = z S_j(z) - z A_{j-1}(z)$$

Substituting this expression for $A_{j-1}^R(z)$ into Eq. (5.146) we have

$$\begin{aligned} A_j(z) &= A_{j-1}(z) + \Gamma_j z^{-1} \Big[z S_j(z) - z A_{j-1}(z) \Big] \\ &= (1 - \Gamma_j) A_{j-1}(z) + \Gamma_j S_j(z) \end{aligned} \tag{5.147}$$

Now, as we saw in Eq. (5.145), $A_j(z)$ may be written in terms of $S_j(z)$ and $S_{j+1}(z)$. Therefore, multiplying both sides of Eq. (5.147) by $(1-z)$

$$(1-z)A_j(z) = (1-\Gamma_j)(1-z)A_{j-1}(z) + \Gamma_j(1-z)S_j(z)$$

and using Eq. (5.145) to eliminate $(1-z)A_j(z)$ and $(1-z)A_{j-1}(z)$ we have, after combining together common terms,

$$S_{j+1}(z) = (1+z^{-1})S_j(z) - \delta_j z^{-1} S_{j-1}(z) \tag{5.148}$$

where we have defined

$$\delta_j = (1-\Gamma_j)(1+\Gamma_{j-1}) \tag{5.149}$$

In the time-domain, Eq. (5.148) becomes

$$\boxed{s_{j+1}(i) = s_j(i) + s_j(i-1) - \delta_j s_{j-1}(i-1)} \tag{5.150}$$

which is referred to as the *three-term recurrence*.[12] Expressed in terms of the singular predictor vectors, this three-term recurrence is

$$\begin{bmatrix} s_{j+1}(0) \\ s_{j+1}(1) \\ \vdots \\ s_{j+1}(j) \\ s_{j+1}(j+1) \end{bmatrix} = \begin{bmatrix} s_j(0) \\ s_j(1) \\ \vdots \\ s_j(j) \\ 0 \end{bmatrix} + \begin{bmatrix} 0 \\ s_j(0) \\ \vdots \\ s_j(j-1) \\ s_j(j) \end{bmatrix} - \delta_j \begin{bmatrix} 0 \\ s_{j-1}(0) \\ \vdots \\ s_{j-1}(j-1) \\ 0 \end{bmatrix} \tag{5.151}$$

Equation (5.148) or, equivalently, Eq. (5.151), gives us a method for computing the singular predictor polynomial $S_{j+1}(z)$ given the previous two polynomials, $S_j(z)$ and $S_{j-1}(z)$, and the parameter δ_j. What is needed to complete the recursion is a method for computing δ_j. Since the reflection coefficients Γ_j are not known, we cannot use Eq. (5.149) directly. However, using Eq. (5.139) we see that

$$\frac{\tau_j}{\tau_{j-1}} = \frac{\epsilon_j}{1+\Gamma_j}\frac{1+\Gamma_{j-1}}{\epsilon_{j-1}}$$

However, since $\epsilon_j = \epsilon_{j-1}(1-\Gamma_j^2)$, then

$$\frac{\tau_j}{\tau_{j-1}} = \frac{(1-\Gamma_j^2)(1+\Gamma_{j-1})}{1+\Gamma_j} = (1-\Gamma_j)(1+\Gamma_{j-1}) \tag{5.152}$$

which, from Eq. (5.149), is equal to δ_j. Therefore, δ_j may be evaluated using

$$\boxed{\delta_j = \frac{\tau_j}{\tau_{j-1}}} \tag{5.153}$$

[12]This relation is called a three-term recurrence since there are three terms that are used to update the singular predictor vector. By contrast, note that the Levinson-Durbin order update equation is a *two-term recurrence*.

along with Eq. (5.140). Equations (5.140), (5.150), and (5.153) constitute the split Levinson recursion for generating the singular predictor polynomials. The initial conditions required to begin the recursion are

$$\begin{aligned} S_0(z) &= \frac{1}{1+\Gamma_0}\Big[A_0(z) + A_0^R(z)\Big] = 2 \\ S_1(z) &= A_0(z) + z^{-1}A_0^R(z) = 1 + z^{-1} \end{aligned} \tag{5.154}$$

Now that we have a recursion for $S_j(z)$, all that is left is to show how to derive the prediction error filter $A_p(z)$ from the singular predictor polynomials. From Eq. (5.145) with $j = p$ we have

$$(1-z)A_p(z) = (1+\Gamma_p)S_p(z) - zS_{p+1}(z) \tag{5.155}$$

Multiplying both sides by $-z^{-1}$ and solving for $A_p(z)$ gives

$$A_p(z) = z^{-1}A_p(z) + S_{p+1}(z) - z^{-1}(1+\Gamma_p)S_p(z) \tag{5.156}$$

which, in the coefficient domain becomes

$$\boxed{a_p(j) = a_p(j-1) + s_{p+1}(j) - (1+\Gamma_p)s_p(j-1)} \tag{5.157}$$

With the initial condition $a_p(0) = 1$ we thus have a recursion for the coefficients $a_p(j)$. All that is missing is an expression for $(1+\Gamma_p)$. However, this may be computed from $\mathbf{s}_{p+1}$ and $\mathbf{s}_p$ as follows. Setting $z = 1$ in Eq. (5.155) we have

$$(1+\Gamma_p)S_p(z)\Big|_{z=1} - S_{p+1}(z)\Big|_{z=1} = 0 \tag{5.158}$$

Therefore, since

$$S_p(z)\Big|_{z=1} = \sum_{i=0}^{p} s_p(i) \quad \text{and} \quad S_{p+1}(z)\Big|_{z=1} = \sum_{i=0}^{p+1} s_{p+1}(i)$$

then

$$\boxed{(1+\Gamma_p) = \frac{\sum_{i=0}^{p+1} s_{p+1}(i)}{\sum_{i=0}^{p} s_p(i)}} \tag{5.159}$$

The complete *Split Levinson Recursion* is summarized in Table 5.7. It is presented, however, in a form that does not take full advantage of the symmetries in the polynomials $S_j(z)$. In its most efficient form, for example, the evaluation of the coefficients τ_j would be accomplished as follows

$$\tau_j = \begin{cases} \displaystyle\sum_{i=0}^{(j-1)/2} \Big[r_x(i) + r_x(j-i)\Big]s_j(i) & ; \quad j \text{ odd} \\ \displaystyle\sum_{i=0}^{(j-1)/2} \Big[r_x(i) + r_x(j-i)\Big]s_j(i) + r_x(j/2)s_j(j/2) & ; \quad j \text{ even} \end{cases}$$

Table 5.7 ***The Split Levinson Recursion***

1. Initialize the recursion
   (a) $\mathbf{s}_0 = 2$ and $\mathbf{s}_1 = [1, 1]^T$
   (b) $\tau_0 = r_x(0)$
2. For $j = 1, 2, \ldots, p$
   (a) $\tau_j = \sum_{i=0}^{j} r_x(i) s_j(i)$
   (b) $\delta_j = \tau_j / \tau_{j-1}$
   (c) $s_{j+1}(0) = s_{j+1}(j+1) = 1$
   (d) For $k = 1, 2, \ldots, j$
   $$s_{j+1}(k) = s_j(k) + s_j(k-1) - \delta_j s_{j-1}(k-1)$$
3. Set $(1 + \Gamma_p) = \sum_{i=0}^{p+1} s_{p+1}(i) / \sum_{i=0}^{p} s_p(i)$
4. For $j = 1, 2, \ldots, p$
   (a) $a_p(j) = a_p(j-1) + s_{p+1}(j) - (1 + \Gamma_p) s_p(j-1)$
5. Done

In addition, due to the symmetry of s_j, only half of the coefficients need to be evaluated, thereby reducing the number of multiplications and additions by approximately a factor of two.

Although not included as a part of the split Levinson recursion in Table 5.7, it is also possible to compute the reflection coefficient recursively from the coefficients δ_j. Specifically, solving Eq. (5.152) for Γ_j we find that

$$\Gamma_j = 1 - \frac{\delta_j}{1 + \Gamma_{j-1}} \tag{5.160}$$

The initial condition for this recursion is $\Gamma_0 = 0$.

Example 5.4.1 ***The Split Levinson Recursion***

Let us use the split Levinson recursion to find the predictor polynomial corresponding to the autocorrelation sequence

$$\mathbf{r} = \begin{bmatrix} 4, & 1, & 1, & -2 \end{bmatrix}^T$$

In order to make the discussion easier to follow, we will not employ the symmetry of $\mathbf{s}_j$ in our calculations.

1. Initialization of the recursion.

$$\begin{aligned} \tau_0 &= r(0) = 4 \\ \mathbf{s}_0 &= 2 \\ \mathbf{s}_1 &= \begin{bmatrix} 1, & 1 \end{bmatrix}^T \end{aligned}$$

2. For $j = 1$,

$$\tau_1 = \begin{bmatrix} 4, & 1 \end{bmatrix} \begin{bmatrix} 1 \\ 1 \end{bmatrix} = 5 \quad ; \quad \delta_1 = \tau_1/\tau_0 = 5/4$$

Now, using Eq.(5.151) we have

$$\mathbf{s}_2 = \begin{bmatrix} 1 \\ 1 \\ 0 \end{bmatrix} + \begin{bmatrix} 0 \\ 1 \\ 1 \end{bmatrix} - 5/4 \begin{bmatrix} 0 \\ 2 \\ 0 \end{bmatrix} = \begin{bmatrix} 1 \\ -1/2 \\ 1 \end{bmatrix}$$

3. For $j = 2$ we have

$$\tau_2 = \begin{bmatrix} 4, & 1, & 1 \end{bmatrix} \begin{bmatrix} 1 \\ -1/2 \\ 1 \end{bmatrix} = 9/2 \quad ; \quad \delta_2 = \tau_2/\tau_1 = (9/2)(1/5) = 9/10$$

Again using Eq.(5.151) to find $\mathbf{s}_3$ we have

$$\mathbf{s}_3 = \begin{bmatrix} 1 \\ -1/2 \\ 1 \\ 0 \end{bmatrix} + \begin{bmatrix} 0 \\ 1 \\ -1/2 \\ 1 \end{bmatrix} - 9/10 \begin{bmatrix} 0 \\ 1 \\ 1 \\ 0 \end{bmatrix} = \begin{bmatrix} 1 \\ -2/5 \\ -2/5 \\ 1 \end{bmatrix}$$

4. Finally, for $j = 3$ we have

$$\tau_3 = \begin{bmatrix} 4, & 1, & 1, & -2 \end{bmatrix} \begin{bmatrix} 1 \\ -2/5 \\ -2/5 \\ 1 \end{bmatrix} = 6/5$$

$$\delta_3 = \tau_3/\tau_2 = (6/5)(2/9) = 4/15$$

Again using Eq.(5.151) to find $\mathbf{s}_4$ we have

$$\mathbf{s}_4 = \begin{bmatrix} 1 \\ -2/5 \\ -2/5 \\ 1 \\ 0 \end{bmatrix} + \begin{bmatrix} 0 \\ 1 \\ -2/5 \\ -2/5 \\ 1 \end{bmatrix} - 4/15 \begin{bmatrix} 0 \\ 1 \\ -1/2 \\ 1 \\ 0 \end{bmatrix} = \begin{bmatrix} 1 \\ 1/3 \\ -2/3 \\ 1/3 \\ 1 \end{bmatrix}$$

Finally, we compute the predictor polynomial with the help of Eqs. (5.157) and (5.159).

First, using Eq. (5.159) we have

$$(1+\Gamma_3) = \frac{\sum_{i=0}^{4} s_4(i)}{\sum_{i=0}^{3} s_3(i)} = 5/3$$

Then, from Eq. (5.157) we have

$$\mathbf{a}_3 = \begin{bmatrix} 1 \\ a_3(1) \\ a_3(2) \\ a_3(3) \end{bmatrix} = \begin{bmatrix} 0 \\ 1 \\ a_3(1) \\ a_3(2) \end{bmatrix} + \begin{bmatrix} 1 \\ s_4(1) \\ s_4(2) \\ s_4(3) \end{bmatrix} - (1+\Gamma_3) \begin{bmatrix} 0 \\ 1 \\ s_3(1) \\ s_3(2) \end{bmatrix}$$

Incorporating the values for the singular predictor vectors gives

$$\begin{bmatrix} 1 \\ a_3(1) \\ a_3(2) \\ a_3(3) \end{bmatrix} = \begin{bmatrix} 0 \\ 1 \\ a_3(1) \\ a_3(2) \end{bmatrix} + \begin{bmatrix} 1 \\ -4/3 \\ 0 \\ 1 \end{bmatrix}$$

Solving this equation recursively for the coefficients $a_3(i)$ we find

$$\mathbf{a}_3 = \begin{bmatrix} 1 \\ -1/3 \\ -1/3 \\ 2/3 \end{bmatrix}$$

5.5 SUMMARY

In this chapter, we have looked at several different algorithms for recursively solving a set of Toeplitz equations. Each of these algorithms exploits the symmetries and redundancies of the Toeplitz matrix. This chapter began with a derivation of the Levinson-Durbin recursion for solving a set of Hermitian Toeplitz equations in which the right-hand side is a unit vector,

$$\mathbf{R}_p \mathbf{a}_p = \epsilon_p \mathbf{u}_1 \tag{5.161}$$

Compared with $O(p^3)$ operations that are required to solve a set of p linear equations using Gaussian elimination, the Levinson-Durbin recursion requires $O(p^2)$ operations. Using the Levinson-Durbin recursion we were able to establish some important properties of the solution $\mathbf{a}_p$ to these equations. It was shown, for example, that $\mathbf{a}_p$ generates a minimum phase polynomial if and only if the Toeplitz matrix $\mathbf{R}_p$ is positive definite and, in this case, the reflection coefficients Γ_j that are generated by the Levinson-Durbin recursion are bounded by one in magnitude, $|\Gamma_j| < 1$. As a result, we were able to establish that the all-pole models that are formed using Prony's method and the autocorrelation method are guaranteed to be stable. We were also able to demonstrate that if $\mathbf{R}_p > 0$, then $r_x(k)$ for $k = 0, 1, \ldots, p$ represents a valid partial autocorrelation sequence and may be extended for $k > p$. This extrapolation may be performed by setting $|\Gamma_k| < 1$ for $k > p$ and finding the corresponding autocorrelation sequence using the inverse Levinson-Durbin recursion.

In addition to the Levinson-Durbin recursion, we derived a number of other "Levinson-like" recursions. For example, with the view that the Levinson-Durbin recursion is a mapping from a set of autocorrelations $r_x(k)$ to a set of filter coefficients $a_p(k)$ and a set of reflection coefficients Γ_j, we saw that it was possible to derive recursions to find the filter coefficients from the reflection coefficients (step-up recursion), the reflection coefficients from the filter coefficients (step-down recursion), and the autocorrelation sequence from the reflection coefficients and the modeling error ϵ_p (inverse Levinson-Durbin recursion). Two additional recursions that were developed for solving equations of the form given in Eq. (5.161) are the split Levinson recursion and the Schur recursion. The split Levinson recursion reduces the computational requirements of the Levinson-Durbin recursion by approximately a factor of two by introducing the set of singular vectors $\mathbf{s}_j$. The Schur recursion, on the other hand, achieves some modest computational savings by generating only the reflection coefficients from the autocorrelation sequence. The main advantage of the Schur recursion over the Levinson-Durbin and split Levinson recursions, however, is that it is suitable for parallel processing in a multiprocessor environment. Finally, we derived the general Levinson recursion for solving a general set of Toeplitz equations of the form

$$\mathbf{R}_p\mathbf{a}_p = \mathbf{b} \tag{5.162}$$

where $\mathbf{b}$ is an arbitrary vector. As with the Levinson-Durbin recursion, the general Levinson recursion requires $O(p^2)$ operations.

In addition to the recursions developed in this chapter, there are other *fast* algorithms that may be of interest. With respect to the problem of signal modeling, one particular algorithm of interest is the *fast covariance algorithm* for solving the covariance normal equations given in Eq. (4.127). As with the Levinson and the Levinson-Durbin recursions, the fast covariance algorithm requires on the order of p^2 operations. Although the approach that is used to derive the fast covariance algorithm is similar in style to that used for the Levinson recursion, it is considerably more involved. For example, the fast covariance algorithm requires a time-update recursion in addition to the *order-update recursion* that is found in the Levinson recursion. This time-update recursion relates the solution to the covariance normal equations over the interval $[L, U]$ to the solution that is obtained over the intervals $[L+1, U]$ and $[L, U-1]$. The details of this recursion may be found in [18, 19]. Another recursion of interest is the Trench algorithm, which generalizes the Levinson recursion by allowing the Toeplitz matrix to be non-symmetric [10, 25]. The Trench algorithm therefore may be used to solve the Padé equations and the modified Yule-Walker equations. In its most general form, the Trench algorithm gives both the inverse matrix along with the solution $\mathbf{x}_p$ to Eq. (5.162). Finally, it should be noted that, in addition to the "Levinson-like" recursions, there are other *fast algorithms* for solving linear equations that have other types of structured matrices such as Hankel and Vandermond matrices [8, 15].

References

1. Y. Bistritz, H. Lev-Ari, and T. Kailath, "Complexity reduced lattice filters for digital speech processing," *Proc. 1987 Int. Conf. on Acoust., Speech, Sig. Proc.*, pp. 21–24, April 1987.
2. Y. Bistritz, H. Lev-Ari, and T. Kailath, "Immittance-type three-term Schur and Levinson recursions for quasi-Toeplitz and complex Hermitian matrices," *SIAM Journal on Matrix Analysis and Applications*, vol. 12, no. 3, pp. 497–520, July 1991.
3. R. V. Churchill and J. W. Brown, *Complex Variables and Applications*, McGraw-Hill, New York, 1990.
4. J. R. Deller, J. G. Proakis, and J. H. L. Hansen, *Discrete-time Processing of Speech Signals*, MacMillan, New York, 1993.
5. P. Delsarte and Y. V. Genin, "The split Levinson algorithm," *IEEE Trans. Acoust., Speech, Sig. Proc.*, vol. ASSP-34, no. 3, pp. 470–478, June 1986.
6. P. Delsarte and Y. V. Genin, "On the splitting of classical algorithms in linear prediction theory," *IEEE Trans. Acoust., Speech, Sig. Proc.*, vol. ASSP-35, no. 5, pp. 645–653, May 1987.
7. J. Durbin, "The fitting of time series models," *Rev. Internat. Statist. Inst.*, vol. 23, pp. 233–244, 1960.
8. I. Gohberg, T. Kailath, and I. Koltracht, "Efficient solution of linear systems of equations with recursive structure," *Lin. Alg. Appl.*, vol. 80, no. 81, 1986.
9. I. Gohberg, ed., *I. Schur Methods in Operator Theory and Signal Processing*, vol. 18, Birkhäuser, Boston, 1986.
10. G. H. Golub and C. F. Van Loan, *Matrix Computations*, Johns Hopkins University Press, Baltimore, 1989.
11. M. H. Hayes and M. A. Clements, "An efficient algorithm for computing Pisarenko's harmonic decomposition using Levinson's recursion," *IEEE Trans. Acoust., Speech, Sig. Proc.*, vol. ASSP-34, no. 3, pp. 485–491, June 1986.
12. H. Krishna and S. D. Morgera, "The Levinson recurrence and fast algorithms for solving Toeplitz systems of linear equations," *IEEE Trans. Acoust., Speech, Sig. Proc.*, vol. ASSP-35, no. 6, pp. 839–848, June 1987.
13. S. Y. Kung and Y. H. Hu, "A highly concurrent algorithm and pipelined architecture for solving Toeplitz systems," *IEEE Trans. Acoust., Speech, Sig. Proc.*, vol. ASSP-31, pp. 66–76, Feb. 1983.
14. S. Lang and J. McClellan, "A simple proof of stability for all-pole linear prediction models," *Proc. IEEE*, vol. 67, pp. 860–861, May 1979.
15. H. Lev-Ari and T. Kailath, "Triangular factorization of structured Hermitian matrices," in *Schur Methods in Operator Theory and Signal Processing*, I. Gohberg, Ed., vol. 18, Boston, Birkhäuser, 1986.
16. N. Levinson, "The Wiener rms error criterion in filter design and prediction," *J. Math. Phys.*, vol. 25, pp. 261–278, 1947.
17. J. Makhoul, "On the eigenvalues of symmetric Toeplitz matrices," *IEEE Trans. Acoust., Speech, Sig. Proc.*, vol. ASSP-29, no. 4, pp. 868–872, Aug. 1981.
18. J. H. McClellan, "Parametric Signal Modeling," Chapter 1 in *Advanced Topics in Signal Processing*, Prentice-Hall, NJ, 1988.
19. M. Morf, B. W. Dickenson, T. Kailath, and A. C. G. Vieira, "Efficient solution of covariance equations for linear prediction," *IEEE Trans. Acoust., Speech, Sig. Proc.*, vol. ASSP-25, pp. 429–433, Oct. 1977.
20. L. Pakula and S. Kay, "Simple proofs of the minimum phase property of the prediction error filter," *IEEE Trans. Acoust., Speech, Sig. Proc.*, vol. ASSP-31, no. 2, pp. 501, April 1983.
21. R. A. Roberts and C. T. Mullis, *Digital Signal Processing*, Addison Wesley, Reading, MA. 1987.
22. I. Schur, "On power series which are bounded in the interior of the unit circle," *J. Reine Angew. Math.*, vol. 147, pp. 205–232, 1917, vol. 148, pp. 122–125, 1918. (Translation of Parts I and II may be found in I. Gohberg, pp. 31–59 and 61–88).

23. F. K. Soong and B.-H. Juang, "Line spectrum pair (LSP) and speech compression," *Proc. 1984 Int. Conf. on Acoust., Speech, Sig. Proc.*, pp. 1.10.1–1.10.4, March 1984.
24. S. Treitel and T. J. Ulrych, "A new proof of the minimum-phase property of the unit prediction error operator," *IEEE Trans. Acoust., Speech, Sig. Proc.*, vol. ASSP-27, no. 1, pp. 99–100, February 1979.
25. W. F. Trench, "An algorithm for the inversion of finite Toeplitz matrices," *J. SIAM*, vol. 12, no. 3, pp. 512–522, 1964.
26. R. A. Wiggins and E. A. Robinson, "Recursive solution to the multichannel filtering problem," *J. Geophys. Research*, vol. 70, no. 8, pp. 1885–1891, April 1965.
27. A. Vieira and T. Kailath, "On another approach to the Schur-Cohn criterion," *IEEE Trans. Circ. Syst*, CAS-24, no. 4, pp. 218–220, 1977.

5.6 PROBLEMS

5.1. Given the autocorrelation sequence

$$r_x(0) = 1, \qquad r_x(1) = 0.8, \qquad r_x(2) = 0.5, \qquad r_x(3) = 0.1$$

find the reflection coefficients, Γ_j, the model parameters, $a_j(k)$, and the modeling errors, ϵ_j, for $j = 1, 2, 3$.

5.2. Let $A_{p-1}(z)$ be a polynomial of order $(p-1)$ of the form

$$A_{p-1}(z) = 1 + \sum_{k=1}^{p-1} a_{p-1}(k) z^{-k} \tag{P5.2-1}$$

and let $\Gamma_1, \Gamma_2, \ldots, \Gamma_{p-1}$ be the reflection coefficients that are generated by the Levinson-Durbin recursion. A pth-order polynomial is formed via the Levinson update equation as follows

$$A_p(z) = A_{p-1}(z) + \Gamma_p z^{-p} A^*_{p-1}(1/z^*)$$

(a) If $|\Gamma_j| < 1$ for $j = 1, \ldots, p-1$ and if $|\Gamma_p| = 1$, what can be said about the location of the zeros of $A_p(z)$?

(b) Suppose $A_p(z)$ may be factored as

$$A_p(z) = \prod_{k=1}^{p} (1 - \alpha_k z^{-1})$$

i.e., the zeros of $A_p(z)$ are at $\alpha_1, \alpha_2, \ldots, \alpha_p$. If $\widetilde{A}_p(z)$ is a polynomial with reflection coefficients $\widetilde{\Gamma}_j$ where

$$\begin{aligned} \widetilde{\Gamma}_j &= \Gamma_j \quad \text{for} \quad j = 1, 2, \ldots, p-1 \\ \widetilde{\Gamma}_p &= 1/\Gamma_p^* \end{aligned}$$

How are the zeros of $\widetilde{A}_p(z)$ related to those of $A_p(z)$?

(c) If we consider $A_p(z)$ in (P5.2-1) to be a continuous function of the reflection coefficient Γ_p, using your results derived in parts (a) and (b), describe how the zeros move in the z-plane as Γ_p is varied in a continuous fashion from some number, say ϵ, to its reciprocal, $1/\epsilon$.

5.3. Let $a_p(k)$ be the filter coefficients corresponding to the reflection coefficients Γ_k, for $k = 1, 2, \ldots p$.

(a) Prove that if the reflection coefficients are modulated by $(-1)^k$

$$\hat{\Gamma}_k = (-1)^k \Gamma_k$$

then the new set of filter coefficients $\hat{a}_p(k)$ are

$$\hat{a}_p(k) = (-1)^k a_p(k)$$

(b) Can you generalize the result in part (a) to the case in which

$$\hat{\Gamma}_k = \alpha^k \Gamma_k$$

where α is a complex number with $|\alpha| = 1$? What about if $|\alpha| < 1$?

5.4. For the reflection coefficient sequence

$$\Gamma_k = \alpha^k \quad ; \quad k = 1, 2, \ldots$$

with $|\alpha| < 1$, prove that the zeros of the polynomials $A_p(k)$ lie on a circle of radius α for every $p \geq 1$.

5.5. Without factoring any polynomials, determine whether or not a linear shift-invariant filter with system function

$$H(z) = \frac{1 + 0.8z^{-1} - 0.9z^{-2} + 0.3z^{-3}}{1 - 0.9z^{-1} + 0.8z^{-2} - 0.5z^{-3}}$$

is minimum phase, i.e., all the poles and zeros are inside the unit circle.

5.6. Consider the signal

$$x(n) = \delta(n) + b\delta(n-1)$$

Suppose that we observe $x(n)$ for $n = 0, 1, \ldots, N$.

(a) Using the autocorrelation method, find the 2nd-order all-pole model for $x(n)$.

(b) Suppose we want to find a p-pole model for $x(n)$. Let Γ_j denote the *jth* reflection coefficient of the lattice filter implementation of the all-pole model. Find a recursion for the reflection coefficients that expresses Γ_j in terms of Γ_{j-1} and Γ_{j-2}.

5.7. Suppose we have a data sequence whose z-transform is of the form:

$$X(z) = \frac{G}{1 + \sum_{k=1}^{p} a_p(k) z^{-k}}$$

although the value of p is unknown. The coefficients of the model are computed using Levinson's recursion. How can the value of p be determined by looking at the sequence of reflection coefficients Γ_j for $j = 1, 2, \ldots$?

5.8. Let $r_x(k)$ be a complex autocorrelation sequence given by

$$\mathbf{r}_x = \left[2,\ 0.5(1+j),\ 0.5j\right]^T$$

Use the Levinson-Durbin recursion to solve the autocorrelation normal equations for a second-order all-pole model.

5.9. Determine whether the following statements are *True* or *False*.

(a) If $r_x(k)$ is an autocorrelation sequence with $r_x(k) = 0$ for $|k| > p$ then $\Gamma_k = 0$ for $|k| > p$.

(b) Given an autocorrelation sequence, $r_x(k)$ for $k = 0, \ldots, p$, if the $(p+1) \times (p+1)$ Toeplitz matrix

$$\mathbf{R}_p = \text{Toep}\{r_x(0), r_x(1), \ldots, r_p(p)\}$$

is positive definite, then

$$\mathbf{r}_x = [r_x(0), \ldots, r_x(p), 0, 0, \ldots]^T$$

will *always* be a valid autocorrelation sequence, i.e., extending $r_x(k)$ with zeros is a valid autocorrelation extension.

(c) If $r_x(k)$ is periodic then Γ_j will be periodic with the same period.

5.10. In our discussions of the Levinson-Durbin recursion, we demonstrated the equivalence between the following three sets of parameters:

- $r_x(0), r_x(1), \ldots, r_x(p)$
- $a_p(1), a_p(2), \ldots, a_p(p), b(0)$
- $\Gamma_1, \Gamma_2, \ldots, \Gamma_p, \epsilon_p$

For each of the following signal transformations, determine which of these parameters change and, if possible, describe how.

(a) The signal $x(n)$ is scaled by a constant C,

$$x'(n) = Cx(n)$$

(b) The signal $x(n)$ is modulated by $(-1)^n$,

$$x'(n) = (-1)^n x(n)$$

5.11. The autocorrelation of a signal consisting of a random phase sinusoid in noise is

$$r_x(k) = P\cos(k\omega_0) + \sigma_w^2 \delta(k)$$

where ω_0 is the frequency of the sinusoid, P is the power, and σ_w^2 is the variance of the noise. Suppose that we fit an AR(2) model to the data.

(a) Find the coefficients $\mathbf{a}_2 = [1,\ a_2(1),\ a_2(2)]^T$ of the AR(2) model as a function of ω_0, σ_w^2, and P.

(b) Find the reflection coefficients, Γ_1 and Γ_2, corresponding to the AR(2) model.

(c) What are the limiting values of the AR(2) parameters and the reflection coefficients as $\sigma_w^2 \to 0$?

5.12. Given that $r_x(0) = 1$ and that the first three reflection coefficients are $\Gamma_1 = 0.5$, $\Gamma_2 = 0.5$, and $\Gamma_3 = 0.25$,

(a) Find the corresponding autocorrelation sequence $r_x(1), r_x(2), r_x(3)$.

(b) Find the autocorrelation sequence $r_x(1), r_x(2), r_x(3)$ for the case in which the reflection coefficient Γ_3 is a free variable, i.e., solve for the autocorrelation values as a function of Γ_3.

(c) Repeat part (b) when both Γ_2 and Γ_3 are free parameters.

5.13. Using the autocorrelation method, an all-pole model of the form

$$H(z) = \frac{b(0)}{1 + a(1)z^{-1} + a(2)z^{-2} + a(3)z^{-3}}$$

has been found for a signal $x(n)$. The constant in the numerator has been chosen so that the autocorrelation of the signal matches the autocorrelation of the model, i.e.,

$$r_x(k) = r_h(k)$$

where $r_h(k)$ is the autocorrelation of $h(n)$. In addition, you know the following about the signal and the model:

1. $r_x(0) = 4$
2. $\Gamma_3 = 0.5$
3. $\Gamma_1 > 0$ and $\Gamma_2 > 0$
4. $x(0) = 1$
5. $\epsilon_3 = 11/16$, where ϵ_3 is the third-order modeling error.
6. $\det(\mathbf{R}_2) = 11$, where $\mathbf{R}_2 = \text{Toep}\{r_x(0), r_x(1), r_x(2)\}$.

Find the values for the model parameters

$$a(1), \quad a(2), \quad a(3), \quad \text{and} \quad b(0)$$

5.14. The first seven values of the unit sample response, $h(n)$, of a 3rd-order linear shift-invariant filter

$$H(z) = A\frac{1 + b(1)z^{-1} + b(2)z^{-2} + b(3)z^{-3}}{1 + a(1)z^{-1} + a(2)z^{-2} + a(3)z^{-3}}$$

are given by

$$\mathbf{h} = \left[1,\ 1/4,\ 1/2,\ 1,\ 0,\ 0,\ 7/8\right]^T$$

Determine whether or not the filter is stable. If more information is needed, state what is required and make the necessary assumptions.

5.15. The extendibility problem in power spectrum estimation concerns the issue of whether or not a finite length sequence,

$$r_x(1), r_x(2), r_x(3), \ldots, r_x(p)$$

may be extended (extrapolated) into a legitimate autocorrelation sequence so that

$$P_x(e^{j\omega}) = \sum_{k=-\infty}^{\infty} r_x(k)e^{-jk\omega} \qquad \text{(P5.15-1)}$$

is a valid power spectrum. In other words, is it possible to find values of $r_x(k)$ for $|k| > p$ such that $P_x(e^{j\omega})$ in (P5.15-1) is a non-negative real function of ω?

(a) Develop a procedure that uses Levinson's recursion to test the extendibility of a sequence.

(b) Use your procedure developed in (a) to determine the constraints on a and b that are necessary and sufficient in order for the sequence

$$r_x(0) = 1, \qquad r_x(1) = a, \qquad r_x(2) = b$$

to be an extendible sequence.

(c) Assuming that the sequence in (b) is extendible, find two different legitimate extensions.

5.16. Which of the following autocorrelation sequences are extendible? For those that are extendible, find an extension and determine whether or not the extension is unique.

(a) $\mathbf{r}_x = \left[1.0,\ 0.6,\ 0.6\right]^T$

(b) $\mathbf{r}_x = \left[1.0,\ 0.6,\ -0.6\right]^T$

(c) $\mathbf{r}_x = \left[1.0,\ 0.0,\ 1.0\right]^T$

5.17. Let $\mathbf{R}_3$ be the symmetric Toeplitz matrix formed from the autocorrelation sequence $r_x(0)$, $r_x(1)$, $r_x(2)$, and $r_x(3)$. If the reflection coefficients that result from applying the Levinson-Durbin recursion to $\mathbf{R}_3$ are

$$\Gamma_1 = \tfrac{1}{2} \qquad \Gamma_2 = \tfrac{1}{3} \qquad \Gamma_3 = \tfrac{1}{4}$$

and if $r_x(0) = 1$, find the determinant of $\mathbf{R}_3$.

5.18. Let $r_x(k) = \sigma_x^2\delta(k) + 1$. Find the reflection coefficients Γ_k for all $k \geq 0$ and find the all-pole models $A_p(k)$ for all $p \geq 1$.

5.19. A pth-order all-pole model for a signal $x(n)$ is parameterized by the $p+1$ parameters ϵ_p and $a_p(k)$. Since the reflection coefficients Γ_k for a stable model are bounded by one in magnitude, they are automatically scaled. In speech processing, therefore, there has been an interest in coding a speech waveform in terms of its reflection coefficient sequence. The relationship between a reflection coefficient sequence and the spectrum, however, is not easily discernable. Consider, for example, the following three reflection coefficient sequences

1. $\Gamma_k = \dfrac{1}{k+1}$
2. $\Gamma_k = -\dfrac{1}{k+1}$
3. $\Gamma_k = \dfrac{(-1)^k}{k+1}$

Although the only difference between these reflection coefficient sequences is in terms of the sign of Γ_k, the power spectra are quite different. For each of these sequences, find the corresponding autocorrelation sequence, $r_x(k)$, and power spectrum, $P_x(e^{j\omega})$, in closed form.

5.20. Let $x(n)$ be a random process with autocorrelation sequence

$$r_x(k) = (0.2)^{|k|}$$

(a) Find the reflection coefficients Γ_1 and Γ_2 for a second-order predictor and draw the lattice filter network.

(b) Suppose that uncorrelated white noise with a variance of $\sigma_w^2 = 0.1$ is added to $x(n)$,

$$y(n) = x(n) + w(n)$$

How do the reflection coefficients change?

(c) Can you make any general statements about the effect on the reflection coefficients when white noise is added to a process?

5.21. The reflection coefficients corresponding to a third-order all-pole model are

$$\Gamma_1 = 0.25 \qquad \Gamma_2 = 0.50 \qquad \Gamma_3 = 0.25$$

and the modeling error is given by

$$\epsilon_3 = (15/16)^2$$

(a) Find the direct form filter coefficients, $a_3(k)$, for this third-order model.

(b) Find the autocorrelation values $r_x(0)$, $r_x(1)$, $r_x(2)$, and $r_x(3)$ that led to this model.

(c) If a fourth-order model were found for $x(n)$, what value or values for $r_x(4)$ result in the minimum model error, ϵ_4?

(d) Repeat part (c) and find the value or values for $r_x(4)$ that produce the maximum error, ϵ_4.

5.22. The reflection coefficients for a two-pole model of a signal $x(n)$ are $\Gamma_1 = 0.25$ and $\Gamma_2 = 0.25$ and the "modeling error" is $\epsilon_2 = 9$.

(a) If $r_x(3) = 1$ find the modeling error, ϵ_3, for a three-pole model.

(b) If the signal values, $x(n)$, are multiplied by $1/2$, i.e., $y(n) = 0.5x(n)$, find the reflection coefficients and the modeling error for a two-pole model of $y(n)$.

5.23. You are given the following sequence of autocorrelation values

$$\mathbf{r}_x = \left[10,\ -1,\ 0.1,\ -1\right]^T$$

(a) Use the Schur recursion to find the reflection coefficient sequence $\Gamma_1, \Gamma_2, \Gamma_3$.

(b) What is the final generator matrix, $\mathbf{G}_3$, equal to?

(c) Find the modeling error, ϵ_3.

5.24. Let $r_x(0), r_x(1), \ldots, r_x(p)$ be a set of autocorrelation values and let $\mathbf{R}_p$ be the corresponding $(p+1) \times (p+1)$ autocorrelation matrix. Show that

$$1 - |\Gamma_p|^2 = \frac{\det \mathbf{R}_p \det \mathbf{R}_{p-2}}{\left[\det \mathbf{R}_{p-1}\right]^2}$$

where Γ_p is the pth reflection coefficient.

5.25. If $|\Gamma_j| < 1$, derive a bound for the coefficients δ_j in the split Levinson recursion.

5.26. Let $h_{n_0}(n)$ be the FIR least squares inverse filter of length N with delay n_0 for a sequence $g(n)$, i.e.,

$$h_{n_0}(n) * g(n) \approx \delta(n - n_0)$$

The coefficients $h_{n_0}(n)$ are the solution to the Toeplitz equations (see p. 174 in Chapter 4)

$$\mathbf{R}_g \mathbf{h}_{n_0} = \mathbf{g}_{n_0} \qquad \text{(P5.26-1)}$$

which may be solved efficiently using the Levinson recursion. Since the value for the delay n_0 that produces the smallest least squares error is typically unknown, to find the optimum value for n_0 these equations must be solved for each value of n_0, beginning with $n_0 = 0$. Instead of using the Levinson recursion to solve these equations repeatedly, it is possible to take advantage of the relationship between $\mathbf{g}_{n_0}$ and $\mathbf{g}_{n_0+1}$,

$$\mathbf{g}_{n_0} = \begin{bmatrix} g^*(n_0) \\ g^*(n_0-1) \\ \vdots \\ g^*(0) \\ 0 \\ 0 \\ \vdots \\ 0 \end{bmatrix} \quad ; \quad \mathbf{g}_{n_0+1} = \begin{bmatrix} g^*(n_0+1) \\ g^*(n_0) \\ \vdots \\ g^*(1) \\ g^*(0) \\ 0 \\ \vdots \\ 0 \end{bmatrix}$$

to derive a recursion for $h_{n_0}(n)$. In this problem we derive this recursion which is known as the *Simpson sideways recursion* [26].

(a) The solution to Eq. (P5.26-1) for $n_0 = 0$ may be found using the Levinson-Durbin recursion. Show how to generate the solution for $n_0 = 1$ from the solution for $n_0 = 0$ in less than $4N$ multiplications and divisions where N is the length of the inverse filter $\mathbf{h}_{n_0}$. Note that any information generated in the Levinson-Durbin recursion (for $n_0 = 0$) can be used to construct the new solution.

(b) Generalize the result of part (a) to obtain a recursion that will successively construct the solution for all $n_0 > 0$. Again your method should have less than $4N$ multiplications and divisions at each step.

(c) Write an expression for the error $\mathcal{E}_{n_0}$ at the n_0th step of the recursion in terms of the coefficients $g(n)$ and the coefficients of the least squares inverse filter $h_{n_0}(n)$.

(d) Write a MATLAB program that implements the Simpson sideways recursion.

(e) How can this recursion be used to find the inverse of a Toeplitz matrix?

Computer Exercises

C5.1. Modify the m-file `rtog.m` to find the Cholesky (LDU) decomposition of a Hermitian Toeplitz matrix $\mathbf{R}_x$. Compare the efficiency of your m-file with the MATLAB Cholesky decomposition program `chol.m`.

C5.2. The inverse Levinson-Durbin recursion is a mapping from the sequence of reflection coefficients Γ_j to the autocorrelation sequence $r_x(k)$. Beginning with $r_x(0)$, the autocorrelations $r_x(k)$ for $k > 0$ are determined recursively. In this problem, we derive and implement another approach for the recovery of the autocorrelation sequence.

(a) Beginning with the normal equations

$$\mathbf{R}_p \mathbf{a}_p = \epsilon_p \mathbf{u}_1$$

rewrite these equations in the form

$$\mathbf{A}_p \mathbf{r}_p = \epsilon_p \mathbf{u}_1$$

where $\mathbf{A}_p$ is a matrix containing the pth-order model coefficients $a_p(k)$ and $\mathbf{r}_p$ is the vector containing the autocorrelation values that are to be determined (assume that the data is complex). What is the structure of the matrix $\mathbf{A}_p$ and how may it be generated?

(b) Write a MATLAB program that will set up these equations and solve for the autocorrelation sequence.

(c) Compare the complexity of this solution to the inverse Levinson-Durbin recursion.

C5.3. The Schur recursion is a mapping from the autocorrelation sequence $r_x(k)$ to a set of reflection coefficients.

(a) Write a MATLAB program that implements the Schur recursion. Using the `flops` command, compare the efficiency of the Schur recursion to the Levinson-Durbin recursion to find the reflection coefficients Γ_j from an autocorrelation sequence $r_x(k)$.

(b) Derive the *inverse Schur recursion* that produces the autocorrelation sequence $r_x(k)$ from the reflection coefficients Γ_j, and write a MATLAB program to implement this recursion. Compare the efficiency of the inverse Schur recursion to the inverse Levinson-Durbin recursion to find the autocorrelation sequence from the reflection coefficients.

C5.4. In the derivation of the split Levinson recursion, we defined a new set of coefficients, δ_j. As we will see in Chapter 6, these coefficients are used in the implementation of a split lattice filter.

(a) Show that these two sets of coefficients are equivalent in the sense that one set may be derived from the other,

$$\left\{\Gamma_1,\ \Gamma_2,\ \ldots,\ \Gamma_p\right\} \longleftrightarrow \left\{\delta_1,\ \delta_2,\ \ldots,\ \delta_p\right\}$$

(b) Write a MATLAB program `gtod.m` that will convert the reflection coefficients Γ_j into the split Levinson coefficients δ_j.

(c) Write a MATLAB program `dtog.m` that will convert the split Levinson coefficients δ_j into a set of reflection coefficients Γ_j.

(d) Use these m-files to study the relationship between the coefficients Γ_j and δ_j. For example, what does the constraint $|\Gamma_j| < 1$ imply about δ_j? What happens if $\Gamma_p = \pm 1$? Is it possible to determine whether or not $|\Gamma_j| < 1$ by simply looking at δ_j?

C5.5. The *line spectral pair* (LSP) coefficients were introduced in the 1980's as an alternative to the filter coefficients $a_p(k)$ and the reflection coefficients Γ_j in the representation of an all-pole model for a signal $x(n)$. Given the prediction error filter $A_p(z)$, the LSP representation is as follows. With $S_j(z)$ and $S_j^\star(z)$ the singular predictor polynomials that were introduced in the derivation of the split Levinson recursion, note that $A_p(z)$ may be represented in terms of $S_{p+1}(z)$ and $S_{p+1}^\star(z)$ as follows,

$$A_p(z) = \frac{1}{2}\Big[S_{p+1}(z) + S_{p+1}^\star(z)\Big]$$

Since the roots of $S_{p+1}(z)$ and $S_{p+1}^\star(z)$ lie on the unit circle, the angles of these zeros uniquely define the singular predictor polynomials and, thus, the prediction error filter $A_p(z)$.

(a) Generate several different singular predictor polynomials, $S_{p+1}(z)$ and $S_{p+1}^\star(z)$, and look at the locations of the roots of these polynomials. What relationship do you observe between the locations of the roots of $S_{p+1}(z)$ and the roots of $S_{p+1}^\star(z)$?

(b) Each zero of $A_p(z)$ maps into one zero in each of the polynomials $S_{p+1}(z)$ and $S^{\star}_{p+1}(z)$. Investigate what happens when a zero of $S_{p+1}(z)$ is close to a zero of $S^{\star}_{p+1}(z)$. Does this imply anything about the spectrum $A_p(e^{j\omega})$?

(c) Let θ_k denote the angles of the roots of the polynomials $S_{p+1}(z)$ and $S^{\star}_{p+1}(z)$ that lie within the interval $[0, \pi]$. Assume that the angles are ordered so that $\theta_{k+1} \geq \theta_k$, and let $\Delta\theta_k = \theta_{k+1} - \theta_k$. Write a MATLAB program `lsp.m` that will produce the prediction error filter $A_p(z)$ from the angles $\Delta\theta_k$.

(d) Generate a number of different AR(12) processes by filtering unit variance white noise with a 12th-order all-pole filter. Place the poles of the filter close to the unit circle, at approximately the same radius, and at angles that are approximately harmonically related to each other. Compare the accuracy in the model for $x(n)$ when quantizing the model coefficients $a_p(k)$ to 16 bits and quantizing the LSP coefficients $\Delta\theta_k$ to 16 bits.

LATTICE FILTERS 6

6.1 INTRODUCTION

In Chapter 5 we discovered how the reflection coefficients from the Levinson-Durbin recursion may be used in a lattice filter implementation of the *inverse filter* $A_p(z)$. This structure has a number of interesting and important properties including modularity, low sensitivity to parameter quantization effects, and a simple test for ensuring that $A_p(z)$ is minimum phase. As a result, lattice filters have found their way into many different and important signal processing applications and have been used both for signal analysis and synthesis. In this chapter, we look again at the lattice filter structure, derive some alternative lattice filter forms, and explore how these filters may be used for signal modeling.

This chapter begins with a derivation of the FIR lattice filter structure in Section 6.2 where we explore the relationship between the lattice filter and the problems of forward and backward linear prediction. In Section 6.3 we then develop the split lattice filter which is based on the singular predictor polynomials and the split Levinson recursion. All-pole lattice filters, allpass lattice filters, and pole-zero lattice filters are then considered in Section 6.4. In Section 6.5 we then turn our attention to the use of lattice filters for all-pole signal modeling. We derive three approaches that are based on the sequential estimation of the reflection coefficients. These methods are the forward covariance method, the backward covariance method, and Burg's method. We also present a nonsequential method known as the modified covariance method. Finally, in Section 6.6 we look at how the lattice methods of signal modeling may be used to model stochastic signals.

6.2 THE FIR LATTICE FILTER

To derive the FIR lattice filter structure, we will proceed exactly as we did in Section 5.2.2, beginning with the problem of all-pole signal modeling. Therefore, let $x(n)$ be a signal that is to be modeled as the unit sample response of an all-pole filter of the form

$$H(z) = \frac{b(0)}{A_p(z)}$$

As we saw in Section 4.4.3 of Chapter 4, with Prony's method the coefficients of $A_p(z)$ are found by minimizing the squared error,[1]

$$\mathcal{E}_p = \sum_{n=0}^{\infty} |e_p(n)|^2$$

where

$$e_p(n) = x(n) * a_p(n) = x(n) + \sum_{k=1}^{p} a_p(k)x(n-k)$$

with $a_p(0) = 1$. Note that by minimizing $\mathcal{E}_p$ we are solving for the coefficients $a_p(k)$ that minimize the difference between $x(n)$ and what may be considered to be an estimate or *prediction* of $x(n)$,

$$\hat{x}(n) = -\sum_{k=1}^{p} a_p(k)x(n-k)$$

This estimate is formed by taking a linear combination of the previous p values of $x(n)$ and is referred to as the forward predictor of $x(n)$ and

$$e_p(n) = x(n) - \hat{x}(n)$$

is referred to as the pth-order *forward prediction error*. In order to distinguish between the *forward prediction error* and the *backward prediction error*, which will be introduced shortly, we will use the following notation,

$$\boxed{e_p^+(n) = x(n) + \sum_{k=1}^{p} a_p(k)x(n-k)} \tag{6.1}$$

In addition, we will use $\mathcal{E}_p^+$ to denote the sum of the squares of $e_p^+(n)$,

$$\boxed{\mathcal{E}_p^+ = \sum_{n=0}^{\infty} |e_p^+(n)|^2} \tag{6.2}$$

Equation (6.1) allows us to write the pth-order forward prediction error in the z-domain as follows

$$E_p^+(z) = A_p(z)X(z) \tag{6.3}$$

where

$$A_p(z) = 1 + \sum_{k=1}^{p} a_p(k)z^{-k} \tag{6.4}$$

Therefore, $e_p^+(n)$ may be generated by filtering $x(n)$ with the FIR filter $A_p(z)$ which is referred to as the *forward prediction error filter*. This relationship between $e_p^+(n)$ and $x(n)$ is illustrated in Fig. 6.1a.

As we saw in Section 4.4.3, the coefficients of the forward prediction error filter are the solution to the normal equations

$$\mathbf{R}_p\mathbf{a}_p = \epsilon_p\mathbf{u}_1 \tag{6.5}$$

where $\mathbf{R}_p$ is a Hermitian Toeplitz matrix of autocorrelations. As we have seen in Chapter 5,

[1]Note that we have added a subscript p to $e(n)$ to indicate the order of the model.

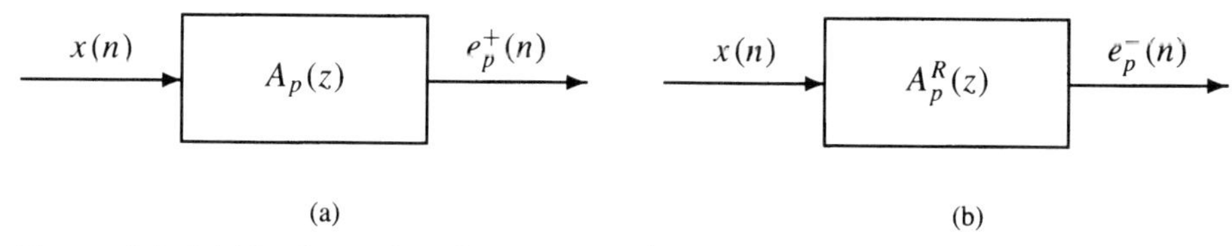

Figure 6.1 *(a) The forward prediction error, $e_p^+(n)$, expressed as the output of the forward prediction error filter, $A_p(z)$ and (b) the backward prediction error, $e_p^-(n)$, expressed as the output of the backward prediction error filter, $A_p^R(z)$.*

these equations may be solved recursively using the Levinson-Durbin recursion. Thus, the coefficients of the $(j+1)$st order prediction error filter, $a_{j+1}(i)$, are related to the coefficients of the jth order prediction error filter through the model-order update equation given in Eq. (5.16) which is repeated below for convenience,

$$a_{j+1}(i) = a_j(i) + \Gamma_{j+1} a_j^*(j-i+1) \tag{6.6}$$

Expressing Eq. (6.6) in the z-domain, we may relate the prediction error filter $A_{j+1}(z)$ to $A_j(z)$ as follows

$$A_{j+1}(z) = A_j(z) + \Gamma_{j+1}\left[z^{-(j+1)} A_j^*(1/z^*)\right] \tag{6.7}$$

Multiplying both sides of Eq. (6.7) by $X(z)$ and using Eq. (6.3) we obtain the following relationship between $E_{j+1}^+(z)$ and $E_j^+(z)$,

$$E_{j+1}^+(z) = E_j^+(z) + z^{-1}\Gamma_{j+1} E_j^-(z) \tag{6.8}$$

where we have defined $E_j^-(z)$ as follows

$$E_j^-(z) = z^{-j} X(z) A_j^*(1/z^*) \tag{6.9}$$

If, as in Chapter 5, we introduce the notation $A_j^R(z) = z^{-j} A_j^*(1/z^*)$, then $E_j^-(z)$ may be written as follows

$$\boxed{E_j^-(z) = A_j^R(z) X(z)} \tag{6.10}$$

Taking the inverse z-transform of both sides of Eq. (6.8), we obtain the time-domain recursion

$$\boxed{e_{j+1}^+(n) = e_j^+(n) + \Gamma_{j+1} e_j^-(n-1)} \tag{6.11}$$

that relates $e_{j+1}^+(n)$ to $e_j^+(n)$ and $e_j^-(n)$.

A signal processing interpretation of the signal $e_j^-(n)$ may be derived from Eq. (6.9) by taking the inverse z-transform as follows,

$$e_j^-(n) = x(n-j) + \sum_{k=1}^{j} a_j^*(k) x(n-j+k) \tag{6.12}$$

As we did with the forward prediction error, $e_j^-(n)$ may be expressed as the difference between $x(n-j)$ and what may be considered to be the estimate or *prediction* of $x(n-j)$ that is formed by taking a linear combination of the j signal values $x(n), x(n-1), \ldots,$ $x(n-j+1)$, i.e.,

$$e_j^-(n) = x(n-j) - \hat{x}(n-j)$$

where

$$\hat{x}(n-j) = -\sum_{k=1}^{j} a_j^*(k)x(n-j+k)$$

What is interesting about this interpretation is that if we minimize the sum of the squares of $e_j^-(n)$,

$$\mathcal{E}_j^- = \sum_{n=0}^{\infty} \left|e_j^-(n)\right|^2$$

then we find that the coefficients that minimize $\mathcal{E}_j^-$ are the same as those that minimize $\mathcal{E}_j^+$ and therefore are found by solving the normal equations given in Eq. (6.5). Thus, $e_j^-(n)$ is referred to as the jth-order *backward prediction error* and $A_j^R(z)$ is known as the *backward prediction error filter*. The relationship between $x(n)$ and $e_j^-(n)$ is illustrated in Fig. 6.1*b*.

Equation (6.11) provides a recursion for the $(j+1)$st-order forward prediction error in terms of the jth-order forward and backward prediction errors. A similar recursion may be derived for the backward prediction error as follows. Taking the complex conjugates of both sides of Eq. (6.6) and substituting $j-i+1$ for i we have

$$a_{j+1}^*(j-i+1) = a_j^*(j-i+1) + \Gamma_{j+1}^* a_j(i) \tag{6.13}$$

Expressing Eq. (6.13) in the z-domain we find that

$$z^{-(j+1)} A_{j+1}^*(1/z^*) = z^{-(j+1)} A_j^*(1/z^*) + \Gamma_{j+1}^* A_j(z) \tag{6.14}$$

Multiplying both sides of Eq. (6.14) by $X(z)$ and using the definitions for $E_j^+(z)$ and $E_j^-(z)$ in Eqs. (6.3) and (6.9), respectively, yields

$$E_{j+1}^-(z) = z^{-1}E_j^-(z) + \Gamma_{j+1}^* E_j^+(z) \tag{6.15}$$

Finally, taking the inverse z-transform we obtain the desired recursion

$$\boxed{e_{j+1}^-(n) = e_j^-(n-1) + \Gamma_{j+1}^* e_j^+(n)} \tag{6.16}$$

Equations (6.11) and (6.16) represent a pair of coupled difference equations that correspond to the two-port network shown in Fig. 6.2*a*. With a cascade of two-port networks having reflection coefficients Γ_j, we have the pth-order FIR lattice filter as shown in Fig. 6.2*b*. Note that since

$$e_0^+(n) = e_0^-(n) = x(n) \tag{6.17}$$

then the two inputs to the first stage of the lattice are the same.

It is interesting to note that with a forward prediction error filter of the form

$$A_p(z) = \prod_{i=1}^{p}\left(1 - \alpha_i z^{-1}\right) \tag{6.18}$$

the system function of the backward prediction error filter is

$$A_p^R(z) = z^{-p}A_p^*(1/z^*) = \prod_{i=1}^{p}\left(z^{-1} - \alpha_i^*\right) \tag{6.19}$$

Therefore, the zeros of $A_p(z)$ are at the conjugate reciprocal locations of those of $A_p^R(z)$, i.e., if there is a zero at $z = \alpha_i$ in $A_p(z)$ then there is a zero at $z = 1/\alpha_i^*$ in $A_p^R(z)$. Consequently,

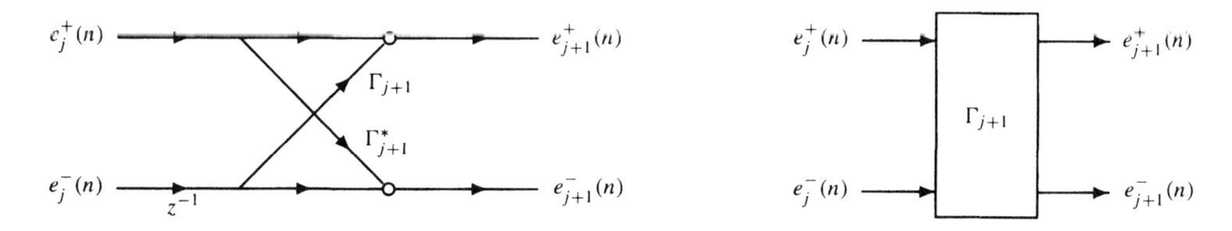

(a) Single stage of an FIR lattice filter.

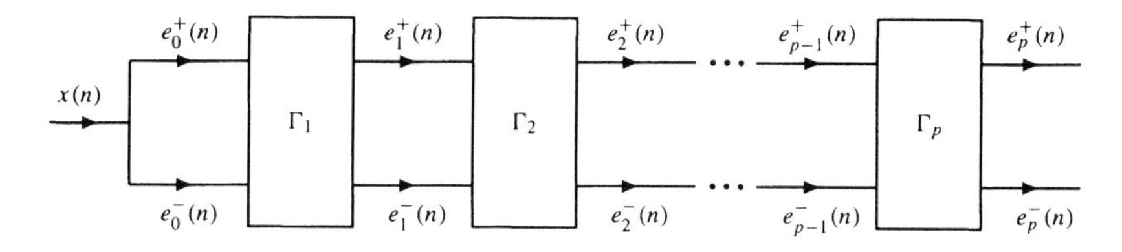

(b) A pth-order FIR lattice filter.

Figure 6.2 *The FIR lattice filter.*

since the forward prediction error filter that is obtained from the autocorrelation method is minimum phase (assuming $\mathbf{R}_p > 0$) then the backward prediction error filter is *maximum phase*. Another interesting relationship to note is that since

$$E_p^+(z) = A_p(z)X(z)$$

and

$$E_p^-(z) = A_p^R(z)X(z)$$

then

$$E_p^-(z) = \frac{A_p^R(z)}{A_p(z)}E_p^+(z) = \prod_{i=1}^{p}\left[\frac{z^{-1} - \alpha_i^*}{1 - \alpha_i z^{-1}}\right]E_p^+(z) = H_{ap}(z)E_p^+(z) \tag{6.20}$$

where

$$H_{ap}(z) = \prod_{i=1}^{p}\frac{z^{-1} - \alpha_i^*}{1 - \alpha_i z^{-1}} \tag{6.21}$$

is an allpass filter. Thus, as illustrated in Fig. 6.3, the backward prediction error may be generated by filtering the forward prediction error with an allpass filter.[2] Note that when $A_p(z)$ is minimum phase, all of the poles of the allpass filter are inside the unit circle and all of the zeros are outside. Therefore, $H_{ap}(z)$ is a stable and causal filter.

There are a number of advantages of a lattice filter over a direct form filter that often make it a popular structure to use in signal processing applications. The first is the modularity of the filter. It is this modularity that allows one to increase or decrease the order of a lattice filter in a linear prediction or all-pole signal modeling application without having to recompute the reflection coefficients. By contrast, when the prediction error filter is implemented in direct form and if the filter order is increased or decreased then all of the filter coefficients, in general, will change. Lattice filters also have the advantage of being easy to determine whether or not the filter is minimum phase (all of the roots inside the unit circle). Specifically, the filter will be minimum phase if and only if the reflection

[2]An allpass lattice filter for computing $e_p^-(n)$ from $e_p^+(n)$ will be derived in Section 6.4.1.

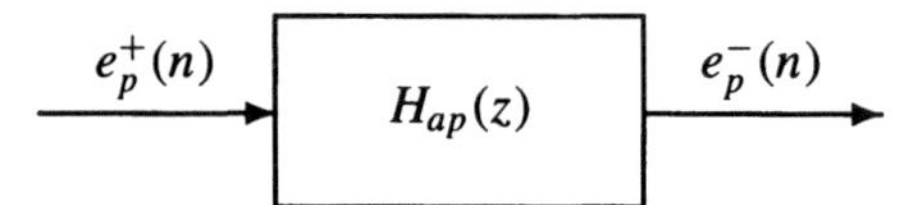

Figure 6.3 *Allpass filter for generating the backward prediction error $e_p^-(n)$ from the forward prediction error $e_p^+(n)$.*

coefficients are bounded by one in magnitude. In the case of IIR lattice filters, which will be discussed in the following section, this allows one to easily ensure that a filter is stable by simply imposing the constraint that $|\Gamma_j| < 1$. This is particularly useful in an adaptive filtering environment in which the reflection coefficients are changing as a function of time (See Chapter 9). This minimum phase condition thus allows one to easily update the filter coefficients in time in such a way that, for each value of n, the roots of the filter remain inside the unit circle. Finally, compared to other filter structures, the lattice filter tends to be less sensitive to parameter quantization effects [13]. Therefore, a lattice filter is frequently used when it is important to minimize the parameter quantization effects or when a signal is to be *coded* in terms of the filter coefficients of the model that provides the best approximation to the signal. There are other forms of FIR lattice filter, including the one-multiplier and the normalized lattice filters [7]. In the following section, we derive the split lattice filter.

6.3 SPLIT LATTICE FILTER*

In Section 5.4 of Chapter 5, the singular predictor polynomials were introduced as a way of imposing symmetry in the prediction error filters $A_j(z)$, thereby allowing for an increase in the efficiency of the Levinson-Durbin recursion. These polynomials were defined to be

$$S_j(z) = A_{j-1}(z) + z^{-1}A_{j-1}^R(z)$$

and they were shown to satisfy the 3-term recurrence

$$S_{j+1}(z) = (1 + z^{-1})S_j(z) - \delta_j z^{-1} S_{j-1}(z)$$

We now show how this recurrence leads to a *split lattice filter* structure [4]. As in Section 5.4 where we derived the split Levinson algorithm, here we restrict ourselves to the case of real signals.

Paralleling our development of the FIR lattice filter in the previous section, note that if we were to filter $x(n)$ with $S_j(z)$, then the filter's response

$$\varepsilon_j(n) = s_j(n) * x(n)$$

will be the forward prediction error that would result if, at the jth stage of the Levinson-Durbin recursion, $\Gamma_j = 1$. Therefore, we will refer to $\varepsilon_j(n)$ as the jth-order *singular forward prediction error*. Using the 3-term recurrence for $S_j(z)$, it follows that $\varepsilon_j(n)$ satisfies the difference equation

$$\varepsilon_{j+1}(n) = \varepsilon_j(n) + \varepsilon_j(n-1) - \delta_j \varepsilon_{j-1}(n-1)$$

This relationship, shown in block diagram form in Fig. 6.4*a*, represents one stage of a split lattice filter. The next stage, which takes the two inputs $\varepsilon_{j+1}(n)$ and $\varepsilon_j(n-1)$ and produces the outputs $\varepsilon_{j+2}(n)$ and $\varepsilon_{j+1}(n-1)$, is defined by the equation

$$\varepsilon_{j+2}(n) = \varepsilon_{j+1}(n) + \varepsilon_{j+1}(n-1) - \delta_{j+1}\varepsilon_j(n-1)$$

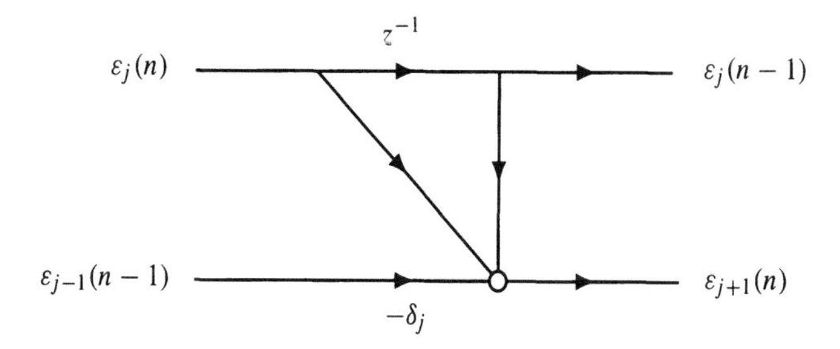

(a) A network corresponding to the 3-term recurrence for the singular predictor polynomials.

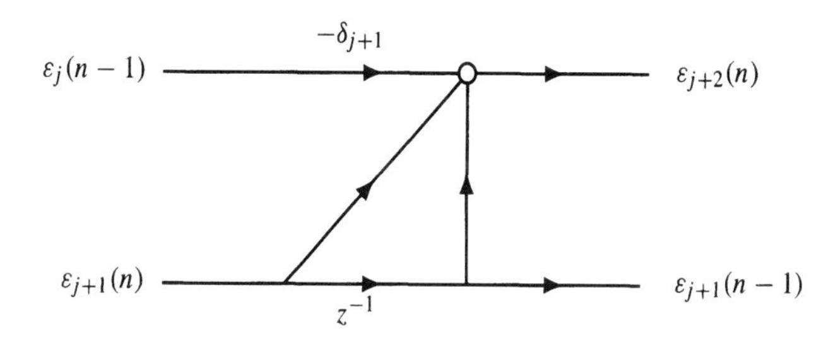

(b) A network that produces $\varepsilon_{j+2}(n)$ from $\varepsilon_{j+1}(n)$ and $\varepsilon_j(n-1)$.

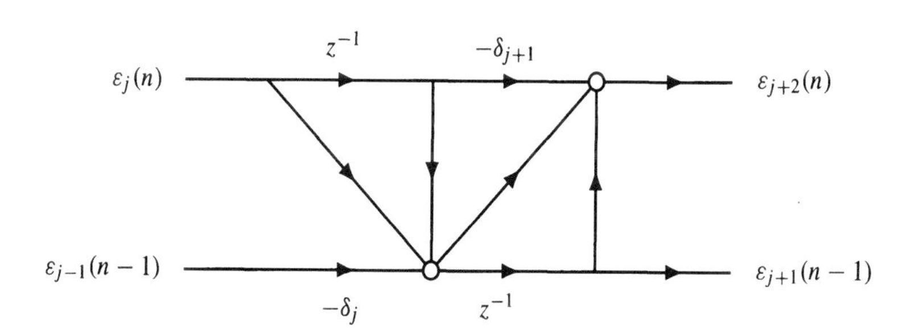

(c) A second-order split lattice module that is formed by cascading the networks in (a) and (b).

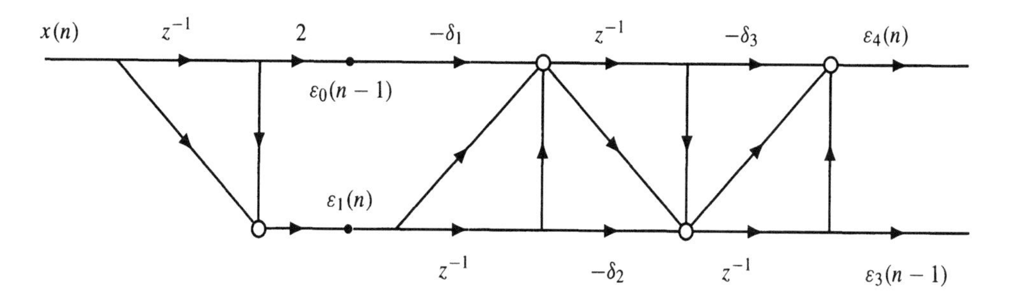

(d) A 3rd-order split lattice filter that produces the singular prediction errors $\varepsilon_4(n)$ and $\varepsilon_3(n-1)$.

Figure 6.4 *A split lattice filter.*

and is shown in block diagram form in Fig. 6.4*b*. Note that this network is the mirror image of the previous one. Cascading these two sections together produces the second-order split lattice module shown in Fig. 6.4*c*. What is needed to complete the split lattice filter is an initial section to transform $x(n)$ into $\varepsilon_0(n)$ and $\varepsilon_1(n)$, and a final section to convert the singular prediction errors $\varepsilon_{p+1}(n)$ and $\varepsilon_p(n)$ into the forward prediction error $e_p^+(n)$. The initial conditions in the split Levinson recursion are $S_0(z) = 2$ and $S_1(z) = 1 + z^{-1}$ (see p. 273), which require that we form the signals

$$\begin{aligned}\varepsilon_0(n) &= 2x(n)\\ \varepsilon_1(n) &= x(n) + x(n-1)\end{aligned}$$

These signals may be generated as illustrated in Fig. 6.4*d* which shows a 3rd-order split lattice filter for producing the singular prediction errors $\varepsilon_4(n)$ and $\varepsilon_3(n-1)$. Note that a multiplication may be saved if we combine the factor of 2 with δ_1.

Finally, to convert the singular prediction errors $\varepsilon_{p+1}(n)$ and $\varepsilon_p(n)$ at the output of the split lattice filter into the forward prediction error $e_p^+(n)$, recall that $A_p(z)$ is related to the singular predictor polynomials as follows

$$A_p(z) = z^{-1}A_p(z) + S_{p+1}(z) - z^{-1}(1+\Gamma_p)S_p(z)$$

Therefore, the pth-order forward prediction error may be generated from the singular prediction errors $\varepsilon_p(n)$ and $\varepsilon_{p+1}(n)$ as follows,

$$e_p^+(n) = e_p^+(n-1) + \varepsilon_{p+1}(n) - (1+\Gamma_p)\varepsilon_p(n-1) \tag{6.22}$$

The following example illustrates the procedure.

Example 6.3.1 *Split Lattice Filter*

Consider the third-order all-zero filter

$$H(z) = 1 - \frac{1}{3}z^{-1} - \frac{1}{3}z^{-2} + \frac{2}{3}z^{-3}$$

In order to implement this filter using a split lattice filter structure, it is necessary that we derive the coefficients δ_j from the coefficients $a(k)$. This may be done in a straightforward manner as follows. First, we convert the coefficients $a(k)$ into reflection coefficients Γ_j. Using the m-file `atog.m` we find

$$\mathbf{\Gamma} = \left[-1/4,\ -1/5,\ 2/3\right]^T$$

Second, using the relationship between Γ_j and δ_j given in Eq. (5.149), i.e.,

$$\delta_j = (1-\Gamma_j)(1+\Gamma_{j-1}) \quad ; \quad \Gamma_0 = 0$$

we have

$$\begin{aligned}\delta_1 &= 1 - \Gamma_1 = 5/4\\ \delta_2 &= (1-\Gamma_2)(1+\Gamma_1) = 9/10\\ \delta_3 &= (1-\Gamma_3)(1+\Gamma_2) = 4/15\end{aligned}$$

Finally, with $\Gamma_3 = \frac{2}{3}$, using the termination specified in Eq. (6.22) we have the split lattice filter shown in Fig. 6.5.

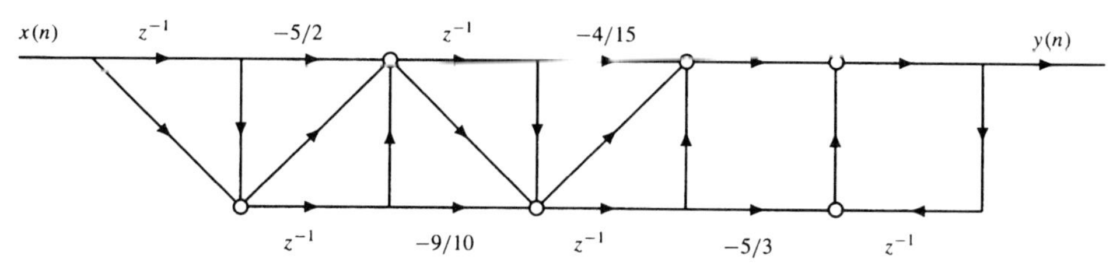

Figure 6.5 *A third-order split lattice filter.*

6.4 IIR LATTICE FILTERS

In Section 6.2 we saw how the reflection coefficients Γ_j could be used to implement an FIR filter $A_p(z)$ using a lattice filter structure. In this section we develop a lattice filter structure for IIR filters. We begin with a derivation of the "standard" two-multiplier all-pole lattice filter for implementing the system function $1/A_p(z)$. At the same time we show how this all-pole lattice may also be used as an allpass filter $H_{ap}(z) = A_p^R(z)/A_p(z)$. We then look at some other structures including the Kelly-Lochbaum lattice filter, the normalized all-pole lattice, and the one-multiplier all-pole lattice. As with the FIR lattice filter, these structures enjoy the same advantages of modularity, simple tests for stability, and decreased sensitivity to parameter quantization effects. Finally, we will look at how the all-pole lattice filter may be modified to implement a general pole-zero filter of the form $H(z) = B_q(z)/A_p(z)$.

6.4.1 All-pole Filter

A pth-order FIR lattice filter is shown in Fig. 6.2*b*. The input to the filter is the zeroth-order forward prediction error $e_0^+(n)$ and the outputs are the pth-order forward and backward prediction errors, $e_p^+(n)$ and $e_p^-(n)$, respectively. The system function of this filter is

$$A_p(z) = \frac{E_p^+(z)}{E_0^+(z)} = 1 + \sum_{k=1}^{p} a_p(k)z^{-k}$$

The all-pole filter

$$\frac{1}{A_p(z)} = \frac{E_0^+(z)}{E_p^+(z)} = \frac{1}{1 + \sum_{k=1}^{p} a_p(k)z^{-k}}$$

on the other hand, would produce a response of $e_0^+(n)$ to the input $e_p^+(n)$. Therefore, as illustrated in Fig. 6.6, whereas the FIR lattice filter builds up the prediction errors $e_j^+(n)$ and $e_j^-(n)$ from $e_0^+(n)$, the all-pole lattice filter produces the lower order prediction errors, $e_j^+(n)$ and $e_j^-(n)$, from $e_p^+(n)$. Thus, in order to implement a pth-order all-pole lattice filter, we must determine how the lower-order prediction errors $e_j^+(n)$ and $e_j^-(n)$ may be generated from $e_p^+(n)$. Note that if we solve Eq. (6.11) for $e_j^+(n)$ we have, along with Eq. (6.16), the following pair of coupled difference equations

$$e_j^+(n) = e_{j+1}^+(n) - \Gamma_{j+1}e_j^-(n-1) \tag{6.23}$$

$$e_{j+1}^-(n) = e_j^-(n-1) + \Gamma_{j+1}^* e_j^+(n) \tag{6.24}$$

These two equations define the two-port network shown in Fig. 6.7*a* and represent a single stage of an all-pole lattice filter. With a cascade of p such sections we have the pth-order

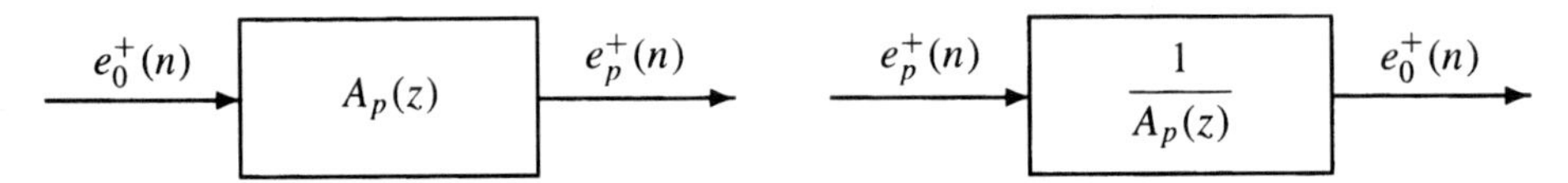

(a) A pth-order all-zero filter for generating $e_p^+(n)$ from $e_0^+(n)$.

(b) A pth-order all-pole filter for generating $e_0^+(n)$ from $e_p^+(n)$.

Figure 6.6 *Generating the pth-order forward prediction error $e_p^+(n)$ from $e_0^+(n)$ and vice versa.*

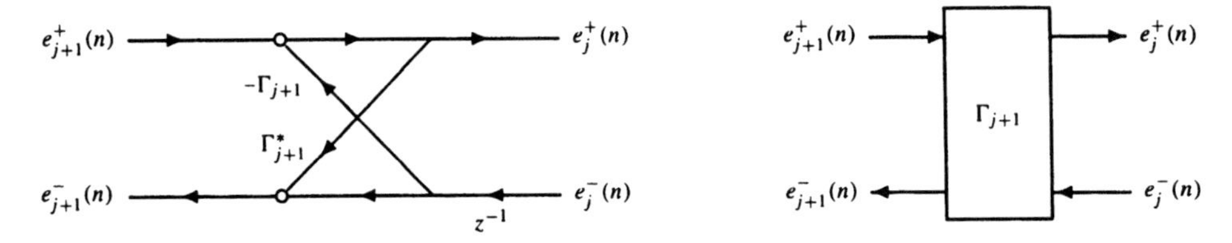

(a) Single stage of an all-pole lattice filter.

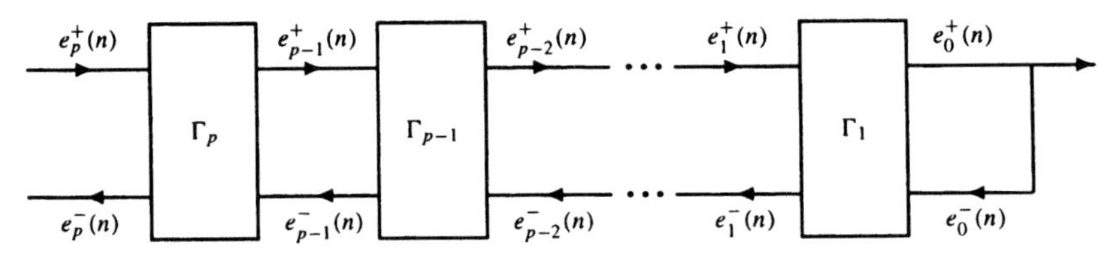

(b) A pth-order all-pole lattice filter and a pth-order allpass filter that is formed by cascading p modules of the form given in (a).

Figure 6.7 *A lattice filter for implementing all-pole and allpass filters.*

all-pole filter shown in Fig. 6.7*b*. Recall that since $e_0^+(n) = e_0^-(n) = x(n)$ then the lattice filter is terminated by feeding $e_0^+(n)$ back into the lower input of the last module. While the system function relating $e_0^+(n)$ to $e_p^+(n)$ is the all-pole filter

$$H(z) = \frac{E_0^+(z)}{E_p^+(z)} = \frac{1}{A_p(z)} \tag{6.25}$$

note that, as we saw in Section 6.2, the system function relating $e_p^+(n)$ to $e_p^-(n)$ is the allpass filter

$$H_{ap}(z) = \frac{E_p^-(z)}{E_p^+(z)} = z^{-p}\frac{A_p^*(1/z^*)}{A_p(z)} = \frac{A_p^R(z)}{A_p(z)} \tag{6.26}$$

Therefore, this structure may also be used to implement allpass systems.

Example 6.4.1 *All-pole Lattice Filter*

Let us implement the third-order all-pole filter

$$H(z) = \frac{0.328}{1 - 0.8z^{-1} + 0.64z^{-2} - 0.512z^{-3}}$$

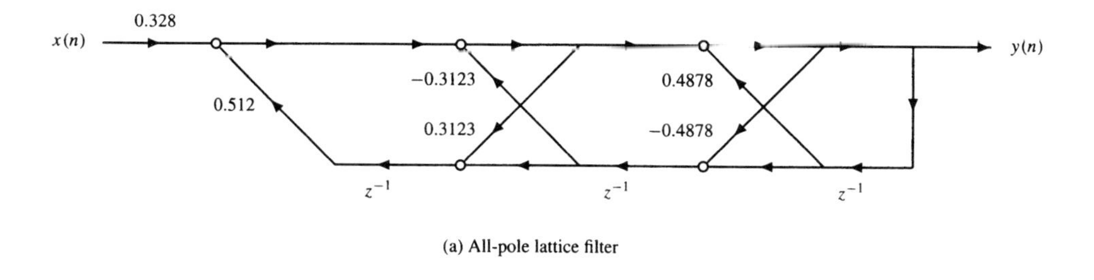

(a) All-pole lattice filter

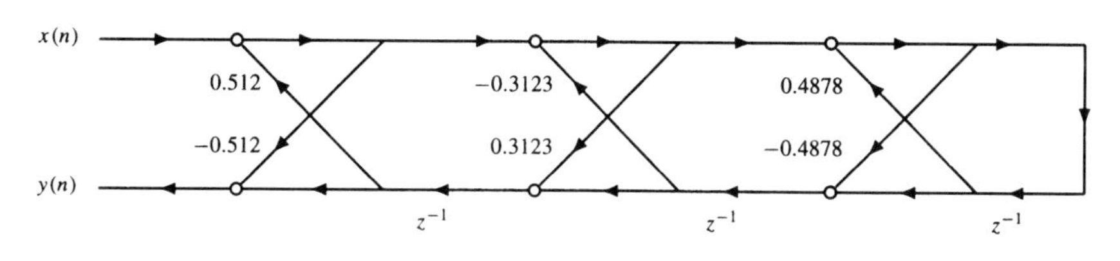

(b) An allpass lattice filter

Figure 6.8 *All-pole and allpass lattice filter implementations.*

using a lattice filter structure. With the step-down recursion we find that the reflection coefficients corresponding to the filter coefficients $a_p(k)$ are

$$\mathbf{\Gamma} = \left[-0.4878,\ 0.3123,\ -0.512\right]^T$$

Thus, the all-pole lattice implementation of $H(z)$ is as illustrated in Fig. 6.8*a*. This structure may also be used to implement the allpass filter

$$H_{ap}(z) = \frac{-0.512 + 0.64z^{-1} - 0.8z^{-2} + z^{-3}}{1 - 0.8z^{-1} + 0.64z^{-2} - 0.512z^{-3}}$$

as illustrated in Fig. 6.8*b*.

6.4.2 Other All-pole Lattice Structures*

Beginning with the all-pole lattice filter in Fig. 6.7, we may develop some additional all-pole lattice filter structures. One such structure may be derived from Eqs. (6.23) and (6.24) as follows. Note that Eq. (6.23) expresses the forward prediction error $e_j^+(n)$ in terms of the two lattice filter inputs $e_{j+1}^+(n)$ and $e_j^-(n-1)$ whereas Eq. (6.24) expresses the backward prediction error $e_{j+1}^-(n)$ in terms of the output $e_j^+(n)$ and the input $e_j^-(n-1)$. However, we may also express $e_{j+1}^-(n)$ in terms of the two inputs $e_{j+1}^+(n)$ and $e_j^-(n-1)$ by substituting Eq. (6.23) for $e_j^+(n)$ into Eq. (6.24) as follows

$$\begin{aligned} e_{j+1}^-(n) &= e_j^-(n-1) + \Gamma_{j+1}^* e_j^+(n) \\ &= e_j^-(n-1) + \Gamma_{j+1}^*\left[e_{j+1}^+(n) - \Gamma_{j+1} e_j^-(n-1)\right] \\ &= \left[1 - |\Gamma_{j+1}|^2\right] e_j^-(n-1) + \Gamma_{j+1}^* e_{j+1}^+(n) \end{aligned} \tag{6.27}$$

Equation (6.27) along with (6.23) define the two-port network shown in Fig. 6.9. Note that, unlike the all-pole lattice in Fig. 6.7, this implementation requires three multiplies per stage.

Although not a particularly useful structure, through a sequence of flowgraph manipulations described below we may derive some interesting and important all-pole lattice structures. The first is the Kelly-Lochbaum model that has been used in speech modeling as well as models for the propagation of waves through a layered media [11,12]. Next, we derive the normalized all-pole lattice filter which, although requiring four multiplications per section, has an advantage in terms of computational accuracy using fixed-point arithmetic [10]. Finally, we derive the one-multiplier lattice which is the most efficient structure when measured in terms of the number of multiplications per section.

Beginning with the three-multiplier lattice module in Fig. 6.9, suppose we scale the input to the upper branch by a constant $1/\alpha_{j+1}$ and scale the outputs of the branch point by the inverse of this scale factor, α_{j+1}. This produces the equivalent two-port network shown in Fig. 6.10*a*. Next, if we divide the two inputs to the adder in the lower branch by α_{j+1} and multiply the output of the adder by α_{j+1} then we have the equivalent network shown in Fig. 6.10*b*. Now, recall that both of the outputs $e_j^+(n)$ and $e_{j+1}^-(n)$ are formed by taking a linear combination of the two inputs $e_{j+1}^+(n)$ and $e_j^-(n)$. Therefore, if we scale the two inputs by α_{j+1} and rescale the outputs by $1/\alpha_{j+1}$ then the network will remain the same. The net effect, as shown in Fig. 6.10*c*, is to bring the two scale factors on the left of the two-port network over to the right side. Forming a cascade of these modules, moving all of the scale factors to the end of the network, results in the equivalent all-pole lattice filter shown in Fig. 6.10*d* where

$$\alpha = \prod_{j=1}^{p} \alpha_j$$

Since $e_0^+(n) = e_0^-(n)$ then the network may be terminated by feeding the output of the final stage of the lattice back into the input. With the scale factors of α and $1/\alpha$ canceling each other at the input to the last module, we are then left with only a scale factor of $1/\alpha$ at the output that is applied to $e_0^+(n)$. This leads to the structure shown in Fig. 6.11 consisting of a cascade of *modified all-pole lattice modules*. Having dropped the final scale factor $1/\alpha$ from the network, the system function between the input $x(n)$ and the output $y(n)$ is

$$H(z) = \frac{Y(z)}{X(z)} = \frac{\prod_{j=1}^{p} \alpha_j}{A_p(z)} \tag{6.28}$$

and the system function relating $x(n)$ to $w(n)$ is the allpass filter

$$H(z) = \frac{W(z)}{X(z)} = \frac{A_p^R(z)}{A_p(z)} \tag{6.29}$$

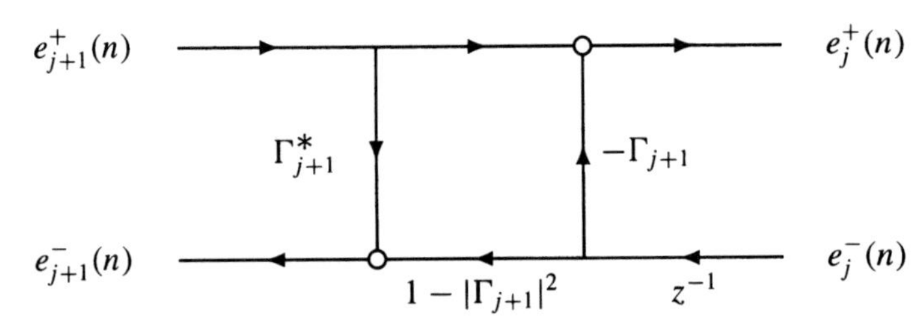

Figure 6.9 *A three multiplier all-pole lattice filter module.*

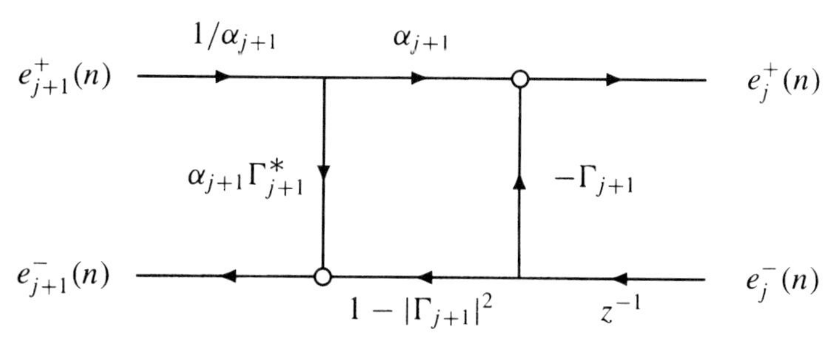

(a) Scaling the input in the upper branch of the all-pole lattice module by $1/\alpha_{j+1}$ and rescaling the outputs of the first branch point by α_{j+1}.

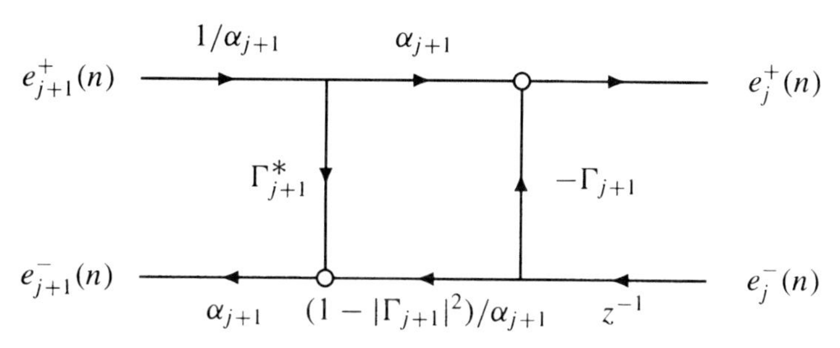

(b) Scaling the inputs to the adder in the lower branch by $1/\alpha_{j+1}$ and rescaling the output by α_{j+1}.

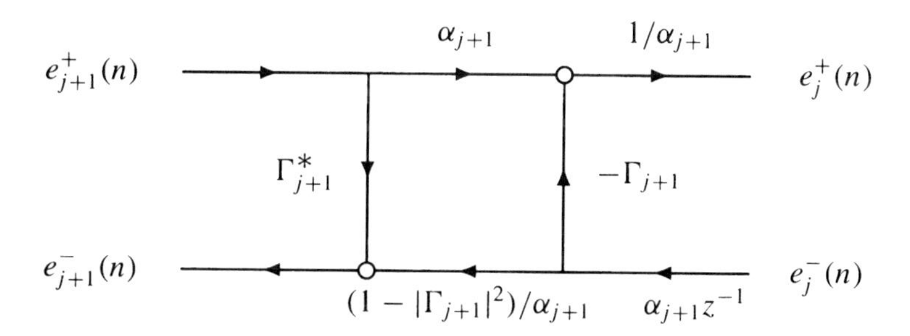

(c) Moving the scale factors α_{j+1} and $1/\alpha_{j+1}$ on the left to the right.

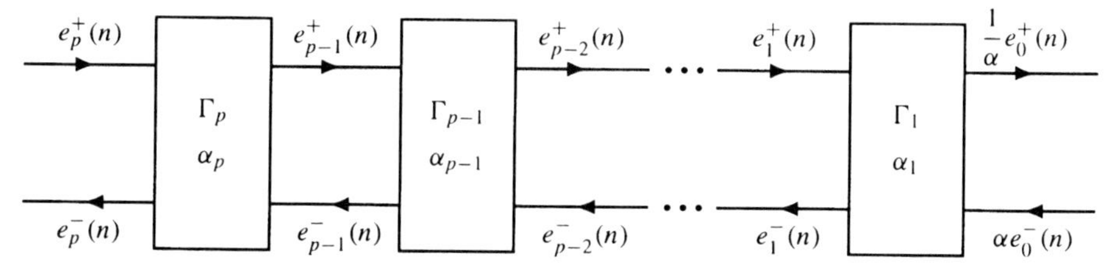

(d) A cascade of lattice filter modules of the form given in (c) with all of the scale factors moved to the final stage.

Figure 6.10 *A sequence of lattice filter flowgraph manipulations that lead to equivalent all-pole lattice filter structures.*

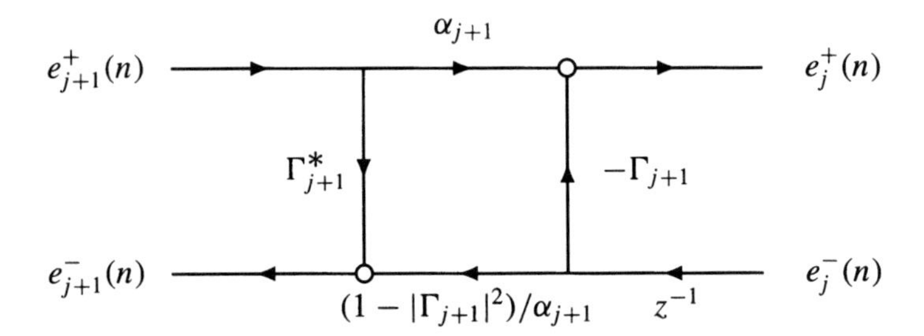

(a) The modified all-pole lattice module.

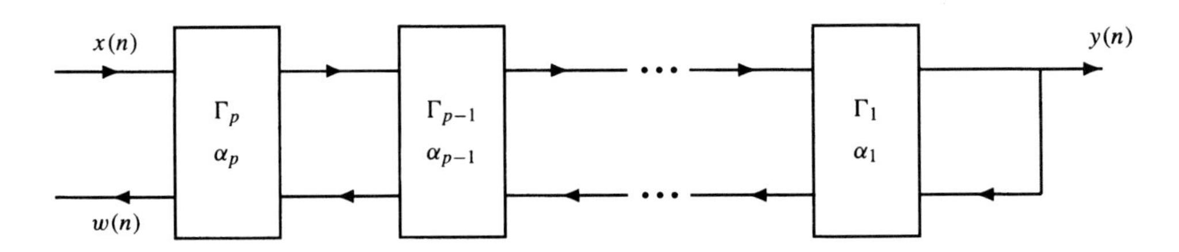

(b) A cascade of modified all-pole lattice modules.

Figure 6.11 *A modified all-pole lattice filter built-up as a cascade of modified all-pole lattice modules shown in (a).*

Given the all-pole module in Fig. 6.11, we may now derive three important all-pole lattice filter structures. In each case, we assume that the reflection coefficients are real-valued. The first is based on the factorization $(1 - \Gamma^2_{j+1}) = (1 + \Gamma_{j+1})(1 - \Gamma_{j+1})$ and is obtained by setting

$$\alpha_{j+1} = 1 + \Gamma_{j+1} \tag{6.30}$$

This results in the lattice filter module shown in Fig. 6.12*a* which is known as the Kelly-Lochbaum model. This structure has been used as a model for speech production in which the vocal tract is approximated as an interconnection of cylindrical acoustic tubes of equal length and differing cross-sectional areas where the reflection coefficients Γ_j correspond to the amount of acoustic energy that is reflected at the interface between two adjacent tubes having a differing acoustic impedance [3,6]. The difference equations for this structure are

$$\begin{aligned} e^+_j(n) &= \left[1 + \Gamma_{j+1}\right] e^+_{j+1}(n) - \Gamma_{j+1} e^-_j(n-1) \\ e^-_{j+1}(n) &= \left[1 - \Gamma_{j+1}\right] e^-_j(n-1) + \Gamma_{j+1} e^+_{j+1}(n) \end{aligned} \tag{6.31}$$

and the system function is

$$H(z) = \frac{\prod_{j=1}^{p}(1 + \Gamma_j)}{A_p(z)} \tag{6.32}$$

This structure requires four multiplications, two additions, and one delay for each stage.

The next all-pole lattice structure is derived by introducing a new set of variables, θ_{j+1}, where θ_{j+1} is related to the reflection coefficients Γ_{j+1} as follows

$$\theta_{j+1} = \sin^{-1} \Gamma_{j+1} \tag{6.33}$$

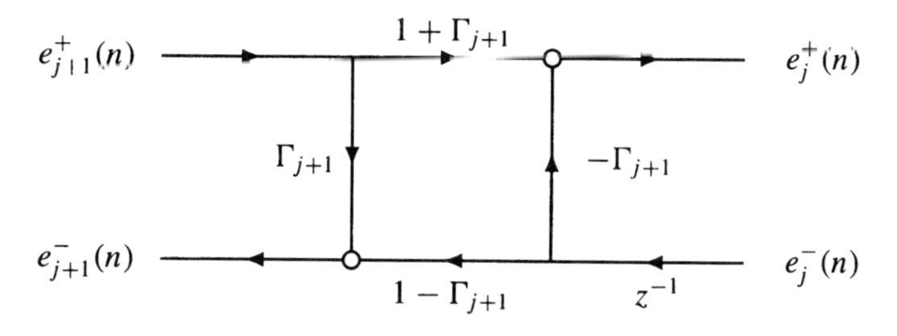

(a) Kelly-Lochbaum lattice filter.

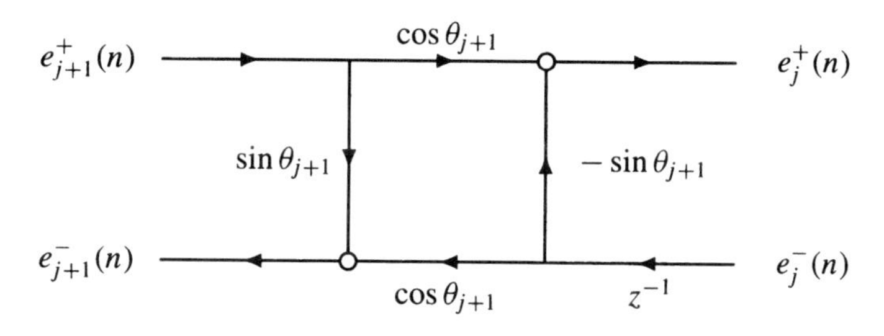

(b) Normalized lattice filter.

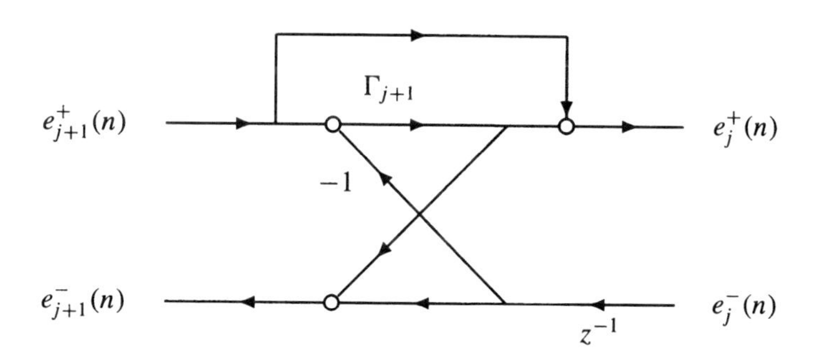

(c) One-multiplier lattice filter.

Figure 6.12 *Three all-pole lattice filter structures with real-valued filter coefficients. Each of these lattice modules is equivalent to within a scale factor.*

Assuming that the filter is stable so that $|\Gamma_{j+1}| < 1$, this transformation is well-defined. With

$$\Gamma_{j+1} = \sin\theta_{j+1} \tag{6.34}$$

and

$$1 - \Gamma^2_{j+1} = \cos^2\theta_{j+1} \tag{6.35}$$

let us define the scale factor α_{j+1} by

$$\alpha_{j+1} = \cos\theta_{j+1} \tag{6.36}$$

This leads to the *normalized all-pole lattice filter* shown in Fig. 6.12*b*. The difference equations for this network are

$$\begin{aligned} e_j^+(n) &= \big[\cos\theta_{j+1}\big]e_{j+1}^+(n) - \big[\sin\theta_{j+1}\big]e_j^-(n-1) \\ e_{j+1}^-(n) &= \big[\cos\theta_{j+1}\big]e_j^-(n-1) + \big[\sin\theta_{j+1}\big]e_{j+1}^+(n) \end{aligned} \tag{6.37}$$

and the system function is

$$H(z) = \frac{\prod_{j=1}^{p} \cos\theta_j}{A_p(z)} \tag{6.38}$$

In this implementation, there are four multiplications, two additions, and one delay for each stage as in the Kelly-Lochbaum structure. However, what makes this structure interesting is the fact that it requires only one complex multiplication. Specifically, note that if we form the complex signal $e_{i+1}^+(n) + je_i^-(n-1)$ from the two input signals, and the complex signal $e_i^+(n) + je_{i+1}^-(n)$ from the two output signals, then these two signals are related as follows[3]

$$e_i^+(n) + je_{i+1}^-(n) = e^{j\theta_{i+1}}\Big[e_{i+1}^+(n) + je_i^-(n-1)\Big] \tag{6.39}$$

In particular, note that Eq. (6.37) corresponds to the real and imaginary parts of both sides of this equation. This relationship also demonstrates that the complex signals are *normalized* in the sense that their magnitudes do not change from one section to the next,

$$\big|e_i^+(n) + je_{i+1}^-(n)\big| = \big|e_{i+1}^+(n) + je_i^-(n-1)\big| \tag{6.40}$$

Thus, this structure is referred to as the normalized lattice. One of the advantages of the normalized lattice filter is its robustness to round-off errors when using fixed-point arithmetic [10].

The last structure is the one-multiplier lattice which, as its name implies, requires only one real multiplication per module [5]. This structure may be derived by re-writing the Kelly-Lochbaum equations as follows,

$$e_j^+(n) = \Gamma_{j+1}\big[e_{j+1}^+(n) - e_j^-(n-1)\big] + e_{j+1}^+(n) \tag{6.41}$$

$$e_{j+1}^-(n) = \Gamma_{j+1}\big[e_{j+1}^+(n) - e_j^-(n-1)\big] + e_j^-(n-1) \tag{6.42}$$

Since the reflection coefficient Γ_{j+1} multiplies the same signal in both equations, forming the sum $e_{j+1}^+(n) - e_j^-(n-1)$ prior to multiplying by Γ_{j+1} leads to a structure requiring only one multiplication per stage. The one-multiplier lattice structure is shown in Fig. 6.12*c*. In addition to one multiply, this structure requires three additions and one delay.

6.4.3 Lattice Filters having Poles and Zeros

In the previous section, several different all-pole lattice structures were derived from the basic all-pole lattice given in Fig. 6.7. From this structure we may also develop a lattice filter that realizes a general rational system function of the form

$$H(z) = \frac{B_q(z)}{A_p(z)} = \frac{b_q(0) + b_q(1)z^{-1} + \cdots + b_q(q)z^{-q}}{1 + a_p(1)z^{-1} + \cdots + a_p(p)z^{-p}} \tag{6.43}$$

[3]Note that we have changed the subscript indices on the forward and backward prediction errors from j to i in order to avoid confusion with the imaginary number $j = \sqrt{-1}$.

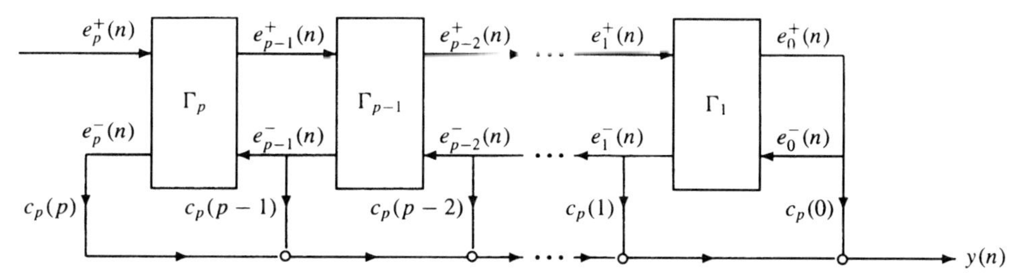

Figure 6.13 *A lattice filter having p poles and p zeros.*

where $q \leq p$. To see how this may be done, note that $H(z)$ may be implemented as a cascade of an all-pole filter $1/A_p(z)$ with an all-zero filter $B_q(z)$. The pair of difference equations corresponding to this cascade are

$$w(n) = x(n) - \sum_{k=1}^{p} a_p(k)w(n-k)$$

where $w(n)$ is the output of the all-pole filter $1/A_p(z)$ and

$$y(n) = \sum_{k=0}^{q} b_q(k)w(n-k)$$

is the output of the FIR filter $B_q(z)$. Since the zeros of $H(z)$ are introduced by taking a linear combination of delayed outputs of the all-pole filter, with the all-pole lattice filter in Fig. 6.7, a rational system function may similarly be realized with a linear combination of the signals $e_0^+(n-k)$ for $k = 0, 1, \ldots, q$. However, a more efficient way is to take a linear combination of the backward prediction errors $e_j^-(n)$

$$y(n) = \sum_{j=0}^{q} c_q(j)e_j^-(n) \tag{6.44}$$

as shown in Fig. 6.13 (in this figure we assume that $q = p$). To show that this filter does, in fact, have a rational system function, we begin by expressing Eq. (6.44) in the z-domain as follows

$$Y(z) = \sum_{j=0}^{q} c_q(j)E_j^-(z) \tag{6.45}$$

With

$$E_j^-(z) = A_j^R(z)E_0^+(z) = \frac{A_j^R(z)}{A_p(z)}E_p^+(z)$$

it follows that

$$Y(z) = \sum_{j=0}^{q} c_q(j)\frac{A_j^R(z)}{A_p(z)}E_p^+(z)$$

Therefore, the system function relating the input $E_p^+(z)$ to the output $Y(z)$ is

$$H(z) = \frac{\displaystyle\sum_{j=0}^{q} c_q(j)A_j^R(z)}{A_p(z)} \tag{6.46}$$

which has p poles and q zeros. Note that the zeros of $H(z)$ are the roots of the polynomial

$$B_q(z) = \sum_{j=0}^{q} c_q(j) A_j^R(z) \tag{6.47}$$

which is a function of not only the tap weights $c_q(j)$ but also the reflection coefficients Γ_j.

To see how the coefficients $b_q(j)$ are related to the coefficients $c_q(j)$, we substitute the following expression for $A_j^R(z)$,

$$A_j^R(z) = z^{-j} A_j^*(1/z^*) = \sum_{m=0}^{j} a_j^*(m) z^{m-j}$$

into Eq. (6.47) as follows

$$B_q(z) = \sum_{j=0}^{q} c_q(j) \sum_{m=0}^{j} a_j^*(m) z^{m-j} \tag{6.48}$$

With the substitution $k = j - m$, Eq. (6.48) becomes

$$B_q(z) = \sum_{j=0}^{q} \sum_{k=0}^{j} c_q(j) a_j^*(j-k) z^{-k}$$

Interchanging the order of the summations and making the appropriate changes in the indexing we have

$$B_q(z) = \sum_{k=0}^{q} \left[\sum_{j=k}^{q} c_q(j) a_j^*(j-k) \right] z^{-k} = \sum_{k=0}^{q} b_q(k) z^{-k} \tag{6.49}$$

Finally, equating powers of z on both sides of Eq. (6.49) gives the desired expression

$$\boxed{b_q(k) = \sum_{j=k}^{q} c_q(j) a_j^*(j-k)} \tag{6.50}$$

This equation shows how the coefficients $b_q(k)$ may be found from the coefficients $c_q(k)$ and Γ_j. Specifically, from the reflection coefficients Γ_j of the all-pole filter we may use the step-up recursion to find the all-pole coefficients for model orders $j = 0, 1, \ldots, q$. With $a_j(k)$ along with the coefficients $c_q(k)$, Eq. (6.50) may then be used to determine the coefficients $b_q(k)$. Equation (6.50) may also be used to find the coefficients $c_q(k)$ required to implement a given rational system function with a numerator $B_q(z)$. To do this, Eq. (6.50) is rewritten as follows

$$b_q(k) = c_q(k) + \sum_{j=k+1}^{q} c_q(j) a_j^*(j-k) \tag{6.51}$$

(here we have used the fact that $a_j(0) = 1$). Thus, the coefficients $c_q(k)$ may be computed recursively as follows

$$\boxed{c_q(k) = b_q(k) - \sum_{j=k+1}^{q} c_q(j) a_j^*(j-k)} \tag{6.52}$$

for $k = q-1, q-2, \ldots, 0$. This recursion is initialized by setting

$$c_q(q) = b_q(q) \tag{6.53}$$

The following example illustrates the procedure.

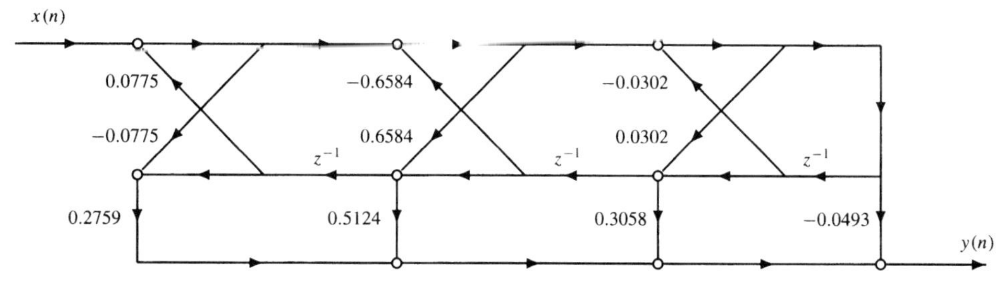

Figure 6.14 *Lattice filter realization of a 4th-order low-pass elliptic filter.*

Example 6.4.2 ***Lattice Filter Implementation of a Low-Pass Elliptic Filter***

A third-order low-pass elliptic filter with a cutoff frequency of $\omega = 0.5\pi$ has a system function given by

$$H(z) = \frac{0.2759 + 0.5121z^{-1} + 0.5121z^{-2} + 0.2759z^{-3}}{1 - 0.0010z^{-1} + 0.6546z^{-2} - 0.0775z^{-3}}$$

To implement this filter using a lattice filter structure, we first transform the denominator coefficients into reflection coefficients. Using the step-down recursion we find

$$\mathbf{\Gamma} = \begin{bmatrix}0.0302, & 0.6584, & -0.0775\end{bmatrix}^T$$

with the second-order coefficients given by

$$\mathbf{a}_2 = \begin{bmatrix}1, & 0.0501, & 0.6584\end{bmatrix}^T$$

and the first-order coefficients equal to

$$\mathbf{a}_1 = \begin{bmatrix}1, & 0.0302\end{bmatrix}^T$$

Next, we find the coefficients $c_3(k)$ that correspond to the given numerator using the recursion in Eq.(6.52). Beginning with

$$c_3(3) = b_3(3) = 0.2759$$

we then find

$$\begin{aligned}
c_3(2) &= b_3(2) - c_3(3)a_3(1) = 0.5124 \\
c_3(1) &= b_3(1) - c_3(2)a_2(1) - c_3(3)a_3(2) = 0.3058 \\
c_3(0) &= b_3(0) - c_3(1)a_1(1) - c_3(2)a_2(2) - c_3(3)a_3(3) = -0.0493
\end{aligned}$$

This leads to the lattice filter implementation illustrated in Fig. 6.14.

6.5 LATTICE METHODS FOR ALL-POLE SIGNAL MODELING

In Chapter 4 we developed several different approaches for finding an all-pole model for a signal $x(n)$. In each case, we first defined an error that was to be minimized and then solved the minimization problem for the optimum set of transversal filter coefficients $a_p(k)$. Since the lattice filter provides an alternate parameterization of the all-pole filter, i.e., in terms of its reflection coefficients, we may also consider formulating the all-pole signal modeling problem as one of finding the reflection coefficients that minimize some error. In this section we look at several such *lattice methods* for signal modeling including the *forward covariance method*, the *backward covariance method*, *Burg's method*, and the *modified covariance method*. Except for the modified covariance method, each of these methods is a *sequential* optimization procedure. That is to say, given an error that is to be minimized, such as the sum of the squares of the forward prediction error, the first reflection coefficient, Γ_1, is found that minimizes this error. Then, with Γ_1 fixed, the optimum value of Γ_2 is determined. Thus, in general, with $\Gamma_1, \ldots, \Gamma_{j-1}$ held constant, the value of Γ_j is found that minimizes some error. We begin, in the following section, with the forward covariance method.

6.5.1 The Forward Covariance Method

Let $x(n)$ for $n = 0, 1, \ldots, N$ be a signal, either real or complex, that is to be modeled as the unit sample response of an all-pole filter of order p. For an all-pole lattice filter, it is necessary to determine the reflection coefficients which, for reasons soon to become apparent, we will denote by Γ_j^+. In the *forward covariance method* the reflection coefficients are computed *sequentially* with the jth reflection coefficient, Γ_j^+, determined by minimizing the *covariance-type* error

$$\mathcal{E}_j^+ = \sum_{n=j}^{N} \left|e_j^+(n)\right|^2 \tag{6.54}$$

Note that since $\mathcal{E}_j^+$ is a function only of the signal values $x(n)$ over the interval $[0, N]$, as with the covariance method described in Section 4.6.2, it is not necessary to window the signal or make any assumptions about the values of $x(n)$ outside the given interval.

To begin the sequential minimization of $\mathcal{E}_j^+$ using the forward covariance method, we must first find the reflection coefficient Γ_1^+ that minimizes the first-order error

$$\mathcal{E}_1^+ = \sum_{n=1}^{N} \left|e_1^+(n)\right|^2 \tag{6.55}$$

This minimization may be performed by setting the derivative of $\mathcal{E}_1^+$ with respect to $(\Gamma_1^+)^*$ equal to zero as follows

$$\frac{\partial}{\partial(\Gamma_1^+)^*}\mathcal{E}_1^+ = \sum_{n=1}^{N} e_1^+(n)\frac{\partial}{\partial(\Gamma_1^+)^*}\left[e_1^+(n)\right]^* = 0 \tag{6.56}$$

Since

$$e_1^+(n) = e_0^+(n) + \Gamma_1^+ e_0^-(n-1) \tag{6.57}$$

then

$$\frac{\partial\left[e_1^+(n)\right]^*}{\partial(\Gamma_1^+)^*} = \left[e_0^-(n-1)\right]^*$$

and Eq. (6.56) becomes

$$\sum_{n=1}^{N} e_1^+(n)\left[e_0^-(n-1)\right]^* = 0 \tag{6.58}$$

Note that if we consider $e_1^+(n)$ and $e_0^-(n-1)$ for $n = 1, 2, \ldots, N$ to be N-dimensional vectors,

$$\begin{aligned} \mathbf{e}_1^+ &= \left[e_1^+(1),\ e_1^+(2), \cdots,\ e_1^+(N)\right]^T \\ \mathbf{e}_0^- &= \left[e_0^-(0),\ e_0^-(1), \cdots,\ e_0^-(N-1)\right]^T \end{aligned} \tag{6.59}$$

then Eq. (6.58) states that $\mathcal{E}_1^+$ will be minimized when $\mathbf{e}_1^+$ and $\mathbf{e}_0^-$ are *orthogonal*,

$$\langle \mathbf{e}_1^+,\ \mathbf{e}_0^- \rangle = 0$$

Substituting Eq. (6.57) into Eq. (6.58) and solving for Γ_1^+ leads to the following expression for Γ_1^+,

$$\Gamma_1^+ = -\frac{\displaystyle\sum_{n=1}^{N} e_0^+(n)\left[e_0^-(n-1)\right]^*}{\displaystyle\sum_{n=1}^{N} \left|e_0^-(n-1)\right|^2} \tag{6.60}$$

which, in terms of the vectors $\mathbf{e}_0^+$ and $\mathbf{e}_0^-$, may be expressed as

$$\Gamma_1^+ = -\frac{\langle \mathbf{e}_0^+,\ \mathbf{e}_0^- \rangle}{\|\mathbf{e}_0^-\|^2}$$

Continuing with this process, sequentially minimizing $\mathcal{E}_j^+$ while holding $\Gamma_1^+, \ldots, \Gamma_{j-1}^+$ fixed, the value of Γ_j^+ is found by setting the derivative of $\mathcal{E}_j^+$ with respect to $(\Gamma_j^+)^*$ equal to zero,

$$\frac{\partial}{\partial(\Gamma_j^+)^*}\mathcal{E}_j^+ = \sum_{n=j}^{N} e_j^+(n)\frac{\partial}{\partial(\Gamma_j^+)^*}\left[e_j^+(n)\right]^* = 0 \tag{6.61}$$

Using the forward prediction error update equation

$$e_j^+(n) = e_{j-1}^+(n) + \Gamma_j^+ e_{j-1}^-(n-1) \tag{6.62}$$

it follows that the partial derivative of $\left[e_j^+(n)\right]^*$ with respect to $(\Gamma_j^+)^*$ is $\left[e_{j-1}^-(n-1)\right]^*$. Therefore, Eq. (6.61) becomes

$$\boxed{\sum_{n=j}^{N} e_j^+(n)\left[e_{j-1}^-(n-1)\right]^* = 0} \tag{6.63}$$

Equation (6.63) states that the vectors $\mathbf{e}_j^+$ and $\mathbf{e}_{j-1}^-$ are orthogonal,

$$\boxed{\langle \mathbf{e}_j^+,\ \mathbf{e}_{j-1}^- \rangle = 0} \tag{6.64}$$

where

$$\begin{aligned} \mathbf{e}_j^+ &= \left[e_j^+(j),\ e_j^+(j+1), \ldots,\ e_j^+(N)\right]^T \\ \mathbf{e}_{j-1}^- &= \left[e_{j-1}^-(j-1),\ e_{j-1}^-(j), \ldots,\ e_{j-1}^-(N-1)\right]^T \end{aligned} \tag{6.65}$$

are vectors of length $N - j + 1$. Substituting Eq. (6.62) for $e_j^+(n)$ into Eq. (6.63) and solving for Γ_j^+ we find that the jth reflection coefficient is

The Forward Covariance Method

```
function [gamma,err] = fcov(x,p)
%
x = x(:);
N=length(x);
eplus  = x(2:N);
eminus = x(1:N-1);
N=N-1;
for j=1:p;
    gamma(j)  = -eminus'*eplus/(eminus'*eminus);
    temp1     =  eplus  + gamma(j)*eminus;
    temp2     =  eminus + conj(gamma(j))*eplus;
    err(j)    =  temp1'*temp1;
    eplus     =  temp1(2:N);
    eminus    =  temp2(1:N-1);
    N=N-1;
    end;
```

Figure 6.15 *A* MATLAB *program for finding the reflection coefficients for a pth-order all-pole model of a signal $x(n)$ using the forward covariance method.*

$$\Gamma_j^+ = -\frac{\displaystyle\sum_{n=j}^{N} e_{j-1}^+(n)\big[e_{j-1}^-(n-1)\big]^*}{\displaystyle\sum_{n=j}^{N} \big|e_{j-1}^-(n-1)\big|^2} = -\frac{\langle \mathbf{e}_{j-1}^+, \mathbf{e}_{j-1}^-\rangle}{\|\mathbf{e}_{j-1}^-\|^2} \tag{6.66}$$

From a computational point of view, the forward covariance algorithm works as follows. Given the first $j-1$ reflection coefficients, $\mathbf{\Gamma}_{j-1}^+ = \big[\Gamma_1^+, \Gamma_2^+, \ldots, \Gamma_{j-1}^+\big]^T$, and given the forward and backward prediction errors $e_{j-1}^+(n)$ and $e_{j-1}^-(n)$, the jth reflection coefficient is found by evaluating Eq. (6.66). Then, using the lattice filter, the $(j-1)$st-order forward and backward prediction errors are updated to form the jth-order errors $e_j^+(n)$ and $e_j^-(n)$ and the process is repeated. A MATLAB program for finding a pth-order all-pole model for a complex signal $x(n)$ using the forward covariance method is given in Fig. 6.15.

Example 6.5.1 *Forward Covariance Method*

Given the signal $x(n) = \beta^n u(n)$ for $n = 0, \ldots, N$, let us find the pth-order all-pole model for $x(n)$ using the forward covariance method. We begin by initializing the forward and backward prediction errors as follows

$$e_0^+(n) = e_0^-(n) = x(n) = \beta^n \quad ; \quad n = 0, 1, \ldots, N$$

Next, we evaluate the norm of $e_0^-(n-1)$,

$$\|\mathbf{e}_0^-\|^2 = \sum_{n=1}^{N} \big[e_0^-(n-1)\big]^2 = \sum_{n=0}^{N-1} \beta^{2n} = \frac{1-\beta^{2N}}{1-\beta^2}$$

and the inner product between $e_0^+(n)$ and $e_0^-(n-1)$,

$$\langle \mathbf{e}_0^+, \mathbf{e}_0^- \rangle = \sum_{n=1}^{N} e_0^+(n) e_0^-(n-1) = \beta \sum_{n=0}^{N-1} \beta^{2n} = \beta \frac{1-\beta^{2N}}{1-\beta^2}$$

Then, using Eq. (6.60), we find for the first reflection coefficient,

$$\Gamma_1^+ = -\frac{\langle \mathbf{e}_0^+, \mathbf{e}_0^- \rangle}{\|\mathbf{e}_0^-\|^2} = -\beta$$

Updating the forward prediction error using Eq. (6.62) we have

$$e_1^+(n) = e_0^+(n) + \Gamma_1^+ e_0^-(n-1) = \beta^n u(n) - \beta(\beta)^{n-1} u(n-1) = \delta(n)$$

Therefore, the first-order modeling error is zero

$$\mathcal{E}_1^+ = \sum_{n=1}^{N} \left[e_1^+(n)\right]^2 = 0$$

The first-order backward prediction error, on the other hand, is not equal to zero. In fact, using Eq. (6.16) we see that

$$e_1^-(n) = e_0^-(n-1) + \Gamma_1^+ e_0^+(n) = \beta^{n-1} u(n-1) - \beta^{n+1} u(n)$$

For the second reflection coefficient, since $e_1^+(n) = 0$ for $n > 0$ then

$$\Gamma_2^+ = -\frac{\langle \mathbf{e}_1^+, \mathbf{e}_1^- \rangle}{\|\mathbf{e}_1^-\|^2} = 0$$

With the second reflection coefficient equal to zero it follows that $e_2^+(n) = e_1^+(n) = \delta(n)$ and that $\Gamma_3^+ = 0$. Continuing, we see that $\Gamma_j^+ = 0$ for all $j > 1$,

$$\boldsymbol{\Gamma}^+ = \left[-\beta,\ 0,\ 0, \ldots\right]^T$$

Lest we be misled by the results of the first example, as another example let us consider the signal $x(n)$ shown in Fig. 6.16*a* which is the unit sample response of the third-order filter

$$H(z) = \frac{1}{A_3(z)} = \frac{1}{1 - 0.12z^{-1} - 0.456z^{-2} + 0.6z^{-3}}$$

The lattice filter coefficients for this filter are

$$\boldsymbol{\Gamma} = \left[0.6,\ -0.6,\ 0.6\right]^T$$

Using the MATLAB program for the forward covariance method given in Fig. 6.15 with $p = 3$ and $N = 60$ we find that the reflection coefficients are

$$\boldsymbol{\Gamma}^+ = \left[0.5836,\ -0.4962,\ 0.5603\right]^T$$

and the sequence of squared errors is

$$\underline{\mathcal{E}}^+ = \left[1.4179,\ 0.4128,\ 0.0117\right]^T$$

The system function corresponding to these reflection coefficients is

$$\hat{H}_3(z) = \frac{1}{\hat{A}_3(z)} = \frac{1}{1 + 0.0160z^{-1} - 0.3315z^{-2} + 0.5603z^{-3}}$$

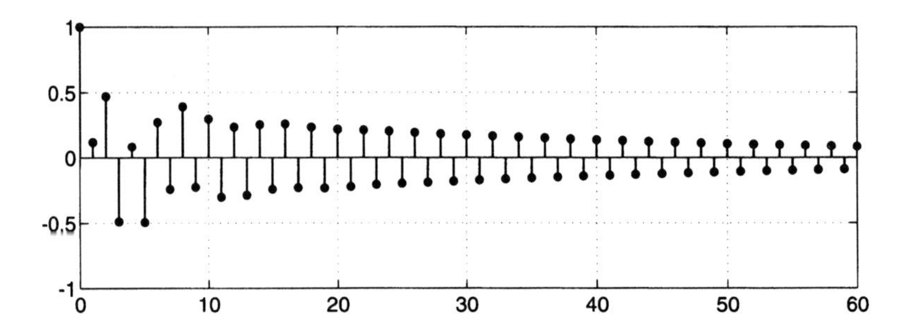

(a) A third-order all-pole signal that is to modeled.

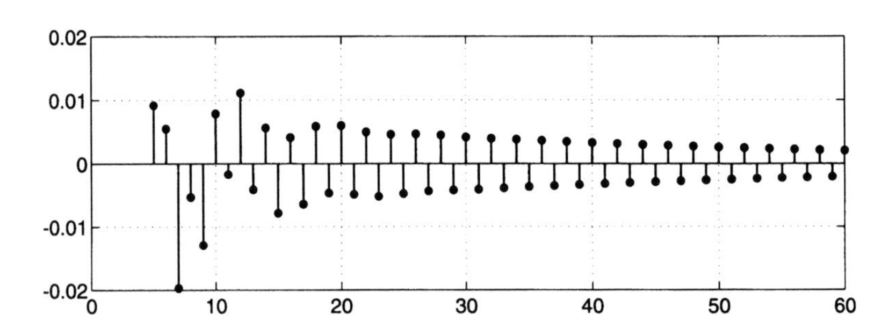

(b) The forward covariance error, $e_5^+(n)$, using a model order of $p = 5$.

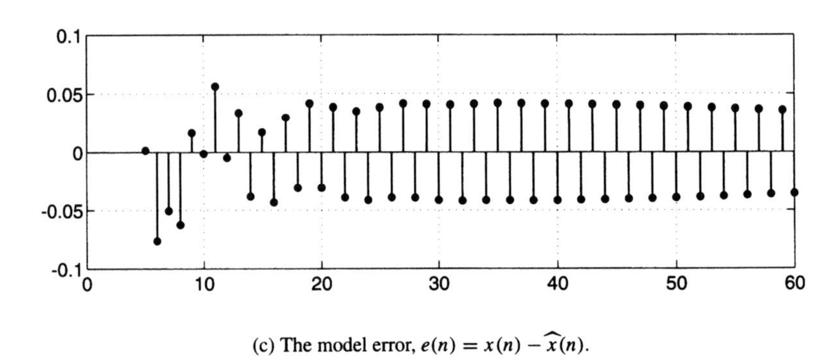

(c) The model error, $e(n) = x(n) - \widehat{x}(n)$.

Figure 6.16 *Using the forward covariance method to model a signal that is the unit sample response of a third-order all-pole filter.*

Note that although $x(n)$ is the unit sample response of an all-pole filter, the forward covariance method is unable to correctly model $x(n)$. If we increase the model order to $p = 5$, the next two reflection coefficients are

$$\Gamma_4^+ = 0.0483 \quad ; \quad \Gamma_5^+ = 0.1067$$

and the corresponding modeling errors are

$$\mathcal{E}_4^+ = 0.0087 \quad ; \quad \mathcal{E}_5^+ = 0.0016$$

Thus, extra poles may be used to further decrease the forward prediction error. Shown in Fig. 6.16*b* is the forward covariance error $e_5^+(n)$ over the interval $[5, N]$, the interval over which $e_5^+(n)$ is minimized. Finally, in Fig. 6.16c is the error, $e(n) = x(n) - \hat{x}(n)$. If we compute the squared model error we find

$$\mathcal{E}_{LS} = \sum_{n=5}^{N} \left[x(n) - \hat{x}(n)\right]^2 = 0.0889$$

The fact that $\mathcal{E}_{LS}$ and $\mathcal{E}_5^+$ are *different* should not be surprising since they are defined differently. It is important to keep in mind that in the forward covariance method $\mathcal{E}_p^+$ is the sum of the squares of the forward prediction error,

$$e_j^+(n) = a_j(n) * x(n)$$

whereas $\mathcal{E}_{LS}$ is the sum of the squares of the modeling error

$$e(n) = x(n) - \hat{x}(n)$$

Finally, note that since $x(n)$ is the output of a third-order all-pole filter, if we were to use the covariance method then $x(n)$ would be modeled exactly and the third-order modeling error $\mathcal{E}_3^C$ would be zero.

As illustrated in the previous example, the model that is produced with the forward covariance method is, in general, different from that obtained using the covariance method. Therefore, it is not necessarily true that the forward covariance method will produce an exact model for a signal $x(n)$, even if it is the output of an all-pole filter. In addition, as we observed when modeling the signal $x(n) = \beta^n$, the forward covariance method does not guarantee a stable model.

6.5.2 The Backward Covariance Method

As we saw in the previous section, using the forward covariance method the reflection coefficients of the lattice filter are found by sequentially minimizing the sum of the squares of the forward prediction error. Since a lattice filter generates both forward and backward prediction errors, we may just as easily minimize the sum of the squares of the backward prediction error. These two approaches are not entirely independent of each other, however, since minimizing the backward prediction error is the same as *time reversing* $x(n)$ and minimizing the forward prediction error. One may argue, in fact, that since $x(n)$ and $x^*(N-n)$ are statistically equivalent in the sense that they have identical deterministic autocorrelation sequences, there is no inherent reason why one error should be preferred over the other. Therefore, in this section we will look at the *backward covariance method* which performs a sequential minimization of the sum of the squares of the backward prediction error,

$$\mathcal{E}_j^- = \sum_{n=j}^{N} \left|e_j^-(n)\right|^2 \tag{6.67}$$

In order to distinguish this solution from that obtained using the forward covariance algorithm, we will denote the reflection coefficients in the backward covariance method by Γ_j^-. Proceeding as we did for the forward covariance method, setting the derivative of $\mathcal{E}_j^-$ with respect to $(\Gamma_j^-)^*$ equal to zero, we find that the reflection coefficient Γ_j^- that minimizes the backward prediction error $\mathcal{E}_j^-$ is given by

$$\Gamma_j^- = -\frac{\sum_{n=j}^{N} e_{j-1}^+(n)\left[e_{j-1}^-(n-1)\right]^*}{\sum_{n=j}^{N} \left|e_{j-1}^+(n)\right|^2} = -\frac{\langle \mathbf{e}_{j-1}^+, \mathbf{e}_{j-1}^- \rangle}{\|\mathbf{e}_{j-1}^+\|^2} \tag{6.68}$$

From a computational point of view, the backward covariance method works in the same way as the forward covariance method. Given the first $j-1$ reflection coefficients, $\mathbf{\Gamma}^- = \left[\Gamma_1^-, \Gamma_2^-, \ldots, \Gamma_{j-1}^-\right]^T$, and given the forward and backward prediction errors $e_{j-1}^+(n)$ and $e_{j-1}^-(n)$, the jth reflection coefficient that minimizes $\mathcal{E}_j^-$ is computed using Eq. (6.68). Then, using the lattice filter the $(j-1)$st-order forward and backward prediction errors are updated to form the jth-order errors $e_j^+(n)$ and $e_j^-(n)$ and the process is repeated. A MATLAB program for the backward covariance method is easily derived by modifying `fcov.m` in Fig. 6.15. Changes are required only in the lines that compute the reflection coefficient Γ_j^- and the error $\mathcal{E}_j^-$.

Example 6.5.2 *Backward Covariance Method*

Let us look again at the modeling problem considered in Example 6.5.1, this time using the backward covariance method. As before, we begin by initializing the forward and backward errors

$$e_0^+(n) = e_0^-(n) = \beta^n u(n) \quad ; \quad n = 0, 1, \ldots, N$$

Evaluating the norm

$$\|\mathbf{e}_0^+\|^2 = \sum_{n=1}^{N} \left[e_0^+(n)\right]^2 = \sum_{n=1}^{N} \beta^{2n} = \beta^2 \frac{1-\beta^{2N}}{1-\beta^2}$$

and the inner product

$$\langle \mathbf{e}_0^+, \mathbf{e}_0^- \rangle = \sum_{n=1}^{N} e_0^+(n) e_0^-(n-1) = \beta \frac{1-\beta^{2N}}{1-\beta^2}$$

we find that the first reflection coefficient is

$$\Gamma_1^- = -\frac{1}{\beta}$$

Updating the backward prediction error

$$e_1^-(n) = e_0^-(n-1) + \Gamma_1^- e_0^+(n) = \beta^{n-1} u(n-1) - \frac{1}{\beta}\beta^n u(n) = -\frac{1}{\beta}\delta(n)$$

it follows that the first-order modeling error is zero

$$\mathcal{E}_1^- = \sum_{n=1}^{N} \left[e_1^-(n)\right]^2 = 0$$

The first-order forward prediction error, however, is nonzero

$$e_1^+(n) = e_0^+(n) + \Gamma_1^- e_0^-(n-1) = \beta^n u(n) - \beta^{-1}\beta^{n-1} u(n-1)$$

Nevertheless, since $e_1^-(n) = 0$ for $n > 0$ then $\langle \mathbf{e}_1^-, \mathbf{e}_1^+ \rangle = 0$ and the second reflection coefficient is equal to zero, $\Gamma_2 = 0$. In fact, as with the forward covariance method, all succeeding reflection coefficients will also be equal to zero,

$$\mathbf{\Gamma}^- = \left[-1/\beta,\ 0,\ 0, \ldots\right]^T$$

As another example, let us again consider the signal $x(n)$ from Example 6.5.1 that is the unit sample response of the third-order filter

$$H(z) = \frac{1}{1 - 0.12z^{-1} - 0.456z^{-2} + 0.6z^{-3}}$$

which has lattice filter coefficients

$$\mathbf{\Gamma} = \left[0.6,\ -0.6,\ 0.6\right]$$

Using the backward covariance method with $p = 3$ and $N = 60$, the reflection coefficient sequence is

$$\mathbf{\Gamma}^- = \left[0.8011,\ -1.1787,\ 0.6796\right]^T$$

which corresponds to a filter having a system function

$$\hat{H}_3(z) = \frac{1}{1 - 0.9441z^{-1} - 1.2760z^{-2} + 0.6796z^{-3}}$$

Furthermore, the sequence of squared errors is

$$\underline{\mathcal{E}}^- = \left[3.6556,\ 1.9460,\ 0.9147\right]^T$$

Note that, unlike the solution to the forward covariance method, the model is unstable. If we increase the model order to $p = 5$, the next two reflection coefficients are

$$\Gamma_4^- = -1.1255 \quad ; \quad \Gamma_5^- = -0.4135$$

and the corresponding errors are

$$\mathcal{E}_4^- = 0.5189 \quad ; \quad \mathcal{E}_5^- = 0.1100$$

As illustrated in the previous example, the reflection coefficients in the backward covariance method are, in general, different from those obtained using the forward covariance method. As with the forward covariance method, it is not necessarily true that the backward covariance method will produce an exact model for a signal $x(n)$, even if it is the unit sample response of an all-pole filter. In addition, the backward covariance method may lead to an unstable model.

6.5.3 Variations

One of the properties that we have discovered about the forward and the backward covariance methods is that the reflection coefficients are not guaranteed to be less than one in magnitude. As a result, it is possible for either of these methods to produce an unstable model. Since ensuring a stable model may be necessary in applications such as speech synthesis or signal extrapolation, there is an interest in modeling techniques that are guaranteed to produce a stable model. There are several ways that we may *combine* the models generated by the forward and backward covariance methods to produce a stable model. One such approach proposed by Itakura [5] is to model a signal using reflection coefficients that are given by

$$\Gamma_j^I = -\frac{\sum_{n=j}^{N} e_{j-1}^+(n)\big[e_{j-1}^-(n-1)\big]^*}{\sqrt{\sum_{n=j}^{N} \big|e_{j-1}^+(n)\big|^2 \sum_{n=j}^{N} \big|e_{j-1}^-(n-1)\big|^2}} = -\frac{\langle \mathbf{e}_{j-1}^+, \mathbf{e}_{j-1}^- \rangle}{\|\mathbf{e}_{j-1}^+\| \, \|\mathbf{e}_{j-1}^-\|} \tag{6.69}$$

If we compare Γ_j^I with the reflection coefficients used in the forward and backward covariance methods, we see that $|\Gamma_j^I|$ is the *geometric mean* of $|\Gamma_j^+|$ and $|\Gamma_j^-|$,

$$|\Gamma_j^I| = \sqrt{|\Gamma_j^+| \, |\Gamma_j^-|} \tag{6.70}$$

We may show that $|\Gamma_j^I| \le 1$ by applying the *Cauchy-Schwartz inequality* to the vectors $\mathbf{e}_{j-1}^-$ and $\mathbf{e}_{j-1}^+$ as follows,

$$\big|\langle \mathbf{e}_{j-1}^+, \mathbf{e}_{j-1}^- \rangle\big| \le \|\mathbf{e}_{j-1}^+\| \, \|\mathbf{e}_{j-1}^-\| \tag{6.71}$$

where equality holds if and only if $\mathbf{e}_{j-1}^+ = \alpha\, \mathbf{e}_{j-1}^-$ for some constant α. Therefore, it follows that $|\Gamma_j^I| \le 1$.

There are other ways to ensure that the reflection coefficients are bounded by one in magnitude [8]. For example, it follows from the properties of the geometric mean that

$$\min\big\{|\Gamma_j^+|, |\Gamma_j^-|\big\} \le |\Gamma_j^I| \le \max\big\{|\Gamma_j^+|, |\Gamma_j^-|\big\}$$

Since $|\Gamma_j^I| \le 1$, it follows that if $|\Gamma_j^+|$ is greater than one in magnitude then $|\Gamma_j^-|$ will be less than one in magnitude, and vice versa. Therefore, if we set

$$|\Gamma_j^{\min}| = \min\big\{|\Gamma_j^+|, |\Gamma_j^-|\big\}$$

then $|\Gamma_j^{\min}|$ is guaranteed to be bounded by one in magnitude.

Although both the geometric mean and the minimum method produce reflection coefficients that are bounded by one in magnitude and thus represent a stable model, neither approach corresponds to a solution that results from the minimization of an error. In the next section we look at Burg's method which uses a sequential approach to find the reflection coefficients and does so in a way that guarantees that the reflection coefficients are bounded by one in magnitude and, therefore, guarantees that the model is stable.

6.5.4 Burg's Method

In Chapter 4 we saw that the covariance method for all-pole signal modeling, while more accurate than the autocorrelation method since it does not apply a window to the data, has the property that it does not always lead to a stable model. In addition, as we have seen in the previous two sections, sequentially minimizing either the forward or the backward prediction error using a "covariance-like" error may also lead to an unstable model. In the 1960s, Burg developed a method for spectrum estimation known as the maximum entropy method [1]. As a part of this method, which involves finding an all-pole model for the data, he proposed that the reflection coefficients be computed sequentially by minimizing the *sum* of the forward and backward prediction errors [2],

$$\mathcal{E}_j^B = \mathcal{E}_j^+ + \mathcal{E}_j^- = \sum_{n=j}^{N} |e_j^+(n)|^2 + \sum_{n=j}^{N} |e_j^-(n)|^2 \tag{6.72}$$

This *blending* of the forward and backward prediction errors places an equal emphasis on $\mathcal{E}_j^+$ and $\mathcal{E}_j^-$, and minimizing this error may be justified based on the statistical equivalence of $x(n)$ and $x^*(N-n)$ as pointed out in Section 6.5.2. As we will soon see, what makes the Burg error interesting and attractive is the fact that sequentially minimizing $\mathcal{E}_j^B$ guarantees that the reflection coefficients will be bounded by one in magnitude and thus, the model will be stable.

As we have done many times before, we may find the value of the reflection coefficient, Γ_j^B, that minimizes $\mathcal{E}_j^B$ by setting the derivative of $\mathcal{E}_j^B$ with respect to $(\Gamma_j^B)^*$ equal to zero as follows

$$\begin{aligned}\frac{\partial}{\partial(\Gamma_j^B)^*}\mathcal{E}_j^B &= \frac{\partial}{\partial(\Gamma_j^B)^*}\sum_{n=j}^{N}\Big\{|e_j^+(n)|^2 + |e_j^-(n)|^2\Big\} \\ &= \sum_{n=j}^{N}\Big\{e_j^+(n)\big[e_{j-1}^-(n-1)\big]^* + \big[e_j^-(n)\big]^* e_{j-1}^+(n)\Big\} = 0\end{aligned} \tag{6.73}$$

Substituting the error update equations for $e_j^+(n)$ and $[e_j^-(n)]^*$, which are similar to those given for $e_{j+1}^+(n)$ and $e_{j+1}^-(n)$ in Eqs. (6.11) and (6.16), and solving for Γ_j^B we find that the value of Γ_j^B that minimizes $\mathcal{E}_j^B$ is

$$\Gamma_j^B = -\frac{2\sum_{n=j}^{N} e_{j-1}^+(n)\big[e_{j-1}^-(n-1)\big]^*}{\sum_{n=j}^{N}\Big\{|e_{j-1}^+(n)|^2 + |e_{j-1}^-(n-1)|^2\Big\}} = -\frac{2\langle \mathbf{e}_{j-1}^+, \mathbf{e}_{j-1}^-\rangle}{\|\mathbf{e}_{j-1}^+\|^2 + \|\mathbf{e}_{j-1}^-\|^2} \tag{6.74}$$

To show that $|\Gamma_j^B| \le 1$ we use the inequality given in Eq. (2.26) on p. 23 which states that, for any two vectors **a** and **b**,

$$2|\langle \mathbf{a}, \mathbf{b}\rangle| \le \|\mathbf{a}\|^2 + \|\mathbf{b}\|^2 \tag{6.75}$$

with equality holding if and only if $\mathbf{a} = \pm\mathbf{b}$. With $\mathbf{a} = \mathbf{e}_{j-1}^-$ and $\mathbf{b} = \mathbf{e}_{j-1}^+$ it follows immediately that $|\Gamma_j^B| \le 1$ with equality if and only if $e_{j-1}^+(n) = e_{j-1}^-(n-1)$.

In the process of deriving the Levinson-Durbin recursion we saw that the modeling errors ϵ_j and ϵ_{j-1} for the autocorrelation method are related by

$$\epsilon_j = \epsilon_{j-1}\big[1 - |\Gamma_j|^2\big] \tag{6.76}$$

A similar relationship exists between the Burg errors $\mathcal{E}_j^B$ and $\mathcal{E}_{j-1}^B$. This relationship may be derived by incorporating the lattice filter update equations into the Burg error given in Eq. (6.72) and simplifying.[4] Specifically, we have

$$\begin{aligned}\mathcal{E}_j^B &= \sum_{n=j}^{N}\Big\{|e_j^+(n)|^2 + |e_j^-(n)|^2\Big\} \\ &= \sum_{n=j}^{N}\big|e_{j-1}^+(n) + \Gamma_j^B e_{j-1}^-(n-1)\big|^2 + \sum_{n=j}^{N}\big|e_{j-1}^-(n-1) + (\Gamma_j^B)^* e_{j-1}^+(n)\big|^2\end{aligned} \tag{6.77}$$

[4] Although not difficult, this derivation is a bit messy due to the plethora of complex conjugates and inner products.

Expanding the squares and combining together common terms we may write Eq. (6.77) as follows

$$\mathcal{E}_j^B = \left[1 + |\Gamma_j^B|^2\right] \sum_{n=j}^{N} \left\{ |e_{j-1}^+(n)|^2 + |e_{j-1}^-(n-1)|^2 \right\}$$

$$+ 2(\Gamma_j^B)^* \sum_{n=j}^{N} e_{j-1}^+(n) \left[e_{j-1}^-(n-1)\right]^* + 2\Gamma_j^B \sum_{n=j}^{N} e_{j-1}^-(n-1) \left[e_{j-1}^+(n)\right]^* \qquad (6.78)$$

Now, note from Eq. (6.74) that the inner product between $e_{j-1}^+(n)$ and $e_{j-1}^-(n-1)$ may be written as

$$2 \sum_{n=j}^{N} e_{j-1}^+(n) \left[e_{j-1}^-(n-1)\right]^* = -\Gamma_j^B \sum_{n=j}^{N} \left\{ |e_{j-1}^+(n)|^2 + |e_{j-1}^-(n-1)|^2 \right\}$$

Substituting this expression and its conjugate into Eq. (6.78) and simplifying we have

$$\mathcal{E}_j^B = \left[1 - |\Gamma_j^B|^2\right] \sum_{n=j}^{N} \left\{ |e_{j-1}^+(n)|^2 + |e_{j-1}^-(n-1)|^2 \right\} \qquad (6.79)$$

Finally, since

$$\begin{aligned} \sum_{n=j}^{N} \left\{ |e_{j-1}^+(n)|^2 + |e_{j-1}^-(n-1)|^2 \right\} &= \sum_{n=j-1}^{N} \left\{ |e_{j-1}^+(n)|^2 + |e_{j-1}^-(n)|^2 \right\} \\ &\quad - |e_{j-1}^+(j-1)|^2 - |e_{j-1}^-(N)|^2 \\ &= \mathcal{E}_{j-1}^B - |e_{j-1}^+(j-1)|^2 - |e_{j-1}^-(N)|^2 \end{aligned} \qquad (6.80)$$

we have the following recursion for the Burg error

$$\boxed{\mathcal{E}_j^B = \left\{ \mathcal{E}_{j-1}^B - |e_{j-1}^+(j-1)|^2 - |e_{j-1}^-(N)|^2 \right\} \left[1 - |\Gamma_j^B|^2\right]} \qquad (6.81)$$

which is initialized with

$$\mathcal{E}_0^B = \sum_{n=0}^{N} \left\{ |e_0^+(n)|^2 + |e_0^-(n)|^2 \right\} = 2 \sum_{n=0}^{N} |x(n)|^2 \qquad (6.82)$$

From a computational point of view, Burg's method works in the same way as both the forward and the backward covariance methods. Specifically, given the first $j-1$ reflection coefficients, $\mathbf{\Gamma}_{j-1}^B = \left[\Gamma_1^B, \ldots, \Gamma_{j-1}^B\right]^T$, and given the forward and backward prediction errors $e_{j-1}^+(n)$ and $e_{j-1}^-(n)$, the jth reflection coefficient is computed using Eq. (6.74). Then, using the lattice filter, the $(j-1)$st-order forward and backward prediction errors are updated to form the jth-order errors $e_j^+(n)$ and $e_j^-(n)$. A MATLAB program that implements Burg's method, sometimes referred to as the Burg recursion, is given in Fig. 6.17.

The primary computational requirements in the Burg recursion come from the evaluation of the inner product and the norms in Eq. (6.74) that are necessary to evaluate Γ_j^B. It is possible, however, to derive a recursion for updating the denominator of Eq. (6.74), thereby reducing the amount of computation. This recursion may be derived by noting that the denominator, which we will denote by D_j, is equal to the term on the left hand side of

The Burg Algorithm

```
function [gamma,err] = burg(x,p)
%
x=x(:);
N=length(x);
eplus  = x(2:N);
eminus = x(1:N-1);
N=N-1;
for j=1:p;
    gamma(j) = -2*eminus'*eplus/(eplus'*eplus+eminus'*eminus);
    temp1    = eplus  + gamma(j)*eminus;
    temp2    = eminus + conj(gamma(j))*eplus;
    err(j)   = temp1'*temp1+temp2'*temp2;
    eplus    = temp1(2:N);
    eminus   = temp2(1:N-1);
    N=N-1;
    end;
```

Figure 6.17 *A* MATLAB *program for finding the reflection coefficients corresponding to a pth-order model for a signal $x(n)$ using Burg's method.*

Eq. (6.80) and, therefore,

$$D_j = \mathcal{E}_{j-1}^B - |e_{j-1}^+(j-1)|^2 - |e_{j-1}^-(N)|^2 \tag{6.83}$$

This, however, is equivalent to the term in braces in Eq. (6.81). Thus, substituting Eq. (6.83) into Eq. (6.81) we have

$$\mathcal{E}_j^B = D_j\Big[1 - |\Gamma_j^B|^2\Big] \tag{6.84}$$

Since

$$D_{j+1} = \mathcal{E}_j^B - |e_j^+(j)|^2 - |e_j^-(N)|^2 \tag{6.85}$$

incorporating Eq. (6.84) into Eq. (6.85) we have the desired recursion,

$$\boxed{D_{j+1} = D_j\Big[1 - |\Gamma_j^B|^2\Big] - |e_j^+(j)|^2 - |e_j^-(N)|^2} \tag{6.86}$$

The equations for the Burg recursion are summarized in Table 6.1. From this table we may evaluate the computational requirements of the Burg algorithm. First, note that in order to initialize the recursion we need N multiplications and $N-1$ additions (assuming the multiplication by 2 can be done with a data shift). Since the evaluation of the error $\mathcal{E}_j$ at each stage of the recursion is not necessary, at the jth step of the recursion there are $3(N-j)+7$ multiplications, $3(N-j)+5$ additions, and one division. Finally, since step $2(a)$ is the only calculation that needs to be performed during the last stage of the recursion, $(j=p)$, then the total number of multiplications is

$$N + \sum_{j=1}^{p-1}\big[3(N-j)+7\big] + N - p + 1 = N(3p-1) - \frac{3}{2}p(p-1) + 6(p-1) \tag{6.87}$$

Table 6.1 ***The Burg Recursion***

1. Initialize the recursion

 (a) $e_0^+(n) = e_0^-(n) = x(n)$

 (b) $D_1 = 2\sum_{n=1}^{N}\left\{|x(n)|^2 - |x(n-1)|^2\right\}$

2. For $j = 1$ to p

 a) $\Gamma_j^B = -\dfrac{2}{D_j}\sum_{n=j}^{N} e_{j-1}^+(n)\left[e_{j-1}^-(n-1)\right]^*$

 b) For $n = j$ to N

$$e_j^+(n) = e_{j-1}^+(n) + \Gamma_j^B e_{j-1}^-(n-1)$$
$$e_j^-(n) = e_{j-1}^-(n-1) + (\Gamma_j^B)^* e_{j-1}^+(n)$$

 c) $D_{j+1} = D_j\left(1 - |\Gamma_j^B|^2\right) - |e_j^+(j)|^2 - |e_j^-(N)|^2$

 d) $\mathcal{E}_j^B = D_j\left[1 - |\Gamma_j^B|^2\right]$

and a similar number of additions. Therefore, assuming that $N \gg p$, the Burg recursion requires on the order of $3Np$ multiplications and additions. The autocorrelation method, on the other hand, requires only $p^2 + 2p$ multiplications and p^2 additions to solve the autocorrelation normal equations using the Levinson-Durbin recursion. In addition to this, however, is the requirement for computing the $p+1$ autocorrelations $r_x(k)$ for $k = 0, 1, \ldots, p$. For a sequence of length N this requires approximately an additional $N(p+1)$ multiplications and additions. Both approaches, therefore, require on the order of Np multiplies and adds.

Example 6.5.3 *The Burg Recursion*

Let us compute the reflection coefficients for the exponential signal

$$x(n) = \beta^n u(n) \quad ; \quad n = 0, 1, \ldots, N$$

using Burg's method. Using the norms and inner products derived in Examples 6.5.1 and 6.5.2 we find for the first-order Burg model

$$\Gamma_1^B = -2\frac{\langle \mathbf{e}_0^+, \mathbf{e}_0^- \rangle}{\|\mathbf{e}_0^+\|^2 + \|\mathbf{e}_0^-\|^2} = -\frac{2\beta}{1+\beta^2}$$

Note that $|\Gamma_1^B| \le 1$ for any value of β. Updating the forward and backward errors,

$$\begin{aligned} e_1^+(n) &= e_0^+(n) + \Gamma_1^B e_0^-(n-1) &= \beta^n u(n) + \Gamma_1^B \beta^{n-1} u(n-1) \\ e_1^-(n) &= e_0^-(n-1) + \Gamma_1^B e_0^+(n) &= \beta^{n-1} u(n-1) + \Gamma_1^B \beta^n u(n) \end{aligned}$$

Computing the error norms and inner products for $e_1^+(n)$ and $e_1^-(n)$ we have, after a lot of algebra,

$$\|\mathbf{e}_1^+\|^2 = \sum_{n=2}^{N}[e_1^+(n)]^2 = \beta^4(1-\beta^2)\frac{1-\beta^{2(N-1)}}{(1+\beta^2)^2}$$

$$\|\mathbf{e}_1^-\|^2 = \sum_{n=2}^{N}[e_1^-(n-1)]^2 = (1-\beta^2)\frac{1-\beta^{2(N-1)}}{(1+\beta^2)^2}$$

$$\langle \mathbf{e}_1^-, \mathbf{e}_1^+\rangle = \sum_{n=2}^{N} e_1^+(n)e_1^-(n-1) = -\beta^2(1-\beta^2)\frac{1-\beta^{2(N-1)}}{(1+\beta^2)^2}$$

Thus, the second reflection coefficient is equal to

$$\Gamma_2^B = -2\frac{\langle \mathbf{e}_1^+, \mathbf{e}_1^-\rangle}{\|\mathbf{e}_1^+\|^2 + \|\mathbf{e}_1^-\|^2} = 2\frac{\beta^2(1-\beta^2)}{\beta^4(1-\beta^2)+(1-\beta^2)} = \frac{2\beta^2}{1+\beta^4}$$

Let us now evaluate the Burg error for this model using the recursion given in Eq. (6.81). Assuming that $N \gg 1$ and $|\beta| < 1$, for the zeroth-order error we have

$$\mathcal{E}_0^B = 2\sum_{n=0}^{N} x^2(n) = 2\frac{1-\beta^{2(N+1)}}{1-\beta^2} \approx \frac{2}{1-\beta^2}$$

For the first-order error,

$$\begin{aligned}
\mathcal{E}_1^B &= \left\{\mathcal{E}_0^B - \left[e_0^+(0)\right]^2 - \left[e_0^-(N)\right]^2\right\}\left[1-|\Gamma_1^B|^2\right] \\
&= \left\{\mathcal{E}_0^B - 1 - \beta^{2N}\right\}\left[1-|\Gamma_1^B|^2\right] \\
&\approx \frac{1-\beta^2}{1+\beta^2}
\end{aligned}$$

Finally, for the second-order error we find, after some algebra,

$$\begin{aligned}
\mathcal{E}_2^B &= \left\{\mathcal{E}_1^B - \left[e_1^+(1)\right]^2 - \left[e_1^-(N)\right]^2\right\}\left[1-|\Gamma_2^B|^2\right] \\
&\approx \frac{(1-\beta^2)^3}{1+\beta^4}
\end{aligned}$$

As a final example, we again consider the signal in Example 6.5.1 which is the unit sample response of the second-order system

$$H(z) = \frac{1}{1-0.12z^{-1}-0.456z^{-2}+0.6z^{-3}}$$

which has lattice filter coefficients

$$\mathbf{\Gamma} = \left[0.6,\ -0.6,\ 0.6\right]^T$$

Using the MATLAB program for the Burg algorithm we find, with $p = 3$ and $N = 60$, the reflection coefficient sequence

$$\mathbf{\Gamma}^B = \left[0.6753,\ -0.6653,\ 0.8920\right]^T$$

which corresponds to a filter having a system function

$$\hat{H}_3(z) = \frac{1}{1-0.3675z^{-1}-0.4637z^{-2}+0.8920z^{-3}}$$

Furthermore, the sequence of squared errors is

$$\underline{\mathcal{E}}^B = \left[3.4371,\ 1.5627,\ 0.3134\right]^T$$

Again, if we increase the model order to $p = 5$, the next two reflection coefficients are

$$\Gamma_4^B = -0.5324 \quad ; \quad \Gamma_5^B = 0.0390$$

and the corresponding errors are

$$\mathcal{E}_4^B = 0.2027 \quad ; \quad \mathcal{E}_5^B = 0.2022$$

An interesting property of the Burg algorithm is that Γ_j^B is the *harmonic mean* of the reflection coefficients Γ_j^+ and Γ_j^-. Specifically, note that if we are given the forward and backward prediction errors $e_j^+(n)$ and $e_j^-(n)$, and if we were to compute the jth reflection coefficient using the forward covariance method, the backward covariance method, and Burg's method, then these reflection coefficients would be related to each other as follows[5]

$$\frac{2}{\Gamma_j^B} = \frac{1}{\Gamma_j^+} + \frac{1}{\Gamma_j^-} \tag{6.88}$$

It is important to be careful in using this equation, however, because it does not imply that we may use the forward and backward covariance methods to compute Γ_j^+ and Γ_j^- and then use Eq. (6.88) to find Γ_j^B. This relationship between the reflection coefficients is only valid when Γ_j^+ and Γ_j^- are computed using the forward and backward prediction errors that are produced with the Burg algorithm, i.e., using the lattice filter with reflection coefficients Γ_j^B.

6.5.5 Modified Covariance Method

In the previous section we derived the Burg recursion, which finds the reflection coefficients for an all-pole model by sequentially minimizing the sum of the squares of the forward and backward prediction errors. In this section we look at the *modified covariance method* or *forward-backward algorithm* for all-pole signal modeling. As with the Burg algorithm, the modified covariance method minimizes the sum of the squares of the forward and backward prediction errors,

$$\mathcal{E}_p^M = \mathcal{E}_p^+ + \mathcal{E}_p^-$$

The difference, however, between the two approaches is that, in the modified covariance method, the minimization is not performed sequentially. In other words, for a pth-order model, the modified covariance method finds the set of reflection coefficients or, equivalently, the set of transversal filter coefficients $a_p(k)$, that minimize $\mathcal{E}_p^M$.

To find the filter coefficients that minimize $\mathcal{E}_p^M$ we set the derivative of $\mathcal{E}_p^M$ with respect to $a_p^*(l)$ equal to zero for $l = 1, 2, \ldots, p$ as we did in the covariance method. Since

$$e_p^+(n) = x(n) + \sum_{k=1}^{p} a_p(k)x(n-k) \tag{6.89}$$

[5]This relationship follows directly from the definitions given in Eqs. (6.74), (6.66) and (6.68).

and

$$e_p^-(n) = x(n-p) + \sum_{k=1}^{p} a_p^*(k)x(n-p+k) \tag{6.90}$$

then

$$\begin{aligned}\frac{\partial \mathcal{E}_p^M}{\partial a_p^*(l)} &= \sum_{n=p}^{N}\Big[e_p^+(n)\frac{\partial\left[e_p^+(n)\right]^*}{\partial a_p^*(l)} + \left[e_p^-(n)\right]^*\frac{\partial e_p^-(n)}{a_p^*(l)}\Big] \\ &= \sum_{n=p}^{N}\Big[e_p^+(n)x^*(n-l) + \left[e_p^-(n)\right]^* x(n-p+l)\Big] = 0\end{aligned} \tag{6.91}$$

Substituting Eqs. (6.89) and (6.90) into Eq. (6.91) and simplifying we find that the normal equations for the modified covariance method are given by

$$\sum_{k=1}^{p}\Big[r_x(l,k) + r_x(p-k,p-l)\Big]a_p(k) = -\big[r_x(l,0) + r_x(p,p-l)\big]; \quad l = 1,\ldots,p \tag{6.92}$$

where

$$r_x(l,k) = \sum_{n=p}^{N} x(n-k)x^*(n-l) \tag{6.93}$$

Recall that if we were to use the covariance method and minimize $\mathcal{E}_p^+$, then the normal equations would be

$$\sum_{k=1}^{p} r_x(l,k)a_p(k) = -r_x(l,0) \tag{6.94}$$

On the other hand, if we were to minimize $\mathcal{E}_p^-$, then the normal equations would be

$$\sum_{k=1}^{p} r_x(p-k,p-l)a_p(k) = -r_x(p,p-l) \tag{6.95}$$

Therefore, the normal equations for the modified covariance method is a combination of these two sets of equations.

For the modified covariance error,

$$\mathcal{E}_p^M = \sum_{n=p}^{N}\Big\{\left|e_p^+(n)\right|^2 + \left|e_p^-(n)\right|^2\Big\}$$

we may use the orthogonality condition in Eq. (6.91) to express $\mathcal{E}_p^M$ as follows

$$\mathcal{E}_p^M = \sum_{n=p}^{N}\left[e_p^+(n)x^*(n) + \left[e_p^-(n)\right]^* x(n-p)\right]$$

Substituting the expressions given in Eqs. (6.89) and (6.90) for $e_p^+(n)$ and $e_p^-(n)$ and simplifying, we have

The Modified Covariance Method

```
function [a,err] = mcov(x,p)
%
x   = x(:);
N   = length(x);
if p>=length(x), error('Model order too large'), end
X   = toeplitz(x(p+1:N),flipud(x(1:p+1)));
R   = X'*X;
R1  = R(2:p+1,2:p+1);
R2  = flipud(fliplr(R(1:p,1:p)));
b1  = R(2:p+1,1);
b2  = flipud(R(1:p,p+1));
a   = [1 ; -(R1+R2)\(b1+b2)];
err = R(1,:)*a+fliplr(R(p+1,:))*a;
end;
```

Figure 6.18 *A* MATLAB *program for finding a pth-order all-pole model for a signal $x(n)$ using the modified covariance method.*

$$\mathcal{E}_p^M = r_x(0,0) + r_x(p,p) + \sum_{k=1}^{p} a(k)\big[r_x(0,k) + r_x(p,p-k)\big] \tag{6.96}$$

A MATLAB program for finding a pth-order all-pole model for $x(n)$ is given in Fig. 6.18.

Example 6.5.4 *The Modified Covariance Method*

Let us use the modified covariance method to find a second-order all-pole model for

$$x(n) = \beta^n u(n)$$

which is assumed to be known for $n = 0, 1, 2, \ldots, N$. We begin by evaluating the autocorrelations given in Eq. (6.93). With $p = 2$ we have

$$\begin{aligned} r_x(k,l) &= \sum_{n=2}^{N} \beta^{n-k}\beta^{n-l} = \beta^{-k-l}\sum_{n=2}^{N}\beta^{2n} \\ &= \beta^4 \frac{1-\beta^{2(N-1)}}{1-\beta^2}\beta^{-k-l} \equiv \lambda\beta^{4-k-l} \end{aligned}$$

where we have defined, for convenience,

$$\lambda = \frac{1-\beta^{2(N-1)}}{1-\beta^2}$$

The coefficients of the second-order model, $a(1)$ and $a(2)$, are the solution to the linear equations

$$\begin{bmatrix} r_x(1,1)+r_x(1,1) & r_x(1,2)+r_x(0,1) \\ r_x(2,1)+r_x(1,0) & r_x(2,2)+r_x(0,0) \end{bmatrix}\begin{bmatrix} a(1) \\ a(2) \end{bmatrix} = -\begin{bmatrix} r_x(1,0)+r_x(2,1) \\ r_x(2,0)+r_x(2,0) \end{bmatrix}$$

Inserting the computed values for $r_x(k, l)$ we have

$$\lambda \begin{bmatrix} 2\beta^2 & \beta + \beta^3 \\ \beta + \beta^3 & 1 + \beta^4 \end{bmatrix} \begin{bmatrix} a(1) \\ a(2) \end{bmatrix} = -\lambda \begin{bmatrix} \beta + \beta^3 \\ 2\beta^2 \end{bmatrix}$$

Solving these equations for $a(1)$ and $a(2)$ we find that $a(1) = -(1 + \beta^2)/\beta$ and $a(2) = 1$ which leads to an all-pole model with

$$A(z) = 1 - \frac{1 + \beta^2}{\beta} z^{-1} + z^{-2}$$

Note that since the roots of $A(z)$ are at $z = \beta$ and $z = 1/\beta$ then the model is *unstable* for all values of β. We find an interesting result, however, when when we examine the error. Inserting the given values for the autocorrelations and the computed values for $a(k)$ into Eq. (6.96) we have

$$\begin{aligned} \mathcal{E}_2^M &= \lambda \left[\beta^4 + 1 + a(1) \left(\beta + \beta^3 \right) + a(2) \left(2\beta^2 \right) \right] \\ &= \lambda \left[\beta^4 + 1 - \frac{1 + \beta^2}{\beta} \beta (1 + \beta^2) + 2\beta^2 \right] = 0 \end{aligned}$$

Thus, the sum of the squares of the forward and backward prediction errors is equal to zero!

6.6 STOCHASTIC MODELING

In Section 6.5 we considered several different approaches for all-pole modeling of deterministic signals. As discussed in Section 4.7 of Chapter 4, in some applications it is necessary to find models for stochastic processes. We may easily adapt the lattice methods developed in this chapter to stochastic processes by replacing the least squares errors with mean-square errors. For example, we may use the forward covariance method to model a stochastic process $x(n)$ as the output of an all-pole filter driven by white noise by sequentially minimizing the mean-square forward prediction error,

$$\xi_j^+ = E\left\{ \left| e_j^+(n) \right|^2 \right\}$$

As we did in the deterministic case, the value for Γ_j^+ that minimizes ξ_j^+ is found by setting the derivative of ξ_j^+ with respect to $(\Gamma_j^+)^*$ equal to zero and solving for Γ_j^+. The result, not too surprisingly, is

$$\boxed{\Gamma_j^+ = -\frac{E\left\{ e_{j-1}^+(n) \left[e_{j-1}^-(n-1) \right]^* \right\}}{E\left\{ |e_{j-1}^-(n-1)|^2 \right\}}} \tag{6.97}$$

which is related to the reflection coefficients in the deterministic case by replacing the sums in Eq. (6.66) with expectations. Instead of ξ_j^+, if we minimize the mean-square backward prediction error,

$$\xi_j^- = E\left\{ \left| e_j^-(n) \right|^2 \right\}$$

then we find that the jth reflection coefficient is given by

$$\Gamma_j^- = -\frac{E\left\{e_{j-1}^+(n)\left[e_{j-1}^-(n-1)\right]^*\right\}}{E\left\{|e_{j-1}^+(n)|^2\right\}} \tag{6.98}$$

On the other hand, minimizing the mean-square Burg error

$$\xi_j^B = E\left\{\left|e_j^+(n)\right|^2 + \left|e_j^-(n)\right|^2\right\}$$

results in reflection coefficients given by

$$\Gamma_j^B = -2\frac{E\left\{e_{j-1}^+(n)\left[e_{j-1}^-(n-1)\right]^*\right\}}{E\left\{|e_{j-1}^+(n)|^2\right\} + E\left\{|e_{j-1}^-(n-1)|^2\right\}} \tag{6.99}$$

Of course, in any practical application these ensemble averages are unknown and it is necessary to estimate them from the given data. Replacing the expected values with estimates that are formed by taking time averages, however, takes us back to the deterministic techniques discussed in Section 6.5. For example, if we use the estimates

$$\hat{E}\left\{e_{j-1}^+(n)\left[e_{j-1}^-(n-1)\right]^*\right\} = \sum_{n=j}^{N} e_{j-1}^+(n)\left[e_{j-1}^-(n-1)\right]^*$$

and

$$\hat{E}\left\{\left|e_j^-(n)\right|^2\right\} = \sum_{n=j}^{N}\left|e_j^-(n)\right|^2$$

in Eq. (6.97), then we have the expression for Γ_j^+ given in Eq. (6.66).

Using the stochastic formulation of the lattice all-pole signal modeling techniques, we may establish some useful and interesting orthogonality relations. In the deterministic case these relations may be assumed to be approximately true. One orthogonality relation that we will find particularly useful in our discussions of adaptive lattice filters (see Section 9.2.8) is the condition that the backward prediction errors $e_i^-(n)$ and $e_j^-(n)$ are orthogonal when $i \neq j$,

$$E\left\{e_i^-(n)\left[e_j^-(n)\right]^*\right\} = 0 \quad ; \quad i \neq j$$

To establish this orthogonality relation, recall from Eq. (6.12) that the backward prediction errors $e_j^-(n)$ are generated by filtering $x(n)$ with the backward prediction error filter $A_j^R(z)$ as follows

$$e_j^-(n) = x(n) * a_j^R(n) = x(n-j) + \sum_{k=1}^{j} a_j^*(k)x(n-j+k)$$

Writing these equations in matrix form for $j = 0, 1, \ldots, p$ we have

$$\begin{bmatrix} e_0^-(n) \\ e_1^-(n) \\ e_2^-(n) \\ \vdots \\ e_p^-(n) \end{bmatrix} = \begin{bmatrix} 1 & 0 & 0 & \cdots & 0 \\ a_1^*(1) & 1 & 0 & \cdots & 0 \\ a_2^*(2) & a_2^*(1) & 1 & \cdots & 0 \\ \vdots & \vdots & \vdots & & \vdots \\ a_p^*(p) & a_p^*(p-1) & a_p^*(p-2) & \cdots & 1 \end{bmatrix} \begin{bmatrix} x(n) \\ x(n-1) \\ x(n-2) \\ \vdots \\ x(n-p) \end{bmatrix} \tag{6.100}$$

which may be written in vector form as follows

$$\mathbf{e}^-(n) = \mathbf{A}_p^T \mathbf{x}(n)$$

where $\mathbf{A}_p$ is the matrix of predictor coefficients defined in Eq. (5.84) (see p. 250). Therefore, it follows that the correlation matrix for the backward prediction errors is

$$E\{[\mathbf{e}^-(n)]^* [\mathbf{e}^-(n)]^T\} = E\{\mathbf{A}_p^H \mathbf{x}^*(n)\mathbf{x}^T(n)\mathbf{A}_p\} = \mathbf{A}_p^H \mathbf{R}_p \mathbf{A}_p$$

where $\mathbf{R}_p$ is the autocorrelation matrix for $x(n)$. However, as we saw in Section 5.2.7, the matrix $\mathbf{A}_p$ diagonalizes the autocorrelation matrix $\mathbf{R}_p$. Therefore

$$E\{[\mathbf{e}^-(n)]^* [\mathbf{e}^-(n)]^T\} = \mathbf{D}_p = \text{diag}\{\epsilon_0,\ \epsilon_1,\ \ldots,\ \epsilon_p\}$$

and we have

$$E\Big\{e_i^-(n)\big[e_j^-(n)\big]^*\Big\} = \begin{cases} \epsilon_j & ; \quad i = j \\ 0 & ; \quad i \neq j \end{cases} \tag{6.101}$$

and the backward prediction errors are orthogonal. Thus, the lattice filter transforms the input sequence, $x(n), x(n-1), \ldots, x(n-p)$ into an orthogonal sequence, $e_0^-(n), e_1^-(n), \ldots, e_p^-(n)$. This orthogonalization results in a *decoupling* of the successive errors form from each other.

6.7 SUMMARY

In the first part of this chapter we derived a number of different lattice filter structures for all-zero, all-pole, and pole-zero filters. These lattice filters are important in signal processing applications due their modularity, robustness to finite precision effects, and ease in ensuring filter stability. Using the Levinson-Durbin recursion, we derived a two-multiplier lattice filter and showed how the forward and backward prediction errors, $e_j^+(n)$ and $e_j^-(n)$, respectively, are generated by this filter. Another FIR lattice filter structure, one that is based on the three-term recurrence for the singular predictor polynomials, $S_j(z)$, was then derived. This structure, known as the split lattice filter, is parameterized in terms of the coefficients δ_j produced by the split Levinson recursion. Next, we developed several different lattice filter structures for all-pole filters. Beginning with the two-multiplier all-pole lattice that follows directly from the FIR lattice filter, it was shown how this structure could be used to implement an allpass filter. Through a sequence of flowgraph manipulations, we then derived the Kelly-Lochbaum lattice filter, the normalized lattice filter, and the one-multiplier lattice filter. Finally, a lattice filter structure was presented for implementing a filter that has both poles and zeros. This structure consists of an all-pole lattice along with a set of tap weights $c_q(k)$ that form a linear combination of the backward prediction errors $e_j^-(n)$ to produce the filter output.

The second part of the chapter was concerned with the application of lattice filters to signal modeling. Since a lattice filter is parameterized in terms of its reflection coefficients, Γ_j, or some set of parameters derived from these coefficients, we considered a number of different ways to optimally select the reflection coefficients. The first approach, known as the forward covariance method, involved the sequential minimization of the forward prediction error using a covariance-type error, i.e., one that does not require knowledge of the signal outside the interval $[0, N]$. The second, the backward covariance method, performed a sequential minimization of the backward prediction errors. Since neither the forward nor the backward covariance methods produce a model that is guaranteed to be stable, we then considered the sequential minimization of the Burg error, which is the *sum* of the forward and backward prediction errors. It was shown that the sequential minimization of the Burg error leads to a model that is guaranteed to be stable. Finally, we consided the modified covariance method, which drops the requirement that the minimization of the Burg error be done sequentially, and finds the global minimum. Although optimum in the sense of minimizing the Burg error, the model produced with this method is not guaranteed to be stable.

In the last section of this chapter, we looked briefly at how the lattice methods for modeling deterministic signals may be generalized to model stochastic processes. What we saw was that a stochastic signal model may be derived using the approaches used for deterministic signals by simply replacing the inner product for deterministic signals with an inner product that is appropriate for stochastic processes, i.e., the expected value of the product of two processes.

References

1. J. Burg, "Maximum entropy spectral analysis," *Proc. 37th Meeting of Soc. of Exploration Geophysicists*, Oklahoma City, OK., October 1967 (reprinted in *Modern Spectrum Estimation*, D.G. Childers, Ed., IEEE Press, New York).
2. J. Burg, "Maximum entropy spectral analysis," Ph.D. dissertation, Stanford University, Stanford, CA, May, 1975.
3. J. R. Deller Jr., J. G. Proakis, and J. H. L. Hansen, *Discrete-time Processing of Speech Signals*, MacMillan, New York, 1993.
4. P. Delsarte and Y. V. Genin, "On the splitting of classical algorithms in linear prediction theory," *IEEE Trans. Acoust., Speech, Sig. Proc.*, vol. ASSP-35, no. 5, pp. 645–653, May 1987.
5. F. Itakura and S. Saito, "Digital filtering techniques for speech analysis and synthesis," *7th Int. Conf. Acoustics*, Budapest, 1971, Paper 25-C-1.
6. J. L. Kelly and C. C. Lochbaum, "Speech synthesis," *Proc. 4th Int. Congress on Acoust.*, vol. G42, pp. 1–4, 1962.
7. J. Makhoul, "A class of all-zero lattice digital filters: Properties and applications," *IEEE Trans. Acoust., Speech, Sig. Proc.*, vol. ASSP-26, no. 4, pp. 304–314, Aug. 1978.
8. J. Makhoul, "Stable and efficient lattice methods for linear prediction," *IEEE Trans. Acoust., Speech, Sig. Proc.*, vol. ASSP-25, no. 5, pp. 423–428, October 1977.
9. J. D. Markel and A. H. Gray, Jr., *Linear Prediction of Speech*, Springer-Verlag, New York, 1976.
10. J. D. Markel and A. H. Gray, Jr., "Roundoff noise characteristics of a class of orthogonal polynomial structures," *IEEE Trans. Acoust., Speech, Sig. Proc.*, vol. ASSP-23, no. 5, pp. 473–486, October 1975.
11. S. J. Orfanidis, *Optimal Signal Processing: An Introduction*, 2nd Ed., Macmillan, New York, 1988.
12. E. A. Robinson and S. Treitel, *Geophysical Signal Analysis*, Prentice-Hall, Englewood Cliffs, N. J., 1980.

13. R. Viswanathan and J. Makhoul, "Quantization properties of transmission parameters in linear predictive systems," *IEEE Trans. Acoust., Speech, Sig. Proc.*, vol. ASSP-22, no. 3, pp. 309–321, June 1975.

6.8 PROBLEMS

6.1. Design a two-pole lattice filter that has poles at $re^{j\theta}$ and $re^{-j\theta}$ and draw a carefully labeled flowgraph of your filter.

6.2. Consider the all-pole filter

$$H(z) = \frac{1}{1 - 0.2z^{-1} + 0.9z^{-2} + 0.6z^{-3}}$$

Draw the flowgraph for a lattice filter implementation of $H(z)$ using

(a) A Kelly-Lochbaum lattice filter.

(b) A normalized lattice filter.

(c) A one-muliplier lattice filter.

For each structure, determine the number of multiplies, adds, and delays required to implement the filter and compare them to a direct-form realization of $H(z)$. Based solely on computational considerations, which structure is the most efficient?

6.3. Find the system function $H(z)$ for the lattice filter given in the figure below.

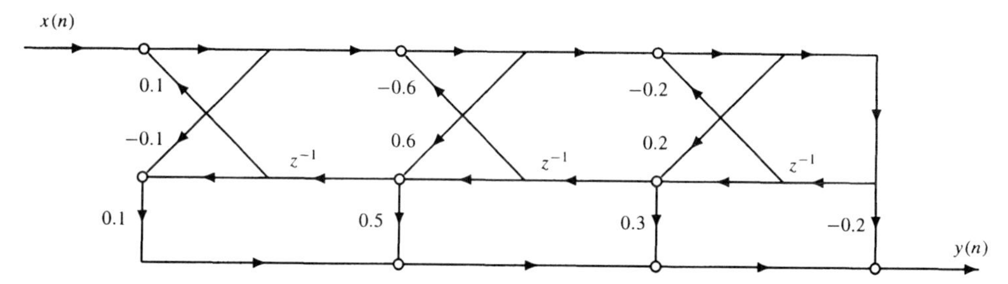

6.4. Sketch a lattice filter structure for each of the following system functions.

(a) $H(z) = \dfrac{2 - z^{-1}}{1 + 0.7z^{-1} + 0.49z^{-2}}$

(b) $H(z) = \dfrac{1 + 1.3125z^{-1} + 0.75z^{-2}}{1 + 0.875z^{-1} + 0.75z^{-2}}$

(c) $H(z) = \dfrac{0.75 + 0.875z^{-1} + z^{-2}}{1 + 0.875z^{-1} + 0.75z^{-2}}$

6.5. Determine the system function of the lattice filter shown in the figure below.

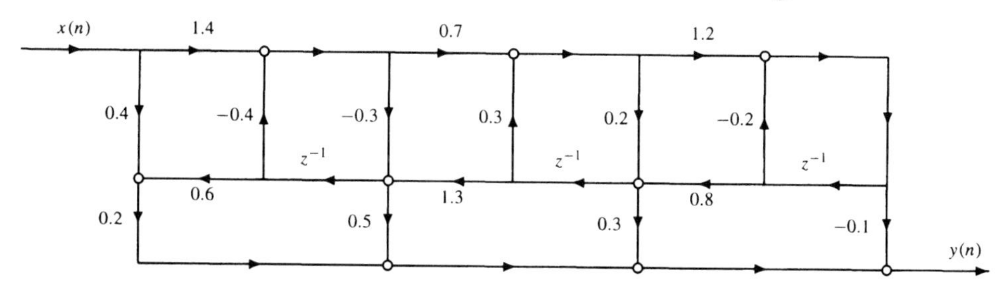

6.6. Shown in the figure below is a split lattice filter.

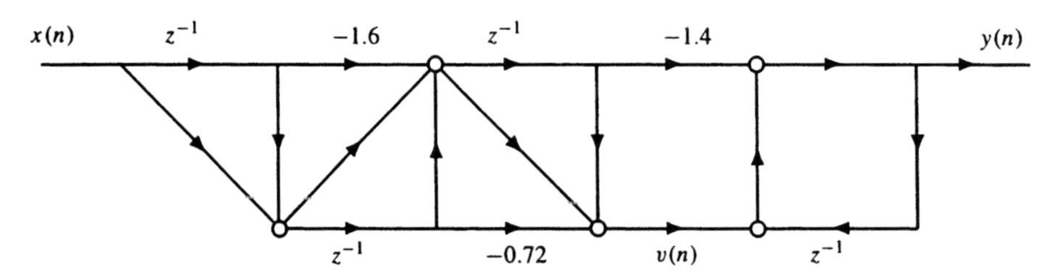

(a) What is the order of the filter (number of poles and zeros)?

(b) Does this filter have minimum phase?

(c) Find the system function $H(z) = Y(z)/X(z)$.

(d) What is the transfer function $V(z)/X(z)$ between $x(n)$ and $v(n)$?

(e) How would you modify this structure to add a zero in $H(z)$ at $z = -1$?

6.7. *True* or *False*: Let $H(z)$ be the system function of a linear shift-invariant filter with coefficients $a_p(k)$ and $b_q(k)$, and let Γ_k and $c_q(k)$ be the coefficients in the lattice filter realization of $H(z)$. If $a_p(k)$ and $b_q(k)$ are modified as follows

$$\widetilde{a}_p(k) = (-1)^k a_p(k) \quad ; \quad \widetilde{b}_p(k) = (-1)^k b_p(k)$$

then the coefficients in the lattice filter are modified in the same way, i.e.,

$$\widetilde{\Gamma}_k = (-1)^k \Gamma_k \quad ; \quad \widetilde{c}_q(k) = (-1)^k c_q(k)$$

6.8. As shown in Fig. 6.2*b*, a lattice filter may be used to generate the forward and backward prediction errors, $e_p^+(n)$ and $e_p^-(n)$, respectively.

(a) What is the relationship between the magnitudes of the discrete-time Fourier transforms of $e_p^+(n)$ and $e_p^-(n)$?

(b) Is it possible to design a realizable filter (causal and stable) that will produce the response $e_p^-(n)$ to the input $e_p^+(n)$? If so, describe how and, if not, state why not.

(c) Is it possible to design a realizable filter that will produce the response $e_p^+(n)$ to the input $e_p^-(n)$? If so, describe how and, if not, state why not.

6.9. The all-pole lattice filter in Fig. 6.7*b* may be used to generate the all-pole approximation $\hat{x}(n)$ to a signal $x(n)$. In this problem, we investigate another use for this filter. Suppose $e_0^+(n)$ is initialized to one at time $n = 0$, and all of the remaining states are set equal to zero. Determine the output of the all-pole filter for $n = 1, 2, \ldots, N$. Hint: Consider the expression for γ_j given in Eq. (5.10) of Chapter 5.

6.10. Consider the following modification to the Burg error

$$\mathcal{E}_j^w = \sum_{n=j}^{N} w_j(n) \left\{ \left|e_j^+(n)\right|^2 + \left|e_j^-(n)\right|^2 \right\}$$

where $w_j(n)$ is a window that is applied to the forward and backward prediction errors.

(a) Derive an expression that defines the value for the reflection coefficient Γ_j^w that minimizes the modified Burg error $\mathcal{E}_j^w$.

(b) What conditions, if any, are necessary in order to guarantee that the reflection coefficients are bounded by one in magnitude?

6.11. Consider the sequence of reflection coefficients Γ_j^B, Γ_j^I, and $\Gamma_j^{\min}$ where Γ_j^I is the reflection coefficient proposed by Itakura and $\Gamma_j^{\min}$ is the reflection coefficient that is formed using the minimum method (see Section 6.5.3).

(a) Establish the following relationship between these reflection coefficients:

$$\left|\Gamma_j^{\min}\right| \le \left|\Gamma_j^B\right| \le \left|\Gamma_j^I\right|$$

(b) Are there any conditions under which all three reflection coefficients will be the same for all j?

(c) Let Γ_j^M be the set of reflection coefficients corresponding to the all-pole model that is derived from the modified covariance method. Is it possible to upper or lower bound these coefficients in terms of Γ_j^B, Γ_j^I, or $\Gamma_j^{\min}$?

6.12. In Section 6.6 it was shown that the backward prediction errors are orthogonal, i.e.,

$$E\left\{e_i^-(n)\left[e_j^-(n)\right]^*\right\} = \begin{cases} \epsilon_j & ; \quad i = j \\ 0 & ; \quad i \neq j \end{cases}$$

Establish the following orthogonality conditions:

(a) $E\left\{e_i^+(n)x^*(n-k)\right\} = 0 \quad ; \quad 1 \le k \le i$

(b) $E\left\{e_i^-(n)x^*(n-k)\right\} = 0 \quad ; \quad 0 \le k \le i-1$

(c) $E\left\{e_i^+(n)x^*(n)\right\} = E\left\{e_i^-(n)x^*(n-i)\right\} = \epsilon_i$

6.13. In this problem it will be shown that the prediction error filters, $A_j(z)$, are *orthogonal* on the unit circle. Specifically, let $P_x(e^{j\omega})$ be the power spectrum of a zero mean random process $x(n)$ and let $A_j(z)$ be the system function of the jth-order prediction error filter. Show that these polynomials satisfy the orthogonality property

$$\int_{-\pi}^{\pi} P_x(e^{j\omega})A_j(e^{j\omega})A_k^*(e^{j\omega})d\omega = \lambda_k \delta_{jk}$$

and find an expression for the constant λ_k. Polynomials that satisfy this orthogonality condition are called Szegö polynomials.

Computer Exercises

C6.1. One approach for estimating the frequency of a sinusoid from noisy samples is to use a constrained lattice filter. Specifically, consider the following signal

$$y(n) = x(n) + w(n)$$

where

$$x(n) = A\cos(n\omega_0 + \phi)$$

and $w(n)$ is white Gaussian noise with a variance of σ_w^2. Consider the constrained second-order lattice filter shown in the figure below where the second reflection coefficient has been set equal to one.

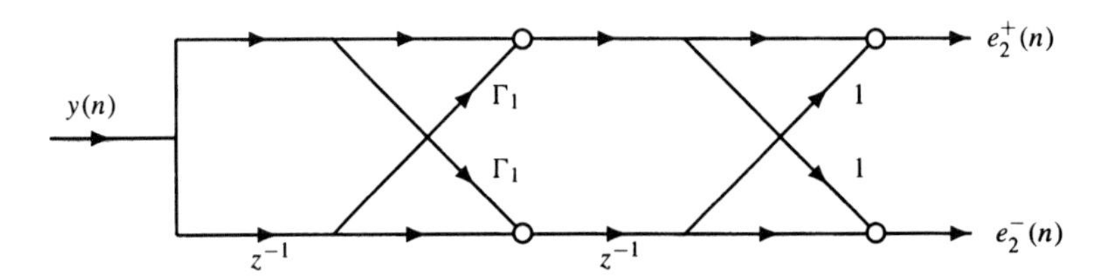

The outputs of the lattice filter, $e_2^+(n)$ and $e_2^-(n)$, are the second-order forward and backward prediction errors, respectively.

(a) Find the location of the zeros of the lattice filter as a function of the reflection coefficient Γ_1.

(b) Given $y(n)$ for $n = 0, 1, \ldots, N-1$, suppose that we would like to estimate the frequency of the sinusoid. One way to accomplish this is to find the value of the reflection coefficient Γ_1 in the constrained lattice filter that minimizes some error that depends upon $e_2^+(n)$ and $e_2^-(n)$, and to use this reflection coefficient to estimate the frequency. Derive an expression for the value of Γ_1 that minimizes the error

$$\mathcal{E}_2^B = \sum_{n=2}^{N-1} \left\{ [e_2^+(n)]^2 + [e_2^-(n)]^2 \right\}$$

Your expression should be expressed only in terms of the observations $y(n)$.

(c) Describe how you would use Γ_1 found in part (b) to estimate the frequency ω_0.

(d) Evaluate the performance of your constrained lattice filter in estimating the frequency of the sinusoid. Consider the effect of the data record length, signal to noise ratio, A^2/σ_w^2, and frequency, ω_0. Compare the performance of the constrained lattice filter to frequency estimates derived from the Burg algorithm, the modified covariance method, and the autocorrelation method.

(e) How may this frequency estimation technique be modified to estimate the frequency of a complex harmonic process, i.e., if

$$y(n) = x(n) + w(n) \qquad \text{for} \quad n = 0, 1, \ldots, N-1$$

where

$$x(n) = Ae^{j(n\omega_0 + \phi)}$$

and $w(n)$ is complex white Gaussian noise.

(f) How may the frequency estimation technique in part (e) be modified to estimate the frequencies of two complex-valued harmonic processes:

$$y(n) = A_1 e^{j(n\omega_1 + \phi_1)} + A_2 e^{j(n\omega_2 + \phi_2)} + w(n)$$

C6.2. In Section 6.4.3 we presented a recursion for converting the direct-form coefficients $b_q(k)$ into the lattice filter coefficients $c_q(k)$ and a recursion for converting the coefficients $c_q(k)$ into $b_q(k)$.

(a) Create an m-file `btoc.m` for converting the coefficients $b_q(k)$ into $c_q(k)$.

(b) Create an m-file `ctob.m` for converting the coefficients $c_q(k)$ into $b_q(k)$.

C6.3. The m-file `fcov.m` given in Fig. 6.15 finds a model for a signal $x(n)$ using the forward covariance method. Modify this program and create an m-file `bcov.m` that will find a model for $x(n)$ using the backwards covariance method.

C6.4. Consider the signal $x(n) = (n+1)u(n)$ for $n = 0, 1, \ldots, N$.

(a) Find the first-order model and the minimum error using the covariance method, the forward and backward covariance methods, the Burg algorithm, and the modified covariance method. Comment on the differences in the models and compare $x(n)$ to the model, $\hat{x}(n)$.

(b) Repeat part (a) using a second-order model. What behavior do you observe for the forward and backward covariance methods?

(c) Examine what happens using the Burg algorithm for model orders $p > 2$. Can you suggest a way to estimate the model order by looking at the sequence of errors $\mathcal{E}_j^B$?

(d) Repeat parts (a) and (b) for the signal $x(n) = n(0.9)^n$. Add noise to $x(n)$ and discuss the sensitivity of your model to the noise.

C6.5. Generate a signal, $x(n)$, of the form

$$x(n) = \alpha^n + \beta^n$$

for $n = 0, 1, 2, \ldots, 15$.

(a) With $\alpha = 0.8$ and $\beta = 1.25$, find a second order all-pole model for $x(n)$ using the modified covariance method. What is the modeling error? What can you say about the stability of the model?

(b) Repeat part (a) using Burg's method, the covariance method, and the autocorrelation method. Discuss the differences that you observe between each of the modeling techniques.

(c) Repeat parts (a) and (b) with $\alpha = 0.75$ and $\beta = 2$.

C6.6. Write an m-file for finding the model for a signal using the method proposed by Itakura (see Section 6.5.3). Compare the effectiveness of Itakura's method to the forward covariance method and Burg's method on a number of different signals.

C6.7. Modify the m-file for the Burg algorithm given in Fig. 6.17 to incorporate the recursion for the denominator given in Eq. (6.86). How much savings is there in terms of the number of multiplies and adds? Do you notice any degradation in the algorithm resulting from numerical errors introduced by finite precision effects?

OPTIMUM FILTERS 7

7.1 INTRODUCTION

The estimation of one signal from another is one of the most important problems in signal processing and it embraces a wide range of interesting applications. In many of these applications the desired signal, whether it is speech, a radar signal, an EEG, or an image, is not available or observed directly. Instead, for a variety of reasons, the desired signal may be noisy and distorted. For example, the equipment used to measure the signal may be of limited resolution, the signal may be observed in the presence of noise or other interfering signals, or it may be distorted due to the propagation of the signal from the source to the receiver as in a digital communication system. In very simple and idealized environments, it may be possible to design a classical filter such as a lowpass, highpass, or bandpass filter, to *restore* the desired signal from the measured data. Rarely, however, will these filters be *optimum* in the sense of producing the *best* estimate of the signal. Therefore, in this chapter we consider the design of *optimum digital filters*, which include the digital Wiener filter and the discrete Kalman filter [2].

In the 1940s, driven by important applications in communication theory, Norbert Wiener pioneered research in the problem of designing a filter that would produce the optimum estimate of a signal from a noisy measurement or observation. The discrete form of the Wiener filtering problem, shown in Fig. 7.1, is to design a filter to recover a signal $d(n)$ from noisy observations

$$x(n) = d(n) + v(n)$$

Assuming that both $d(n)$ and $v(n)$ are wide-sense stationary random processes, Wiener considered the problem of designing the filter that would produce the minimum mean-square error estimate of $d(n)$. Thus, with

$$\xi = E\left\{|e(n)|^2\right\}$$

where

$$e(n) = d(n) - \hat{d}(n)$$

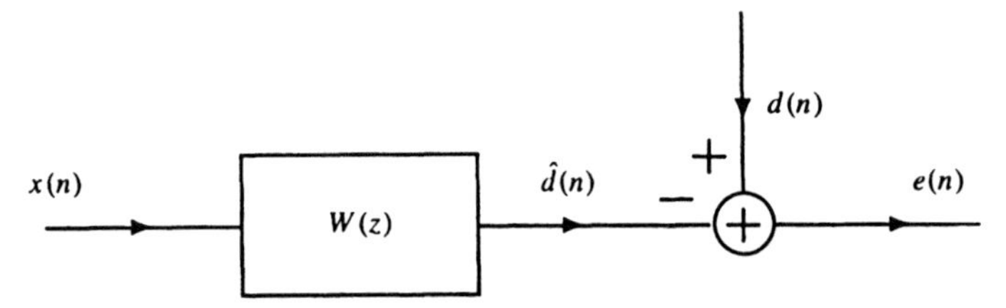

Figure 7.1 *Illustration of the general Wiener filtering problem. Given two wide-sense stationary processes, $x(n)$ and $d(n)$, that are statistically related to each other, the filter $W(z)$ is to produce the minimum mean-square error estimate, $\hat{d}(n)$, of $d(n)$.*

the problem is to find the filter that minimizes ξ. We begin this chapter by considering the general problem of Wiener filtering in which a linear shift-invariant filter, $W(z)$, is to be designed that will filter a given signal, $x(n)$, to produce the minimum mean square estimate, $\hat{d}(n)$, of another signal, $d(n)$. Depending upon how the signals $x(n)$ and $d(n)$ are related to each other, a number of different and important problems may be cast into a Wiener filtering framework. Some of the problems that will be considered in this chapter include:

1. **Filtering**. This is the classic problem considered by Wiener in which we are given $x(n) = d(n) + v(n)$ and the goal is to estimate $d(n)$ using a causal filter, i.e., to estimate $d(n)$ from the current and past values of $x(n)$.
2. **Smoothing**. This is the same as the *filtering* problem except that the filter is allowed to be noncausal. A Wiener smoothing filter, for example, may be designed to estimate $d(n)$ from $x(n) = d(n) + v(n)$ using all of the available data.
3. **Prediction**. If $d(n) = x(n+1)$ and $W(z)$ is a causal filter, then the Wiener filter becomes a linear predictor. In this case, the filter is to produce a prediction (estimate) of $x(n+1)$ in terms of a linear combination of previous values of $x(n)$.
4. **Deconvolution**. When $x(n) = d(n) * g(n) + v(n)$ with $g(n)$ being the unit sample response of a linear shift-invariant filter, the Wiener filter becomes a deconvolution filter.

First, we consider the design of FIR Wiener filters in Section 7.2. The main result here will be the derivation of the discrete form of the Wiener-Hopf equations which specify the filter coefficients of the optimum (minimum mse) filter. Solutions to the Wiener-Hopf equations are then given for the cases of filtering, smoothing, prediction, and noise cancellation. In Section 7.3 we then consider the design of IIR Wiener filters. First, in Section 7.3.1 we solve the noncausal Wiener filtering problem. Then, in Section 7.3.2, the problem of designing a causal Wiener filter is considered. Unlike the noncausal Wiener filter, solving for the optimum causal Wiener filter is a nonlinear problem that requires a spectral factorization of the power spectrum of the input process, $x(n)$. The design of a causal Wiener filter is illustrated with examples of Wiener filtering, Wiener prediction, and Wiener deconvolution. Finally, in Section 7.4 we consider recursive approaches to signal estimation and derive what is known as the discrete Kalman filter. Unlike the Wiener filter, which is a linear shift-invariant filter for estimating stationary processes, the Kalman filter is shift-varying and applicable to nonstationary as well as stationary processes.

7.2 THE FIR WIENER FILTER

In this section we consider the design of an FIR Wiener filter that produces the minimum mean-square estimate of a given process $d(n)$ by filtering a set of observations of a statistically related process $x(n)$. It is assumed that $x(n)$ and $d(n)$ are jointly wide-sense stationary with known autocorrelations, $r_x(k)$ and $r_d(k)$, and known cross-correlation $r_{dx}(k)$. Denoting the unit sample response of the Wiener filter by $w(n)$, and assuming a $(p-1)$st-order filter, the system function is

$$W(z) = \sum_{n=0}^{p-1} w(n) z^{-n}$$

With $x(n)$ the input to the filter, the output, which we denote by $\hat{d}(n)$, is the convolution of $w(n)$ with $x(n)$,

$$\hat{d}(n) = \sum_{l=0}^{p-1} w(l) x(n-l) \tag{7.1}$$

The Wiener filter design problem requires that we find the filter coefficients, $w(k)$, that minimize the mean-square error[1]

$$\xi = E\{|e(n)|^2\} = E\{|d(n) - \hat{d}(n)|^2\} \tag{7.2}$$

As discussed in Section 2.3.10 of Chapter 2, in order for a set of filter coefficients to minimize ξ it is necessary and sufficient that the derivative of ξ with respect to $w^*(k)$ be equal to zero for $k = 0, 1, \ldots, p-1$,

$$\frac{\partial \xi}{\partial w^*(k)} = \frac{\partial}{\partial w^*(k)} E\{e(n)e^*(n)\} = E\left\{e(n)\frac{\partial e^*(n)}{\partial w^*(k)}\right\} = 0 \tag{7.3}$$

With

$$e(n) = d(n) - \sum_{l=0}^{p-1} w(l) x(n-l) \tag{7.4}$$

it follows that

$$\frac{\partial e^*(n)}{\partial w^*(k)} = -x^*(n-k)$$

and Eq. (7.3) becomes

$$\boxed{E\{e(n)x^*(n-k)\} = 0 \quad ; \quad k = 0, 1, \ldots, p-1} \tag{7.5}$$

which is known as the *orthogonality principle* or the *projection theorem*.[2] Substituting Eq. (7.4) into Eq. (7.5) we have

$$E\{d(n)x^*(n-k)\} - \sum_{l=0}^{p-1} w(l) E\{x(n-l)x^*(n-k)\} = 0 \tag{7.6}$$

[1] Note that our wide-sense stationarity assumption implies that the mean-square error does not depend upon n.

[2] Compare this with the orthogonality principle in Chapter 4, p. 146.

Finally, since $x(n)$ and $d(n)$ are jointly WSS then $E\{x(n-l)x^*(n-k)\} = r_x(k-l)$ and $E\{d(n)x^*(n-k)\} = r_{dx}(k)$ and Eq. (7.6) becomes

$$\boxed{\sum_{l=0}^{p-1} w(l) r_x(k-l) = r_{dx}(k) \quad ; \quad k = 0, 1, \ldots, p-1} \tag{7.7}$$

which is a set of p linear equations in the p unknowns $w(k)$, $k = 0, 1, \ldots, p-1$. In matrix form, using the fact that the autocorrelation sequence is conjugate symmetric, $r_x(k) = r_x^*(-k)$, Eq. (7.7) becomes

$$\begin{bmatrix} r_x(0) & r_x^*(1) & \cdots & r_x^*(p-1) \\ r_x(1) & r_x(0) & \cdots & r_x^*(p-2) \\ r_x(2) & r_x(1) & \cdots & r_x^*(p-3) \\ \vdots & \vdots & & \vdots \\ r_x(p-1) & r_x(p-2) & \cdots & r_x(0) \end{bmatrix} \begin{bmatrix} w(0) \\ w(1) \\ w(2) \\ \vdots \\ w(p-1) \end{bmatrix} = \begin{bmatrix} r_{dx}(0) \\ r_{dx}(1) \\ r_{dx}(2) \\ \vdots \\ r_{dx}(p-1) \end{bmatrix} \tag{7.8}$$

which is the matrix form of the *Wiener-Hopf equations*. Equation (7.8) may be written more concisely as

$$\boxed{\mathbf{R}_x \mathbf{w} = \mathbf{r}_{dx}} \tag{7.9}$$

where $\mathbf{R}_x$ is a $p \times p$ Hermitian Toeplitz matrix of autocorrelations, $\mathbf{w}$ is the vector of filter coefficients, and $\mathbf{r}_{dx}$ is the vector of cross-correlations between the desired signal $d(n)$ and the observed signal $x(n)$.

The minimum mean-square error in the estimate of $d(n)$ may be evaluated from Eq. (7.2) as follows. With

$$\begin{aligned} \xi &= E\left\{|e(n)|^2\right\} = E\left\{e(n)\left[d(n) - \sum_{l=0}^{p-1} w(l)x(n-l)\right]^*\right\} \\ &= E\left\{e(n)d^*(n)\right\} - \sum_{l=0}^{p-1} w^*(l) E\left\{e(n)x^*(n-l)\right\} \end{aligned} \tag{7.10}$$

recall that if $w(k)$ is the solution to the Wiener-Hopf equations, then, it follows from Eq. (7.5) that $E\{e(n)x^*(n-k)\} = 0$. Therefore, the second term in Eq. (7.10) is equal to zero and

$$\xi_{\min} = E\left\{e(n)d^*(n)\right\} = E\left\{\left[d(n) - \sum_{l=0}^{p-1} w(l)x(n-l)\right]d^*(n)\right\}$$

Finally, taking expected values we have

$$\boxed{\xi_{\min} = r_d(0) - \sum_{l=0}^{p-1} w(l) r_{dx}^*(l)} \tag{7.11}$$

or, using vector notation,

$$\xi_{\min} = r_d(0) - \mathbf{r}_{dx}^H \mathbf{w} \tag{7.12}$$

Alternatively, since

$$\mathbf{w} = \mathbf{R}_x^{-1} \mathbf{r}_{dx}$$

Table 7.1 *The Wiener-Hopf Equations for the FIR Wiener Filter and the Minimum Mean-Square Error*

Wiener-Hopf equations	$\sum_{l=0}^{p-1} w(l) r_x(k-l) = r_{dx}(k) \quad ; \quad k = 0, 1, \ldots, p-1$
Correlations	$r_x(k) = E\{x(n)x^*(n-k)\}$ $r_{dx}(k) = E\{d(n)x^*(n-k)\}$
Minimum error	$\xi_{\min} = r_d(0) - \sum_{l=0}^{p-1} w(l) r_{dx}^*(l)$

the minimum error may also be written explicitly in terms of the autocorrelation matrix $\mathbf{R}_x$ and the cross-correlation vector $\mathbf{r}_{dx}$ as follows:

$$\boxed{\xi_{\min} = r_d(0) - \mathbf{r}_{dx}^H \mathbf{R}_x^{-1} \mathbf{r}_{dx}} \tag{7.13}$$

The FIR Wiener filtering equations are summarized in Table 7.1.

We now look at some Wiener filtering applications that illustrate how to formulate the Wiener filtering problem, set up the Wiener-Hopf equations (7.9), and solve for the filter coefficients.

7.2.1 Filtering

In the filtering problem, a signal $d(n)$ is to be estimated from a noise corrupted observation

$$x(n) = d(n) + v(n)$$

Filtering, or noise reduction, is an extremely important and pervasive problem that is found in many applications such as the transmission of speech in a noisy environment and the reception of data across a noisy channel. It is also important in the detection and location of targets using sensor arrays, the restoration of old recordings, and the enhancement of images.

Using the results in the previous section, the optimum FIR Wiener filter may be easily derived. It will be assumed that the noise has zero mean and that it is uncorrelated with $d(n)$. Therefore, $E\{d(n)v^*(n-k)\} = 0$ and the cross-correlation between $d(n)$ and $x(n)$ becomes

$$\begin{aligned} r_{dx}(k) &= E\{d(n)x^*(n-k)\} \\ &= E\{d(n)d^*(n-k)\} + E\{d(n)v^*(n-k)\} \\ &= r_d(k) \end{aligned} \tag{7.14}$$

Next, since

$$\begin{aligned} r_x(k) &= E\{x(n+k)x^*(n)\} \\ &= E\{[d(n+k) + v(n+k)][d(n) + v(n)]^*\} \end{aligned} \tag{7.15}$$

with $v(n)$ and $d(n)$ uncorrelated processes it follows that

$$r_x(k) = r_d(k) + r_v(k)$$

Therefore, with $\mathbf{R}_d$ the autocorrelation matrix for $d(n)$, $\mathbf{R}_v$ the autocorrelation matrix for $v(n)$, and $\mathbf{r}_{dx} = \mathbf{r}_d = [r_d(0), \ldots, r_d(p-1)]^T$, the Wiener-Hopf equations become

$$\big[\mathbf{R}_d + \mathbf{R}_v\big]\mathbf{w} = \mathbf{r}_d \tag{7.16}$$

In order to simplify these equations any further, however, specific information about the statistics of the signal and noise are required.

Example 7.2.1 *Filtering*

Let $d(n)$ be an AR(1) process with an autocorrelation sequence

$$r_d(k) = \alpha^{|k|}$$

with $0 < \alpha < 1$, and suppose that $d(n)$ is observed in the presence of uncorrelated white noise, $v(n)$, that has a variance of σ_v^2,

$$x(n) = d(n) + v(n)$$

Let us design a first-order FIR Wiener filter to reduce the noise in $x(n)$. With

$$W(z) = w(0) + w(1)z^{-1}$$

the Wiener-Hopf equations are

$$\begin{bmatrix} r_x(0) & r_x(1) \\ r_x(1) & r_x(0) \end{bmatrix} \begin{bmatrix} w(0) \\ w(1) \end{bmatrix} = \begin{bmatrix} r_{dx}(0) \\ r_{dx}(1) \end{bmatrix}$$

Since $d(n)$ and $v(n)$ are assumed to be uncorrelated, then

$$r_{dx}(k) = r_d(k) = \alpha^{|k|}$$

and

$$r_x(k) = r_d(k) + r_v(k) = \alpha^{|k|} + \sigma_v^2\delta(k)$$

Thus, the Wiener-Hopf equations become

$$\begin{bmatrix} 1+\sigma_v^2 & \alpha \\ \alpha & 1+\sigma_v^2 \end{bmatrix} \begin{bmatrix} w(0) \\ w(1) \end{bmatrix} = \begin{bmatrix} 1 \\ \alpha \end{bmatrix}$$

Solving for $w(0)$ and $w(1)$ we have

$$\begin{bmatrix} w(0) \\ w(1) \end{bmatrix} = \frac{1}{(1+\sigma_v^2)^2 - \alpha^2} \begin{bmatrix} 1+\sigma_v^2 - \alpha^2 \\ \alpha\sigma_v^2 \end{bmatrix}$$

Therefore, the Wiener filter is

$$W(z) = \frac{1}{(1+\sigma_v^2)^2 - \alpha^2}\left[(1+\sigma_v^2 - \alpha^2) + \alpha\sigma_v^2 z^{-1}\right]$$

which has a zero at $z = -\alpha\sigma_v^2/(1+\sigma_v^2-\alpha^2)$. As a specific example, let $\alpha = 0.8$ and $\sigma_v^2 = 1$. In this case the Wiener filter becomes

$$W(e^{j\omega}) = 0.4048 + 0.2381e^{-j\omega}$$

which is a lowpass filter that has a magnitude response as shown in Fig. 7.2*a*. The fact that $W(e^{j\omega})$ is a lowpass filter should not be surprising. Specifically, note that the power spectrum of $d(n)$ is

$$P_d(e^{j\omega}) = \frac{1-\alpha^2}{(1+\alpha^2) - 2\alpha\cos\omega}$$

which, for $\alpha = 0.8$ becomes

$$P_d(e^{j\omega}) = \frac{0.36}{1.64 - 1.6\cos\omega}$$

as shown in Fig. 7.2*b*. Therefore, since $P_d(e^{j\omega})$ decreases with increasing ω and since the power spectrum of the noise is constant for all ω, then the signal-to-noise ratio decreases with increasing ω. Thus, it follows that the filter should have a frequency response that has a magnitude that *decreases* with increasing ω. For the mean-square error, we have

$$\xi_{\min} = E\{|e(n)|^2\} = r_d(0) - w(0)r_{dx}^*(0) - w(1)r_{dx}^*(1) = \sigma_v^2 \frac{1+\sigma_v^2-\alpha^2}{(1+\sigma_v^2)^2-\alpha^2}$$

which, for the case in which $\alpha = 0.8$ and $\sigma_v^2 = 1$, is $\xi_{\min} = 0.4048$.

Let us evaluate how much the signal-to-noise ratio is increased by using the Wiener filter. Prior to filtering, since $r_d(0) = \sigma_d^2 = 1$ and $\sigma_v^2 = 1$, then the power in $d(n)$ is equal to the power in $v(n)$, $E\{|d(n)|^2\} = E\{|v(n)|^2\} = 1$, and the signal to noise ratio (SNR)

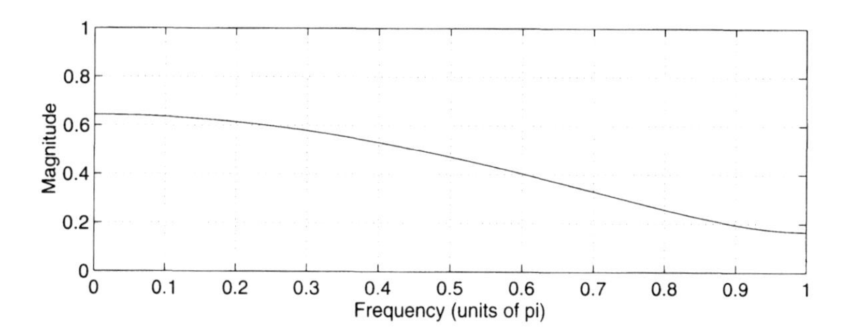

(a) The magnitude of the frequency response of the Wiener filter.

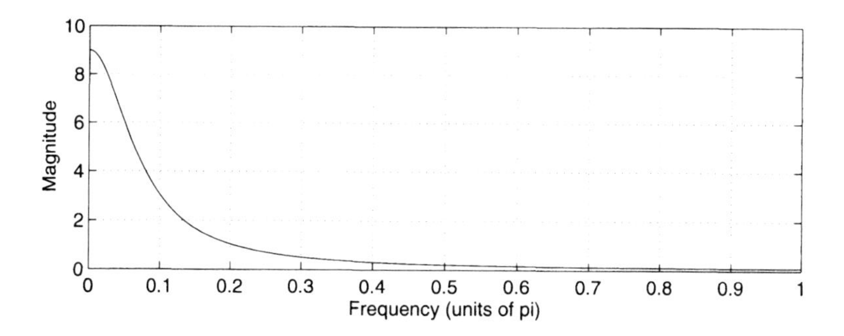

(b) The power spectrum of an AR(1) process with an autocorrelation of $r_d(k) = 0.8^{|k|}$.

Figure 7.2 *FIR Wiener filter for filtering an AR(1) process in white noise.*

is 0 dB. After filtering, it follows from Eq. (3.90) on p. 101 that the power in the signal $d'(n) = w(n) * d(n)$ is

$$E\{|d'(n)|^2\} = \mathbf{w}^T \mathbf{R}_d \mathbf{w} = \begin{bmatrix} w(0) & w(1) \end{bmatrix} \begin{bmatrix} 1 & \alpha \\ \alpha & 1 \end{bmatrix} \begin{bmatrix} w(0) \\ w(1) \end{bmatrix} = 0.3748$$

and the noise power is

$$E\{|v'(n)|^2\} = \mathbf{w}^T \mathbf{R}_v \mathbf{w} = \begin{bmatrix} w(0) & w(1) \end{bmatrix} \begin{bmatrix} w(0) \\ w(1) \end{bmatrix} = 0.2206$$

Therefore, the SNR at the output of the Wiener filter is

$$\text{SNR} = 10 \log_{10} \frac{0.3748}{0.2206} = 2.302 \text{ dB}$$

Thus, the Wiener filter increases the SNR by more than 2dB.

7.2.2 Linear Prediction

As discussed in previous chapters, linear prediction is an important problem in many signal processing applications. With noise-free observations, linear prediction is concerned with the estimation (prediction) of $x(n+1)$ in terms of a linear combination of the current and previous values of $x(n)$ as shown in Fig. 7.3. Thus, an FIR linear predictor of order $p-1$ has the form

$$\hat{x}(n+1) = \sum_{k=0}^{p-1} w(k) x(n-k)$$

where $w(k)$ for $k = 0, 1, \ldots, p-1$ are the coefficients of the prediction filter. The linear predictor may be cast into a Wiener filtering problem by setting $d(n) = x(n+1)$ in Fig. 7.1. To set up the Wiener-Hopf equations, all that is needed is to evaluate the cross-correlation between $d(n)$ and $x(n)$. Since

$$r_{dx}(k) = E\{d(n)x^*(n-k)\} = E\{x(n+1)x^*(n-k)\} = r_x(k+1)$$

then the Wiener-Hopf equations for the optimum linear predictor are

$$\begin{bmatrix} r_x(0) & r_x^*(1) & r_x^*(2) & \cdots & r_x^*(p-1) \\ r_x(1) & r_x(0) & r_x^*(1) & \cdots & r_x^*(p-2) \\ r_x(2) & r_x(1) & r_x(0) & \cdots & r_x^*(p-3) \\ \vdots & \vdots & \vdots & & \vdots \\ r_x(p-1) & r_x(p-2) & r_x(p-3) & \cdots & r_x(0) \end{bmatrix} \begin{bmatrix} w(0) \\ w(1) \\ w(2) \\ \vdots \\ w(p-1) \end{bmatrix} = \begin{bmatrix} r_x(1) \\ r_x(2) \\ r_x(3) \\ \vdots \\ r_x(p) \end{bmatrix} \qquad (7.17)$$

and the mean-square error is

$$\xi_{\min} = r_x(0) - \sum_{k=0}^{p-1} w(k) r_x^*(k+1)$$

Comparing Eq. (7.17) with the Prony all-pole normal equations given in Eq. (4.79) we see that, except for the fact that $r_x(k)$ is a deterministic autocorrelation sequence in the Prony all-pole normal equations and $r_x(k)$ is a stochastic autocorrelation in Eq. (7.17), the two sets of equations are the same (the minus sign simply changes the sign of the coefficients).

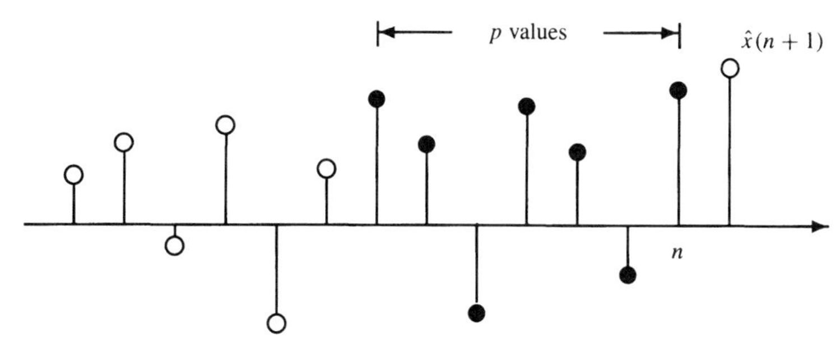

Figure 7.3 *Linear prediction is the problem of finding the minimum mean-square estimate of $x(n+1)$ using a linear combination of the p signal values from $x(n)$ to $x(n-p+1)$.*

Example 7.2.2 ***Linear Prediction***

In this example we find the optimum linear predictor for an AR(1) process $x(n)$ that has an autocorrelation sequence given by

$$r_x(k) = \alpha^{|k|}$$

With a first-order predictor of the form

$$\hat{x}(n+1) = w(0)x(n) + w(1)x(n-1)$$

the Wiener-Hopf equations are

$$\begin{bmatrix} 1 & \alpha \\ \alpha & 1 \end{bmatrix} \begin{bmatrix} w(0) \\ w(1) \end{bmatrix} = \begin{bmatrix} \alpha \\ \alpha^2 \end{bmatrix}$$

Solving for the predictor coefficients we find

$$\begin{bmatrix} w(0) \\ w(1) \end{bmatrix} = \frac{1}{1-\alpha^2} \begin{bmatrix} 1 & -\alpha \\ -\alpha & 1 \end{bmatrix} \begin{bmatrix} \alpha \\ \alpha^2 \end{bmatrix} = \begin{bmatrix} \alpha \\ 0 \end{bmatrix}$$

Therefore, the predictor for $x(n+1)$ is

$$\hat{x}(n+1) = \alpha x(n)$$

and the value of $x(n-1)$ is never used in the prediction of $x(n+1)$. This result may be explained intuitively as follows. Since $x(n)$ is an AR(1) process, it satisfies a first-order difference equation of the form

$$x(n) = \alpha x(n-1) + v(n)$$

or, equivalently,

$$x(n+1) = \alpha x(n) + v(n+1)$$

Since $v(n)$ is a white noise process, it cannot be predicted from previous values of either $x(n)$ or $v(n)$. Therefore, the best estimate of $x(n+1)$ in terms of $x(n)$ and $x(n-1)$ is obtained by replacing $v(n)$ with its expected value as follows

$$\hat{x}(n+1) = \alpha x(n) + E\{v(n+1)\}$$

Since $E\{v(n+1)\} = 0$ then $\hat{x}(n+1) = \alpha x(n)$ as before.

The mean-square linear prediction error is

$$\xi_{\min} = r_x(0) - w(0)r_x(1) - w(1)r_x(2) = 1 - \alpha^2$$

Note that as α increases, the correlation between successive samples of $x(n)$ increases, and the mean-square prediction error decreases. For an uncorrelated process, $\alpha = 0$ and

$$\xi_{\min} = 1$$

which is equal to the variance of $x(n)$, and the optimum predictor is

$$\hat{x}(n+1) = 0$$

i.e., the mean value of the process.

Thus far, we have assumed that $x(n)$ is measured in the absence of noise. Unfortunately, this is typically not the case. A more realistic model for linear prediction is the one shown in Fig. 7.4 in which the signal that is to be predicted is measured in the presence of noise. With the input to the Wiener filter given by

$$y(n) = x(n) + v(n)$$

the goal is to design a filter that will estimate $x(n+1)$ in terms of a linear combination of p previous values of $y(n)$

$$\hat{x}(n+1) = \sum_{k=0}^{p-1} w(k)y(n-k) = \sum_{k=0}^{p-1} w(k)\big[x(n-k) + v(n-k)\big]$$

The Wiener-Hopf equations are

$$\mathbf{R}_y\mathbf{w} = \mathbf{r}_{dy}$$

If the noise $v(n)$ is uncorrelated with the signal $x(n)$, then $\mathbf{R}_y$, the autocorrelation matrix for $y(n)$, is

$$r_y(k) = E\big\{y(n)y^*(n-k)\big\} = r_x(k) + r_v(k)$$

and $\mathbf{r}_{dy}$, the vector of cross-correlations between $d(n)$ and $y(n)$, is

$$r_{dy}(k) = E\big\{d(n)y^*(n-k)\big\} = E\big\{x(n+1)y^*(n-k)\big\} = r_x(k+1)$$

Thus, the only difference between linear prediction with and without noise is in the autocorrelation matrix for the input signal where, in the case of noise that is uncorrelated with $x(n)$, $\mathbf{R}_x$ is replaced with $\mathbf{R}_y = \mathbf{R}_x + \mathbf{R}_v$.

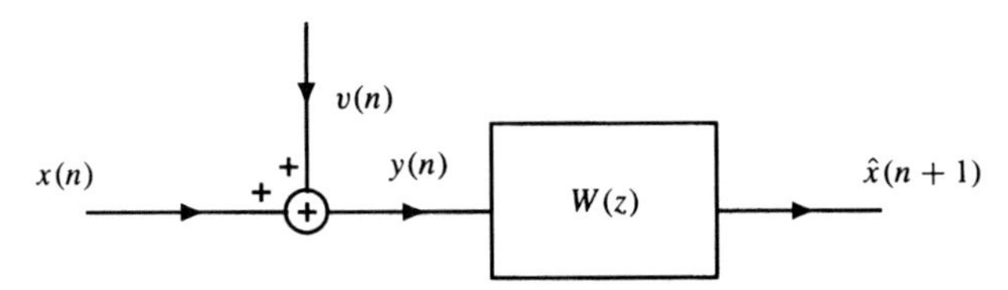

Figure 7.4 *Wiener prediction in a noisy environment.*

Example 7.2.3 *Linear Prediction in Noise*

Let us reconsider the linear prediction problem in Example 7.2.2 when the measurement of $x(n)$ is noisy. Suppose that

$$y(n) = x(n) + v(n)$$

where $v(n)$ is zero mean white noise with a variance of σ_v^2. Assuming that $v(n)$ is uncorrelated with $x(n)$, the Wiener-Hopf equations are

$$\left[\mathbf{R}_x + \sigma_v^2\mathbf{I}\right]\mathbf{w} = \mathbf{r}_{dy}$$

where

$$r_{dy}(k) = r_x(k+1)$$

If $x(n)$ is an AR(1) process with an autocorrelation sequence

$$r_x(k) = \alpha^{|k|}$$

for a first-order predictor

$$W(z) = w(0) + w(1)z^{-1}$$

the Wiener-Hopf equations are

$$\begin{bmatrix} 1+\sigma_v^2 & \alpha \\ \alpha & 1+\sigma_v^2 \end{bmatrix}\begin{bmatrix} w(0) \\ w(1) \end{bmatrix} = \begin{bmatrix} \alpha \\ \alpha^2 \end{bmatrix}$$

Solving for the predictor coefficients, we find

$$\begin{bmatrix} w(0) \\ w(1) \end{bmatrix} = \frac{\alpha}{(1+\sigma_v^2)^2 - \alpha^2}\begin{bmatrix} 1+\sigma_v^2-\alpha^2 \\ \alpha\sigma_v^2 \end{bmatrix} \tag{7.18}$$

Note that as $\sigma_v^2 \rightarrow 0$ the predictor coefficients approach the solution in the noise-free case, i.e., $\mathbf{w} = [\alpha, 0]^T$.

The previous two examples considered the problem of predicting $x(n+1)$ in terms of a linear combination of the current and previous values of $x(n)$. This problem, sometimes referred to as *one-step linear prediction*, may be generalized to the problem of *multistep prediction* shown in Fig. 7.5 in which $x(n+\alpha)$ is to be predicted in terms of a linear combination of the p values $x(n), x(n-1), \ldots, x(n-p+1)$,

$$\hat{x}(n+\alpha) = \sum_{k=0}^{p-1} w(k)x(n-k)$$

where α is a positive integer. In setting up the Wiener-Hopf equations for the multistep predictor, the only term that changes from the one-step linear predictor is the cross-correlation vector $\mathbf{r}_{dx}$. In multistep prediction, since $d(n) = x(n+\alpha)$, then

$$r_{dx}(k) = E\left\{x(n+\alpha)x^*(n-k)\right\} = r_x(\alpha+k)$$

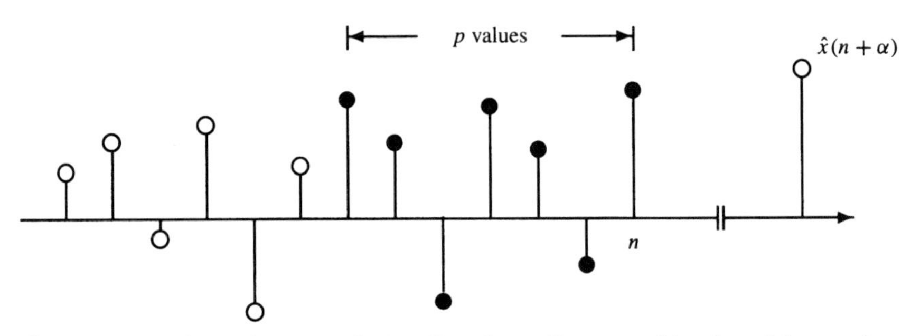

Figure 7.5 *Multistep linear prediction. Based on a linear combination of the p values $x(n), x(n-1), \ldots, x(n-p+1)$, the minimum mean-square estimate of $x(n+\alpha)$ is to be found.*

and the Wiener-Hopf equations become

$$\begin{bmatrix} r_x(0) & r_x^*(1) & r_x^*(2) & \cdots & r_x^*(p-1) \\ r_x(1) & r_x(0) & r_x^*(1) & \cdots & r_x^*(p-2) \\ r_x(2) & r_x(1) & r_x(0) & \cdots & r_x^*(p-3) \\ \vdots & \vdots & \vdots & & \vdots \\ r_x(p-1) & r_x(p-2) & r_x(p-3) & \cdots & r_x(0) \end{bmatrix} \begin{bmatrix} w(0) \\ w(1) \\ w(2) \\ \vdots \\ w(p-1) \end{bmatrix} = \begin{bmatrix} r_x(\alpha) \\ r_x(\alpha+1) \\ r_x(\alpha+2) \\ \vdots \\ r_x(\alpha+p-1) \end{bmatrix} \tag{7.19}$$

Equation (7.19) may be written in matrix form as

$$\mathbf{R}_x \mathbf{w} = \mathbf{r}_\alpha$$

where $\mathbf{r}_\alpha$ is the vector of autocorrelations beginning with $r_x(\alpha)$. Finally, for the minimum mean-square error it follows that

$$\xi_{\min} = r_x(0) - \sum_{k=0}^{p-1} w(k) r_x^*(k+\alpha) = r_x(0) - \mathbf{r}_\alpha^H \mathbf{w}$$

We now look at an example of multistep linear prediction.

Example 7.2.4 *Multistep Linear Prediction*

Let us consider the problem of multistep prediction for a random process having an autocorrelation sequence of the form

$$r_x(k) = \delta(k) + (0.9)^{|k|} \cos(\pi k/4)$$

Thus, the first eight autocorrelation values are

$$\mathbf{r}_x = [2.0,\ 0.6364,\ 0,\ -0.5155,\ -0.6561,\ -0.4175,\ 0,\ 0.3382]^T$$

We will begin with a first-order one-step linear predictor, which is the solution to the following set of linear equations,

$$\begin{bmatrix} r_x(0) & r_x(1) \\ r_x(1) & r_x(0) \end{bmatrix} \begin{bmatrix} w(0) \\ w(1) \end{bmatrix} = \begin{bmatrix} r_x(1) \\ r_x(2) \end{bmatrix}$$

Using the given autocorrelation values and solving for the predictor coefficients, we find

$$\begin{bmatrix} w(0) \\ w(1) \end{bmatrix} = \begin{bmatrix} 0.3540 \\ -0.1127 \end{bmatrix}$$

Thus, the optimum one-step predictor is

$$\hat{x}(n+1) = 0.3540x(n) - 0.1127x(n-1)$$

and the minimum mean-square error, which we will denote by $\{\xi_1\}_{\min}$, is

$$\{\xi_1\}_{\min} = r_x(0) - w(0)r_x(1) - w(1)r_x(2) = 1.7747$$

Now, let us consider a three-step predictor. In this case, the Wiener-Hopf equations become

$$\begin{bmatrix} r_x(0) & r_x(1) \\ r_x(1) & r_x(0) \end{bmatrix} \begin{bmatrix} w(0) \\ w(1) \end{bmatrix} = \begin{bmatrix} r_x(3) \\ r_x(4) \end{bmatrix}$$

which leads to the following set of predictor coefficients,

$$\begin{bmatrix} w(0) \\ w(1) \end{bmatrix} = \begin{bmatrix} -0.1706 \\ -0.2738 \end{bmatrix}$$

i.e.,

$$\hat{x}(n+3) = -0.1706x(n) - 0.2738x(n-1)$$

For the mean-square error, $\{\xi_3\}_{\min}$, we have

$$\{\xi_3\}_{\min} = r_x(0) - w(0)r_x(3) - w(1)r_x(4) = 1.7324$$

which is smaller than the mean-square prediction error using a one-step predictor. Therefore, we have an interesting case in which it is easier to predict three samples ahead than it is to predict the next sample. Although perhaps at first surprising, if we compare the values of the autocorrelation sequence at lags $k = 1, 2$ (used in the one-step predictor), with the values at lags $k = 3, 4$ (used in the three-step predictor), we find that there is a higher degree of correlation between $x(n+3)$ and the two signal values that are used in the predictor, $x(n)$ and $x(n-1)$, than there is between $x(n+1)$ and the same two signal values. It is for this reason that the minimum mean-square error is smaller for the three-step predictor than it is for the one-step predictor.

Another way to look at the multistep predictor is in terms of a one-step predictor that uses a linear combination of the values of $x(n)$ over the interval extending from $(n-\alpha-p+2)$ to $(n-\alpha+1)$. Thus, as illustrated in Fig. 7.6, the multistep predictor may be expressed in the form

$$\hat{x}(n+1) = \sum_{k=0}^{p-1} w(k)x(n-k-\alpha+1) \tag{7.20}$$

This interpretation also suggests the following interesting variation to the linear prediction problem. With a linear predictor of the form given in Eq. (7.20), suppose that the delay α is a free parameter and that the problem is to find the coefficients, $w(k)$, along with the delay α that minimize the mean-square prediction error. This design problem, summarized in Fig. 7.7, is considered in the following example.

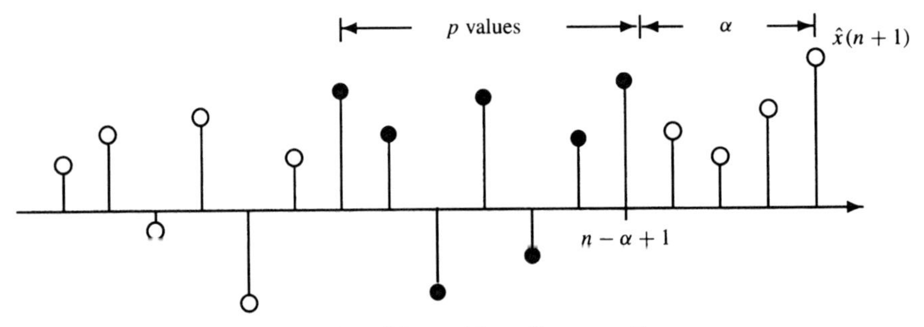

Figure 7.6 *Another interpretation of the multistep linear predictor.*

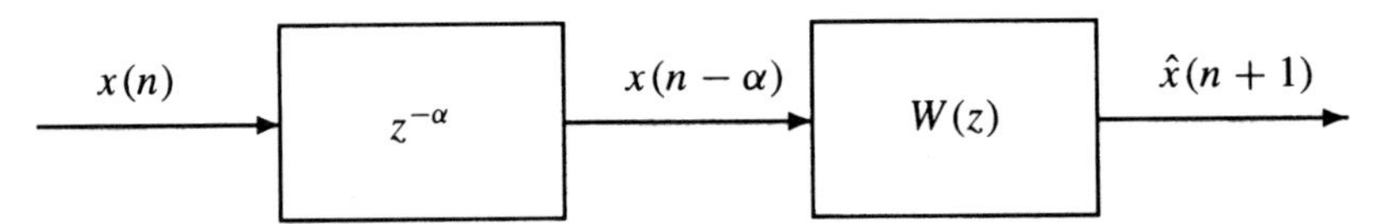

Figure 7.7 *Linear prediction with a delay.*

Example 7.2.5 ***Linear Prediction with the Optimum Delay***

Suppose we reconsider Example 7.2.4 and allow the delay α to be a free variable. To design the optimum linear predictor it is necessary to find the value for α that gives the smallest mean-square prediction error. This search may be simplified by writing the mean-square error in the form

$$\{\xi_\alpha\}_{\min} = r_x(0) - \mathbf{r}_\alpha^H \mathbf{w}$$

and using the fact that the vector of predictor coefficients may be written as

$$\mathbf{w} = \mathbf{R}_x^{-1}\mathbf{r}_\alpha$$

Therefore, the minimum mean-square error is given by

$$\{\xi_\alpha\}_{\min} = r_x(0) - \mathbf{r}_\alpha^H \mathbf{R}_x^{-1}\mathbf{r}_\alpha$$

and, by precomputing $\mathbf{R}_x^{-1}$, the errors may be easily evaluated. Using this approach, the first six values of $\{\xi_\alpha\}_{\min}$ are found to be

$$\begin{aligned}
\{\xi_1\}_{\min} &= 1.7747\\
\{\xi_2\}_{\min} &= 1.8522\\
\{\xi_3\}_{\min} &= 1.7324\\
\{\xi_4\}_{\min} &= 1.7605\\
\{\xi_5\}_{\min} &= 1.9030\\
\{\xi_6\}_{\min} &= 1.9364
\end{aligned}$$

Thus, for values of α between 1 and 6, $\alpha = 3$ gives the minimum mean-square error (this also turns out to be the global minimum over all values of α).

7.2.3 Noise Cancellation

Another application of Wiener filtering is the problem referred to as *noise cancellation*. As in the filtering problem, the goal of a noise canceller is to estimate a signal $d(n)$ from a noise corrupted observation

$$x(n) = d(n) + v_1(n)$$

that is recorded by a primary sensor. However, unlike the filtering problem, which requires that the autocorrelation of the noise be known, with a noise canceller this information is obtained from a secondary sensor that is placed within the noise field as illustrated in Fig. 7.8. Although the noise measured by this secondary sensor, $v_2(n)$, will be *correlated* with the noise in the primary sensor, the two processes will not be equal. There may be a number of reasons for this, such as differences in the sensor characteristics, differences in the propagation paths from the noise source to the two sensors, and leakage of the signal $d(n)$ into the measurements made by the secondary sensor. Since $v_1(n) \neq v_2(n)$ it is not possible to estimate $d(n)$ by simply subtracting $v_2(n)$ from $x(n)$. Instead, the noise canceller consists of a Wiener filter that is designed to estimate the noise $v_1(n)$ from the signal received by the secondary sensor. This estimate, $\hat{v}_1(n)$, is then subtracted from the primary signal $x(n)$, to form an estimate of $d(n)$, which is given by

$$\hat{d}(n) = x(n) - \hat{v}_1(n)$$

An example of where such a system may be useful is in air-to-air communications between pilots in fighter aircraft or in air-to-ground communications between a pilot and the control tower. Since there is often a large amount of engine and wind noise within the cockpit of the fighter aircraft, communication is often a difficult problem. However, if a secondary microphone is placed within the cockpit of an aircraft, then one may estimate the noise that is transmitted when the pilot speaks into the microphone, and subtract this estimate from the transmitted signal, thereby increasing the signal-to-noise ratio.

The Wiener-Hopf equations for the noise cancellation system in Fig. 7.8 may be derived as follows. With $v_2(n)$ the input to the Wiener filter that is used to estimate the noise $v_1(n)$, the Wiener-Hopf equations are

$$\mathbf{R}_{v_2}\mathbf{w} = \mathbf{r}_{v_1 v_2}$$

where $\mathbf{R}_{v_2}$ is the autocorrelation matrix of $v_2(n)$ and $\mathbf{r}_{v_1 v_2}$ is the vector of cross-correlations between the desired signal $v_1(n)$ and Wiener filter input, $v_2(n)$. For the cross-correlation

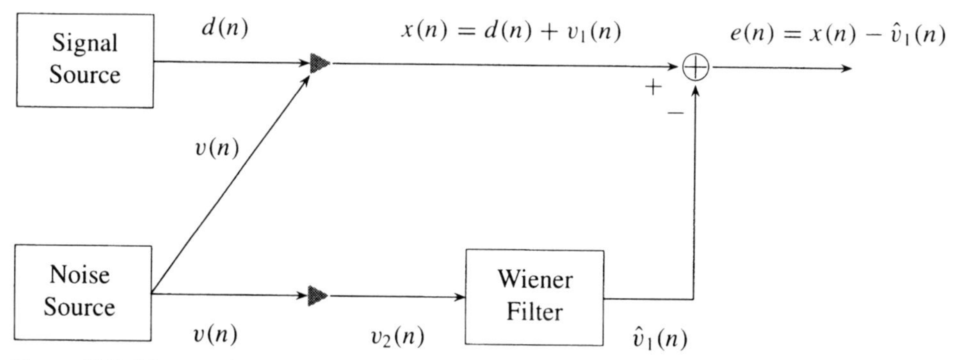

Figure 7.8 *Wiener noise cancellation using a secondary sensor to measure the additive noise* $v_1(n)$.

between $v_1(n)$ and $v_2(n)$ we have

$$r_{v_1v_2}(k) = E\{v_1(n)v_2^*(n-k)\} = E\{[x(n)-d(n)]v_2^*(n-k)\}$$

$$= E\{x(n)v_2^*(n-k)\} - E\{d(n)v_2^*(n-k)\} \tag{7.21}$$

If we assume that $v_2(n)$ is uncorrelated with $d(n)$, then the second term is zero and the cross-correlation becomes

$$r_{v_1v_2}(k) = E\{x(n)v_2^*(n-k)\} = r_{xv_2}(k) \tag{7.22}$$

Therefore, the Wiener-Hopf equations are

$$\mathbf{R}_{v_2}\mathbf{w} = \mathbf{r}_{xv_2} \tag{7.23}$$

We will now look at a specific example.

Example 7.2.6 *Noise Cancellation*

Suppose that the desired signal $d(n)$ in Fig. 7.8 is a sinusoid

$$d(n) = \sin(n\omega_0 + \phi)$$

with $\omega_0 = 0.05\pi$, and that the noise sequences $v_1(n)$ and $v_2(n)$ are AR(1) processes that are generated by the first-order difference equations

$$v_1(n) = 0.8v_1(n-1) + g(n)$$

$$v_2(n) = -0.6v_2(n-1) + g(n)$$

where $g(n)$ is zero-mean, unit variance white noise that is uncorrelated with $d(n)$. Shown in Fig. 7.9*a* is a plot of 200 samples of $x(n) = d(n) + v_1(n)$ with the desired signal, $d(n)$, indicated by the dashed line, and shown in Fig. 7.9*b* is the reference signal $v_2(n)$ that is used to estimate $v_1(n)$. Estimating $r_{v_2}(k)$ using the sample autocorrelation

$$\hat{r}_{v_2}(k) = \frac{1}{N}\sum_{n=0}^{N-1} v_2(n)v_2(n-k)$$

and $r_{xv_2}(k)$ using the sample cross-correlation

$$\hat{r}_{xv_2}(k) = \frac{1}{N}\sum_{n=0}^{N-1} x(n)v_2(n-k)$$

FIR Wiener filters of orders $p = 6$ and $p = 12$ were found by solving Eq. (7.23). Using these filters to estimate $v_1(n)$, the sinusoid $d(n)$ was then estimated by subtracting $\hat{v}_1(n)$ from $x(n)$. The results are shown in Figs. 7.9*c* and *d*.

In typical applications, $d(n)$ and $v_1(n)$ are often found to be non-stationary processes. Therefore, the use of a linear shift-invariant Wiener filter will not be optimum. However, as we will see in Chapter 9, an adaptive Wiener filter that has filter coefficients that are allowed to vary as a function of time may provide effective noise cancellation in nonstationary environments.

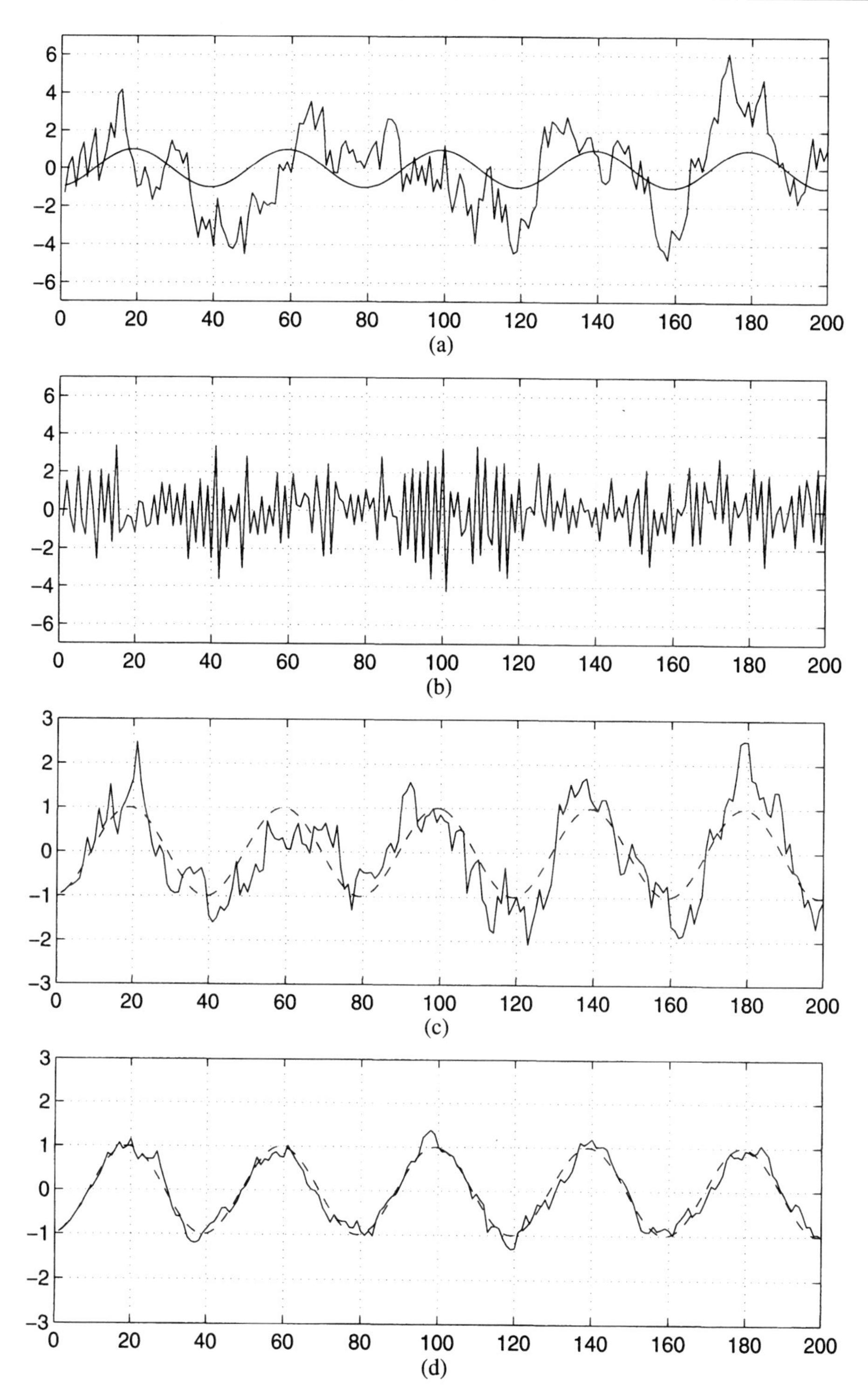

Figure 7.9 *Noise cancellation example. (a) Noise corrupted sinusoid, (b) Reference signal used by secondary sensor, (c) Output of sixth-order Wiener noise canceller, (d) Output of twelth-order Wiener noise canceller.*

7.2.4 Lattice Representation for the FIR Wiener Filter

In this section, we derive a lattice filter implementation for an FIR Wiener filter. We begin by noting that the output of the Wiener filter may be written in vector form as follows:

$$\hat{d}(n) = \mathbf{x}^T(n)\mathbf{w}$$

where

$$\mathbf{x}(n) = \left[x(n),\ x(n-1),\ \ldots\ ,\ x(n-p)\right]^T$$

and where $\mathbf{w}$ is the solution to the Wiener-Hopf equations given in Eq. (7.9). Using the *triangular decomposition* for the inverse of $\mathbf{R}_x$ given by Eq. (5.102)

$$\mathbf{R}_x^{-1} = \mathbf{A}_{p-1}\mathbf{D}_{p-1}^{-1}\mathbf{A}_{p-1}^H$$

where $\mathbf{D}_{p-1} = \text{diag}\left\{\epsilon_0, \epsilon_1, \ldots, \epsilon_{p-1}\right\}$ is a diagonal matrix of forward prediction errors and where

$$\mathbf{A}_{p-1} = \begin{bmatrix} 1 & a_1^*(1) & a_2^*(2) & \cdots & a_{p-1}^*(p-1) \\ 0 & 1 & a_2^*(1) & \cdots & a_{p-1}^*(p-2) \\ 0 & 0 & 1 & \cdots & a_{p-1}^*(p-3) \\ \vdots & \vdots & \vdots & & \vdots \\ 0 & 0 & 0 & \cdots & 1 \end{bmatrix} \tag{7.24}$$

is the matrix of prediction error filters as defined in Eq. (5.84). Therefore, the output of the Wiener filter may be written as

$$\hat{d}(n) = \mathbf{x}^T(n)\mathbf{R}_x^{-1}\mathbf{r}_{dx} = \mathbf{x}^T(n)\mathbf{A}_{p-1}\mathbf{D}_{p-1}^{-1}\mathbf{A}_{p-1}^H\mathbf{r}_{dx} \tag{7.25}$$

Recall, however, from Eq. (6.101) that

$$\mathbf{A}_{p-1}^T\mathbf{x}(n) = \mathbf{e}^-(n)$$

where

$$\mathbf{e}^-(n) = \begin{bmatrix} e_0^-(n) \\ e_1^-(n) \\ \vdots \\ e_{p-1}^-(n) \end{bmatrix}$$

is the vector of backward prediction errors at time n. With

$$\boxed{\mathbf{b} = \mathbf{D}_{p-1}^{-1}\mathbf{A}_{p-1}^H\mathbf{r}_{dx}} \tag{7.26}$$

it follows therefore, that $\hat{d}(n) = \mathbf{b}^T\mathbf{e}^-(n)$, i.e.,

$$\hat{d}(n) = \sum_{k=0}^{p-1} b(k)e_k^-(n) \tag{7.27}$$

Therefore, the Wiener estimate $\hat{d}(n)$ is formed by taking a linear combination of the backward prediction errors using the filter coefficients $b(k)$, as shown in Fig. 7.10. This Wiener lattice filter, sometimes called *joint process estimator*, may be viewed as two filters operating jointly. The first is a *lattice predictor* that transforms the sequence of correlated

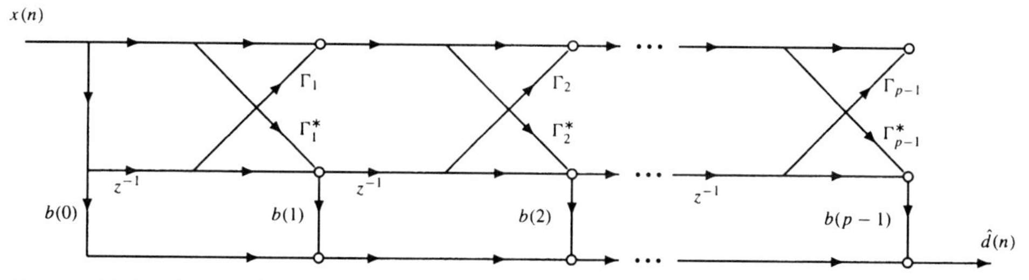

Figure 7.10 *A lattice filter implementation of an FIR Wiener filter.*

input samples, $x(n), x(n-1), \ldots, x(n-p)$, into a sequence of uncorrelated backward prediction errors, $e_0^-(n), e_1^-(n), \ldots, e_p^-(n)$ (see the discussion leading up to Eq. (6.102) in Section 6.6 in Chapter 6). The second filter, called a *multiple regression filter* [5] which is characterized by the weights $b(k)$, uses the backward prediction errors $e_k^-(n)$ as inputs to produce an estimate of the desired signal $d(n)$.

7.3 THE IIR WIENER FILTER

Having solved the FIR Wiener filter design problem and having explored a few Wiener filtering applications, we now consider the design of an IIR digital Wiener filter. As was the case for an FIR Wiener filter, given a process $x(n)$ our goal is to design a filter $h(n)$ that produces an output $y(n) = x(n) * h(n)$ that is as close as possible in the mean-square sense to a desired process, $d(n)$. Although the problem formulation is the same for both the FIR and the IIR Wiener filters, there is an important difference that significantly changes the solution. Specifically, for the FIR Wiener filter there are only a finite number of filter coefficients that must be determined, whereas for the IIR Wiener filter there are an infinite number of unknowns, i.e., the values of $h(n)$ for all n. In this section, we will consider two cases. First, in Section 7.3.1, we will solve the Wiener filter design problem for the case in which there are no constraints placed on the solution. What we will find is that the optimum filter will, in general, be noncausal and therefore unrealizable. Then, in Section 7.3.2, we will constrain the solution to be causal by forcing $h(n)$ to be zero for $n \leq 0$. For the noncausal Wiener filter, we will find a simple closed form expression for the frequency response, whereas for the causal Wiener filter, we will only be able to specify the system function implicitly in terms of a spectral factorization.

7.3.1 Noncausal IIR Wiener Filter

For a noncausal (unconstrained) IIR Wiener filter, the problem is to find the unit sample response, $h(n)$, of the IIR filter

$$H(z) = \sum_{n=-\infty}^{\infty} h(n) z^{-n}$$

that minimizes the mean-square error

$$\xi = E\{|e(n)|^2\} \tag{7.28}$$

where $e(n)$ is the difference between the desired process $d(n)$ and the output of the Wiener filter, $\hat{d}(n)$,

$$e(n) = d(n) - \hat{d}(n) = d(n) - \sum_{l=-\infty}^{\infty} h(l)x(n-l) \tag{7.29}$$

This problem may be solved in exactly the same way that we solved the FIR Wiener filtering problem, i.e., by differentiating ξ with respect to $h^*(k)$ for each k and setting the derivatives equal to zero. Performing this differentiation we have

$$\frac{\partial \xi}{\partial h^*(k)} = -E\{e(n)x^*(n-k)\} = 0 \quad ; \quad -\infty < k < \infty$$

or

$$\boxed{E\{e(n)x^*(n-k)\} = 0 \quad ; \quad -\infty < k < \infty} \tag{7.30}$$

Equation (7.30) is referred to as the *orthogonality principle*, and it is identical to the orthogonality principle for the FIR Wiener filter given in Eq. (7.5) except that here the equality must hold for all k. Substituting Eq. (7.29) for $e(n)$ into Eq. (7.30) and rearranging terms gives

$$\sum_{l=-\infty}^{\infty} h(l)E\{x(n-l)x^*(n-k)\} = E\{d(n)x^*(n-k)\} \quad ; \quad -\infty < k < \infty \tag{7.31}$$

Note that the expectation on the left is the autocorrelation of $x(n)$, and the expectation on the right is the cross-correlation between $x(n)$ and $d(n)$. Therefore, Eq. (7.31) may be written as

$$\boxed{\sum_{l=-\infty}^{\infty} h(l)r_x(k-l) = r_{dx}(k) \quad ; \quad -\infty < k < \infty} \tag{7.32}$$

which are the Wiener-Hopf equations of the noncausal IIR Wiener filter. Comparing the Wiener-Hopf equations for the FIR Wiener filter and the noncausal IIR Wiener filter, Eqs. (7.7) and (7.32), respectively, we see that the only difference is in the limits on the summation and the range of values for which the equations must hold. Although Eq. (7.32) corresponds to an infinite set of linear equations with an infinite number of unknowns, finding the solution to these equations is straightforward if we write the left side of the equation as the convolution of $h(k)$ with $r_x(k)$,

$$h(k) * r_x(k) = r_{dx}(k) \tag{7.33}$$

In the frequency domain, Eq. (7.33) becomes

$$H(e^{j\omega})P_x(e^{j\omega}) = P_{dx}(e^{j\omega}) \tag{7.34}$$

Therefore, it follows that the frequency response of the IIR Wiener filter is

$$\boxed{H(e^{j\omega}) = \frac{P_{dx}(e^{j\omega})}{P_x(e^{j\omega})}} \tag{7.35}$$

and the system function is

$$H(z) = \frac{P_{dx}(z)}{P_x(z)} \tag{7.36}$$

Note that the denominator of $H(z)$ is a power spectral density, $P_x(z)$, and that the numerator, $P_{dx}(z)$, is a cross-power spectral density.

Having found the noncausal IIR Wiener filter, we now evaluate the mean-square error. Following the steps that were taken for an FIR Wiener filter, we find that the mean-square error is

$$\xi_{\min} = r_d(0) - \sum_{l=-\infty}^{\infty} h(l) r_{dx}^*(l) \tag{7.37}$$

Note that the only difference between this and the mean-square error for the FIR Wiener filter given in Eq. (7.11) is in the limits on the summation. Using Parseval's theorem this error may be expressed in the frequency domain as follows

$$\boxed{\xi_{\min} = r_d(0) - \frac{1}{2\pi}\int_{-\pi}^{\pi} H(e^{j\omega}) P_{dx}^*(e^{j\omega}) d\omega} \tag{7.38}$$

Since

$$r_d(0) = \frac{1}{2\pi}\int_{-\pi}^{\pi} P_d(e^{j\omega}) d\omega$$

then we also have

$$\xi_{\min} = \frac{1}{2\pi}\int_{-\pi}^{\pi} \left[P_d(e^{j\omega}) - H(e^{j\omega}) P_{dx}^*(e^{j\omega})\right] d\omega \tag{7.39}$$

This error may also be expressed in terms of the complex variable z as follows:

$$\xi_{\min} = r_d(0) - \frac{1}{2\pi j}\oint_C H(z) P_{dx}^*(1/z^*) z^{-1} dz \tag{7.40}$$

or, equivalently,

$$\xi_{\min} = \frac{1}{2\pi j}\oint_C \left[P_d(z) - H(z) P_{dx}^*(1/z^*)\right] z^{-1} dz \tag{7.41}$$

where the contour, C, may be taken to be the unit circle. The noncausal Wiener filtering equations are summarized in Table 7.2.

We conclude this section with a derivation of the Wiener smoothing filter for producing the minimum mean-square estimate of a process $d(n)$ using the noisy observations

$$x(n) = d(n) + v(n)$$

for all n. Since the optimum noncausal IIR Wiener filter is given by Eq. (7.35), all that must be done is to find $P_x(e^{j\omega})$ and $P_{dx}(e^{j\omega})$. Assuming that $d(n)$ and $v(n)$ are uncorrelated zero mean random processes, the autocorrelation of $x(n)$ is

$$r_x(k) = r_d(k) + r_v(k)$$

and the power spectrum is

$$P_x(e^{j\omega}) = P_d(e^{j\omega}) + P_v(e^{j\omega})$$

Table 7.2 ***The Frequency Response and Minimum Error for a Noncausal Wiener Filter***

Wiener-Hopf equations	$H(e^{j\omega}) = \dfrac{P_{dx}(e^{j\omega})}{P_x(e^{j\omega})}$
Correlations	$r_x(k) = E\{x(n)x^*(n-k)\}$ $r_{dx}(k) = E\{d(n)x^*(n-k)\}$
Minimum error	$\xi_{\min} = r_d(0) - \sum_{l=-\infty}^{\infty} h(l) r_{dx}^*(l)$ $= \dfrac{1}{2\pi}\int_{-\pi}^{\pi}\left[P_d(e^{j\omega}) - H(e^{j\omega})P_{dx}^*(e^{j\omega})\right] d\omega$

Furthermore, the cross-correlation, $r_{dx}(k)$, is

$$r_{dx}(k) = E\{d(n)x^*(n-k)\} = E\{d(n)d^*(n-k)\} + E\{d(n)v^*(n-k)\} = r_d(k)$$

Therefore,

$$P_{dx}(e^{j\omega}) = P_d(e^{j\omega})$$

and the IIR Wiener smoothing filter is

$$H(e^{j\omega}) = \frac{P_d(e^{j\omega})}{P_d(e^{j\omega}) + P_v(e^{j\omega})} \tag{7.42}$$

Note that for those values of ω for which $P_d(e^{j\omega}) \gg P_v(e^{j\omega})$, the signal-to-noise ratio is high and $|H(e^{j\omega})| \approx 1$. Therefore, over those frequency bands where the signal dominates, the filter passes the process with little attenuation. On the other hand, for those values of ω for which the signal-to-noise ratio is small, $P_d(e^{j\omega}) \ll P_v(e^{j\omega})$, the frequency response is small, $|H(e^{j\omega})| \approx 0$. Therefore, over those frequency bands where the noise dominates, $H(e^{j\omega})$ is small in order to filter out or suppress the noise. Finally, since

$$P_{dx}(e^{j\omega}) = P_d(e^{j\omega})$$

if we evaluate the mean-square error using Eq. (7.39) and use the fact that $P_d(e^{j\omega})$ is real we have

$$\xi_{\min} = \frac{1}{2\pi}\int_{-\pi}^{\pi}\left[P_d(e^{j\omega}) - H(e^{j\omega})P_{dx}^*(e^{j\omega})\right] d\omega = \frac{1}{2\pi}\int_{-\pi}^{\pi} P_d(e^{j\omega})\left[1 - H(e^{j\omega})\right] d\omega$$

Substituting Eq. (7.42) for the frequency response of the Wiener smoothing filter we find that the minimum mean-square error is

$$\xi_{\min} = \frac{1}{2\pi}\int_{-\pi}^{\pi} P_d(e^{j\omega})\frac{P_v(e^{j\omega})}{P_d(e^{j\omega}) + P_v(e^{j\omega})} d\omega = \frac{1}{2\pi}\int_{-\pi}^{\pi} P_v(e^{j\omega})H(e^{j\omega}) d\omega \tag{7.43}$$

which, if expressed in the z-domain, becomes

$$\xi_{\min} = \frac{1}{2\pi j}\oint_C P_v(z)H(z)z^{-1}dz \tag{7.44}$$

Example 7.3.1 *Noncausal Wiener Smoothing*

Let $d(n)$ be a real-valued AR(1) process with power spectrum

$$P_d(z) = \frac{b^2(0)}{(1 - \alpha z^{-1})(1 - \alpha z)}$$

and suppose that $d(n)$ is observed in the presence of zero mean white noise with a variance σ_v^2,

$$x(n) = d(n) + v(n)$$

Assuming that $v(n)$ is uncorrelated with $d(n)$, we will design a noncausal IIR Wiener smoothing filter for estimating $d(n)$ from $x(n)$ and find the mean-square error in the estimation of $d(n)$.

From Eq. (7.42) it follows that the system function of the noncausal Wiener smoothing filter is

$$H(z) = \frac{P_d(z)}{P_d(z) + P_v(z)}$$

Substituting the given expression for $P_d(z)$ into $H(z)$ and setting $P_v(z) = \sigma_v^2$, we have

$$H(z) = \frac{b^2(0)}{b^2(0) + \sigma_v^2(1 - \alpha z^{-1})(1 - \alpha z)}$$

Evaluating the minimum error using Eq. (7.43), with $P_v(z) = \sigma_v^2$ we have the remarkably simple result

$$\xi_{\min} = \frac{1}{2\pi}\int_{-\pi}^{\pi} P_v(e^{j\omega})H(e^{j\omega})d\omega = \sigma_v^2 \frac{1}{2\pi}\int_{-\pi}^{\pi} H(e^{j\omega})\, d\omega = \sigma_v^2 h(0)$$

Now consider the specific case in which $b^2(0) = 0.25$, $\alpha = 0.5$, and $\sigma_v^2 = 0.25$. For the Wiener filter we have

$$H(z) = \frac{0.25}{0.25 + 0.25(1 - 0.5z^{-1})(1 - 0.5z)} = \frac{2(0.2344)}{(1 - 0.2344z^{-1})(1 - 0.2344z)}$$

Using the z-transform pair

$$\alpha^{|n|} \longleftrightarrow \frac{1 - \alpha^2}{(1 - \alpha z^{-1})(1 - \alpha z)}$$

we have for the unit sample response,

$$h(n) = 0.4960\ (0.2344)^{|n|}$$

which, clearly, is noncausal. For the minimum mean-square error we have

$$\xi_{\min} = \sigma_v^2 h(0) = (0.25)(0.4960) = 0.1240$$

Finally, it is interesting to see how much the error is reduced as a result of filtering $x(n)$ with a Wiener filter. Without a filter, setting $\hat{d}(n) = x(n)$ the mean-square error is

$$E\left\{|e(n)|^2\right\} = E\left\{|v(n)|^2\right\} = 0.25$$

Thus, the noncausal Wiener filter reduces the mean-square error by approximately a factor of two.

7.3.2 The Causal IIR Wiener Filter

In the previous section, we considered the design of an IIR digital Wiener filter, and placed no constraints on the form of the solution. In this section, we reconsider the design for the case in which the Wiener filter is constrained to be causal. For a causal filter, the unit sample response is zero for $n < 0$ and the estimate of $d(n)$ takes the form

$$\hat{d}(n) = x(n) * h(n) = \sum_{k=0}^{\infty} h(k)x(n-k)$$

To find the filter coefficients that minimize the mean-square error, we proceed in exactly the same way that we did for the noncausal Wiener filter. Specifically, differentiating ξ with respect to $h^*(k)$ for $k \geq 0$ and setting the derivatives to zero we find

$$\boxed{\sum_{l=0}^{\infty} h(l)r_x(k-l) = r_{dx}(k) \quad ; \quad 0 \leq k < \infty} \tag{7.45}$$

which are the Wiener-Hopf equations for the causal IIR Wiener filter. Note that the only differences between Eqs. (7.32) and (7.45) are in the limits on the summation and the values of k for which the equations must hold. This restriction on k is important since it implies that $r_{dx}(k)$ may no longer be expressed as the convolution of $h(k)$ and $r_x(k)$. As a result, solving Eq. (7.45) for the coefficients $h(l)$ of the Wiener filter is much more difficult than solving Eq. (7.32).

To solve the Wiener-Hopf equations, we begin by looking at the special case in which the input to the filter is unit variance white noise, $\epsilon(n)$. Denoting the coefficients of the Wiener filter by $g(n)$, the Wiener-Hopf equations are

$$\sum_{l=0}^{\infty} g(l)r_\epsilon(k-l) = r_{d\epsilon}(k) \quad ; \quad 0 \leq k < \infty \tag{7.46}$$

With $r_\epsilon(k) = \delta(k)$, the left side of Eq. (7.46) reduces to $g(k)$. Therefore, $g(k) = r_{d\epsilon}(k)$ for $k \geq 0$ and, since the Wiener filter is causal, $g(k) = 0$ for $n < 0$. Thus, the causal Wiener filter for a white noise input $\epsilon(n)$ is

$$g(n) = r_{d\epsilon}(n)u(n) \tag{7.47}$$

where $u(n)$ is the unit step function. We will express this solution in the z-domain as follows:

$$\boxed{G(z) = [P_{d\epsilon}(z)]_+} \tag{7.48}$$

where the subscript "+" is used to indicate the "positive-time part" of the sequence whose z-transform is contained within the brackets.

In a typical Wiener filtering application, it is unlikely that the input to the Wiener filter will be white noise. Therefore, suppose that $x(n)$ is a random process with a rational power spectrum that has no poles or zeros on the unit circle. We may then perform a spectral factorization and write $P_x(z)$ as follows (see p. 105)

$$P_x(z) = \sigma_0^2 Q(z)Q^*(1/z^*) \tag{7.49}$$

where $Q(z)$ is minimum phase and of the form

$$Q(z) = 1 + q(1)z^{-1} + q(2)z^{-2} + \cdots = \frac{N(z)}{D(z)}$$

with $N(z)$ and $D(z)$ minimum phase monic polynomials. If $x(n)$ is filtered with a filter having a system function of the form (see Fig. 7.11)

$$F(z) = \frac{1}{\sigma_0 Q(z)} \tag{7.50}$$

then the power spectrum of the output process, $\epsilon(n)$, will be

$$P_\epsilon(z) = P_x(z)F(z)F^*(1/z^*) = 1$$

Therefore, $\epsilon(n)$ is white noise and $F(z)$ is referred to as a *whitening filter*. Note that since $Q(z)$ is minimum phase, then $F(z)$ is stable and causal and has a stable and causal inverse, $F^{-1}(z)$. As a result, $x(n)$ may be recovered from $\epsilon(n)$ by filtering with the inverse filter, $F^{-1}(z)$. In other words, there is no loss of information in the linear transformation that produces the white noise process from $x(n)$.

With this background, we are now in a position to derive the optimum causal Wiener filter when the input to the filter $x(n)$ has a rational power spectrum. Let $H(z)$ be the causal Wiener filter that produces the minimum mean-square estimate of $d(n)$ from $x(n)$, and suppose that $x(n)$ is filtered with a cascade of three filters, $F(z)$, $F^{-1}(z)$, and $H(z)$ as shown in Fig. 7.12, where $F(z)$ is the causal whitening filter for $x(n)$ and $F^{-1}(z)$ is the causal inverse. The cascade

$$G(z) = F^{-1}(z)H(z)$$

is the causal Wiener filter that produces the minimum mean-square estimate of $d(n)$ from the white noise process $\epsilon(n)$. The causality of $G(z)$ follows from the fact that both $F^{-1}(z)$ and $H(z)$ are causal. The optimality of the filter follows from the observation that if there were another filter, $G'(z)$, that produced an estimate of $d(n)$ having a smaller mean-square error

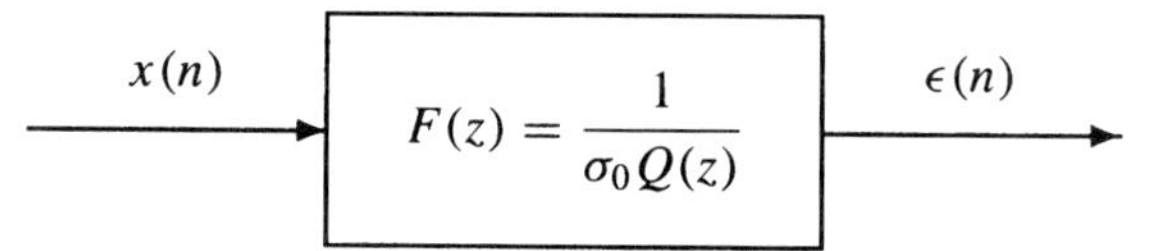

Figure 7.11 *A whitening filter that produces white noise with power spectrum $P_\epsilon(e^{j\omega}) = 1$ when the input, $x(n)$, has a power spectrum $P_x(z) = \sigma_0^2 Q(z)Q^*(1/z^*)$.*

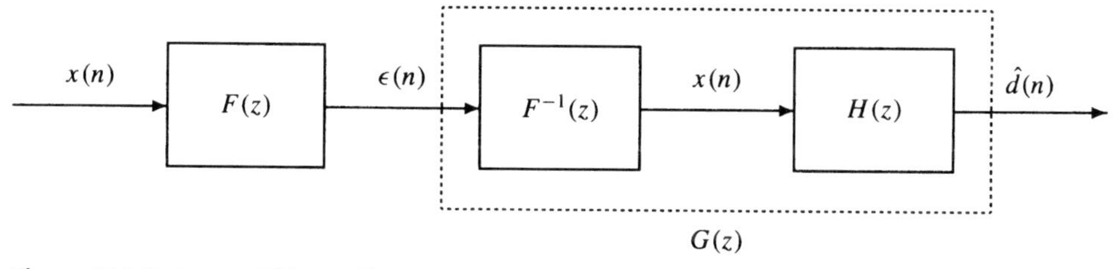

Figure 7.12 *A causal Wiener filter, $G(z)$, operating on a whitened input, $\epsilon(n)$, with $H(z)$ the causal Wiener filter for estimating $d(n)$ from $x(n)$.*

than $G(z)$, then $H'(z) = F(z)G'(z)$ would produce an estimate of $d(n)$ with a smaller mean-square error than $H(z)$, the causal Wiener filter. This, however, contradicts the assumed optimality of $H(z)$.

With $\epsilon(n)$ a white noise process, we see from Eq. (7.48) that $G(z)$, the causal IIR Wiener filter for estimating $d(n)$ from $\epsilon(n)$, is $G(z) = [P_{d\epsilon}(z)]_+$. Since $\epsilon(n)$ is formed by filtering $x(n)$ with the whitening filter $f(n)$, then the cross-correlation between $d(n)$ and $\epsilon(n)$ is

$$\begin{aligned} r_{d\epsilon}(k) &= E\{d(n)\epsilon^*(n-k)\} = E\Big\{d(n)\Big[\sum_{l=-\infty}^{\infty} f(l)x(n-k-l)\Big]^*\Big\} \\ &= \sum_{l=-\infty}^{\infty} f^*(l)r_{dx}(k+l) \end{aligned} \tag{7.51}$$

Therefore, the cross-power spectral density, $P_{d\epsilon}(z)$, is

$$P_{d\epsilon}(z) = P_{dx}(z)F^*(1/z^*) = \frac{P_{dx}(z)}{\sigma_0 Q^*(1/z^*)}$$

and the causal Wiener filter for estimating $d(n)$ from $\epsilon(n)$ is

$$G(z) = \frac{1}{\sigma_0}\left[\frac{P_{dx}(z)}{Q^*(1/z^*)}\right]_+ \tag{7.52}$$

The design is completed by recalling that the causal Wiener filter for estimating $d(n)$ from $x(n)$ is the cascade of $F(z)$ and $G(z)$,

$$H(z) = F(z)G(z)$$

Thus, combining Eqs. (7.50) and (7.52) leads to the desired solution

$$\boxed{H(z) = \frac{1}{\sigma_0^2 Q(z)}\left[\frac{P_{dx}(z)}{Q^*(1/z^*)}\right]_+} \tag{7.53}$$

In the case of real processes, $h(n)$ is real and the causal Wiener filter takes the form

$$H(z) = \frac{1}{\sigma_0^2 Q(z)}\left[\frac{P_{dx}(z)}{Q(z^{-1})}\right]_+ \tag{7.54}$$

Finally, as with the noncausal IIR Wiener filter, the mean-square error for the causal IIR Wiener filter is

$$\xi_{\min} = r_d(0) - \sum_{l=0}^{\infty} h(l)r_{dx}^*(l) \tag{7.55}$$

where the sum now extends only over the interval $0 \le l < \infty$ since $h(l) = 0$ for $l < 0$. In the frequency domain, this error may be written as

$$\xi_{\min} = \frac{1}{2\pi}\int_{-\pi}^{\pi} \left[P_d(e^{j\omega}) - H(e^{j\omega})P_{dx}^*(e^{j\omega})\right] d\omega \tag{7.56}$$

or, equivalently,

$$\xi_{\min} = \frac{1}{2\pi j}\oint_C \left[P_d(z) - H(z)P_{dx}^*(1/z^*)\right] z^{-1} dz \tag{7.57}$$

The causal Wiener filtering equations are summarized in Table 7.3.

Table 7.3 ***The System Function and Minimum Error for a Causal Wiener Filter***

System function

$$H(z) = \frac{1}{\sigma_0^2 Q(z)} \left[\frac{P_{dx}(z)}{Q^*(1/z^*)} \right]_+$$

Spectral Factorization

$$P_x(z) = \sigma_0^2 Q(z) Q^*(1/z^*)$$

Minimum error

$$\xi_{\min} = r_d(0) - \sum_{l=0}^{\infty} h(l) r_{dx}^*(l)$$

$$= \frac{1}{2\pi} \int_{-\pi}^{\pi} \left[P_d(e^{j\omega}) - H(e^{j\omega}) P_{dx}^*(e^{j\omega}) \right] d\omega$$

An interesting interpretation of the causal Wiener filter follows if we compare Eq. (7.53) to the noncausal Wiener filter given in Eq. (7.36). Denoting the noncausal Wiener filter by $H_{\text{nc}}(z)$ and using the spectral factorization of $P_x(z)$ given in Eq. (7.49), we see that the noncausal Wiener filter may be written as

$$H_{\text{nc}}(z) = \frac{P_{dx}(z)}{\sigma_0^2 Q(z) Q^*(1/z^*)} \tag{7.58}$$

Viewed as a cascade of two filters, the noncausal Wiener filter is

$$\boxed{H_{\text{nc}}(z) = \frac{1}{\sigma_0^2 Q(z)} \left[\frac{P_{dx}(z)}{Q^*(1/z^*)} \right]} \tag{7.59}$$

as shown in Fig. 7.13*a*. Note that the first filter is the causal whitening filter that generates the white noise process $\epsilon(n)$ from $x(n)$, and the second filter is the *noncausal* filter that produces the minimum mean-square estimate of $d(n)$ from the whitened signal. Comparing Eq. (7.59) to Eq. (7.53), note that the causal IIR Wiener filter shown in Fig. 7.13*b* is formed by taking the causal part of $[P_{dx}(z)/Q^*(1/z^*)]$. We now look at some applications of causal Wiener filtering.

7.3.3 Causal Wiener Filtering

In Section 7.3.1, we considered the problem of noncausal Wiener smoothing for estimating a process $d(n)$ from the noisy observations

$$x(n) = d(n) + v(n)$$

The system function of the causal Wiener filter for estimating $d(n)$ is given in Eq. (7.53). If the noise $v(n)$ is uncorrelated with $d(n)$ then $P_{dx}(z) = P_d(z)$ and the causal Wiener filter becomes

$$H(z) = \frac{1}{\sigma_0^2 Q(z)} \left[\frac{P_d(z)}{Q^*(1/z^*)} \right]_+ \tag{7.60}$$

where

$$P_x(z) = P_d(z) + P_v(z) = \sigma_0^2 Q(z) Q^*(1/z^*) \tag{7.61}$$

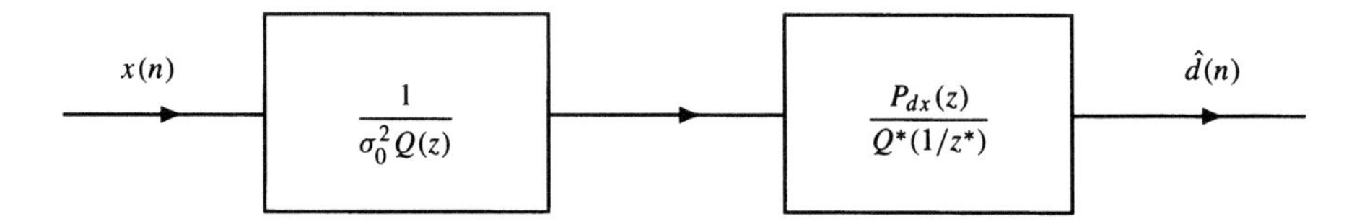

(a) Noncausal Wiener filter.

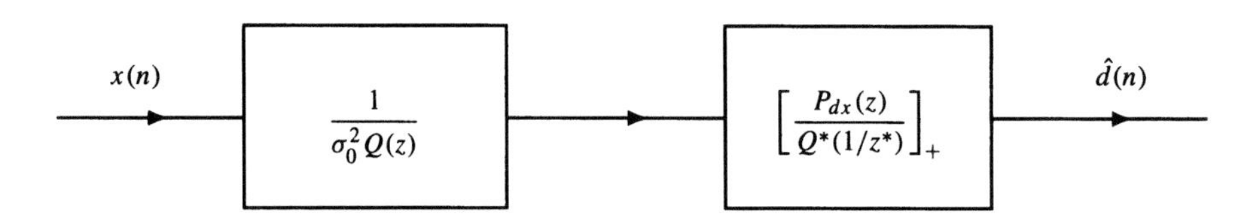

(b) Causal Wiener filter

Figure 7.13 *(a) A noncausal Wiener filter expressed in terms of a cascade of a causal filter with a noncausal filter. (b) The causal Wiener filter that is formed by taking the causal part of the noncausal factor in (a).*

However, without specific expressions for the power spectral densities $P_d(z)$ and $P_v(z)$, Eq. (7.60) cannot be simplified any further. In the following example we design a causal Wiener filter for estimating an AR(1) process that is measured in white noise.

Example 7.3.2 *Causal Wiener Filtering*

Suppose that we wish to estimate a signal $d(n)$ from the noisy observation

$$x(n) = d(n) + v(n)$$

where $v(n)$ is unit variance white noise that is uncorrelated with $d(n)$. The signal $d(n)$ is an AR(1) process that is generated by the difference equation

$$d(n) = 0.8d(n-1) + w(n)$$

where $w(n)$ is white noise with variance $\sigma_w^2 = 0.36$. Therefore $r_d(k) = (0.8)^{|k|}$.

To find the optimum *causal* Wiener filter for estimating $d(n)$ from $x(n)$ we begin by noting that

$$\begin{aligned} P_{dx}(z) &= P_d(z) \\ P_x(z) &= P_d(z) + P_v(z) = P_d(z) + 1 \end{aligned}$$

Therefore, with

$$P_d(z) = \frac{0.36}{(1-0.8z^{-1})(1-0.8z)}$$

the power spectrum of $x(n)$ is

$$P_x(z) = 1 + \frac{0.36}{(1-0.8z^{-1})(1-0.8z)} = 1.6\frac{(1-0.5z^{-1})(1-0.5z)}{(1-0.8z^{-1})(1-0.8z)}$$

and, since $x(n)$ is real, then $P_x(z) = \sigma_0^2 Q(z)Q(z^{-1})$ with

$$\sigma_0^2 = 1.6 \quad \text{and} \quad Q(z) = \frac{(1-0.5z^{-1})}{(1-0.8z^{-1})}$$

Since the causal Wiener filter is

$$H(z) = \frac{1}{\sigma_0^2 Q(z)}\left[\frac{P_{dx}(z)}{Q(z^{-1})}\right]_+$$

then we have

$$\begin{aligned}\frac{P_{dx}(z)}{Q(z^{-1})} &= \frac{0.36}{(1-0.8z^{-1})(1-0.8z)}\frac{(1-0.8z)}{(1-0.5z)} \\ &= \frac{0.36z^{-1}}{(1-0.8z^{-1})(z^{-1}-0.5)} \\ &= \frac{0.6}{1-0.8z^{-1}} + \frac{0.3}{z^{-1}-0.5}\end{aligned}$$

Therefore,

$$\left[\frac{P_{dx}(z)}{Q(z^{-1})}\right]_+ = \frac{0.6}{1-0.8z^{-1}}$$

and the Wiener filter is

$$H(z) = \frac{1}{1.6}\frac{(1-0.8z^{-1})}{(1-0.5z^{-1})}\frac{0.6}{(1-0.8z^{-1})} = \frac{0.375}{1-0.5z^{-1}}$$

or,

$$h(n) = 0.375\left(\tfrac{1}{2}\right)^n u(n)$$

Since $\hat{D}(z) = H(z)X(z)$, the estimate of $d(n)$ may be computed recursively as follows

$$\hat{d}(n) = 0.5\hat{d}(n-1) + 0.375x(n)$$

Finally, for the mean-square error in the estimate of $d(n)$ we have

$$\xi_{\min} = E\left\{\left[d(n) - \hat{d}(n)\right]^2\right\} = r_d(0) - \sum_{l=0}^{\infty} h(l)r_{dx}(l) = 1 - \tfrac{3}{8}\sum_{l=0}^{\infty}\left(\tfrac{1}{2}\right)^l (0.8)^l = 0.3750$$

Note that, for the second-order FIR Wiener filter designed in Example 7.2.1, the mean-square error was $\xi_{\min} = 0.4048$. Therefore, using *all* of the previous observations of $x(n)$ only slightly improves the performance of the Wiener filter. For another comparison, let us find the *noncausal* Wiener filter and evaluate the mean-square error. The system function of the noncausal Wiener filter is

$$H(z) = \frac{P_{dx}(z)}{P_x(z)} = \frac{P_d(z)}{P_x(z)} = \frac{0.36/1.6}{(1-0.5z^{-1})(1-0.5z)}$$

and the unit sample response is

$$h(n) = \tfrac{3}{10}\left(\tfrac{1}{2}\right)^{|n|}$$

Computing the mean-square error we find

$$\xi_{\min} = r_d(0) - \sum_{l=-\infty}^{\infty} h(l) r_{dx}(l) = 1 - 2\sum_{k=0}^{\infty} \tfrac{3}{10}\left(\tfrac{1}{2}\right)^k (0.8)^k + \tfrac{3}{10} = 0.3$$

which is smaller, as it should be. Alternatively, as we saw in Example 7.3.1,

$$\xi_{\min} = \sigma_u^2 h(c)$$

Therefore, $\xi_{\min} = 0.3$ as above.

In the previous example we looked at the problem of estimating a process $d(n)$ from noisy measurements

$$x(n) = d(n) + v(n) \tag{7.62}$$

What we found was that if $d(n)$ is generated by the difference equation

$$d(n) = 0.8d(n-1) + w(n) \tag{7.63}$$

then the estimate of $d(n)$, when computed recursively, is given by

$$\hat{d}(n) = 0.5\hat{d}(n-1) + 0.375x(n)$$

What is interesting to note is that this recursive estimator may also be rewritten in the form

$$\hat{d}(n) = 0.8\hat{d}(n-1) + 0.375\Big[x(n) - 0.8\hat{d}(n-1)\Big] \tag{7.64}$$

The interpretation of this equation is as follows. The quantity $\hat{d}(n)$ is the minimum mean-square estimate of $d(n)$ that is based on all observations of $x(n)$ up to time n. Similarly, $\hat{d}(n-1)$ is the minimum mean-square estimate of $d(n-1)$ that is formed from all of the observations of $x(n)$ up to time $n-1$. Given $\hat{d}(n-1)$, before the next value of $x(n)$ is observed, we may "predict" what the next value of $d(n)$ should be. In light of Eq. (7.63), since $w(n)$ is a zero mean process then this estimate should be $0.8\hat{d}(n-1)$. Given that $x(n) = d(n)+v(n)$ we may then use this prediction to "predict" what the next measurement should be,

$$\hat{x}(n) = 0.8\hat{d}(n-1)$$

Finally, with the arrival of the new measurement, $x(n)$, since the prediction will not be perfect, there will be an error

$$\alpha(n) = x(n) - \hat{x}(n)$$

This error, called the *innovations* process, represents the "new information" that is brought with the observation $x(n)$. In other words, $\alpha(n)$ corresponds to that part of $x(n)$ that cannot be predicted. Therefore, the estimate $\hat{d}(n)$ is modified by adding a correction, which is the innovations process after it has been scaled by a *gain*, K, referred to as the *Kalman Gain*. As we will soon discover in Section 7.4, Eq. (7.64) is the steady-state Kalman filter for estimating the stationary process $d(n)$.

7.3.4 Causal Linear Prediction

In Section 7.2.2, we derived the FIR Wiener filter for linear prediction. In this section, we look at the design of the optimum causal linear predictor

$$\hat{x}(n+1) = \sum_{k=0}^{\infty} h(k)x(n-k)$$

that produces the best estimate of $x(n+1)$ given $x(k)$ for all $k \le n$. Since the infinite past is now being used to predict the next value of $x(n)$, we expect a smaller mean-square prediction error than for an FIR linear predictor.

For linear prediction, $d(n) = x(n+1)$ and the cross-correlation between $x(n)$ and $d(n)$ is

$$r_{dx}(k) = E\{d(n)x^*(n-k)\} = E\{x(n+1)x^*(n-k)\} = r_x(k+1)$$

Therefore, $P_{dx}(z) = zP_x(z)$ and the causal Wiener predictor is

$$H(z) = \frac{1}{\sigma_0^2 Q(z)}\left[\frac{zP_x(z)}{Q^*(1/z^*)}\right]_+ \tag{7.65}$$

However, since $P_x(z) = \sigma_0^2 Q(z)Q^*(1/z^*)$, then Eq. (7.65) may be simplified as follows

$$H(z) = \frac{1}{Q(z)}\Big[z\,Q(z)\Big]_+ \tag{7.66}$$

Now, recall that $Q(z)$ is a monic polynomial

$$Q(z) = 1 + q(1)z^{-1} + q(2)z^{-2} + q(3)z^{-3} + \cdots$$

Therefore, the positive-time part of $zQ(z)$ is

$$\begin{aligned}\Big[zQ(z)\Big]_+ &= \Big[z + q(1) + q(2)z^{-1} + q(3)z^{-2} + \cdots\Big]_+ \\ &= q(1) + q(2)z^{-1} + q(3)z^{-2} + \cdots \\ &= z\Big[Q(z) - 1\Big]\end{aligned} \tag{7.67}$$

Substituting Eq. (7.67) into Eq. (7.66) the causal linear predictor becomes

$$\boxed{H(z) = \frac{1}{Q(z)}z\Big[Q(z) - 1\Big] = z\Big[1 - \frac{1}{Q(z)}\Big]} \tag{7.68}$$

Finally, for the minimum mean-square error, we have

$$\xi_{\min} = \frac{1}{2\pi j}\oint_C \Big[P_d(z) - H(z)P_{dx}^*(1/z^*)\Big]z^{-1}dz \tag{7.69}$$

Since $P_d(z) = P_x(z)$ and $P_{dx}(z) = zP_x(z)$, it follows that the minimum mean-square error is

$$\xi_{\min} = \frac{1}{2\pi j}\oint_C \Big[P_x(z) - z^{-1}H(z)P_x^*(1/z^*)\Big]z^{-1}dz$$

Using the symmetry of the power spectrum, $P_x(z) = P_x^*(1/z^*)$, this becomes

$$\xi_{\min} = \frac{1}{2\pi j}\oint_C P_x(z)\Big[1 - z^{-1}H(z)\Big]z^{-1}dz$$

Substituting the system function for the causal Wiener linear predictor, Eq. (7.68), into this expression yields

$$\xi_{\min} = \frac{1}{2\pi j}\oint_C P_x(z)\left[1 - \left(1 - \frac{1}{Q(z)}\right)\right] z^{-1}dz$$

$$= \frac{1}{2\pi j}\oint_C \frac{P_x(z)}{Q(z)} z^{-1}dz = \frac{1}{2\pi j}\oint_C \sigma_0^2 Q^*(1/z^*) z^{-1}dz$$

$$= \sigma_0^2 q(0) \tag{7.70}$$

Thus, since $Q(z)$ is monic with $q(0) = 1$ we have the remarkably simple result

$$\boxed{\xi_{\min} = \sigma_0^2} \tag{7.71}$$

An interesting relationship between the minimum mean-square error and the power spectrum follows from Eq. (3.103) on p. 105. Specifically,

$$\boxed{\xi_{\min} = \exp\left\{\frac{1}{2\pi}\int_{-\pi}^{\pi} \ln P_x(e^{j\omega})d\omega\right\}} \tag{7.72}$$

which is known as the *Kolmogorov-Szegö formula*.

Let us now apply these results to the linear prediction of an autoregressive process. If $x(n)$ is an AR(p) process with a power spectrum

$$P_x(z) = \frac{\sigma_0^2}{A(z)A^*(1/z^*)}$$

where

$$A(z) = 1 + \sum_{k=1}^{p} a(k)z^{-k}$$

is a minimum phase polynomial having all of its roots *inside* the unit circle, then the optimum linear predictor is

$$H(z) = z\Big[1 - A(z)\Big] = -a(1) - a(2)z^{-1} - \cdots - a(p)z^{-p+1} \tag{7.73}$$

which is an FIR filter. Therefore, given the entire past history of $x(n)$, only the last p values of $x(n)$ are used in the prediction of $x(n+1)$. However, this should not be surprising if we recall that an AR(p) random process satisfies a difference equation of the form

$$x(n) = -a(1)x(n-1) - a(2)x(n-2) - \cdots - a(p)x(n-p) + w(n)$$

where $w(n)$ is white noise. Since $w(n+1)$ cannot be predicted from $x(n)$ or previous values of $x(n)$ then the best that can be done in predicting $x(n+1)$ is to use the model for $x(n)$ and ignore the noise

$$\hat{x}(n+1) = -a(1)x(n) - a(2)x(n-1) - \cdots - a(p)x(n-p+1)$$

This predictor, of course, is the same as the one given in Eq. (7.73).

Example 7.3.3 ***Causal Linear Prediction for an AR Process***

Consider the real-valued AR(2) process

$$x(n) = 0.9x(n-1) - 0.2x(n-2) + w(n) \tag{7.74}$$

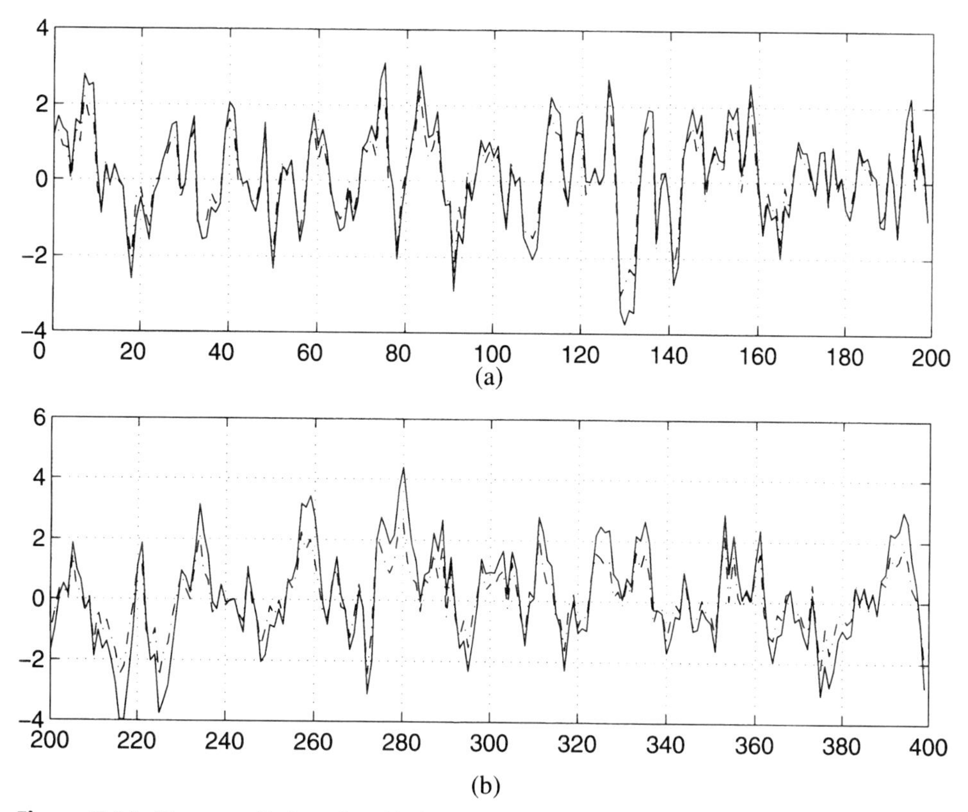

Figure 7.14 *Linear prediction of an AR(2) process using an IIR Wiener filter. The process is represented by the solid line and the prediction by a dotted line. (a) Prediction assuming that the model parameters are known. (b) Prediction using estimated model parameters.*

where $w(n)$ is unit variance zero mean white noise. Since

$$P_x(z) = \frac{1}{A(z)A(z^{-1})}$$

where

$$A(z) = 1 - 0.9z^{-1} + 0.2z^{-2}$$

then the optimum linear predictor is

$$H(z) = z\Big[1 - A(z)\Big] = z\Big[0.9z^{-1} - 0.2z^{-2}\Big] = 0.9 - 0.2z^{-1}$$

and the prediction of $x(n)$ is formed as follows

$$\hat{x}(n+1) = 0.9x(n) - 0.2x(n-1) \tag{7.75}$$

As a specific example, an AR(2) process $x(n)$ that is generated according to Eq. (7.74) is shown in Fig. 7.14*a*. Also shown is the prediction of $x(n)$ using the predictor given in Eq. (7.75), which has an average squared prediction error of

$$\xi = \frac{1}{N}\sum_{n=0}^{N-1}\big[x(n+1) - \hat{x}(n+1)\big]^2 = 0.0324$$

This example, however, is somewhat contrived since, in practice, we cannot expect to have prior knowledge of the statistics of $x(n)$. Therefore, a more realistic example is the following. Using the given data, $x(n)$, in Fig. 7.14*a* we first estimate the AR parameters using the autocorrelation method. With

$$\hat{r}_x(k) = \frac{1}{N}\sum_{n=0}^{N-1} x(n)x(n-k)$$

where $N = 200$ we find

$$r_x(0) = 2.1904, \quad r_x(1) = 1.5462, \quad \text{and} \quad r_x(2) = 0.8670$$

Thus, the normal equations are

$$\begin{bmatrix} 2.1904 & 1.5462 \\ 1.5462 & 2.1904 \end{bmatrix}\begin{bmatrix} a(1) \\ a(2) \end{bmatrix} = -\begin{bmatrix} 1.5462 \\ 0.8670 \end{bmatrix}$$

Solving for $a(1)$ and $a(2)$ we find

$$\begin{bmatrix} a(1) \\ a(2) \end{bmatrix} = \begin{bmatrix} -0.8500 \\ 0.2042 \end{bmatrix}$$

Therefore, the predictor becomes

$$\hat{x}(n+1) = 0.85x(n) - 0.2042x(n-1)$$

Now, instead of using the predictor on the data that is used to estimate the predictor coefficients, we apply the predictor to the next sequence of 200 data values. The results are shown in Fig. 7.14*b* and we see that the performance of the predictor is similar to that in Fig. 7.14*a*.

In the previous example, we looked at linear prediction for an autoregressive random process. Although these predictors are FIR filters, as we see in the next example, the Wiener predictor for an ARMA processes is an IIR filter.

Example 7.3.4 *Causal Linear Prediction for an ARMA Process*

In this example, we consider the design of the optimum causal linear predictor for a process, $x(n)$, that has a power spectrum

$$P_x(z) = \frac{(1 - 0.6z^{-1} + 0.36z^{-2})(1 - 0.6z + 0.36z^2)}{(1 - 0.8z^{-1})(1 - 0.8z)}$$

The optimum predictor is

$$H(z) = z\left[1 - \frac{1}{Q(z)}\right]$$

where $Q(z)$ is the minimum phase spectral factor of $P_x(z)$. Since

$$Q(z) = \frac{1 - 0.6z^{-1} + 0.36z^{-2}}{1 - 0.8z^{-1}}$$

then the system function of the linear predictor is

$$H(z) = z\left[1 - \frac{1 - 0.8z^{-1}}{1 - 0.6z^{-1} + 0.36z^{-2}}\right] = z\left[\frac{0.2z^{-1} + 0.36z^{-2}}{1 - 0.6z^{-1} + 0.36z^{-2}}\right]$$

$$= \frac{0.2 + 0.36z^{-1}}{1 - 0.6z^{-1} + 0.36z^{-2}}$$

which is an IIR filter. Therefore, the prediction of $x(n+1)$ is computed using the recursive equation

$$\hat{x}(n+1) = 0.6\hat{x}(n) - 0.36\hat{x}(n-1) + 0.2x(n) + 0.36x(n-1)$$

7.3.5 Wiener Deconvolution

We conclude this section on Wiener filtering with a discussion of an extremely difficult and important problem that arises in many applications. This problem, known as *deconvolution*, is concerned with the recovery of a signal $d(n)$ that has been convolved with a filter $g(n)$ that may not be precisely known,

$$x(n) = d(n) * g(n) \tag{7.76}$$

Convolutional distortion is often introduced in the process of measuring or recording data. For example, a camera that is slightly out of focus and a band-limited communication channel may be modeled as systems that introduce convolutional distortion. In some applications, convolutional distortion may be inadvertently introduced as would be the case if a camera were in motion during photographic exposure [1]. If the "blurring" function, $g(n)$, is known perfectly and if $g(n)$ has an inverse $g^{-1}(n)$ so that

$$g(n) * g^{-1}(n) = \delta(n)$$

then there is, in theory, no difficulty in recovering $d(n)$ from $x(n)$ since the inverse filter would perform the desired restoration, i.e.,

$$d(n) = x(n) * g^{-1}(n) \tag{7.77}$$

or, in the frequency domain,

$$D(e^{j\omega}) = \frac{X(e^{j\omega})}{G(e^{j\omega})}$$

In practice, however, precise knowledge of $g(n)$ or $G(e^{j\omega})$ is generally not available. Another problem is that the frequency response, $G(e^{j\omega})$, is often equal to zero at one or more frequencies or is very small over a band of frequencies. This results in either a noninvertible $G(e^{j\omega})$ or one that is ill-conditioned. In addition, since noise is invariably introduced in the measurement process, a more accurate model for the observed signal would be

$$x(n) = d(n) * g(n) + w(n)$$

where $w(n)$ is additive noise that is often assumed to be uncorrelated with $d(n)$. In this case, even if the inverse filter $G^{-1}(e^{j\omega})$ exists and is well-behaved, when the inverse filter is applied to $x(n)$ the restored signal is

$$\hat{D}(e^{j\omega}) = D(e^{j\omega}) + \underbrace{\frac{W(e^{j\omega})}{G(e^{j\omega})}}_{\text{noise}} = D(e^{j\omega}) + V(e^{j\omega})$$

which, in the time domain, becomes

$$\hat{d}(n) = d(n) + v(n) \tag{7.78}$$

Thus, $\hat{d}(n)$ is the sum of the original signal $d(n)$ and a filtered noise term $v(n)$. The difficulty with this solution is that, if $G(e^{j\omega}) \approx 0$ over some range of frequencies, then the inverse filter, $1/G(e^{j\omega})$, becomes large and the noise $w(n)$ will be amplified.

Another approach to the deconvolution problem is to design a Wiener filter that produces the minimum mean-square estimate of $d(n)$ from $x(n)$. Thus, let $h(n)$ be an IIR linear shift-invariant filter and let $\hat{d}(n)$ be the estimate of $d(n)$ that is produced by filtering $x(n)$ with $h(n)$

$$\hat{d}(n) = x(n) * h(n) = \sum_{l=-\infty}^{\infty} h(l)x(n-l)$$

Note that we are assuming that the filter is noncausal. The filter coefficients $h(n)$ that minimize the mean-square error

$$\xi = E\left\{|d(n) - \hat{d}(n)|^2\right\}$$

are the solution to the Wiener-Hopf equations, which, in the frequency domain, becomes

$$H(e^{j\omega}) = \frac{P_{dx}(e^{j\omega})}{P_x(e^{j\omega})} \tag{7.79}$$

Therefore, all that needs to be done in the design of $H(e^{j\omega})$ is to find the power spectral densities $P_{dx}(e^{j\omega})$ and $P_x(e^{j\omega})$. Since $w(n)$ is assumed to be uncorrelated with $d(n)$, then $w(n)$ will also be uncorrelated with $d(n) * g(n)$. As a result, the power spectral density of $x(n)$ is the sum of the power spectrum of $d(n) * g(n)$ and the power spectrum of $w(n)$,

$$P_x(e^{j\omega}) = P_d(e^{j\omega})|G(e^{j\omega})|^2 + P_w(e^{j\omega}) \tag{7.80}$$

In addition, the cross-power spectral density $P_{dx}(e^{j\omega})$ is

$$P_{dx}(e^{j\omega}) = P_d(e^{j\omega})G^*(e^{j\omega}) \tag{7.81}$$

Therefore, substituting Eqs. (7.80) and (7.81) into Eq. (7.79) we find that the optimum Wiener filter for deconvolution is given by

$$H(e^{j\omega}) = \frac{P_d(e^{j\omega})G^*(e^{j\omega})}{P_d(e^{j\omega})|G(e^{j\omega})|^2 + P_w(e^{j\omega})}$$

It is interesting to note that if we assume that there are no values of ω for which $G(e^{j\omega})$ is equal to zero, and if we factor out the inverse filter $1/G(e^{j\omega})$ from $H(e^{j\omega})$, then the noncausal Wiener filter may be written as

$$H(e^{j\omega}) = \frac{1}{G(e^{j\omega})}\left[\frac{P_d(e^{j\omega})}{P_d(e^{j\omega}) + P_w(e^{j\omega})/|G(e^{j\omega})|^2}\right] \tag{7.82}$$

Since the power spectrum of the filtered noise, $v(n)$, in Eq. (7.78) is

$$P_v(e^{j\omega}) = P_w(e^{j\omega})/|G(e^{j\omega})|^2$$

it follows that the term in brackets in Eq. (7.82) may be written as

$$F(e^{j\omega}) = \frac{P_d(e^{j\omega})}{P_d(e^{j\omega}) + P_v(e^{j\omega})} \tag{7.83}$$

Therefore, comparing Eq. (7.83) with Eq. (7.42) we see that $F(e^{j\omega})$ is the noncausal IIR Wiener smoothing filter for estimating $d(n)$ from

$$y(n) = d(n) + v(n) \tag{7.84}$$

Figure 7.15 *Wiener deconvolution realized as the cascade of an inverse filter with a smoothing filter* $F(e^{j\omega})$.

Thus, as shown in Fig. 7.15, the Wiener filter $H(e^{j\omega})$ may be viewed as a cascade of the inverse filter $1/G(e^{j\omega})$ followed by a noncausal Wiener smoothing filter for reducing the filtered noise.

7.4 DISCRETE KALMAN FILTER

In Section 7.3.2 we considered the problem of designing a causal Wiener filter to estimate a process $d(n)$ from a set of noisy observations $x(n) = d(n) + v(n)$. The primary limitation with the solution that was derived is that it requires that $d(n)$ and $x(n)$ be jointly wide-sense stationary processes. Since most processes encountered in practice are nonstationary, this constraint limits the usefulness of the Wiener filter. Therefore, in this section we re-examine this estimation problem within the context of nonstationary processes and derive what is known as the *discrete Kalman filter*.

To begin, let us look briefly once again at the causal Wiener filter for estimating a process $x(n)$ from the noisy measurements[3]

$$y(n) = x(n) + v(n) \tag{7.85}$$

In Example 7.3.2 we considered the specific problem of estimating an AR(1) process of the form[4]

$$x(n) = a(1)x(n-1) + w(n)$$

from

$$y(n) = x(n) + v(n)$$

where $w(n)$ and $v(n)$ are uncorrelated white noise processes. What we discovered was that the optimum linear estimate of $x(n)$, using *all* of the measurements, $y(k)$, for $k \leq n$, could be computed with a recursion of the form

$$\hat{x}(n) = a(1)\hat{x}(n-1) + K\Big[y(n) - a(1)\hat{x}(n-1)\Big] \tag{7.86}$$

where K is a constant, referred to as the *Kalman gain*, that minimizes the mean-square error $E\{|x(n) - \hat{x}(n)|^2\}$. However, there are two problems with this solution that need to be

[3]Here we have introduced a slight change in notation in order to be consistent with the notation that is commonly used in the Kalman filtering literature. Instead of $d(n)$, the signal that is to be estimated will be denoted by $x(n)$ and the noisy observations denoted by $y(n)$.

[4]In the literature on Kalman filtering, the process $x(n)$ is usually assumed to be generated according to the difference equation

$$x(n) = a(1)x(n-1) + w(n-1)$$

This difference is not significant, however, since both processes have the same autocorrelation.

addressed. First is the requirement that $x(n)$ and $y(n)$ be jointly wide-sense stationary processes. For example, Eq. (7.86) is not the optimum linear estimate if $x(n)$ is a nonstationary process, such as the one that is generated by the time-varying difference equation

$$x(n) = a_{n-1}(1)x(n-1) + w(n)$$

Nevertheless, what we will discover is that the optimum estimate may be written as

$$\hat{x}(n) = a_{n-1}(1)\hat{x}(n-1) + K(n)\Big[y(n) - a_{n-1}(1)\hat{x}(n-1)\Big] \tag{7.87}$$

where $K(n)$ is a suitably chosen (time-varying) gain. The second problem with the Wiener solution is that it does not allow the filter to be "turned on" at time $n = 0$. In other words, implicit in the development of the causal Wiener filter is the assumption that the observations $y(k)$ are available for all $k \leq n$. Again, however, we will find that this problem is addressed with an estimate of the form given in Eq. (7.87).

Although the discussion above is only concerned with the estimation of an AR(1) process from noisy measurements, using state variables we may easily extend these results to more general processes. For example, let $x(n)$ be an AR(p) process that is generated according to the difference equation

$$x(n) = \sum_{k=1}^{p} a(k)x(n-k) + w(n) \tag{7.88}$$

and suppose that $x(n)$ is measured in the presence of additive noise

$$y(n) = x(n) + v(n) \tag{7.89}$$

If we let $\mathbf{x}(n)$ be the p-dimensional state vector

$$\mathbf{x}(n) = \begin{bmatrix} x(n) \\ x(n-1) \\ \vdots \\ x(n-p+1) \end{bmatrix}$$

then Eqs. (7.88) and (7.89) may be written in terms of $\mathbf{x}(n)$ as follows

$$\mathbf{x}(n) = \begin{bmatrix} a(1) & a(2) & \cdots & a(p-1) & a(p) \\ 1 & 0 & \cdots & 0 & 0 \\ 0 & 1 & \cdots & 0 & 0 \\ \vdots & \vdots & \cdots & & \vdots \\ 0 & 0 & \cdots & 1 & 0 \end{bmatrix} \mathbf{x}(n-1) + \begin{bmatrix} 1 \\ 0 \\ 0 \\ \vdots \\ 0 \end{bmatrix} w(n) \tag{7.90}$$

and

$$y(n) = \big[1,\ 0,\ \ldots,\ 0\big]\mathbf{x}(n) + v(n) \tag{7.91}$$

Using matrix notation to simplify these equations we have

$$\begin{aligned} \mathbf{x}(n) &= \mathbf{A}\mathbf{x}(n-1) + \mathbf{w}(n) \\ y(n) &= \mathbf{c}^T\mathbf{x}(n) + v(n) \end{aligned} \tag{7.92}$$

where $\mathbf{A}$ is a $p \times p$ *state transition matrix*, $\mathbf{w}(n) = \big[w(n),\ 0,\ \ldots,\ 0\big]^T$ is a vector noise process, and $\mathbf{c}$ is a unit vector of length p. As in Eq. (7.87) for the case of an AR(1) process, the optimum estimate of the state vector $\mathbf{x}(n)$, using all of the measurements up to time n, may be expressed in the form

$$\hat{\mathbf{x}}(n) = \mathbf{A}\hat{\mathbf{x}}(n-1) + \mathbf{K}\Big[y(n) - \mathbf{c}^T\mathbf{A}\hat{\mathbf{x}}(n-1)\Big] \tag{7.93}$$

where $\mathbf{K}$ is a Kalman gain vector.

Although only applicable to stationary AR(p) processes, Eq. (7.92) may be easily generalized to nonstationary processes as follows. Let $\mathbf{x}(n)$ be a state vector of dimension p that evolves according to the difference equation

$$\boxed{\mathbf{x}(n) = \mathbf{A}(n-1)\mathbf{x}(n-1) + \mathbf{w}(n)} \tag{7.94}$$

where $\mathbf{A}(n-1)$ is a time-varying $p \times p$ state transition matrix and $\mathbf{w}(n)$ is a vector of zero-mean white noise processes with

$$E\Big\{\mathbf{w}(n)\mathbf{w}^H(k)\Big\} = \begin{cases} \mathbf{Q}_w(n) & ; \quad k = n \\ \mathbf{0} & ; \quad k \neq n \end{cases} \tag{7.95}$$

In addition, let $\mathbf{y}(n)$ be a *vector* of observations that are formed according to

$$\boxed{\mathbf{y}(n) = \mathbf{C}(n)\mathbf{x}(n) + \mathbf{v}(n)} \tag{7.96}$$

where $\mathbf{y}(n)$ is a vector of length q, $\mathbf{C}(n)$ is a time-varying $q \times p$ matrix, and $\mathbf{v}(n)$ is a vector of zero mean white noise processes that are statistically independent of $\mathbf{w}(n)$ with

$$E\Big\{\mathbf{v}(n)\mathbf{v}^H(k)\Big\} = \begin{cases} \mathbf{Q}_v(n) & ; \quad k = n \\ \mathbf{0} & ; \quad k \neq n \end{cases} \tag{7.97}$$

Generalizing the result given in Eq. (7.93), we expect the optimum *linear* estimate of $\mathbf{x}(n)$ to be expressible in the form

$$\hat{\mathbf{x}}(n) = \mathbf{A}(n-1)\hat{\mathbf{x}}(n-1) + \mathbf{K}(n)\Big[\mathbf{y}(n) - \mathbf{C}(n)\mathbf{A}(n-1)\hat{\mathbf{x}}(n-1)\Big]$$

With the appropriate *Kalman gain matrix* $\mathbf{K}(n)$, this recursion corresponds to the *discrete Kalman filter*. We will now show that the optimum *linear* recursive estimate of $\mathbf{x}(n)$ has this form and derive the optimum Kalman gain $\mathbf{K}(n)$ that minimizes the mean-square estimation error. In the following discussion it is assumed that $\mathbf{A}(n)$, $\mathbf{C}(n)$, $\mathbf{Q}_w(n)$, and $\mathbf{Q}_v(n)$ are known.

In our development of the discrete Kalman filter, we will let $\hat{\mathbf{x}}(n|n)$ denote the best linear estimate of $\mathbf{x}(n)$ at time n given the observations $\mathbf{y}(i)$ for $i = 1, 2, \ldots, n$, and we will let $\hat{\mathbf{x}}(n|n-1)$ denote the best estimate given the observations up to time $n-1$. With $\mathbf{e}(n|n)$ and $\mathbf{e}(n|n-1)$ the corresponding state estimation errors,

$$\begin{aligned} \mathbf{e}(n|n) &= \mathbf{x}(n) - \hat{\mathbf{x}}(n|n) \\ \mathbf{e}(n|n-1) &= \mathbf{x}(n) - \hat{\mathbf{x}}(n|n-1) \end{aligned}$$

and $\mathbf{P}(n|n)$ and $\mathbf{P}(n|n-1)$ the error covariance matrices,

$$\begin{aligned} \mathbf{P}(n|n) &= E\big\{\mathbf{e}(n|n)\mathbf{e}^H(n|n)\big\} \\ \mathbf{P}(n|n-1) &= E\big\{\mathbf{e}(n|n-1)\mathbf{e}^H(n|n-1)\big\} \end{aligned} \tag{7.98}$$

the problem that we would like to solve is the following. Suppose that matrix we are given an estimate $\hat{\mathbf{x}}(0|0)$ of the state $\mathbf{x}(0)$, and that the error covariance matrix for this estimate, $\mathbf{P}(0|0)$, is known. When the measurement $\mathbf{y}(1)$ becomes available the goal is to update $\hat{\mathbf{x}}(0|0)$ and find the estimate $\hat{\mathbf{x}}(1|1)$ of the state at time $n = 1$ that minimizes the mean-square error

$$\xi(1) = E\{\|\mathbf{e}(1|1)\|^2\} = \text{tr}\{\mathbf{P}(1|1)\} = \sum_{i=0}^{p-1} E\{|e_i(1|1)|^2\} \tag{7.99}$$

After $\hat{\mathbf{x}}(1|1)$ has been determined and the error covariance $\mathbf{P}(1|1)$ found, the estimation is repeated for the next observation $\mathbf{y}(2)$. Thus, for each $n > 0$, given $\hat{\mathbf{x}}(n-1|n-1)$ and $\mathbf{P}(n-1|n-1)$, when a new observation, $y(n)$, becomes available, the problem is to find the minimum mean-square estimate $\hat{\mathbf{x}}(n|n)$ of the state vector $\mathbf{x}(n)$. The solution to this problem will be derived in two steps. First, given $\hat{\mathbf{x}}(n-1|n-1)$ we will find $\hat{\mathbf{x}}(n|n-1)$, which is the best estimate of $\mathbf{x}(n)$ *without* the observation $\mathbf{y}(n)$. Then, given $\mathbf{y}(n)$ and $\hat{\mathbf{x}}(n|n-1)$ we will estimate $\mathbf{x}(n)$.

In the first step, since no new measurements are used to estimate $\mathbf{x}(n)$, all that is known is that $\mathbf{x}(n)$ evolves according to the state equation

$$\mathbf{x}(n) = \mathbf{A}(n-1)\mathbf{x}(n-1) + \mathbf{w}(n)$$

Since $\mathbf{w}(n)$ is a zero mean white noise process (and the values of $\mathbf{w}(n)$ are unknown), then we may predict $\mathbf{x}(n)$ as follows,

$$\boxed{\hat{\mathbf{x}}(n|n-1) = \mathbf{A}(n-1)\hat{\mathbf{x}}(n-1|n-1)} \tag{7.100}$$

which has an estimation error given by

$$\begin{aligned}\mathbf{e}(n|n-1) &= \mathbf{x}(n) - \hat{\mathbf{x}}(n|n-1) \\ &= \mathbf{A}(n-1)\mathbf{x}(n-1) + \mathbf{w}(n) - \mathbf{A}(n-1)\hat{\mathbf{x}}(n-1|n-1) \\ &= \mathbf{A}(n-1)\mathbf{e}(n-1|n-1) + \mathbf{w}(n)\end{aligned} \tag{7.101}$$

Note that since $\mathbf{w}(n)$ has zero mean, if $\hat{\mathbf{x}}(n-1|n-1)$ is an unbiased estimate of $\mathbf{x}(n-1)$, i.e.,

$$E\{\mathbf{e}(n-1|n-1)\} = 0$$

then $\hat{\mathbf{x}}(n|n-1)$ will be an unbiased estimate of $\mathbf{x}(n)$,

$$E\{\mathbf{e}(n|n-1)\} = 0$$

Finally, since the estimation error $\mathbf{e}(n-1|n-1)$ is uncorrelated with $\mathbf{w}(n)$ (a consequence of the fact that $\mathbf{w}(n)$ is a white noise sequence), then

$$\boxed{\mathbf{P}(n|n-1) = \mathbf{A}(n-1)\mathbf{P}(n-1|n-1)\mathbf{A}^H(n-1) + \mathbf{Q}_w(n)} \tag{7.102}$$

where $\mathbf{Q}_w(n)$ is the covariance matrix for the noise process $\mathbf{w}(n)$. This completes the first step of the Kalman filter.

In the second step we incorporate the new measurement $\mathbf{y}(n)$ into the estimate $\hat{\mathbf{x}}(n|n-1)$. A linear estimate of $\mathbf{x}(n)$ that is based on $\hat{\mathbf{x}}(n|n-1)$ and $\mathbf{y}(n)$ is of the form

$$\hat{\mathbf{x}}(n|n) = \mathbf{K}'(n)\hat{\mathbf{x}}(n|n-1) + \mathbf{K}(n)\mathbf{y}(n) \tag{7.103}$$

where $\mathbf{K}(n)$ and $\mathbf{K}'(n)$ are matrices, yet to be specified. The requirement that is imposed on $\hat{\mathbf{x}}(n|n)$ is that it be *unbiased*, $E\{\mathbf{e}(n|n)\} = 0$, and that it minimize the mean-square error, $E\{\|\mathbf{e}(n|n)\|^2\}$. Using Eq. (7.103) we may express $\mathbf{e}(n|n)$ in terms of $\mathbf{e}(n|n-1)$ as follows

$$\begin{aligned} \mathbf{e}(n|n) &= \mathbf{x}(n) - \mathbf{K}'(n)\hat{\mathbf{x}}(n|n-1) - \mathbf{K}(n)\mathbf{y}(n) \\ &= \mathbf{x}(n) - \mathbf{K}'(n)\Big[\mathbf{x}(n) - \mathbf{e}(n|n-1)\Big] - \mathbf{K}(n)\Big[\mathbf{C}(n)\mathbf{x}(n) + \mathbf{v}(n)\Big] \\ &= \Big[\mathbf{I} - \mathbf{K}'(n) - \mathbf{K}(n)\mathbf{C}(n)\Big]\mathbf{x}(n) + \mathbf{K}'(n)\mathbf{e}(n|n-1) - \mathbf{K}(n)\mathbf{v}(n) \end{aligned} \tag{7.104}$$

Since $E\{\mathbf{v}(n)\} = 0$ and $E\{e(n|n-1)\} = 0$, then $\hat{\mathbf{x}}(n|n)$ will be unbiased for any $\mathbf{x}(n)$ only if the term in brackets is zero,

$$\mathbf{K}'(n) = \mathbf{I} - \mathbf{K}(n)\mathbf{C}(n)$$

With this constraint, it follows from Eq. (7.103) that $\hat{\mathbf{x}}(n|n)$ has the form

$$\hat{\mathbf{x}}(n|n) = \Big[\mathbf{I} - \mathbf{K}(n)\mathbf{C}(n)\Big]\hat{\mathbf{x}}(n|n-1) + \mathbf{K}(n)\mathbf{y}(n) \tag{7.105}$$

or

$$\boxed{\hat{\mathbf{x}}(n|n) = \hat{\mathbf{x}}(n|n-1) + \mathbf{K}(n)\Big[\mathbf{y}(n) - \mathbf{C}(n)\hat{\mathbf{x}}(n|n-1)\Big]} \tag{7.106}$$

and the error is

$$\begin{aligned} \mathbf{e}(n|n) &= \mathbf{K}'(n)\mathbf{e}(n|n-1) - \mathbf{K}(n)\mathbf{v}(n) \\ &= \Big[\mathbf{I} - \mathbf{K}(n)\mathbf{C}(n)\Big]\mathbf{e}(n|n-1) - \mathbf{K}(n)\mathbf{v}(n) \end{aligned} \tag{7.107}$$

Since $\mathbf{v}(n)$ is uncorrelated with $\mathbf{w}(n)$, then $\mathbf{v}(n)$ is uncorrelated with $\mathbf{x}(n)$ and, therefore, it is uncorrelated with $\hat{\mathbf{x}}(n|n-1)$. In addition, since $\mathbf{e}(n|n-1) = \mathbf{x}(n) - \hat{\mathbf{x}}(n|n-1)$, then $\mathbf{v}(n)$ is uncorrelated with $\mathbf{e}(n|n-1)$,

$$E\{\mathbf{e}(n|n-1)v(n)\} = 0$$

Thus, the error covariance matrix for $\mathbf{e}(n|n)$ is

$$\mathbf{P}(n|n) = E\{\mathbf{e}(n|n)\mathbf{e}^H(n|n)\} \tag{7.108}$$

$$= \big[\mathbf{I} - \mathbf{K}(n)\mathbf{C}(n)\big]\mathbf{P}(n|n-1)\big[\mathbf{I} - \mathbf{K}(n)\mathbf{C}(n)\big]^H + \mathbf{K}(n)\mathbf{Q}_v(n)\mathbf{K}^H(n) \tag{7.109}$$

Next, we must find the value for the Kalman gain $\mathbf{K}(n)$ that minimizes the mean-square error

$$\xi(n) = \text{tr}\{\mathbf{P}(n|n)\}$$

This may be accomplished in a couple of different ways. Although requiring some special matrix differentiation formulas, we will take the most expedient approach of differentiating $\xi(n)$ with respect to $\mathbf{K}(n)$, setting the derivative to zero, and solving for $\mathbf{K}(n)$. Using the matrix differentiation formulas

$$\frac{d}{d\mathbf{K}}\text{tr}(\mathbf{K}\mathbf{A}) = \mathbf{A}^H \tag{7.110}$$

and

$$\frac{d}{d\mathbf{K}}\text{tr}(\mathbf{K}\mathbf{A}\mathbf{K}^H) = 2\mathbf{K}\mathbf{A} \tag{7.111}$$

we have

$$\frac{d}{d\mathbf{K}}\text{tr}\big\{\mathbf{P}(n|n)\big\} = -2\Big[\mathbf{I} - \mathbf{K}(n)\mathbf{C}(n)\Big]\mathbf{P}(n|n-1)\mathbf{C}^H(n) + 2\mathbf{K}(n)\mathbf{Q}_v(n) = 0 \qquad (7.112)$$

Solving for $\mathbf{K}(n)$ gives the desired expression for the Kalman gain,

$$\boxed{\mathbf{K}(n) = \mathbf{P}(n|n-1)\mathbf{C}^H(n)\Big[\mathbf{C}(n)\mathbf{P}(n|n-1)\mathbf{C}^H(n) + \mathbf{Q}_v(n)\Big]^{-1}} \qquad (7.113)$$

Having found the Kalman gain vector, we may simplify the expression given in Eq. (7.109) for the error covariance. First, we rewrite the expression for $\mathbf{P}(n|n)$ as follows,

$$\begin{aligned}\mathbf{P}(n|n) &= \Big[\mathbf{I} - \mathbf{K}(n)\mathbf{C}(n)\Big]\mathbf{P}(n|n-1) \\ &\quad - \Big\{\big[\mathbf{I} - \mathbf{K}(n)\mathbf{C}(n)\big]\mathbf{P}(n|n-1)\mathbf{C}^H(n) + \mathbf{K}(n)\mathbf{Q}_v(n)\Big\}\mathbf{K}^H(n)\end{aligned}$$

From Eq. (7.112), however, it follows that the second term is equal to zero, which leads to the desired expression for the error covariance matrix

$$\boxed{\mathbf{P}(n|n) = \Big[\mathbf{I} - \mathbf{K}(n)\mathbf{C}(n)\Big]\mathbf{P}(n|n-1)} \qquad (7.114)$$

Thus far we have derived the Kalman filtering equations for recursively estimating the state vector $\mathbf{x}(n)$. All that needs to be done to complete the recursion is to determine how the recursion should be initialized at time $n = 0$. Since the value of the initial state is unknown, in the absence of any observed data at time $n = 0$, the initial estimate is chosen to be

$$\hat{\mathbf{x}}(0|0) = E\big\{\mathbf{x}(0)\big\}$$

and, for the initial value for the error covariance matrix, we have

$$\mathbf{P}(0|0) = E\big\{\mathbf{x}(0)\mathbf{x}^H(0)\big\}$$

This choice for the initial conditions makes $\hat{\mathbf{x}}(0|0)$ an unbiased estimate of $\mathbf{x}(0)$ and ensures that $\hat{\mathbf{x}}(n|n)$ will be unbiased for all n (recall that the Kalman filtering update equations were derived with the constraint that $\hat{\mathbf{x}}(n|n)$ be unbiased). This completes the derivation of the discrete Kalman filter which is summarized in Table 7.4. One interesting property to note about the Kalman filter is that the Kalman gain $\mathbf{K}(n)$ and the error covariance matrix $\mathbf{P}(n|n)$ do not depend on the data $\mathbf{x}(n)$. Therefore, it is possible for both of these terms to be computed off-line prior to any filtering.

Example 7.4.1 *Using a Kalman Filter to Estimate an Unknown Constant*

Let us consider the problem of estimating the value of an (unknown) constant x given measurements that are corrupted by uncorrelated, zero mean white noise $v(n)$ that has a variance σ_v^2. Since the value of x does not change with time n, then the state equation is

$$x(n) = x(n-1)$$

The measurement equation, on the other hand, is

$$y(n) = x(n) + v(n)$$

Therefore, $\mathbf{A}(n) = 1$, $\mathbf{C}(n) = 1$, $\mathbf{Q}_w(n) = 0$, and $\mathbf{Q}_v(n) = \sigma_v^2$. Since $x(n)$ is a scalar, the error covariance is also a scalar and will be denoted by

$$P(n|n) = E\big\{e^2(n|n)\big\}$$

Table 7.4 ***The Discrete Kalman Filter***

State Equation	$\mathbf{x}(n) = \mathbf{A}(n-1)\mathbf{x}(n-1) + \mathbf{w}(n)$
Observation Equation	$\mathbf{y}(n) = \mathbf{C}(n)\mathbf{x}(n) + \mathbf{v}(n)$
Initialization:	$\hat{\mathbf{x}}(0\|0) = E\{\mathbf{x}(0)\}$
	$\mathbf{P}(0\|0) = E\{\mathbf{x}(0)\mathbf{x}^H(0)\}$
Computation:	For $n = 1, 2, \ldots$ compute
	$\hat{\mathbf{x}}(n\|n-1) = \mathbf{A}(n-1)\hat{\mathbf{x}}(n-1\|n-1)$
	$\mathbf{P}(n\|n-1) = \mathbf{A}(n-1)\mathbf{P}(n-1\|n-1)\mathbf{A}^H(n-1) + \mathbf{Q}_w(n)$
	$\mathbf{K}(n) = \mathbf{P}(n\|n-1)\mathbf{C}^H(n)\Big[\mathbf{C}(n)\mathbf{P}(n\|n-1)\mathbf{C}^H(n) + \mathbf{Q}_v(n)\Big]^{-1}$
	$\hat{\mathbf{x}}(n\|n) = \hat{\mathbf{x}}(n\|n-1) + \mathbf{K}(n)\Big[\mathbf{y}(n) - \mathbf{C}(n)\hat{\mathbf{x}}(n\|n-1)\Big]$
	$\mathbf{P}(n\|n) = \Big[\mathbf{I} - \mathbf{K}(n)\mathbf{C}(n)\Big]\mathbf{P}(n\|n-1)$

where $e(n|n) = x(n) - \hat{x}(n|n)$. From Eq. (7.102) it follows that $P(n|n-1)$ and $P(n-1|n-1)$ are equal,

$$P(n|n-1) = P(n-1|n-1)$$

Therefore, in order to simplify notation we will use $P(n-1)$ to denote both $P(n-1|n-1)$ and $P(n|n-1)$. From the expression for the Kalman gain in Eq. (7.113) we have

$$K(n) = P(n-1)\Big[P(n-1) + \sigma_v^2\Big]^{-1} \tag{7.115}$$

Thus, from Eq. (7.114) it follows that the update for $P(n|n)$ is

$$\begin{aligned} P(n) &= \Big[1 - K(n)\Big]P(n-1) \\ &= \left[1 - \frac{P(n-1)}{P(n-1) + \sigma_v^2}\right]P(n-1) \\ &= \frac{P(n-1)\sigma_v^2}{P(n-1) + \sigma_v^2} \end{aligned}$$

We may solve this difference equation recursively as follows

$$\begin{aligned} P(1) &= \frac{P(0)\sigma_v^2}{P(0) + \sigma_v^2} \\ P(2) &= \frac{P(1)\sigma_v^2}{P(1) + \sigma_v^2} = \frac{P(0)\sigma_v^2}{2P(0) + \sigma_v^2} \\ P(3) &= \frac{P(2)\sigma_v^2}{P(2) + \sigma_v^2} = \frac{P(0)\sigma_v^2}{3P(0) + \sigma_v^2} \\ &\vdots \end{aligned}$$

Thus, for $P(n)$ we have, in general,

$$P(n) = \frac{P(0)\sigma_v^2}{nP(0) + \sigma_v^2}$$

Incorporating this into the expression for the Kalman gain given in Eq. (7.115) we have

$$K(n) = \frac{P(n-1)}{P(n-1)+\sigma_v^2} = \frac{P(0)}{nP(0)+\sigma_v^2}$$

Finally, for the discrete Kalman filter we have

$$\hat{x}(n) = \hat{x}(n-1) + \frac{P(0)}{nP(0)+\sigma_v^2}\Big[y(n) - \hat{x}(n-1)\Big]$$

Note that as $n \to \infty$ then $K(n) \to 0$ and $\hat{x}(n)$ approaches a steady state value.

There are some special cases of this Kalman filter that are of interest. First, note that if $\sigma_v^2 \to \infty$, which implies that the measurements are completely unreliable, then the Kalman gain goes to zero and the estimate becomes

$$\hat{x}(n) = \hat{x}(n-1)$$

Thus, the measurements are ignored and $\hat{x}(n) = \hat{x}(0)$, the initial estimate, which has an error variance $P(0)$. Second, suppose that $\hat{x}(0) = 0$ and $P(0) \longrightarrow \infty$. This corresponds to the case of no a priori information about x. In this case $K(n) = 1/n$ and the estimate becomes

$$\hat{x}(n) = \hat{x}(n-1) + \frac{1}{n}\big[y(n) - \hat{x}(n-1)\big] = \frac{n-1}{n}\hat{x}(n-1) + \frac{1}{n}y(n)$$

Note, however, that this is simply a recursive implementation of the sample mean,

$$\hat{x}(n) = \frac{1}{n}\sum_{k=1}^{n} y(k)$$

In the next example, we consider the use of a Kalman filter to solve the filtering problem considered in Example 7.3.2. As we will see, the Kalman filter is time-varying due to the fact that, unlike the Wiener filter, the filter begins operation at time $n = 0$. However, since the processes are stationary, after the initial transients have died out, the Kalman filter settles down into its steady-state behavior, which is equivalent to the causal Wiener filter solution.

Example 7.4.2 ***Using a Kalman Filter to Estimate an AR(1) Process***

Let $x(n)$ be the AR(1) process

$$x(n) = 0.8x(n-1) + w(n)$$

where $w(n)$ is white noise with a variance $\sigma_w^2 = 0.36$, and let

$$y(n) = x(n) + v(n)$$

be noisy measurements of $x(n)$ where $v(n)$ is unit variance white noise that is uncorrelated with $w(n)$. Thus, with $\mathbf{A}(n) = 0.8$ and $\mathbf{C}(n) = 1$ the Kalman filter state estimation equation is

$$\hat{x}(n) = 0.8\hat{x}(n-1) + K(n)\big[y(n) - 0.8\hat{x}(n-1)\big]$$

Since the state vector is a scalar, the equations for computing the Kalman gain are scalar equations,

$$\begin{aligned} P(n|n-1) &= (0.8)^2 P(n-1|n-1) + 0.36 \\ K(n) &= P(n|n-1)\big[P(n|n-1)+1\big]^{-1} \\ P(n|n) &= \big[1 - K(n)\big]P(n|n-1) \end{aligned}$$

With $\hat{x}(0) = E\{x(0)\} = 0$ and $P(0|0) = E\{|x(0)|^2\} = 1$, the Kalman gain and the

Table 7.5 ***The Kalman Gain and Error Covariances***

n	$P(n\|n-1)$	$K(n)$	$P(n\|n)$
1	1.0000	0.5000	0.5000
2	0.6800	0.4048	0.4048
3	0.6190	0.3824	0.3824
4	0.6047	0.3768	0.3768
5	0.6012	0.3755	0.3755
6	0.6003	0.3751	0.3751
⋮	⋮	⋮	⋮
∞	0.6000	0.3750	0.3750

error covariances for the first few values of n are shown in Table 7.5. Note that after a few iterations the Kalman filter settles down into its steady state solution

$$\hat{d}(n) = 0.8\hat{d}(n-1) + 0.375\Big[x(n) - 0.8\hat{d}(n-1)\Big]$$

with a final mean-square error of $\xi = 0.375$ which, as we see from our discussion following Example 7.3.2 (p. 364), is identical to the causal Wiener filter.

The goal of the discrete Kalman filter is to use the measurements, $\mathbf{y}(n)$, to estimate the state $\mathbf{x}(n)$ of a dynamic system. The Kalman filter is a remarkably versatile and powerful recursive estimation algorithm that has found applications in a wide variety of different areas including spacecraft orbit determination, radar tracking, estimation and prediction of target trajectories, adaptive equalization of telephone channels, adaptive equalization of fading dispersive channels, and adaptive antenna arrays. In this section our intent was only to provide a brief introduction to the problem of recursive estimation. Since a detailed study on the use, application, and numerical properties of the Kalman filter would fill an entire textbook, we have only touched upon the problem. A more detailed treatment of the discrete Kalman filter may be found in many excellent references such as [2,3,5,6]. In addition, some signal processing applications of Kalman filtering may be found in [4,5,8], and a historical perspective of Kalman filtering is given in [9] which traces its roots back to the invention of least squares theory by Gauss in 1809. Finally, an interesting collection of papers on the theory and application of Kalman filtering may be found in [10].

7.5 SUMMARY

In this chapter, we considered the problem of designing the optimum filter for estimating a process $d(n)$ in terms of measurements of a related process $x(n)$. The first problem that we considered was the design of the optimum FIR filter that minimizes the mean-square estimation error $\xi = E\big\{|d(n) - \hat{d}(n)|^2\big\}$. Assuming that $d(n)$ and $x(n)$ are jointly wide-sense stationary processes with known autocorrelation $r_x(k)$ and cross-correlation $r_{dx}(k)$, the solution is given by the Wiener-Hopf equations, which are a set of linear Toeplitz equations. These equations are a generalization of the linear Toeplitz equations derived in Chapter 4 for all-pole signal modeling and, as we saw in Chapter 5, may be solved using the general Levinson recursion. Since the Wiener-Hopf equations apply to the estimation of *any* process $d(n)$, we then considered the special cases of filtering, linear prediction, and noise cancellation. It was then shown how the FIR Wiener filter could be implemented in lattice filter form.

Next, we considered the design of an IIR Wiener filter. Without imposing a causality constraint on the solution, what we discovered was that these filters are generally noncausal and, therefore, unrealizable. As a result, these filters would not generally be appropriate in real-time signal processing applications. Nevertheless, assuming that the power spectrum of $x(n)$ and the cross-power spectral density between $x(n)$ and $d(n)$ are known, these filters are easily designed and may be used to set an upper bound on the performance of an FIR Wiener filter or a causal IIR Wiener filter.

After looking at the use of a noncausal Wiener filter for smoothing, we then considered the design of a causal IIR Wiener filter. What we found was that, by imposing a causality constraint on the filter, it becomes necessary to perform a spectral factorization of the power spectrum of the input process $x(n)$. As a result, compared with the design of an FIR Wiener filter or a noncausal Wiener filter, these filters are generally much more difficult to design. We then looked at specific examples of designing causal Wiener filters, including filtering, linear prediction, and deconvolution.

Finally, we briefly considered the problem of recursive filtering and derived the discrete Kalman filter. Unlike the Wiener filter, the Kalman filter may by used for nonstationary processes as well as stationary ones, and may be initialized to begin operating at time $n = 0$. The next step is to relax the requirement that the statistics of $x(n)$ and $d(n)$ be known. This is the subject of Chapter 9.

References

1. H. C. Andrews and B. R. Hunt, *Digital Image Restoration*, Prentice-Hall, Englewood Cliffs, NJ, 1977.
2. R. G. Brown, *Introduction to Random Signal Analysis and Kalman Filtering*, John Wiley & Sons, New York, 1983.
3. A. Gelb, Ed., *Applied Optimal Estimation*, M.I.T. Press, Cambridge, MA, 1974.
4. D. Godard, "Channel equalization using a Kalman filter for fast data transmission," *IBS J. Res. Dev.*, vol. 18, pp. 267–273, 1974.
5. S. Haykin, *Adaptive Filter Theory*, Prentice-Hall, Englewood Cliffs, NJ, 1986.
6. T. Kailath, "An innovations approach to least-squares estimation – Part I: Linear filtering in additive noise," *IEEE Trans. Autom. Control*, vol. AC-13, pp. 641–655, 1968.
7. R. E. Kalman, "A new approach to linear filtering and prediction problems", *Trans. ASME, J. Basic Eng.*, Ser. 82D, pp. 35–45, March 1960.
8. R. A. Monzingo and T. W. Miller, *Introduction to Adaptive Arrays*, Wiley-Interscience, New York, 1980.
9. H. W. Sorenson, "Least-squares estimation: From Gauss to Kalman," *IEEE Spectrum*, vol. 7, pp. 63–68, July 1970.
10. H. W. Sorenson, ed., *Kalman Filtering: Theory and Application*, IEEE Press, New York, 1985.
11. N. Wiener, *Extrapolation, Interpolation, and Smoothing of Stationary Time Series with Engineering Applications*, MIT Press, Cambridge, MA, 1949.

7.6 PROBLEMS

7.1. A random process $x(n)$ is generated as follows

$$x(n) = \alpha x(n-1) + v(n) + \beta v(n-1)$$

where $v(n)$ is white noise with mean m_v and variance σ_v^2.

(a) Design a first-order linear predictor

$$\hat{x}(n+1) = w(0)x(n) + w(1)x(n-1)$$

that minimizes the mean-square error in the prediction of $x(n+1)$, and find the minimum mean-square error.

(b) Now consider a predictor of the form

$$\hat{x}(n+1) = c + w(0)x(n) + w(1)x(n-1)$$

Find the values for c, $w(0)$, and $w(1)$ that minimize the mean-square error, and compare the mean-square error of this predictor with that found in part (a).

7.2. In this problem we consider the design of a three-step predictor using a first-order filter

$$W(z) = w(0) + w(1)z^{-1}$$

In other words, with $x(n)$ the input to the predictor $W(z)$, the output

$$\hat{x}(n+3) = w(0)x(n) + w(1)x(n-1)$$

is the minimum mean-square estimate of $x(n+3)$.

(a) What are the Wiener-Hopf equations for the Wiener three-step predictor?

(b) If the values of $r_x(k)$ for lags $k = 0$ to $k = 4$ are

$$\mathbf{r}_x = \begin{bmatrix} 1.0, & 0, & 0.1, & -0.2, & -0.9 \end{bmatrix}^T$$

solve the Wiener-Hopf equations and find the optimum three-step predictor.

(c) Does the prediction error filter

$$F(z) = 1 + w(0)z^{-3} + w(1)z^{-4}$$

have minimum phase, i.e., are the zeros of $F(z)$ inside the unit circle? How does this compare to what you know about the prediction error filter for a one-step predictor?

7.3. Repeat Example 7.2.1 using a second-order Wiener filter, and compare the mean-square error for the second-order filter with the mean-square error of the first-order filter.

7.4. Consider the system shown in the figure below for estimating a process $d(n)$ from $x(n)$.

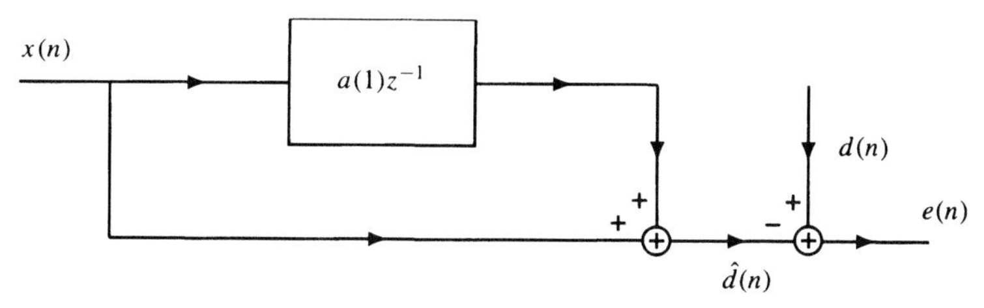

If $\sigma_d^2 = 4$ and

$$\mathbf{r}_x = \begin{bmatrix} 1.0, & 0.5, & 0.25 \end{bmatrix}^T \quad ; \quad \mathbf{r}_{dx} = \begin{bmatrix} -1.0, & 1.0 \end{bmatrix}^T$$

find the value of $a(1)$ that minimizes the mean-square error $\xi = E\{|e(n)|^2\}$, and find the minimum mean-square error.

7.5. In this problem we consider linear prediction in a noisy environment. Suppose that a signal $d(n)$ is corrupted by noise,

$$x(n) = d(n) + w(n)$$

where $r_w(k) = 0.5\delta(k)$ and $r_{dw}(k) = 0$. The signal $d(n)$ is an AR(1) process that satisfies the difference equation

$$d(n) = 0.5d(n-1) + v(n)$$

where $v(n)$ is white noise with variance $\sigma_v^2 = 1$. Assume that $w(n)$ and $v(n)$ are uncorrelated.

(a) Design a first-order FIR linear predictor $W(z) = w(0) + w(1)z^{-1}$ for $d(n)$ and find the mean-square prediction error $\xi = E\{[d(n+1) - \hat{d}(n+1)]^2\}$.

(b) Design a causal Wiener predictor and compare the mean-square prediction error with that found in part (a).

7.6. Suppose that a process $x(n)$ has been recorded, but there is a missing gap of data over the interval $[N_1, N_2]$, i.e., $x(n)$ is unknown over this interval.

(a) Derive the optimum estimate of $x(N_1)$ using the data in the semi-infinite interval $(-\infty, N_1 - 1]$.

(b) Derive the optimum estimate of $x(N_1)$ using the data in the semi-infinite interval $[N_2 + 1, \infty)$.

(c) Derive the optimum estimate of $x(N_1)$ that is formed by combining together the two estimates found in parts (a) and (b).

(d) Generalize your result in part (c) to find the optimum estimate of $x(n)$ at an arbitrary point n in the interval $[N_1, N_2]$.

7.7. In this problem we consider the design of a causal IIR Wiener filter for p-step prediction,

$$\hat{x}(n+p) = \sum_{k=0}^{\infty} h(k)x(n-k)$$

(a) If $x(n)$ is a real-valued random process with power spectral density

$$P_x(z) = \sigma_x^2 Q(z)Q(z^{-1})$$

find the system function of the causal Wiener filter that minimizes the mean-square error

$$\xi = E\big\{|x(n+p) - \hat{x}(n+p)|^2\big\}$$

(b) If $x(n)$ is an AR(1) process with power spectrum

$$P_x(z) = \frac{1-a^2}{(1-az^{-1})(1-az)}$$

find the causal p-step linear predictor and evaluate the mean-square error.

(c) If $x(n)$ is an MA(2) process that is generated by the difference equation

$$x(n) = 4v(n) - 2v(n-1) + v(n-2)$$

where $v(n)$ is zero mean unit variance white noise, find the system function of the two-step ($p = 2$) predictor and evaluate the mean-square error.

(d) Repeat part (c) for a three-step predictor.

7.8. Suppose that we would like to estimate a process $d(n)$ from the noisy observations

$$x(n) = d(n) + v(n)$$

where the noise, $v(n)$, is uncorrelated with $d(n)$. The power spectral densities of $d(n)$ and $v(n)$ are shown in the following figure.

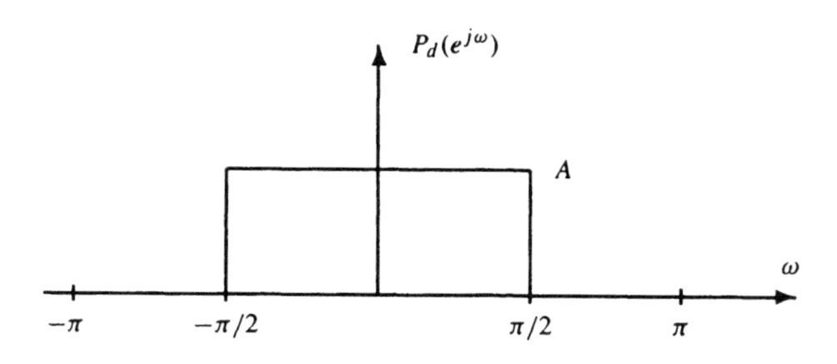

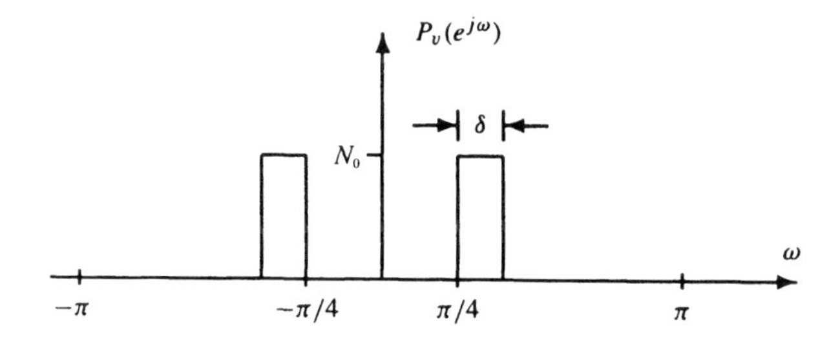

(a) Design a *noncausal* Wiener smoothing filter for estimating $d(n)$ from $x(n)$,

$$\hat{d}(n) = \sum_{k=-\infty}^{\infty} h(k)x(n-k)$$

(b) Compute the mean-square error $E\{|d(n) - \hat{d}(n)|^2\}$ and compare it to the mean-square error that results when $h(n) = \delta(n)$, i.e., with no filtering of $x(n)$.

(c) Design a second-order FIR Wiener filter

$$W(z) = w(0) + w(1)z^{-1} + w(2)z^{-2}$$

for estimating $d(n)$ from $x(n)$, and compare the mean-square error in your estimate to that found in part (b).

7.9. We would like to estimate a process $d(n)$ from noisy observations,

$$x(n) = d(n) + v(n)$$

where $v(n)$ is white noise with variance $\sigma_v^2 = 1$ and $d(n)$ is a wide-sense stationary random process with the first four values of the autocorrelation sequence given by

$$\mathbf{r}_d = \left[1.5,\ 0,\ 1.0,\ 0\right]^T$$

Assume that $d(n)$ and $v(n)$ are uncorrelated. Our goal is to design an FIR filter to reduce the noise in $d(n)$. Hardware constraints, however, limit the filter to only three nonzero coefficients in $W(z)$.

(a) Derive the optimal three-multiplier causal filter

$$W(z) = w(0) + w(1)z^{-1} + w(2)z^{-2}$$

for estimating $d(n)$, and evaluate the mean-square error $E\{|d(n) - \hat{d}(n)|^2\}$.

(b) Repeat part (a) for the noncausal FIR filter

$$W(z) = w(-1)z + w(0) + w(1)z^{-1}$$

(c) Can you suggest a way to reduce the mean-square error below that obtained for the filters designed in parts (a) and (b) without using any more than three filter coefficients?

7.10. Suppose that a signal $x(n)$ is recorded and that, due to measurement errors, there are *outliers* in the data, i.e., for some values of n there is a large error in the measured value of $x(n)$. Instead of eliminating these data values, suppose that we perform a *minimum mean-square interpolation* as follows. Given a "bad" data value at time $n = n_0$, consider an estimate for $x(n_0)$ of the form

$$\hat{x}(n_0) = ax(n_0 - 1) + bx(n_0 + 1)$$

(a) Assuming that $x(n)$ is a wide-sense stationary random process with autocorrelation sequence $r_x(k)$, find the values for a and b that minimize the mean-square error

$$\xi = E\{|x(n_0) - \hat{x}(n_0)|^2\}$$

(b) If $r_x(k) = (0.5)^{|k|}$, evaluate the mean-square error for the interpolator found in part (a).

(c) Discuss when it may be better to use an estimator of the form

$$\hat{x}(n_0) = ax(n_0 - 1) + bx(n_0 - 2)$$

or explain why such an estimator should not be used.

(d) Given an autocorrelation sequence $r_x(k)$, derive the Wiener-Hopf equations that define the optimum filter for interpolating $x(n)$ to produce the best estimate of $x(n_0)$ in terms of the $2p$ data values

$$x(n_0 - 1), x(n_0 - 2), \ldots, x(n_0 - p) \quad \text{and} \quad x(n_0 + 1), x(n_0 + 2), \ldots, x(n_0 + p)$$

(e) Find an expression for the minimum mean-square error for your estimate in part (d).

7.11. In this problem we consider the design of an optimum smoothing filter for estimating a process $d(n)$ from the measurements

$$x(n) = d(n) + v(n)$$

Our goal is to use a noncausal FIR filter that has a system function of the form:

$$W(z) = \sum_{k=-p}^{p} w(k)z^{-k}$$

In other words, we want to produce an estimate of $d(n)$ as follows

$$\hat{d}(n) = \sum_{k=-p}^{p} w(k)x(n-k)$$

(a) Derive the Wiener-Hopf equations that define the set of coefficients that minimize the mean-square error

$$\xi = E\{|d(n) - \hat{d}(n)|^2\}$$

(b) How would the Wiener-Hopf equations derived in part (a) change if we used a causal filter with the same number of coefficients? In other words, if the system function were of the form

$$W(z) = \sum_{k=0}^{2p} w(k) z^{-k}$$

how would you modify your equations in (a)?

(c) State qualitatively when you might prefer the noncausal filter over the causal filter and vice versa. For example, for what types of signals and for what types of noise would you expect a causal filter to be superior to the noncausal filter?

(d) FIR digital filters with linear phase (or zero phase) are important in signal processing applications where frequency dispersion due to nonlinear phase is harmful. An FIR filter with zero phase is characterized by the property that

$$w(n) = w(-n)$$

Thus, the system function may be written as

$$W(z) = w(0) + \sum_{k=1}^{p} w(k)[z^{-k} + z^{k}]$$

Derive the Wiener-Hopf equations that define the optimum zero phase smoothing filter.

(e) With $r_d(k) = 4(0.5)^{|k|}$ and $r_v(k) = \delta(k)$, find the optimum values for the filter coefficients $w(0)$ and $w(1)$ in the zero phase filter

$$W(z) = w(0) + w(1)\left[z + z^{-1}\right]$$

7.12. We observe a signal, $x(n)$, in a noisy and reverberant environment,

$$y(n) = x(n) + 0.8x(n-1) + v(n)$$

where $v(n)$ is white noise with variance $\sigma_w^2 = 1$ that is uncorrelated with $x(n)$. We know that $x(n)$ is a wide-sense stationary AR(1) random process with autocorrelation values

$$\mathbf{r}_x = \left[4,\ 2,\ 1,\ 0.5\right]^T$$

(a) Find the noncausal IIR Wiener filter, $H(z)$, that produces the minimum mean-square estimate of $x(n)$.

(b) Design a causal IIR Wiener filter, $H(z)$, that produces the minimum mean-square estimate of $x(n)$.

7.13. A wide-sense stationary random process has an autocorrelation sequence of the form

$$r_x(k) = \sigma_x^2 \alpha^{|k|}$$

where $|\alpha| < 1$. Over a given time interval, $[n_A, n_B]$, the process $x(n)$ is only known at the end points, i.e., the only given data is $x(n_A)$ and $x(n_B)$. Based on these two observations, determine the optimum estimate

$$\hat{x}(n) = a(n)x(n_A) + b(n)x(n_B)$$

of $x(n)$ over each of the following intervals

(a) $n > n_B$.

(b) $n < n_A$.

(c) $n_A < n < n_B$.

7.14. As shown in Figure 7.12, the Wiener filter may be viewed as a cascade of a whitening filter with a causal filter that produces the minimum mean-square estimate of $d(n)$ from $\epsilon(n)$. For real processes, the system function of the cascade is

$$H(z) = F(z)G(z) = \frac{1}{\sigma_0^2 Q(z)} \left[\frac{P_{dx}(z)}{Q(z^{-1})} \right]_+$$

and the mean-square error is

$$\xi_{\min} = r_d(0) - \sum_{l=0}^{\infty} h(l) r_{dx}(l)$$

(a) If $r_{d\epsilon}(k) = \delta(k)$ and

$$P_x z = \frac{4}{(1 - 0.5z^{-1})(1 - 0.5z)}$$

find the unit sample response, $h(n)$, of the causal Wiener filter.

(b) Derive an expression for the mean-square error that expresses $\xi_{\min}$ in terms of the cross-correlation, $r_{d\epsilon}(k)$, and evaluate the mean-square error when

$$r_{d\epsilon}(k) = (\tfrac{1}{2})^k u(k) + (\tfrac{1}{3})^{-k} u(-k-1)$$

and

$$E\{d^2(n)\} = 4$$

7.15. Let $x(n)$ be an AR(1) process of the following form

$$x(n) = a(1)x(n-1) + b(0)w(n)$$

where $w(n)$ is unit variance white noise, and let $y(n)$ be noisy measurements

$$y(n) = x(n) + v(n)$$

where $v(n)$ is unit variance white noise that is uncorrelated with $w(n)$. We have seen that the causal Wiener filter for estimating $x(n)$ from $y(n)$ has the form

$$\hat{x}(n) = a(1)\hat{x}(n-1) + K\Big[y(n) - a(1)\hat{x}(n-1)\Big]$$

Find the value of K in terms of $a(1)$ and $b(0)$ that minimizes the mean-square error

$$E\Big\{[x(n) - \hat{x}(n)]^2\Big\}$$

7.16. In the derivation of the Kalman filtering equations, we made use of the following matrix differentiation formulas

$$\frac{d}{d\mathbf{K}} \mathrm{tr}\big(\mathbf{KA}\big) = \mathbf{A}^H$$

and

$$\frac{d}{d\mathbf{K}} \mathrm{tr}\big(\mathbf{KAK}^H\big) = 2\mathbf{KA}$$

(a) Show that these matrix differentiation formulas are valid.

(b) Derive an expression for the derivative

$$\frac{d}{d\mathbf{K}}\operatorname{tr}(\mathbf{AK}^H)$$

(c) Use these matrix differentiation formulas to derive the expression for the Kalman gain given in Eq. (7.113).

7.17. Consider a system consisting of two sensors, each making a single measurement of an unknown constant x. Each measurement is noisy and may be modeled as follows

$$\begin{aligned} y(1) &= x + v(1) \\ y(2) &= x + v(2) \end{aligned}$$

where $v(1)$ and $v(2)$ are zero mean uncorrelated random variables with variance σ_1^2 and σ_2^2, respectively.

(a) In the absence of any other information, we seek the best linear estimate of x of the form

$$\hat{x} = k_1 y(1) + k_2 y(2)$$

Find the values for k_1 and k_2 that produce an unbiased estimate of x that minimizes the mean-square error, $E\{[x - \hat{x}]^2\}$.

(b) Repeat part (a) for the case where the measurement errors are correlated,

$$E\{v(1)v(2)\} = \rho\sigma_1\sigma_2$$

where ρ is the correlation coefficient.

(c) Repeat part (a) within the framework of Kalman filtering, treating the measurements $y(1)$ and $y(2)$ sequentially.

7.18. An autoregressive process of order 1 is described by the difference equation

$$x(n) = 0.5x(n-1) + w(n)$$

where $w(n)$ is zero-mean white noise with a variance $\sigma_w^2 = 0.64$. The observed process $y(n)$ is given by

$$y(n) = x(n) + v(n)$$

where $v(n)$ is zero-mean white noise with a variance $\sigma_v^2 = 1$.

(a) Write the Kalman filtering equations to find the minimum mean-square estimate, $\hat{x}(n|n)$, of $x(n)$ given the observations $y(i)$, $i = 1, \ldots, n$. The initial conditions are $\hat{x}(0|0) = 0$ and $E\{\epsilon^2(0|0)\} = 1$ where $\epsilon(0|0) = x(0) - \hat{x}(0|0)$.

(b) Assuming that the filter reaches a steady state solution, find the steady state Kalman gain and the limiting form of the estimation equation for $\hat{x}(n|n)$.

7.19. In many cases, the error covariance matrix $\mathbf{P}(n|n-1)$ will converge to a steady-state value $\mathbf{P}$ as $n \to \infty$. Assume that $\mathbf{C}$, $\mathbf{Q}_w$, and $\mathbf{Q}_v$ are the limiting values of $\mathbf{C}(n)$, $\mathbf{Q}_w(n)$, and $\mathbf{Q}_v(n)$, respectively.

(a) For $\mathbf{A}(n) = \mathbf{I}$, show that if $\mathbf{P}(n|n-1)$ converges to a steady state value $\mathbf{P}$, then the limiting value satisfies the *algebraic Ricatti equation*

$$\mathbf{P}\mathbf{C}^H\left(\mathbf{C}\mathbf{P}\mathbf{C}^H + \mathbf{Q}_v\right)^{-1}\mathbf{C}\mathbf{P} - \mathbf{Q}_w = \mathbf{0}$$

(b) Derive the Ricatti equation for a general state transition matrix $\mathbf{A}(n)$ that has a limiting value of $\mathbf{A}$.

7.20. In Example 7.4.1 we derived the Kalman filter for estimating an unknown constant from noisy measurements. The estimate at time n was shown to be

$$\hat{x}(n) = \hat{x}(n-1) + \frac{P(0)}{nP(0) + \sigma_v^2}\left[y(n) - \hat{x}(n-1)\right]$$

(a) Solve this difference equation and find a closed-form expression for $\hat{x}(n)$ in terms of $\hat{x}(0)$ and the measurements $y(0), y(1), \ldots, y(n)$.

(b) What does $\hat{x}(n)$ converge to as $n \to \infty$?

7.21. In this problem we will derive the following expression for the Kalman gain,

$$\mathbf{K}(n) = \mathbf{P}(n|n)\mathbf{C}^H(n)\mathbf{Q}_v^{-1}(n) \tag{P7.21-1}$$

(a) By substituting Eq. (7.113) for the Kalman gain into Eq. (7.114), show that the error covariance matrix $\mathbf{P}(n|n)$ may be written as

$$\begin{aligned}\mathbf{P}(n|n) = \mathbf{P}(n|n-1) - &\mathbf{P}(n|n-1)\mathbf{C}^H(n) \\ &\times\left[\mathbf{C}(n)\mathbf{P}(n|n-1)\mathbf{C}^H(n) + \mathbf{Q}_v(n)\right]^{-1}\mathbf{C}(n)\mathbf{P}(n|n-1)\end{aligned}$$

(b) Using the matrix inversion lemma given in Eq. (2.28) on p. 29, show that the inverse covariance matrix may be written as

$$\mathbf{P}^{-1}(n|n) = \mathbf{P}^{-1}(n|n-1) + \mathbf{C}^H(n)\mathbf{Q}_v^{-1}(n)\mathbf{C}(n)$$

(c) By multiplying the expression for the Kalman gain given in Eq. (7.113) on the left by $\mathbf{P}(n|n)\mathbf{P}^{-1}(n|n)$, use your results in part (b) to derive the expression for $\mathbf{K}(n)$ given in Eq. (P7.21-1).

7.22. Consider the ARMA(1,1) process $y(n)$ given by[5]

$$y(n) + ay(n-1) = v(n) + bv(n-1)$$

where $v(n)$ is a zero-mean white-noise process with a variance σ_v^2.

(a) Show that the state-space representation for this process may be written as

$$\mathbf{x}(n) = \begin{bmatrix} -a & 1 \\ 0 & 0 \end{bmatrix}\mathbf{x}(n-1) + \begin{bmatrix} 1 \\ b \end{bmatrix} w(n)$$

$$y(n) = \begin{bmatrix} 1 & 0 \end{bmatrix}\mathbf{x}(n)$$

where $\mathbf{x}(n)$ is a two-dimensional state vector.

(b) Assuming that the error covariance $\mathbf{P}(n|n)$ converges to a steady state value of $\mathbf{P}$ and is a solution of the Ricatti equation given in Prob. 7.19, show that

$$\mathbf{P} = \sigma_v^2\begin{bmatrix} 1+c & b \\ b & b^2 \end{bmatrix}$$

[5]This problem is adapted from Haykin [5]

where c is a scalar that satisfies the second-order equation

$$c = (b-a)^2 + a^2c - \frac{(b-a-ac)^2}{1+c}$$

and find the two values of c that satisfy this equation. For each of these values, find the corresponding values for $\mathbf{P}$.

(c) Find the steady-state Kalman gain, and determine the values for $\mathbf{K}$ that correspond to the solutions for c found in part (b).

Computer Exercises

C7.1. Let $d(n)$ be an AR(1) process with an autocorrelation

$$r_d(k) = \alpha^{|k|}$$

with $0 < \alpha < 1$. Suppose that $d(n)$ is observed in the presence of zero-mean uncorrelated white noise $v(n)$ with a variance σ_v^2,

$$x(n) = d(n) + v(n)$$

In Example 7.2.1 we considered the design of a first-order FIR Wiener filter for estimating $d(n)$ from $x(n)$ and in Example 7.3.2 we looked at the design of a causal Wiener filter.

(a) Write a MATLAB program to design a pth-order FIR Wiener filter to estimate $d(n)$ and evaluate the mean-square error.

(b) With $\alpha = 0.8$, evaluate the error for filters of order $p = 1, 2, \ldots, 20$ and compare these errors to the causal Wiener filter. Explain your results.

(c) With $p = 10$, plot the mean-square error versus α for $\alpha = 0.1, 0.2, \ldots, 0.9$. Explain your results.

C7.2. In this exercise we look at the noise cancellation problem considered in Example 7.2.6. Let

$$x(n) = d(n) + g(n)$$

where $d(n)$ is the harmonic process

$$d(n) = \sin(n\omega_0 + \phi)$$

with $\omega_0 = 0.05\pi$ and ϕ is a random variable that is uniformly distributed between $-\pi$ and π. Assume that $g(n)$ is unit variance white noise. Suppose that a noise process $v_2(n)$ that is correlated with $g(n)$ is measured by a secondary sensor. The noise $v_2(n)$ is related to $g(n)$ by a filtering operation,

$$v_2(n) = 0.8v_2(n) + g(n)$$

(a) Using MATLAB, generate 500 samples of the processes $x(n)$ and $v_2(n)$.

(b) Derive the Wiener-Hopf equations that define the optimum pth-order FIR filter for estimating $g(n)$ from $v_2(n)$.

(c) Using filters of order $p = 2, 4$, and 6, design and implement the Wiener noise cancellation filters. Make plots of the estimated process $\hat{g}(n)$ and compare the average squared errors for each filter.

(d) In some situations, the desired signal may leak into the secondary sensor. In this case, the performance of the Wiener filter may be severely compromised. To see what effect this has, suppose the input to the Wiener filter is

$$v_0(n) = v_2(n) + \alpha d(n)$$

where $v_2(n)$ is the filtered noise defined above. Evaluate the performance of the Wiener noise canceller for several different values of α for filter orders of $p = 2$, 4, and 6. Comment on your observations.

C7.3. The state estimation equation in the discrete Kalman filter is

$$\hat{\mathbf{x}}(n|n) = \mathbf{A}(n-1)\hat{\mathbf{x}}(n-1|n-1) + \mathbf{K}(n)\Big[\mathbf{y}(n) - \mathbf{C}(n)\mathbf{A}(n-1)\hat{\mathbf{x}}(n-1|n-1)\Big]$$

Thus, given the state transition matrix $\mathbf{A}(n)$ and the observation matrix $\mathbf{C}(n)$, all that is required is the Kalman gain $\mathbf{K}(n)$. Since the Kalman gain does not depend upon the state $\mathbf{x}(n)$ or the observations $\mathbf{y}(n)$, the Kalman gain may be computed off-line prior to filtering.

(a) Write a MATLAB program `gain.m` to compute the Kalman gain $\mathbf{K}(n)$ for a stationary process with

$$\begin{aligned}\mathbf{x}(n) &= \mathbf{A}\mathbf{x}(n-1) + \mathbf{w}(n) \\ \mathbf{y}(n) &= \mathbf{C}\mathbf{x}(n) + \mathbf{v}(n)\end{aligned}$$

(b) Suppose that $x(n)$ is a third-order autoregressive process

$$x(n) = -0.1x(n-1) - 0.09x(n-2) + 0.648x(n-3) + w(n)$$

where $w(n)$ is unit variance white noise, and that the observations are

$$y(n) = x(n) + v(n)$$

where $v(n)$ is white noise with a variance $\sigma_v^2 = 0.64$. What initialization should you use for $P(0|0)$? Using this initialization, find the Kalman gain $\mathbf{K}(n)$ for $n = 0$ to $n = 10$.

(c) What is the steady-state value for the Kalman gain? How is it affected by the initialization $P(0|0)$?

(d) Generate the processes $x(n)$ and $y(n)$ in part (b) and use your Kalman filter to estimate $x(n)$ from $y(n)$. Plot your estimate and compare it to $x(n)$.

(e) Repeat parts (b) and (d) for the process

$$x(n) = -0.95x(n-1) - 0.9025x(n-2) + w(n) - w(n-1)$$

where $w(n)$ is unit variance white noise, and the observations are

$$y(n) = x(n) + v(n)$$

where $v(n)$ is white noise with a variance $\sigma_v^2 = 0.8$.

SPECTRUM ESTIMATION

8

8.1 INTRODUCTION

In this chapter we consider the problem of estimating the power spectral density of a wide-sense stationary random process. As discussed in Chapter 3, the power spectrum is the Fourier transform of the autocorrelation sequence. Therefore, estimating the power spectrum is equivalent to estimating the autocorrelation. For an autocorrelation ergodic process, recall that

$$\lim_{N\to\infty}\left\{\frac{1}{2N+1}\sum_{n=-N}^{N} x(n+k)x^*(n)\right\} = r_x(k) \tag{8.1}$$

Thus, if $x(n)$ is known for all n, estimating the power spectrum is straightforward, in theory, since all that must be done is to determine the autocorrelation sequence $r_x(k)$ using Eq. (8.1), and then compute its Fourier transform. However, there are two difficulties with this approach that make spectrum estimation both an interesting and a challenging problem. First, the amount of data that one has to work with is never unlimited and, in many cases, it may be very small. Such a limitation may be an inherent characteristic of the data collection process as is the case, for example, in the analysis of seismic data from an earthquake in which the signal persists for only a short period of time. It is also possible, however, that a limited data set is imposed by the requirement that the spectral characteristics of the process remain constant over the duration of the data record. In speech, for example, a stationarity requirement will restrict the length of time over which the signal may be assumed to be approximately stationary to a few milliseconds or less. The second difficulty is that the data is often corrupted by noise or contaminated with an interfering signal. Thus, spectrum estimation is a problem that involves estimating $P_x(e^{j\omega})$ from a finite number of noisy measurements of $x(n)$. In some applications, however, estimating the power spectrum may be facilitated by having prior knowledge about how the process is generated. It may be known, for example, that $x(n)$ is an autoregressive process or that it consists of one or more sinusoids in noise. This type of information may then allow one to parametrically estimate the power spectrum or, perhaps, to extrapolate the data or its autocorrelation in order to improve the performance of a spectrum estimation algorithm.

Spectrum estimation is a problem that is important in a variety of different fields and applications. It was shown in Chapter 7, for example, that the frequency response of a noncausal Wiener smoothing filter is

$$H(e^{j\omega}) = \frac{P_d(e^{j\omega})}{P_d(e^{j\omega}) + P_v(e^{j\omega})}$$

where $P_d(e^{j\omega})$ is the power spectrum of $d(n)$, the desired output of the Wiener filter, and $P_v(e^{j\omega})$ is the power spectrum of the noise, $v(n)$. Therefore, before a Wiener smoothing filter can be designed and implemented, the power spectrum of both $d(n)$ and $v(n)$ must be determined. Since these power spectral densities are not generally known a priori, one is faced with the problem of estimating them from measurements. Another application in which spectrum estimation plays an important role is signal detection and tracking. Suppose, for example, that a sonar array is placed on the ocean floor to listen for the narrow-band acoustic signals that are generated by the rotating machinery or propellers of a ship. Once a narrow-band signal is detected, the problem of interest is to estimate its center frequency in order to determine the ships direction or velocity. Since these narrow-band signals are typically recorded in a very noisy environment, signal detection and frequency estimation are nontrivial problems that require robust, high-resolution spectrum estimation techniques. Other applications of spectrum estimation include harmonic analysis and prediction, time series extrapolation and interpolation, spectral smoothing, bandwidth compression, beam-forming and direction finding [25,34].

The approaches for spectrum estimation may be generally categorized into one of two classes. The first includes the *classical* or *nonparametric* methods that begin by estimating the autocorrelation sequence $r_x(k)$ from a given set of data. The power spectrum is then estimated by Fourier transforming the estimated autocorrelation sequence. The second class includes the nonclassical or parametric approaches, which are based on using a model for the process in order to estimate the power spectrum. For example, if it is known that $x(n)$ is a pth-order autoregressive process, then measured values of $x(n)$ may be used to estimate the parameters of the all-pole model, $a_p(k)$, and these estimated model parameters, $\hat{a}_p(k)$, may then, in turn, be used to estimate the power spectrum as follows:

$$\hat{P}_x(e^{j\omega}) = \frac{1}{\left|\sum_{k=0}^{p} \hat{a}_p(k) e^{-jk\omega}\right|^2}$$

We begin this chapter with the classical or nonparametric spectrum estimation techniques. These methods, which are described in Section 8.2, include the periodogram, the modified periodogram, Bartlett's method, Welch's method, and the Blackman-Tukey method. The minimum variance method is then considered in Section 8.3. This technique involves the design of a narrow-band filter bank to generate a set of narrow-band random processes. The power spectrum at the center frequency of each bandpass filter is then estimated by measuring the power in the narrow-band process and dividing by the filter bandwidth. Next, in Section 8.4, the maximum entropy method (MEM) is presented, and it is shown that MEM is equivalent to spectrum estimation using an all-pole model. Then, in Section 8.5, we look at power spectrum estimation techniques that are based on a parametric model for the data. These models include moving average (MA), autoregressive (AR), and autoregressive moving average (ARMA). Next, in Section 8.6, we consider frequency estimation algorithms for harmonic processes that consist of a sum of sinusoids or complex exponentials in noise. These methods, sometimes referred to as noise subspace methods, include the Pisarenko harmonic decomposition, MUSIC, the eigenvector method, and the minimum norm algorithm. Finally, in Section 8.7, we look at principal components frequency estimation. This approach, which also assumes that the process is harmonic, forms a low-rank approximation to the autocorrelation matrix, which is then incorporated into a spectrum estimation algorithm such as the minimum variance method or MEM.

8.2 NONPARAMETRIC METHODS

In this section, we consider nonparametric techniques of spectrum estimation. These methods are based on the idea of estimating the autocorrelation sequence of a random process from a set of measured data, and then taking the Fourier transform to obtain an estimate of the power spectrum. We begin with the *periodogram*, a nonparametric method first introduced by Schuster in 1898 in his study of periodicities in sunspot numbers [47,49]. As we will see, although the periodogram is easy to compute, it is limited in its ability to produce an accurate estimate of the power spectrum, particularly for short data records. We will then examine a number of modifications to the periodogram that have been proposed to improve its statistical properties. These include the modified periodogram, Bartlett's method, Welch's method, and the Blackman-Tukey method.

8.2.1 The Periodogram

The power spectrum of a wide-sense stationary random process is the Fourier transform of the autocorrelation sequence,

$$P_x(e^{j\omega}) = \sum_{k=-\infty}^{\infty} r_x(k)e^{-jk\omega}$$

Therefore, spectrum estimation is, in some sense, an autocorrelation estimation problem. For an *autocorrelation ergodic* process and an unlimited amount of data, the autocorrelation sequence may, in theory, be determined using the time-average

$$r_x(k) = \lim_{N\to\infty} \frac{1}{2N+1} \sum_{n=-N}^{N} x(n+k)x^*(n) \tag{8.2}$$

However, if $x(n)$ is only measured over a finite interval, say $n = 0, 1, \ldots, N-1$, then the autocorrelation sequence must be estimated using, for example, Eq. (8.2) with a finite sum,

$$\hat{r}_x(k) = \frac{1}{N} \sum_{n=0}^{N-1} x(n+k)x^*(n) \tag{8.3}$$

In order to ensure that the values of $x(n)$ that fall outside the interval $[0, N-1]$ are excluded from the sum, Eq. (8.3) will be rewritten as follows:

$$\hat{r}_x(k) = \frac{1}{N} \sum_{n=0}^{N-1-k} x(n+k)x^*(n) \quad ; \quad k = 0, 1, \ldots, N-1 \tag{8.4}$$

with the values of $\hat{r}_x(k)$ for $k < 0$ defined using conjugate symmetry, $\hat{r}_x(-k) = \hat{r}_x^*(k)$, and with $\hat{r}_x(k)$ set equal to zero for $|k| \geq N$. Taking the discrete-time Fourier transform of

$\hat{r}_x(k)$ leads to an estimate of the power spectrum known as the *periodogram*,

$$\hat{P}_{per}(e^{j\omega}) = \sum_{k=-N+1}^{N-1} \hat{r}_x(k)e^{-jk\omega} \tag{8.5}$$

Although defined in terms of the estimated autocorrelation sequence $\hat{r}_x(k)$, it will be more convenient to express the periodogram directly in terms of the process $x(n)$. This may be done as follows. Let $x_N(n)$ be the finite length signal of length N that is equal to $x(n)$ over the interval $[0, N-1]$, and is zero otherwise,

$$x_N(n) = \begin{cases} x(n) & ; \quad 0 \le n < N \\ 0 & ; \quad \text{otherwise} \end{cases} \tag{8.6}$$

Thus, $x_N(n)$ is the product of $x(n)$ with a rectangular window $w_R(n)$,

$$x_N(n) = w_R(n)x(n) \tag{8.7}$$

In terms of $x_N(n)$, the estimated autocorrelation sequence may be written as follows:

$$\hat{r}_x(k) = \frac{1}{N}\sum_{n=-\infty}^{\infty} x_N(n+k)x_N^*(n) = \frac{1}{N}x_N(k) * x_N^*(-k) \tag{8.8}$$

Taking the Fourier transform and using the convolution theorem, the periodogram becomes

$$\boxed{\hat{P}_{per}(e^{j\omega}) = \frac{1}{N}X_N(e^{j\omega})X_N^*(e^{j\omega}) = \frac{1}{N}\left|X_N(e^{j\omega})\right|^2} \tag{8.9}$$

where $X_N(e^{j\omega})$ is the discrete-time Fourier transform of the N-point data sequence $x_N(n)$,

$$X_N(e^{j\omega}) = \sum_{n=-\infty}^{\infty} x_N(n)e^{-jn\omega} = \sum_{n=0}^{N-1} x(n)e^{-jn\omega} \tag{8.10}$$

Thus, the periodogram is proportional to the squared magnitude of the DTFT of $x_N(n)$, and may be easily computed using a DFT as follows:

$$x_N(n) \xrightarrow{\text{DFT}} X_N(k) \longrightarrow \frac{1}{N}|X_N(k)|^2 = \hat{P}_{per}(e^{j2\pi k/N})$$

A MATLAB program for the periodogram is given in Fig. 8.1.

The Periodogram

```
function Px = periodogram(x,n1,n2)
%
   x    = x(:);
   if nargin == 1
      n1 = 1;   n2 = length(x);   end;
   Px = abs(fft(x(n1:n2),1024)).^2/(n2-n1+1);
   Px(1)=Px(2);
   end;
```

Figure 8.1 *A MATLAB program for computing the periodogram of $x(n)$.*

Example 8.2.1 *Periodogram of White Noise*

If $x(n)$ is white noise with a variance of σ_x^2, then $r_x(k) = \sigma_x^2\delta(k)$ and the power spectrum is a constant,

$$P_x(e^{j\omega}) = \sigma_x^2$$

Shown in Fig. 8.2*a* is a sample realization of unit variance white noise of length $N = 32$. The autocorrelation sequence that is estimated using Eq. (8.3) is shown in Fig. 8.2*b*. Note that

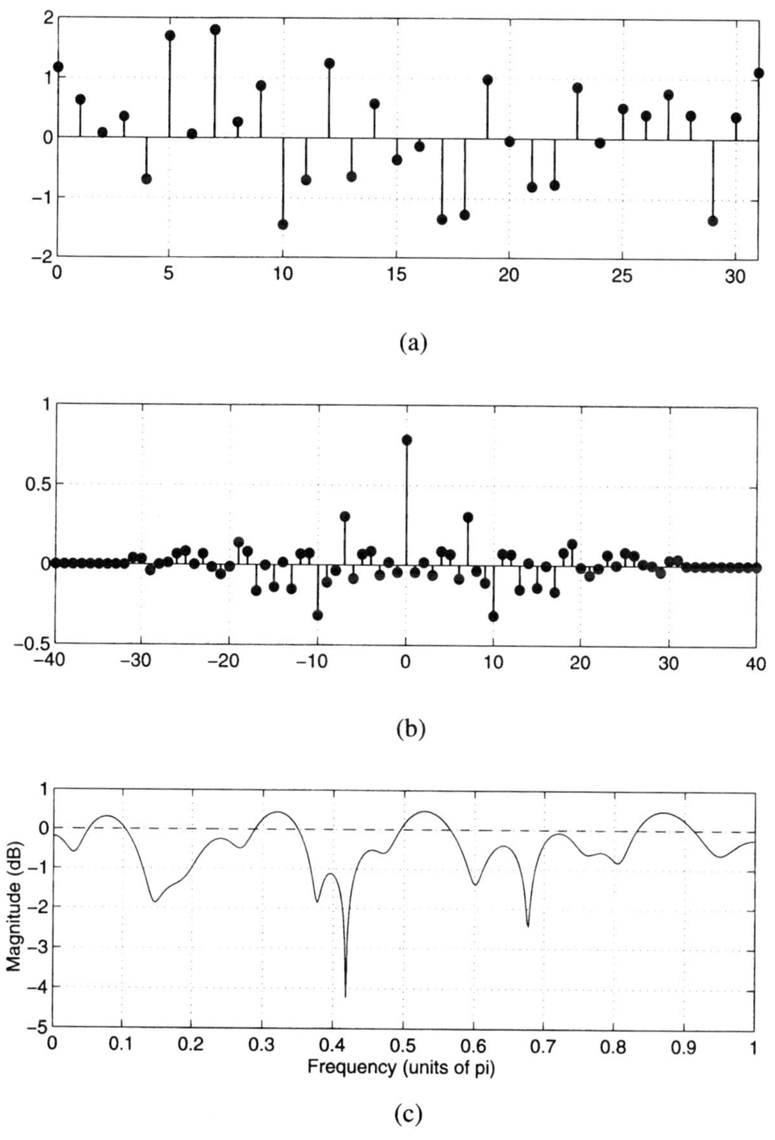

Figure 8.2 *(a) A sample realization of unit variance white noise of length $N = 32$. (b) The estimated autocorrelation sequence. (c) The periodogram along with the true power spectrum, $P_x(e^{j\omega}) = 1$, which is indicated by the dotted line.*

although $\hat{r}_x(k)$ is zero for $|k| \geq 32$, it is nonzero for all other values of k. The periodogram, which is the Fourier transform of $\hat{r}_x(k)$, is shown in Fig. 8.2c along with the true power spectrum. Note that although the periodogram is approximately equal to $P_x(e^{j\omega})$, on the average, we see that there is a considerable amount of variation in $\hat{P}_{per}(e^{j\omega})$ as ω varies. As we will see, such variations are not uncommon with the periodogram.

As we now show, the periodogram has an interesting interpretation in terms of filter banks. Let $h_i(n)$ be an FIR filter of length N that is defined as follows:

$$h_i(n) = \frac{1}{N} e^{jn\omega_i} w_R(n) = \begin{cases} \frac{1}{N} e^{jn\omega_i} & ; \quad 0 \leq n < N \\ 0 & ; \quad \text{otherwise} \end{cases} \tag{8.11}$$

The frequency response of this filter is

$$H_i(e^{j\omega}) = \sum_{n=0}^{N-1} h_i(n) e^{-jn\omega} = e^{-j(\omega-\omega_i)(N-1)/2} \frac{\sin[N(\omega - \omega_i)/2]}{N \sin[(\omega - \omega_i)/2]} \tag{8.12}$$

which, as illustrated in Fig. 8.3, is a bandpass filter with a center frequency ω_i, and a bandwidth that is approximately equal to $\Delta\omega = 2\pi/N$. If a WSS random process $x(n)$ is filtered with $h_i(n)$, then the output process is

$$y_i(n) = x(n) * h_i(n) = \sum_{k=n-N+1}^{n} x(k) h_i(n-k) = \frac{1}{N} \sum_{k=n-N+1}^{n} x(k) e^{j(n-k)\omega_i} \tag{8.13}$$

Since $|H_i(e^{j\omega})|_{\omega=\omega_i} = 1$, then the power spectrum of $x(n)$ and $y(n)$ are equal at frequency ω_i,

$$P_x(e^{j\omega_i}) = P_y(e^{j\omega_i})$$

Furthermore, if the bandwidth of the filter is small enough so that the power spectrum of $x(n)$ may be assumed to be approximately constant over the passband of the filter, then the

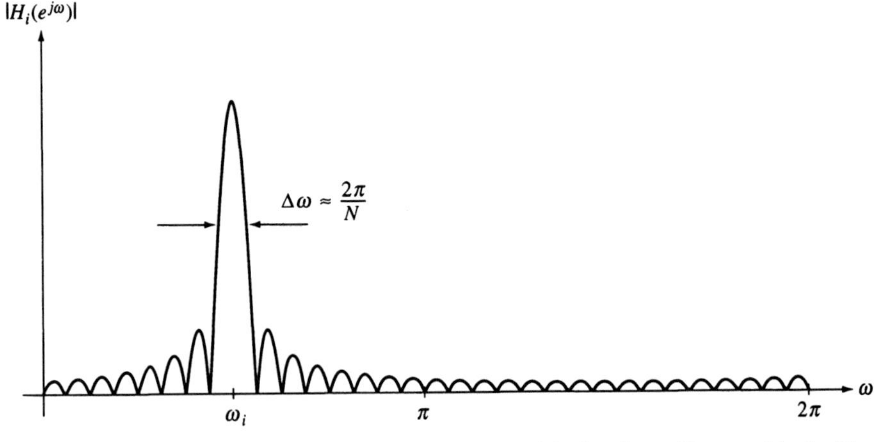

Figure 8.3 *The magnitude of the frequency response of the bandpass filter used in the filter bank interpretation of the periodogram.*

power in $y_i(n)$ will be approximately[1]

$$E\{|y_i(n)|^2\} = \frac{1}{2\pi}\int_{-\pi}^{\pi} P_x(e^{j\omega})|H_i(e^{j\omega})|^2 d\omega \approx \frac{\Delta\omega}{2\pi}P_x(e^{j\omega_i}) = \frac{1}{N}P_x(e^{j\omega_i})$$

and, therefore,

$$P_x(e^{j\omega_i}) \approx NE\{|y_i(n)|^2\} \tag{8.14}$$

Thus, if we are able to estimate the power in $y_i(n)$, then the power spectrum at frequency ω_i may be estimated as follows:

$$\hat{P}_x(e^{j\omega_i}) = N\hat{E}\{|y_i(n)|^2\} \tag{8.15}$$

One simple yet very crude way to estimate the power is to use a one-point sample average,

$$\hat{E}\{|y_i(n)|^2\} = |y_i(N-1)|^2$$

From Eq. (8.13) we see that this is equivalent to

$$|y_i(N-1)|^2 = \frac{1}{N^2}\left|\sum_{k=0}^{N-1} x(k)e^{-jk\omega_i}\right|^2 \tag{8.16}$$

Therefore,

$$\hat{P}_x(e^{j\omega_i}) = N|y_i(N-1)|^2 = \frac{1}{N}\left|\sum_{k=0}^{N-1} x(k)e^{-jk\omega_i}\right|^2 \tag{8.17}$$

which is equivalent to the periodogram. Thus, the periodogram may be viewed as the estimate of the power spectrum that is formed using a filter bank of bandpass filters as illustrated in Fig. 8.4, with $\hat{P}_{per}(e^{j\omega_i})$ being derived from a one-point sample average of the power in the filtered process $y_i(n)$. Of course, the periodogram has such a filter bank "built into it" so that it is not necessary to implement the filter bank. However, in Section 8.3

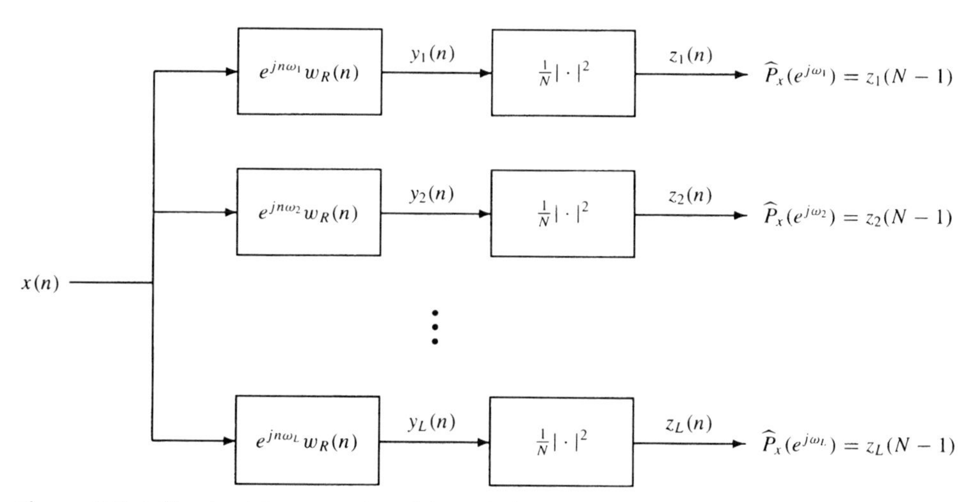

Figure 8.4 *A filter bank interpretation of the periodogram.*

[1] See the discussion leading up to Eq. (3.96) on p. 103.

we will consider a generalization of this filter bank idea that allows the filters to be *data dependent* and to have a frequency response that varies with the center frequency ω_i.

8.2.2 Performance of the Periodogram

In the previous section, it was shown that the periodogram is proportional to the squared magnitude of the DTFT of the finite length sequence $x_N(n)$. Therefore, from a computational point of view, the periodogram is simple to evaluate. In this section, we look at the performance of the periodogram. Ideally, as the length of the data record increases, the periodogram should converge to the power spectrum of the process, $P_x(e^{j\omega})$. However, we must be careful when discussing the convergence of the periodogram to $P_x(e^{j\omega})$. Specifically, since $\hat{P}_{per}(e^{j\omega})$ is a function of the random variables $x(0), \ldots, x(N)$, it is necessary to consider convergence in a statistical sense. Therefore, in this section, we will look at mean-square convergence of the periodogram [33,40], i.e., we will be interested in whether or not

$$\lim_{N\to\infty} E\left\{\left[\hat{P}_{per}(e^{j\omega}) - P_x(e^{j\omega})\right]^2\right\} = 0$$

In order for the periodogram to be mean-square convergent, it is necessary that it be asymptotically unbiased

$$\lim_{N\to\infty} E\left\{\hat{P}_{per}(e^{j\omega})\right\} = P_x(e^{j\omega}) \tag{8.18}$$

and have a variance that goes to zero as the data record length N goes to infinity,

$$\lim_{N\to\infty} \mathrm{Var}\left\{\hat{P}_{per}(e^{j\omega})\right\} = 0 \tag{8.19}$$

In other words, $\hat{P}_{per}(e^{j\omega})$ must be a *consistent* estimate of the power spectrum (see Section 3.2.8). First, we consider the bias of the periodogram.

Periodogram Bias. To compute the bias of the periodogram, we begin by finding the expected value of $\hat{r}_x(k)$. From Eq. (8.4) it follows that the expected value of $\hat{r}_x(k)$ for $k = 0, 1, \ldots N-1$ is

$$E\{\hat{r}_x(k)\} = \frac{1}{N}\sum_{n=0}^{N-1-k} E\left\{x(n+k)x^*(n)\right\} = \frac{1}{N}\sum_{n=0}^{N-1-k} r_x(k) = \frac{N-k}{N} r_x(k)$$

and, for $k \geq N$, the expected value is zero. Using the conjugate symmetry of $\hat{r}_x(k)$ we have

$$E\{\hat{r}_x(k)\} = w_B(k) r_x(k) \tag{8.20}$$

where

$$w_B(k) = \begin{cases} \dfrac{N-|k|}{N} & ; \quad |k| \leq N \\ 0 & ; \quad |k| > N \end{cases} \tag{8.21}$$

is a Bartlett (triangular) window.[2] Therefore, $\hat{r}_x(k)$ is a biased estimate of the autocorrelation.

[2]Since the Bartlett window is applied to the autocorrelation sequence, $w_B(k)$ is referred to as a *lag window*. This is in contrast to a *data window* that is applied to $x(n)$.

Using Eq. (8.20) it follows that the expected value of the periodogram is

$$E\left\{\hat{P}_{per}(e^{j\omega})\right\} = E\left\{\sum_{k=-N+1}^{N-1} \hat{r}_x(k)e^{-jk\omega}\right\} = \sum_{k=-N+1}^{N-1} E\left\{\hat{r}_x(k)\right\} e^{-jk\omega}$$

$$= \sum_{k=-\infty}^{\infty} r_x(k)w_B(k)e^{-jk\omega} \tag{8.22}$$

Since $E\{\hat{P}_{per}(e^{j\omega})\}$ is the Fourier transform of the product $r_x(k)w_B(k)$, using the frequency convolution theorem we have

$$E\left\{\hat{P}_{per}(e^{j\omega})\right\} = \frac{1}{2\pi}P_x(e^{j\omega}) * W_B(e^{j\omega}) \tag{8.23}$$

where $W_B(e^{j\omega})$ is the Fourier transform of the Bartlett window, $w_B(k)$,

$$W_B(e^{j\omega}) = \frac{1}{N}\left[\frac{\sin(N\omega/2)}{\sin(\omega/2)}\right]^2 \tag{8.24}$$

Thus, *the expected value of the periodogram is the convolution of the power spectrum* $P_x(e^{j\omega})$ *with the Fourier transform of a Bartlett window* and, therefore, the periodogram is a biased estimate. However, since $W_B(e^{j\omega})$ converges to an impulse as N goes to infinity, the periodogram is *asymptotically unbiased*

$$\lim_{N\to\infty} E\left\{\hat{P}_{per}(e^{j\omega})\right\} = P_x(e^{j\omega}) \tag{8.25}$$

To illustrate the effect of the lag window, $w_B(k)$, on the expected value of the periodogram, consider a random process consisting of a random phase sinusoid in white noise

$$x(n) = A\sin(n\omega + \phi) + v(n)$$

where ϕ is a random variable that is uniformly distributed over the interval $[-\pi, \pi]$, and $v(n)$ is white noise with a variance σ_v^2. The power spectrum of $x(n)$ is

$$P_x(e^{j\omega}) = \sigma_v^2 + \frac{1}{2}\pi A^2\Big[u_0(\omega - \omega_0) + u_0(\omega + \omega_0)\Big]$$

Therefore, it follows from Eq. (8.23) that the expected value of the periodogram is

$$E\left\{\hat{P}_{per}(e^{j\omega})\right\} = \frac{1}{2\pi}P_x(e^{j\omega}) * W_B(e^{j\omega})$$

$$= \sigma_v^2 + \frac{1}{4}A^2\left[W_B(e^{j(\omega-\omega_0)}) + W_B(e^{j(\omega+\omega_0)})\right] \tag{8.26}$$

The power spectrum $P_x(e^{j\omega})$ and the expected value of the periodogram are shown in Fig. 8.5 for $N = 64$. There are two effects that should be noted in this example. First, is the spectral smoothing that is produced by $W_B(e^{j\omega})$, which leads to a spreading of the power in the sinusoid over a band of frequencies that has a bandwidth of approximately $4\pi/N$. The second effect is the power leakage through the sidelobes of the window, which creates secondary spectral peaks at frequencies $\omega_k \approx \omega_0 \pm \frac{2\pi}{N}k$. As we will see in Section 8.2.3, it is possible for these sidelobes to mask low-level narrowband components.

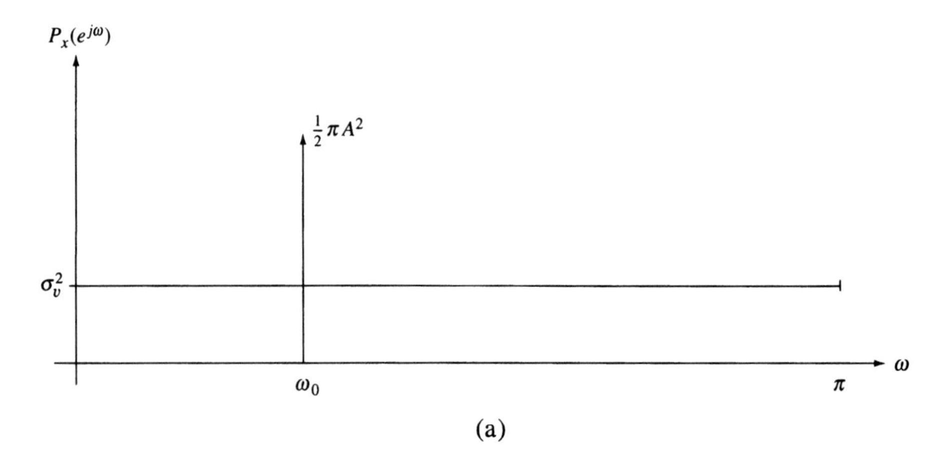

(a)

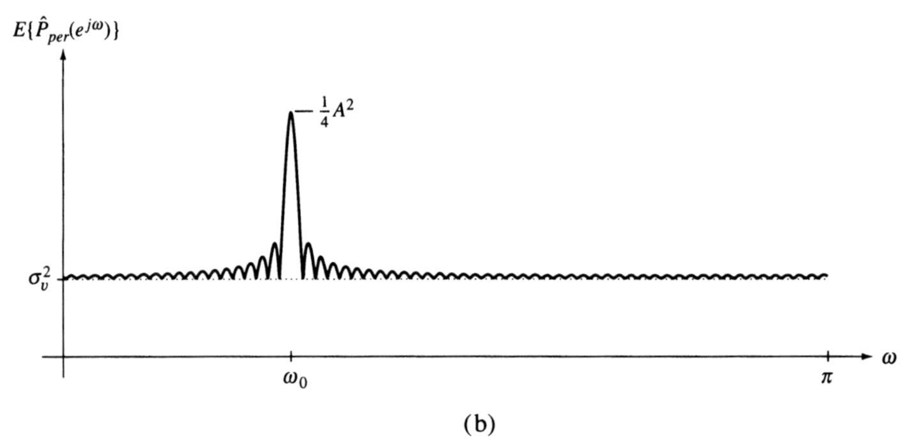

(b)

Figure 8.5 *(a) The power spectrum of a single sinusoid in white noise and (b) the expected value of the periodogram.*

Example 8.2.2 ***Periodogram of a Sinusoid in Noise***

Let $x(n)$ be a wide-sense stationary process consisting of a random phase sinusoid in unit variance white noise

$$x(n) = A\sin(n\omega_0 + \phi) + v(n)$$

With $A = 5$, $\omega_0 = 0.4\pi$, and $N = 64$, fifty different realizations of this process were generated and the periodogram of each was computed. Shown in Fig. 8.6*a* is an overlay of the 50 periodograms. Note that although each periodogram has a peak at approximately $\omega = 0.4\pi$, there is a considerable amount of variation from one periodogram to the next. In Fig. 8.6*b* the average of all 50 periodograms is shown. This average is approximately equal to the expected value given in Eq. (8.26). By increasing the number of data values to $N = 256$, we obtain the periodograms shown in Fig. 8.6*c* and *d*. Note that with the additional data, the power in the sinusoid is spread out over a much narrower band of frequencies.

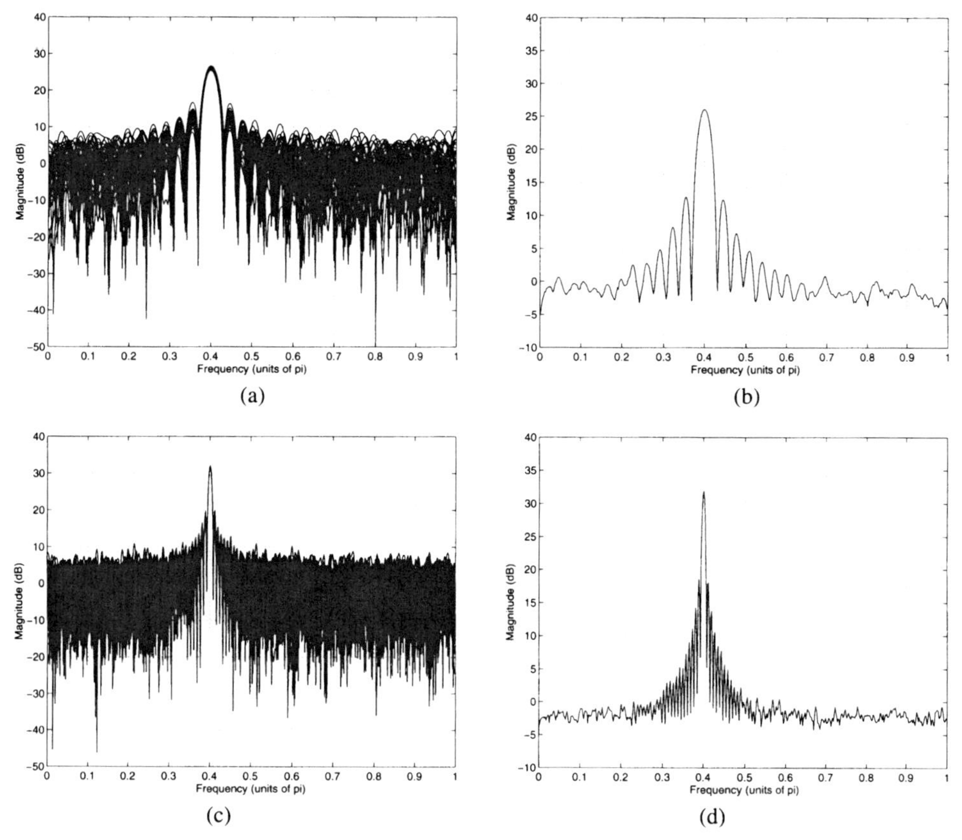

Figure 8.6 *The periodogram of a sinusoid in white noise. (a) Overlay plot of 50 periodograms using $N = 64$ data values and (b) the periodogram average. (c) Overlay plot of 50 periodograms using $N = 256$ data values and (d) the periodogram average.*

In addition to biasing the periodogram, the smoothing that is introduced by the Bartlett window also limits the ability of the periodogram to resolve closely-spaced narrowband components in $x(n)$. Consider, for example, a random process consisting of two sinusoids in white noise

$$x(n) = A_1 \sin(n\omega_1 + \phi_1) + A_2 \sin(n\omega_2 + \phi_2) + v(n)$$

where ϕ_1 and ϕ_2 are uncorrelated uniformly distributed random variables and where $v(n)$ is white noise with a variance of σ_v^2. The power spectrum of $x(n)$ is

$$P_x(e^{j\omega}) = \sigma_v^2 + \frac{1}{2}\pi A_1^2\Big[u_0(\omega - \omega_1) + u_0(\omega + \omega_1)\Big] + \frac{1}{2}\pi A_2^2\Big[u_0(\omega - \omega_2) + u_0(\omega + \omega_2)\Big]$$

and the expected value of the periodogram is

$$\begin{aligned} E\left\{\hat{P}_{per}(e^{j\omega})\right\} &= \frac{1}{2\pi} P_x(e^{j\omega}) * W_B(e^{j\omega}) \\ &= \sigma_v^2 + \frac{1}{4}A_1^2\left[W_B(e^{j(\omega-\omega_1)}) + W_B(e^{j(\omega+\omega_1)})\right] \\ &\quad + \frac{1}{4}A_2^2\left[W_B(e^{j(\omega-\omega_2)}) + W_B(e^{j(\omega+\omega_2)})\right] \end{aligned} \tag{8.27}$$

which is shown in Fig. 8.7 for $A_1 = A_2$ and $N = 64$. Since the width of the main lobe of $W_B(e^{j\omega})$ increases as the data record length decreases, for a given data record length, N, there is a limit on how closely two sinusoids or two narrowband processes may be located before they can no longer be resolved. One way to define this resolution limit is to set $\Delta\omega$ equal to the width of the main lobe of the spectral window, $W_B(e^{j\omega})$, at its "half-power" or 6 dB point. For the Bartlett window, $\Delta\omega = 0.89(2\pi/N)$, which means that the resolution of the periodogram is

$$\boxed{\text{Res}\left[\hat{P}_{per}(e^{j\omega})\right] = 0.89\frac{2\pi}{N}} \tag{8.28}$$

It is important to note, however, that Eq. (8.28) is nothing more than a *rule of thumb* that should only be used as a guideline in determining the amount of data that is necessary for a given resolution. There is nothing sacred, for example, with the proportionality constant $0.89(2\pi)$. On the other hand, what is important is the fact that the resolution is inversely proportional to the amount of data, N. Nevertheless, although the definition of $\Delta\omega$ given in Eq. (8.28) is somewhat arbitrary, one generally finds that it is difficult to resolve details in the spectrum that are much finer than this.

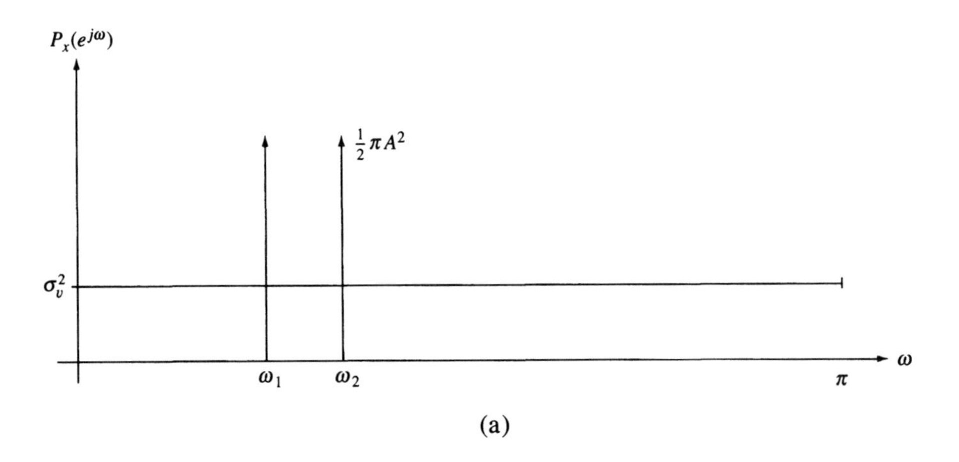

(a)

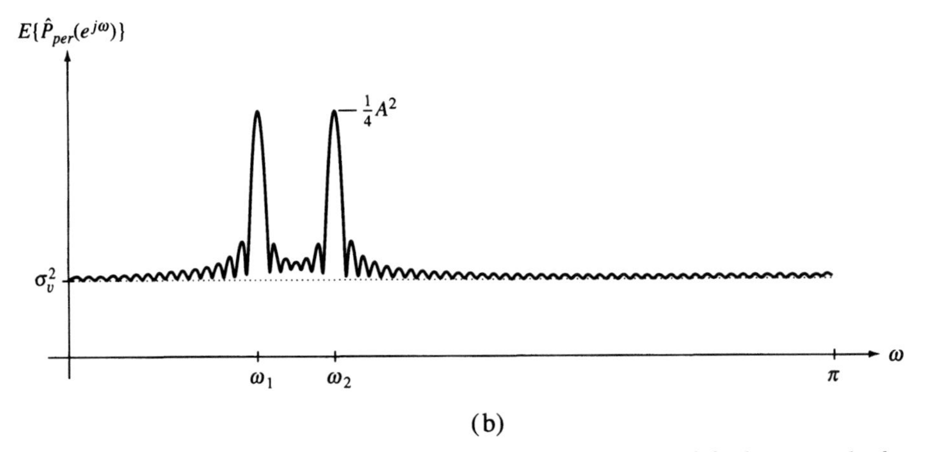

(b)

Figure 8.7 *(a) The power spectrum of two sinusoids in white noise and (b) the expected value of the periodogram.*

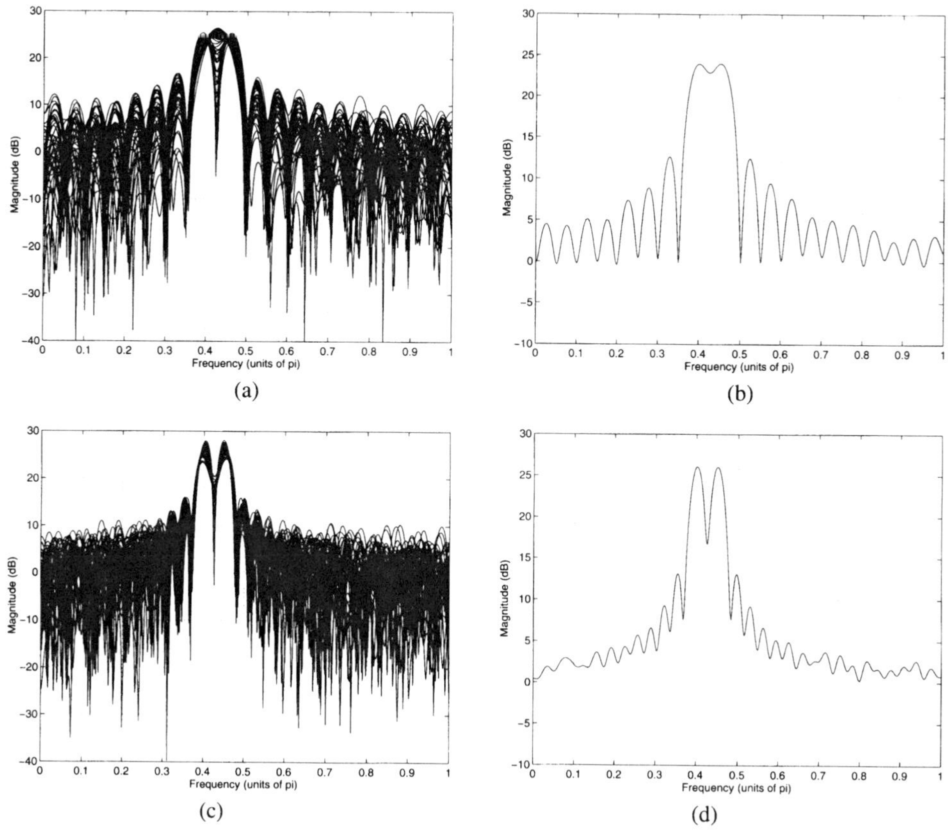

Figure 8.8 *The periodogram of two sinusoids in white noise with $\omega_1 = 0.4\pi$ and $\omega_2 = 0.45\pi$. (a) Overlay plot of 50 periodograms using $N = 40$ data values and (b) the ensemble average. (c) Overlay plot of 50 periodograms using $N = 64$ data values and (d) the ensemble average.*

Example 8.2.3 *Periodogram Resolution*

Let $x(n)$ be a random process consisting of two equal amplitude sinusoids in unit variance white noise

$$x(n) = A\sin(n\omega_1 + \phi_1) + A\sin(n\omega_2 + \phi_2) + v(n)$$

where $\omega_1 = 0.4\pi$, $\omega_2 = 0.45\pi$, and $A = 5$. Using Eq. (8.28), with $\Delta\omega = 0.05\pi$ we see that in order to resolve the two narrowband components, a data record length on the order of $N = 36$ is required. Using $N = 40$, an overlay of 50 periodograms is shown in Fig. 8.8*a*. It is clear from this figure that it is not always possible to resolve the two sinusoidal components when $N = 40$. The average of the 50 periodograms, shown in 8.8*b*, illustrates the overlap of the two Bartlett windows. However, as shown in Figures 8.8*c* and *d*, if $N = 64$ then the two sinusoids are clearly resolved.

Variance of the Periodogram. We have seen that the periodogram is an asymptotically unbiased estimate of the power spectrum. In order for it to be a consistent estimate, it is necessary that the variance go to zero as $N \to \infty$. Unfortunately, it is difficult to evaluate the variance of the periodogram for an arbitrary process $x(n)$ since the variance depends on the fourth-order moments of the process. However, as we show next, the variance may be evaluated in the special case of white Gaussian noise.

Let $x(n)$ be a Gaussian white noise process with variance σ_x^2. Using Eq. (8.9) the periodogram may be expressed as follows:

$$\hat{P}_{per}(e^{j\omega}) = \frac{1}{N}\left|\sum_{k=0}^{N-1} x(k)e^{-jk\omega}\right|^2 = \frac{1}{N}\left\{\sum_{k=0}^{N-1} x(k)e^{-jk\omega}\right\}\left\{\sum_{l=0}^{N-1} x^*(l)e^{jl\omega}\right\}$$

$$= \frac{1}{N}\sum_{k=0}^{N-1}\sum_{l=0}^{N-1} x(k)x^*(l)e^{-j(k-l)\omega} \tag{8.29}$$

Therefore, the second-order moment of the periodogram is

$$E\left\{\hat{P}_{per}(e^{j\omega_1})\hat{P}_{per}(e^{j\omega_2})\right\}$$

$$= \frac{1}{N^2}\sum_{k=0}^{N-1}\sum_{l=0}^{N-1}\sum_{m=0}^{N-1}\sum_{n=0}^{N-1} E\big\{x(k)x^*(l)x(m)x^*(n)\big\}e^{-j(k-l)\omega_1}e^{-j(m-n)\omega_2} \tag{8.30}$$

which depends on the fourth-order moments of $x(n)$. Since $x(n)$ is Gaussian, we may use the moment factoring theorem to simplify these moments [17,44]. For complex Gaussian random variables, the moment factoring theorem is[3]

$$\begin{aligned} E\big\{x(k)x^*(l)x(m)x^*(n)\big\} &= E\big\{x(k)x^*(l)\big\}E\big\{x(m)x^*(n)\big\} \\ &\quad + E\big\{x(k)x^*(n)\big\}E\big\{x(m)x^*(l)\big\} \end{aligned} \tag{8.31}$$

Substituting Eq. (8.31) into Eq. (8.30), the second-order moment of the periodogram becomes a sum of two terms. The first term contains products of $E\big\{x(k)x^*(l)\big\}$ with $E\big\{x(m)x^*(n)\big\}$. For white noise, these terms are equal to σ_x^4 when $k = l$ and $m = n$, and they are equal to zero otherwise. Thus, the first term simplifies to

$$\frac{1}{N^2}\sum_{k=0}^{N-1}\sum_{m=0}^{N-1}\sigma_x^4 = \sigma_x^4 \tag{8.32}$$

The second term, on the other hand, contains products of $E\big\{x(k)x^*(n)\big\}$ with $E\big\{x(m)x^*(l)\big\}$. Again, for white noise, these terms are equal to σ_x^4 when $k = n$ and $l = m$, and they are equal to zero otherwise. Therefore, the second term becomes

$$\begin{aligned} \frac{1}{N^2}\sum_{k=0}^{N-1}\sum_{l=0}^{N-1}\sigma_x^4 e^{-j(k-l)\omega_1}e^{j(k-l)\omega_2} &= \frac{\sigma_x^4}{N^2}\sum_{k=0}^{N-1} e^{-jk(\omega_1-\omega_2)}\sum_{l=0}^{N-1} e^{jl(\omega_1-\omega_2)} \\ &= \frac{\sigma_x^4}{N^2}\left[\frac{1-e^{-jN(\omega_1-\omega_2)}}{1-e^{-j(\omega_1-\omega_2)}}\right]\left[\frac{1-e^{jN(\omega_1-\omega_2)}}{1-e^{j(\omega_1-\omega_2)}}\right] \\ &= \sigma_x^4\left[\frac{\sin N(\omega_1-\omega_2)/2}{N\sin(\omega_1-\omega_2)/2}\right]^2 \end{aligned} \tag{8.33}$$

[3]Note that this is different from the moment factoring theorem for real Gaussian random variables, which contains three terms instead of two.

Combining Eq. (8.32) and Eq. (8.33) it follows that

$$E\left\{\hat{P}_{per}(e^{j\omega_1})\hat{P}_{per}(e^{j\omega_2})\right\} = \sigma_x^4\left\{1 + \left[\frac{\sin N(\omega_1-\omega_2)/2}{N\sin(\omega_1-\omega_2)/2}\right]^2\right\} \tag{8.34}$$

Since

$$\text{Cov}\left\{\hat{P}_{per}(e^{j\omega_1})\hat{P}_{per}(e^{j\omega_2})\right\} = E\left\{\hat{P}_{per}(e^{j\omega_1})\hat{P}_{per}(e^{j\omega_2})\right\} - E\left\{\hat{P}_{per}(e^{j\omega_1})\right\}E\left\{\hat{P}_{per}(e^{j\omega_2})\right\}$$

and $E\left\{\hat{P}_{per}(e^{j\omega})\right\} = \sigma_x^2$, then the covariance of the periodogram is

$$\text{Cov}\left\{\hat{P}_{per}(e^{j\omega_1})\hat{P}_{per}(e^{j\omega_2})\right\} = \sigma_x^4\left[\frac{\sin N(\omega_1-\omega_2)/2}{N\sin(\omega_1-\omega_2)/2}\right]^2 \tag{8.35}$$

Finally, setting $\omega_1 = \omega_2$ we have, for the variance,

$$\text{Var}\left\{\hat{P}_{per}(e^{j\omega})\right\} = \sigma_x^4 \tag{8.36}$$

Thus, the variance does not go to zero as $N \to \infty$, and the periodogram is *not a consistent estimate* of the power spectrum. In fact, since $P_x(e^{j\omega}) = \sigma_x^2$ then the variance of the periodogram of white Gaussian noise is proportional to the square of the power spectrum,

$$\text{Var}\left\{\hat{P}_{per}(e^{j\omega})\right\} = P_x^2(e^{j\omega}) \tag{8.37}$$

Example 8.2.4 ***Periodogram of White Noise***

Let $x(n)$ be white Gaussian noise with

$$P_x(e^{j\omega}) = 1$$

From Eq. (8.23) it follows that the expected value of the periodogram is equal to one,

$$E\{\hat{P}_{per}(e^{j\omega})\} = 1$$

and from Eq. (8.36) it follows that the variance is also equal to one,

$$\text{Var}\left\{\hat{P}_{per}(e^{j\omega})\right\} = 1$$

Thus, although the periodogram is unbiased, the variance is equal to a constant that is independent of the data record length, N. In Fig. 8.9*a*, *c*, and *e* are overlay plots of 50 periodograms of white noise that were generated using data records of length $N = 64$, 128, and 256, respectively. In Fig. 8.9*b*, *d*, and *f*, on the other hand, are the periodogram averages. What we observe is that, although the average value of the periodogram is approximately equal to $\sigma_x^2 = 1$, the variance does not decrease as the amount of data increases.

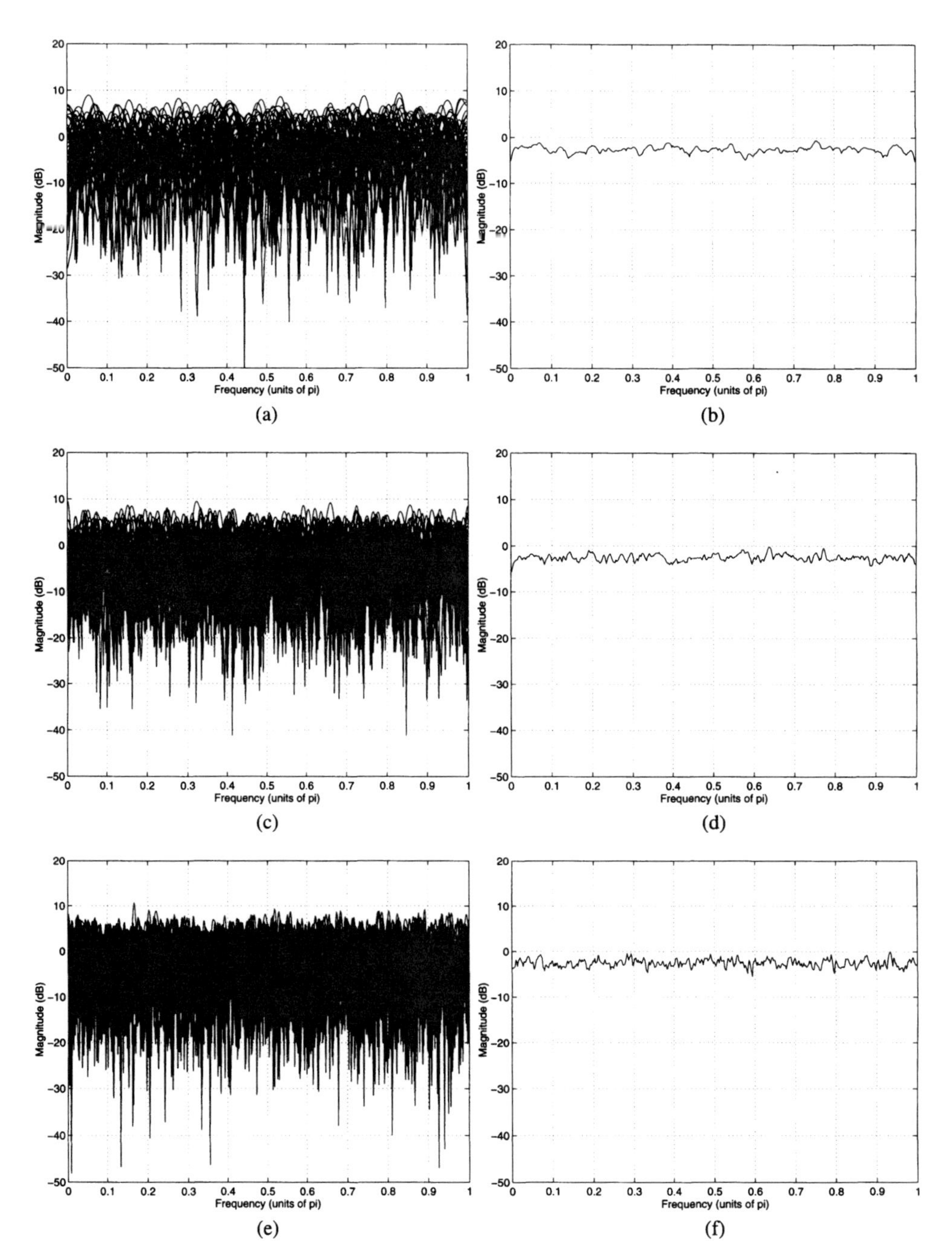

Figure 8.9 *The periodogram of unit variance white Gaussian noise. (a) Overlay plot of 50 periodograms with* $N = 64$ *data values and (b) the periodogram average. (c) Overlay plot of 50 periodograms with* $N = 128$ *data values and (d) the periodogram average. (e) Overlay plot of 50 periodograms with* $N = 256$ *data values and (f) the periodogram average.*

The analysis given above for the variance of the periodogram assumes that $x(n)$ is white Gaussian noise. Although the statistical analysis of a nonwhite Gaussian process is much more difficult, we may derive an approximate expression for the variance as follows. Recall that a random process $x(n)$ with power spectrum $P_x(e^{j\omega})$ may be generated by filtering unit variance white noise $v(n)$ with a linear shift-invariant filter $h(n)$ that has a frequency response $H(e^{j\omega})$ with

$$|H(e^{j\omega})|^2 = P_x(e^{j\omega}) \tag{8.38}$$

As defined in Eq. (8.7), if $x_N(n)$ and $v_N(n)$ are the sequences of length N that are formed by windowing $x(n)$ and $v(n)$, respectively, then the periodograms of these processes are

$$\hat{P}_{per}^{(x)}(e^{j\omega}) = \frac{1}{N}\left|X_N(e^{j\omega})\right|^2 \tag{8.39}$$

$$\hat{P}_{per}^{(v)}(e^{j\omega}) = \frac{1}{N}\left|V_N(e^{j\omega})\right|^2 \tag{8.40}$$

Although $x_N(n)$ is not equal to the convolution of $v_N(n)$ with $h(n)$, if N is large compared to the length of $h(n)$ so that the transient effects are small, then

$$x_N(n) \approx h(n) * v_N(n)$$

Since

$$\left|X_N(e^{j\omega})\right|^2 \approx \left|H(e^{j\omega})\right|^2 \left|V_N(e^{j\omega})\right|^2 = P_x(e^{j\omega}) \left|V_N(e^{j\omega})\right|^2 \tag{8.41}$$

substituting Eqs. (8.39) and (8.40) into Eq. (8.41) we have

$$\hat{P}_{per}^{(x)}(e^{j\omega}) \approx P_x(e^{j\omega}) \hat{P}_{per}^{(v)}(e^{j\omega})$$

Therefore,

$$\text{Var}\left\{\hat{P}_{per}^{(x)}(e^{j\omega})\right\} \approx P_x^2(e^{j\omega}) \text{Var}\left\{\hat{P}_{per}^{(v)}(e^{j\omega})\right\}$$

and, since the variance of the periodogram of $v(n)$ is equal to one, then we have

$$\text{Var}\left\{\hat{P}_{per}^{(x)}(e^{j\omega})\right\} \approx P_x^2(e^{j\omega}) \tag{8.42}$$

Thus, assuming that N is large, the variance of the periodogram of a Gaussian random process is proportional to the square of its power spectrum.

The second-order moment and covariance of the periodogram may be similarly generalized for nonwhite Gaussian noise. For the second-order moment, Eq. (8.34) becomes

$$E\left\{\hat{P}_{per}(e^{j\omega_1})\hat{P}_{per}(e^{j\omega_2})\right\} \approx P_x(e^{j\omega_1})P_x(e^{j\omega_2})\left\{1 + \left[\frac{\sin N(\omega_1 - \omega_2)/2}{N\sin(\omega_1 - \omega_2)/2}\right]^2\right\} \tag{8.43}$$

and, for the covariance, Eq. (8.35) becomes

$$\text{Cov}\left\{\hat{P}_{per}(e^{j\omega_1})\hat{P}_{per}(e^{j\omega_2})\right\} \approx P_x(e^{j\omega_1})P_x(e^{j\omega_2})\left[\frac{\sin N(\omega_1 - \omega_2)/2}{N\sin(\omega_1 - \omega_2)/2}\right]^2 \tag{8.44}$$

Note that for large N, the term in brackets is approximately equal to zero provided $\omega_1 - \omega_2 \gg 2\pi/N$, which implies that there is little correlation between one frequency and another. The properties of the periodogram are summarized in Table 8.1.

Table 8.1 *Properties of the Periodogram*

$$\hat{P}_{per}(e^{j\omega}) = \frac{1}{N}\left|\sum_{n=0}^{N-1} x(n)e^{-jn\omega}\right|^2$$

Bias	$E\left\{\hat{P}_{per}(e^{j\omega})\right\} = \frac{1}{2\pi} P_x(e^{j\omega}) * W_B(e^{j\omega})$
Resolution	$\Delta\omega = 0.89\frac{2\pi}{N}$
Variance	$\text{Var}\left\{\hat{P}_{per}(e^{j\omega})\right\} \approx P_x^2(e^{j\omega})$

8.2.3 The Modified Periodogram

In Section 8.2.1 we saw that the periodogram is proportional to the squared magnitude of the Fourier transform of the windowed signal $x_N(n) = x(n)w_R(n)$,

$$\hat{P}_{per}(e^{j\omega}) = \frac{1}{N}|X_N(e^{j\omega})|^2 = \frac{1}{N}\left|\sum_{n=-\infty}^{\infty} x(n)w_R(n)e^{-jn\omega}\right|^2 \tag{8.45}$$

Instead of applying a rectangular window to $x(n)$, Eq. (8.45) suggests the possibility of using other data windows. Would there be any benefit, for example, in replacing the rectangular window with a triangular (Bartlett) window? To answer this question, let us examine the effect of the data window on the bias of the periodogram.

Using Eq. (8.45), the expected value of the periodogram is

$$\begin{aligned}
E\left\{\hat{P}_{per}(e^{j\omega})\right\} &= \frac{1}{N}E\left\{\left[\sum_{n=-\infty}^{\infty} x(n)w_R(n)e^{-jn\omega}\right]\left[\sum_{m=-\infty}^{\infty} x(m)w_R(m)e^{-jm\omega}\right]^*\right\} \\
&= \frac{1}{N}E\left\{\sum_{m=-\infty}^{\infty}\sum_{n=-\infty}^{\infty} x(n)x^*(m)w_R(m)w_R(n)e^{-j(n-m)\omega}\right\} \\
&= \frac{1}{N}\sum_{m=-\infty}^{\infty}\sum_{n=-\infty}^{\infty} r_x(n-m)w_R(m)w_R(n)e^{-j(n-m)\omega}
\end{aligned} \tag{8.46}$$

With the change of variables, $k = n - m$, Eq. (8.46) becomes

$$\begin{aligned}
E\left\{\hat{P}_{per}(e^{j\omega})\right\} &= \frac{1}{N}\sum_{k=-\infty}^{\infty}\sum_{n=-\infty}^{\infty} r_x(k)w_R(n)w_R(n-k)e^{-jk\omega} \\
&= \frac{1}{N}\sum_{k=-\infty}^{\infty} r_x(k)\left[\sum_{n=-\infty}^{\infty} w_R(n)w_R(n-k)\right]e^{-jk\omega} \\
&= \frac{1}{N}\sum_{k=-\infty}^{\infty} r_x(k)w_B(k)e^{-jk\omega}
\end{aligned} \tag{8.47}$$

where

$$w_B(k) = w_R(k) * w_R(-k) = \sum_{n=-\infty}^{\infty} w_R(n)w_R(n-k)$$

is a Bartlett window. Using the frequency convolution theorem, it follows that the expected value of the periodogram is

$$E\left\{\hat{P}_{per}(e^{j\omega})\right\} = \frac{1}{2\pi N} P_x(e^{j\omega}) * |W_R(e^{j\omega})|^2 \tag{8.48}$$

where

$$W_R(e^{j\omega}) = \frac{\sin(N\omega/2)}{\sin(\omega/2)} e^{-j(N-1)\omega/2}$$

is the Fourier transform of the rectangular data window, $w_R(n)$.[4] Therefore, the amount of smoothing in the periodogram is determined by the window that is applied to the data. Although a rectangular window has a narrow main lobe compared to other windows and, therefore, produces the least amount of spectral smoothing, it has relatively large sidelobes that may lead to masking of weak narrowband components. Consider, for example, a random process consisting of two sinusoids in white noise

$$x(n) = 0.1 \sin(n\omega_1 + \phi_1) + \sin(n\omega_2 + \phi_2) + v(n)$$

With $\omega_1 = 0.2\pi$, $\omega_2 = 0.3\pi$, and $N = 128$, the expected value of the periodogram is shown in Fig. 8.10*a*. What we observe is that the sinusoid at frequency ω_1 is almost completely masked by the sidelobes of the window at frequency ω_2. However. if the rectangular window is replaced with a Hamming window, then the sinusoid at frequency ω_1 is clearly visible as illustrated in Fig. 8.10*b*. This is due to the smaller sidelobes of a Hamming window, which are down about 30 dB compared to the sidelobes of a rectangular window. On the

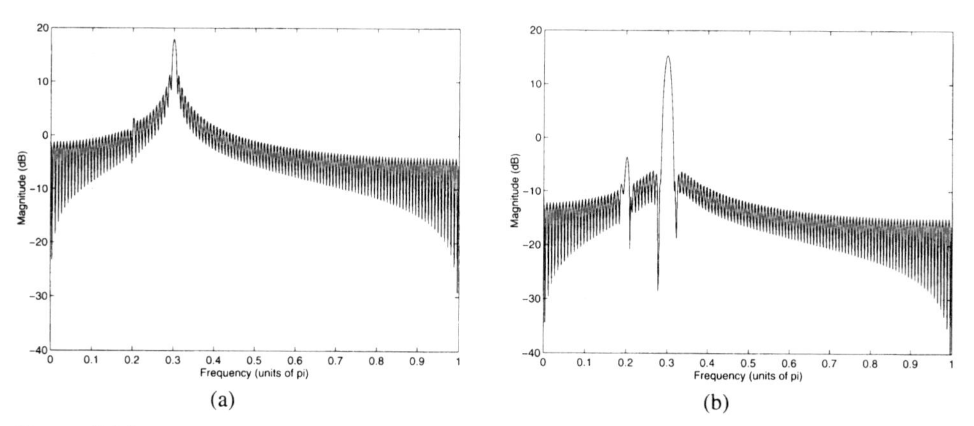

Figure 8.10 *Spectral analysis of two sinusoids in white noise with sinusoidal frequencies of* $\omega_1 = .2\pi$ *and* $\omega_2 = .3\pi$ *and a data record length of* $N = 128$ *points. (a) The expected value of the periodogram. (b) The expected value of the modified periodogram using a Hamming data window.*

[4]Note the equivalence of Eq. (8.48) and Eq. (8.23).

other hand, this reduction in the sidelobe amplitude comes at the expense of an increase in the width of the mainlobe which, in turn, affects the resolution. The periodogram of a process that is windowed with a general window $w(n)$ is called a *modified periodogram* and is given by

$$P_M(e^{j\omega}) = \frac{1}{NU}\left|\sum_{n=-\infty}^{\infty} x(n)w(n)e^{-jn\omega}\right|^2 \tag{8.49}$$

where N is the length of the window and

$$U = \frac{1}{N}\sum_{n=0}^{N-1} |w(n)|^2 \tag{8.50}$$

is a constant that, as we will see, is defined so that $\hat{P}_M(e^{j\omega})$ will be asymptotically unbiased. A MATLAB program for the modified periodogram is given in Fig. 8.11.

Let us now evaluate the performance of the modified periodogram. It follows from the derivation of Eq. (8.48) that the expected value of $\hat{P}_M(e^{j\omega})$ is

$$E\left\{\hat{P}_M(e^{j\omega})\right\} = \frac{1}{2\pi NU} P_x(e^{j\omega}) * |W(e^{j\omega})|^2 \tag{8.51}$$

where $W(e^{j\omega})$ is the Fourier transform of the data window. Note that with

$$U = \frac{1}{N}\sum_{n=0}^{N-1} |w(n)|^2 = \frac{1}{2\pi N}\int_{-\pi}^{\pi} |W(e^{j\omega})|^2 d\omega \tag{8.52}$$

The Modified Periodogram

```
function Px = mper(x,win,n1,n2)
%
   x  = x(:);
   if nargin == 2
      n1 = 1;  n2 = length(x);  end;
   N  = n2 - n1 + 1;
   w  = ones(N,1);
   if (win == 2) w = hamming(N);
      elseif (win == 3) w = hanning(N);
      elseif (win == 4) w = bartlett(N);
      elseif (win == 5) w = blackman(N);
      end;
   xw = x(n1:n2).*w/norm(w);
   Px = N*periodogram(xw);
   end;
```

Figure 8.11 *A MATLAB program for computing the modified periodogram of $x(n)$.*

then

$$\frac{1}{2\pi NU}\int_{-\pi}^{\pi}|W(e^{j\omega})|^2 d\omega = 1$$

and, with an appropriate window, $|W(e^{j\omega})|^2/NU$ will converge to an impulse of unit area as $N \to \infty$, and the modified periodogram will be asymptotically unbiased. (Note that if $w(n)$ is a rectangular window, then $U = 1$ and the modified periodogram reduces to the periodogram).

Since the modified periodogram is simply the periodogram of a windowed data sequence, the variance of $\hat{P}_M(e^{j\omega})$ will be approximately the same as that for the periodogram, i.e.,

$$\text{Var}\left\{\hat{P}_M(e^{j\omega})\right\} \approx P_x^2(e^{j\omega}) \tag{8.53}$$

Therefore, the modified periodogram is not a consistent estimate of the power spectrum and the data window offers no benefit in terms of reducing the variance. What the window does provide, however, is a trade-off between spectral resolution (main lobe width) and spectral masking (sidelobe amplitude). For example, with the resolution of $\hat{P}_M(e^{j\omega})$ defined to be the 3 dB bandwidth of the data window,[5]

$$\text{Res}\left[\hat{P}_M(e^{j\omega})\right] = (\Delta\omega)_{3\text{dB}} \tag{8.54}$$

we see from Table 8.2 that a 43 dB reduction in the sidelobe amplitude that comes from using a Hamming data window instead of a rectangular window results in a reduction in the spectral resolution of about 50%. The properties of the modified periodogram are summarized in Table 8.3. An extensive list of windows and window characteristics may be found in [14].

Table 8.2 ***Properties of a Few Commonly Used Windows. Each Window is Assumed to be of Length N.***

Window	Sidelobe Level (dB)	3 dB BW $(\Delta\omega)_{3\text{dB}}$
Rectangular	−13	$0.89(2\pi/N)$
Bartlett	−27	$1.28(2\pi/N)$
Hanning	−32	$1.44(2\pi/N)$
Hamming	−43	$1.30(2\pi/N)$
Blackman	−58	$1.68(2\pi/N)$

[5] Note that this is consistent with the definition of resolution given in Eq. (8.28) for the periodogram. Specifically, although the periodogram resolution is defined to be the 6 dB bandwidth of $W_B(e^{j\omega})$, since $W_B(e^{j\omega}) = |W_R(e^{j\omega})|^2$, this is equivalent to the 3 dB bandwidth of $W_R(e^{j\omega})$.

Table 8.3 Properties of the Modified Periodogram

$$\hat{P}_M(e^{j\omega}) = \frac{1}{NU}\left|\sum_{n=-\infty}^{\infty} w(n)x(n)e^{-jn\omega}\right|^2$$

$$U = \frac{1}{N}\sum_{n=0}^{N-1}|w(n)|^2$$

Bias	$E\left\{\hat{P}_M(e^{j\omega})\right\} = \frac{1}{2\pi NU}P_x(e^{j\omega}) * \|W(e^{j\omega})\|^2$
Resolution	Window dependent
Variance	$\text{Var}\left\{\hat{P}_M(e^{j\omega})\right\} \approx P_x^2(e^{j\omega})$

8.2.4 Bartlett's Method: Periodogram Averaging

In this section, we look at Bartlett's method of periodogram averaging, which, unlike either the periodogram or the modified periodogram, produces a consistent estimate of the power spectrum [5]. The motivation for this method comes from the observation that the expected value of the periodogram converges to $P_x(e^{j\omega})$ as the data record length N goes to infinity,

$$\lim_{N\to\infty} E\{\hat{P}_{per}(e^{j\omega})\} = P_x(e^{j\omega}) \tag{8.55}$$

Therefore, if we can find a consistent estimate of the mean, $E\{\hat{P}_{per}(e^{j\omega})\}$, then this estimate will be a consistent estimate of $P_x(e^{j\omega})$.

In our discussion of the sample mean in Section 3.2.8, we saw how averaging a set of uncorrelated measurements of a random variable x yields a consistent estimate of the mean, $E\{x\}$. This suggests that we consider estimating the power spectrum of a random process by periodogram averaging. Thus, let $x_i(n)$ for $i = 1, 2, \ldots, K$ be K uncorrelated realizations of a random process $x(n)$ over the interval $0 \le n < L$. With $\hat{P}_{per}^{(i)}(e^{j\omega})$ the periodogram of $x_i(n)$,

$$\hat{P}_{per}^{(i)}(e^{j\omega}) = \frac{1}{L}\left|\sum_{n=0}^{L-1} x_i(n)e^{-jn\omega}\right|^2 \quad ; \quad i = 1, 2, \ldots, K \tag{8.56}$$

the average of these periodograms is

$$\hat{P}_x(e^{j\omega}) = \frac{1}{K}\sum_{i=1}^{K}\hat{P}_{per}^{(i)}(e^{j\omega}) \tag{8.57}$$

Evaluating the expected value of $\hat{P}_x(e^{j\omega})$ we have

$$E\left\{\hat{P}_x(e^{j\omega})\right\} = E\left\{\hat{P}_{per}^{(i)}(e^{j\omega})\right\} = \frac{1}{2\pi}P_x(e^{j\omega}) * W_B(e^{j\omega}) \tag{8.58}$$

where $W_B(e^{j\omega})$ is the Fourier transform of a Bartlett window, $w_B(k)$, that extends from $-L$

to L. Therefore, as with the periodogram, $\hat{P}_x(e^{j\omega})$ is asymptotically unbiased. In addition, with our assumption that the data records are uncorrelated, it follows that the variance of $\hat{P}_x(e^{j\omega})$ is

$$\text{Var}\left\{\hat{P}_x(e^{j\omega})\right\} = \frac{1}{K}\text{Var}\left\{\hat{P}_{per}^{(i)}(e^{j\omega})\right\} \approx \frac{1}{K}P_x^2(e^{j\omega}) \tag{8.59}$$

which goes to zero as K goes to infinity. Therefore, $\hat{P}_x(e^{j\omega})$ is a consistent estimate of the power spectrum provided that both K and L are allowed to go to infinity. However, the difficulty with this approach is that uncorrelated realizations of a process are generally not available. Instead, one typically only has a single realization of length N. Therefore, Bartlett proposed that $x(n)$ be partitioned into K nonoverlapping sequences of length L where $N = KL$ as illustrated in Fig. 8.12. The Bartlett estimate is then computed as in Eq. (8.56) and Eq. (8.57) with

$$x_i(n) = x(n + iL) \qquad \begin{array}{l} n = 0, 1, \ldots, L-1 \\ i = 0, 1, \ldots, K-1 \end{array}$$

Thus, the Bartlett estimate is

$$\boxed{\hat{P}_B(e^{j\omega}) = \frac{1}{N}\sum_{i=0}^{K-1}\left|\sum_{n=0}^{L-1} x(n + iL)e^{-jn\omega}\right|^2} \tag{8.60}$$

A MATLAB program to compute the Bartlett estimate is given in Fig. 8.13.

Based on our analysis of the periodogram and the modified periodogram, we may easily evaluate the performance of Bartlett's method as follows. First, as in Eq. (8.58), the expected value of Bartlett's estimate is

$$\boxed{E\left\{\hat{P}_B(e^{j\omega})\right\} = \frac{1}{2\pi}P_x(e^{j\omega}) * W_B(e^{j\omega})} \tag{8.61}$$

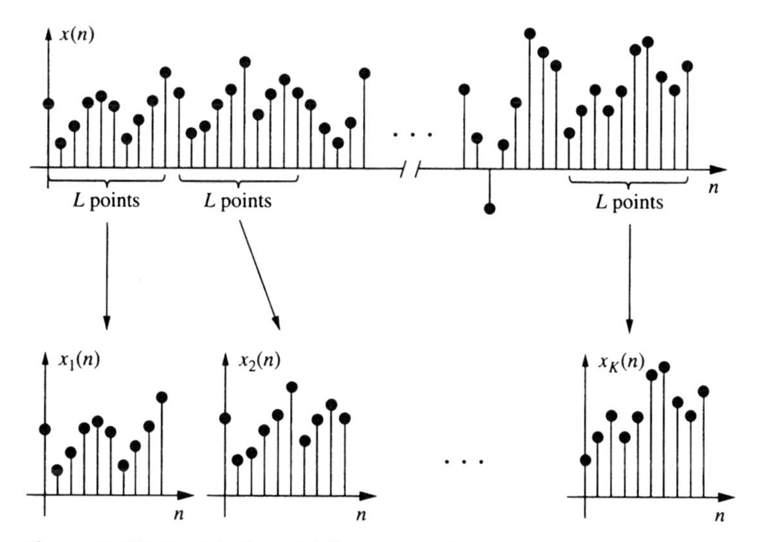

Figure 8.12 *Partitioning $x(n)$ into nonoverlapping subsequences.*

Bartlett's Method

```
function Px = bart(x,nsect)
%
   L  = floor(length(x)/nsect);
   Px = 0;
   n1 = 1;
   for i=1:nsect
       Px = Px + periodogram(x(n1:n1+L-1))/nsect;
       n1 = n1 + L;
       end;
```

Figure 8.13 *A* MATLAB *program for estimating the power spectrum using Bartlett's method of averaging periodograms. Note that this m-file calls* `periodogram.m`.

Therefore, $\hat{P}_B(e^{j\omega})$ is asymptotically unbiased. Second, since the periodograms used in $\hat{P}_B(e^{j\omega})$ are computed using sequences of length L, then the resolution is

$$\text{Res}\left[\hat{P}_B(e^{j\omega})\right] = 0.89\frac{2\pi}{L} = 0.89K\frac{2\pi}{N} \tag{8.62}$$

which is K times larger (worse) than the periodogram. Finally, since the sequences $x_i(n)$ are generally correlated with one another, unless $x(n)$ is white noise, then the variance reduction will not be as large as that given in Eq. (8.59). However, the variance will be inversely proportional to K and, assuming that the data sequences are approximately uncorrelated, for large N the variance is approximately

$$\text{Var}\left\{\hat{P}_B(e^{j\omega})\right\} \approx \frac{1}{K}\text{Var}\left\{\hat{P}_{per}^{(i)}(e^{j\omega})\right\} \approx \frac{1}{K}P_x^2(e^{j\omega}) \tag{8.63}$$

Thus, if both K and L are allowed to go to infinity as $N \to \infty$, then $\hat{P}_B(e^{j\omega})$ will be a consistent estimate of the power spectrum. In addition, for a given value of N, Bartlett's method allows one to trade a reduction in spectral resolution for a reduction in variance by simply changing the values of K and L. Table 8.4 summarizes the properties of Bartlett's method.

Table 8.4 ***Properties of Bartlett's Method***

$$\hat{P}_B(e^{j\omega}) = \frac{1}{N}\sum_{i=0}^{K-1}\left|\sum_{n=0}^{L-1} x(n+iL)e^{-jn\omega}\right|^2$$

Bias

$$E\left\{\hat{P}_B(e^{j\omega})\right\} = \frac{1}{2\pi}P_x(e^{j\omega}) * W_B(e^{j\omega})$$

Resolution

$$\Delta\omega = 0.89K\frac{2\pi}{N}$$

Variance

$$\text{Var}\left\{\hat{P}_B(e^{j\omega})\right\} \approx \frac{1}{K}P_x^2(e^{j\omega})$$

Example 8.2.5 *Bartlett's Method*

In Example 8.2.4, we considered using the periodogram to estimate the power spectrum of white noise. What we observed was that the variance of the estimate did not decrease as we increased the length N of the data sequence. Shown in Fig. 8.14*a* are the periodograms ($K = 1$) of 50 different unit variance white noise sequences of length $N = 512$ and in Fig. 8.14*b* is the ensemble average. The Bartlett estimates for these sequences with $K = 4$ sections of length $L = 128$ are shown in Fig. 8.14*c* with the average given in Fig. 8.14*d*. Similarly, the Bartlett estimates of these 50 processes using $K = 16$ sections of length $L = 32$ are shown in Fig. 8.14*e* and the average in Fig. 8.14*f*. What we observe in these examples is that the variance of the estimate decreases in proportion to the number of sections K.

As another example, let $x(n)$ be a process consisting of two sinusoids in unit variance white noise,

$$x(n) = A\sin(n\omega_1 + \phi_1) + \sin(n\omega_2 + \phi_2) + v(n)$$

where $\omega_1 = 0.2\pi$, $\omega_2 = 0.25\pi$, and $A = \sqrt{10}$. With $N = 512$, an overlay plot of 50 periodograms is shown in Fig. 8.15*a* and the ensemble average of these periodograms is given in Fig. 8.15*b*. Using Bartlett's method with $K = 4$ and $K = 16$ sections, the overlay plots and ensemble averages are shown in Fig. 8.15*c-f*. Note that although the variance of the estimate decreases with K, there is a corresponding decrease in the resolution as evidenced by the broadening of the spectral peaks.

8.2.5 Welch's Method: Averaging Modified Periodograms

In 1967, Welch proposed two modifications to Bartlett's method [59]. The first is to allow the sequences $x_i(n)$ to overlap, and the second is to allow a data window $w(n)$ to be applied to each sequence, thereby producing a set of *modified periodograms* that are to be averaged. Assuming that successive sequences are offset by D points and that each sequence is L points long then the ith sequence is given by

$$x_i(n) = x(n + iD) \quad ; \quad n = 0, 1, \ldots, L-1$$

Thus, the amount of overlap between $x_i(n)$ and $x_{i+1}(n)$ is $L - D$ points, and if K sequences cover the entire N data points, then

$$N = L + D(K-1).$$

For example, with no overlap ($D = L$) we have $K = N/L$ sections of length L as in Bartlett's method. On the other hand, if the sequences are allowed to overlap by 50% ($D = L/2$) then we may form

$$K = 2\frac{N}{L} - 1$$

sections of length L, thus maintaining the same resolution (section length) as Bartlett's method while doubling the number of modified periodograms that are averaged, thereby reducing the variance. However, with a 50% overlap we could also form

$$K = \frac{N}{L} - 1$$

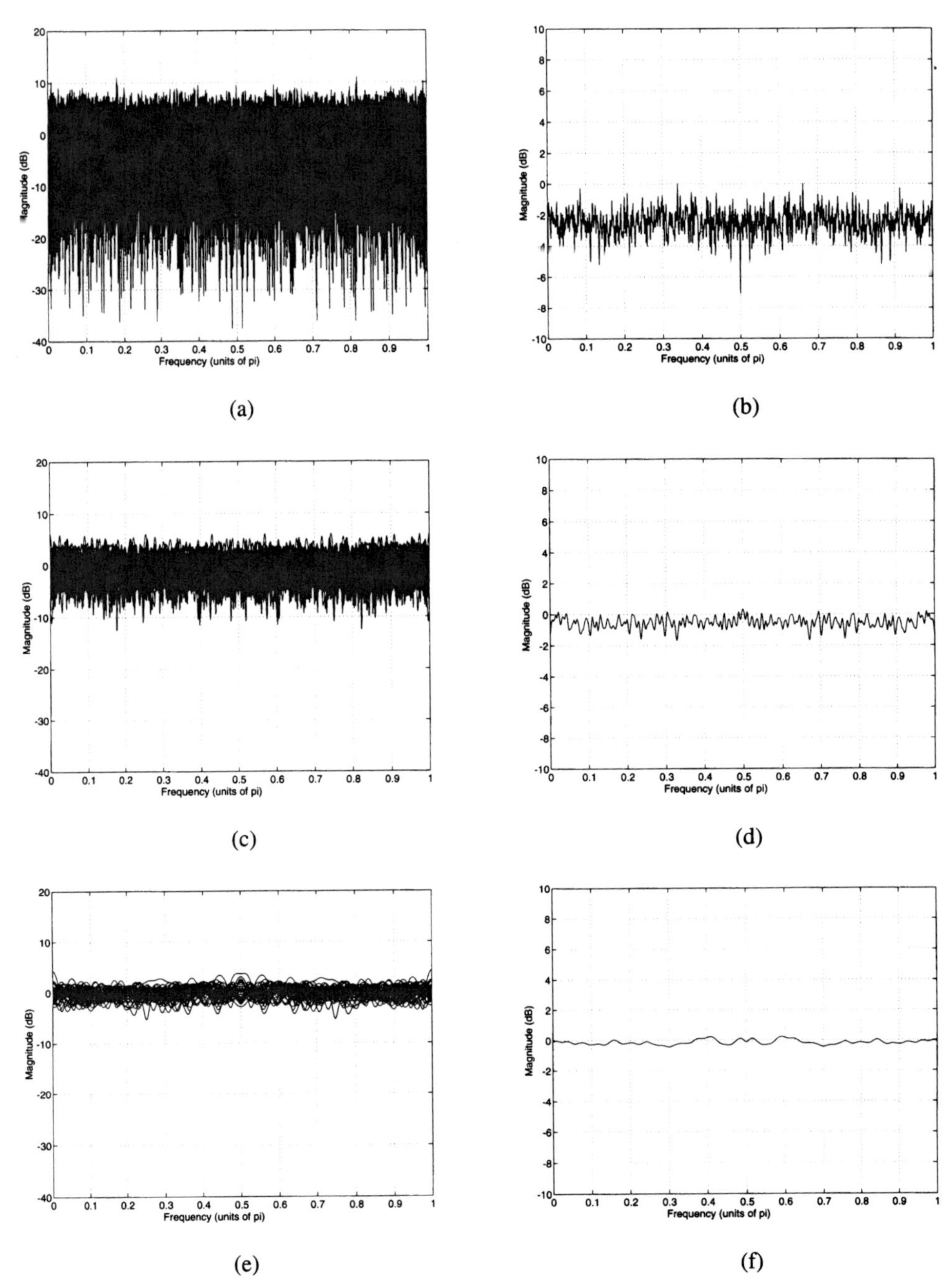

Figure 8.14 *Spectrum estimation of unit variance white Gaussian noise. (a) Overlay plot of 50 periodograms with $N = 512$ and (b) the ensemble average. (c) Overlay plot of 50 Bartlett estimates with $K = 4$ and $L = 128$ and (d) the ensemble average. (e) Overlay plot of 50 Bartlett estimates with $K = 8$ and $L = 64$ and (f) the ensemble average.*

sequences of length $2L$, thus increasing the resolution while maintaining the same variance as Bartlett's method. Therefore, by allowing the sequences to overlap it is possible to increase the number and/or length of the sequences that are averaged, thereby trading a

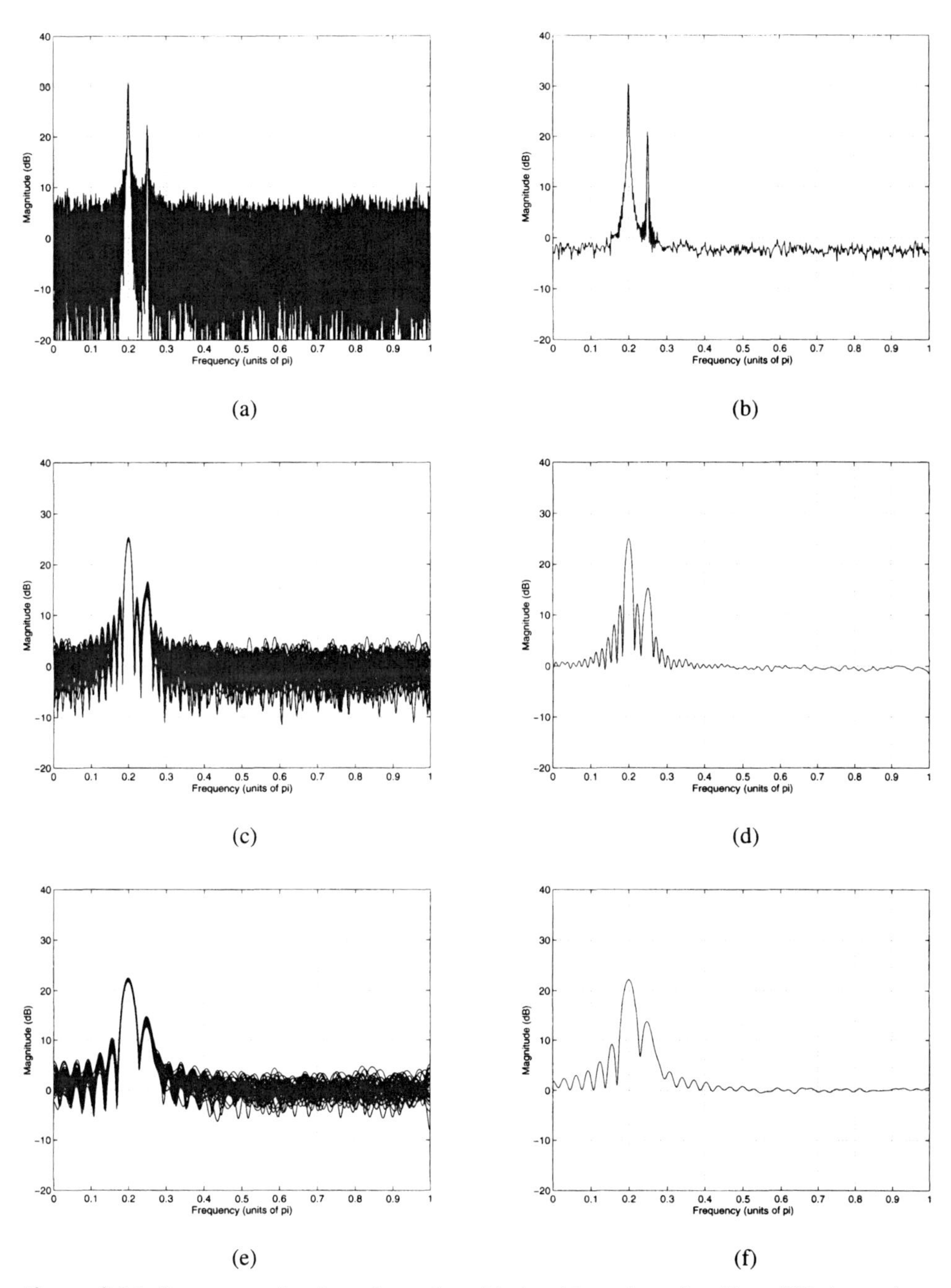

Figure 8.15 *Spectrum estimation of two sinusoids in white noise using* $N = 512$ *data values. (a) Overlay plot of 50 periodograms and (b) the ensemble average. (c) Overlay plot of 50 Bartlett estimates with* $K = 4$ *and* $L = 128$ *and (d) the ensemble average. (e) Overlay plot of 50 Bartlett estimates with* $K = 8$ *and* $L = 64$ *and (f) the ensemble average.*

reduction in variance for a reduction in resolution. A MATLAB program for estimating the power spectrum using Welch's method is given in Fig. 8.16.

Let us now evaluate the performance of Welch's method. Note that the estimate produced

Welch's Method

```
function Px = welch(x,L,over,win)
%
   if (over >= 1) | (over < 0)
      error('Overlap is invalid'), end
   n1 = 1;
   n0 = (1-over)*L;
   nsect=1+floor((length(x)-L)/(n0));
   Px=0;
   for i=1:nsect
       Px = Px + mper(x,win,n1,n1+L-1)/nsect;
       n1 = n1 + n0;
       end;
```

Figure 8.16 *A* MATLAB *program for estimating the power spectrum using Welch's method of averaging modified periodograms. Note that this m-file calls* `mper.m`.

with Welch's method may be written explicitly in terms of $x(n)$ as follows

$$\hat{P}_W(e^{j\omega}) = \frac{1}{KLU}\sum_{i=0}^{K-1}\left|\sum_{n=0}^{L-1} w(n)x(n+iD)e^{-jn\omega}\right|^2 \tag{8.64}$$

or, more succinctly, in terms of modified periodograms as

$$\hat{P}_W(e^{j\omega}) = \frac{1}{K}\sum_{i=0}^{K-1}\hat{P}_M^{(i)}(e^{j\omega}) \tag{8.65}$$

Therefore, the expected value of Welch's estimate is

$$\boxed{E\{\hat{P}_W(e^{j\omega})\} = E\{\hat{P}_M(e^{j\omega})\} = \frac{1}{2\pi LU}P_x(e^{j\omega}) * |W(e^{j\omega})|^2} \tag{8.66}$$

where $W(e^{j\omega})$ is the Fourier transform of the L-point data window, $w(n)$, used in Eq. (8.64) to form the modified periodograms. Thus, as with each of the previous periodogram-based methods, Welch's method is an asymptotically unbiased estimate of the power spectrum. The resolution, however, depends on the data window. As with the modified periodogram, the resolution is defined to be the 3 dB bandwidth of the data window (see Table 8.2). The variance, on the other hand, is more difficult to compute since, with overlapping sequences, the modified periodograms cannot be assumed to be uncorrelated. Nevertheless, it has been shown that, with a Bartlett window and a 50% overlap, the variance is approximately [59]

$$\text{Var}\{\hat{P}_W(e^{j\omega})\} \approx \frac{9}{8K}P_x^2(e^{j\omega}) \tag{8.67}$$

Comparing Eqs. (8.67) and (8.63) we see that, for a given number of sections K, the variance of the estimate using Welch's method is larger than that for Bartlett's method by a factor of $9/8$. However, for a fixed amount of data, N, and a given resolution (sequence length L), with a 50% overlap twice as many sections may be averaged in Welch's method. Expressing

Table 8.5 ***Properties of Welch's Method***

$$\hat{P}_W(e^{j\omega}) = \frac{1}{KLU}\sum_{i=0}^{K-1}\left|\sum_{n=0}^{L-1} w(n)x(n+iD)e^{-jn\omega}\right|^2$$

$$U = \frac{1}{L}\sum_{n=0}^{L-1}|w(n)|^2$$

Bias

$$E\left\{\hat{P}_W(e^{j\omega})\right\} = \frac{1}{2\pi LU}P_x(e^{j\omega}) * |W(e^{j\omega})|^2$$

Resolution Window dependent

Variance†

$$\text{Var}\left\{\hat{P}_W(e^{j\omega})\right\} \approx \frac{9}{16}\frac{L}{N}P_x^2(e^{j\omega})$$

† Assuming 50% overlap and a Bartlett window.

the variance in terms of L and N, we have (assuming a 50% overlap)

$$\boxed{\text{Var}\left\{\hat{P}_W(e^{j\omega})\right\} \approx \frac{9}{16}\frac{L}{N}P_x^2(e^{j\omega})} \tag{8.68}$$

Since N/L is the number of sections that are used in Bartlett's method, it follows from Eq. (8.63) that

$$\text{Var}\left\{\hat{P}_W(e^{j\omega})\right\} \approx \frac{9}{16}\text{Var}\{\hat{P}_B(e^{j\omega})\}$$

Although it is possible to average more sequences for a given amount of data by increasing the amount of overlap, the computational requirements increase in proportion with K. In addition, since an increase in the amount of overlap increases the correlation between the sequences $x_i(n)$, there are diminishing returns when increasing K for a fixed N. Therefore the amount of overlap is typically either 50% or 75%. The properties of Welch's method are summarized in Table 8.5.

Example 8.2.6 *Welch's Method*

Consider the process defined in Example 8.2.5 consisting of two sinusoids in unit variance white noise. Using Welch's method with $N = 512$, a section length $L = 128$, a 50% overlap (7 sections), and a Hamming window, an overlay plot of the spectrum estimates for 50 different realizations of the process are shown in Fig. 8.17*a* and the ensemble average is shown in Fig. 8.17*b*. Comparing these estimates with those shown in Fig. 8.15*e* and *f* of Example 8.2.5, we see that, since the number of sections used in both examples are about the same (7 versus 8), then the variance of the two estimates are approximately the same. In addition, although the width of the main lobe of the Hamming window used in Welch's method is 1.46 times the width of the rectangular window used in Bartlett's method, the resolution is about the same. The reason for this is due to the fact that the 50% overlap that

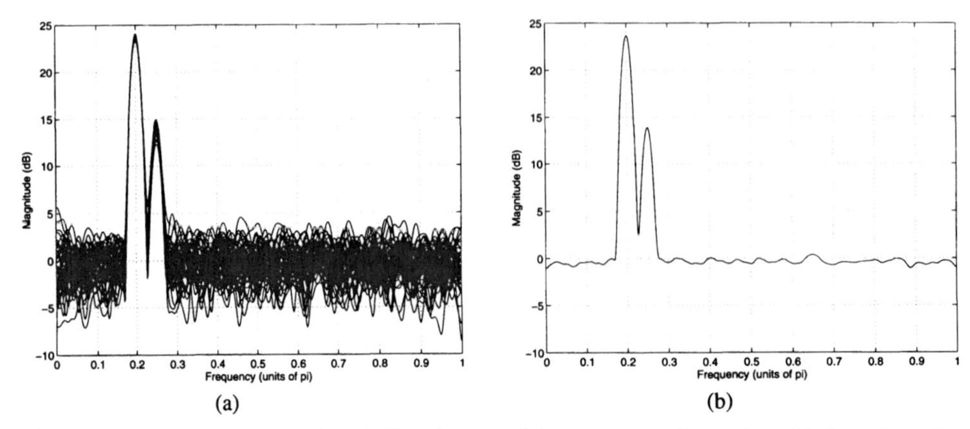

Figure 8.17 *(a) An overlay plot of 50 estimates of the spectrum of two sinusoids in noise using Welch's method with $N = 512$, a section length of $L = 128$, 50% overlap (7 sections), and a Hamming window. (b) The average of the estimates in (a).*

is used in Welch's method allows for the section length to be twice the length of that used in Bartlett's method. So what do we gain with Welch's method? We gain a reduction in the amount of spectral leakage that takes place through the sidelobes of the data window.

8.2.6 Blackman-Tukey Method: Periodogram Smoothing

The methods of Bartlett and Welch are designed to reduce the variance of the periodogram by averaging periodograms and modified periodograms, respectively. Another method for decreasing the statistical variability of the periodogram is periodogram smoothing, often referred to as the Blackman-Tukey method after the pioneering work of Blackman and Tukey in spectrum analysis [6]. To see how periodogram smoothing may reduce the variance of the periodogram, recall that the periodogram is computed by taking the Fourier transform of a consistent estimate of the autocorrelation sequence. However, for any finite data record of length N, the variance of $\hat{r}_x(k)$ will be large for values of k that are close to N. For example, note that the estimate of $r_x(k)$ at lag $k = N - 1$ is

$$\hat{r}_x(N-1) = \frac{1}{N}x(N-1)x(0)$$

Since there is little averaging that goes into the formation of the estimates of $r_x(k)$ for $|k| \approx N$, no matter how large N becomes, these estimates will always be unreliable. Consequently, the only way to reduce the variance of the periodogram is to reduce the variance of these estimates or to reduce the contribution that they make to the periodogram. In the methods of Bartlett and Welch, the variance of the periodogram is decreased by reducing the variance of the autocorrelation estimate by averaging. In the Blackman-Tukey method, the variance of the periodogram is reduced by applying a window to $\hat{r}_x(k)$ in order to decrease the contribution of the unreliable estimates to the periodogram. Specifically, the Blackman-Tukey spectrum estimate is

$$\hat{P}_{BT}(e^{j\omega}) = \sum_{k=-M}^{M} \hat{r}_x(k)w(k)e^{-jk\omega} \tag{8.69}$$

where $w(k)$ is a *lag window* that is applied to the autocorrelation estimate. For example, if $w(k)$ is a rectangular window extending from $-M$ to M with $M < N-1$, then the estimates of $r_x(k)$ having the largest variance are set to zero and, consequently, the power spectrum estimate will have a smaller variance. What is traded for this reduction in variance, however, is a reduction in resolution since a smaller number of autocorrelation estimates are used to form the estimate of the power spectrum.

Using the frequency convolution theorem, the Blackman-Tukey spectrum may be written in the frequency domain as follows:

$$\hat{P}_{BT}(e^{j\omega}) = \frac{1}{2\pi}\hat{P}_{per}(e^{j\omega}) * W(e^{j\omega}) = \frac{1}{2\pi}\int_{-\pi}^{\pi} \hat{P}_{per}(e^{ju})W(e^{j(\omega-u)})du \tag{8.70}$$

Therefore, the Blackman-Tukey estimate smoothes the periodogram by convolving with the Fourier transform of the autocorrelation window, $W(e^{j\omega})$. Although there is considerable flexibility in the choice of the window that may be used, $w(k)$ should be conjugate symmetric so that $W(e^{jw})$ is real-valued, and the window should have a nonnegative Fourier transform, $W(e^{j\omega}) \geq 0$, so that $\hat{P}_{BT}(e^{j\omega})$ is guaranteed to be nonnegative. A MATLAB program for generating the Blackman-Tukey spectrum estimate is given in Fig. 8.18.

To analyze the performance of the Blackman-Tukey method, we will evaluate the bias and the variance (the resolution is window-dependent). The bias may be computed by taking the expected value of Eq. (8.70) as follows:

$$E\{\hat{P}_{BT}(e^{j\omega})\} = \frac{1}{2\pi}E\{\hat{P}_{per}(e^{j\omega})\} * W(e^{j\omega}) \tag{8.71}$$

Blackman-Tukey Method

```
function Px = per_smooth(x,win,M,n1,n2)
%
   x   = x(:);
   if nargin == 3
       n1 = 1;   n2 = length(x);   end;
   R   = covar(x(n1:n2),M);
   r   = [fliplr(R(1,2:M)),R(1,1),R(1,2:M)];
   M   = 2*M-1;
   w   = ones(M,1);
   if (win == 2) w = hamming(M);
       elseif (win == 3) w = hanning(M);
       elseif (win == 4) w = bartlett(M);
       elseif (win == 5) w = blackman(M);
       end;
   r = r'.*w;
   Px = abs(fft(r,1024));
   Px(1)=Px(2);
   end;
```

Figure 8.18 A MATLAB *program for estimating the power spectrum using the Blackman-Tukey method (periodogram smoothing).*

Substituting Eq. (8.23) for the expected value of the periodogram we have

$$E\{\hat{P}_{BT}(e^{j\omega})\} = \frac{1}{2\pi} P_x(e^{j\omega}) * W_B(e^{j\omega}) * W(e^{j\omega}) \tag{8.72}$$

or, equivalently,

$$E\{\hat{P}_{BT}(e^{j\omega})\} = \sum_{k=-M}^{M} r_x(k) w_B(k) w(k) e^{-jk\omega} \tag{8.73}$$

If we let $w_{BT}(k) = w_B(k)w(k)$ be the combined window that is applied to the autocorrelation sequence $r_x(k)$, using the frequency convolution theorem we have

$$E\{\hat{P}_{BT}(e^{j\omega})\} = \frac{1}{2\pi} P_x(e^{j\omega}) * W_{BT}(e^{j\omega}) \tag{8.74}$$

If we assume that $M \ll N$ so that $w_B(k)w(k) \approx w(k)$, then we have

$$\boxed{E\{\hat{P}_{BT}(e^{j\omega})\} \approx \frac{1}{2\pi} P_x(e^{j\omega}) * W(e^{j\omega})} \tag{8.75}$$

where $W(e^{j\omega})$ is the Fourier transform of the lag window $w(k)$.

Evaluating the variance of the Blackman-Tukey spectrum estimate requires a bit more work.[6] Since

$$\text{Var}\left\{\hat{P}_{BT}(e^{j\omega})\right\} = E\left\{\hat{P}_{BT}^2(e^{j\omega})\right\} - E^2\left\{\hat{P}_{BT}(e^{j\omega})\right\} \tag{8.76}$$

we begin by finding the mean-square value $E\{\hat{P}_{BT}^2(e^{j\omega})\}$. From Eq. (8.70) we have

$$\hat{P}_{BT}^2(e^{j\omega}) = \frac{1}{4\pi^2} \int_{-\pi}^{\pi} \int_{-\pi}^{\pi} \hat{P}_{per}(e^{ju}) \hat{P}_{per}(e^{jv}) W(e^{j(\omega-u)}) W(e^{j(\omega-v)}) du dv$$

Therefore, the mean-square value is

$$E\left\{\hat{P}_{BT}^2(e^{j\omega})\right\} = \frac{1}{4\pi^2} \int_{-\pi}^{\pi} \int_{-\pi}^{\pi} E\left\{\hat{P}_{per}(e^{ju}) \hat{P}_{per}(e^{jv})\right\} W(e^{j(\omega-u)}) W(e^{j(\omega-v)}) du dv$$

Using the approximation given in Eq. (8.43) for $E\{\hat{P}_{per}(e^{ju})\hat{P}_{per}(e^{jv})\}$ leads to an expression for the mean-square value that contains two terms. The first term is

$$\begin{aligned} &\frac{1}{4\pi^2} \int_{-\pi}^{\pi} \int_{-\pi}^{\pi} P_x(e^{ju}) P_x(e^{jv}) W(e^{j(\omega-u)}) W(e^{j(\omega-v)}) du dv \\ &\qquad = \left[\frac{1}{2\pi} \int_{-\pi}^{\pi} P_x(e^{ju}) W(e^{j(\omega-u)}) du\right]^2 = E^2\left\{\hat{P}_{BT}(e^{j\omega})\right\} \end{aligned} \tag{8.77}$$

which is cancelled by the second term in Eq. (8.76). Therefore, the variance is

$$\begin{aligned} \text{Var}\left\{\hat{P}_{BT}(e^{j\omega})\right\} = \frac{1}{4\pi^2} &\int_{-\pi}^{\pi} \int_{-\pi}^{\pi} P_x(e^{ju}) P_x(e^{jv}) \\ &\times \left[\frac{\sin N(u-v)/2}{N \sin(u-v)/2}\right]^2 W(e^{j(\omega-u)}) W(e^{j(\omega-v)}) du dv \end{aligned} \tag{8.78}$$

[6] The reader may jump to the final result given in Eq. (8.79) without any loss in continuity.

Since

$$W_B(e^{j\omega}) = \frac{1}{N}\left[\frac{\sin(N\omega/2)}{\sin(\omega/2)}\right]^2$$

is the discrete-time Fourier transform of a Bartlett window, then $w_B(k)$ approaches a constant as $N \to \infty$ and $W_B(e^{j\omega})$ converges to an impulse. Therefore, if N is large then the term in brackets may be approximated by an impulse of area $2\pi/N$,

$$\left[\frac{\sin N(u-v)/2}{N\sin(u-v)/2}\right]^2 \approx \frac{2\pi}{N}u_0(u-v)$$

Thus, for large N, the variance of the Blackman-Tukey estimate is approximately

$$\text{Var}\left\{\hat{P}_{BT}(e^{j\omega})\right\} \approx \frac{1}{2\pi N}\int_{-\pi}^{\pi} P_x^2(e^{ju})W^2(e^{j(\omega-u)})du$$

If M is large enough so that we may assume that $P_x(e^{j\omega})$ is constant across the main lobe of $W(e^{j(\omega-u)})$, then $P_x^2(e^{j\omega})$ may be pulled out of the integral,

$$\text{Var}\left\{\hat{P}_{BT}(e^{j\omega})\right\} \approx \frac{1}{2\pi N}P_x^2(e^{j\omega})\int_{-\pi}^{\pi} W^2(e^{j(\omega-u)})du$$

Finally, using Parseval's theorem we have

$$\boxed{\text{Var}\left\{\hat{P}_{BT}(e^{j\omega})\right\} \approx P_x^2(e^{j\omega})\frac{1}{N}\sum_{k=-M}^{M} w^2(k)} \qquad (8.79)$$

provided $N \gg M \gg 1$. Thus, from Eqs. (8.75) and (8.79) we again see the trade-off between bias and variance. For a small bias, M should be large in order to minimize the width of the main lobe of $W(e^{jw})$ whereas M should be small in order to minimize the sum in Eq. (8.79). Generally, it is recommended that M have a maximum value of $M = N/5$ [26]. The properties of the Blackman-Tukey method are summarized in Table 8.6.

Table 8.6 ***Properties of the Blackman-Tukey Method***

	$\hat{P}_{BT}(e^{j\omega}) = \sum_{k=-M}^{M} \hat{r}_x(k)w(k)e^{-jk\omega}$
Bias	$E\left\{\hat{P}_{BT}(e^{j\omega})\right\} \approx \frac{1}{2\pi}P_x(e^{j\omega}) * W(e^{j\omega})$
Resolution	Window dependent
Variance	$\text{Var}\left\{\hat{P}_{BT}(e^{j\omega})\right\} \approx P_x^2(e^{j\omega})\frac{1}{N}\sum_{k=-M}^{M} w^2(k)$

8.2.7 Performance Comparisons

An important issue in the selection of a spectrum estimation technique is the performance of the estimator. In the previous sections, we have seen that, in comparing one nonparametric method to another, there is a trade-off between resolution and variance. This section summarizes the performance of each nonparametric technique in terms of two criteria. The first is the *variability* of the estimate,

$$\mathcal{V} = \frac{\text{Var}\left\{\hat{P}_x(e^{j\omega})\right\}}{E^2\left\{\hat{P}_x(e^{j\omega})\right\}}$$

which is a *normalized variance*. The second measure is an overall *figure of merit* that is defined as the product of the variability and the resolution,

$$\mathcal{M} = \mathcal{V}\Delta\omega$$

This figure of merit should be as small as possible. However, as we will soon discover, this figure of merit is approximately the same for all of the nonparametric methods that we have considered.

Periodogram. In Section 8.2.2 it was shown that the periodogram is asymptotically unbiased and that, for large N, the variance is approximately equal to $P_x^2(e^{j\omega})$. Asymptotically, therefore, the variability of the periodogram is equal to one

$$\mathcal{V}_{per} = \frac{P_x^2(e^{j\omega})}{P_x^2(e^{j\omega})} = 1$$

Thus, since the resolution of the periodogram is

$$\Delta\omega = 0.89\frac{2\pi}{N}$$

then the overall figure of merit is

$$\mathcal{M}_{per} = 0.89\frac{2\pi}{N}$$

which is inversely proportional to the data record length, N.

Bartlett's Method. In Bartlett's method, a reduction in variance is achieved by averaging periodograms. With $N = KL$, if N is large then the variance is approximately

$$\text{Var}\left\{\hat{P}_B(e^{j\omega})\right\} \approx \frac{1}{K}P_x^2(e^{j\omega})$$

and the variability is

$$\mathcal{V}_B = \frac{1}{K}\frac{P_x^2(e^{j\omega})}{P_x^2(e^{j\omega})} = \frac{1}{K}$$

Since the resolution is $\Delta\omega = 0.89(2\pi K/N)$ then the figure of merit is

$$\mathcal{M}_B = 0.89\frac{2\pi}{N}$$

which is the same as the figure of merit for the periodogram.

Welch's Method. The statistical properties of Welch's method depend on the amount of overlap that is used and on the type of data window. With a 50% overlap and a Bartlett window, the variability for large N is approximately

$$\mathcal{V}_W = \frac{9}{8}\frac{1}{K} = \frac{9}{16}\frac{L}{N}$$

Since the 3 dB bandwidth of a Bartlett window of length L is $1.28(2\pi/L)$ the resolution is

$$\Delta\omega = 1.28\frac{2\pi}{L}$$

and the figure of merit becomes

$$\mathcal{M}_W = 0.72\frac{2\pi}{N}$$

Blackman-Tukey Method. Since the variance and resolution of the Blackman-Tukey method depend on the window that is used, suppose $w(k)$ is a Bartlett window of length $2M$ that extends from $k = -M$ to $k = M$. Assuming that $N \gg M \gg 1$, the variance of the Blackman-Tukey estimate is approximately

$$\text{Var}\left\{\hat{P}_{BT}(e^{j\omega})\right\} \approx P_x^2(e^{j\omega})\frac{1}{N}\sum_{k=-M}^{M}\left(1-\frac{|k|}{M}\right)^2 \approx P_x^2(e^{j\omega})\frac{2M}{3N}$$

(here we have used the series given in Table 2.3 on p. 16 to evaluate the sum). Therefore, the variability is

$$\mathcal{V}_{BT} = \frac{2M}{3N}$$

Since the 3 dB bandwidth of a Bartlett window of length $2M$ is equal to $1.28(2\pi/2M)$, then the resolution is

$$\Delta\omega = 0.64\frac{2\pi}{M}$$

and the figure of merit becomes

$$\mathcal{M}_{BT} = 0.43\frac{2\pi}{N}$$

which is slightly smaller than the figure of merit for Welch's method.

Summary. Table 8.7 provides a summary of the performance measures presented above for the periodogram-based spectrum estimation techniques discussed in this section. What is apparent from this table is that each technique has a figure of merit that is approximately the same, and that these figures of merit are inversely proportional to the length of the data sequence, N. Therefore, although each method differs in its resolution and variance, the overall performance is fundamentally limited by the amount of data that is available. In the following sections, we will look at some entirely different approaches to spectrum estimation in the hope of being able to find a high-resolution spectrum estimate with a small variance that works well on short data records.

Table 8.7 *Performance Measures for the Nonparametric Methods of Spectrum Estimation*

	Variability $\mathcal{V}$	Resolution $\Delta\omega$	Figure of Merit $\mathcal{M}$
Periodogram	1	$0.89\frac{2\pi}{N}$	$0.89\frac{2\pi}{N}$
Bartlett	$\frac{1}{K}$	$0.89K\frac{2\pi}{N}$	$0.89\frac{2\pi}{N}$
Welch†	$\frac{9}{8}\frac{1}{K}$	$1.28\frac{2\pi}{L}$	$0.72\frac{2\pi}{N}$
Blackman-Tukey	$\frac{2}{3}\frac{M}{N}$	$0.64\frac{2\pi}{M}$	$0.43\frac{2\pi}{N}$

† 50% overlap and a Bartlett window.

8.3 MINIMUM VARIANCE SPECTRUM ESTIMATION

Up to this point, we have been considering nonparametric techniques for estimating the power spectrum of a random process. Relying on the DTFT of an estimated autocorrelation sequence, the performance of these methods is limited by the length of the data record. In this section, we develop the Minimum Variance (MV) method of spectrum estimation, which is an adaptation of the Maximum Likelihood Method (MLM) developed by Capon for the analysis of two-dimensional power spectral densities [8]. In the MV method, the power spectrum is estimated by filtering a process with a bank of narrowband bandpass filters. The motivation for this approach may be seen by looking, once again, at the effect of filtering a WSS random process with a narrowband bandpass filter. Therefore, let $x(n)$ be a zero mean wide-sense stationary random process with a power spectrum $P_x(e^{j\omega})$ and let $g_i(n)$ be an ideal bandpass filter with a bandwidth Δ and center frequency ω_i,

$$|G_i(e^{j\omega})| = \begin{cases} 1 & ; \quad |\omega - \omega_i| < \Delta/2 \\ 0 & ; \quad \text{otherwise} \end{cases}$$

If $x(n)$ is filtered with $g_i(n)$, then the power spectrum of the output process, $y_i(n)$, is

$$P_i(e^{j\omega}) = P_x(e^{j\omega})|G_i(e^{j\omega})|^2$$

and the power in $y_i(n)$ is

$$\begin{aligned} E\left\{|y_i(n)|^2\right\} &= \frac{1}{2\pi}\int_{-\pi}^{\pi} P_i(e^{j\omega})d\omega = \frac{1}{2\pi}\int_{-\pi}^{\pi} P_x(e^{j\omega})|G_i(e^{j\omega})|^2 d\omega \\ &= \frac{1}{2\pi}\int_{\omega_i-\Delta/2}^{\omega_i+\Delta/2} P_x(e^{j\omega})d\omega \end{aligned} \tag{8.80}$$

If Δ is small enough so that $P_x(e^{j\omega})$ is approximately constant over the passband of the filter, then the power in $y_i(n)$ is approximately

$$E\left\{|y_i(n)|^2\right\} \approx P_x(e^{j\omega_i})\frac{\Delta}{2\pi} \tag{8.81}$$

Therefore, it is possible to estimate the power spectral density of $x(n)$ at frequency $\omega = \omega_i$ from the filtered process by estimating the power in $y_i(n)$ and dividing by the normalized filter bandwidth, $\Delta/2\pi$,

$$\hat{P}_x(e^{j\omega_i}) = \frac{E\left\{|y_i(n)|^2\right\}}{\Delta/2\pi} \tag{8.82}$$

As we saw in Section 8.2.1, the periodogram produces an estimate of the power spectrum in a similar fashion. Specifically, $x(n)$ is filtered with a bank of bandpass filters, $h_i(n)$, where

$$|H_i(e^{j\omega})| = \frac{\sin[N(\omega - \omega_i)/2]}{N\sin[(\omega - \omega_i)/2]} \tag{8.83}$$

and the power in each of the filtered signals is estimated using a one-point sample average,

$$\hat{E}\left\{|y_i(n)|^2\right\} = |y_i(N-1)|^2$$

The periodogram is then formed by dividing this power estimate by the filter bandwidth, $\Delta = 2\pi/N$. Since each filter in the filter bank for the periodogram is the same, differing only in the center frequency, these filters are *data independent*. As a result, when a random process contains a significant amount of power in frequency bands within the sidelobes of the bandpass filter, leakage through the sidelobes will lead to significant distortion in the power estimates. Therefore, a better approach would be to allow each filter in the filter bank to be *data adaptive* so that each filter may be designed to be "optimum" in the sense of rejecting as much out-of-band signal power as possible. The minimum variance spectrum estimation technique described in this section is based on this idea and involves the following steps:

1. Design a bank of bandpass filters $g_i(n)$ with center frequency ω_i so that each filter rejects the maximum amount of out-of-band power while passing the component at frequency ω_i with no distortion.
2. Filter $x(n)$ with each filter in the filter bank and estimate the power in each output process $y_i(n)$.
3. Set $\hat{P}_x(e^{j\omega_i})$ equal to the power estimated in step (2) divided by the filter bandwidth.

To derive the minimum variance spectrum estimate, we begin with the design of the bandpass filter bank.

Suppose that we would like to estimate the power spectral density of $x(n)$ at frequency ω_i. Let $g_i(n)$ be a complex-valued FIR bandpass filter of order p. To ensure that the filter does not alter the power in the input process at frequency ω_i, $G_i(e^{j\omega})$ will be constrained to have a gain of one at $\omega = \omega_i$,

$$G_i(e^{j\omega_i}) = \sum_{n=0}^{p} g_i(n)e^{-jn\omega_i} = 1 \tag{8.84}$$

Let $\mathbf{g}_i$ be the vector of filter coefficients $g_i(n)$,

$$\mathbf{g}_i = \left[g_i(0), g_i(1), \ldots, g_i(p)\right]^T$$

and let $\mathbf{e}_i$ be the vector of complex exponentials $e^{jk\omega_i}$,

$$\mathbf{e}_i = \left[1, e^{j\omega_i}, \ldots, e^{jp\omega_i}\right]^T$$

The constraint on the frequency response given in Eq. (8.84) may be written in vector form as follows[7]

$$\mathbf{g}_i^H \mathbf{e}_i = \mathbf{e}_i^H \mathbf{g}_i = 1 \tag{8.85}$$

Now, in order for the power spectrum of $x(n)$ at frequency ω_i to be measured as accurately as possible, the bandpass filter should reject as much out-of-band power as possible. Therefore, the criterion that will be used to design the bandpass filter will be to minimize the power in the output process subject to the linear constraint given in Eq. (8.85). Since the power in $y_i(n)$ may be expressed in terms of the autocorrelation matrix $\mathbf{R}_x$ as follows (see Eq. (3.90) on p. 101)

$$E\{|y_i(n)|^2\} = \mathbf{g}_i^H \mathbf{R}_x \mathbf{g}_i \tag{8.86}$$

the filter design problem becomes one of minimizing Eq. (8.86) subject to the linear constraint given in Eq. (8.85). As we saw in Section 2.3.10, the solution to this problem is

$$\mathbf{g}_i = \frac{\mathbf{R}_x^{-1}\mathbf{e}_i}{\mathbf{e}_i^H \mathbf{R}_x^{-1} \mathbf{e}_i} \tag{8.87}$$

where the minimum value of $E\{|y_i(n)|^2\}$ is equal to

$$\min_{\mathbf{g}_i} E\{|y_i(n)|^2\} = \frac{1}{\mathbf{e}_i^H \mathbf{R}_x^{-1} \mathbf{e}_i} \tag{8.88}$$

Thus, Eq. (8.87) defines the optimum filter for estimating the power in $x(n)$ at frequency ω_i, and Eq. (8.88) gives the power in $y_i(n)$, which is used as the estimate, $\hat{\sigma}_x^2(\omega_i)$, of the power in $x(n)$ at frequency ω_i. Note that although these equations were derived for a specific frequency ω_i, since this frequency was arbitrary, then these equations are valid for all ω. Thus, the optimum filter for estimating the power in $x(n)$ at frequency ω is

$$\boxed{\mathbf{g} = \frac{\mathbf{R}_x^{-1}\mathbf{e}}{\mathbf{e}^H \mathbf{R}_x^{-1} \mathbf{e}}} \tag{8.89}$$

and the power estimate is

$$\boxed{\hat{\sigma}_x^2(\omega) = \frac{1}{\mathbf{e}^H \mathbf{R}_x^{-1} \mathbf{e}}} \tag{8.90}$$

where $\mathbf{e} = \left[1, e^{j\omega}, \ldots, e^{jp\omega}\right]^T$.

Example 8.3.1 *White Noise*

Let us consider using Eq. (8.90) to estimate the power in white noise that has a variance of σ_x^2. Since the autocorrelation matrix is $\mathbf{R}_x = \sigma_x^2 \mathbf{I}$, then the minimum variance bandpass filter is

$$\mathbf{g} = \frac{\mathbf{R}_x^{-1}\mathbf{e}}{\mathbf{e}^H \mathbf{R}_x^{-1} \mathbf{e}} = \frac{\sigma_x^{-2}\mathbf{e}}{\sigma_x^{-2}\mathbf{e}^H \mathbf{e}} = \frac{1}{p+1}\mathbf{e}$$

[7] The first equality follows from the fact that $\mathbf{g}_i^H \mathbf{e}_i$ is constrained to be equal to one and, therefore, is real-valued.

which is the same, to within a constant, as the filter in Eq. (8.11) that appears in the filter bank interpretation of the periodogram. From Eq. (8.90) it follows that the estimate of the power in $x(n)$ at frequency ω is

$$\hat{\sigma}_x^2(\omega) = \frac{1}{\mathbf{e}^H \mathbf{R}_x^{-1} \mathbf{e}} = \frac{1}{p+1}\sigma_x^2$$

which is independent of ω. Therefore, the distribution of power in $x(n)$ is a constant. In addition, note that the estimated power decreases as the filter order, p, increases. This is a result of the fact that, as the order of the bandpass filter increases, the bandwidth decreases and less power is allowed to pass through the filter. Since the power spectral density of white noise is constant, the total power over a frequency band of width $\Delta\omega$ is $\sigma_x^2 \Delta\omega$, which goes to zero as $\Delta\omega$ goes to zero. As we will see, what must be done in order to obtain an estimate of the power spectrum is to divide the power estimate by the bandwidth of the filter.

Having designed the bandpass filter bank and estimated the distribution of power in $x(n)$ as a function of frequency, we may now estimate the power spectrum by dividing the power estimate by the bandwidth of the bandpass filter. Although there are several different criteria that may be used to define bandwidth, the simplest is to use the value of Δ that produces the correct power spectral density for white noise. Since the minimum variance estimate of the power in white noise is $E\left\{|y_i(n)|^2\right\} = \sigma_x^2/(p+1)$, it follows from Eq. (8.82) that the spectrum estimate is

$$\hat{P}_x(e^{j\omega_i}) = \frac{E\left\{|y_i(n)|^2\right\}}{\Delta/2\pi} = \frac{\sigma_x^2}{p+1}\frac{2\pi}{\Delta} \tag{8.91}$$

Therefore, if we set

$$\Delta = \frac{2\pi}{p+1} \tag{8.92}$$

then $\hat{P}_x(e^{j\omega}) = \sigma_x^2$. Using Eq. (8.92) as the bandwidth of the filter $g(n)$, the power spectrum estimate becomes, in general,

$$\boxed{\hat{P}_{MV}(e^{j\omega}) = \frac{p+1}{\mathbf{e}^H \mathbf{R}_x^{-1} \mathbf{e}}} \tag{8.93}$$

which is the *minimum variance spectrum estimate*. Note that $\hat{P}_{MV}(e^{j\omega})$ is defined in terms of the autocorrelation matrix $\mathbf{R}_x$, of $x(n)$. If, as is normally the case, the autocorrelation matrix is unknown, then $\mathbf{R}_x$ may be replaced with an estimated autocorrelation matrix, $\hat{\mathbf{R}}_x$. A MATLAB program for computing the MV spectrum estimate in given in Fig. 8.19.

From a computational point of view, the minimum variance spectrum estimate requires the inversion of the autocorrelation matrix $\mathbf{R}_x$ (or $\hat{\mathbf{R}}_x$). Since $\mathbf{R}_x$ is Toeplitz, the inverse may be found using either the Levinson recursion or the Cholesky decomposition. Once the inverse has been found, the quadratic form $\mathbf{e}^H \mathbf{R}_x^{-1} \mathbf{e}$ must be evaluated, which may be done

The Minimum Variance Method

```
function Px = minvar(x,p)
%
   x = x(:);
   R = covar(x,p);
   [v,d]=eig(R);
   U = diag(inv(abs(d)+eps));
   V  = abs(fft(v,1024)).^2;
   Px = 10*log10(p)-10*log10(V*U);
   end;
```

Figure 8.19 *A* MATLAB *program for estimating the power spectrum using the minimum variance method.*

efficiently as follows. Let $v_x(k, l)$ denote the (k, l)th entry in the inverse of $\mathbf{R}_x$

$$\mathbf{R}_x^{-1} = \begin{bmatrix} v_x(0,0) & v_x(0,1) & \cdots & v_x(0,p) \\ v_x(1,0) & v_x(1,1) & \cdots & v_x(1,p) \\ \vdots & \vdots & & \vdots \\ v_x(p,0) & v_x(p,1) & \cdots & v_x(p,p) \end{bmatrix}$$

The quadratic form is

$$\mathbf{e}^H \mathbf{R}_x^{-1} \mathbf{e} = \sum_{k=0}^{p} \sum_{l=0}^{p} e^{-jk\omega} v_x(k,l) e^{jl\omega} = \sum_{k=0}^{p} \sum_{l=0}^{p} v_x(k,l) e^{j(l-k)\omega} \tag{8.94}$$

which may be written as

$$\mathbf{e}^H \mathbf{R}_x^{-1} \mathbf{e} = \sum_{n=-p}^{p} \left[\sum_{k=\max(0,n)}^{\min(p,p+n)} v_x(k, k-n) \right] e^{-jn\omega}$$

Note that the expression in brackets,

$$q(n) = \sum_{k=\max(0,n)}^{\min(p,p+n)} v_x(k, k-n)$$

is the sequence that is formed by summing the terms in the inverse matrix along the diagonals, i.e., $q(0)$ is the sum along the main diagonal, $q(1)$ is the sum along the first diagonal below the main diagonal, $q(-1)$ is the sum along the first diagonal above the main diagonal, and so on. Therefore,

$$\mathbf{e}^H \mathbf{R}_x^{-1} \mathbf{e} = \sum_{n=-p}^{p} q(n) e^{-jn\omega}$$

which is the discrete-time Fourier transform of the sequence $q(n)$. Since $\mathbf{R}_x^{-1}$ is Hermitian, then $q(n) = q^*(-n)$ and we also have

$$\mathbf{e}^H \mathbf{R}_x^{-1} \mathbf{e} = q(0) + 2 \sum_{n=1}^{p} \mathrm{Re}\left\{ q(n) e^{-jn\omega} \right\}$$

or, if $r_x(k)$ is real,

$$\mathbf{e}^H\mathbf{R}_x^{-1}\mathbf{e} = q(0) + 2\sum_{n=1}^{p} q(n)\cos(n\omega)$$

In the following example, we consider the minimum variance estimate of an AR(1) process.

Example 8.3.2 ***The MV Estimate of an AR(1) Process***

Let $x(n)$ be an AR(1) random process with autocorrelation

$$r_x(k) = \frac{1}{1-\alpha^2}\alpha^{|k|}$$

where $|\alpha| < 1$. The power spectrum of $x(n)$ is

$$P_x(e^{j\omega}) = \frac{1}{|1-\alpha e^{-j\omega}|^2} = \frac{1}{1+\alpha^2-2\alpha\cos\omega}$$

Given $r_x(k)$ for $|k| \le p$, the pth-order minimum variance spectrum estimate is

$$\hat{P}_{MV}(e^{j\omega}) = \frac{p+1}{\mathbf{e}^H\mathbf{R}_x^{-1}\mathbf{e}}$$

where $\mathbf{R}_x = \dfrac{1}{1-\alpha^2}\text{Toep}\{1,\ \alpha,\ \ldots\ ,\ \alpha^p\}$. From Example 5.2.11 (p. 258) it follows that the inverse of $\mathbf{R}_x$ is

$$\mathbf{R}_x^{-1} = \begin{bmatrix} 1 & -\alpha & 0 & \cdots & 0 & 0 \\ -\alpha & 1+\alpha^2 & -\alpha & \cdots & 0 & 0 \\ 0 & -\alpha & 1+\alpha^2 & \cdots & 0 & 0 \\ \vdots & \vdots & \vdots & & \vdots & \vdots \\ 0 & 0 & 0 & \cdots & 1+\alpha^2 & -\alpha \\ 0 & 0 & 0 & \cdots & -\alpha & 1 \end{bmatrix}$$

Therefore,

$$q(0) = \left[2+(p-1)(1+\alpha^2)\right]$$

and

$$q(1) = -\alpha p$$

so the minimum variance estimate is

$$\hat{P}_{MV}(e^{j\omega}) = \frac{(p+1)}{2+(p-1)(1+\alpha^2)-2\alpha p\cos\omega}$$

Note that $\hat{P}_{MV}(e^{j\omega})$ converges to $P_x(e^{j\omega})$ as $p \to \infty$.

In a number of applications it is important to be able to estimate the frequency of a sinusoid or a complex exponential in noise. The following example considers the use of the minimum variance method for frequency estimation.

Example 8.3.3 *The MV Estimate of a Complex Exponential in Noise*

Let $x(n)$ be a random phase complex exponential in white noise

$$x(n) = A_1 e^{jn\omega_1} + w(n)$$

where $A_1 = |A_1|e^{j\phi}$ with ϕ a random variable that is uniformly distributed over the interval $[-\pi, \pi]$. If the variance of $w(n)$ is σ_w^2, then the autocorrelation of $x(n)$ is

$$r_x(k) = P_1 e^{jk\omega_1} + \sigma_w^2 \delta(k)$$

where $P_1 = |A_1|^2$. Therefore, we may write the autocorrelation matrix as follows:

$$\mathbf{R}_x = P_1 \mathbf{e}_1 \mathbf{e}_1^H + \sigma_w^2 \mathbf{I}$$

where $\mathbf{e}_1 = \left[1, e^{j\omega_1}, \ldots, e^{jp\omega_1}\right]^T$. Using Woodbury's identity (see p. 29) it follows that the inverse of $\mathbf{R}_x$ is

$$\mathbf{R}_x^{-1} = \frac{1}{\sigma_w^2}\mathbf{I} - \frac{\dfrac{1}{\sigma_w^4} P_1 \mathbf{e}_1 \mathbf{e}_1^H}{1 + \dfrac{P_1}{\sigma_w^2}\mathbf{e}_1^H \mathbf{e}_1} = \frac{1}{\sigma_w^2}\left[\mathbf{I} - \frac{P_1}{\sigma_w^2 + (p+1)P_1}\mathbf{e}_1 \mathbf{e}_1^H\right]$$

(also see Example 2.3.9 on p. 45). Thus,

$$\hat{P}_{MV}(e^{j\omega}) = \frac{p+1}{\mathbf{e}^H \mathbf{R}_x^{-1} \mathbf{e}} = \frac{p+1}{\dfrac{1}{\sigma_w^2}\mathbf{e}^H\left[\mathbf{I} - \dfrac{P_1}{\sigma_w^2 + (p+1)P_1}\mathbf{e}_1 \mathbf{e}_1^H\right]\mathbf{e}}$$

$$= \frac{\sigma_w^2}{1 - \dfrac{P_1/(p+1)}{\sigma_w^2 + (p+1)P_1}|\mathbf{e}^H \mathbf{e}_1|^2} \tag{8.95}$$

Since

$$\mathbf{e}^H \mathbf{e}_1 = \sum_{k=0}^{p} e^{-jk\omega} e^{jk\omega_1} = \sum_{k=0}^{p} e^{-jk(\omega-\omega_1)} = W_R(e^{j(\omega-\omega_1)})$$

where $W_R(e^{j\omega})$ is the discrete-time Fourier transform of a rectangular window that extends from $k = 0$ to $k = p$, then the minimum variance estimate is

$$\hat{P}_{MV}(e^{j\omega}) = \frac{\sigma_w^2}{1 - \dfrac{P_1/(p+1)}{\sigma_w^2 + (p+1)P_1}\left|W_R(e^{j(\omega-\omega_1)})\right|^2}$$

From this expression we see that the minimum variance estimate attains its maximum at $\omega = \omega_1$ with

$$\hat{P}_{MV}(e^{j\omega})\Big|_{\omega=\omega_1} = \frac{\sigma_w^2}{1 - \dfrac{(p+1)P_1}{\sigma_w^2 + (p+1)P_1}} = \sigma_w^2 + (p+1)P_1$$

Therefore, the estimate of the *power* in $x(n)$ at frequency $\omega = \omega_1$ is

$$\hat{\sigma}_x^2(\omega_1) = \frac{1}{p+1}\hat{P}_{MV}(e^{j\omega_1}) = \frac{\sigma_w^2}{p+1} + P_1$$

Thus, if the signal-to-noise ratio is large, $P_1 \gg \sigma_w^2$, then the minimum variance estimate of the power at $\omega = \omega_1$ becomes

$$\hat{\sigma}_x^2(\omega_1) = P_1$$

Furthermore, if $p \gg 1$ and $\omega \neq \omega_1$, then $W_R(e^{j(\omega-\omega_1)}) \approx 0$ and the minimum variance estimate of the power spectrum becomes

$$\hat{P}_{MV}(e^{j\omega}) \approx \sigma_w^2$$

One issue that we have not yet addressed concerns the question of how to select the filter order, p. Clearly, the larger the order of the filter the better the filter will be in rejecting out-of-band power. Therefore, with this in mind, the order of the filter should be as large as possible. From a practical point of view, however, there is a limit on how large the filter order may be. Specifically, note that for a pth-order filter the MV spectrum estimate requires the evaluation of the inverse of a $(p+1) \times (p+1)$ autocorrelation matrix $\mathbf{R}_x$. However, in order to be able to invert this matrix, it is necessary that $r_x(k)$ be known or estimated for $k = 0, 1, \ldots, p$. Therefore, for a fixed data record length, N, since it is only possible to estimate $r_x(k)$ for $k = 0, 1, \ldots, N-1$, the filter order is limited to $p \leq N$. In practice, however, the filter order will generally be much smaller than this due to the large variance of the autocorrelation estimates for values of k that are close to N.

In addition to the approach described above, there are other forms of minimum variance spectrum estimates that differ either in the way that the filter is designed or in the way that the power in the output signal is computed [12,31,54]. For example, in the Data Adaptive Spectrum Estimation (DASE) algorithm [12], $G_i(e^{j\omega})$ is constrained to have unit energy within a passband of width Δ that is centered at frequency ω_i,

$$\frac{1}{\Delta}\int_{\omega_i-\Delta/2}^{\omega_i+\Delta/2} \left|G_i(e^{j\omega})\right|^2 d\omega = 1$$

Solving the constrained minimization problem in this case leads to a generalized eigenvalue problem.

8.4 THE MAXIMUM ENTROPY METHOD

One of the limitations with the classical approach to spectrum estimation is that, for a data record of length N, the autocorrelation sequence can only be estimated for lags $|k| < N$. As a result, $\hat{r}_x(k)$ is set to zero for $|k| \geq N$. Since many signals of interest have autocorrelations that are nonzero for $|k| \geq N$, this windowing may significantly limit the resolution and accuracy of the estimated spectrum. This is particularly true, for example, in the case of narrowband processes that have autocorrelations that decay slowly with k. Since the classical methods effectively extrapolate the autocorrelation sequence with zeros, if it were possible to perform a more accurate extrapolation of the autocorrelation sequence, then the effects of the window could be mitigated and a more accurate estimate of the spectrum could be found. A difficult question to answer, however, is the following: How should this extrapolation be performed? The maximum entropy method (MEM) that is developed in this section suggests one possible way to perform this extrapolation.

Given the autocorrelation $r_x(k)$ of a WSS process for lags $|k| \le p$, the problem that we wish to address, illustrated in Fig. 8.20, is how to extrapolate $r_x(k)$ for $|k| > p$. Denoting the extrapolated values by $r_e(k)$, it is clear that some constraints should be placed on $r_e(k)$. For example, if

$$P_x(e^{j\omega}) = \sum_{k=-p}^{p} r_x(k)e^{-jk\omega} + \sum_{|k|>p} r_e(k)e^{-jk\omega} \tag{8.96}$$

then $P_x(e^{j\omega})$ should correspond to a valid power spectrum, i.e., $P_x(e^{j\omega})$ should be real-valued and nonnegative for all ω. In general, however, only constraining $P_x(e^{j\omega})$ to be real and nonnegative is not sufficient to guarantee a unique extrapolation. Therefore, some additional constraints must be imposed on the set of allowable extrapolations.[8] One such constraint, proposed by Burg [7], is to perform the extrapolation in such a way so as to maximize the entropy of the process.[9] Since entropy is a measure of randomness or uncertainty, a maximum entropy extrapolation is equivalent to finding the sequence of autocorrelations, $r_e(k)$, that make $x(n)$ as *white* (random) as possible. In some sense, such an extrapolation places as few constraints as possible or the least amount of structure on $x(n)$. In terms of the power spectrum, this corresponds to the constraint that $P_x(e^{j\omega})$ be "as flat as possible" (see Fig. 8.21).

For a Gaussian random process with power spectrum $P_x(e^{j\omega})$, the entropy is [11]

$$H(x) = \frac{1}{2\pi}\int_{-\pi}^{\pi} \ln P_x(e^{j\omega})d\omega \tag{8.97}$$

Therefore, for Gaussian processes with a given partial autocorrelation sequence, $r_x(k)$ for $|k| \le p$, the maximum entropy power spectrum is the one that maximizes Eq. (8.97) subject to the constraint that the inverse discrete-time Fourier transform of $P_x(e^{j\omega})$ equals the given set of autocorrelations for $|k| \le p$,

$$\frac{1}{2\pi}\int_{-\pi}^{\pi} P_x(e^{j\omega})e^{jk\omega}d\omega = r_x(k) \quad ; \quad |k| \le p \tag{8.98}$$

The values of $r_e(k)$ that maximize the entropy may be found by setting the derivative of $H(x)$ with respect to $r_e^*(k)$ equal to zero as follows

$$\frac{\partial H(x)}{\partial r_e^*(k)} = \frac{1}{2\pi}\int_{-\pi}^{\pi} \frac{1}{P_x(e^{j\omega})}\frac{\partial P_x(e^{j\omega})}{\partial r_e^*(k)}d\omega = 0 \quad ; \quad |k| > p \tag{8.99}$$

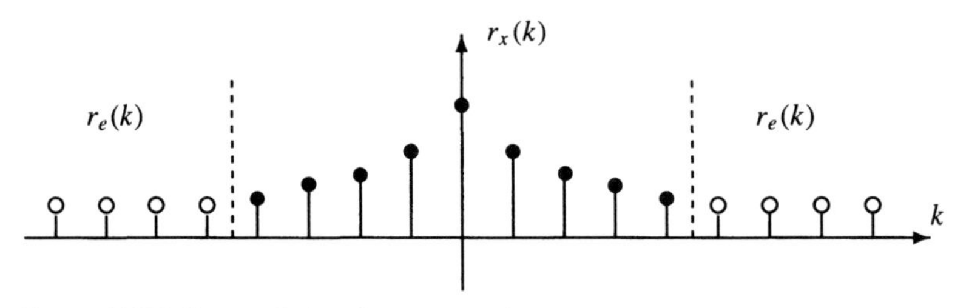

Figure 8.20 *Extrapolating the autocorrelation sequence.*

[8]It is assumed that the given partial autocorrelation sequence is *extendible* so that there is at least one extrapolation that produces a valid power spectrum. See Section 5.2.8 for a discussion of the conditions under which an autocorrelation sequence is extendible.

[9]Entropy has its origins in information theory and is a subject that is well outside the scope of this book. For a treatment of entropy the reader may consult [4, 11, 40].

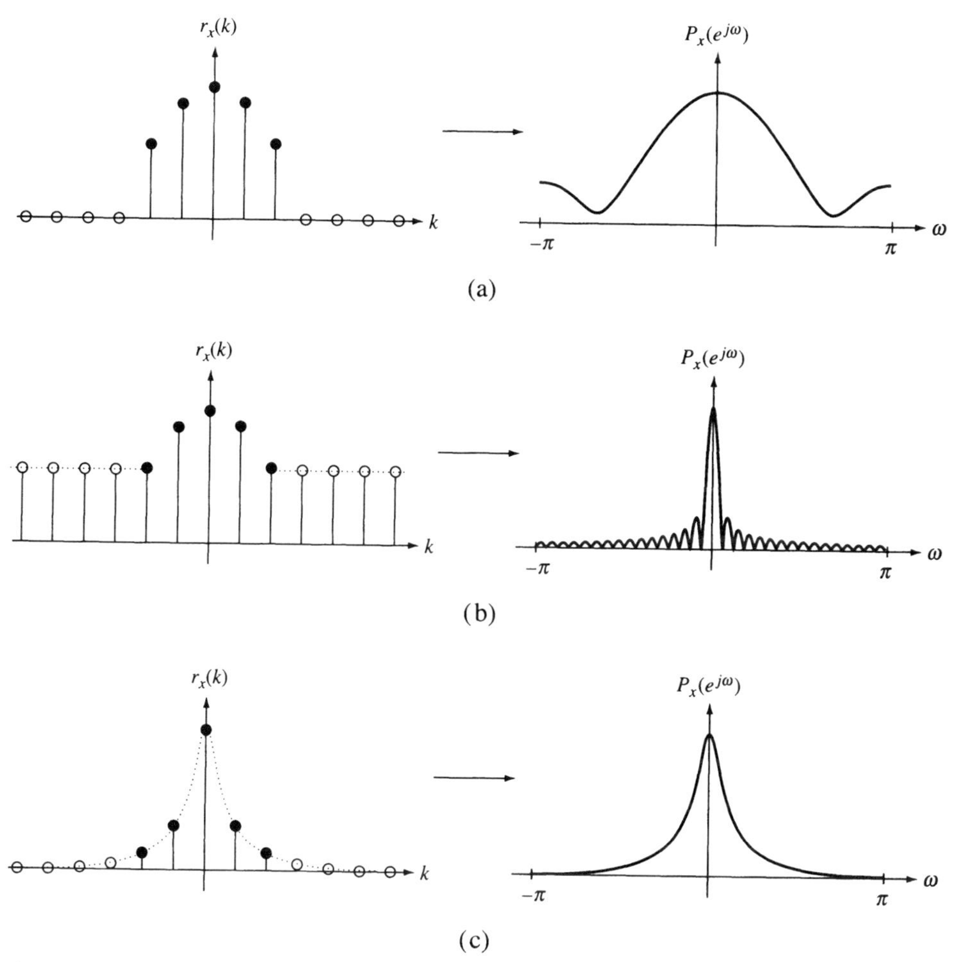

Figure 8.21 *Different extrapolations of a partial autocorrelation sequence and the corresponding power spectral densities. The problem is to find the extrapolation that produces a spectrum that is as flat as possible.*

From Eq. (8.96) we see that[10]

$$\frac{\partial P_x(e^{j\omega})}{\partial r_e^*(k)} = e^{jk\omega}$$

which, when substituted into Eq. (8.99), yields

$$\frac{1}{2\pi}\int_{-\pi}^{\pi} \frac{1}{P_x(e^{j\omega})} e^{jk\omega} d\omega = 0 \quad ; \quad |k| > p \tag{8.100}$$

Defining $Q_x(e^{j\omega}) = 1/P_x(e^{j\omega})$, Eq. (8.100) states that the inverse discrete-time Fourier transform of $Q_x(e^{j\omega})$ is a finite-length sequence that is equal to zero for $|k| > p$,

$$q_x(k) = \frac{1}{2\pi}\int_{-\pi}^{\pi} Q_x(e^{j\omega}) e^{jk\omega} d\omega = 0 \quad ; \quad |k| > p \tag{8.101}$$

[10]Recall that $r_x(k)$ is conjugate symmetric so $r_x(-k) = r_x^*(k)$.

Therefore,

$$Q_x(e^{j\omega}) = \frac{1}{P_x(e^{j\omega})} = \sum_{k=-p}^{p} q_x(k) e^{-jk\omega}$$

and it follows that the maximum entropy power spectrum for a Gaussian process, which we will denote by $\hat{P}_{mem}(e^{j\omega})$, is an all-pole power spectrum,

$$\hat{P}_{mem}(e^{j\omega}) = \frac{1}{\sum_{k=-p}^{p} q_x(k) e^{-jk\omega}} \tag{8.102}$$

Using the spectral factorization theorem, it follows that Eq. (8.102) may be expressed as

$$\hat{P}_{mem}(e^{j\omega}) = \frac{|b(0)|^2}{A_p(e^{j\omega}) A_p^*(e^{j\omega})} = \frac{|b(0)|^2}{\left|1 + \sum_{k=1}^{p} a_p(k) e^{-jk\omega}\right|^2}$$

Alternatively, in terms of the vectors $\mathbf{a}_p = [1, a_p(1), \ldots, a_p(p)]^T$ and $\mathbf{e} = [1, e^{j\omega}, \ldots, e^{jp\omega}]^T$, the MEM spectrum may be written as

$$\hat{P}_{mem}(e^{j\omega}) = \frac{|b(0)|^2}{|\mathbf{e}^H \mathbf{a}_p|^2} \tag{8.103}$$

Having determined the form of the MEM spectrum, all that remains is to find the coefficients $a_p(k)$ and $b(0)$. Due to the constraint given in Eq. (8.98), these coefficients must be chosen in such a way that the inverse discrete-time Fourier transform of $\hat{P}_{mem}(e^{j\omega})$ produces an autocorrelation sequence that matches the given values of $r_x(k)$ for $|k| \le p$. As we saw in Section 5.2.3, if the coefficients $a_p(k)$ are the solution to the autocorrelation normal equations

$$\begin{bmatrix} r_x(0) & r_x^*(1) & \cdots & r_x^*(p) \\ r_x(1) & r_x(0) & \cdots & r_x^*(p-1) \\ \vdots & \vdots & & \vdots \\ r_x(p) & r_x(p-1) & \cdots & r_x(0) \end{bmatrix} \begin{bmatrix} 1 \\ a_p(1) \\ \vdots \\ a_p(p) \end{bmatrix} = \epsilon_p \begin{bmatrix} 1 \\ 0 \\ \vdots \\ 0 \end{bmatrix} \tag{8.104}$$

and if

$$|b(0)|^2 = r_x(0) + \sum_{k=1}^{p} a_p(k) r_x^*(k) = \epsilon_p \tag{8.105}$$

then the autocorrelation matching constraint given in Eq. (8.98) will be satisfied. Thus, the MEM spectrum is

$$\boxed{\hat{P}_{mem}(e^{j\omega}) = \frac{\epsilon_p}{|\mathbf{e}^H \mathbf{a}_p|^2}} \tag{8.106}$$

where $\mathbf{a}_p$ is the solution to Eq. (8.104). In summary, given a sequence of autocorrelations, $r_x(k)$ for $k = 0, 1, \ldots, p$, the MEM spectrum is computed as follows. First, the autocorrelation normal equations (8.104) are solved for the all-pole coefficients $a_p(k)$ and ϵ_p. Then, the MEM spectrum is formed by incorporating these parameters into Eq. (8.106). Note that since $\hat{P}_{mem}(e^{j\omega})$ is an all-pole power spectrum, then $r_x(k)$ satisfies the Yule-Walker

The Maximum Entropy Method

```
function Px = mem(x,p)
%
   [a,e] = acm(x,p);
   Px    = 20*(log10(e)-log10(abs(fft(a,1024))));
   end;
```

Figure 8.22 *A* MATLAB *program for estimating the power spectrum using the maximum entropy method. Note that this m-file calls* `acm.m`.

equations

$$r_x(l) = -\sum_{k=1}^{p} a_p(k) r_x(k-l) \qquad \text{for } \ l > 0 \tag{8.107}$$

Therefore, the maximum entropy method extrapolates the autocorrelation sequence according to this recursion. A MATLAB program for computing the MEM spectrum is given in Fig. 8.22.

The properties of the maximum entropy method have been studied extensively and, as a spectrum analysis tool, the maximum entropy method is subject to different interpretations [10,29]. It may be argued, for example, that in the absence of any information or constraints on a process $x(n)$, given a set of autocorrelation values, $r_x(0), \ldots, r_x(p)$, the best way to estimate the power spectrum is to Fourier Transform the autocorrelation sequence formed from the given values along with an extrapolation that imposes the least amount of structure on the data, i.e., perform a maximum entropy extrapolation. This would seem to be preferable to an extrapolation that somewhat arbitrarily sets $r_x(k) = 0$ for $|k| > p$ as in the classical approach. On the other hand, it may also be argued that since the maximum entropy extrapolation imposes an all-pole model on the data, unless the process is known to be consistent with this model, then the estimated spectrum may not be very accurate. Whether or not the MEM estimate is "better" than the minimum variance method or an estimate produced using a classical approach depends critically on what type of process is being analyzed and on how closely the process may be modeled as an autoregressive process.

Example 8.4.1 *The MEM Estimate of a Complex Exponential in Noise*

In Example 8.3.3, we found the minimum variance estimate of a complex exponential in white noise

$$x(n) = A_1 e^{jn\omega_1} + w(n)$$

where $A_1 = |A_1|e^{j\phi}$ with ϕ being a random variable that is uniformly distributed over the interval $[-\pi, \pi]$ and with the variance of $w(n)$ being equal to σ_w^2. To find the pth-order MEM spectrum, we must solve the autocorrelation normal equations (8.104) for the AR coefficients $a_p(k)$. The $(p+1) \times (p+1)$ autocorrelation matrix for $x(n)$ is

$$\mathbf{R}_x = P_1 \mathbf{e}_1 \mathbf{e}_1^H + \sigma_w^2 \mathbf{I}$$

where $\mathbf{e}_1 = [1, e^{j\omega_1}, \ldots, e^{jp\omega_1}]^T$ and $P_1 = |A_1|^2$. As we saw in Example 8.3.3, the inverse of $\mathbf{R}_x$ is

$$\mathbf{R}_x^{-1} = \frac{1}{\sigma_w^2}\left[\mathbf{I} - \frac{P_1}{\sigma_w^2 + (p+1)P_1}\mathbf{e}_1\mathbf{e}_1^H\right]$$

Therefore,

$$\mathbf{a}_p = \epsilon_p \mathbf{R}_x^{-1}\mathbf{u}_1 = \frac{\epsilon_p}{\sigma_w^2}\left[\mathbf{u}_1 - \frac{P_1}{\sigma_w^2 + (p+1)P_1}\mathbf{e}_1\right] \tag{8.108}$$

where $\mathbf{u}_1 = \left[1,\ 0,\ \ldots,\ 0\right]^T$, and the MEM spectrum is

$$\hat{P}_{mem}(e^{j\omega}) = \frac{\epsilon_p}{|\mathbf{e}^H \mathbf{a}_p|^2} = \frac{\epsilon_p}{\left(\dfrac{\epsilon_p}{\sigma_w^2}\right)^2 \left|\mathbf{e}^H\left(\mathbf{u}_1 - \dfrac{P_1}{\sigma_w^2 + (p+1)P_1}\mathbf{e}_1\right)\right|^2}$$

$$= \frac{\epsilon_p}{\left(\dfrac{\epsilon_p}{\sigma_w^2}\right)^2 \left|1 - \dfrac{P_1}{\sigma_w^2 + (p+1)P_1}\mathbf{e}^H\mathbf{e}_1\right|^2}$$

Since

$$\mathbf{e}^H\mathbf{e}_1 = \sum_{k=0}^{p} e^{-jk(\omega-\omega_1)} = W_R(e^{j(\omega-\omega_1)})$$

where $W_R(e^{j\omega})$ is the discrete-time Fourier transform of a rectangular window, then the MEM spectrum becomes

$$\hat{P}_{mem}(e^{j\omega}) = \frac{\epsilon_p}{\left(\dfrac{\epsilon_p}{\sigma_w^2}\right)^2 \left|1 - \dfrac{P_1}{\sigma_w^2 + (p+1)P_1}W_R(e^{j(\omega-\omega_1)})\right|^2}$$

To determine the value of ϵ_p, note that since $a_p(0) = 1$, then it follows from Eq. (8.108) that

$$a_p(0) = \frac{\epsilon_p}{\sigma_w^2}\left[1 - \frac{P_1}{\sigma_w^2 + (p+1)P_1}\right] = 1$$

Solving for ϵ_p we have

$$\epsilon_p = \sigma_w^2\left[1 + \frac{P_1}{\sigma_w^2 + pP_1}\right]$$

and the MEM spectrum becomes

$$\hat{P}_{mem}(e^{j\omega}) = \frac{\sigma_w^2\left[1 - \dfrac{P_1}{\sigma_w^2 + (p+1)P_1}\right]}{\left|1 - \dfrac{P_1}{\sigma_w^2 + (p+1)P_1}W_R(e^{j(\omega-\omega_1)})\right|^2}$$

As with the minimum variance method, the peak of the MEM spectrum occurs at frequency $\omega = \omega_1$, and

$$\hat{P}_{mem}(e^{j\omega})\Big|_{\omega=\omega_1} = \frac{\sigma_w^2\left[1 - \dfrac{P_1}{\sigma_w^2 + (p+1)P_1}\right]}{\left|1 - \dfrac{(p+1)P_1}{\sigma_w^2 + (p+1)P_1}\right|^2} = \frac{1}{\sigma_w^2}\left[\sigma_w^2 + pP_1\right]\left[\sigma_w^2 + (p+1)P_1\right]$$

If the signal-to-noise ratio is large, $P_1 \gg \sigma_w^2$, then the MEM spectrum estimate at $\omega = \omega_1$ is approximately

$$\hat{P}_{mem}(e^{j\omega})\Big|_{\omega=\omega_1} \approx p^2 \frac{P_1^2}{\sigma_w^2}$$

Thus, the peak in the MEM spectrum is proportional to the *square* of the power in the complex exponential.

There is an interesting relationship that exists between the MEM and MV spectrum estimates. This relationship states that the pth-order MV estimate is the harmonic mean of the MEM estimates of orders $k = 0, 1, \ldots, p$. To derive this relationship, we will use the recursion given in Eq. (5.114) for computing the inverse of a Toeplitz matrix. This recursion provides the following expression for the inverse of the $(p+1) \times (p+1)$ Toeplitz matrix $\mathbf{R}_p$,

$$\mathbf{R}_p^{-1} = \left[\begin{array}{c|ccc} 0 & 0 & \cdots & 0 \\ \hline 0 & & & \\ \vdots & & \mathbf{R}_{p-1}^{-1} & \\ 0 & & & \end{array}\right] + \frac{1}{\epsilon_p}\mathbf{a}_p\mathbf{a}_p^H \tag{8.109}$$

For a WSS process with an autocorrelation matrix $\mathbf{R}_p$, the pth-order MV estimate is

$$\hat{P}_{MV}(e^{j\omega}) = \frac{p+1}{\mathbf{e}^H\mathbf{R}_p^{-1}\mathbf{e}} \tag{8.110}$$

Multiplying $\mathbf{R}_p^{-1}$ on the left by $\mathbf{e}^H$ and on the right by $\mathbf{e}$ we have

$$\mathbf{e}^H\mathbf{R}_p^{-1}\mathbf{e} = \mathbf{e}^H \left[\begin{array}{c|ccc} 0 & 0 & \cdots & 0 \\ \hline 0 & & & \\ \vdots & & \mathbf{R}_{p-1}^{-1} & \\ 0 & & & \end{array}\right] \mathbf{e} + \frac{1}{\epsilon_p}\mathbf{e}^H\mathbf{a}_p\mathbf{a}_p^H\mathbf{e} \tag{8.111}$$

or,[11]

$$\mathbf{e}^H\mathbf{R}_p^{-1}\mathbf{e} = \mathbf{e}^H\mathbf{R}_{p-1}^{-1}\mathbf{e} + \frac{1}{\epsilon_p}|\mathbf{e}^H\mathbf{a}_p|^2 \tag{8.112}$$

Note that the first two terms in Eq. (8.112) are proportional to the inverse of the minimum variance estimates of order p and $p-1$, respectively, and the last term is the reciprocal of the pth-order MEM spectrum. Therefore, we have

$$\frac{p+1}{\hat{P}_{MV}^{(p)}(e^{j\omega})} = \frac{p}{\hat{P}_{MV}^{(p-1)}(e^{j\omega})} + \frac{1}{\hat{P}_{mem}^{(p)}(e^{j\omega})}$$

which is a recursion for the pth-order MV estimate in terms of the $(p-1)$st-order MV estimate and the pth-order MEM spectrum. Solving this recursion for $\hat{P}_{MV}^{(p)}(e^{j\omega})$ we find

$$\boxed{\frac{1}{\hat{P}_{MV}^{(p)}(e^{j\omega})} = \frac{1}{p+1}\sum_{k=0}^{p}\frac{1}{\hat{P}_{mem}^{(k)}(e^{j\omega})}} \tag{8.113}$$

[11]Note that, in order to keep the notation as simple as possible, we are using $\mathbf{e}$ to denote a vector of complex exponentials $e^{jk\omega}$ of varying lengths. Thus, the number of elements in $\mathbf{e}$ is determined by the size of the vector or matrix that it is multiplied by.

Therefore, the MV estimate is the harmonic mean of the MEM spectra from the low order to the high order estimates. As a result of this smoothing, for WSS processes consisting of narrowband components in noise, the MEM spectrum generally provides a higher resolution spectrum estimate than the minimum variance method.

8.5 PARAMETRIC METHODS

One of the limitations of the nonparametric methods of spectrum estimation presented in Sections 8.2 and 8.3 is that they are not designed to incorporate information that may be available about the process into the estimation procedure. In some applications this may be an important limitation, particularly when some knowledge is available about how the data samples are generated. In speech processing, for example, an acoustic tube model for the vocal tract imposes an autoregressive model on the speech waveform [43]. As a result, for intervals of time over which the speech waveform may be assumed to be approximately stationary, the spectrum is of the form,

$$P_x(e^{j\omega}) = \frac{|b(0)|^2}{\left|1 + \sum_{k=1}^{p} a_p(k)e^{-jk\omega}\right|^2}$$

However, with a nonparametric approach such as the periodogram, the power spectrum has a form that is consistent with a moving average process,

$$\hat{P}_{per}(e^{j\omega}) = |X_N(e^{j\omega})|^2 = \sum_{k=-N}^{N} \hat{r}_x(k)e^{-jk\omega}$$

Therefore, one would hope that if it were possible to incorporate a model for the process directly into the spectrum estimation algorithm, then a more accurate and higher resolution estimate could be found. This may be easily done using a parametric approach to spectrum estimation. With a parametric approach, the first step is to select an appropriate model for the process. This selection may be based on a priori knowledge about how the process is generated or, perhaps, on experimental results indicating that a particular model "works well." Models that are commonly used include autoregressive (AR), moving average (MA), autoregressive moving average (ARMA), and harmonic (complex exponentials in noise). Once a model has been selected, the next step is to estimate the model parameters from the given data. The final step is to estimate the power spectrum by incorporating the estimated parameters into the parametric form for the spectrum. For example, with an autoregressive moving average model, with estimates $\hat{b}_q(k)$ and $\hat{a}_p(k)$ of the model parameters, the spectrum estimate would be

$$\hat{P}_x(e^{j\omega}) = \frac{\left|\sum_{k=0}^{q} \hat{b}_q(k)e^{-jk\omega}\right|^2}{\left|1 + \sum_{k=1}^{p} \hat{a}_p(k)e^{-jk\omega}\right|^2}$$

Although it is possible to significantly improve the resolution of the spectrum estimate with a parametric method, it is important to realize that, unless the model that is used is appropriate for the process that is being analyzed, inaccurate or misleading estimates may be obtained. Consider, for example, the two estimates shown in Fig. 8.23*a* that were formed from $N = 64$ values of a process consisting of two sinusoids in unit variance white noise

$$x(n) = 5\sin(0.45\pi n + \phi_1) + 5\sin(0.55\pi n + \phi_2) + w(n)$$

The spectrum estimate shown by the solid line was derived assuming an all-pole model for $x(n)$ whereas the estimate shown by the dashed line was formed using the Blackman-Tukey method. Clearly, the estimate produced with the model-based approach provides much better resolution than the Blackman-Tukey method. Fig. 8.23*b*, on the other hand, shows the spectrum estimates that are formed using the same two approaches for the MA(2) process

$$x(n) = w(n) - w(n-2)$$

where $w(n)$ is unit variance white noise. The power spectrum of $x(n)$ is

$$P_x(e^{j\omega}) = 2 - 2\cos 2\omega$$

In this case, the model-based approach inaccurately represents the underlying spectrum, indicating that the process may contain two narrowband components. The Blackman-Tukey method, on the other hand, which makes no assumptions about the process, produces a more accurate estimate of the power spectrum.

In this section, we consider power spectrum estimation using AR, MA, and ARMA models. The problem of frequency estimation for processes consisting of complex exponentials in noise will be considered in Section 8.6.

8.5.1 Autoregressive Spectrum Estimation

An autoregressive process, $x(n)$, may be represented as the output of an all-pole filter that is driven by unit variance white noise. The power spectrum of a pth-order autoregressive

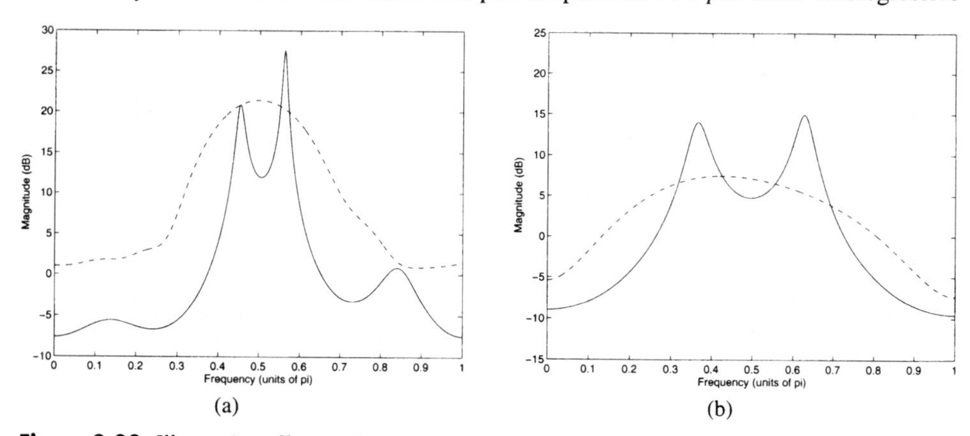

Figure 8.23 *Illustration of how an inappropriate model may lead to an inaccurate spectrum estimate. Using a spectrum estimation technique that assumes an all-pole model for the process (solid line) and the Blackman-Tukey method (dashed line), the spectrum estimates are shown for (a) a process consisting of two sinusoids in noise and (b) a second-order moving average process.*

process is

$$P_x(e^{j\omega}) = \frac{|b(0)|^2}{\left|1 + \sum_{k=1}^{p} a_p(k)e^{-jk\omega}\right|^2} \tag{8.114}$$

Therefore, if $b(0)$ and $a_p(k)$ can be estimated from the data, then an estimate of the power spectrum may be formed using

$$\hat{P}_{AR}(e^{j\omega}) = \frac{|\hat{b}(0)|^2}{\left|1 + \sum_{k=1}^{p} \hat{a}_p(k)e^{-jk\omega}\right|^2} \tag{8.115}$$

Clearly, the accuracy of $\hat{P}_{AR}(e^{j\omega})$ will depend on how accurately the model parameters may be estimated and, even more importantly, on whether or not an autoregressive model is consistent with the way in which the data is generated. If, for example, Eq. (8.115) is applied to a moving average process, then one should expect the estimate to be poor.

Since autoregressive spectrum estimation requires that an all-pole model be found for the process, there is a variety of techniques that may be used to estimate the all-pole parameters. However, once the all-pole parameters have been estimated, each method generates an estimate of the power spectrum in exactly the same way, i.e., using Eq. (8.115). In the following subsections, we briefly review the AR modeling techniques and describe some of the properties of these techniques as they apply to spectrum estimation.

The Autocorrelation Method. In the autocorrelation method of all-pole modeling, the AR coefficients $a_p(k)$ are found by solving the autocorrelation normal equations

$$\begin{bmatrix} r_x(0) & r_x^*(1) & r_x^*(2) & \cdots & r_x^*(p) \\ r_x(1) & r_x(0) & r_x^*(1) & \cdots & r_x^*(p-1) \\ r_x(2) & r_x(1) & r_x(0) & \cdots & r_x^*(p-2) \\ \vdots & \vdots & \vdots & & \vdots \\ r_x(p) & r_x(p-1) & r_x(p-2) & \cdots & r_x(0) \end{bmatrix} \begin{bmatrix} 1 \\ a_p(1) \\ a_p(2) \\ \vdots \\ a_p(p) \end{bmatrix} = \epsilon_p \begin{bmatrix} 1 \\ 0 \\ 0 \\ \vdots \\ 0 \end{bmatrix} \tag{8.116}$$

where

$$r_x(k) = \frac{1}{N} \sum_{n=0}^{N-1-k} x(n+k)x^*(n) \quad ; \quad k = 0, 1, \ldots, p \tag{8.117}$$

(for simplicity we are suppressing the hats over the autocorrelations $r_x(k)$ and all-pole parameters $a_p(k)$). Solving Eq. (8.116) for the coefficients $a_p(k)$, setting

$$|b(0)|^2 = \epsilon_p = r_x(0) + \sum_{k=1}^{p} a_p(k) r_x^*(k) \tag{8.118}$$

and incorporating these parameters into Eq. (8.115) produces an estimate of the power spectrum, which is sometimes referred to as the *Yule-Walker method*.[12] Note that the Yule-Walker method is equivalent to the maximum entropy method. In fact, the only difference between the two methods lies in the assumptions that are made about the process $x(n)$.

[12]This name comes from the fact that the autocorrelation normal equations are equivalent to the Yule-Walker equations for autoregressive processes (see Section 3.6.2).

Specifically, with the Yule-Walker method it is assumed that $x(n)$ is an autoregressive process, whereas with the maximum entropy method it is assumed that $x(n)$ is Gaussian.

Since the autocorrelation matrix $\mathbf{R}_x$ in the autocorrelation normal equations is Toeplitz, the Levinson-Durbin recursion may be used to solve these equations for $a_p(k)$. Furthermore, if $\mathbf{R}_x > 0$, then the roots of $A_p(z)$ will lie inside the unit circle. However, because the autocorrelation method effectively applies a rectangular window to the data when estimating the autocorrelation sequence using Eq. (8.117), the data is effectively extrapolated with zeros. Therefore, the autocorrelation method generally produces a lower resolution estimate than the approaches that do not window the data, such as the covariance method and Burg's method. Consequently, for short data records the autocorrelation method is not generally used.

An artifact that may be observed with the autocorrelation method is *spectral line splitting*. This artifact involves the splitting of a single spectral peak into two separate and distinct peaks. Typically, spectral line splitting occurs when $x(n)$ is *overmodeled*, i.e., when p is too large. An example of spectral line splitting is shown in Fig. 8.24 for an AR(2) process that is generated according to the difference equation

$$x(n) = -0.9x(n-2) + w(n)$$

where $w(n)$ is unit variance white noise. Using model orders of $p = 4$ and $p = 12$ and a data record of length $N = 64$, spectrum estimates are shown for the autocorrelation method. What we observe is that, although the true spectrum has a single spectral peak at $\omega = \pi/2$, when $p = 12$ this peak is split into two peaks.[13]

Since the autocorrelation estimate in Eq. (8.117) is biased, a variation of the autocorrelation method is to use the unbiased estimate

$$\hat{r}_x(k) = \frac{1}{N-k} \sum_{n=0}^{N-1-k} x(n+k)x^*(n) \quad ; \quad k = 0, 1, \ldots, p \tag{8.119}$$

In this case, however, the autocorrelation matrix is not guaranteed to be positive definite and, as a result, the variance of the spectrum estimate tends to become large when $\hat{\mathbf{R}}_x$

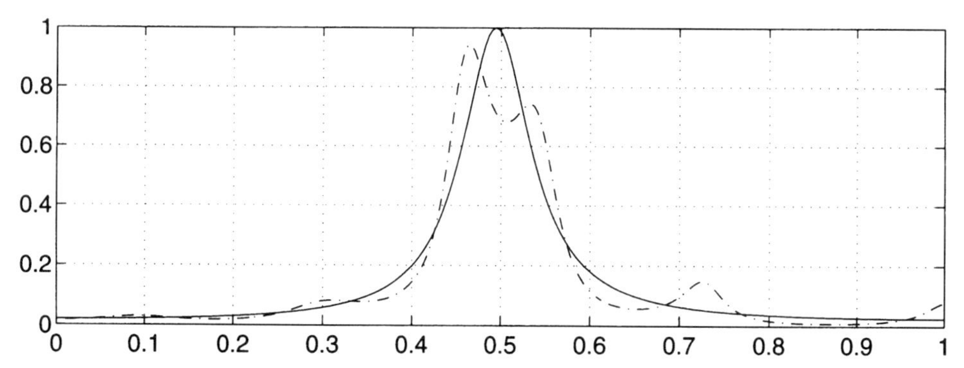

Figure 8.24 *Spectral line splitting of an AR(2) process. Two all-pole spectrum estimates were computed using the autocorrelation method with orders $p = 4$ (solid line) and $p = 12$ (dash-dot line).*

[13]Note that for a process such as this, spectral line splitting will not *always* be observed. Whether or not line splitting occurs depends on the specific white noise process that generates $x(n)$.

is ill-conditioned or singular [26,37]. Therefore, the biased estimate of $r_x(k)$ is generally preferred over the unbiased estimate.

The Covariance Method. Another approach for estimating the AR parameters is the covariance method. The covariance method requires finding the solution to the set of linear equations,

$$\begin{bmatrix} r_x(1,1) & r_x(2,1) & \cdots & r_x(p,1) \\ r_x(1,2) & r_x(2,2) & \cdots & r_x(p,2) \\ \vdots & \vdots & & \vdots \\ r_x(1,p) & r_x(2,p) & \cdots & r_x(p,p) \end{bmatrix} \begin{bmatrix} a_p(1) \\ a_p(2) \\ \vdots \\ a_p(p) \end{bmatrix} = - \begin{bmatrix} r_x(0,1) \\ r_x(0,2) \\ \vdots \\ r_x(0,p) \end{bmatrix} \tag{8.120}$$

where

$$r_x(k,l) = \sum_{n=p}^{N-1} x(n-l)x^*(n-k) \tag{8.121}$$

Unlike the linear equations in the autocorrelation method, these equations are not Toeplitz. However, the advantage of the covariance method over the autocorrelation method is that no windowing of the data is required in the formation of the autocorrelation estimates, $r_x(k,l)$. Therefore, for short data records the covariance method generally produces higher resolution spectrum estimates than the autocorrelation method. However, as the data record length increases and becomes large compared to the model order, $N \gg p$, the effect of the data window becomes small and the difference between the two approaches becomes negligible.

The Modified Covariance Method. The modified covariance method is similar to the covariance method in that no window is applied to the data. However, instead of finding the autoregressive model that minimizes the sum of the squares of the forward prediction error, the modified covariance method minimizes the sum of the squares of the forward and backward prediction errors.[14] As a result, the autoregressive parameters in the modified covariance method are found by solving a set of linear equations of the form given in Eq. (8.120) with

$$r_x(k,l) = \sum_{n=p}^{N-1} \Big[x(n-l)x^*(n-k) + x(n-p+l)x^*(n-p+k) \Big]$$

replacing the estimate in Eq. (8.121). As with the covariance method, the autocorrelation matrix is not Toeplitz.

In contrast to other AR spectrum estimation techniques, the modified covariance method appears to give statistically stable spectrum estimates with high resolution [37,51]. Furthermore, in the spectral analysis of sinusoids in white noise, although the modified covariance method is characterized by a shifting of the peaks from their true locations duc to additive noise, this shifting appears to be less pronounced than with other autoregressive estimation techniques [53]. In addition, the peak locations tend to be less sensitive to phase [9,57]. Finally, unlike the previous methods, it appears that the modified covariance method is not subject to spectral line splitting [27].

[14]The modified covariance method has also been referred to as the *Forward-Backward Method* [37] and as the *Least Squares Method* [57].

The Burg Algorithm. As with the modified covariance algorithm, the Burg algorithm finds a set of all-pole model parameters that minimizes the sum of the squares of the forward and backward prediction errors. However, in order to assure that the model is stable, this minimization is performed sequentially with respect to the reflection coefficients. Although less accurate than the modified covariance method, since the Burg algorithm does not apply a window to the data, the estimates of the autoregressive parameters are more accurate than those obtained with the autocorrelation method. In the analysis of sinusoids in noise, the Burg algorithm is subject to spectral line splitting and the peak locations are highly dependent upon the phases of the sinusoids [27,35].

Examples. It is difficult to provide a complete set of examples to illustrate the properties of the AR spectrum estimation techniques for different types of processes and to compare them to other spectrum estimation algorithms. It would be interesting, for example, to compare the effectiveness of each approach in estimating the spectra of narrowband and wideband autoregressive processes of various orders using different data record lengths. It would also be interesting to look at what happens when these techniques are applied to other types of processes such as MA processes, ARMA processes, and harmonic processes. Rather than attempting to cover all of the possibilities, in the following we only consider the use of AR spectrum estimation techniques to analyze a short data record that is derived from a fourth-order narrowband autoregressive process. Additional experiments with the AR techniques are left to the computer exercises.

Consider the AR(4) process that is generated by the difference equation

$$\begin{aligned} x(n) = 2.7377x(n-1) - 3.7476x(n-2) + 2.6293x(n-3) \\ - 0.9224x(n-4) + w(n) \end{aligned} \tag{8.122}$$

where $w(n)$ is unit variance white Gaussian noise. The filter that generates $x(n)$ has a pair of complex poles at $z = 0.98e^{\pm j(0.2\pi)}$ and a pair of complex poles at $z = 0.98e^{\pm j(0.3\pi)}$. Using data records of length $N = 128$, an ensemble of 50 spectrum estimates were computed using the Yule-Walker method, the covariance method, the modified covariance method, and Burg's method. Shown in part *a* of Figs. 8.25 to 8.28 are overlay plots of these estimates and in part *b* the ensemble average is plotted along with the true power spectrum. What we observe from these plots is that, for this narrowband process, all of the estimates, except the Yule-Walker method, appear to be unbiased and to have a comparable variance. The Yule-Walker method, on the other hand, is unable to resolve the spectral peaks and has a larger variance.

Selecting the Model Order. A question that remains to be answered in the use of an AR spectrum estimation method is how to select the model order p of the AR process. If the model order that is used is too small, then the resulting spectrum will be smoothed and will have poor resolution. If, on the other hand, the model order is too large, then the spectrum may contain spurious peaks and, as illustrated in Fig. 8.24, may lead to spectral line splitting. Therefore, it would be useful to have a criterion that indicates the appropriate model order to use for a given set of data. One approach would be to increase the model order until the modeling error is minimized. The difficulty with this, however, is that the error is a monotonically nonincreasing function of the model order p. This problem may be overcome by incorporating a penalty function that increases with the model order p. Several criteria have been proposed that include a penalty term that increases *linearly* with p

$$C(p) = N \log \mathcal{E}_p + f(N)p \tag{8.123}$$

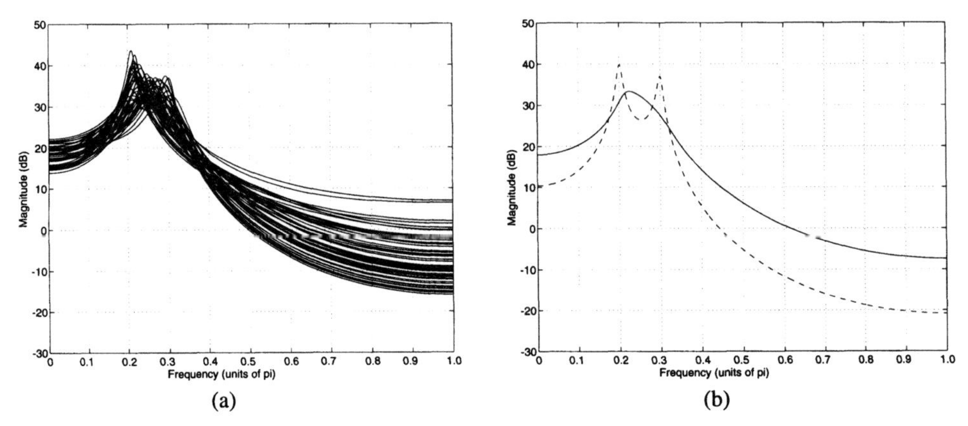

Figure 8.25 *Spectrum estimation of a fourth-order autoregressive process using the Yule-Walker method. (a) Overlay plot of 50 spectrum estimates. (b) The average of the estimates in (a) with the true power spectrum indicated by the dashed line.*

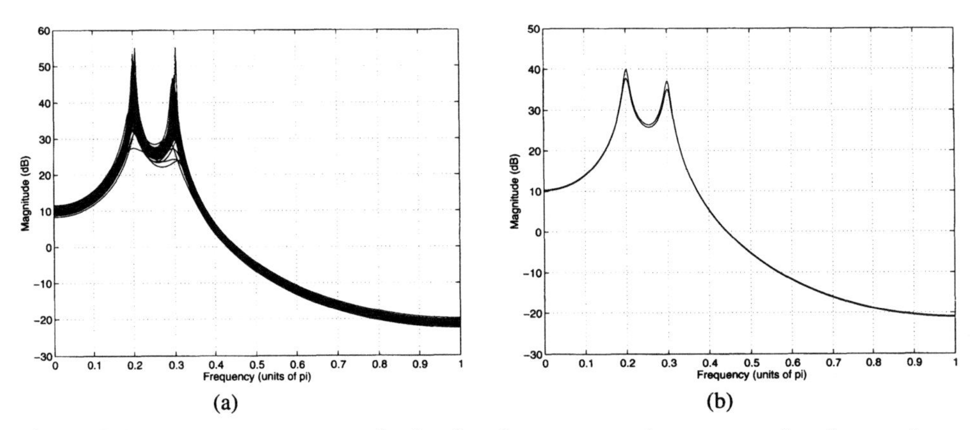

Figure 8.26 *Spectrum estimation of a fourth-order autoregressive process using the covariance method. (a) Overlay plot of 50 spectrum estimates. (b) The average of the estimates in (a) with the true power spectrum indicated by the dashed line.*

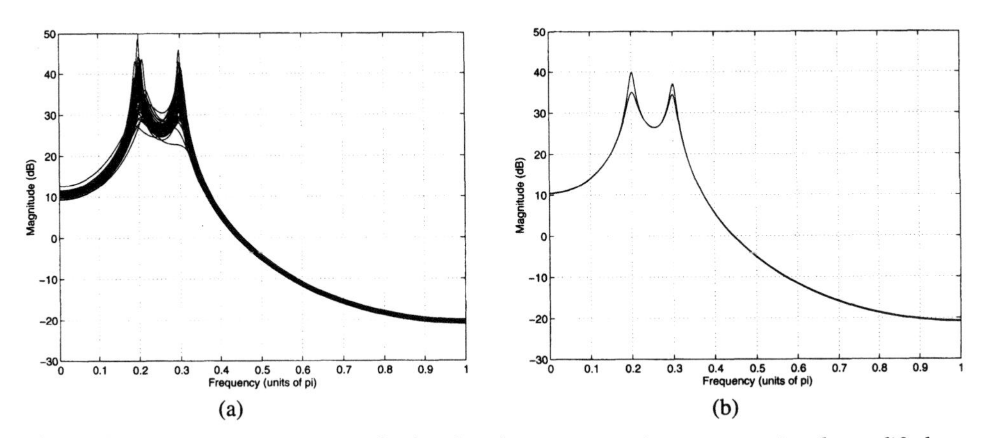

Figure 8.27 *Spectrum estimation of a fourth-order autoregressive process using the modified covariance method. (a) Overlay plot of 50 spectrum estimates. (b) The average of the estimates in (a) with the true power spectrum indicated by the dashed line.*

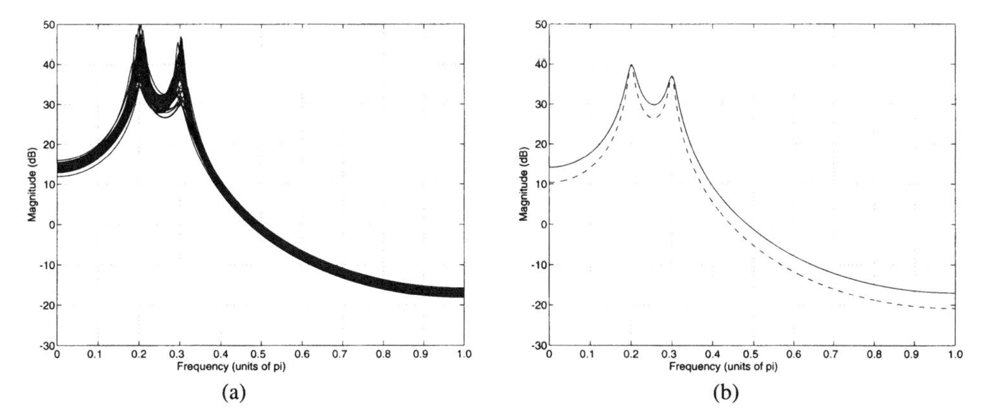

Figure 8.28 *Spectrum estimation of a fourth-order autoregressive process using Burg's method. (a) Overlay plot of 50 spectrum estimates. (b) The average of the estimates in (a) with the true power spectrum indicated by the dashed line.*

Here, $\mathcal{E}_p$ is the modeling error, N is the data record length, and $f(N)$ is a constant that may depend upon N. The idea, then, is to select the value of p that minimizes $C(p)$. Two criteria that are of this form are the Akaike Information Criterion [1,2]

$$\text{AIC}(p) = N \log \mathcal{E}_p + 2p \tag{8.124}$$

and the minimum description length proposed by Rissanen [45],

$$\text{MDL}(p) = N \log \mathcal{E}_p + (\log N)p \tag{8.125}$$

The AIC was derived by minimizing an information theoretic function and includes the penalty $2p$ for any extra AR coefficients that do not significantly reduce the prediction error. It has been observed that the AIC gives an estimate for the order p that is too small when applied to nonautoregressive processes and that it tends to overestimate the order as N increases [23,24,55]. The MDL, on the other hand, contains the penalty term $(\log N)p$, which increases with the data record length N and the model order p. It has been shown that the MDL is a consistent model-order estimator in the sense that it converges to the true order as the number of observations, N, increases [45,58]. Two other model order selection criteria that are often used are Akaike's Final Prediction Error [2],

$$\text{FPE}(p) = \mathcal{E}_p \frac{N + p + 1}{N - p - 1} \tag{8.126}$$

and Parzen's Criterion Autoregressive Transfer function [39]

$$\text{CAT}(p) = \left[\frac{1}{N} \sum_{j=1}^{p} \frac{N - j}{N\mathcal{E}_j} \right] - \frac{N - p}{N\mathcal{E}_p} \tag{8.127}$$

The success of each of these criteria in estimating model orders has been mixed. In fact, for short data sequences, none of the criteria tend to work particularly well [56]. Therefore, these criteria should only be used as "indicators" of the model order. It should also be pointed out that since each of these criteria depends upon the prediction error, $\mathcal{E}_p$, the model order will depend on the modeling technique that is used, e.g. the predicted model order may not be the same when using the autocorrelation method and the Burg algorithm. Table 8.8 summarizes these model order criteria.

Table 8.8 ***Model Order Selection Criteria***

$$\mathrm{AIC}(p) = N \log \mathcal{E}_p + 2p$$

$$\mathrm{MDL}(p) = N \log \mathcal{E}_p + (\log N)p$$

$$\mathrm{FPE}(p) = \mathcal{E}_p \frac{N+p+1}{N-p-1}$$

$$\mathrm{CAT}(p) = \left[\frac{1}{N}\sum_{j=1}^{p} \frac{N-j}{N\mathcal{E}_j}\right] - \frac{N-p}{N\mathcal{E}_p}$$

8.5.2 Moving Average Spectrum Estimation

A moving average process may be generated by filtering unit variance white noise, $w(n)$, with an FIR filter as follows:

$$x(n) = \sum_{k=0}^{q} b_q(k) w(n-k) \tag{8.128}$$

As discussed in Section 3.6.3, the relationship between the power spectrum of a moving average process and the coefficients $b_q(k)$ is

$$P_x(e^{j\omega}) = \left|\sum_{k=0}^{q} b_q(k) e^{-jk\omega}\right|^2 \tag{8.129}$$

Equivalently, the power spectrum may be written in terms of the autocorrelation sequence $r_x(k)$ as

$$P_x(e^{j\omega}) = \sum_{k=-q}^{q} r_x(k) e^{-jk\omega} \tag{8.130}$$

where $r_x(k)$ is related to the filter coefficients $b_q(k)$ through the Yule-Walker equations (see Section 3.6.3)

$$r_x(k) = \sum_{l=0}^{q-k} b_q(l+k) b_q^*(l) \quad ; \quad k = 0, 1, \ldots, q \tag{8.131}$$

with $r_x(-k) = r_x^*(k)$ and $r_x(k) = 0$ for $|k| > q$.

With a moving average model, the spectrum may be estimated in one of two ways. The first approach is to take advantage of the fact that the autocorrelation sequence of a moving average process is finite in length. Specifically, since $r_x(k) = 0$ for $|k| > q$, then a natural estimate to use is

$$\boxed{\hat{P}_{MA}(e^{j\omega}) = \sum_{k=-q}^{q} \hat{r}_x(k) e^{-jk\omega}} \tag{8.132}$$

where $\hat{r}_x(k)$ is a suitable estimate of the autocorrelation sequence. Note that although $\hat{P}_{MA}(e^{j\omega})$ is equivalent to the Blackman-Tukey estimate using a rectangular window, there is a subtle difference in the assumptions that are behind these two estimates. In particular, since Eq. (8.132) assumes that $x(n)$ is a moving average process of order q, then the true autocorrelation sequence is zero for $|k| > q$. Thus, if an unbiased estimate of the autocorrelation sequence is used for $|k| \leq q$, then

$$E\left\{\hat{P}_{MA}(e^{j\omega})\right\} = P_x(e^{j\omega})$$

i.e., $\hat{P}_{MA}(e^{j\omega})$ is unbiased. The Blackman-Tukey method, on the other hand, makes no assumptions about $x(n)$ and may be applied to any type of process. Therefore, due to the windowing of the autocorrelation sequence, unless $x(n)$ is a moving average process, the Blackman-Tukey spectrum will be biased.

The second approach is to estimate the moving average parameters, $b_q(k)$, from $x(n)$ and then substitute these estimates into Eq. (8.129) as follows:

$$\boxed{\hat{P}_{MA}(e^{j\omega}) = \left|\sum_{k=0}^{q} \hat{b}_q(k)e^{-jk\omega}\right|^2} \tag{8.133}$$

For example, $b_q(k)$ may be estimated using the two-stage approach developed by Durbin (see Section 4.7.3).

As with autoregressive spectrum estimation, it is useful to have a criterion for estimating the order of the MA model that should be used for a given process $x(n)$. A discussion of some of these estimation methods may be found in [25].

Examples. As we did with the autoregressive methods, here we present only a few examples of MA spectrum estimation rather than attempting to be complete in covering all of the interesting possibilities. Specifically, we consider a fourth-order moving average process that is generated by the difference equation

$$\begin{aligned} x(n) = w(n) &- 1.5857w(n-1) + 1.9208w(n-2) \\ &- 1.5229w(n-3) + 0.9224w(n-4) \end{aligned} \tag{8.134}$$

where $w(n)$ is unit variance white Gaussian noise. The filter that generates $x(n)$ has a pair of complex zeroes at $z = 0.98e^{\pm j(0.2\pi)}$ and a pair of complex zeroes at $z = 0.98e^{\pm j(0.5\pi)}$. Using data records of length $N = 128$, an ensemble of 50 spectrum estimates were computed using the Blackman-Tukey method with a rectangular window that extends from $k = -4$ to $k = 4$. Shown in Fig. 8.29*a* is an overlay plot of these estimates and in *b* is the ensemble average along with the true power spectrum. These plots are repeated in Fig. (8.30) using Durbin's method with $q = 4$ and $p = 32$ (recall that p is the order of the all-pole model that is used to model $x(n)$ prior to estimating the moving average coefficients).

8.5.3 Autoregressive Moving Average Spectrum Estimation

An autoregressive moving average process has a power spectrum of the form

$$P_x(e^{j\omega}) = \frac{\left|\sum_{k=0}^{q} b_q(k)e^{-jk\omega}\right|^2}{\left|1 + \sum_{k=1}^{p} a_p(k)^{-jk\omega}\right|^2} \tag{8.135}$$

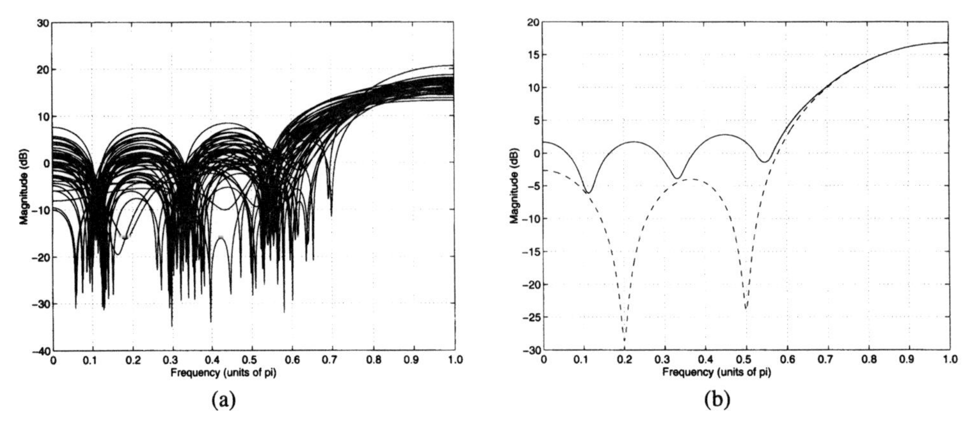

Figure 8.29 *Spectrum estimation of a fourth-order moving average process using the Blackman-Tukey method. (a) Overlay plot of 50 spectrum estimates. (b) The average of the estimates in (a) with the true power spectrum indicated by the dashed line.*

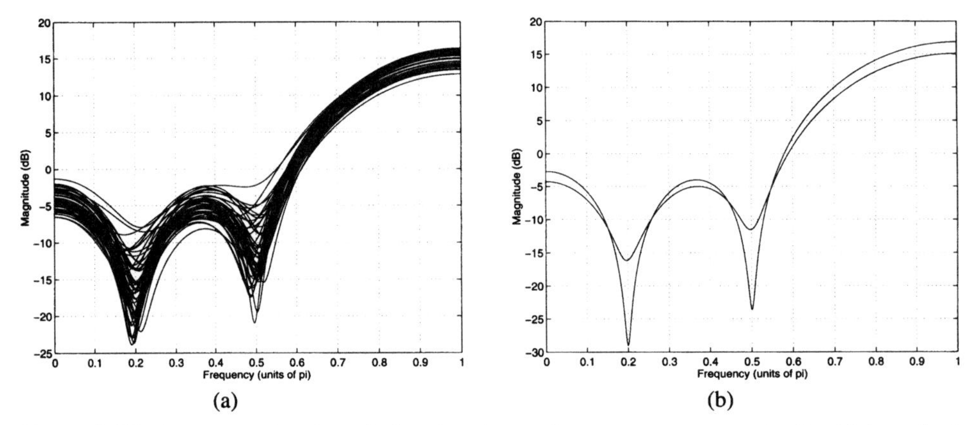

Figure 8.30 *Spectrum estimation of a fourth-order moving average process using Durbin's method. (a) Overlay plot of 50 spectrum estimates. (b) The average of the estimates in (a) with the true power spectrum indicated by the dashed line.*

which may be generated by filtering unit variance white noise with a filter having both poles and zeros,

$$H(z) = \frac{B_q(z)}{A_p(z)} = \frac{\sum_{k=0}^{q} b_q(k) z^{-k}}{1 + \sum_{k=1}^{p} a_p(k) z^{-k}}$$

Following the approach used for AR(p) and MA(q) spectrum estimation, the spectrum of an ARMA(p, q) process may be estimated from Eq. (8.135) using estimates of the model parameters

$$\hat{P}_{ARMA}(e^{j\omega}) = \frac{\left| \sum_{k=0}^{q} \hat{b}_q(k) e^{-jk\omega} \right|^2}{\left| 1 + \sum_{k=1}^{p} \hat{a}_p(k) e^{-jk\omega} \right|^2} \tag{8.136}$$

As discussed in Section 4.7.1, the AR model parameters may be estimated from the modified Yule-Walker equations either directly or by using a least squares approach. Once the coefficients $\hat{a}_p(k)$ have been estimated, a moving average modeling technique such as Durbin's method may be used to estimate the moving average parameters $b_q(k)$. Examples may be found in [25].

8.6 FREQUENCY ESTIMATION

In the previous section, we considered the problem of estimating the power spectrum of a WSS random process that could be modeled as the output of a linear shift-invariant filter that is driven by white noise. Another model of importance is one in which $x(n)$ is a sum of complex exponentials in white noise,

$$x(n) = \sum_{i=1}^{p} A_i e^{jn\omega_i} + w(n) \tag{8.137}$$

It is assumed that the amplitudes A_i are complex,

$$A_i = |A_i| e^{j\phi_i}$$

with ϕ_i uncorrelated random variables that are uniformly distributed over the interval $[-\pi, \pi]$. Although the frequencies and magnitudes of the complex exponentials, ω_i and $|A_i|$, respectively, are not random, they are assumed to be unknown. Thus, the power spectrum of $x(n)$ consists of a set of p impulses of area $2\pi|A_i|$ at frequency ω_i for $i = 1, 2, \ldots, p$, plus the power spectrum of the additive noise $w(n)$. Signals of this form are found in a number of applications such as sonar signal processing and speech processing. Typically, the complex exponentials are the "information bearing" part of the signal and it is the estimation of the frequencies and amplitudes that is of interest rather than the estimation of the power spectrum itself. In the case of sonar signals, for example, the frequencies ω_i may represent bearing or velocity information, whereas for speech signals they would correspond to the formant frequencies [42]. Although it is possible to estimate the frequencies of the complex exponentials from the peaks of the spectrum that is estimated using any of the techniques discussed in the previous sections, this approach would not fully exploit the assumed parametric form of the process. Therefore, in this section we consider *frequency estimation* algorithms that take into account the known properties of the process. The methods that we will be considering are based on an eigendecomposition of the autocorrelation matrix into two subspaces, a *signal* subspace and a *noise* subspace. Once these subspaces have been determined, a frequency estimation function is then used to extract estimates of the frequencies. Before looking at these *subspace methods*, we begin with a discussion of the eigendecomposition of the autocorrelation matrix.

8.6.1 Eigendecomposition of the Autocorrelation Matrix

In order to motivate the use of an eigendecomposition of the autocorrelation matrix as an approach that may be used for frequency estimation, consider the first-order process

$$x(n) = A_1 e^{jn\omega_1} + w(n)$$

that consists of a single complex exponential in white noise. The amplitude of the complex exponential is $A_1 = |A_1| e^{j\phi_1}$ where ϕ_1 is a uniformly distributed random variable, and

$w(n)$ is white noise that has a variance of σ_w^2. As shown in Section 3.6.4, the autocorrelation sequence of $x(n)$ is

$$r_x(k) = P_1 e^{jk\omega_1} + \sigma_w^2 \delta(k)$$

where $P_1 = |A_1|^2$ is the *power* in the complex exponential. Therefore, the $M \times M$ autocorrelation matrix for $x(n)$ is a sum of an autocorrelation matrix due to the signal, $\mathbf{R}_s$, and an autocorrelation matrix due to the noise, $\mathbf{R}_n$,

$$\boxed{\mathbf{R}_x = \mathbf{R}_s + \mathbf{R}_n} \tag{8.138}$$

where the signal autocorrelation matrix is

$$\mathbf{R}_s = P_1 \begin{bmatrix} 1 & e^{-j\omega_1} & e^{-j2\omega_1} & \cdots & e^{-j(M-1)\omega_1} \\ e^{j\omega_1} & 1 & e^{-j\omega_1} & \cdots & e^{-j(M-2)\omega_1} \\ e^{j2\omega_1} & e^{j\omega_1} & 1 & \cdots & e^{-j(M-3)\omega_1} \\ \vdots & \vdots & \vdots & & \vdots \\ e^{j(M-1)\omega_1} & e^{j(M-2)\omega_1} & e^{j(M-3)\omega_1} & \cdots & 1 \end{bmatrix} \tag{8.139}$$

and has a rank of one, and the noise autocorrelation matrix is diagonal,

$$\mathbf{R}_n = \sigma_w^2 \mathbf{I}$$

and has full rank. Note that if we define

$$\mathbf{e}_1 = [1, e^{j\omega_1}, e^{j2\omega_1}, \ldots, e^{j(M-1)\omega_1}]^T \tag{8.140}$$

then $\mathbf{R}_s$ may be written in terms of $\mathbf{e}_1$ as follows:

$$\mathbf{R}_s = P_1 \mathbf{e}_1 \mathbf{e}_1^H$$

Since the rank of $\mathbf{R}_s$ is equal to one, then $\mathbf{R}_s$ has only one nonzero eigenvalue. With

$$\mathbf{R}_s \mathbf{e}_1 = P_1 \left(\mathbf{e}_1 \mathbf{e}_1^H\right) \mathbf{e}_1 = P_1 \mathbf{e}_1 \left(\mathbf{e}_1^H \mathbf{e}_1\right) = M P_1 \mathbf{e}_1$$

it follows that the nonzero eigenvalue is equal to $M P_1$, and that $\mathbf{e}_1$ is the corresponding eigenvector. In addition, since $\mathbf{R}_s$ is Hermitian then the remaining eigenvectors, $\mathbf{v}_2, \mathbf{v}_3, \ldots, \mathbf{v}_M$, will be orthogonal to $\mathbf{e}_1$,[15]

$$\mathbf{e}_1^H \mathbf{v}_i = 0 \quad ; \quad i = 2, 3, \ldots, M \tag{8.141}$$

Finally, note that if we let λ_i^s be the eigenvalues of $\mathbf{R}_s$, then

$$\mathbf{R}_x \mathbf{v}_i = \left(\mathbf{R}_s + \sigma_w^2 \mathbf{I}\right) \mathbf{v}_i = \lambda_i^s \mathbf{v}_i + \sigma_w^2 \mathbf{v}_i = (\lambda_i^s + \sigma_w^2) \mathbf{v}_i \tag{8.142}$$

Therefore, the eigenvectors of $\mathbf{R}_x$ are the same as those of $\mathbf{R}_s$, and the eigenvalues of $\mathbf{R}_x$ are

$$\lambda_i = \lambda_i^s + \sigma_w^2$$

As a result, the largest eigenvalue of $\mathbf{R}_x$ is

$$\lambda_{\max} = M P_1 + \sigma_w^2$$

and the remaining $M - 1$ eigenvalues are equal to σ_w^2. Thus, it is possible to extract all of the parameters of interest about $x(n)$ from the eigenvalues and eigenvectors of $\mathbf{R}_x$ as follows:

[15]Recall that for a Hermitian matrix the eigenvectors corresponding to distinct eigenvalues are orthogonal. See Property 4 on p. 43 of Chapter 2.

1. Perform an eigendecomposition of the autocorrelation matrix, $\mathbf{R}_x$. The largest eigenvalue will be equal to $MP_1 + \sigma_w^2$ and the remaining eigenvalues will be equal to σ_w^2.
2. Use the eigenvalues of $\mathbf{R}_x$ to solve for the power P_1 and the noise variance as follows:

$$\begin{aligned} \sigma_w^2 &= \lambda_{\min} \\ P_1 &= \frac{1}{M}\left(\lambda_{\max} - \lambda_{\min}\right) \end{aligned}$$

3. Determine the frequency ω_1 from the eigenvector $\mathbf{v}_{\max}$ that is associated with the largest eigenvalue using, for example, the second coefficient of $\mathbf{v}_{\max}$,[16]

$$\omega_i = \arg\Big\{ v_{\max}(1) \Big\}$$

The following example illustrates the procedure.

Example 8.6.1 ***Eigendecomposition of a Complex Exponential in Noise***

Let $x(n)$ be a first-order harmonic process consisting of a single complex exponential in white noise,

$$x(n) = A_1 e^{jn\omega_1} + w(n)$$

with a 2×2 autocorrelation matrix given by

$$\mathbf{R}_x = \begin{bmatrix} 3 & 2(1-j) \\ 2(1+j) & 3 \end{bmatrix}$$

The eigenvalues of $\mathbf{R}_x$ are

$$\lambda_{1,2} = 3 \pm \big|2(1+j)\big| = 3 \pm 2\sqrt{2}$$

and the eigenvectors are

$$\mathbf{v}_{1,2} = \begin{bmatrix} 1 \\ \pm\dfrac{\sqrt{2}}{2}(1+j) \end{bmatrix}$$

Therefore, the variance of the white noise is

$$\sigma_w^2 = \lambda_{\min} = 3 - 2\sqrt{2}$$

and the power in the complex exponential is

$$P_1 = \frac{1}{M}\big(\lambda_{\max} - \lambda_{\min}\big) = 2\sqrt{2}$$

Finally, the frequency of the complex exponential is

$$\omega_1 = \arg\left\{ \frac{\sqrt{2}}{2}(1+j) \right\} = \pi/4$$

It should be pointed out that, for a first-order harmonic process, finding the parameters of the process is actually a bit easier than this. For example, note that

$$r_x(1) = 2(1+j) = 2\sqrt{2}e^{j\pi/4} = P_1 e^{j\omega_1}$$

[16] The components of the eigenvectors are numbered from $v_i(0)$ to $v_i(M-1)$.

Therefore, we see immediately that $P_1 = 2\sqrt{2}$ and $\omega_1 = \pi/4$. Furthermore, once P_1 is known, then the variance of the white noise may be determined as follows:

$$\sigma_w^2 = r_x(0) - P_1 = 3 - 2\sqrt{2}$$

In practice, the approach described above is of limited value in frequency estimation since it requires that the autocorrelation matrix be known exactly. Although an estimated autocorrelation matrix could be used in place of $\mathbf{R}_x$, if this is done then the largest eigenvalue will only be approximately equal to $P_1 + \sigma_w^2$ and the corresponding eigenvector will only be an approximation to $\mathbf{e}_1$. Since the eigenvalues and eigenvectors may be quite sensitive to small errors in $r_x(k)$, instead of estimating the frequency of the complex exponential from a single eigenvector, we may consider using a weighted average as follows. Let $\mathbf{v}_i$ be a *noise eigenvector* of $\mathbf{R}_x$, i.e., one that has an eigenvalue of σ_w^2, and let $v_i(k)$ be the kth component of $\mathbf{v}_i$. If we compute the discrete-time Fourier transform of the coefficients in $\mathbf{v}_i$,

$$V_i(e^{j\omega}) = \sum_{k=0}^{M-1} v_i(k)e^{-jk\omega} = \mathbf{e}^H\mathbf{v}_i \tag{8.143}$$

then the orthogonality condition given in Eq. (8.141) implies that $V_i(e^{j\omega})$ will be equal to zero at $\omega = \omega_1$, the frequency of the complex exponential. Therefore, if we form the *frequency estimation function*

$$\hat{P}_i(e^{j\omega}) = \frac{1}{\left|\sum_{k=0}^{M-1} v_i(k)e^{-jk\omega}\right|^2} = \frac{1}{|\mathbf{e}^H\mathbf{v}_i|^2} \tag{8.144}$$

then $\hat{P}_i(e^{j\omega})$ will be large (in theory, infinite) at $\omega = \omega_1$. Thus, the location of the peak of this frequency estimation function may be used to estimate the frequency of the complex exponential. However, since Eq. (8.144) uses only a single eigenvector and, therefore, may be sensitive to errors in the estimation of $\mathbf{R}_x$, we may consider using a weighted average of all of the noise eigenvectors as follows:

$$\hat{P}(e^{j\omega}) = \frac{1}{\sum_{i=2}^{M} \alpha_i |\mathbf{e}^H\mathbf{v}_i|^2} \tag{8.145}$$

where α_i are some appropriately chosen constants.

Example 8.6.2 ***Eigendecomposition of a Complex Exponential in Noise (cont.)***

Let $x(n)$ be a WSS process consisting of a single complex exponential in unit variance white noise,

$$x(n) = 4\, e^{j(n\pi/4+\phi)} + w(n)$$

where ϕ is a uniformly distributed random variable. Using $N = 64$ values of $x(n)$, a 6×6 autocorrelation matrix was estimated and an eigendecomposition performed. Shown in Fig. 8.31a is a plot of the frequency estimation function defined in Eq. (8.145) with $\alpha_i = 1$. The peak of $\hat{P}(e^{j\omega})$ occurs at frequency $\omega = 0.2539\pi$ and the minimum eigenvalue is

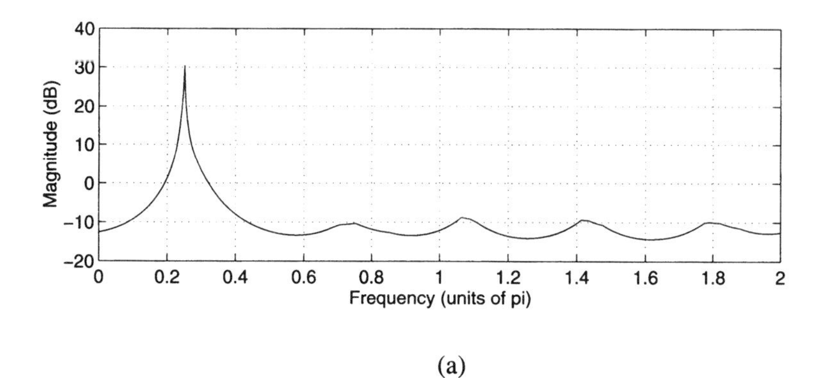

(a)

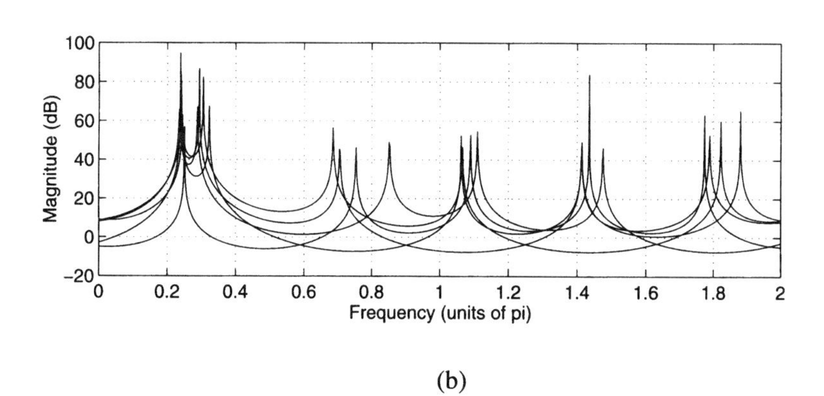

(b)

Figure 8.31 *Frequency estimation functions of a single complex exponential in white noise. (a) The frequency estimation function that uses all of the noise eigenvectors with a weighting $\alpha_i = 1$. (b) An overlay plot of the frequency estimation functions $V_i(e^{j\omega}) = 1/|\mathbf{e}^H \mathbf{v}_i|^2$ that are derived from each noise eigenvector.*

$\lambda_{\min} = 1.08$, which is close to the variance of the white noise. By contrast, shown in Fig. 8.31*b* is an overlay plot of the frequency estimation functions $V_i(e^{j\omega})$ for each of the noise eigenvectors as defined in Eq. (8.143). As we see from this figure, although each plot has a peak that is close to $\omega = 0.25\pi$, with a single plot it is difficult to distinguish the correct peak from a spurious one.

Let us now consider what happens in the case of two complex exponentials in white noise,

$$x(n) = A_1 e^{jn\omega_1} + A_2 e^{jn\omega_2} + w(n)$$

where $A_i = |A_i| e^{j\phi_i}$ for $i = 1, 2$ are the amplitudes of the complex exponentials, and ω_1 and ω_2 are the frequencies with $\omega_1 \neq \omega_2$. If the variance of $w(n)$ is σ_w^2 then the autocorrelation of $x(n)$ is

$$r_x(k) = P_1 e^{jk\omega_1} + P_2 e^{jk\omega_2} + \sigma_w^2 \delta(k)$$

where $P_1 = |A_1|^2$ and $P_2 = |A_2|^2$. Thus, the autocorrelation matrix may again be written as a sum

$$\mathbf{R}_x = P_1\mathbf{e}_1\mathbf{e}_1^H + P_2\mathbf{e}_2\mathbf{e}_2^H + \sigma_w^2\mathbf{I}$$

where

$$\mathbf{R}_s = P_1\mathbf{e}_1\mathbf{e}_1^H + P_2\mathbf{e}_2\mathbf{e}_2^H$$

is a rank two matrix representing the component of $\mathbf{R}_x$ that is due to the signal, and

$$\mathbf{R}_n = \sigma_w^2\mathbf{I}$$

is a diagonal matrix that is due to the noise. A more concise way to express this decomposition is to write $\mathbf{R}_x$ as follows:

$$\mathbf{R}_x = \mathbf{E}\mathbf{P}\mathbf{E}^H + \sigma_w^2\mathbf{I} \tag{8.146}$$

where

$$\mathbf{E} = \big[\mathbf{e}_1,\ \mathbf{e}_2\big]$$

is an $M \times 2$ matrix containing the two signal vectors $\mathbf{e}_1$ and $\mathbf{e}_2$ and $\mathbf{P} = \text{diag}\{P_1,\ P_2\}$ is a diagonal matrix containing the signal powers.

In addition to decomposing $\mathbf{R}_x$ into a sum of two autocorrelation matrices as in Eq. (8.146), we may also perform an eigendecomposition of $\mathbf{R}_x$ as follows. Let $\mathbf{v}_i$ and λ_i be the eigenvectors and eigenvalues of $\mathbf{R}_x$, respectively, with the eigenvalues arranged in decreasing order,

$$\lambda_1 \geq \lambda_2 \geq \cdots \geq \lambda_M$$

Since $\mathbf{R}_x = \mathbf{R}_s + \sigma_w^2\mathbf{I}$ then

$$\lambda_i = \lambda_i^s + \sigma_w^2$$

where λ_i^s are the eigenvalues of $\mathbf{R}_s$. Since the rank of $\mathbf{R}_s$ is equal to two, then $\mathbf{R}_s$ has only two nonzero eigenvalues, and both of these are greater than zero ($\mathbf{R}_s$ is nonnegative definite). Therefore, the first two eigenvalues of $\mathbf{R}_x$ are greater than σ_w^2 and the remaining eigenvalues are equal to σ_w^2. Thus, the eigenvalues and eigenvectors of $\mathbf{R}_x$ may be divided into two groups. The first group, consisting of the two eigenvectors that have eigenvalues greater than σ_w^2, are referred to as *signal eigenvectors* and span a two-dimensional subspace called the *signal subspace*. The second group, consisting of those eigenvectors that have eigenvalues equal to σ_w^2, are referred to as the *noise eigenvectors* and span an $(M - 2)$-dimensional subspace called the *noise subspace*.[17] Since $\mathbf{R}_x$ is Hermitian, the eigenvectors $\mathbf{v}_i$ form an orthonormal set (see the discussion following Property 4 on p. 43). Therefore, the signal and noise subspaces are orthogonal. That is to say, for any vector $\mathbf{u}$ in the signal subspace and for any vector $\mathbf{v}$ in the noise subspace, $\mathbf{u}^H\mathbf{v} = 0$. The geometry of these subspaces is illustrated in Fig. 8.32.

Unlike the case for a single complex exponential, with a sum of two complex exponentials in noise, the signal eigenvectors will generally not be equal to $\mathbf{e}_1$ and $\mathbf{e}_2$.[18] Nevertheless, $\mathbf{e}_1$ and $\mathbf{e}_2$ will lie in the signal subspace that is spanned by the signal eigenvectors $\mathbf{v}_1$ and $\mathbf{v}_2$, and since the signal subspace is orthogonal to the noise subspace, then $\mathbf{e}_1$ and $\mathbf{e}_2$ will be

[17] Note that this term is a bit misleading since the noise has components in both the noise and signal subspaces.

[18] This may be seen easily by noting that the signal vectors will not, in general, be orthogonal whereas the eigenvectors will always be orthogonal.

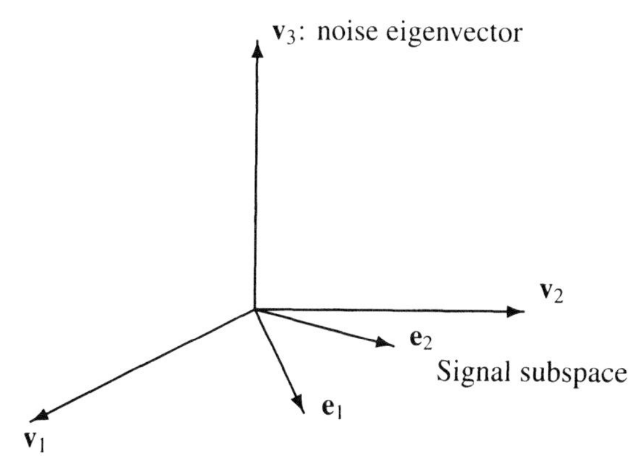

Figure 8.32 *Geometrical interpretation of the orthogonality of the signal and noise subspaces. The one-dimensional noise subspace is spanned by the noise eigenvector* $\mathbf{v}_3$ *and the signal subspace which contains the signal vectors* $\mathbf{e}_1$ *and* $\mathbf{e}_2$ *is spanned by the signal eigenvectors* $\mathbf{v}_1$ *and* $\mathbf{v}_2$.

orthogonal to the noise eigenvectors $\mathbf{v}_i$, i.e.,

$$\begin{aligned} \mathbf{e}_1^H \mathbf{v}_i &= 0 \quad ; \quad i = 3, 4, \ldots, M \\ \mathbf{e}_2^H \mathbf{v}_i &= 0 \quad ; \quad i = 3, 4, \ldots, M \end{aligned} \tag{8.147}$$

Therefore, as in the case of one complex exponential, the complex exponential frequencies, ω_1 and ω_2, may be estimated using a frequency estimation function of the form

$$\hat{P}(e^{j\omega}) = \frac{1}{\sum_{i=3}^{M} \alpha_i |\mathbf{e}^H \mathbf{v}_i|^2}.$$

Let us now consider the general case of a wide-sense stationary process consisting of p *distinct* complex exponentials in white noise. The $M \times M$ autocorrelation sequence is

$$r_x(k) = \sum_{i=1}^{p} P_i e^{jk\omega_i} + \sigma_w^2 \delta(k)$$

where $P_i = |A_i|^2$ is the *power* in the ith component. Therefore, the autocorrelation matrix may be written as

$$\mathbf{R}_x = \mathbf{R}_s + \mathbf{R}_n = \sum_{i=1}^{p} P_i \mathbf{e}_i \mathbf{e}_i^H + \sigma_w^2 \mathbf{I} \tag{8.148}$$

where

$$\mathbf{e}_i = [1,\ e^{j\omega_i},\ e^{j2\omega_i},\ \ldots,\ e^{j(M-1)\omega_i}]^T \quad ; \quad i = 1, 2, \ldots, p$$

is a set of p linearly independent vectors. As was the case for two complex exponentials, Eq. (8.148) may be written concisely as follows:

$$\boxed{\mathbf{R}_x = \mathbf{E}\mathbf{P}\mathbf{E}^H + \sigma_w^2 \mathbf{I}} \tag{8.149}$$

where $\mathbf{E} = \left[\mathbf{e}_1, \ \ldots, \ \mathbf{e}_p\right]$ is an $M \times p$ matrix containing the p signal vectors, $\mathbf{e}_i$, and $\mathbf{P} = \text{diag}\left\{P_1, \ \ldots, \ P_p\right\}$ is a diagonal matrix of signal powers. Since the eigenvalues of $\mathbf{R}_x$ are $\lambda_i = \lambda_i^s + \sigma_w^2$ where λ_i^s are the eigenvalues of $\mathbf{R}_s$, and since $\mathbf{R}_s$ is a matrix of rank p, then the first p eigenvalues of $\mathbf{R}_x$ will be greater than σ_w^2 and the last $M - p$ eigenvalues will be equal to σ_w^2. Therefore, the eigenvalues and eigenvectors of $\mathbf{R}_x$ may again be divided into two groups: the signal eigenvectors $\mathbf{v}_1, \ldots, \mathbf{v}_p$ that have eigenvalues greater than σ_w^2, and the noise eigenvectors $\mathbf{v}_{p+1}, \ldots, \mathbf{v}_M$ that have eigenvalues equal to σ_w^2. Assuming that the eigenvectors have been normalized to have unit norm, we may use the spectral theorem (p. 44) to decompose $\mathbf{R}_x$ as follows:

$$\mathbf{R}_x = \sum_{i=1}^{p}(\lambda_i^s + \sigma_w^2)\mathbf{v}_i\mathbf{v}_i^H + \sum_{i=p+1}^{M} \sigma_w^2\mathbf{v}_i\mathbf{v}_i^H$$

In matrix notation, this decomposition may be written as

$$\boxed{\mathbf{R}_x = \mathbf{V}_s\mathbf{V}_s^H + \mathbf{V}_n\mathbf{V}_n^H} \tag{8.150}$$

where

$$\mathbf{V}_s = \left[\mathbf{v}_1, \ \mathbf{v}_2, \ \ldots, \ \mathbf{v}_p\right]$$

is the $M \times p$ matrix of signal eigenvectors and

$$\mathbf{V}_n = \left[\mathbf{v}_{p+1}, \ \mathbf{v}_{p+2}, \ \ldots, \ \mathbf{v}_M\right]$$

is the $M \times (M - p)$ matrix of noise eigenvectors, and where ${}_s$ and ${}_n$ are diagonal matrices that contain the eigenvalues $\lambda_i = \lambda_i^s + \sigma_w^2$ and $\lambda_i = \sigma_w^2$, respectively. Later, we will be interested in projecting a vector onto either the signal subspace or the noise subspace. The projection matrices $\mathbf{P}_s$ and $\mathbf{P}_n$ that will perform these projections onto the signal and noise subspaces, respectively, are (see p. 34)

$$\boxed{\mathbf{P}_s = \mathbf{V}_s\mathbf{V}_s^H \quad ; \quad \mathbf{P}_n = \mathbf{V}_n\mathbf{V}_n^H} \tag{8.151}$$

As was the case for one and two complex exponentials in white noise, the orthogonality of the signal and noise subspaces may be used to estimate the frequencies of the complex exponentials. Specifically, since each signal vector $\mathbf{e}_1, \ldots, \mathbf{e}_p$ lies in the signal subspace, this orthogonality implies that $\mathbf{e}_i$ will be orthogonal to each of the noise eigenvectors,

$$\mathbf{e}_i^H\mathbf{v}_k = 0 \quad ; \quad \begin{array}{l} i = 1, 2, \ldots, p \\ k = p+1, p+2, \ldots, M \end{array}$$

Therefore, the frequencies may be estimated using a frequency estimation function such as

$$\hat{P}(e^{j\omega}) = \frac{1}{\displaystyle\sum_{i=p+1}^{M} \alpha_i|\mathbf{e}^H\mathbf{v}_i|^2} \tag{8.152}$$

In the following sections, we will develop several different types of frequency estimation algorithms that are based on Eq. (8.152). We begin with the Pisarenko harmonic decomposition, which uses a frequency estimator of this form with $M = p + 1$ and $\alpha_M = 1$.

8.6.2 Pisarenko Harmonic Decomposition

In 1973, V. Pisarenko considered the problem of estimating the frequencies of a sum of complex exponentials in white noise [41]. Based on a theorem of Carathedory, he demonstrated that the frequencies could be derived from the eigenvector corresponding to the minimum eigenvalue of the autocorrelation matrix. Although the resulting technique, referred to as the *Pisarenko harmonic decomposition*, is somewhat limited in its usefulness due to its sensitivity to noise, this decomposition is of theoretical interest, has led to important insights into the frequency estimation problem, and has provided the stimulus for the development of other eigenvalue decomposition methods that are more robust.

In the Pisarenko harmonic decomposition, it is assumed that $x(n)$ is a sum of p complex exponentials in white noise, and that the number of complex exponentials, p, is known. It is also assumed that $p+1$ values of the autocorrelation sequence are either known or have been estimated. With a $(p+1) \times (p+1)$ autocorrelation matrix, the dimension of the noise subspace is equal to one, and it is spanned by the eigenvector corresponding to the minimum eigenvalue, $\lambda_{\min} = \sigma_w^2$. Denoting this noise eigenvector by $\mathbf{v}_{\min}$, it follows that $\mathbf{v}_{\min}$ will be orthogonal to each of the signal vectors, $\mathbf{e}_i$,

$$\mathbf{e}_i^H \mathbf{v}_{\min} = \sum_{k=0}^{p} v_{\min}(k) e^{-jk\omega_i} = 0 \quad ; \quad i = 1, 2, \ldots, p \tag{8.153}$$

Therefore,

$$V_{\min}(e^{j\omega}) = \sum_{k=0}^{p} v_{\min}(k) e^{-jk\omega}$$

is equal to zero at each of the complex exponential frequencies ω_i for $i = 1, 2, \ldots, p$. Consequently, the z-transform of the noise eigenvector, referred to as an *eigenfilter*, has p zeros on the unit circle,

$$\boxed{V_{\min}(z) = \sum_{k=0}^{p} v_{\min}(k) z^{-k} = \prod_{k=1}^{p} \left(1 - e^{j\omega_k} z^{-1}\right)} \tag{8.154}$$

and the frequencies of the complex exponentials may be extracted from the roots of the eigenfilter. As an alternative to rooting $V_{\min}(z)$, we may also form the *frequency estimation function*

$$\boxed{\hat{P}_{PHD}(e^{j\omega}) = \frac{1}{|\mathbf{e}^H \mathbf{v}_{\min}|^2}} \tag{8.155}$$

which is a special case of Eq. (8.152) with $M = p+1$ and $\alpha_{p+1} = 1$. Since $\hat{P}_{PHD}(e^{j\omega})$ will be large (in theory, infinite) at the frequencies of the complex exponentials, the locations of the peaks in $\hat{P}_{PHD}(e^{j\omega})$ may be used as frequency estimates. Although written in the form of a power spectrum, $\hat{P}_{PHD}(e^{j\omega})$ is called a *pseudospectrum* (sometimes referred to as an *eigenspectrum*) since it does not contain any information about the power in the complex exponentials, nor does it contain a component due to the noise.

Once the frequencies of the complex exponentials have been determined, the powers P_i may be found from the eigenvalues of $\mathbf{R}_x$ as follows. Let us assume that the signal subspace eigenvectors $\mathbf{v}_1, \mathbf{v}_2, \ldots, \mathbf{v}_p$, have been normalized so that $\mathbf{v}_i^H \mathbf{v}_i = 1$. With

$$\mathbf{R}_x \mathbf{v}_i = \lambda_i \mathbf{v}_i \quad ; \quad i = 1, 2, \ldots, p \tag{8.156}$$

if both sides of Eq. (8.156) are multiplied on the left by $\mathbf{v}_i^H$, then

$$\mathbf{v}_i^H \mathbf{R}_x \mathbf{v}_i = \lambda_i \mathbf{v}_i^H \mathbf{v}_i = \lambda_i \quad ; \quad i = 1, 2, \ldots, p \tag{8.157}$$

Substituting the expression for $\mathbf{R}_x$ given in Eq. (8.148) into Eq. (8.157) we have

$$\mathbf{v}_i^H \mathbf{R}_x \mathbf{v}_i = \mathbf{v}_i^H \left\{ \sum_{k=1}^{p} P_k \mathbf{e}_k \mathbf{e}_k^H + \sigma_w^2 \mathbf{I} \right\} \mathbf{v}_i = \lambda_i$$

which may be simplified to

$$\boxed{\sum_{k=1}^{p} P_k |\mathbf{e}_k^H \mathbf{v}_i|^2 = \lambda_i - \sigma_w^2 \quad ; \quad i = 1, 2, \ldots, p} \tag{8.158}$$

Note that the terms $|\mathbf{e}_k^H \mathbf{v}_i|^2$ in the sum correspond to the squared magnitude of the DTFT of the signal subspace eigenvector $\mathbf{v}_i$ evaluated at frequency ω_k,

$$|\mathbf{e}_k^H \mathbf{v}_i|^2 = \left| V_i(e^{j\omega_k}) \right|^2$$

where

$$V_i(e^{j\omega}) = \sum_{l=0}^{p} v_i(l) e^{-jl\omega}$$

Therefore, Eq. (8.158) may also be written as

$$\sum_{k=1}^{p} P_k |V_i(e^{j\omega_k})|^2 = \lambda_i - \sigma_\omega^2 \quad ; \quad i = 1, 2, \ldots, p \tag{8.159}$$

Equation (8.159) is a set of p linear equations in the p unknowns, P_k,

$$\begin{bmatrix} |V_1(e^{j\omega_1})|^2 & |V_1(e^{j\omega_2})|^2 & \cdots & |V_1(e^{j\omega_p})|^2 \\ |V_2(e^{j\omega_1})|^2 & |V_2(e^{j\omega_2})|^2 & \cdots & |V_2(e^{j\omega_p})|^2 \\ \vdots & \vdots & & \vdots \\ |V_p(e^{j\omega_1})|^2 & |V_p(e^{j\omega_2})|^2 & \cdots & |V_p(e^{j\omega_p})|^2 \end{bmatrix} \begin{bmatrix} P_1 \\ P_2 \\ \vdots \\ P_p \end{bmatrix} = \begin{bmatrix} \lambda_1 - \sigma_w^2 \\ \lambda_2 - \sigma_w^2 \\ \vdots \\ \lambda_p - \sigma_w^2 \end{bmatrix} \tag{8.160}$$

which may be solved for the powers P_k. Thus, as summarized in Table 8.9, the Pisarenko

Table 8.9 ***Pisarenko's Method for Frequency Estimation***

Step 1: Given that a process consists of p complex exponentials in white noise, find the minimum eigenvalue $\lambda_{\min}$ and the corresponding eigenvector $\mathbf{v}_{\min}$ of the $(p+1) \times (p+1)$ autocorrelation matrix $\mathbf{R}_x$.

Step 2: Set the white noise power equal to the minimum eigenvalue, $\lambda_{\min} = \sigma_w^2$, and set the frequencies equal to the angles of the roots of the eigenfilter

$$V_{\min}(z) = \sum_{k=0}^{p} v_{\min}(k) z^{-k}$$

or the location of the peaks in the frequency estimation function

$$\hat{P}_{PHD}(e^{j\omega}) = \frac{1}{|\mathbf{e}^H \mathbf{v}_{\min}|^2}$$

Step 3: Compute the powers of the complex exponentials by solving the linear equations (8.160).

The Pisarenko Harmonic Decomposition

```
function [vmin,sigma] = phd(x,p)
%
   x = x(:);
   R = covar(x,p+1);
   [v,d]=eig(R);
   sigma=min(diag(d));
   index=find(diag(d)==sigma);
   vmin = v(:,index);
   end;
```

Figure 8.33 *A* MATLAB *program to estimate the frequencies of p complex exponentials in white noise using Pisarenko's method.*

harmonic decomposition estimates the complex exponential frequencies either from the roots of the eigenfilter $V_{\min}(z)$ or from the locations of the peaks in the frequency estimation function $\hat{P}_{PHD}(e^{j\omega})$, and then solves the linear equations (8.160) for the powers of the complex exponentials. A MATLAB program for the Pisarenko decomposition is given in Fig 8.33.

Example 8.6.3 *Pisarenko's Method for Two Complex Exponentials in Noise*

Suppose that the first three autocorrelations of a random process consisting of two complex exponentials in white noise are

$$\begin{aligned} r_x(0) &= 6 \\ r_x(1) &= 1.92705 + j4.58522 \\ r_x(2) &= -3.42705 + j3.49541 \end{aligned}$$

Let us use Pisarenko's method to find the frequencies and powers of the complex exponentials. Since $p = 2$, we must perform an eigendecomposition of the 3×3 autocorrelation matrix

$$\mathbf{R}_x = \begin{bmatrix} 6 & 1.92705 - j4.58522 & -3.42705 - j3.49541 \\ 1.92705 + j4.58522 & 6 & 1.92705 - j4.58522 \\ -3.42705 + j3.49541 & 1.92705 + j4.58522 & 6 \end{bmatrix}$$

The eigenvectors are

$$\mathbf{V} = \begin{bmatrix} \mathbf{v}_1, & \mathbf{v}_2, & \mathbf{v}_3 \end{bmatrix} = \begin{bmatrix} 0.5763 - 0.0000j & -0.2740 + 0.6518j & -0.2785 - 0.3006j \\ 0.2244 + 0.5342j & 0.0001 + 0.0100j & -0.3209 + 0.7492j \\ -0.4034 + 0.4116j & 0.2830 + 0.6480j & 0.4097 - 0.0058j \end{bmatrix}$$

and the eigenvalues are

$$\lambda_1 = 15.8951 \quad ; \quad \lambda_2 = 1.1049 \quad ; \quad \lambda_3 = 1.0000$$

Therefore, the minimum eigenvalue is $\lambda_{\min} = 1$ and the corresponding eigenvector is

$$\mathbf{v}_{\min} = \begin{bmatrix} -0.2785 - 0.3006j \\ -0.3209 + 0.7492j \\ 0.4097 - 0.0058j \end{bmatrix}$$

Since the roots of the eigenfilter are

$$z_1 = 0.5 + j0.8660 = e^{j\pi/3} \quad ; \quad z_2 = 0.3090 + j0.9511 = e^{j2\pi/5}$$

then the frequencies of the complex exponentials are

$$\omega_1 = \pi/3 \quad ; \quad \omega_2 = 2\pi/5$$

To find the powers in the complex exponentials, we must compute the squared magnitude of the DTFT of the signal eigenvectors, $\mathbf{v}_1$ and $\mathbf{v}_2$, at the complex exponential frequencies ω_1 and ω_2. With

$$|V_1(e^{j\omega_1})|^2 = 2.9685 \quad ; \quad |V_1(e^{j\omega_2})|^2 = 2.9861$$

and

$$|V_2(e^{j\omega_1})|^2 = 0.0315 \quad ; \quad |V_2(e^{j\omega_2})|^2 = 0.0139$$

then Eq. (8.160) becomes

$$\begin{bmatrix} 2.9685 & 2.9861 \\ 0.0315 & 0.0139 \end{bmatrix} \begin{bmatrix} P_1 \\ P_2 \end{bmatrix} = \begin{bmatrix} \lambda_1 - \sigma_w^2 \\ \lambda_2 - \sigma_w^2 \end{bmatrix}$$

where $\sigma_w^2 = \lambda_{\min} = 1$. Solving for P_1 and P_2 we find

$$P_1 = 2 \quad ; \quad P_2 = 3$$

In the previous example, we used Pisarenko's method to estimate the frequencies of two complex exponentials in white noise. In the following example we consider the problem of estimating the frequency of a single sinusoid in white noise. Although a sinusoid is a sum of two complex exponentials, the two frequencies are constrained to be negatives of each other, $\omega_2 = -\omega_1$. This constraint results in an autocorrelation sequence that is real-valued, which forces the eigenvectors to be real and, therefore, constrains the roots of the eigenfilter to occur in complex conjugate pairs.

Example 8.6.4 ***Pisarenko's Method for One Sinusoid***

Let $x(n)$ be a random phase sinusoid in white noise

$$x(n) = A\sin(n\omega_0 + \phi) + w(n)$$

with

$$r_x(0) = 2.2 \quad ; \quad r_x(1) = 1.3 \quad ; \quad r_x(2) = 0.8$$

The eigenvalues of the 3×3 autocorrelation matrix $\mathbf{R}_x = \text{toep}\{2.2, 1.3, 0.8\}$ are

$$\lambda_1 = 4.4815 \quad ; \quad \lambda_2 = 1.4 \quad ; \quad \lambda_3 = 0.7185$$

Therefore, the white noise power is

$$\sigma_w^2 = \lambda_{\min} = 0.7185$$

The eigenvectors of $\mathbf{R}_x$ are

$$\mathbf{v}_1 = \begin{bmatrix} 0.5506 \\ 0.6275 \\ 0.5506 \end{bmatrix} ; \quad \mathbf{v}_2 = \begin{bmatrix} -0.7071 \\ 0 \\ 0.7071 \end{bmatrix} ; \quad \mathbf{v}_3 = \begin{bmatrix} 0.4437 \\ -0.7787 \\ 0.4437 \end{bmatrix}$$

Note the symmetry of the eigenvectors (See Problem 2.9). To estimate the frequency of the sinusoid we find the roots of the eigenfilter $V_{\min}(z)$, which is the z-transform of $\mathbf{v}_3$,

$$V_{\min}(z) = 0.4437\big(1 - 1.755z^{-1} + z^{-2}\big)$$

Rooting the eigenfilter we find that $V_{\min}(z)$ has roots at $z = e^{\pm j\omega_0}$ where

$$2\cos\omega_0 = 1.755$$

or

$$\omega_0 = 0.159\pi$$

Finally, the power in the sinusoid may be estimated using Eq. (8.160). However, the computation may be simplified by taking into account the fact that $x(n)$ contains a single sinusoid. Specifically, since the autocorrelation sequence for a single sinusoid in white noise is (see Section 3.6.4)

$$r_x(k) = \frac{1}{2}A^2\cos(k\omega_0) + \sigma_w^2\delta(k)$$

then, for $k = 0$,

$$r_x(0) = \frac{1}{2}A^2 + \sigma_w^2$$

With $r_x(0) = 2.2$ and $\sigma_w^2 = 0.7185$ it follows that

$$A^2 = 2.963$$

In spite of the mathematical elegance of the Pisarenko harmonic decomposition, it is not commonly used in practice. One of the reasons for this is that it requires that the number of complex exponentials be known. In the ideal case of exact autocorrelations, the number of complex exponentials may be determined by overestimating the number of exponentials and examining the multiplicity of the minimum eigenvalue. However, with estimated autocorrelations, the multiplicity rule will not work since the minimum eigenvalue will rarely have a multiplicity greater than one. Another limitation with the Pisarenko decomposition is that it assumes that the additive noise is white. In practice, this will generally not be the case and the frequency estimates will be biased. Although Pisarenko's method may be modified to account for nonwhite noise, it is necessary that the power spectrum of the additive noise be known [46].

From a computational point of view, the Pisarenko harmonic decomposition requires finding the minimum eigenvalue and eigenvector of the signal autocorrelation matrix. For high-order problems, this may be computationally time consuming. It is possible, however, to find the minimum eigenvalue and eigenvector efficiently using an iterative algorithm based on the Levinson-Durbin recursion [13,15,19].

8.6.3 MUSIC

In 1979, an improvement to the Pisarenko harmonic decomposition known as the MUltiple SIgnal Classification method (MUSIC) was presented by Schmidt [48]. Like Pisarenko's method, the MUSIC algorithm is a frequency estimation technique. To see how the MUSIC algorithm works, assume that $x(n)$ is a random process consisting of p complex exponentials in white noise with a variance of σ_w^2, and let $\mathbf{R}_x$ be the $M \times M$ autocorrelation matrix of $x(n)$ with $M > p + 1$.[19] If the eigenvalues of $\mathbf{R}_x$ are arranged in decreasing order, $\lambda_1 \geq \lambda_2 \geq \ldots \geq \lambda_M$, and if $\mathbf{v}_1, \mathbf{v}_2, \ldots, \mathbf{v}_M$ are the corresponding eigenvectors, then we may divide these eigenvectors into two groups: the p signal eigenvectors corresponding to the p largest eigenvalues, and the $M - p$ noise eigenvectors that, ideally, have eigenvalues equal to σ_w^2. However, with inexact autocorrelations the smallest $M - p$ eigenvalues will only be approximately equal to σ_w^2. Although we could consider estimating the white noise variance by averaging the $M - p$ smallest eigenvalues

$$\hat{\sigma}_w^2 = \frac{1}{M-p} \sum_{k=p+1}^{M} \lambda_k \tag{8.161}$$

estimating the frequencies of the complex exponentials is a bit more difficult. Since the eigenvectors of $\mathbf{R}_x$ are of length M, each of the noise subspace eigenfilters

$$V_i(z) = \sum_{k=0}^{M-1} v_i(k) z^{-k} \quad ; \quad i = p+1, \ldots, M$$

will have $M - 1$ roots (zeros). Ideally, p of these roots will lie on the unit circle at the frequencies of the complex exponentials, and the eigenspectrum

$$|V_i(e^{j\omega})|^2 = \frac{1}{\left| \sum_{k=0}^{M-1} v_i(k) e^{-jk\omega} \right|^2}$$

associated with the noise eigenvector $\mathbf{v}_i$ will exhibit sharp peaks at the frequencies of the complex exponentials. However, the remaining $(M - p - 1)$ zeros may lie anywhere and, in fact, some may lie close to the unit circle, giving rise to spurious peaks in the eigenspectrum. Furthermore, with inexact autocorrelations, the zeros of $V_i(z)$ that are on the unit circle may not remain on the unit circle. Therefore, when only one noise eigenvector is used to estimate the complex exponential frequencies, there may be some ambiguity in distinguishing the desired peaks from the spurious ones. As we saw in Example 8.6.2, for one complex exponential in white noise, the spurious peaks that are introduced from each of the noise subspace eigenfilters tend to occur at different frequencies. Therefore, in the MUSIC algorithm, the effects of these spurious peaks are reduced by averaging, using the frequency estimation function

$$\hat{P}_{MU}(e^{j\omega}) = \frac{1}{\sum_{i=p+1}^{M} \left| \mathbf{e}^H \mathbf{v}_i \right|^2} \tag{8.162}$$

The frequencies of the complex exponentials are then taken as the locations of the p largest peaks in $\hat{P}_{MU}(e^{j\omega})$. Once the frequencies have been determined the power of each complex exponential may be found using Eq. (8.160).

Instead of searching for the peaks of $\hat{P}_{MU}(e^{j\omega})$, an alternative is to use a method called root MUSIC, which involves rooting a polynomial. Since the z-transform equivalent of

[19]Recall that in the Pisarenko decomposition, $M = p + 1$. Thus, here we are using additional autocorrelations. As we will see, if $M = p + 1$, then the MUSIC algorithm is equivalent to Pisarenko's method.

Eq. (8.162) is

$$\hat{P}_{MU}(z) = \frac{1}{\sum_{i=p+1}^{M} V_i(z)V_i^*(1/z^*)}$$

then the frequency estimates may be taken to be the angles of the p roots of the polynomial

$$D(z) = \sum_{i=p+1}^{M} V_i(z)V_i^*(1/z^*) \tag{8.163}$$

that are closest to the unit circle. A MATLAB program for the MUSIC algorithm is given in Fig 8.34.

8.6.4 Other Eigenvector Methods

In addition to the Pisarenko and Music algorithms, a number of other eigenvector methods have been proposed for estimating the frequencies of complex exponentials in noise. One of these, the EigenVector (EV) method [20], is closely related to the MUSIC algorithm. Specifically, the EV method estimates the exponential frequencies from the peaks of the eigenspectrum

$$\hat{P}_{EV}(e^{j\omega}) = \frac{1}{\sum_{i=p+1}^{M} \frac{1}{\lambda_i}|\mathbf{e}^H\mathbf{v}_i|^2} \tag{8.164}$$

where λ_i is the eigenvalue associated with the eigenvector $\mathbf{v}_i$. Note that if $w(n)$ is white noise and if the autocorrelation sequence $r_x(k)$ is known exactly for $k = 0, 1, \ldots, M-1$,

The MUSIC Algorithm

```
function Px = music(x,p,M)
%
   x   = x(:);
   if M<p+1 | length(x)<M, error('Size of R is inappropriate'), end
   R = covar(x,M);
   [v,d]=eig(R);
   [y,i]=sort(diag(d));
   Px=0;
   for j=1:M-p
       Px=Px+abs(fft(v(:,i(j)),1024));
       end;
   Px=-20*log10(Px);
   end;
```

Figure 8.34 A MATLAB *program for estimating the frequencies of p complex exponentials in white noise using the MUSIC algorithm.*

then the eigenvalues in Eq. (8.164) are equal to the white noise variance, $\lambda_i = \sigma_v^2$, and the EV eigenspectrum will be the same as the MUSIC pseudospectrum to within a constant. However, with estimated autocorrelations the eigenvector method differs from the MUSIC algorithm and appears to produce fewer spurious peaks [20,34]. A MATLAB program for finding the eigenspectrum using the eigenvector method is given in Fig 8.35.

Another eigendecomposition-based method of interest is the *minimum norm algorithm* [29]. Instead of forming an eigenspectrum that uses all of the noise eigenvectors as in the MUSIC and eigenvector algorithms, the minimum norm algorithm uses a single vector **a** that is constrained to lie in the noise subspace, and the complex exponential frequencies are estimated from the peaks of the frequency estimation function

$$\hat{P}_{MN}(e^{j\omega}) = \frac{1}{|\mathbf{e}^H\mathbf{a}|^2} \tag{8.165}$$

With **a** constrained to lie in the noise subspace, if the autocorrelation sequence is known exactly, then $|\mathbf{e}^H\mathbf{a}|^2$ will have nulls at the frequencies of each complex exponential. Therefore, the z-transform of the coefficients in **a** may be factored as follows:

$$A(z) = \sum_{k=0}^{M-1} a(k)z^{-k} = \prod_{k=1}^{p}\left(1 - e^{j\omega_k}z^{-1}\right) \prod_{k=p+1}^{M-1}\left(1 - z_k z^{-1}\right)$$

where z_k for $k = p+1, \ldots, M-1$ are the spurious roots that do not, in general, lie on the unit circle. The problem then is to determine which vector in the noise subspace minimizes the effects of the spurious zeros on the peaks of $\hat{P}_{MN}(e^{j\omega})$. The approach that is used in the minimum norm algorithm is to find the vector **a** that satisfies the following three constraints:

1. The vector **a** lies in the noise subspace.
2. The vector **a** has minimum norm.
3. The first element of **a** is unity.

The first constraint ensures that p roots of $A(z)$ lie on the unit circle. The second constraint ensures that the spurious roots of $A(z)$ lie *inside* the unit circle, i.e., $|z_k| < 1$. Finally, the third constraint ensures that the minimum norm solution is not the zero vector.

The Eigenvector Method

```
function Px = ev(x,p,M)
%
   x   = x(:);
   if M<p+1, error('Specified size of R is too small'), end
   R=covar(x,M);
   [v,d]=eig(R);
   [y,i]=sort(diag(d));
   Px=0;
   for j=1:M-p
       Px=Px+abs(fft(v(:,i(j)),1024)).^2/abs(y(j));
       end;
   Px=-10*log10(Px);
   end;
```

Figure 8.35 A MATLAB *program for estimating the frequencies of p complex exponentials in white noise using the eigenvector method.*

To solve this constrained minimization problem, we begin by noting that the constraint that $\mathbf{a}$ lies in the noise subspace may be written as

$$\mathbf{a} = \mathbf{P}_n \mathbf{v} \tag{8.166}$$

where $\mathbf{P}_n = \mathbf{V}_n \mathbf{V}_n^H$ is the projection matrix that projects an arbitrary vector $\mathbf{v}$ onto the noise subspace (see Eq. (8.151) and p. 34). The third constraint may be expressed as follows:

$$\mathbf{a}^H \mathbf{u}_1 = 1 \tag{8.167}$$

where $\mathbf{u}_1 = [1,\ 0,\ \ldots\ ,\ 0]^T$. This constraint may be combined with the constraint given in Eq. (8.166) as follows,

$$\mathbf{v}^H \big(\mathbf{P}_n^H \mathbf{u}_1\big) = 1 \tag{8.168}$$

Using Eq. (8.166) the norm of $\mathbf{a}$ may be written as

$$\|\mathbf{a}\|^2 = \|\mathbf{P}_n \mathbf{v}\|^2 = \mathbf{v}^H \big(\mathbf{P}_n^H \mathbf{P}_n\big) \mathbf{v}$$

Since $\mathbf{P}_n$ is a projection matrix then it is Hermitian, $\mathbf{P}_n = \mathbf{P}_n^H$, and idempotent, $\mathbf{P}_n^2 = \mathbf{P}_n$. Therefore,

$$\|\mathbf{a}\|^2 = \mathbf{v}^H \mathbf{P}_n \mathbf{v}$$

and it follows that minimizing the norm of $\mathbf{a}$ is equivalent to finding the vector $\mathbf{v}$ that minimizes the quadratic form $\mathbf{v}^H \mathbf{P}_n \mathbf{v}$. We may now reformulate the constrained minimization problem as follows:

$$\boxed{\min \mathbf{v}^H \mathbf{P}_n \mathbf{v} \quad \text{subject to} \quad \mathbf{v}^H \big(\mathbf{P}_n^H \mathbf{u}_1\big) = 1} \tag{8.169}$$

Once the solution to Eq. (8.169) has been found, the minimum norm solution is formed by projecting $\mathbf{v}$ onto the noise subspace using Eq. (8.166).

In Section 2.3.10 we considered the problem of solving constrained minimization problems of the form given in Eq. (8.169). What we found was that the solution is

$$\mathbf{v} = \lambda \mathbf{P}_n^{-1} \big(\mathbf{P}_n^H \mathbf{u}_1\big) = \lambda \mathbf{u}_1$$

where

$$\lambda = \frac{1}{\mathbf{u}_1^H \mathbf{P}_n \mathbf{u}_1}$$

Therefore, the minimum norm solution is

$$\boxed{\mathbf{a} = \mathbf{P}_n \mathbf{v} = \lambda \mathbf{P}_n \mathbf{u}_1 = \frac{\mathbf{P}_n \mathbf{u}_1}{\mathbf{u}_1^H \mathbf{P}_n \mathbf{u}_1}} \tag{8.170}$$

which is simply the projection of the unit vector onto the noise subspace, normalized so that the first coefficient is equal to one. In terms of the eigenvectors of the autocorrelation matrix, the minimum norm solution may be written as

$$\boxed{\mathbf{a} = \frac{(\mathbf{V}_n \mathbf{V}_n^H)\mathbf{u}_1}{\mathbf{u}_1^H (\mathbf{V}_n \mathbf{V}_n^H)\mathbf{u}_1}} \tag{8.171}$$

A MATLAB program for the minimum norm algorithm is given in Fig 8.36. The frequency estimation algorithms are summarized in Table 8.10.

The Minimum Norm Algorithm

```
function Px = min_norm(x,p,M)
%
    x   = x(:);
    if M<p+1, error('Specified size of R is too small'), end
    R=covar(x,M);
    [v,d]=eig(R);
    [y,i]=sort(diag(d));
    for j=1:M-p
        V=[V,v(:,i(j))];
        end;
    a=V*V(1,:)';
    Px=-20*log10(abs(fft(a,1024)));
    end;
```

Figure 8.36 *A* MATLAB *program for estimating the frequencies of p complex exponentials in white noise using the minimum norm algorithm.*

Table 8.10 ***Noise Subspace Methods for Frequency Estimation***

Pisarenko	$\hat{P}_{PHD}(e^{j\omega}) = \dfrac{1}{\lvert\mathbf{e}^H\mathbf{v}_{\min}\rvert^2}$
MUSIC	$\hat{P}_{MU}(e^{j\omega}) = \dfrac{1}{\sum_{i=p+1}^{M} \lvert\mathbf{e}^H\mathbf{v}_i\rvert^2}$
Eigenvector Method	$\hat{P}_{EV}(e^{j\omega}) = \dfrac{1}{\sum_{i=p+1}^{M} \frac{1}{\lambda_i}\lvert\mathbf{e}^H\mathbf{v}_i\rvert^2}$
Minimum Norm	$\hat{P}_{MN}(e^{j\omega}) = \dfrac{1}{\lvert\mathbf{e}^H\mathbf{a}\rvert^2}$; $\mathbf{a} = \lambda\mathbf{P}_n\mathbf{u}_1$

Example 8.6.5 ***A Comparison of Frequency Estimation Methods***

Let $x(n)$ be a process consisting of a sum of four complex exponentials in white noise

$$x(n) = \sum_{k=1}^{4} A_k e^{j(n\omega_k + \phi_k)} + w(n)$$

where the amplitudes A_k are equal to one, the frequencies ω_k are 0.2π, 0.3π, 0.8π, and 1.2π, the phases are uncorrelated random variables that are uniformly distributed over the interval $[0, 2\pi]$, and the variance of the white noise is $\sigma_w^2 = 0.5$. Using ten different realizations of $x(n)$ with $N = 64$ values, overlay plots of the frequency estimation functions using Pisarenko's method, the MUSIC algorithm, the eigenvector method, and the minimum norm algorithm are shown in Fig. 8.37. For Pisarenko's method, the frequency estimation

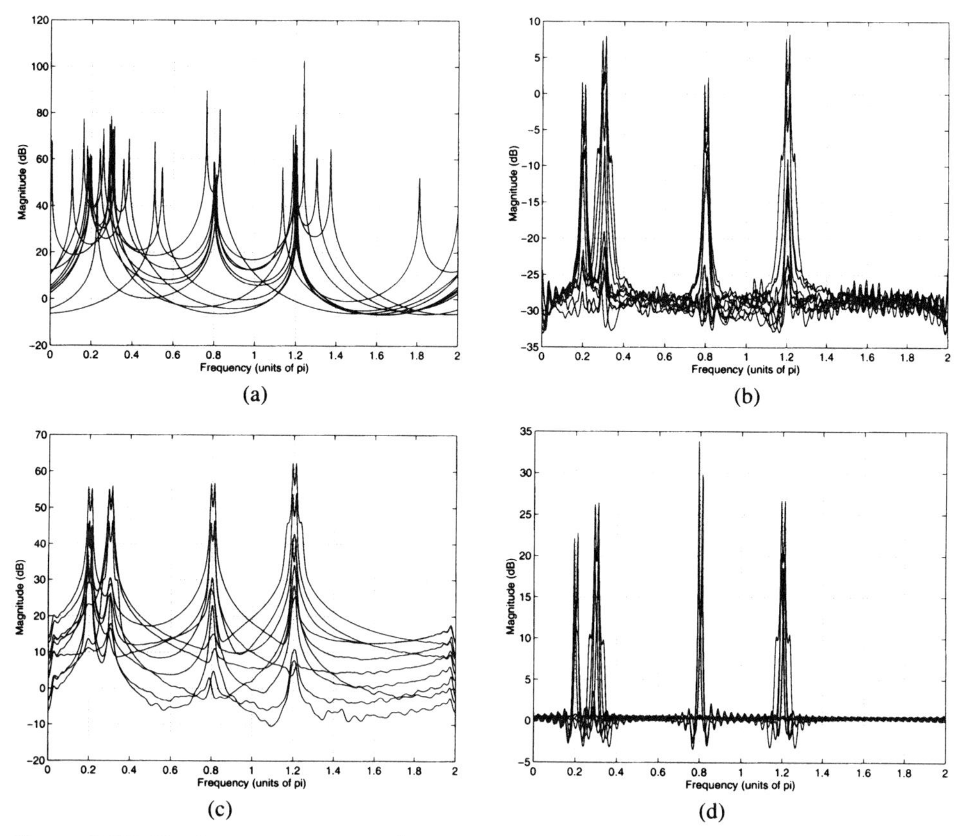

Figure 8.37 *The frequency estimation functions for a process consisting of four complex exponentials in white noise using (a) the Pisarenko harmonic decomposition, (b) the MUSIC algorithm, (c) the eigenvector method and (d) the minimum norm algorithm.*

function was derived from the 5×5 autocorrelation matrix that was estimated from the 64 values of $x(n)$. For the MUSIC, eigenvector, and minimum norm algorithms, the frequency estimation functions were formed from the 64×64 autocorrelation matrix that was estimated from $x(n)$, i.e., $M = 64$. Except for Pisarenko's method, the frequency estimation functions for this process produce accurate estimates of the exponential frequencies, with the most well-defined peaks being produced with the minimum norm algorithm. However, it is important to point out that not all of the frequency estimation functions shown in these overlay plots have four well-defined peaks. In some cases, for example, only two or three peaks are observed.

8.7 PRINCIPAL COMPONENTS SPECTRUM ESTIMATION

In the previous section, we saw how the orthogonality of the signal and noise subspaces could be used to estimate the frequencies of p complex exponentials in white noise. Since these methods only use vectors that lie in the noise subspace, they are often referred to as *noise subspace methods*. In this section, we consider another set of algorithms that use vectors that lie in the signal subspace. These methods are based on a principal components analysis of the autocorrelation matrix and are referred to as *signal subspace methods*. The basic idea of these methods is as follows. Let $\mathbf{R}_x$ be an $M \times M$ autocorrelation matrix of a process that consists of p complex exponentials in white noise. With an eigendecomposition of $\mathbf{R}_x$ we have

$$\mathbf{R}_x = \sum_{i=1}^{M} \lambda_i \mathbf{v}_i \mathbf{v}_i^H = \sum_{i=1}^{p} \lambda_i \mathbf{v}_i \mathbf{v}_i^H + \sum_{i=p+1}^{M} \lambda_i \mathbf{v}_i \mathbf{v}_i^H \tag{8.172}$$

where it is assumed that the eigenvalues have been arranged in decreasing order, $\lambda_1 \geq \lambda_2 \geq \cdots \geq \lambda_M$. Since the second term in Eq. (8.172) is due only to the noise, we may form a *reduced rank* approximation to the signal autocorrelation matrix, $\mathbf{R}_s$, by retaining only the principal eigenvectors of $\mathbf{R}_x$,

$$\hat{\mathbf{R}}_s = \sum_{i=1}^{p} \lambda_i \mathbf{v}_i \mathbf{v}_i^H \tag{8.173}$$

This principal components approximation may then be used in the place of $\mathbf{R}_x$ in a spectral estimator such as the minimum variance method or the maximum entropy method. The net effect of this approach is to *filter out* a portion of the noise, thereby enhancing the estimate of the spectral component due to the signal alone, i.e., the complex exponentials. Another way to view this approach is in terms of a constraint that is being placed on the autocorrelation matrix. Specifically, given that a process consists of p complex exponentials in noise, since the rank of the autocorrelation matrix due to the signal, $\mathbf{R}_s$, is p, then a principal components representation simply imposes this rank-p constraint on $\mathbf{R}_x$. In the following subsections, we discuss how a principal components analysis of the autocorrelation matrix may be used in conjunction with the Blackman-Tukey method, the minimum variance method, and the maximum entropy method to form a principal components spectrum estimate.

8.7.1 Blackman-Tukey Frequency Estimation

The Blackman-Tukey estimate of the power spectrum is formed by taking the discrete-time Fourier transform of a windowed autocorrelation sequence

$$\hat{P}_{BT}(e^{j\omega}) = \sum_{k=-M}^{M} \hat{r}_x(k) w(k) e^{-jk\omega}$$

If $w(k)$ is a Bartlett window, then the Blackman-Tukey estimate may be written in terms of the autocorrelation matrix $\mathbf{R}_x$ as follows:

$$\hat{P}_{BT}(e^{j\omega}) = \frac{1}{M} \sum_{k=-M}^{M} (M - |k|) \hat{r}_x(k) e^{-jk\omega} = \frac{1}{M} \mathbf{e}^H \mathbf{R}_x \mathbf{e} \tag{8.174}$$

With an eigendecomposition of the autocorrelation matrix, $\hat{P}_{BT}(e^{j\omega})$ may therefore be expressed as

$$\hat{P}_{BT}(e^{j\omega}) = \frac{1}{M} \sum_{i=1}^{M} \lambda_i |\mathbf{e}^H \mathbf{v}_i|^2$$

If $x(n)$ is known to consist of p complex exponentials in white noise, and if the eigenvalues of $\mathbf{R}_x$ are arranged in decreasing order, $\lambda_1 \geq \lambda_2 \geq \cdots \geq \lambda_M$, then a principal components version of this spectrum estimate is

$$\hat{P}_{PC-BT}(e^{j\omega}) = \frac{1}{M}\mathbf{e}^H\hat{\mathbf{R}}_s\mathbf{e} = \frac{1}{M}\sum_{i=1}^{p}\lambda_i|\mathbf{e}^H\mathbf{v}_i|^2 \tag{8.175}$$

A MATLAB program for the principal components Blackman-Tukey spectrum estimate is given in Fig. 8.38.

8.7.2 Minimum Variance Frequency Estimation

Given the autocorrelation sequence $r_x(k)$ of a process $x(n)$ for lags $|k| \leq M$, the Mth-order minimum variance spectrum estimate is

$$\hat{P}_{MV}(e^{j\omega}) = \frac{M}{\mathbf{e}^H\mathbf{R}_x^{-1}\mathbf{e}} \tag{8.176}$$

With an eigendecomposition of the autocorrelation matrix, the inverse of $\mathbf{R}_x$ is

$$\mathbf{R}_x^{-1} = \sum_{i=1}^{p}\frac{1}{\lambda_i}\mathbf{v}_i\mathbf{v}_i^H + \sum_{i=p+1}^{M}\frac{1}{\lambda_i}\mathbf{v}_i\mathbf{v}_i^H \tag{8.177}$$

where p is the number of complex exponentials. Retaining only the first p principal components of $\mathbf{R}_x^{-1}$ leads to the principal components minimum variance estimate

$$\hat{P}_{PC-MV}(e^{j\omega}) = \frac{M}{\displaystyle\sum_{i=1}^{p}\frac{1}{\lambda_i}|\mathbf{e}^H\mathbf{v}_i|^2} \tag{8.178}$$

It is interesting to compare Eq. (8.178) with the eigenvector method given in Eq. (8.164).

Principal Components Frequency Estimation
Using the Blackman-Tukey Method

```
function Px = bt_pc(x,p,M)
%
   x   = x(:);
   if M<p+1, error('Specified size of R is too small'), end
   R=covar(x,M);
   [v,d]=eig(R);
   [y,i]=sort(diag(d));
   Px=0;
   for j=M-p+1:M
       Px=Px+abs(fft(v(:,i(j)),1024))*sqrt(real(y(j)));
       end;
   Px=20*log10(Px)-10*log10(M);
   end;
```

Figure 8.38 *A* MATLAB *program for estimating the frequencies of p complex exponentials in white noise using a principal components analysis with the Blackman-Tukey method.*

Whereas $\hat{P}_{PC-MV}(e^{j\omega})$ is based on the first term in the decomposition given in Eq. (8.177), the EV method uses the second. Recall, however, that the EV algorithm produces a frequency estimation function rather than an estimate of the spectrum. Therefore, whereas $\hat{P}_{EV}(e^{j\omega})$ simply produces peaks in the pseudospectrum at the complex exponential frequencies, the minimum variance method provides an estimate of the power spectrum.

8.7.3 Autoregressive Frequency Estimation

Autoregressive spectrum estimation using the autocorrelation, covariance, or modified covariance algorithms involves finding the solution to a set of linear equations of the form

$$\mathbf{R}_x \mathbf{a}_M = \epsilon_M \mathbf{u}_1 \tag{8.179}$$

where $\mathbf{R}_x$ is an $(M+1) \times (M+1)$ autocorrelation matrix. From the solution to these equations,

$$\mathbf{a}_M = \epsilon_M \mathbf{R}_x^{-1} \mathbf{u}_1$$

an estimate of the spectrum is formed as follows:

$$\hat{P}_{AR}(e^{j\omega}) = \frac{|b(0)|^2}{|\mathbf{e}^H \mathbf{a}_M|^2}$$

where $b(0)$ is some appropriately chosen constant such as $|b(0)|^2 = \epsilon_M$. However, if $x(n)$ is known to consist of p complex exponentials in noise, then we may form a principal components solution to Eq. (8.179) as follows:

$$\mathbf{a}_{pc} = \epsilon_M \left(\sum_{i=1}^{p} \frac{1}{\lambda_i} \mathbf{v}_i \mathbf{v}_i^H \right) \mathbf{u}_1$$

or,

$$\mathbf{a}_{pc} = \epsilon_M \sum_{i=1}^{p} \frac{v_i^*(0)}{\lambda_i} \mathbf{v}_i = \epsilon_M \sum_{i=1}^{p} \alpha_i \mathbf{v}_i$$

where $v_i(0)$ is the first element of the normalized eigenvector $\mathbf{v}_i$ and $\alpha_i = v_i^*(0)/\lambda_i$. Therefore, if we set $|b(0)|^2 = \epsilon_M$, then the principal components autoregressive spectrum estimate becomes

$$\hat{P}_{PC-AR}(e^{j\omega}) = \frac{1}{\left| \sum_{i=1}^{p} \alpha_i \mathbf{e}^H \mathbf{v}_i \right|^2}$$

Note that as the number of autocorrelations is increased, only p principal eigenvectors and eigenvalues are used in $\hat{P}_{PC-AR}(e^{j\omega})$. This allows for an increase in the model order without a corresponding increase in the spurious peaks that are due to the noise eigenvectors of the autocorrelation matrix.

We conclude this section with a summary of the signal subspace frequency estimation algorithms, which is given in Table 8.11. As we have seen, each technique is based on using the p principal eigenvectors of the autocorrelation matrix. The techniques differ, however, in the weighting function that is applied to the eigenfilters $\mathbf{e}^H \mathbf{v}_k$ and in whether the linear combination of weighted eigenfilters is in the numerator or denominator of the spectrum estimate.

Table 8.11 ***Signal Subspace Methods***

Blackman-Tukey	$\hat{P}_{PC-BT}(e^{j\omega}) = \dfrac{1}{M}\sum_{i=1}^{p}\lambda_i \lvert \mathbf{e}^H \mathbf{v}_i \rvert^2$
Minimum variance	$\hat{P}_{PC-MV}(e^{j\omega}) = \dfrac{M}{\sum_{i=1}^{p}\dfrac{1}{\lambda_i}\lvert \mathbf{e}^H \mathbf{v}_i \rvert^2}$
Autoregressive	$\hat{P}_{PC-AR}(e^{j\omega}) = \dfrac{1}{\left\lvert \sum_{i=1}^{p}\alpha_i \mathbf{e}^H \mathbf{v}_i \right\rvert^2}$

8.8 SUMMARY

In this chapter, we considered many different approaches for estimating the power spectrum of a wide-sense stationary process. We began with a discussion of the nonparametric methods that are based on computing the discrete-time Fourier transform of an estimate of the autocorrelation sequence. The first of these was the periodogram, which is easily evaluated from the DFT of the given values of the process. Unfortunately, however, the periodogram is not a consistent estimate of the power spectrum. Therefore, we looked at several modifications of the periodogram to improve its statistical properties. These included applying a window to the data, periodogram averaging, and periodogram smoothing. Although periodogram averaging and periodogram smoothing provide a consistent estimate of the power spectrum, they generally do not work well for short data records, and are limited in their ability to resolve closely spaced narrowband processes when the number of data samples is limited. An advantage of these methods, on the other hand, is that they do not make any assumptions or place any constraints on the process and, therefore, may be used on any type of process.

Next, we derived the minimum variance method, which may be viewed as a data-adaptive modification to the periodogram. The basic idea of this approach is to design a filter bank of bandpass filters, measure the power in the processes that are produced at the output of each filter, and estimate the power spectrum by dividing this power estimate by the bandwidth of the filter. Generally, the minimum variance spectrum estimate provides higher resolution than the periodogram and Blackman-Tukey methods.

The primary limitation with the nonparametric methods of spectrum estimation is that the estimate of the spectrum is based on a windowed autocorrelation sequence. In an attempt to overcome this limitation and improve the resolution, we then derived the maximum entropy method, which estimates the spectrum using a maximum entropy extrapolation of a given partial autocorrelation sequence. In other words, given $r_x(k)$ for $|k| \leq p$, the values of $r_x(k)$ for $|k| > p$ are found that make the underlying process as white or as random as possible. What we discovered, however, is that this is equivalent to finding an all-pole model for the process that is consistent with the given autocorrelation sequence, and then computing the power spectrum from the all-pole model.

The next set of techniques that were discussed are the parametric methods of spectrum estimation. With a parametric approach, the first step is to select an appropriate model for the process. This selection may be based on a priori knowledge about how the process is generated or, perhaps, on experimental results indicating that a particular model "works well". Once a model has been selected, the next step is to estimate the model parameters from the given data. For example, if $x(n)$ is assumed to be an autoregressive process, then the covariance method or some other algorithm may be used to estimate the all-pole parameters. The final step is to estimate the power spectrum by incorporating the estimated parameters into the parametric form for the spectrum. Although it is possible to significantly improve the performance of the spectrum estimate with a parametric approach, it is important that the model that is used be consistent with the process that is being analyzed. Otherwise inaccurate or misleading spectrum estimates may result.

The final set of techniques that were discussed are those that assume a harmonic model for the process, i.e., that $x(n)$ is a sum of complex exponentials or sinusoids in white noise. For these processes, the goal is generally to estimate the frequencies of the complex exponentials and, possibly, to determine the powers. Two different approaches to the problem of frequency estimation were considered. The first involves defining a frequency estimation function that produces peaks at the frequencies of the complex exponentials. These frequency estimation functions are designed to take advantage of the fact that the signal and noise subspaces are orthogonal. The second set of approaches use a principal components analysis of the autocorrelation matrix. The resulting reduced-rank approximation to the autocorrelation matrix is then used in a spectrum estimation technique such as the minimum variance method or the maximum entropy method.

Although many different approaches to spectrum estimation have been presented in this chapter, the list is by no means complete. Many other methods have been proposed and, under certain conditions or set of assumptions, these approaches may be superior to the methods described here. For a more complete coverage of spectrum estimation, the reader may consult any one of a number of excellent books on the subject [10,16,25,28,34].

References

1. H. Akaike, "Fitting autoregressive models for prediction," *Ann. Inst. Statist. Math*, vol. 21, pp. 243–247, 1969.
2. H. Akaike, "A new look at the statistical model identification," *IEEE Trans. Autom. Control*, vol. AC-19, pp. 716–723, Dec. 1974.
3. N. O. Anderson, "Comments on the performance of the maximum entropy algorithm," *Proc. IEEE*, vol. 66, pp. 1581–1582, Nov. 1978.
4. R. Ash, *Information Theory*, Wiley Interscience, New York, 1965.
5. M. S. Bartlett, "Smoothing periodograms from time series with continuous spectra," *Nature* (London), vol. 161, pp. 686–687, May 1948.
6. R. B. Blackman and J. W. Tukey, *The Measurement of Power Spectra*, Dover, New York, 1958.
7. J. P. Burg, "Maximum entropy spectral analysis," Ph.D. thesis, Stanford University, Stanford, CA., May 1975.
8. J. Capon, "High-resolution frequency-wavenumber spectrum analysis," *Proc. IEEE*, vol. 57, pp. 1408–1418, Aug. 1969.
9. W. Y. Chen and G. R. Stegen, "Experiments with maximum entropy power spectra of sinusoids," *J. Geophysics Rev.*, vol. 79, pp. 3019–3022, July 1974.
10. D. G. Childers, Ed., *Modern Spectrum Analysis*, IEEE Press, New York, 1978.

11. T. M. Cover and J. A. Thomas, *Elements of Information Theory*, John Wiley & Sons, Inc., New York, 1991.
12. R. E. Davis and L. A. Regier, "Methods of estimating directional wave spectra from multi- element arrays," *Journal of Marine Research*, vol. 35, pp. 453–477, 1977.
13. C. Gueguen, Y. Grenier, and F. Giannella, "Factorial linear modeling, algorithms, and applications," *Proc. 1980 Int. Conf. Acoust., Speech, Signal Proc.*, pp. 618–621, 1980.
14. F. J. Harris, "On the use of windows for harmonic analysis with the discrete Fourier transform," *Proc. IEEE*, vol. 66, pp. 51–83, Jan. 1978.
15. M. H. Hayes and M. A. Clements, "An efficient algorithm for computing Pisarenko's harmonic decomposition using Levinson's recursion," *IEEE Trans. on Acoustics, Speech, Sig. Proc.*, vol. ASSP-34, no. 3, pp. 485–491, June 1986.
16. S. Haykin, Ed., *Nonlinear Methods of Spectral Analysis*, Springer-Verlag, New York, 1979.
17. S. Haykin, *Adaptive Filter Theory*, Prentice-Hall, Englewood Cliffs, NJ, 1986.
18. S. Haykin and S. Kesler, "Prediction-error filtering and maximum-entropy spectral estimation," Chapter 2 in *Nonlinear Methods of Spectral Analysis*, S. Haykin, Ed., Springer-Verlag, New York, 1979.
19. Y. Hu and S. Y. Kung, "Highly concurrent Toeplitz eigen-system solver for high-resolution spectral estimation," *Proc. 1983 Int. Conf. Acoust., Speech, Signal Proc.*, pp. 1422–1425, 1983.
20. D. H. Johnson and S. R. DeGraaf, "Improving the resolution of bearing in passive sonar arrays by eigenvalue analysis," *IEEE Trans. Acoust., Speech, Sig. Proc.*, vol. ASSP-30, no. 4, pp. 638–647, Aug. 1982.
21. R. H. Jones, "Autoregression order selection," *Geophysics*, vol. 41, pp. 771–773, Aug. 1976.
22. R. H. Jones, "Identification and autoregressive spectrum estimation," *IEEE Trans. Autom. Control*, vol. AC-19, pp. 894–897, Dec. 1974.
23. R. L. Kashyap, "Inconsistency of the AIC rule for estimating the order of autoregressive models," *IEEE Trans. Autom. Control*, vol. AC-25, pp. 996–998, Oct. 1980.
24. M. Kaveh and S. P. Bruzzone, "Order determination for autoregressive spectral estimation," *Record of the 1979 RADC Spectral Estimation Workshop*, pp. 139–145, 1979.
25. S. Kay, *Modern Spectrum Estimation: Theory and Applications*, Prentice-Hall, Englewood Cliffs, NJ, 1988.
26. S. Kay, "Noise compensation for autoregressive spectral estimates," *IEEE Trans. Acoust., Speech, Sig. Proc.*, vol. ASSP-28, pp. 292–303, June 1980.
27. S. Kay and S. L. Marple Jr., "Sources and remedies for spectral line splitting in autoregressive spectrum analysis," *Proc. 1979 Int. Conf. on Acoust., Speech, Sig. Proc.*, pp. 151–154, 1979.
28. S. B. Kesler, Ed., *Modern Spectrum Analysis, II*, IEEE Press, New York, 1986.
29. R. Kumaresan and D. W. Tufts, "Estimating the angles of arrival of multiple plane waves," *IEEE Trans. on Aerospace and Elec. Syst.*, vol. AES-19, vol. 1, pp. 134–139, Jan. 1983.
30. R. T. Lacoss, "Data adaptive spectral analysis methods," *Geophysics*, vol. 36, pp. 661–675, Aug. 1971.
31. M. A. Lagunas and A. Gasull-Llampalas, "An improved maximum likelihood method for power spectral density estimation," *IEEE Trans. Acoust., Speech, Sig. Proc.*, vol. ASSP-32, no. 1, pp. 170–172, Feb. 1984.
32. T. E. Landers and R. T. Lacoss, "Some geophysical applications of autoregressive spectral estimates," *IEEE Trans. Geosci. Electron.*, vol. GE-15, pp. 26–32, Jan. 1977.
33. H. L. Larson and B. O. Shubert, *Probabilistic Models in Engineering Sciences: Vol. I, Random Variables and Stochastic Processes*, John Wiley & Sons, New York, 1979.
34. S. L. Marple Jr., *Digital Spectral Analysis with Applications*, Prentice-Hall, Englewood Cliffs, NJ, 1987.
35. S. L. Marple Jr., "A new autoregressive spectrum analysis algorithm," *IEEE Trans. Acoust., Speech, Sig. Proc.*, vol. ASSP-28, pp. 441–454, Aug. 1980.

36. T. L. Marzetta and S. W. Lang, "New interpretations for the MLM and DASE spectral estimators," *Proc. 1983 Int. Conf. Acoust., Speech, Sig. Proc.*, pp. 844–846, April 1983.

37. A. H. Nuttall, "Spectral analysis of a univariate process with bad data points via maximum entropy and linear predictive techniques," Tech. Rep. TR-5303, Naval Underwater Systems Center, New London, CT., March 1976.

38. A. V. Oppenheim and R. W. Schafer, *Discrete-Time Signal Processing*, Prentice-Hall, Englewood Cliffs, NJ, 1989.

39. E. Parzen, "Some recent advances in time series modeling," *IEEE Trans. Autom. Control*, vol. AC-19, pp. 723–730, Dec. 1974.

40. A. Papoulis, *Probability, Random Variables, and Stochastic Processes*, McGraw-Hill, New York, 1984.

41. V. F. Pisarenko, "The retrieval of harmonics from a covariance function," *Geophysics J. Roy. Astron. Soc.*, vol. 33, pp. 347–366, 1973.

42. R. J. McAulay and T. F. Quatieri, "Speech analysis/synthesis based on a sinusoidal representation," *IEEE Trans. Acoust., Speech, Sig. Proc.*, vol. ASSP-34, no. 4, pp. 744–754, Aug. 1986.

43. L. R. Rabiner and R. W. Schafer, *Digital Processing of Speech Signals*, Prentice-Hall, Englewood Cliffs, NJ, 1978

44. I. S. Reed, "On a moment theorem for complex Gaussian processes," *Proc. IEEE Trans. Inf. Theory*, vol. IT-8, pp. 194–195, April 1962.

45. J. Rissanen, "Modeling by shortest data description," *Automatica*, vol. 14, pp. 465–471, 1978.

46. R .A. Roberts and C. T. Mullis, *Digital Signal Processing*, Addison Wesley, Reading, MA. 1987.

47. E. A. Robinson, "A historical perspective of spectrum estimation," *Proc. IEEE*, vol. 70, pp. 885–907, Sept., 1982.

48. R. Schmidt, "Multiple emitter location and signal parameter estimation," *Proc. RADC Spectrum Estimation Workshop*, pp. 243–258, 1979.

49. A. Schuster, "On the investigation of hidden periodicities with application to a supposed twenty-six day period of meteorological phenomena," *Terr. Magn.*, vol. 3, no. 1, pp. 13–41, March 1898.

50. G. Schwartz, "Estimating the dimension of a model," *Annals of Statistics*, vol. 6, pp. 461–464, 1978.

51. S. Shon and K. Mehrota, "Performance comparisons of autoregressive estimation methods," *Proc. 1984 Int. Conf. Acoust., Speech, Sig. Proc.*, pp. 14.3.1–14.3.4., 1984.

52. G. Strang, *Linear Algebra and its Applications*, Academic Press, New York, 1980.

53. D. N. Swingler, "A comparison between Burg's maximum entropy method and a non-recursive technique for the spectral analysis of deterministic signals," *J. Geophysics Res.*, vol. 84, pp. 679–685, Feb. 1979.

54. D. M. Thomas and M. H. Hayes, "A novel data-adaptive power spectrum estimation technique," *Proc. 1987 Int. Conf. on Acoustics, Speech, Sig. Proc.*, pp. 1589–1592, April 1987.

55. H. Tong, "Autoregressive model fitting with noisy data by Akaike's information criterion," *IEEE Trans. Inf. Theory*, vol. IT-21, pp. 476–480, July 1975.

56. T. J. Ulrych and M. Ooe, "Autoregressive and mixed autoregressive-moving average models," Chapter 3 in *Nonlinear Methods of Spectral Analysis*, S. Haykin, Ed., Springer-Verlag, New York, 1979.

57. T. J. Ulrych and R. W. Clayton, "Time series modeling and maximum entropy," *Phys. Earth Planet. Inter.*, vol. 12, pp. 188–200, Aug. 1976.

58. M. Wax and T. Kailath, "Detection of signals by information theoretic criteria," *IEEE Trans. Acoust., Speech, Sig. Proc.*, vol. ASSP-33, no. 2, pp. 387–392, April 1985.

59. P. D. Welch, "The use of fast Fourier transform for the estimation of power spectra: A method based on time averaging over short modified periodograms," *IEEE Trans. Audio and Electroacoust.*, vol. AU-15, pp. 70–73, June 1967.

8.9 PROBLEMS

8.1. Given $N = 10,000$ samples of a process $x(n)$, you are asked to compute the periodogram. However, with only a finite amount of memory resources, you are unable to compute a DFT any longer than 1024. Using these 10,000 samples, describe how you would be able to compute a periodogram that has a resolution of

$$\Delta\omega = 0.89\frac{2\pi}{10000}$$

Hint: Consider how the decimation-in-time FFT algorithm works.

8.2. A continuous-time signal $x_a(t)$ is bandlimited to 5 kHz, i.e., $x_a(t)$ has a spectrum $X_a(f)$ that is zero for $|f| > 5$ kHz. Only 10 seconds of the signal has been recorded and is available for processing. We would like to estimate the power spectrum of $x_a(t)$ using the available data in a radix-2 FFT algorithm, and it is required that the estimate have a resolution of at least 10 Hz. Suppose that we use Bartlett's method of periodogram averaging.

(a) If the data is sampled at the Nyquist rate, what is the minimum section length that you may use to get the desired resolution?

(b) Using the minimum section length determined in part (a), with 10 seconds of data, how many sections are available for averaging?

(c) How does your choice of the sampling rate affect the resolution and variance of your estimate? Are there any benefits to sampling above the Nyquist rate?

8.3. Bartlett's method is used to estimate the power spectrum of a process from a sequence of $N = 2000$ samples.

(a) What is the minimum length L that may be used for each sequence if we are to have a resolution of $\Delta f = 0.005$?

(b) Explain why it would not be advantageous to increase L beyond the value found in (a).

(c) The *quality factor* of a spectrum estimate is defined to be the inverse of the variability,

$$Q = 1/\mathcal{V}$$

Using Bartlett's method, what is the minimum number of data samples, N, that are necessary to achieve a resolution of $\Delta f = 0.005$, and a quality factor that is five times larger than that of the periodogram?

8.4. A random process $x(n)$ is generated by filtering unit variance white noise as shown in the figure below

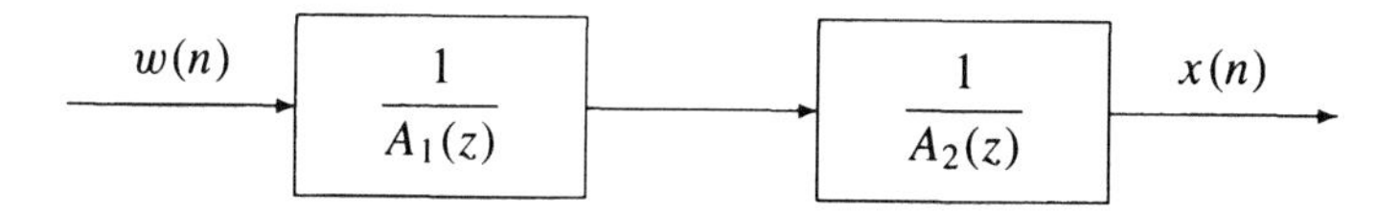

where

$$A_1(z) = 1 + az^{-1} + 0.99z^{-2} \quad ; \quad A_2(z) = 1 - az^{-1} + 0.98z^{-2}$$

(a) Prepare a carefully labeled sketch of the power spectrum of $x(n)$ assuming that a is small, e.g., $0 < a < 0.1$. Pay careful attention to the location and amplitude of the two spectral peaks and the value of $P_x(e^{j\omega})$ at $\omega = \pi/2$.

(b) If $a = 0.1$, determine the section length L required to resolve the spectral peaks of $P_x(e^{j\omega})$ using Bartlett's method. For this value of L, find an approximate value for the bias of the estimate at the peaks of the spectrum. How is the bias related to the area of the spectral peaks?

(c) Consider the method of periodogram smoothing. How many lags of the autocorrelation must be used to obtain a resolution that is comparable to that of Bartlett's estimate considered in part (b)? How much data must be available if the variance of the estimate is to be comparable to that of a four-section Bartlett estimate?

8.5. Many commercial *Fourier analyzers* continuously update the estimate of the power spectrum of a process $x(n)$ by exponential averaging of periodograms as follows,

$$\hat{P}_i(e^{j\omega}) = \alpha \hat{P}_{i-1}(e^{j\omega}) + \frac{1-\alpha}{N}\left|\sum_{n=0}^{N-1} x_i(n)e^{-jn\omega}\right|^2$$

where $x_i(n) = x(n + Ni)$ is the ith sequence of N data values. This update equation is initialized with $\hat{P}_{-1}(e^{j\omega}) = 0$.

(a) Qualitatively describe the philosophy behind this method, and discuss how the value for the weighting factor α should be selected.

(b) Assuming that successive periodograms are uncorrelated and that $0 < \alpha < 1$, find the mean and variance of $\hat{P}_i(e^{j\omega})$ for a Gaussian random process.

(c) Repeat the analysis in part (b) if the periodograms are replaced with modified periodograms.

8.6. The minimum variance method of spectrum estimation constrains the bandpass filter $G_i(e^{j\omega})$ to have a gain of one at frequency $\omega = \omega_i$,

$$G_i(e^{j\omega_i}) = 1$$

Another approach is to constrain the filter to have unit energy over a frequency band that is centered at $\omega = \omega_i$ and has a bandwidth of Δ,

$$\frac{1}{\Delta}\int_{\omega_i-\Delta/2}^{\omega_i+\Delta/2} \left|G_i(e^{j\omega})\right|^2 d\omega = 1$$

With this constraint, the filter coefficients $\mathbf{g}_i = \left[g_i(0), g_i(1), \ldots, g_i(p)\right]^T$ that minimize the power in the filtered process,

$$E\left\{|y_i(n)|^2\right\} = \mathbf{g}_i^H \mathbf{R}_x \mathbf{g}_i$$

may be shown to be the solution to a generalized eigenvalue problem,

$$\mathbf{R}_x \mathbf{g}_i = \lambda(\omega_i, \Delta)\mathbf{T}\mathbf{g}_i$$

where $\mathbf{T}$ is a matrix whose elements depend upon ω_i and Δ. The spectrum estimate, referred to as the DASE estimate, is

$$\hat{P}_{DASE}(e^{j\omega}) = \lambda_{\min}(\omega_i, \Delta)$$

where $\lambda_{\min}(\omega_i, \Delta)$ is the minimum eigenvalue of the generalized eigenvalue problem.

(a) Perform the minimization of $E\left\{|y_i(n)|^2\right\}$ and determine the form of the matrix $\mathbf{T}$.

(b) What happens to the matrix $\mathbf{T}$ in the limit as $\Delta \to 0$? What does the power spectrum estimate correspond to in this case?

(c) Repeat part (b) for $\Delta = 2\pi$.

(d) Find the DASE estimate for white noise.

8.7. Let $x(n)$ be a random process consisting of a single complex exponential in white noise,

$$r_x(k) = Pe^{jk\omega_0} + \sigma_w^2\delta(k)$$

and let $\mathbf{g}_i$ be the minimum variance bandpass filter

$$\mathbf{g}_i = \frac{\mathbf{R}_x^{-1}\mathbf{e}_i}{\mathbf{e}_i^H\mathbf{R}_x^{-1}\mathbf{e}_i}$$

that has a center frequency ω_i with $G(e^{j\omega_i}) = 1$. Assuming that $\omega_i \neq \omega_0$, prove that $G_i(z)$ has a zero that approaches $z = e^{j\omega_0}$ as $\sigma_w^2 \longrightarrow 0$.

8.8. A random process is known to consist of a single sinusoid in white noise,

$$x(n) = A\cos(n\omega_0 + \phi) + w(n)$$

Thus, the autocorrelation sequence for $x(n)$ is

$$r_x(k) = \frac{1}{2}A^2\cos(k\omega_0) + \sigma_w^2\delta(k)$$

(a) If $\omega_0 = \pi/4$, $A = \sqrt{2}$, and $\sigma_w^2 = 1$, find the second-order MEM spectrum, $\hat{P}_{mem}(e^{j\omega})$.

(b) Determine the location of the poles of $\hat{P}_{mem}(z)$.

(c) Does the peak of $\hat{P}_{mem}(e^{j\omega})$ provide an accurate estimate of ω_0? How does this estimate of ω_0 compare to that obtained using the Pisarenko Harmonic decomposition?

8.9. Suppose that we have determined the following values for the autocorrelation sequence of a real-valued random process $x(n)$:

$$r_x(0) = 1 \quad ; \quad r_x(1) = a \quad ; \quad r_x(2) = 0$$

(a) Using the Blackman-Tukey method with a rectangular window, find and make a carefully labeled sketch of the estimated power spectrum, $\hat{P}_{BT}(e^{j\omega})$.

(b) Repeat part (a) for a second-order MEM spectrum estimate, $\hat{P}_{mem}(e^{j\omega})$.

(c) Repeat part (a) for a MV spectrum estimate, $\hat{P}_{MV}(e^{j\omega})$.

(d) What can you say about the autocorrelation sequences that correspond to the spectrum estimates $\hat{P}_{BT}(e^{j\omega})$, $\hat{P}_{mem}(e^{j\omega})$, and $\hat{P}_{MV}(e^{j\omega})$ found in parts (a)-(c)?

8.10. Given that the sixth-order minimum variance spectrum estimate of a process $x(n)$ is

$$\hat{P}_{MV}(e^{j\omega}) = \frac{1}{1 + a\cos 4\omega + 4a\cos 6\omega}$$

and the seventh-order estimate is

$$\hat{P}_{MV}(e^{j\omega}) = \frac{1}{1 - 2a\cos 2\omega - a\cos 7\omega}$$

find the seventh-order maximum entropy spectrum, $\hat{P}_{mem}(e^{j\omega})$.

8.11. The first-order ($p = 1$) minimum variance spectrum estimate of a random process is

$$\hat{P}_{MV}(e^{j\omega}) = \frac{8}{3 - \cos\omega}$$

(a) Find the autocorrelations, $r_x(0)$ and $r_x(1)$, that produced this spectrum estimate.

(b) In general, given the pth-order minimum variance estimate $\hat{P}_{MV}(e^{j\omega})$, is it possible to recover the values of the autocorrelation sequence that produce this estimate?

8.12. The second-order maximum entropy spectrum of a process $x(n)$ is

$$\hat{P}_{mem}(e^{j\omega}) = \frac{2}{\left|1 - 0.5e^{-j\omega} + 0.25e^{-2j\omega}\right|^2}$$

(a) What is the first-order maximum entropy spectrum?

(b) Find the second-order minimum variance spectrum estimate.

8.13. From measurements of a process $x(n)$, we estimate the following values for the autocorrelation sequence:

$$r_x(k) = \alpha^{|k|} \quad ; \quad |k| \leq M$$

where $|\alpha| < 1$. Estimate the power spectrum using

(a) The Blackman-Tukey method with a rectangular window.

(b) The minimum variance method.

(c) The maximum entropy method.

8.14. In Eq. (8.97), the entropy of a Gaussian random process was given as

$$H(x) = \frac{1}{2\pi} \int_{-\pi}^{\pi} \ln P_x(e^{j\omega}) d\omega$$

In this problem, we derive another expression for the entropy. Let $x(n)$ be a real-valued zero mean Gaussian random process, and let $\mathbf{x} = \left[x(0), x(1), \ldots, x(N-1)\right]^T$ be an N-dimensional Gaussian random vector that is formed from samples of this process. The probability density function for this random vector is

$$f_x(\mathbf{x}) = \frac{1}{(2\pi)^{N/2} (\det \mathbf{R}_x)^{1/2}} \exp\left\{-\tfrac{1}{2}\mathbf{x}^T \mathbf{R}_x^{-1} \mathbf{x}\right\}$$

where $\mathbf{R}_x$ is the $N \times N$ autocorrelation matrix of the vector $\mathbf{x}$.

(a) The *average entropy* of a random vector $\mathbf{x}$ is defined as

$$H_N(\mathbf{x}) = -\frac{1}{N} \int f_x(\mathbf{x}) \ln f_x(\mathbf{x}) d\mathbf{x}$$

Show that the average entropy of a zero mean Gaussian random vector is

$$H_N(\mathbf{x}) = \frac{1}{2} \ln(2\pi e) + \frac{1}{2N} \ln \det \mathbf{R}_x$$

(b) Show that the average entropy of a Gaussian random vector may be written as

$$H_N(\mathbf{x}) = \frac{1}{2} \ln(2\pi e) + \frac{1}{N} \sum_{k=0}^{N-1} \ln \epsilon_k$$

where ϵ_k is the prediction error sequence that is generated by the Levinson-Durbin recursion from the autocorrelation sequence $r_x(k)$.

(c) The *entropy rate* of a process $x(n)$ is the limit, as $N \to \infty$, of the average entropy [40],

$$\bar{H}(\mathbf{x}) = \lim_{N \to \infty} H_N(x)$$

Given a partial autocorrelation sequence, $r_x(k)$, for $k = 0, 1, \ldots, N-1$, find the spectrum $P_x(e^{j\omega})$ that maximizes $\bar{H}(x)$ subject to the constraint that the spectrum is consistent with the given autocorrelations.

(d) Let $P_x(e^{j\omega})$ be the power spectrum of a wide-sense stationary process with an $N \times N$ autocorrelation matrix $\mathbf{R}_x$, and let $\lambda_1 \geq \lambda_2 \geq \ldots \geq \lambda_N \geq 0$ be the eigenvalues. *Szegö's theorem* states that if $g(\cdot)$ is a continuous real-valued function then

$$\lim_{N\to\infty} \frac{1}{N}\sum_{k=1}^{N} g(\lambda_k) = \frac{1}{2\pi}\int_{-\pi}^{\pi} g\big[P_x(e^{j\omega})\big]d\omega$$

Use Szegö's theorem to show that

$$H(x) = \frac{1}{2}\ln(2\pi e) + \bar{H}(x)$$

i.e., that the two entropy expressions are equal to within an additive constant.

8.15. In this problem, we examine how the entropy of a process changes with the addition of a harmonic process. Let $y(n)$ be a Gaussian random process with a power spectrum

$$P_y(e^{j\omega}) = P_x(e^{j\omega}) + P_\epsilon(e^{j\omega})$$

where

$$P_\epsilon(e^{j\omega}) = \begin{cases} 1/\epsilon & ; \quad |\omega - \omega_0| < \epsilon \\ 0 & ; \quad \text{otherwise} \end{cases}$$

(a) Find the entropy of $y(n)$.

(b) What is the entropy of this process in the limit as $\epsilon \to 0$?

8.16. Given an autocorrelation sequence $r_x(k)$ for $k = 0, 1, \ldots, p$, the maximum entropy spectrum is

$$\hat{P}_{mem}(e^{j\omega}) = \frac{\epsilon_p}{\left|1 + \sum_{k=1}^{p} a_p(k)e^{-jk\omega}\right|^2}$$

where the coefficients $a_p(k)$ are the solution to the normal equations $\mathbf{R}_x\mathbf{a}_p = \epsilon_p\mathbf{u}_1$. If Γ_k are the reflection coefficients produced by the Levinson-Durbin recursion, show that the MEM spectrum may be upper and lower bounded in terms of Γ_k as follows,

$$r_x(0)\prod_{k=1}^{p}\frac{1-|\Gamma_k|}{1+|\Gamma_k|} \leq \hat{P}_{mem}(e^{j\omega}) \leq r_x(0)\prod_{k=1}^{p}\frac{1+|\Gamma_k|}{1-|\Gamma_k|}$$

Hint: Begin with the frequency domain version of the Levinson order-update equation and use the inequality,

$$\big||a| - |b|\big| \leq \big|a + b\big| \leq |a| + |b|$$

8.17. Let $x(n)$ be a first-order Gaussian autoregressive process with power spectrum

$$P_x(z) = \frac{c}{(1 - az^{-1})(1 - az)}$$

where a and c are real numbers.

(a) With the constraint that the total power in the process is equal to one, find the value or values of a and c that maximize the entropy of $x(n)$.

(b) Repeat part (a) and find the value or values of a and c that minimize the entropy.

8.18. The estimated autocorrelation sequence of a random process $x(n)$ for lags $k = 0, 1, 2, 3, 4$ are

$$r_x(0) = 2 \quad ; \quad r_x(1) = 1 \quad ; \quad r_x(2) = 1 \quad ; \quad r_x(3) = 0.5 \quad ; \quad r_x(4) = 0$$

Estimate the power spectrum of $x(n)$ for each of the following cases.

(a) $x(n)$ is an AR(2) process.

(b) $x(n)$ is an MA(2) process.

(c) $x(n)$ is an ARMA(1,1) process.

(d) $x(n)$ contains a single sinusoid in white noise.

8.19. The first three values of the autocorrelation sequence for a process $x(n)$ are:

$$r_x(0) = 1 \quad ; \quad r_x(1) = 0 \quad ; \quad r_x(2) = -\alpha$$

where $0 < \alpha < 1$. The eigenvalues of the 3×3 autocorrelation matrix that is formed from these autocorrelations are $\lambda_1 = 1 + \alpha$, $\lambda_2 = 1$, and $\lambda_3 = 1 - \alpha$, and the corresponding eigenvectors are

$$\mathbf{v}_1 = \frac{1}{\sqrt{2}} \begin{bmatrix} 1 \\ 0 \\ -1 \end{bmatrix} \quad ; \quad \mathbf{v}_2 = \begin{bmatrix} 0 \\ 1 \\ 0 \end{bmatrix} \quad ; \quad \mathbf{v}_3 = \frac{1}{\sqrt{2}} \begin{bmatrix} 1 \\ 0 \\ 1 \end{bmatrix}$$

(a) Use the Blackman-Tukey method with a rectangular window to estimate the power spectrum of $x(n)$, and make a carefully labeled sketch of your estimate.

(b) Suppose that $x(n)$ is known to consist of two complex exponentials in white noise. Estimate the power spectrum of $x(n)$ and make a carefully labeled sketch of your estimate.

8.20. Suppose that we would like to estimate the power spectrum of an AR(2) process

$$x(n) = a(1)x(n-1) + a(2)x(n-2) + w(n)$$

where $w(n)$ is unit variance white noise. However, our measurements of $x(n)$ are noisy, and what we observe is the process

$$y(n) = x(n) + v(n)$$

where the measurement noise, $v(n)$, is uncorrelated with $x(n)$. It is known that $v(n)$ is a first-order moving average process,

$$v(n) = b(0)q(n) + b(1)q(n-1)$$

where $q(n)$ is white noise. Based on measurements of $v(n)$, the power spectrum of $v(n)$ is estimated to be

$$\hat{P}_v(e^{j\omega}) = 3 + 2\cos\omega$$

From $y(n)$ we estimate the following values of the autocorrelation sequence $r_y(k)$,

$$\hat{r}_y(0) = 5 \quad ; \quad \hat{r}_y(1) = 2 \quad ; \quad \hat{r}_y(2) = 0 \quad ; \quad \hat{r}_y(3) = -1 \quad ; \quad \hat{r}_y(4) = 0.5$$

Using all of the given information, estimate the power spectrum of $x(n)$ using the maximum entropy method.

8.21. Show that for $N \gg 1$, estimating the order of an autoregressive process by minimizing $\text{FPE}(p)$ is equivalent to minimizing $\text{AIC}(p)$. Hint: Show that for large N,

$$N \ln \text{FPE}(p) \approx \text{AIC}(p)$$

and use the fact that, if x is small, then $\ln(1 + x) \approx x$.

8.22. You are given the following values for the autocorrelation sequence of a wide-sense stationary process $x(n)$,

$$r_x(0) = 2 \quad ; \quad r_x(1) = \sqrt{3}/2 \quad ; \quad r_x(2) = 0.5$$

The eigenvalues of the 3×3 Toeplitz autocorrelation matrix are $\lambda_1 = 3.5$, $\lambda_2 = 1.5$, and $\lambda_3 = 1.0$ and the corresponding normalized eigenvectors are

$$\mathbf{v}_1 = \sqrt{2/5}\begin{bmatrix} \sqrt{3}/2 \\ 1 \\ \sqrt{3}/2 \end{bmatrix} \quad ; \quad \mathbf{v}_2 = \frac{1}{\sqrt{2}}\begin{bmatrix} -1 \\ 0 \\ 1 \end{bmatrix} \quad ; \quad \mathbf{v}_3 = \frac{1}{\sqrt{5}}\begin{bmatrix} 1 \\ -\sqrt{3} \\ 1 \end{bmatrix}$$

It is known that $x(n)$ consists of a single sinusoid in white noise.

(a) Estimate the frequency of the sinusoid using the Blackman-Tukey method of frequency estimation.

(b) Use the MUSIC algorithm to estimate the frequency of the sinusoid.

(c) Repeat part (b) using the minimum norm algorithm.

8.23. The Pisarenko harmonic decomposition provides a way to estimate the frequencies of a sum of complex exponentials in white noise. As described in Sect. 8.6.2, the powers of the complex exponentials may be found by solving the set of linear equations given in Eq. (8.160). Another method that may be used is based on the orthogonality of the trigonometric sine and cosine functions. This orthogonality condition implies that

$$\det\begin{bmatrix} \sin\omega_1 & \sin\omega_2 & \cdots & \sin\omega_p \\ \sin 2\omega_1 & \sin 2\omega_2 & \cdots & \sin 2\omega_p \\ \vdots & \vdots & & \vdots \\ \sin p\omega_1 & \sin p\omega_2 & \cdots & \sin p\omega_p \end{bmatrix} \neq 0$$

and

$$\det\begin{bmatrix} 1 & 1 & \cdots & 1 \\ \cos\omega_1 & \cos\omega_2 & \cdots & \cos\omega_p \\ \vdots & \vdots & & \vdots \\ \cos(p-1)\omega_1 & \cos(p-1)\omega_2 & \cdots & \cos(p-1)\omega_p \end{bmatrix} \neq 0$$

provided $0 < \omega_i < \pi$ and $\omega_i \neq \omega_j$.

(a) Given the autocorrelation sequence of a pth-order harmonic process,

$$r_x(k) = \sum_{i=1}^{p} P_i e^{jk\omega_i} + \sigma_w^2 \delta(k)$$

evaluate the imaginary part of $r_x(k)$ and use the orthogonality of the sine functions to derive a set of linear equations that may be solved to find the signal powers P_i.

(b) How would you modify this algorithm if some of the frequencies were equal to zero or π?

(c) How would you modify this approach for a sum of sinusoids in white noise?

8.24. The Pisarenko harmonic decomposition was derived for a process that consists of a sum of complex exponentials in white noise. In this problem we generalize the decomposition to nonwhite noise. To accomplish this, we begin with an alternate derivation of the Pisarenko decomposition for white noise. Let

$$x(n) = \sum_{k=1}^{p} A_k e^{jn\omega_k} + w(n)$$

where $w(n)$ is noise that is uncorrelated with the complex exponentials.

(a) If $w(n)$ is white noise then, as we saw in Eq. (8.149), the autocorrelation matrix for $x(n)$ may be written as

$$\mathbf{R}_x = \mathbf{E}\mathbf{P}\mathbf{E}^H + \sigma_w^2 \mathbf{I}$$

where $\mathbf{E}$ is a matrix of complex exponentials and $\mathbf{P}$ is a diagonal matrix of signal powers. If $x(n)$ is filtered with a pth-order FIR filter $\mathbf{a} = \left[a(0),\ a(1),\ \ldots,\ a(p)\right]^T$, then the power in the output process is

$$\xi = E\left\{|y(n)|^2\right\} = \mathbf{a}^H \mathbf{R}_x \mathbf{a}$$

If $\mathbf{a}$ is constrained to have unit norm, $\mathbf{a}^H\mathbf{a} = 1$, show that the filter that minimizes ξ has p zeros on the unit circle at the frequencies ω_k of the complex exponentials, and show that the minimum value of ξ is equal to σ_w^2.

(b) Now assume that $w(n)$ has an arbitrary power spectrum, $P_w(e^{j\omega})$. If the autocorrelation matrix of $w(n)$ is $\sigma_w^2\mathbf{Q}$, then the autocorrelation matrix for $x(n)$ becomes

$$\mathbf{R}_x = \mathbf{E}\mathbf{P}\mathbf{E}^H + \sigma_w^2 \mathbf{Q}$$

Suppose that $x(n)$ is filtered with a pth-order FIR filter $\mathbf{a} = \left[a(0),\ a(1),\ \ldots,\ a(p)\right]^T$ that is normalized so that

$$\mathbf{a}^H \mathbf{Q} \mathbf{a} = 1$$

Show that the filter that minimizes the power in the filtered process has p zeros on the unit circle at the frequencies ω_k of the complex exponentials, and that the minimum value is equal to σ_w^2.

(c) Show that minimizing $\xi = \mathbf{a}^H\mathbf{R}_x\mathbf{a}$ subject to the constraint $\mathbf{a}^H\mathbf{Q}\mathbf{a} = 1$ is equivalent to solving the generalized eigenvalue problem

$$\mathbf{R}_x \mathbf{a} = \lambda \mathbf{Q} \mathbf{a}$$

for the minimum eigenvalue and eigenvector. Thus, the frequencies of the complex exponentials correspond to the roots of the polynomial that is formed from the minimum eigenvector

$$V_{\min}(z) = \sum_{k=0}^{p} v_{\min}(k) z^{-k}$$

and σ_w^2 corresponds to the minimum eigenvalue.

(d) A random process consists of single sinusoid in nonwhite noise,

$$x(n) = A\sin(n\omega_0 + \phi) + w(n)$$

The first three values of the autocorrelation sequence for $x(n)$ are

$$\mathbf{r}_x = \left[9.515, 7.758, 6.472\right]^T$$

It is known that the additive noise $w(n)$ is a moving average process that is generated by filtering white noise $v(n)$ as follows

$$w(n) = v(n) + 0.1v(n-1)$$

However, the variance of $v(n)$ is unknown. Find the frequency ω_0 and the power, $P = \frac{1}{2}A^2$, of the sinusoid.

8.25. A random process is known to consist of a single sinusoid in white noise

$$x(n) = A\sin(n\omega_0 + \phi) + w(n)$$

where the variance of $w(n)$ is σ_w^2.

(a) Suppose that the first three values of the autocorrelation sequence are estimated and found to be

$$r_x(0) = 1 \quad ; \quad r_x(1) = \beta \quad ; \quad r_x(2) = 0$$

Find and prepare a carefully labeled sketch of the spectrum estimate that is formed using the Blackman-Tukey method with a rectangular window.

(b) With the autocorrelations given in part (a), use the Pisarenko harmonic decomposition to estimate the variance of the white noise, σ_w^2, the frequency of the sinusoid, ω_0, and the sinusoid power, $P = \frac{1}{2}A^2$. How does your estimate of the white noise power and the sinusoid frequency depend upon β? Does the sinusoid power depend upon β?

(c) Suppose that we compute the periodogram $\hat{P}_{per}(e^{j\omega})$ using N samples of $x(n)$. Find and prepare a carefully labeled sketch of the expected value of this spectrum estimate. Is this estimate biased? Is it consistent?

(d) Using the autocorrelations given in part (a), find the second-order MEM power spectrum.

8.26. A random process may be classified in terms of the properties of the prediction error sequence ϵ_k that is produced when fitting an all-pole model to the process. Listed below are five different classifications for the error sequence:

1. $\epsilon_k = c > 0$ for all $k \geq 0$.
2. $\epsilon_k = c > 0$ for all $k \geq k_0$ for some $k_0 > 0$.
3. $\epsilon_k \to c$ as $k \to \infty$ where $c > 0$.
4. $\epsilon_k \to 0$ as $k \to \infty$.
5. $\epsilon_k = 0$ for all $k \geq k_0$ for some $k_0 > 0$.

For each of these classifications, describe as completely as possible the characteristics that may be attributed to the process and its power spectrum.

8.27. In the MUSIC algorithm, finding the peaks of the frequency estimation function

$$\hat{P}_{MU}(e^{j\omega}) = \frac{1}{\sum_{i=p+1}^{M} |\mathbf{e}^H \mathbf{v}_i|^2}$$

is equivalent to finding the minima of the denominator. Show that finding the *minima* of the denominator is equivalent to finding the *maxima* of

$$\sum_{i=1}^{p} |\mathbf{e}^H \mathbf{v}_i|^2$$

Hint: Use the fact that

$$\mathbf{I} = \sum_{i=1}^{M} \mathbf{v}_i \mathbf{v}_i^H$$

8.28. The 3×3 autocorrelation matrix of a harmonic process is

$$\mathbf{R}_x = \begin{bmatrix} 3 & -j & -1 \\ j & 3 & -j \\ -1 & j & 3 \end{bmatrix}$$

(a) Using the Pisarenko harmonic decomposition, find the complex exponential frequencies and the variance of the white noise.

(b) Repeat part (a) using the MUSIC algorithm, the eigenvector method, and the minimum norm method.

8.29. In this problem we prove that the spurious roots in the minimum norm method lie inside the unit circle. Let $x(n)$ be a random process that is a sum of p complex exponentials in white noise, and let $\mathbf{a}$ be an M-dimensional vector that lies in the noise subspace. The z-transform of $\mathbf{a}$ may be factored as follows

$$A(z) = A_0(z)A_1(z)$$

where

$$A_0(z) = \prod_{k=1}^{p}(1 - e^{j\omega_k} z^{-1})$$

is a monic polynomial that has p roots on the unit circle at the frequencies of the complex exponentials in $x(n)$, and $A_1(z)$ is a polynomial that contains the $M - p - 1$ spurious roots.

(a) Show that minimizing $\|\mathbf{a}\|^2$ is equivalent to minimizing

$$\frac{1}{2\pi}\int_{-\pi}^{\pi} |A(e^{j\omega})|^2 d\omega = \frac{1}{2\pi}\int_{-\pi}^{\pi} |A_0(e^{j\omega})|^2 |A_1(e^{j\omega})|^2 d\omega$$

where $A_0(e^{j\omega})$ is fixed and $A_1(e^{j\omega})$ is monic.

(b) From the results of part (a), show that minimizing $\|\mathbf{a}\|^2$ is thus equivalent to using the autocorrelation method to find the prediction error filter $A_1(z)$ for the signal whose z-transform is $A_0(z)$.

(c) From (b), argue that $A_1(z)$ must therefore have all of its roots inside the unit circle.

8.30. In the minimum norm method, the spurious zeros in the polynomial $A(z)$ are separated from those that lie on the unit circle by forcing the spurious roots to lie inside the unit circle. In some applications of eigenvector methods, such as system identification, some of the desired zeros may lie inside the unit circle. In this case, the desired roots cannot be distinguished from the spurious roots. The minimum norm method may be modified, however, to force the spurious zeros to lie *outside* the unit circle. This is done by constraining the *last* element of the vector $\mathbf{a}$ to have a value of one, i.e.,

$$\mathbf{a}^H \mathbf{u}_M = 1$$

where $\mathbf{u}_M = [0, 0, \ldots, 0, 1]^T$ is a unit vector with the last element equal to one.

(a) Derive the *modified minimum norm* algorithm that uses the constant that $\mathbf{a}^H \mathbf{u}_M = 1$ instead of $\mathbf{a}^H \mathbf{u}_1 = 1$ as in the minimum norm algorithm.

(b) The 3×3 autocorrelation matrix for a single complex exponential in white noise is

$$\mathbf{R}_x = \begin{bmatrix} 2 & 1-j & -j\sqrt{2} \\ 1+j & 2 & -j \\ j\sqrt{2} & j & 2 \end{bmatrix}$$

Find the frequency of the complex exponential and the locations of the spurious roots in the minimum norm frequency estimation function.

(c) Repeat part (b) for the modified minimum norm method.

Computer Exercises

C8.1. Consider the broadband AR(4) process $x(n)$ that is produced by filtering unit variance white Gaussian noise with the filter

$$H(z) = \frac{1}{(1 - 0.5z^{-1} + 0.5z^{-2})(1 + 0.5z^{-2})}$$

(a) Generate $N = 256$ samples of this process and estimate the power spectrum using the autocorrelation method with $p = 4$. Make a plot of the estimate and compare it to the true power spectrum.

(b) Repeat part (a) for 20 different realizations of the process $x(n)$. Generate an overlay plot of the 20 estimates and plot the ensemble average. Comment on the variance of the estimate and on how accurately the autocorrelation method is able to estimate the power spectrum.

(c) Repeat part (b) using model orders of $p = 6$, 8, and 12. Describe what happens when the model order becomes too large.

(d) Repeat parts (b) and (c) for the covariance, modified covariance, and Burg algorithms. Which approach seems to be the best for a broadband autoregressive process?

C8.2. Repeat Problem C8.1 for the narrowband AR(4) process that is generated by filtering unit variance white noise with

$$H(z) = \frac{1}{(1 - 1.585z^{-1} + 0.96z^{-2})(1 - 1.152z^{-1} + 0.96z^{-2})}$$

C8.3. Consider the process

$$y(n) = x(n) + w(n)$$

where $w(n)$ is white Gaussian noise with a variance σ_w^2 and $x(n)$ is an AR(2) process that is generated by filtering unit variance white noise with the filter

$$H(z) = \frac{1}{1 - 1.585z^{-1} + 0.96z^{-2}}$$

(a) Plot the power spectrum of $x(n)$ and $y(n)$.

(b) For $\sigma_w^2 = 0.5, 1, 2, 5$, generate $N = 100$ samples of the process $y(n)$, and estimate the power spectrum of $x(n)$ from $y(n)$ using the maximum entropy method with $p = 2$. What is the effect of the noise $w(n)$ on the accuracy of the spectrum estimate?

(c) Repeat part (b) using the maximum entropy method with $p = 5$. Describe your observations.

(d) Since the autocorrelation sequence of $y(n)$ is

$$r_y(k) = r_x(k) + \sigma_w^2 \delta(k)$$

investigate what happens to your estimate in (c) if the autocorrelation sequence is modified by subtracting σ_w^2 from $r_y(0)$ before the maximum entropy spectrum is computed. Does this improve the spectrum estimate?

C8.4. In this exercise, we look at what happens if an autoregressive spectrum estimation technique is used on a moving average process.

(a) Let $x(n)$ be the second-order moving average process that is formed by filtering unit variance white Gaussian noise $w(n)$ as follows,

$$x(n) = w(n) - w(n-2)$$

Examine the spectrum estimates that are produced using an autoregressive technique such as MEM or the covariance method, and discuss your findings. What happens as you let the model order become large?

(b) Repeat part (a) for the MA(3) process that is formed by filtering unit variance white Gaussian noise with the filter

$$H(z) = (1 - 0.98z^{-1})(1 - 0.96z^{-2})$$

C8.5. Write a MATLAB program to estimate the order of an all-pole random process using the Akaike final prediction error, the minimum descriptor length, the Akaike information criterion, and Parzen's CAT. Evaluate the accuracy of the estimates that are produced with these methods if the modeling errors $\mathcal{E}_p$ are derived using the autocorrelation method. Consider both broadband and narrowband AR processes. How much do the estimates change if $\mathcal{E}_p$ is replaced with the Burg error, $\mathcal{E}_p^B$? Discuss your findings.

C8.6. Consider the process $x(n)$ consisting of two sinusoids in noise,

$$x(n) = 2\cos(n\omega_1 + \phi_1) + 2\cos(n\omega_2 + \phi_2) + w(n)$$

where ϕ_1 and ϕ_2 are uncorrelated random variables that are uniformly distributed over the interval $[0, 2\pi]$, and $w(n)$ is a fourth-order moving average process that is generated by filtering unit variance white noise with a linear shift-invariant filter that has a system function

$$H(z) = (1 - z^{-1})(1 + z^{-1})^3$$

Let $\omega_1 = \pi/2$ and $\omega_2 = 1.1\pi/2$.

(a) Plot the power spectrum of $x(n)$.

(b) Find the variance of the moving average process and compare the power in $w(n)$ to the power in each of the sinusoids.

(c) How many values of $x(n)$ are necessary in order to resolve the frequencies of the two sinusoids using the periodogram?

(d) Compare the expected value of the periodogram at the sinusoid frequencies ω_1 and ω_2 with the power spectrum of the noise $P_w(e^{j\omega})$. What record length N is required in order for the expected value of the periodogram at w_1 and w_2 to be twice the value of $P_w(e^{j\omega})$?

(e) Generate a sample realization of $x(n)$ of length $N = 256$ and plot the periodogram. Are the two sinusoidal frequencies clearly visible in the spectrum estimate? Repeat for 20 different realizations of the process and discuss your findings.

(f) Estimate the expected value of the periodogram by averaging the periodograms that are formed from an ensemble of 50 realizations of $x(n)$. Plot the estimate and comment on its resolution and on how closely it estimates the moving average part of the spectrum.

(g) Repeat parts (e) and (f) using Bartlett's method with $K = 2, 4$, and 8 sections.

C8.7. Let $x(n)$ be the process defined in Problem C8.6 consisting of a fourth-order moving average process plus two sinusoids.

(a) Generate a sample realization of $x(n)$ of length $N = 256$, and estimate the spectrum using the maximum entropy method with $p = 4, 8, 16$, and 32. Repeat for 20 different realizations of the process, and generate an overlay plot of the estimates along with the ensemble average. Discuss your findings.

(b) Repeat part (a) using the Burg algorithm and the modified covariance method.

C8.8. One of the methods we have seen for estimating the numerator and denominator coefficients of an ARMA process is iterative prefiltering. In this exercise we consider the use of iterative prefiltering for spectrum estimation.

(a) Generate 100 samples of an ARMA(2,2) process $x(n)$ by filtering unit variance white Gaussian noise with a filter that has a system function of the form

$$H(z) = \frac{1 + 1.5z^{-1} + z^{-2}}{1 + 0.9z^{-2}}$$

Using the method of iterative prefiltering, find the second-order ARMA model for $x(n)$, i.e., $p = q = 2$. Make a plot of the power spectrum that is formed from this model, and compare it to the true power spectrum.

(b) Repeat your experiment in part (a) for 20 different realizations of $x(n)$. How close are the estimated model coefficients to the true model, on the average? How accurate are the estimates of the power spectrum that are generated from these models? What happens if the process is over-modeled by setting $p = q = 3$? What about $p = q = 5$?

C8.9. Let $P_x(e^{j\omega})$ be the power spectrum of a process that has an autocorrelation sequence $r_x(k)$, and let λ_k be the eigenvalues of the $N \times N$ autocorrelation matrix $\mathbf{R}_x$.

(a) Use Szegö's theorem (see Prob. 8.14) to estimate the eigenvalues of the 128 × 128 autocorrelation matrix of the lowpass random process $x(n)$ that has a power spectrum

$$P_x(e^{j\omega}) = \begin{cases} 1 & ; \quad |\omega| < \pi/2 \\ 0 & ; \quad \pi/2 \le |\omega| \le \pi \end{cases}$$

(b) Generate the 128 × 128 autocorrelation matrix for this process, find the eigenvalues, and plot them in increasing order. Given the autocorrelation sequence that is stored in the vector `r`, note that this plot may be generated with the single MATLAB command

```
plot(sort(eig(toeplitz(r)))).
```

Are your results consistent with the estimate found in part (a)?

(c) Repeat parts (a) and (b) for the harmonic process

$$x(n) = A\cos(n\omega_0 + \phi) + w(n)$$

where $w(n)$ is unit variance white noise.

C8.10. It has been said that the minimum variance spectrum estimate exhibits less variance than an AR spectrum estimate for long data records [30]. To investigate this claim, consider the autocorrelation sequence

$$r_x(k) = \cos(0.35\pi k) + \delta(k) + w(k) \quad ; \quad k = 0, 1, \ldots, p$$

where $w(k)$ is uniformly distributed white noise with zero mean and variance σ_w^2. This sequence represents a noisy estimate of the autocorrelation sequence of a random phase sinusoid of frequency $\omega = 0.35\pi$ in unit variance white noise.

(a) Compute the minimum variance spectrum estimate with $p = 10$ and $\sigma_w^2 = 0$.

(b) Generate an overlay plot of 25 minimum variance spectrum estimates with $\sigma_w^2 = 0.1$ and compare these estimates to that generated in part (a).

(c) Repeat parts (a) and (b) using the maximum entropy method and compare your results with the minimum variance estimates. Is the claim substantiated? Suppose the Burg algorithm is used instead of MEM. Are your results any different?

C8.11. In the minimum variance spectrum estimate method, the bandwidths of the bandpass filters are the same for each filter in the filter bank, $\Delta\omega = 2\pi/p$. However, since each filter $\mathbf{g}_i$ has a bandwidth that depends, in general, on the center frequency of the filter, it may be possible to *improve* the power estimate by modifying the definition for the filter bandwidth. For example, consider the "equivalent filter bandwidth," which is defined by

$$\Delta = \mathbf{g}_i^H \mathbf{g}_i$$

(a) Show that the resulting *modified MV spectrum* is given by

$$\hat{P}_x(e^{j\omega}) = \frac{\mathbf{e}^H \mathbf{R}_x^{-1} \mathbf{e}}{\mathbf{e}^H \mathbf{R}_x^{-2} \mathbf{e}}$$

(b) What is the modified MV estimate of white Gaussian noise that has a variance of σ_v^2?

(c) Modify the m-file `minvar.m` so that it finds the modified minimum variance spectrum estimate of a random process $x(n)$.

(d) Let $x(n)$ be a harmonic process consisting of a sum of two sinusoids in white noise. Let the sinusoid frequencies be $\omega_1 = 0.2\pi$ and $\omega_2 = 0.25\pi$, and assume that the signal-to-noise ratio for each sinusoid is 10 dB, i.e.,

$$20 \log_{10} \frac{P}{\sigma_w} = 10$$

where P is the sinusoid power and σ_w^2 is the variance of the white noise. Find the autocorrelation sequence of this process and compute the 10th-order modified MV spectrum estimate. Make a plot of your estimate and compare it to a 10th-order minimum variance estimate, and a 10th-order MEM estimate.

(e) Make a plot of the bandwidths of the bandpass filters versus ω that are used in the modified MV spectrum estimate generated in (d) and compare them to the fixed bandwidth $\Delta\omega = 2\pi/p$ used in the MV spectrum estimate.

(f) Create an ensemble of 25 different harmonic processes of the form given in (d) and generate an overlay plot of the modified MV spectrum estimates. Repeat for the MV spectrum estimate and comment on any differences that you observe.

C8.12. Let $x(n)$ be the harmonic process

$$x(n) = \sum_{i=1}^{3} A_i e^{jn\omega_i} + w(n)$$

where $w(n)$ is unit variance white Gaussian noise. Let $A_1 = 4e^{j\phi_i}$, $A_2 = 3e^{j\phi_2}$, and $A_3 = e^{j\phi_3}$ where ϕ_i are uncorrelated random variables that are uniformly distributed between $-\pi$ and π. In addition, let $\omega_1 = 0.4\pi$, $\omega_2 = 0.45\pi$, and $\omega_3 = 0.8\pi$.

(a) Assuming that it is known that $x(n)$ contains three complex exponentials, use the Pisarenko harmonic decomposition to estimate the frequencies, and discuss the accuracy of your estimates. Repeat for 20 different realizations of $x(n)$. How accurate are the frequency estimates, on the average? How much variance is there in the frequency estimates? What happens if you overestimate the number of complex exponentials? What about if the number of complex exponentials is underestimated?

(b) Write an m-file to estimate the powers in the complex exponentials. Using this m-file, estimate the powers using the frequency estimates derived in part (a). Repeat the estimation of the powers using the correct frequencies.

(c) Repeat part (a) using the MUSIC algorithm, the eigenvector method, and the minimum norm algorithm on 20 different realizations of $x(n)$. Compare the accuracy of the estimates that are produced with each method.

C8.13. If a random process consists of a sum of complex exponentials in white noise, one way to estimate the number of complex exponentials is to perform an eigendecomposition of the autocorrelation matrix and examine the multiplicity of the minimum eigenvalue.

(a) Generate $N = 100$ samples of the random process

$$x(n) = \sum_{i=1}^{3} A_i e^{jn\omega_i} + w(n)$$

where $w(n)$ is white Gaussian noise with a variance σ_w^2. Let $|A_i| = 2$ for $i = 1, 2, 3$, and let the phase of A be a random variable that is uniformly distributed between $-\pi$ and π. If $\omega_1 = 0.4\pi$, $\omega_2 = 0.5\pi$, and $\omega_3 = 1.2\pi$, find the eigenvalues of the 6×6 autocorrelation matrix. Is the number of complex exponentials evident from the eigenvalues?

(b) Repeat part (a) for autocorrelation matrices of size 15×15 and 30×30.

(c) Does the variance of the white noise have any effect on the ability to estimate the number of complex exponentials from the eigenvalues?

(d) Let

$$\alpha(k) = \left[\prod_{i=k+1}^{M} \lambda_i\right]^{1/(M-k)}$$

and let

$$\beta(k) = \frac{1}{M-k}\sum_{i=k+1}^{M} \lambda_i$$

where $\lambda_1 \geq \lambda_2 \geq \cdots \geq \lambda_M$ are the eigenvalues of the $M \times M$ autocorrelation matrix. Thus, $\alpha(k)$ is the harmonic mean of the noise eigenvalues and $\beta(k)$ is the arithmetic mean. The Akaike information criterion for estimating the number of complex exponentials is

$$\text{AIC}(k) = -N(M-k)\log\frac{\alpha(k)}{\beta(k)} + k(2M-k)$$

where N is the length of the data record. An estimate of the number of complex exponentials in a process $x(n)$ is the value of k that minimizes AIC(k). Determine the accuracy of the AIC in estimating the number of complex exponentials by applying it to a number of different harmonic processes.

ADAPTIVE FILTERING

9

9.1 INTRODUCTION

In the previous five chapters, we have considered a variety of different problems including signal modeling, Wiener filtering, and spectrum estimation. In each case, we made an important assumption that the signals that were being analyzed were stationary. In signal modeling, for example, it was assumed that the signal to be modeled could be approximated as the output of a linear *shift-invariant* filter $h(n)$ with an input that is either a unit sample, in the case of deterministic signal modeling, or stationary white noise, in the case of stochastic signal modeling. In Chapter 7, we looked at the problem of designing a linear *shift-invariant* filter that would produce the minimum mean-square estimate of a wide-sense stationary process $d(n)$ and in Chapter 8 we considered the problem of estimating the power spectrum $P_x(e^{j\omega})$ of a wide-sense stationary process $x(n)$. Unfortunately, since the signals that arise in almost every application will be nonstationary, the approaches and techniques that we have been considering thus far would not be appropriate. One way to circumvent this difficulty would be to process these nonstationary processes in blocks, over intervals for which the process may be assumed to be approximately stationary. This approach, however, is limited in its effectiveness for several reasons. First, for rapidly varying processes, the interval over which a process may be assumed to be stationary may be too small to allow for sufficient accuracy or resolution in the estimation of the relevant parameters. Second, this approach would not easily accommodate step changes within the analysis intervals. Third, and perhaps most important, this solution imposes an incorrect model on the data, i.e., piecewise stationary. Therefore, a better approach would be start over and begin with a nonstationarity assumption at the outset.

In order to motivate the approach that we will be considering in this chapter, let us reconsider the Wiener filtering problem within the context of nonstationary processes. Specifically, let $w(n)$ denote the unit sample response of the FIR Wiener filter that produces the minimum mean-square estimate of a desired process $d(n)$,

$$\hat{d}(n) = \sum_{k=0}^{p} w(k)x(n-k) \tag{9.1}$$

As we saw in Chap. 7, if $x(n)$ and $d(n)$ are jointly wide-sense stationary processes, with $e(n) = d(n) - \hat{d}(n)$, then the filter coefficients that minimize the mean-square error

$E\{|e(n)|^2\}$ are found by solving the Wiener-Hopf equations

$$\mathbf{R}_x \mathbf{w} = \mathbf{r}_{dx} \tag{9.2}$$

However, if $d(n)$ and $x(n)$ are nonstationary, then the filter coefficients that minimize $E\{|e(n)|^2\}$ will depend on n, and the filter will be shift-varying, i.e.,

$$\hat{d}(n) = \sum_{k=0}^{p} w_n(k)x(n-k) \tag{9.3}$$

where $w_n(k)$ is the value of the kth filter coefficient at time n. Using vector notation, this estimate may be expressed as

$$\hat{d}(n) = \mathbf{w}_n^T \mathbf{x}(n)$$

where

$$\mathbf{w}_n = \Big[w_n(0),\ w_n(1),\ \ldots,\ w_n(p)\Big]^T$$

is the vector of filter coefficients at time n, and

$$\mathbf{x}(n) = \Big[x(n),\ x(n-1),\ \ldots,\ x(n-p)\Big]^T$$

In many respects, the design of a shift-varying (adaptive) filter is much more difficult than the design of a (shift-invariant) Wiener filter since, for each value of n, it is necessary to find the set of optimum filter coefficients, $w_n(k)$, for $k = 0, 1, \ldots, p$. However, the problem may be simplified considerably if we relax the requirement that $\mathbf{w}_n$ minimize the mean-square error at each time n and consider, instead, a coefficient update equation of the form

$$\boxed{\mathbf{w}_{n+1} = \mathbf{w}_n + \Delta\mathbf{w}_n} \tag{9.4}$$

where $\Delta\mathbf{w}_n$ is a *correction* that is applied to the filter coefficients $\mathbf{w}_n$ at time n to form a new set of coefficients, $\mathbf{w}_{n+1}$, at time $n+1$. This update equation is the heart of the adaptive filters that we will be designing in this chapter. As illustrated in Fig. 9.1, the design of an adaptive filter involves defining how this correction is to be formed. Even for stationary processes, there are several reasons why we may prefer to implement a time-invariant Wiener filter using Eq. (9.4). First, if the order of the filter p is large, then it may be difficult or impractical to solve the Wiener-Hopf equations directly. Second, if $\mathbf{R}_x$ is ill-conditioned (almost singular) then the solution to the Wiener-Hopf equations will be numerically sensitive to round-off errors and finite precision effects. Finally, and perhaps most importantly, is the fact that solving the Wiener-Hopf equations requires that the autocorrelation $r_x(k)$ and the cross-correlation $r_{dx}(k)$ be known. Since these ensemble averages are typically unknown, then it is necessary to estimate them from measurements of the processes. Although we could use estimates such as

$$\begin{aligned}
\hat{r}_x(k) &= \frac{1}{N}\sum_{n=0}^{N-1} x(n)x^*(n-k) \\
\hat{r}_{dx}(k) &= \frac{1}{N}\sum_{n=0}^{N-1} d(n)x^*(n-k)
\end{aligned} \tag{9.5}$$

doing so would result in a delay of N samples. Even more importantly, in an environment for which the ensemble averages are changing in time, these estimates would need to be updated continuously.

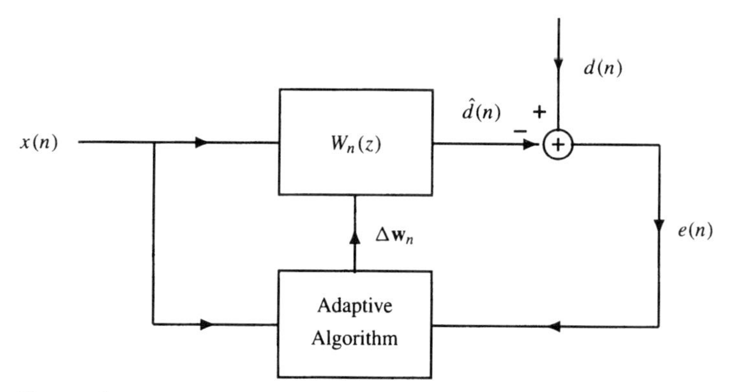

Figure 9.1 *Block diagram of an adaptive filter consisting of a shift-varying filter $W_n(z)$ and an adaptive algorithm for updating the filter coefficients $w_n(k)$.*

The key component of an adaptive filter is the set of rules, or algorithm, that defines how the correction $\Delta\mathbf{w}_n$ is to be formed. Although it is not yet clear what this correction should be, what is clear is that the sequence of corrections should decrease the mean-square error. In fact, whatever algorithm is used, the adaptive filter should have the following properties:

1. In a stationary environment, the adaptive filter should produce a sequence of corrections $\Delta\mathbf{w}_n$ in such a way that $\mathbf{w}_n$ converges to the solution to the Wiener-Hopf equations,
$$\lim_{n\to\infty} \mathbf{w}_n = \mathbf{R}_x^{-1}\mathbf{r}_{dx}$$
2. It should not be necessary to know the signal statistics $r_x(k)$ and $r_{dx}(k)$ in order to compute $\Delta\mathbf{w}_n$. The estimation of these statistics should be "built into" the adaptive filter.
3. For nonstationary signals, the filter should be able to adapt to the changing statistics and "track" the solution as it evolves in time.

An important issue in the implementation of an adaptive filter that is apparent from Fig. 9.1 is the requirement that the error signal, $e(n)$, be available to the adaptive algorithm. Since it is $e(n)$ that allows the filter to *measure* its performance and determine how the filter coefficients should be modified, without $e(n)$ the filter would not be able to adapt. In some applications, acquiring $d(n)$ is straightforward, which makes the evaluation of $e(n)$ easy. Consider, for example, the problem of system identification shown in Fig. 9.2*a* in which a plant produces an output $d(n)$ in response to a known input $x(n)$. The goal is to develop a model for the plant, $W_n(z)$, that produces a response $\hat{d}(n)$ that is as close as possible to $d(n)$. To account for observation noise in the measurement of the plant output, an additive noise source $v(n)$ is shown in the figure. In other applications, the acquisition of $d(n)$ is not as straightforward and some ingenuity is required. For example, consider the problem of interference cancellation in which a signal $d(n)$ is observed in the presence of an interfering signal $v(n)$,

$$x(n) = d(n) + v(n)$$

Since $d(n)$ is unknown, the error sequence cannot be generated directly. In some circumstances, however, the system shown in Fig. 9.2*b* will generate a sequence that may be used by the adaptive filter to estimate $d(n)$. Specifically, suppose that $d(n)$ and $v(n)$ are

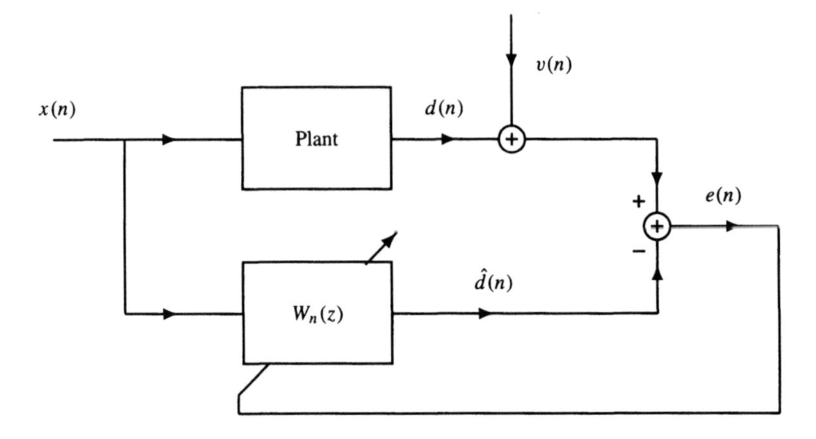

(a) System identification

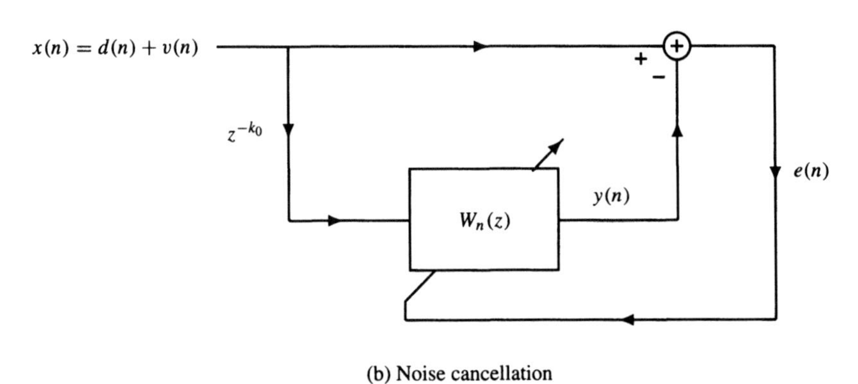

(b) Noise cancellation

Figure 9.2 *Two adaptive filtering applications that illustrate how an error sequence, $e(n)$, may be generated.*

real-valued, uncorrelated zero mean processes. In addition, suppose that $d(n)$ is a narrowband process and that $v(n)$ is a broadband process with an autocorrelation sequence that is approximately zero for lags $k \geq k_0$. The mean-square error that is formed by taking the difference between $x(n)$ and the output of the adaptive filter, $y(n)$, may be expressed as follows:

$$\begin{aligned} E\{e^2(n)\} &= E\Big\{\big[d(n) + v(n) - y(n)\big]^2\Big\} \\ &= E\Big\{v^2(n)\Big\} + E\Big\{\big[d(n) - y(n)\big]^2\Big\} + 2E\Big\{v(n)\big[d(n) - y(n)\big]\Big\} \end{aligned} \tag{9.6}$$

Since $v(n)$ and $d(n)$ are uncorrelated, then $E\{v(n)d(n)\} = 0$ and the last term becomes

$$2E\Big\{v(n)\big[d(n) - y(n)\big]\Big\} = -2E\Big\{v(n)y(n)\Big\}$$

In addition, since the input to the adaptive filter is $x(n - k_0)$, then the output $y(n)$ is

$$y(n) = \sum_{k=0}^{p} w_n(k)x(n - k_0 - k) = \sum_{k=0}^{p} w_n(k)\Big[d(n - k_0 - k) + v(n - k_0 - k)\Big]$$

Thus,

$$E\{v(n)y(n)\} = \sum_{k=0}^{p} w_n(k)\Big[E\{v(n)d(n-k_0-k)\} + E\{v(n)v(n-k_0-k)\}\Big]$$

Finally, since $v(n)$ is uncorrelated with $d(n)$ as well as with $v(n-k_0-k)$, then $E\{v(n)y(n)\} = 0$ and the mean-square error becomes

$$E\{e^2(n)\} = E\{v^2(n)\} + E\{[d(n) - y(n)]^2\}$$

Therefore, minimizing $E\{e^2(n)\}$ is equivalent to minimizing $E\{[d(n)-y(n)]^2\}$, the mean-square error between $d(n)$ and the output of the adaptive filter, $y(n)$. Thus, the output of the adaptive filter is the minimum mean-square estimate of $d(n)$.

In this chapter, we will consider a variety of different methods for designing and implementing adaptive filters. As we will see, the efficiency of the adaptive filter and its performance in estimating $d(n)$ will depend on a number of factors including the type of filter (FIR or IIR), the filter structure (direct form, parallel, lattice, etc.), and the way in which the performance measure is defined (mean-square error, least squares error). The organization of this chapter is as follows. In Section 9.2, we begin with the development of the FIR adaptive filters that are based on the method of steepest descent. Of primary interest will be the LMS adaptive filter. We will consider direct form as well as lattice filter structures. In Section 9.3, we will look briefly at the design of adaptive IIR filters. Being more difficult to characterize in terms of their properties and performance, the focus will be on the LMS adaptive recursive filter. Finally, in Section 9.4, the Recursive Least Squares (RLS) algorithm will be developed. There is a wide variety of applications in which adaptive filters have been successfully used such as linear prediction, echo cancellation, channel equalization, interference cancellation, adaptive notch filtering, adaptive control, system identification, and array processing. Therefore, as we progress through this chapter and look at specific adaptive filtering algorithms, we will briefly introduce some of these applications.

9.2 FIR ADAPTIVE FILTERS

In this section, we begin our study of adaptive filters by looking at the design of FIR (non-recursive) adaptive filters. In contrast to IIR or recursive adaptive filters, FIR filters are routinely used in adaptive filtering applications that range from adaptive equalizers in digital communication systems [26] to adaptive noise control systems [14]. There are several reasons for the popularity of FIR adaptive filters. First, stability is easily controlled by ensuring that the filter coefficients are bounded. Second, there are simple and efficient algorithms for adjusting the filter coefficients. Third, the performance of these algorithms is well understood in terms of their convergence and stability. Finally, FIR adaptive filters very often perform well enough to satisfy the design criteria.

An FIR adaptive filter for estimating a desired signal $d(n)$ from a related signal $x(n)$, as illustrated in Fig. 9.3, is

$$\hat{d}(n) = \sum_{k=0}^{p} w_n(k)x(n-k) = \mathbf{w}_n^T\mathbf{x}(n)$$

Here it is assumed that $x(n)$ and $d(n)$ are nonstationary random process and the goal is to

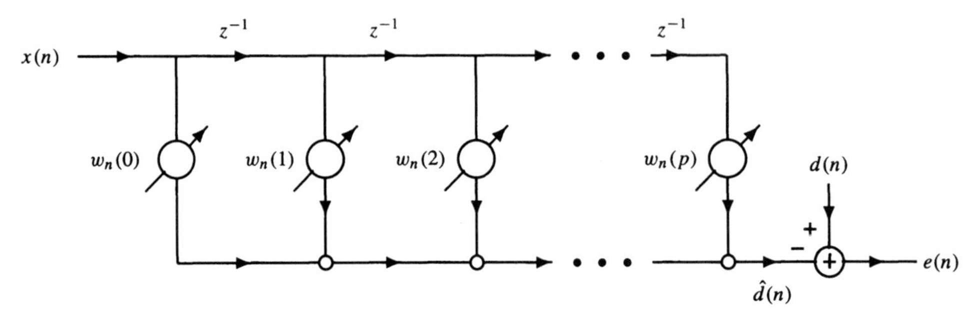

Figure 9.3 *A direct-form FIR adaptive filter.*

find the coefficient vector $\mathbf{w}_n$ at time n that minimizes the mean-square error,

$$\xi(n) = E\left\{|e(n)|^2\right\}$$

where

$$e(n) = d(n) - \hat{d}(n) = d(n) - \mathbf{w}_n^T \mathbf{x}(n) \tag{9.7}$$

As in the derivation of the FIR Wiener filter in Section 7.2, the solution to this minimization problem may be found by setting the derivative of $\xi(n)$ with respect to $w_n^*(k)$ equal to zero for $k = 0, 1, \ldots, p$. The result is

$$E\left\{e(n)x^*(n-k)\right\} = 0 \quad ; \quad k = 0, 1, \ldots, p \tag{9.8}$$

Substituting Eq. (9.7) into Eq. (9.8) we have

$$E\left\{\left[d(n) - \sum_{l=0}^{p} w_n(l)x(n-l)\right]x^*(n-k)\right\} = 0 \quad ; \quad k = 0, 1, \ldots, p$$

which, after rearranging terms, becomes

$$\sum_{l=0}^{p} w_n(l)E\left\{x(n-l)x^*(n-k)\right\} = E\left\{d(n)x^*(n-k)\right\} \quad ; \quad k = 0, 1, \ldots, p \tag{9.9}$$

Equation (9.9) is a set of $p+1$ linear equations in the $p+1$ unknowns $w_n(l)$. However, unlike the case of an FIR Wiener filter where it was assumed that $x(n)$ and $d(n)$ are jointly WSS, the solution to these equations depends on n. We may express these equations in vector form as follows:

$$\boxed{\mathbf{R}_x(n)\mathbf{w}_n = \mathbf{r}_{dx}(n)} \tag{9.10}$$

where

$$\mathbf{R}_x(n) = \begin{bmatrix} E\{x(n)x^*(n)\} & E\{x(n-1)x^*(n)\} & \cdots & E\{x(n-p)x^*(n)\} \\ E\{x(n)x^*(n-1)\} & E\{x(n-1)x^*(n-1)\} & \cdots & E\{x(n-p)x^*(n-1)\} \\ \vdots & \vdots & & \vdots \\ E\{x(n)x^*(n-p)\} & E\{x(n-1)x^*(n-p)\} & \cdots & E\{x(n-p)x^*(n-p)\} \end{bmatrix}$$

is a $(p+1) \times (p+1)$ Hermitian matrix of autocorrelations and

$$\mathbf{r}_{dx}(n) = \Big[E\{d(n)x^*(n)\},\ E\{d(n)x^*(n-1)\},\ \cdots,\ E\{d(n)x^*(n-p)\}\Big]^T \tag{9.11}$$

is a vector of cross-correlations between $d(n)$ and $x(n)$. Note that in the case of jointly WSS processes, Eq. (9.10) reduces to the Wiener-Hopf equations, and the solution $\mathbf{w}_n$ becomes independent of time. Instead of solving Eq. (9.10) for each value of n, which would be impractical in most real-time implementations, in the following section, we consider an iterative approach that is based on the method of steepest descent.

9.2.1 The Steepest Descent Adaptive Filter

In designing an FIR adaptive filter, the goal is to find the vector $\mathbf{w}_n$ at time n that minimizes the quadratic function

$$\xi(n) = E\{|e(n)|^2\}$$

Although the vector that minimizes $\xi(n)$ may be found by setting the derivatives of $\xi(n)$ with respect to $w^*(k)$ equal to zero, another approach is to *search* for the solution using the method of steepest descent. The *method of steepest descent* is an iterative procedure that has been used to find extrema of nonlinear functions since before the time of Newton. The basic idea of this method is as follows. Let $\mathbf{w}_n$ be an estimate of the vector that minimizes the mean-square error $\xi(n)$ at time n. At time $n+1$ a new estimate is formed by adding a correction to $\mathbf{w}_n$ that is designed to bring $\mathbf{w}_n$ closer to the desired solution. The correction involves taking a step of size μ in the direction of *maximum descent* down the quadratic error surface. For example, shown in Fig. 9.4*a* is a three-dimensional plot of a quadratic function of two real-valued coefficients, $w(0)$ and $w(1)$, given by[1]

$$\xi(n) = 6 - 6w(0) - 4w(1) + 6\big[w^2(0) + w^2(1)\big] + 6w(0)w(1) \qquad (9.12)$$

Note that the contours of constant error, when projected onto the $w(0)$-$w(1)$ plane, form a set of concentric ellipses. The direction of steepest descent at any point in the plane is the direction that a marble would take if it were placed on the inside of this quadratic *bowl*. Mathematically, this direction is given by the *gradient*, which is the vector of partial

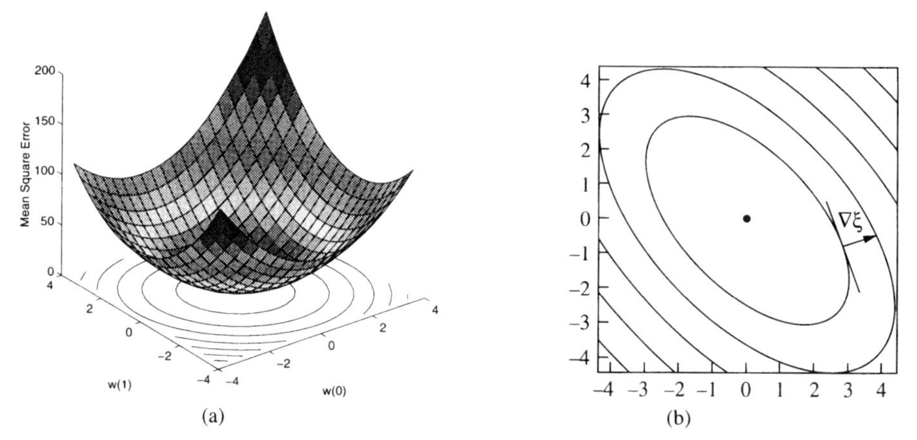

Figure 9.4 *(a) A quadratic function of two weights and (b) the contours of constant error. The gradient vector, which points in the direction of maximum increase in ξ, is orthogonal to the line that is tangent to the contour as illustrated in (b).*

[1] Although this error does not depend on n, we still denote it by $\xi(n)$.

derivatives of $\xi(n)$ with respect to the coefficients $w(k)$. For the quadratic function in Eq. (9.12), the gradient vector is

$$\nabla\xi(n) = \begin{bmatrix} \dfrac{\partial\xi(n)}{\partial w(0)} \\ \dfrac{\partial\xi(n)}{\partial w(1)} \end{bmatrix} = \begin{bmatrix} 12w(0) + 6w(1) - 6 \\ 12w(1) + 6w(0) - 4 \end{bmatrix}$$

As shown in Fig. 9.4*b*, for any vector **w**, the gradient is orthogonal to the line that is tangent to the contour of constant error at **w**. However, since the gradient vector points in the direction of *steepest ascent*, the direction of *steepest descent* points in the negative gradient direction. Thus, the update equation for $\mathbf{w}_n$ is

$$\boxed{\mathbf{w}_{n+1} = \mathbf{w}_n - \mu\nabla\xi(n)}$$

The step size μ affects the rate at which the weight vector moves down the quadratic surface and must be a positive number (a negative value for μ would move the weight vector up the quadratic surface in the direction of maximum ascent and would result in an increase in the error). For very small values of μ, the correction to $\mathbf{w}_n$ is small and the movement down the quadratic surface is slow and, as μ is increased, the rate of descent increases. However, there is an upper limit on how large the step size may be. For values of μ that exceed this limit, the trajectory of $\mathbf{w}_n$ becomes unstable and unbounded. The steepest descent algorithm may be summarized as follows:

1. Initialize the steepest descent algorithm with an initial estimate, $\mathbf{w}_0$, of the optimum weight vector $\mathbf{w}$.
2. Evaluate the gradient of $\xi(n)$ at the current estimate, $\mathbf{w}_n$, of the optimum weight vector.
3. Update the estimate at time n by adding a correction that is formed by taking a step of size μ in the negative gradient direction

$$\mathbf{w}_{n+1} = \mathbf{w}_n - \mu\nabla\xi(n)$$

4. Go back to (2) and repeat the process.

Let us now evaluate the gradient vector $\nabla\xi(n)$. Assuming that **w** is complex, the gradient is the derivative of $E\{|e(n)|^2\}$ with respect to $\mathbf{w}^*$. With

$$\nabla\xi(n) = \nabla E\{|e(n)|^2\} = E\{\nabla|e(n)|^2\} = E\{e(n)\nabla e^*(n)\}$$

and

$$\nabla e^*(n) = -\mathbf{x}^*(n)$$

it follows that

$$\nabla\xi(n) = -E\{e(n)\mathbf{x}^*(n)\}$$

Thus, with a step size of μ, the steepest descent algorithm becomes

$$\boxed{\mathbf{w}_{n+1} = \mathbf{w}_n + \mu E\{e(n)\mathbf{x}^*(n)\}} \qquad (9.13)$$

To see how this steepest descent update equation for $\mathbf{w}_n$ performs, let us consider what happens in the case of stationary processes. If $x(n)$ and $d(n)$ are jointly WSS then

$$\begin{aligned} E\{e(n)\mathbf{x}^*(n)\} &= E\{d(n)\mathbf{x}^*(n)\} - E\{\mathbf{w}_n^T\mathbf{x}(n)\mathbf{x}^*(n)\} \\ &= \mathbf{r}_{dx} - \mathbf{R}_x\mathbf{w}_n \end{aligned}$$

and the steepest descent algorithm becomes

$$\boxed{\mathbf{w}_{n+1} = \mathbf{w}_n + \mu\big(\mathbf{r}_{dx} - \mathbf{R}_x\mathbf{w}_n\big)} \tag{9.14}$$

Note that if $\mathbf{w}_n$ is the solution to the Wiener-Hopf equations, $\mathbf{w}_n = \mathbf{R}_x^{-1}\mathbf{r}_{dx}$, then the correction term is zero and $\mathbf{w}_{n+1} = \mathbf{w}_n$ for all n. Of greater interest, however, is how the weights evolve in time, beginning with an arbitrary initial weight vector $\mathbf{w}_0$. The following property defines what is required for $\mathbf{w}_n$ to converge to $\mathbf{w}$.

Property 1. For jointly wide-sense stationary processes, $d(n)$ and $x(n)$, the steepest descent adaptive filter converges to the solution to the Wiener-Hopf equations

$$\lim_{n\to\infty} \mathbf{w}_n = \mathbf{R}_x^{-1}\mathbf{r}_{dx}$$

if the step size satisfies the condition

$$0 < \mu < \frac{2}{\lambda_{\max}} \tag{9.15}$$

where $\lambda_{\max}$ is the maximum eigenvalue of the autocorrelation matrix $\mathbf{R}_x$.

To establish this property, we begin by rewriting Eq. (9.14) as follows:

$$\mathbf{w}_{n+1} = \big(\mathbf{I} - \mu\mathbf{R}_x\big)\mathbf{w}_n + \mu\mathbf{r}_{dx} \tag{9.16}$$

Subtracting $\mathbf{w}$ from both sides of this equation and using the fact that $\mathbf{r}_{dx} = \mathbf{R}_x\mathbf{w}$ we have

$$\mathbf{w}_{n+1} - \mathbf{w} = \big(\mathbf{I} - \mu\mathbf{R}_x\big)\mathbf{w}_n + \mu\mathbf{R}_x\mathbf{w} - \mathbf{w} = \big[\mathbf{I} - \mu\mathbf{R}_x\big]\big(\mathbf{w}_n - \mathbf{w}\big) \tag{9.17}$$

If we let $\mathbf{c}_n$ be the *weight error vector*,

$$\boxed{\mathbf{c}_n = \mathbf{w}_n - \mathbf{w}} \tag{9.18}$$

then Eq. (9.17) becomes

$$\mathbf{c}_{n+1} = \big(\mathbf{I} - \mu\mathbf{R}_x\big)\mathbf{c}_n \tag{9.19}$$

Note that, unless $\mathbf{R}_x$ is a diagonal matrix, there will be cross-coupling between the coefficients of the weight error vector. However, we may decouple these coefficients by diagonalizing the autocorrelation matrix as follows. Using the spectral theorem (p. 44) the autocorrelation matrix may be factored as

$$\mathbf{R}_x = \mathbf{V}\mathbf{\Lambda}\mathbf{V}^H$$

where $\mathbf{\Lambda}$ is a diagonal matrix containing the eigenvalues of $\mathbf{R}_x$, and $\mathbf{V}$ is a matrix whose columns are the eigenvectors of $\mathbf{R}_x$. Since $\mathbf{R}_x$ is Hermitian and nonnegative definite, the eigenvalues are real and non-negative, $\lambda_k \geq 0$, and the eigenvectors may be chosen to be

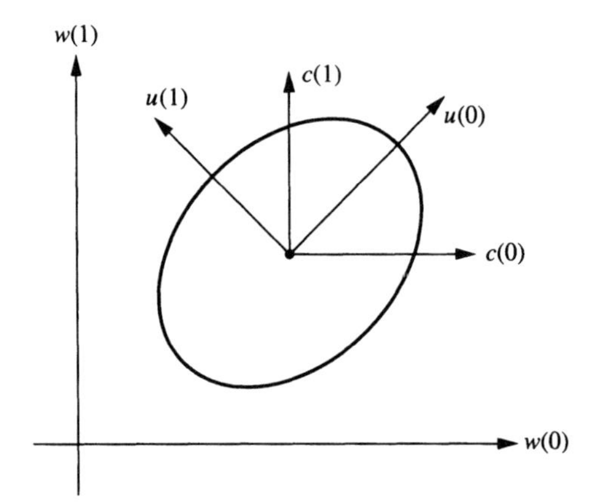

Figure 9.5 *Illustration of the relationships between the vectors* **w**, **c**, *and* **u**.

orthonormal, $\mathbf{V}\mathbf{V}^H = \mathbf{I}$, i.e., $\mathbf{V}$ is *unitary*. Incorporating this factorization into Eq. (9.19) leads to

$$\mathbf{c}_{n+1} = \left(\mathbf{I} - \mu \mathbf{V}\mathbf{\Lambda}\mathbf{V}^H\right)\mathbf{c}_n$$

Using the unitary property of $\mathbf{V}$ we have

$$\mathbf{c}_{n+1} = \left(\mathbf{V}\mathbf{V}^H - \mu \mathbf{V}\mathbf{\Lambda}\mathbf{V}^H\right)\mathbf{c}_n = \mathbf{V}\left(\mathbf{I} - \mu\mathbf{\Lambda}\right)\mathbf{V}^H\mathbf{c}_n$$

Multiplying both sides of the equation by $\mathbf{V}^H$ gives

$$\mathbf{V}^H\mathbf{c}_{n+1} = \left(\mathbf{I} - \mu\mathbf{\Lambda}\right)\mathbf{V}^H\mathbf{c}_n \tag{9.20}$$

If we define

$$\mathbf{u}_n = \mathbf{V}^H\mathbf{c}_n \tag{9.21}$$

then Eq. (9.20) becomes

$$\mathbf{u}_{n+1} = \left(\mathbf{I} - \mu\mathbf{\Lambda}\right)\mathbf{u}_n$$

As illustrated in Fig. 9.5, Eq. (9.21) represents a rotation of the coordinate system for the weight error vector $\mathbf{c}_n$ with the new axes aligned with the eigenvectors $\mathbf{v}_k$ of the autocorrelation matrix. With an initial weight vector $\mathbf{u}_0$, it follows that

$$\mathbf{u}_n = \left(\mathbf{I} - \mu\mathbf{\Lambda}\right)^n\mathbf{u}_0 \tag{9.22}$$

Since $(\mathbf{I} - \mu\mathbf{\Lambda})$ is a diagonal matrix then the kth component of $\mathbf{u}_n$ may be expressed as

$$u_n(k) = (1 - \mu\lambda_k)^n u_0(k) \tag{9.23}$$

In order for $\mathbf{w}_n$ to converge to $\mathbf{w}$ it is necessary that the weight error vector $\mathbf{c}_n$ converge to zero and, therefore, that $\mathbf{u}_n = \mathbf{V}^H\mathbf{c}_n$ converge to zero. This will occur for any $\mathbf{u}_0$ if and only if

$$|1 - \mu\lambda_k| < 1 \quad ; \quad k = 0, 1, \ldots, p$$

which places the following restriction on the step size μ

$$0 < \mu < \frac{2}{\lambda_{\max}}$$

as was to be shown. ■

Having found an expression for the evolution of the vector $\mathbf{u}_n$, we may derive an expression for the evolution of the weight vector $\mathbf{w}_n$. With

$$\mathbf{w}_n = \mathbf{w} + \mathbf{c}_n = \mathbf{w} + \mathbf{V}\mathbf{u}_n = \mathbf{w} + \left[\mathbf{v}_0,\ \mathbf{v}_1,\ \ldots,\ \mathbf{v}_p\right]\begin{bmatrix} u_n(0) \\ u_n(1) \\ \vdots \\ u_n(p) \end{bmatrix}$$

using Eq. (9.23) we have

$$\mathbf{w}_n = \mathbf{w} + \sum_{k=0}^{p} u_n(k)\mathbf{v}_k = \mathbf{w} + \sum_{k=0}^{p}\left(1 - \mu\lambda_k\right)^n u_0(k)\mathbf{v}_k$$

Since $\mathbf{w}_n$ is a linear combination of the eigenvectors $\mathbf{v}_n$, referred to as the *modes* of the filter, then $\mathbf{w}_n$ will converge no faster than the slowest decaying mode. With each mode decaying as $(1 - \mu\lambda_k)^n$, we may define the time constant τ_k to be the time required for the kth mode to reach $1/e$ of its initial value:

$$(1 - \mu\lambda_k)^{\tau_k} = 1/e$$

Taking logarithms we have

$$\boxed{\tau_k = -\frac{1}{\ln(1 - \mu\lambda_k)}} \tag{9.24}$$

If μ is small enough so that $\mu\lambda_k \ll 1$, then the time constant may be approximated by

$$\tau_k \approx \frac{1}{\mu\lambda_k}$$

Defining the overall time constant to be the time that it takes for the slowest decaying mode to converge to $1/e$ of its initial value we have

$$\boxed{\tau = \max\left\{\tau_k\right\} \approx \frac{1}{\mu\lambda_{\min}}} \tag{9.25}$$

Since Property 1 places an upper bound of $2/\lambda_{\max}$ on the step size if $\mathbf{w}_n$ is to converge to $\mathbf{w}$, let us write μ as follows

$$\mu = \alpha\frac{2}{\lambda_{\max}}$$

where α is a normalized step size with $0 < \alpha < 1$. In terms of α, the time constant becomes

$$\tau \approx \frac{1}{2\alpha}\frac{\lambda_{\max}}{\lambda_{\min}} = \frac{1}{2\alpha}\,\chi$$

where $\chi = \lambda_{\max}/\lambda_{\min}$ is the condition number of the autocorrelation matrix. Thus, the rate of convergence is determined by the eigenvalue spread. The reason for this dependence may be explained geometrically as illustrated in Fig. 9.6. In Fig. 9.6*a* are the error contours for a two-dimensional adaptive filter with $\lambda_1 = \lambda_2 = 1$, i.e., $\chi = 1$, along with the trajectory of $\mathbf{w}_n$. Note that the contours are circles and that, at any point, the direction of steepest descent points toward the minimum of the quadratic function. Fig. 9.6*b*, on the other hand, shows the error contours when $\lambda_1 = 3$ and $\lambda_2 = 1$, i.e., $\chi = 3$, along with the trajectory of $\mathbf{w}_n$.

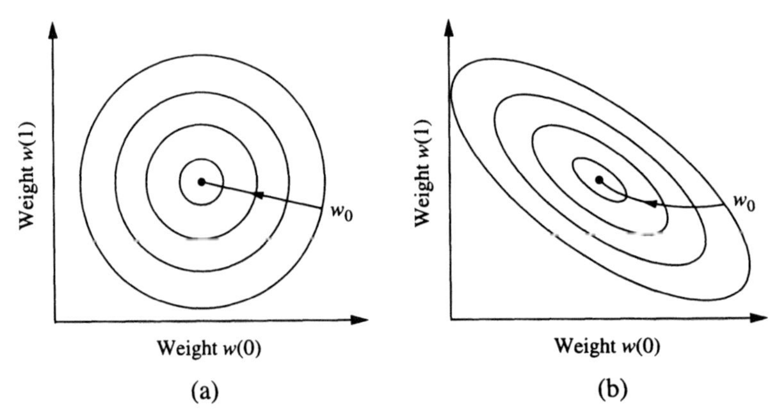

Figure 9.6 *The effect of the condition number on the convergence rate of the steepest descent algorithm. (a) With $\chi = 1$ the direction of steepest descent points toward the minimum of the error surface. (b) With $\chi = 3$ the direction of steepest descent does not, in general, point toward the minimum.*

Note that, unlike the case when $\chi = 1$, the direction of steepest descent is typically not pointing toward the minimum of the quadratic function.

Another important and useful measure of performance is the behavior of the mean-square error as a function of n. From Eq. (7.12) we see that for jointly wide-sense stationary processes the minimum mean-square error is

$$\xi_{\min} = r_d(0) - \mathbf{r}_{dx}^H \mathbf{w} \tag{9.26}$$

For an arbitrary weight vector, $\mathbf{w}_n$, the mean-square error is

$$\begin{aligned} \xi(n) &= E\{|e(n)|^2\} = E\{|d(n) - \mathbf{w}_n^T \mathbf{x}(n)|^2\} \\ &= r_d(0) - \mathbf{r}_{dx}^H \mathbf{w}_n - \mathbf{w}_n^H \mathbf{r}_{dx} + \mathbf{w}_n^H \mathbf{R}_x \mathbf{w}_n \end{aligned} \tag{9.27}$$

Expressing $\mathbf{w}_n$ in terms of the weight error vector $\mathbf{c}_n$ this becomes

$$\xi(n) = r_d(0) - \mathbf{r}_{dx}^H(\mathbf{w} + \mathbf{c}_n) - (\mathbf{w} + \mathbf{c}_n)^H \mathbf{r}_{dx} + (\mathbf{w} + \mathbf{c}_n)^H \mathbf{R}_x (\mathbf{w} + \mathbf{c}_n)$$

Expanding the products and using the fact that $\mathbf{R}_x \mathbf{w} = \mathbf{r}_{dx}$ we have

$$\xi(n) = r_d(0) - \mathbf{r}_{dx}^H \mathbf{w} + \mathbf{c}_n^H \mathbf{R}_x \mathbf{c}_n \tag{9.28}$$

Note that the first two terms are equal to the minimum error, $\xi_{\min}$. Therefore, the error at time n is

$$\xi(n) = \xi_{\min} + \mathbf{c}_n^H \mathbf{R}_x \mathbf{c}_n$$

With the decomposition $\mathbf{R}_x = \mathbf{V} \mathbf{\Lambda} \mathbf{V}^H$ and the definition for $\mathbf{u}_n$ given in Eq. (9.21), we have

$$\boxed{\xi(n) = \xi_{\min} + \mathbf{u}_n^H \mathbf{\Lambda}_x \mathbf{u}_n} \tag{9.29}$$

Expanding the quadratic form and using the expression for $u_n(k)$ given in Eq. (9.23) yields

$$\xi(n) = \xi_{\min} + \sum_{k=0}^{p} \lambda_k |u_n(k)|^2 = \xi_{\min} + \sum_{k=0}^{p} \lambda_k (1 - \mu\lambda_k)^{2n} |u_0(k)|^2 \tag{9.30}$$

Thus, if the step size μ satisfies the condition for convergence given in Eq. (9.15), then $\xi(n)$ decays exponentially to $\xi_{\min}$. A plot of $\xi(n)$ versus n is referred to as the *learning curve* and indicates how rapidly the adaptive filter *learns* the solution to the Wiener-Hopf equations.

Although for stationary processes the steepest descent adaptive filter converges to the solution to the Wiener-Hopf equations when $\mu < 2/\lambda_{\max}$, this algorithm is primarily of theoretical interest and finds little use in adaptive filtering applications. The reason for this is that, in order to compute the gradient vector, it is necessary that $E\{e(n)\mathbf{x}^*(n)\}$ be known. For stationary processes this requires that the autocorrelation matrix of $x(n)$, and the cross-correlation between $d(n)$ and $\mathbf{x}(n)$ be known. In most applications, these ensemble averages are unknown and must be estimated from the data. In the next section, we look at the LMS algorithm, which incorporates an *estimate* of the expectation $E\{e(n)\mathbf{x}^*(n)\}$ into the adaptive algorithm.

9.2.2 The LMS Algorithm

In the previous section, we developed the steepest descent adaptive filter, which has a weight-vector update equation given by

$$\mathbf{w}_{n+1} = \mathbf{w}_n + \mu E\{e(n)\mathbf{x}^*(n)\} \tag{9.31}$$

A practical limitation with this algorithm is that the expectation $E\{e(n)\mathbf{x}^*(n)\}$ is generally unknown. Therefore, it must be replaced with an estimate such as the sample mean

$$\hat{E}\{e(n)\mathbf{x}^*(n)\} = \frac{1}{L}\sum_{l=0}^{L-1} e(n-l)\mathbf{x}^*(n-l) \tag{9.32}$$

Incorporating this estimate into the steepest descent algorithm, the update for $\mathbf{w}_n$ becomes

$$\mathbf{w}_{n+1} = \mathbf{w}_n + \frac{\mu}{L}\sum_{l=0}^{L-1} e(n-l)\mathbf{x}^*(n-l) \tag{9.33}$$

A special case of Eq. (9.33) occurs if we use a one-point sample mean ($L = 1$),

$$\hat{E}\{e(n)\mathbf{x}^*(n)\} = e(n)\mathbf{x}^*(n) \tag{9.34}$$

In this case, the weight vector update equation assumes a particularly simple form

$$\boxed{\mathbf{w}_{n+1} = \mathbf{w}_n + \mu e(n)\mathbf{x}^*(n)} \tag{9.35}$$

and is known as the *LMS algorithm* [36]. The simplicity of the algorithm comes from the fact that the update for the kth coefficient,

$$w_{n+1}(k) = w_n(k) + \mu e(n)x^*(n-k)$$

requires only one multiplication and one addition (the value for $\mu e(n)$ need only be computed once and may be used for all of the coefficients). Therefore, an LMS adaptive filter having $p+1$ coefficients requires $p+1$ multiplications and $(p+1)$ additions to update the filter coefficients. In addition, one addition is necessary to compute the error $e(n) = d(n) - y(n)$ and one multiplication is needed to form the product $\mu e(n)$. Finally, $p+1$ multiplications and p additions are necessary to calculate the output, $y(n)$, of the adaptive filter. Thus, a total of $2p+3$ multiplications and $2p+2$ additions per output point are required. The complete LMS algorithm is summarized in Table 9.1 and a MATLAB program is given in

Table 9.1 ***The LMS Algorithm for a pth-Order FIR Adaptive Filter***

Parameters:	$p =$ Filter order
	$\mu =$ Step size
Initialization:	$\mathbf{w}_0 = \mathbf{0}$
Computation:	For $n = 0, 1, 2, \ldots$
	(a) $y(n) = \mathbf{w}_n^T \mathbf{x}(n)$
	(b) $e(n) = d(n) - y(n)$
	(c) $\mathbf{w}_{n+1} = \mathbf{w}_n + \mu e(n)\mathbf{x}^*(n)$

The LMS Algorithm

```
function [A,E] = lms(x,d,mu,nord,a0)
%
X=convm(x,nord);
[M,N] = size(X);
if nargin < 5,   a0 = zeros(1,N);   end
a0 = a0(:).';
E(1) = d(1) - a0*X(1,:).';
A(1,:) = a0 + mu*E(1)*conj(X(1,:));
if M>1
for k=2:M-nord+1;
    E(k) = d(k) - A(k-1,:)*X(k,:).';
    A(k,:) = A(k-1,:) + mu*E(k)*conj(X(k,:));
    end;
end;
```

Figure 9.7 *A* MATLAB *program to estimate a process $d(n)$ from a related process $x(n)$ using the LMS algorithm.*

Fig. 9.7. Although based on a very crude estimate of $E\{e(n)\mathbf{x}^*(n)\}$, we will see that the LMS adaptive filter often performs well enough to be used successfully in a number of applications. In the following section, we consider the convergence of the LMS adaptive filter.

9.2.3 Convergence of the LMS Algorithm

In estimating the ensemble average $E\{e(n)\mathbf{x}^*(n)\}$ with a one-point sample average $e(n)\mathbf{x}^*(n)$, the LMS algorithm replaces the gradient in the steepest descent algorithm,

$$\nabla\xi(n) = -E\{e(n)\mathbf{x}^*(n)\}$$

with an estimated gradient

$$\hat{\nabla}\xi(n) = -e(n)\mathbf{x}^*(n) \tag{9.36}$$

When this is done, the correction that is applied to $\mathbf{w}_n$ is generally not aligned with the

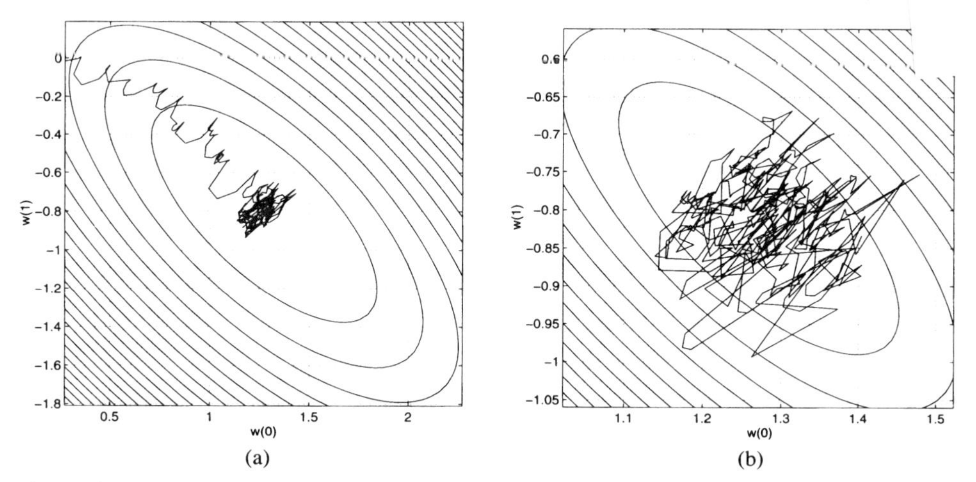

Figure 9.8 *Weight trajectories for a two-coefficient LMS adaptive filter. (a) The trajectory of* $\mathbf{w}_n$ *for 500 iterations with* $\mathbf{w}_0 = \mathbf{0}$. *The solution to the Wiener-Hopf equations is at the center of the ellipses. (b) The trajectory with* $\mathbf{w}_0 = \mathbf{R}_x^{-1}\mathbf{r}_{dx}$.

direction of steepest descent. However, since the gradient estimate is unbiased,

$$E\left\{\hat{\nabla}\xi(n)\right\} = -E\left\{e(n)\mathbf{x}^*(n)\right\} = \nabla\xi(n)$$

then the correction that is applied is, on the average, in the direction of steepest descent. An illustration of the behavior that is typically observed with the LMS algorithm for stationary processes is shown in Fig. 9.8. In Fig. 9.8*a*, with the weight vector initialized to $\mathbf{w}_0 = \mathbf{0}$, we see that the sequence of weights generally moves toward the solution to the Wiener-Hopf equations, $\mathbf{w} = \mathbf{R}_x^{-1}\mathbf{r}_{dx}$. However, as illustrated in Fig. 9.8*b*, if the weight vector is initialized with the solution to the Wiener-Hopf equations then, although the gradient at this point is zero, since a gradient estimate is used, $\mathbf{w}_n$ moves randomly within a neighborhood of this solution. In order to explain these observations, we will now consider the convergence of the LMS adaptive filter.

Since $\mathbf{w}_n$ is a vector of random variables, the convergence of the LMS algorithm must be considered within a statistical framework. Therefore, we will begin by assuming that $x(n)$ and $d(n)$ are jointly wide-sense stationary processes, and will determine when the coefficients $\mathbf{w}_n$ converge in the mean to $\mathbf{w} = \mathbf{R}_x^{-1}\mathbf{r}_{dx}$, i.e.,

$$\lim_{n\to\infty} E\left\{\mathbf{w}_n\right\} = \mathbf{w} = \mathbf{R}_x^{-1}\mathbf{r}_{dx}$$

We begin by substituting Eq. (9.7) into the LMS coefficient update equation as follows:

$$\mathbf{w}_{n+1} = \mathbf{w}_n + \mu\left[d(n) - \mathbf{w}_n^T\mathbf{x}(n)\right]\mathbf{x}^*(n)$$

Taking the expected value, we have

$$E\left\{\mathbf{w}_{n+1}\right\} = E\left\{\mathbf{w}_n\right\} + \mu E\left\{d(n)\mathbf{x}^*(n)\right\} - \mu E\left\{\mathbf{x}^*(n)\mathbf{x}^T(n)\mathbf{w}_n\right\} \tag{9.37}$$

Although the last term in Eq. (9.37) is not easy to evaluate, it may be simplified considerably if we make the following *independence assumption* [20]:

Independence Assumption. The data $\mathbf{x}(n)$ and the LMS weight vector $\mathbf{w}_n$ are statistically independent.[2]

With this assumption, Eq. (9.37) becomes

$$\begin{aligned} E\{\mathbf{w}_{n+1}\} &= E\{\mathbf{w}_n\} + \mu E\{d(n)\mathbf{x}^*(n)\} - \mu E\{\mathbf{x}^*(n)\mathbf{x}^T(n)\}E\{\mathbf{w}_n\} \\ &= (\mathbf{I} - \mu\mathbf{R}_x)E\{\mathbf{w}_n\} + \mu\mathbf{r}_{dx} \end{aligned} \tag{9.38}$$

which is the same as Eq. (9.16) for the weight vector in the steepest descent algorithm. Therefore, the analysis for the steepest descent algorithm is applicable to $E\{\mathbf{w}_{n+1}\}$. In particular, it follows from Eq. (9.22) that

$$E\{\mathbf{u}_n\} = (\mathbf{I} - \mu\mathbf{\Lambda})^n\mathbf{u}_0 \tag{9.39}$$

where

$$\mathbf{u}_n = \mathbf{V}^H[\mathbf{w}_n - \mathbf{w}]$$

Since $\mathbf{w}_n$ will converge in the mean to $\mathbf{w}$ if $E\{\mathbf{u}_n\}$ converges to zero, then we have the following property:

Property 2. For jointly wide-sense stationary processes, the LMS algorithm *converges in the mean* if

$$0 < \mu < \frac{2}{\lambda_{\max}} \tag{9.40}$$

and the independence assumption is satisfied.

Although Eq. (9.40) places a bound on the step size for convergence in the mean, this bound is of limited use for two reasons. First, it is generally acknowledged that the upper bound is too large to ensure stability of the LMS algorithm since it is not sufficient to guarantee that the coefficient vector will remain bounded for all n. For example, although this bound ensures that $E\{\mathbf{w}_n\}$ converges, it places no constraints on how large the variance of $\mathbf{w}_n$ may become. Second, since the upper bound is expressed in terms of the largest eigenvalue of $\mathbf{R}_x$, using this bound requires that $\mathbf{R}_x$ be known. If this matrix is unknown, then it becomes necessary to estimate $\lambda_{\max}$. One way around this difficulty is to use the fact that $\lambda_{\max}$ may be upper bounded by the trace of $\mathbf{R}_x$,

$$\lambda_{\max} \leq \sum_{k=0}^{p} \lambda_k = \text{tr}(\mathbf{R}_x)$$

Therefore, if $x(n)$ is wide-sense stationary, then $\mathbf{R}_x$ is Toeplitz and the trace becomes

$$\text{tr}(\mathbf{R}_x) = (p+1)r_x(0) = (p+1)E\{|x(n)|^2\}$$

[2] Since $\mathbf{w}_n$ depends on the previous input vectors $\mathbf{x}(n-1), \mathbf{x}(n-2), \ldots$, this assumption can only be approximately true. However, experience with the LMS algorithm shows that this assumption leads to convergence properties that are generally in close agreement with experiments and computer simulations.

As a result, Eq. (9.40) may be replaced with the more conservative bound

$$0 < \mu < \frac{2}{(p+1)E\{|x(n)|^2\}} \tag{9.41}$$

Although we have simply replaced one unknown with another, $E\{|x(n)|^2\}$ is more easily estimated since it represents the power in $x(n)$. For example, $E\{|x(n)|^2\}$ could be estimated using an average such as

$$\hat{E}\{|x(n)|^2\} = \frac{1}{N}\sum_{k=0}^{N-1}|x(n-k)|^2$$

In the following example we consider the use of an LMS adaptive filter for linear prediction.

Example 9.2.1 *Adaptive Linear Prediction Using the LMS Algorithm*

Let $x(n)$ be a second-order autoregressive process that is generated according to the difference equation

$$x(n) = 1.2728x(n-1) - 0.81x(n-2) + v(n) \tag{9.42}$$

where $v(n)$ is unit variance white noise. As we saw in Section 7.3.4, the optimum causal linear predictor for $x(n)$ is

$$\hat{x}(n) = 1.2728x(n-1) - 0.81x(n-2)$$

However, in order to design this predictor (i.e., to know that the optimum predictor coefficients are 1.2728 and -0.81) it is necessary to know the autocorrelation sequence of $x(n)$. Therefore, suppose we consider an adaptive linear predictor of the form:

$$\hat{x}(n) = w_n(1)x(n-1) + w_n(2)x(n-2)$$

as shown in Fig. 9.9. With the LMS algorithm, the predictor coefficients $w_n(k)$ are updated as follows:

$$w_{n+1}(k) = w_n(k) + \mu e(n)x^*(n-k)$$

If the step size μ is sufficiently small, then the coefficients $w_n(1)$ and $w_n(2)$ will converge in the mean to their optimum values, $w(1) = 1.2728$ and $w(2) = -0.81$, respectively. Note that the prediction error is

$$e(n) = x(n) - \hat{x}(n) = \left[1.2728 - w_n(1)\right]x(n-1) + \left[-0.81 - w_n(1)\right]x(n-2) + v(n)$$

Therefore, when $w_n(1) = 1.2728$ and $w_n(2) = -0.81$, the error becomes $e(n) = v(n)$, and the minimum mean-square error is[3]

$$\xi_{\min} = \sigma_v^2 = 1$$

Although we might expect the mean-square error $E\{|e(n)|^2\}$ to converge to $\xi_{\min}$ as $\mathbf{w}_n$ converges to $\mathbf{w}$, as we will soon discover, this is not the case.

To see how this adaptive linear predictor behaves in practice, suppose that the weight vector is initialized to zero, $\mathbf{w}_0 = \mathbf{0}$, and that the step size is $\mu = 0.02$. Shown in Fig. 9.10*a*

[3]This may also be shown by evaluating the expression for the minimum mean-square error given in Eq. (9.26).

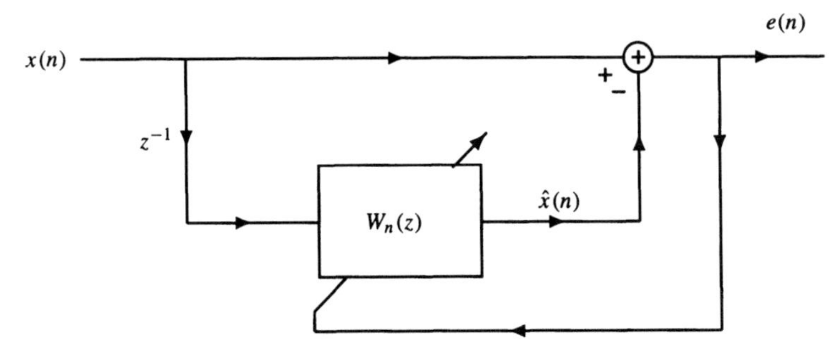

Figure 9.9 *An adaptive filter for linear prediction.*

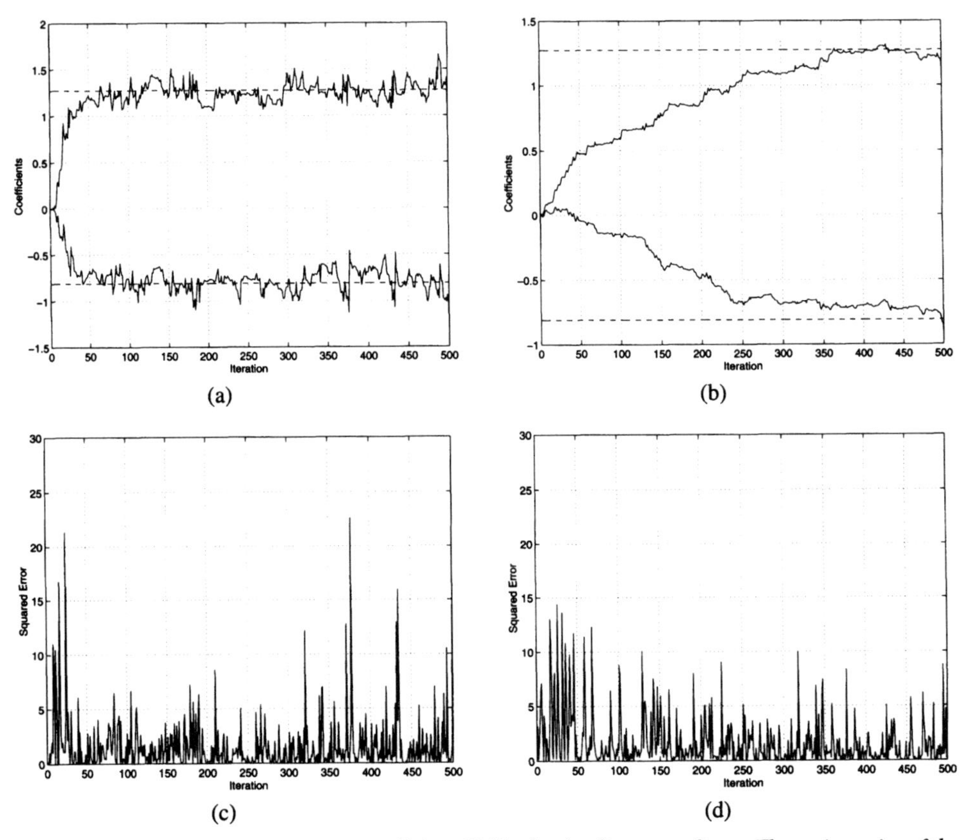

Figure 9.10 *Performance of a two-coefficient LMS adaptive linear predictor. The trajectories of the predictor coefficients are shown for step sizes of (a) $\mu = 0.02$ and (b) $\mu = 0.004$ with the correct values indicated by the dashed lines. A plot of the squared error, $e^2(n)$, is shown in (c) and (d) for step sizes of $\mu = 0.02$ and $\mu = 0.004$, respectively.*

are the trajectories of $w_n(1)$ and $w_n(2)$ versus n. As we see, there is a great deal of fluctuation in the weights, even after they have converged to within a neighborhood of their steady-state values. By contrast, shown in Fig. 9.10*b* are the trajectories of the predictor coefficients for a step size $\mu = 0.004$. Compared to a step size of $\mu = 0.02$, we see that, although the weights take longer to converge, the trajectories are much smoother, illustrating the basic trade-off between rate of convergence and stability (variance) of the final solution. Shown in parts *c* and *d* are plots of the squared error $e^2(n)$ for $\mu = 0.02$ and $\mu = 0.004$, respectively. What is so striking in these plots is the large amount of variation in $e^2(n)$ versus n compared to the learning curves for the steepest descent algorithm.

Before leaving this example, let us compute the maximum step size that is allowed for the adaptive linear predictor to converge in the mean. Given the coefficients of the filter that are used to generate $x(n)$, we may use the inverse Levinson-Durbin recursion to find the autocorrelation sequence $r_x(k)$. Using the m-file `ator.m` we find

$$r_x(0) = 5.7523 \quad ; \quad r_x(1) = 4.0450$$

Therefore, $\lambda_{\max} = r_x(0) + r_x(1) = 9.7973$ and the bound on the step size is

$$0 < \mu < 0.2041$$

Note that the step sizes used in Fig. 9.10 are at least an order of magnitude smaller than the largest value allowed for convergence in the mean. This is typically the case and, in fact, with a step size much larger than $\mu = 0.02$, the coefficients $w_n(k)$ begin to fluctuate wildly and eventually become unstable.

As we observed in the previous example, as the weight vector begins to converge in the mean, the coefficients begin to fluctuate about their optimum values. These fluctuations are due to the noisy gradient vectors that are used to form the corrections to $\mathbf{w}_n$. As a result, the variance of the weight error vector does not go to zero and the mean-square error is larger than the minimum mean-square error by an amount referred to as the *excess mean-square error*. This behavior is illustrated in Fig. 9.8*b* which shows that, as $\mathbf{w}_n$ oscillates about $\mathbf{w} = \mathbf{R}_x^{-1}\mathbf{r}_{dx}$, the corresponding mean-square error $\xi(n)$ has a value that, on the average, exceeds the minimum mean-square error. In order to quantify this excess mean-square error, we will write the error at time n as follows:

$$e(n) = d(n) - \mathbf{w}_n^T\mathbf{x}(n) = d(n) - \left(\mathbf{w} + \mathbf{c}_n\right)^T\mathbf{x}(n) = e_{\min}(n) - \mathbf{c}_n^T\mathbf{x}(n)$$

where $e_{\min}(n)$ is the error that would occur if the optimum filter coefficients were used, i.e.,

$$e_{\min}(n) = d(n) - \mathbf{w}^T\mathbf{x}(n)$$

Assuming that the filter is in the steady-state with $E\{\mathbf{c}_n\} = 0$, the mean-square error may be expressed as

$$\xi(n) = E\left\{|e(n)|^2\right\} = \xi_{\min} + \xi_{\text{ex}}(n)$$

where

$$\xi_{\min} = E\left\{|e_{\min}(n)|^2\right\}$$

is the minimum mean-square error and $\xi_{ex}(n)$ is the *excess mean-square error*, which depends on the statistics of $\mathbf{x}(n)$, $\mathbf{c}_n$, and $d(n)$. Although $\xi_{ex}(n)$ is not easy to evaluate, by invoking the independence assumption, the following property may be established with a fair amount of effort [20].

Property 3. The mean-square error $\xi(n)$ converges to a steady-state value of

$$\xi(\infty) = \xi_{\min} + \xi_{ex}(\infty) = \xi_{\min} \frac{1}{1 - \mu \displaystyle\sum_{k=0}^{p} \frac{\lambda_k}{2 - \mu\lambda_k}} \tag{9.43}$$

and the LMS algorithm is said to *converge in the mean-square* if and only if the step-size μ satisfies the following two conditions:

$$0 < \mu < \frac{2}{\lambda_{\max}} \tag{9.44}$$

$$\mu \sum_{k=0}^{p} \frac{\lambda_k}{2 - \mu\lambda_k} < 1 \tag{9.45}$$

Note that Eq. (9.44) is the condition that is required for the LMS algorithm to converge in the mean, and Eq. (9.45) guarantees that $\xi(\infty)$ is positive. Solving Eq. (9.43) for $\xi_{ex}(\infty)$ we find

$$\xi_{ex}(\infty) = \mu\, \xi_{\min} \frac{\displaystyle\sum_{k=0}^{p} \frac{\lambda_k}{2 - \mu\lambda_k}}{1 - \mu \displaystyle\sum_{k=0}^{p} \frac{\lambda_k}{2 - \mu\lambda_k}} \tag{9.46}$$

If $\mu \ll 2/\lambda_{\max}$, as is typically the case, then $\mu\lambda_k \ll 2$ and Eq. (9.45) may be simplified to

$$\frac{1}{2}\mu \sum_{k=0}^{p} \lambda_k < 1$$

or,

$$\mu < \frac{2}{\text{tr}(\mathbf{R}_x)}$$

When $\mu \ll 2/\lambda_{\max}$ it also follows that

$$\xi(\infty) \approx \xi_{\min} \frac{1}{1 - \frac{1}{2}\mu\, \text{tr}(\mathbf{R}_x)}$$

and the excess mean-square error given in Eq. (9.46) is approximately

$$\xi_{ex}(\infty) \approx \mu\, \xi_{\min} \frac{\frac{1}{2}\text{tr}(\mathbf{R}_x)}{1 - \frac{1}{2}\mu\, \text{tr}(\mathbf{R}_x)} \approx \frac{1}{2}\mu\, \xi_{\min}\, \text{tr}(\mathbf{R}_x) \tag{9.47}$$

Thus, for small μ, the excess mean-square error is proportional to the step size μ. Adaptive filters may be described in terms of their *misadjustment*, which is a normalized mean-square error that is defined as follows.

Definition. The misadjustment $\mathcal{M}$ is the ratio of the steady-state excess mean-square error to the minimum mean-square error,

$$\mathcal{M} = \frac{\xi_{ex}(\infty)}{\xi_{min}}$$

From Eq. (9.47) we see that if the step size is small, $\mu \ll 2/\lambda_{max}$, then the misadjustment is approximately

$$\mathcal{M} \approx \mu \frac{\frac{1}{2}\text{tr}(\mathbf{R}_x)}{1 - \frac{1}{2}\mu\,\text{tr}(\mathbf{R}_x)} \approx \frac{1}{2}\mu\,\text{tr}(\mathbf{R}_x)$$

Example 9.2.2 *LMS Misadjustment*

In this example, we look at the learning curves for the adaptive linear predictor considered in Example 9.2.1, and evaluate the excess mean-square error and the misadjustment for different step sizes. Since the learning curve is a plot of $\xi(n) = E\{|e(n)|^2\}$ versus n, we may approximate the learning curve by averaging plots of $|e(n)|^2$ that are obtained by repeatedly implementing the adaptive predictor. For example, implementing the adaptive predictor K times, and denoting the squared error at time n on the kth trial by $|e_k(n)|^2$, we have

$$\hat{\xi}(n) = \hat{E}\{|e(n)|^2\} = \frac{1}{K}\sum_{k=1}^{K} |e_k(n)|^2$$

With $K = 200$, an initial weight vector of zero, and step sizes of $\mu = 0.02$ and $\mu = 0.004$, these estimates of the learning curves are shown in Fig. 9.11. One property of the LMS algorithm that we are able to observe from these plots is that, when the step size is decreased, the convergence of the adaptive filter to its steady-state value is slower, but the average steady-state squared error is smaller.

We may estimate the steady-state mean-square error from these plots by averaging $\hat{\xi}(n)$ over n after the LMS algorithm has reached steady-state. For example, with

$$\hat{\xi}(\infty) = \frac{1}{100}\sum_{n=901}^{1000} \hat{E}\{|e(n)|^2\}$$

we find

$$\hat{\xi}(\infty) = \begin{cases} 1.1942 & ; \quad \text{for } \mu = 0.02 \\ 1.0155 & ; \quad \text{for } \mu = 0.004 \end{cases}$$

We may compare these results to the theoretical steady-state mean-square error using Eq. (9.43). With $\xi_{min} = 1$, and eigenvalues $\lambda_1 = 9.7924$ and $\lambda_2 = 1.7073$ (see Example 9.2.1), it follows that

$$\xi(\infty) = \begin{cases} 1.1441 & ; \quad \text{for } \mu = 0.02 \\ 1.0240 & ; \quad \text{for } \mu = 0.004 \end{cases}$$

which is in fairly close agreement with the estimated values given above.

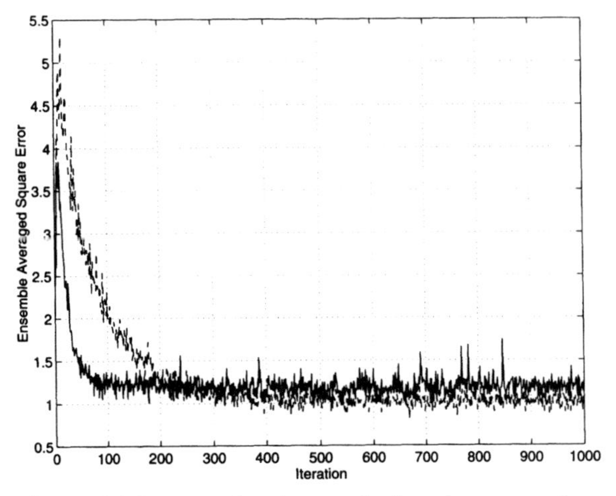

Figure 9.11 *Approximations to the learning curves for a second-order LMS adaptive linear predictor using step sizes of* $\mu = 0.02$ *(solid line) and* $\mu = 0.004$ *(dotted line).*

9.2.4 Normalized LMS

As we have seen, one of the difficulties in the design and implementation of the LMS adaptive filter is the selection of the step size μ. For stationary processes, the LMS algorithm converges in the mean if $0 < \mu < 2/\lambda_{\max}$, and converges in the mean-square if $0 < \mu < 2/\text{tr}(\mathbf{R}_x)$. However, since $\mathbf{R}_x$ is generally unknown, then either $\lambda_{\max}$ or $\mathbf{R}_x$ must be estimated in order to use these bounds. One way around this difficulty is to use the fact that, for stationary processes, $\text{tr}(\mathbf{R}_x) = (p+1)E\{|x(n)|^2\}$. Therefore, the condition for mean-square convergence may be replaced with

$$0 < \mu < \frac{2}{(p+1)E\{|x(n)|^2\}}$$

where $E\{|x(n)|^2\}$ is the power in the process $x(n)$. This power may be estimated using a time average such as[4]

$$\hat{E}\{|x(n)|^2\} = \frac{1}{p+1}\sum_{k=0}^{p}|x(n-k)|^2$$

which leads to the following bound on the step size for mean-square convergence:

$$0 < \mu < \frac{2}{\mathbf{x}^H(n)\mathbf{x}(n)}$$

A convenient way to incorporate this bound into the LMS adaptive filter is to use a (time-varying) step size of the form

$$\mu(n) = \frac{\beta}{\mathbf{x}^H(n)\mathbf{x}(n)} = \frac{\beta}{\|\mathbf{x}(n)\|^2} \tag{9.48}$$

where β is a *normalized step size* with $0 < \beta < 2$. Replacing μ in the LMS weight vector update equation with $\mu(n)$ leads to the Normalized LMS algorithm (NLMS), which is given

[4]Since this estimate uses only those values of $x(n)$ that are within the tapped delay line at time n, no extra memory is required to evaluate this sum.

The Normalized LMS Algorithm

```
function [A,E] = nlms(x,d,beta,nord,a0)
%
X=convm(x,nord);
[M,N] = size(X);
if nargin < 5,    a0 = zeros(1,N);    end
a0 = a0(:).';
E(1) = d(1) - a0*X(1,:).';
DEN=X(1,:)*X(1,:)' + 0.0001;
A(1,:) = a0 + beta/DEN*E(1)*conj(X(1,:));
if M>1
for k=2:M-nord+1;
    E(k) = d(k) - A(k-1,:)*X(k,:).';
    DEN=X(k,:)*X(k,:)' + 0.0001;
    A(k,:) = A(k-1,:) + beta/DEN*E(k)*conj(X(k,:));
    end;
end;
```

Figure 9.12 *A* MATLAB *program for the Normalized LMS algorithm.*

by

$$\mathbf{w}_{n+1} = \mathbf{w}_n + \beta \frac{\mathbf{x}^*(n)}{\|\mathbf{x}(n)\|^2} e(n) \tag{9.49}$$

Note that the effect of the normalization by $\|\mathbf{x}(n)\|^2$ is to alter the magnitude, but not the *direction*, of the estimated gradient vector. Therefore, with the appropriate set of statistical assumptions it may be shown that the normalized LMS algorithm converges in the mean-square if $0 < \beta < 2$ [4,34].

In the LMS algorithm, the correction that is applied to $\mathbf{w}_n$ is proportional to the input vector $\mathbf{x}(n)$. Therefore, when $\mathbf{x}(n)$ is large, the LMS algorithm experiences a problem with *gradient noise amplification*. With the normalization of the LMS step size by $\|\mathbf{x}(n)\|^2$ in the NLMS algorithm, however, this noise amplification problem is diminished. Although the NLMS algorithm bypasses the problem of noise amplification, we are now faced with a similar problem that occurs when $\|\mathbf{x}(n)\|$ becomes too small. An alternative, therefore, is to use the following modification to the NLMS algorithm:

$$\mathbf{w}_{n+1} = \mathbf{w}_n + \beta \frac{\mathbf{x}^*(n)}{\epsilon + \|\mathbf{x}(n)\|^2} e(n) \tag{9.50}$$

where ϵ is some small positive number. A MATLAB program for the normalized LMS algorithm is given in Fig. 9.12.

Compared with the LMS algorithm, the normalized LMS algorithm requires additional computation to evaluate the normalization term $\|\mathbf{x}(n)\|^2$. However, if this term is evaluated recursively as follows

$$\|\mathbf{x}(n+1)\|^2 = \|\mathbf{x}(n)\|^2 + |x(n+1)|^2 - |x(n-p)|^2 \tag{9.51}$$

then the extra computation involves only two squaring operations, one addition, and one substraction.

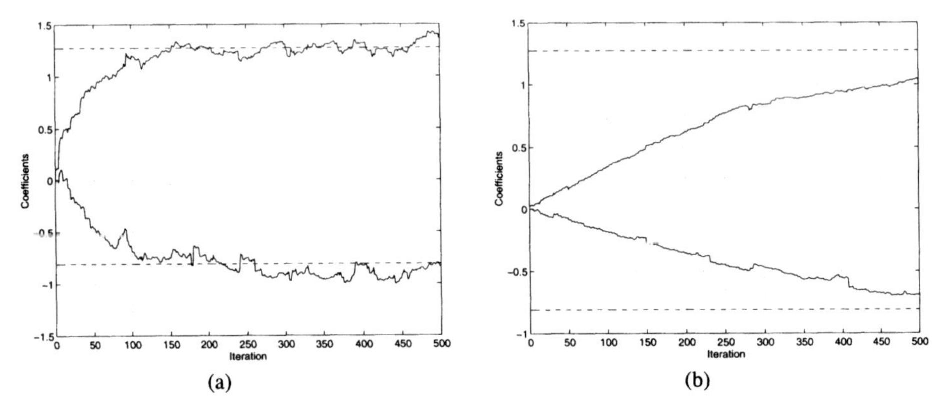

Figure 9.13 *A two-coefficient normalized LMS adaptive linear predictor. The trajectories of the predictor coefficients are shown for step sizes of (a)* $\beta = 0.05$ *and (b)* $\beta = 0.01$ *with the correct values for the coefficients indicated by the dashed lines.*

Example 9.2.3 *Adaptive Linear Prediction Using the NLMS Algorithm*

In this example, we consider once again the problem of adaptive linear prediction. This time, however, we will use the normalized LMS algorithm. As in Example 9.2.1, the process that is to be predicted is the AR(2) process that is generated by the difference equation

$$x(n) = 1.2728x(n-1) - 0.81x(n-2) + v(n)$$

where $v(n)$ is unit variance white noise. With a two-coefficient LMS adaptive predictor we have

$$\hat{x}(n) = w_n(1)x(n-1) + w_n(2)x(n-2)$$

where the predictor coefficients are updated according to

$$w_{n+1}(k) = w_n(k) + \beta \frac{x(n-k)}{\epsilon + x^2(n-1) + x^2(n-2)} e(n) \quad ; \quad k = 1, 2$$

Using normalized step sizes of $\beta = 0.05$ and $\beta = 0.01$, with $\epsilon = 0.0001$ and $w_0(1) = w_0(2) = 0$ we obtain the sequence of predictor coefficients shown in Fig. 9.13*a* and *b*, respectively. Comparing these results to those shown in Fig. 9.10 we see that the trajectories of the coefficients are similar. The difference, however, is that for the NLMS algorithm it is not necessary to estimate $\lambda_{\max}$ in order to select a step size.

9.2.5 Application: Noise Cancellation

In Section 7.2.3, we looked at the problem of *noise cancellation* in which a process $d(n)$ is to be estimated from a noise corrupted observation

$$x(n) = d(n) + v_1(n)$$

Clearly, without any information about $d(n)$ or $v_1(n)$ it is not possible to separate the signal from the noise. However, given a *reference signal*, $v_2(n)$, that is correlated with $v_1(n)$, then this reference signal may be used to estimate the noise $v_1(n)$, and this estimate may then be subtracted from $x(n)$ to form an estimate of $d(n)$,

$$\hat{d}(n) = x(n) - \hat{v}_1(n)$$

For example, if $d(n)$, $v_1(n)$, and $v_2(n)$ are jointly wide-sense stationary processes, and if the autocorrelation $r_{v_2}(k)$ and the cross-correlation $r_{v_1 v_2}(k)$ are known, then a Wiener filter may be designed to find the minimum mean-square estimate of $v_1(n)$ as illustrated in Fig. 9.14*a*. In practice, however, a stationarity assumption is not generally appropriate and, even if it were, the required statistics of $v_2(n)$ and $v_1(n)$ are generally unknown. Therefore, as an alternative to the Wiener filter, let us consider the adaptive noise canceller shown in Fig. 9.14*b*. If the reference signal $v_2(n)$ is uncorrelated with $d(n)$, then it follows from the

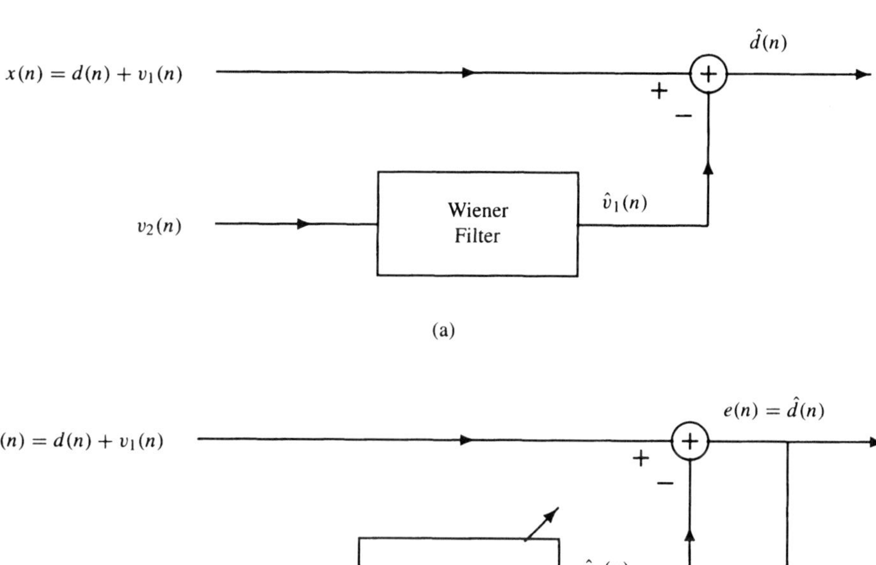

(a)

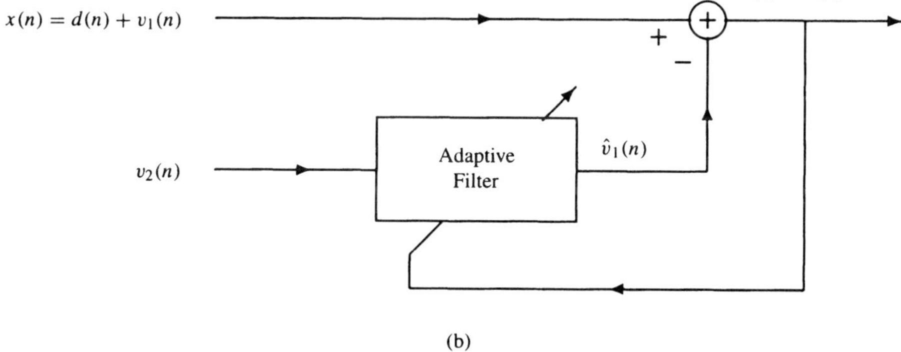

(b)

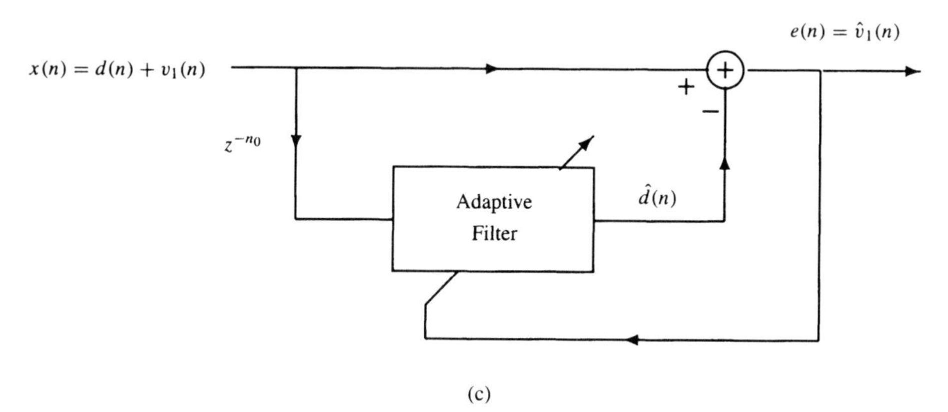

(c)

Figure 9.14 *Noise cancellation using (a) a Wiener filter, (b) an adaptive noise canceller with a reference signal $v(n)$, and (c) adaptive noise cancellation without a reference.*

discussion in Section 9.1 that minimizing the mean-square error $E\{|e(n)|^2\}$ is equivalent to minimizing $E\{|v_1(n) - \hat{v}_1(n)|^2\}$. In other words, the output of the adaptive filter is the minimum mean-square estimate of $v_1(n)$. Basically, if there is no information about $d(n)$ in the reference signal $v_2(n)$, then the best that the adaptive filter can do to minimize $E\{|e(n)|^2\}$ is to remove the part of $e(n)$ that may be estimated from $v_2(n)$, which is $v_1(n)$. Since the output of the adaptive filter is the minimum mean-square estimate of $v_1(n)$, then it follows that $e(n)$ is the minimum mean square estimate of $d(n)$.

As a specific example, let us reconsider the problem presented in Example 7.2.6 in which the signal to be estimated is a sinusoid,

$$d(n) = \sin(n\omega_0 + \phi)$$

with $\omega_0 = 0.05\pi$, and the noise sequences $v_1(n)$ and $v_2(n)$ are generated by the first-order difference equations

$$\begin{aligned} v_1(n) &= 0.8v_1(n-1) + g(n) \\ v_2(n) &= -0.6v_2(n-1) + g(n) \end{aligned} \tag{9.52}$$

where $g(n)$ is a zero-mean, unit variance white noise process that is uncorrelated with $d(n)$. Shown in Fig. 9.15*a* is a plot of 1000 samples of the sinusoid and in Fig. 9.15*b* is the noisy signal $x(n) = d(n) + v_1(n)$. The reference signal $v_2(n)$ that is used to estimate $v_1(n)$ is shown in Fig. 9.15*c*. Using a 12th-order adaptive noise canceller with coefficients that are updated using the normalized LMS algorithm, the estimate of $d(n)$ that is produced with a step size $\beta = 0.25$ is shown in Fig. 9.15*d*. As we see from this figure, after about 100 iterations the adaptive filter is producing a fairly accurate estimate of $d(n)$ and, after about 200 iterations the adaptive filter appears to have settled down into its steady-state behavior. Although not quite as good as the estimate that is produced with a 6th-order Wiener filter as shown in Fig. 7.9*d* on p. 351, the adaptive noise canceller, unlike the Wiener filter, does not require any statistical information.

One of the advantages of this adaptive noise canceller over a Wiener filter is that it may be used when the processes are nonstationary. For example, let $v_1(n)$ and $v_2(n)$ be nonstationary processes that are generated by the first-order difference equations given in Eq. (9.52), where $g(n)$ is nonstationary white noise with a variance that increases linearly from $\sigma_g^2(0) = 0.25$ to $\sigma_g^2(1000) = 6.25$. As before, $d(n)$ is estimated using a 12th-order adaptive noise canceller with coefficients that are updated according to the normalized LMS algorithm. Shown in Fig. 9.16*a* is the desired signal $d(n)$ and in Fig. 9.16*b* is the noisy signal $x(n)$. The increasing variance in the additive noise is clearly evident. The nonstationary reference signal $v_2(n)$ that is used to estimate $v_1(n)$ is shown in Fig. 9.16*c* and in Fig. 9.16*d* is the estimate of $d(n)$ that is produced by the adaptive filter. As we see in this figure, the performance of the adaptive noise canceller is not significantly affected by the nonstationarity of the noise (note that for $n > 250$ the variance of the nonstationary noise is larger than the variance of the noise in Fig. 9.15).

The key to the successful operation of the adaptive noise canceller in Fig. 9.14*b* is the availability of a reference signal, $v_2(n)$, that may be used to estimate the additive noise $v_1(n)$. Unfortunately, in many applications a reference signal is not available and another approach must be considered. In some cases, however, it is possible to derive a reference signal by simply delaying the process $x(n) = d(n) + v_1(n)$. For example, suppose that $d(n)$ is a narrowband process and that $v_1(n)$ is a broadband process with

$$E\{v_1(n)v_1(n-k)\} = 0 \quad ; \quad |k| > k_0$$

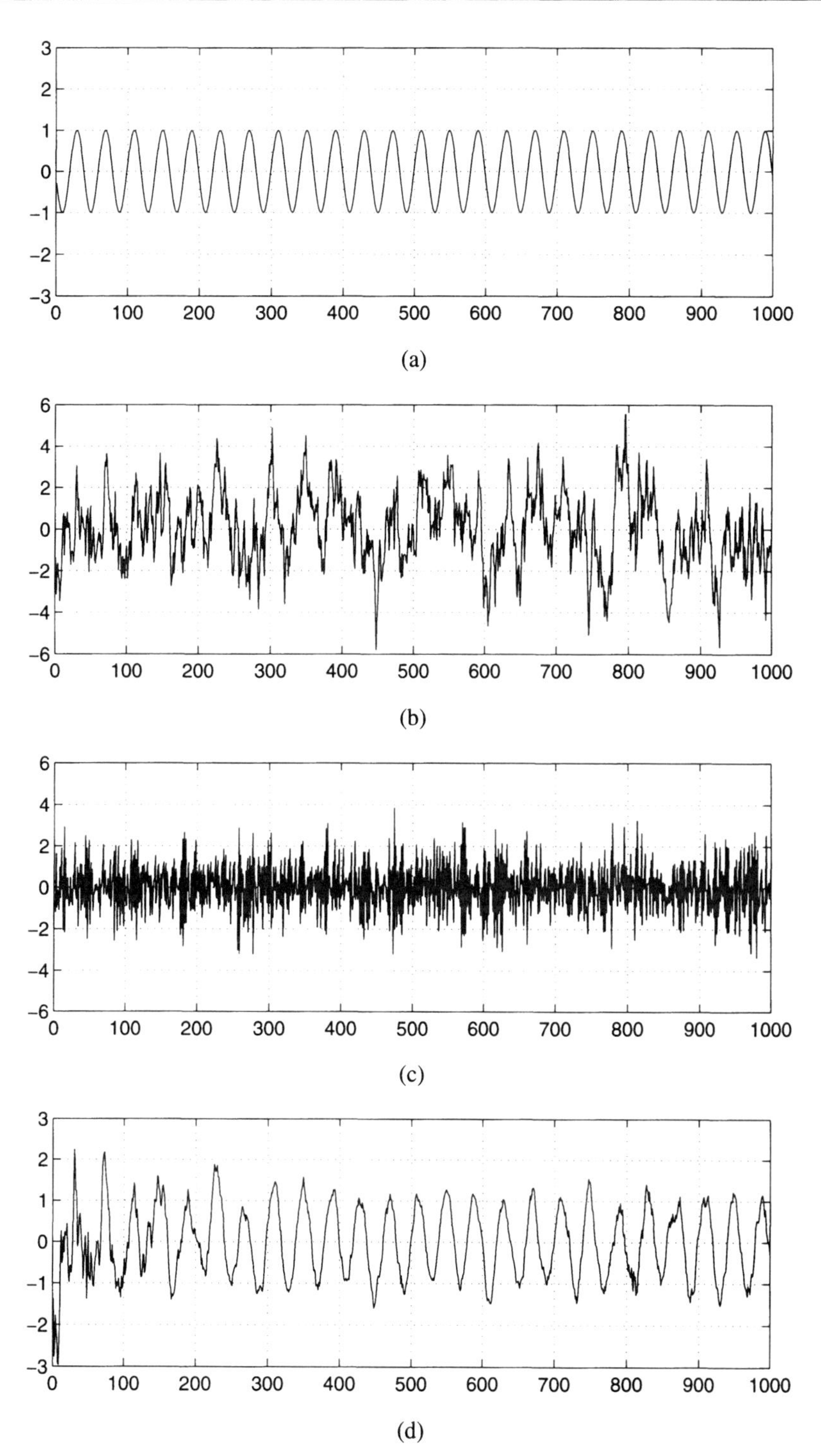

Figure 9.15 *Noise cancellation example. (a) A sinusoid that is to be estimated. (b) Noise corrupted sinusoid. (c) The reference signal used by the secondary sensor. (d) The output of a sixth-order NLMS adaptive noise canceller with a normalized step size* $\beta = 0.25$.

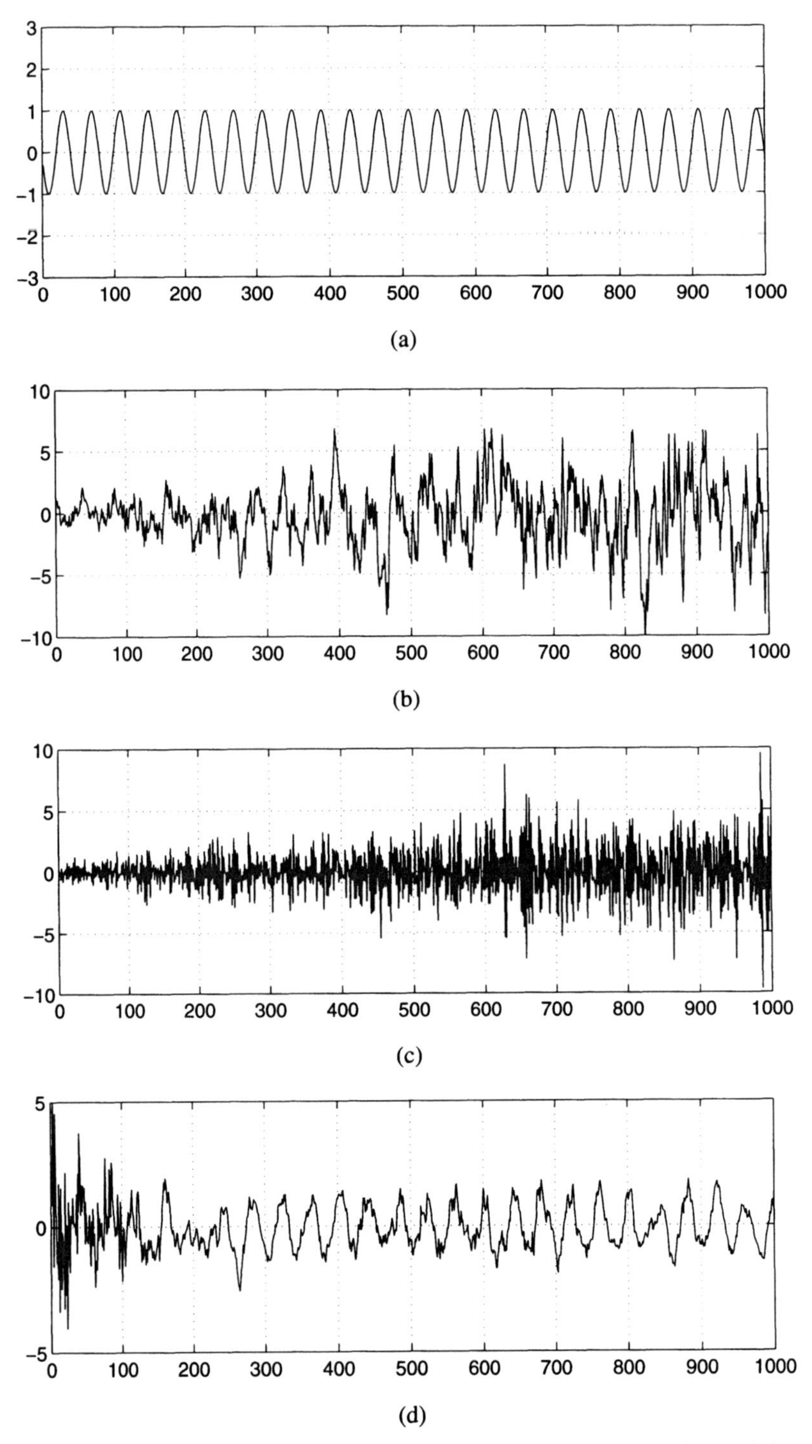

Figure 9.16 *Noise cancellation of nonstationary noise. (a) Desired signal that is to be estimated. (b) Noise corrupted sinusoid. (c) The reference signal used by secondary sensor. (d) The output of a 12th-order NLMS adaptive noise canceller with a normalized step size* $\beta = 0.25$.

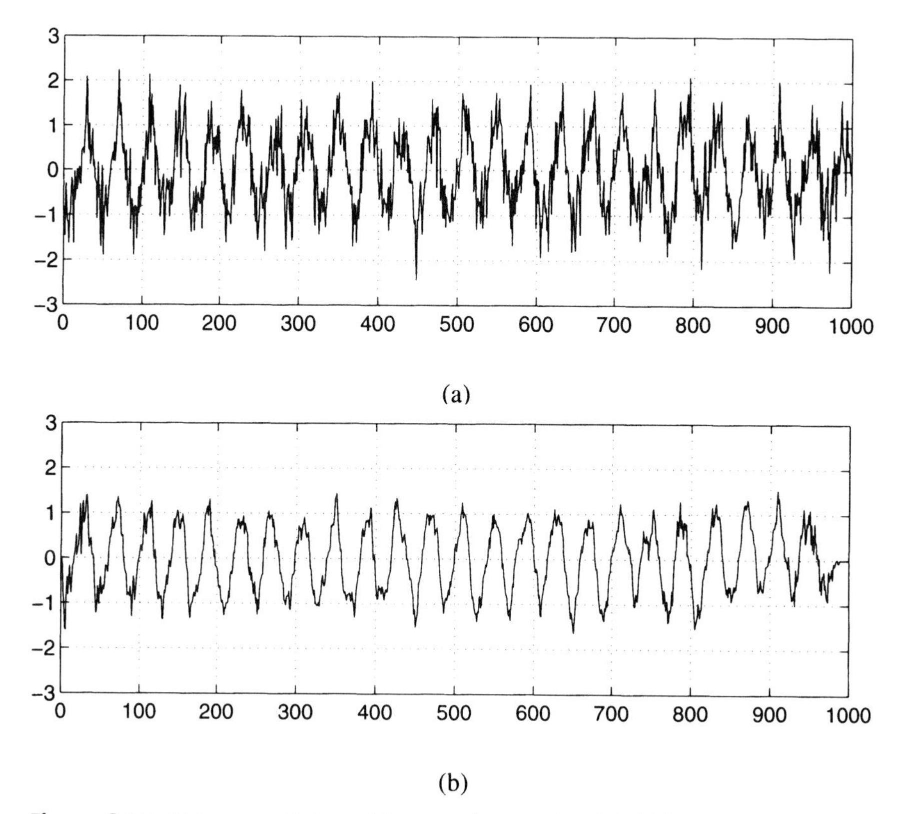

Figure 9.17 *Noise cancellation without a reference signal. (a) The noisy process $x(n)$ and (b) the output of a 12th-order NLMS adaptive noise canceller with $\beta = 0.25$ and $n_0 = 25$.*

If $d(n)$ and $v_1(n)$ are uncorrelated, then

$$E\{v_1(n)x(n-k)\} = E\{v_1(n)d(n-k)\} + E\{v_1(n)v_1(n-k)\} = 0 \quad ; \quad |k| > k_0$$

Therefore, if $n_0 > k_0$ then the delayed process $x(n-n_0)$ will be uncorrelated with the noise $v_1(n)$, and correlated with $d(n)$ (with the assumption that $d(n)$ is a broadband process). Thus, $x(n-n_0)$ may be used as a reference signal to estimate $d(n)$ as illustrated in Fig. 9.14*c*. In contrast to the adaptive noise canceller in Fig. 9.14*b*, note that the adaptive filter in Fig. 9.14*c* produces an estimate of the broadband process, $d(n)$, and the error $e(n)$ corresponds to an estimate of the noise $v_1(n)$. As an example of using an adaptive noise canceller without a reference, let $x(n) = d(n)+v_1(n)$ where $d(n)$ is a sinusoid of frequency $\omega_0 = 0.05\pi$, and let $v_1(n)$ be the AR(1) process defined in Eq. (9.52) where $g(n)$ white noise with a variance of $\sigma_g^2 = 0.25$. Shown in Fig. 9.17*a* is the noisy process $x(n) = d(n) + v_1(n)$ and in Fig. 9.17*b* is the output of the adaptive filter using the NLMS algorithm with a normalized step size of $\beta = 0.25$, and a reference signal that is obtained by delaying $x(n)$ by $n_0 = 25$ samples.

9.2.6 Other LMS-Based Adaptive Filters

In addition to the NLMS algorithm, a number of other modifications to the LMS algorithm have been proposed and studied. Each of these modifications attempts to improve one or more of the properties of the LMS algorithm. In this section, we provide a brief overview of some of these modifications, including the leaky LMS algorithm for dealing with the problems that occur when the autocorrelation matrix is singular, the block LMS algorithm and the sign algorithms for increasing the computational efficiency of the LMS algorithm, and the variable step-size (VS) algorithms for improving the speed of convergence.

The Leaky LMS Algorithm. When the input process to an adaptive filter has an autocorrelation matrix with zero eigenvalues, the LMS adaptive filter has one or more modes that are undriven and undamped. For example, it follows from Eq. (9.39) that if $\lambda_k = 0$, then

$$E\{u_n(k)\} = u_0(k)$$

which does not decay to zero with n. Since it is possible for these undamped modes to become unstable, it is important to stabilize the LMS adaptive filter by forcing these modes to zero [33]. One way to accomplish this is to introduce a leakage coefficient γ into the LMS algorithm as follows:

$$\boxed{\mathbf{w}_{n+1} = (1 - \mu\gamma)\mathbf{w}_n + \mu e(n)\mathbf{x}^*(n)} \tag{9.53}$$

where $0 < \gamma \ll 1$. The effect of this *leakage coefficient* is to force the filter coefficients to zero if either the error $e(n)$ or the input $x(n)$ become zero, and to force any undamped modes of the system to zero.

The properties of the leaky LMS algorithm may be derived by examining the behavior of $E\{\mathbf{w}_n\}$. If we substitute Eq. (9.7) for $e(n)$ into Eq. (9.53), then we have

$$\mathbf{w}_{n+1} = \Big[\mathbf{I} - \mu\Big\{\mathbf{x}^*(n)\mathbf{x}^T(n) + \gamma\mathbf{I}\Big\}\Big]\mathbf{w}_n + \mu d(n)\mathbf{x}^*(n)$$

Taking the expected value and using the independence assumption yields

$$E\{\mathbf{w}_{n+1}\} = \Big[\mathbf{I} - \mu\Big\{\mathbf{R}_x + \gamma\mathbf{I}\Big\}\Big]E\{\mathbf{w}_n\} + \mu\mathbf{r}_{dx} \tag{9.54}$$

Comparing Eq. (9.54) to Eq. (9.38) we see that the autocorrelation matrix $\mathbf{R}_x$ in the LMS algorithm has been replaced with $\mathbf{R}_x + \gamma\mathbf{I}$. Therefore, the coefficient leakage term effectively adds white noise to $x(n)$ by adding γ to the main diagonal of the autocorrelation matrix. Since the eigenvalues of $\mathbf{R}_x + \gamma\mathbf{I}$ are $\lambda_k + \gamma$ and since $\lambda_k \geq 0$, then none of the modes of the leaky LMS algorithm will be undamped. In addition, the constraint on the step size μ for convergence in the mean becomes

$$0 < \mu < \frac{2}{\lambda_{\max} + \gamma}$$

The drawback to the leaky LMS algorithm is that, for stationary processes, the steady-state solution will be biased. Specifically, note that if $\mathbf{w}_n$ converges in the mean then

$$\lim_{n\to\infty} E\{\mathbf{w}_n\} = \big(\mathbf{R}_x + \gamma\mathbf{I}\big)^{-1}\mathbf{r}_{dx}$$

Thus, the leakage coefficient introduces a bias into the steady-state solution.

It is interesting to note that the leaky LMS algorithm may also be derived by using the LMS gradient descent algorithm to minimize the error

$$\xi(n) = |e(n)|^2 + \gamma\,\|\mathbf{w}_n\|^2$$

Specifically, since the gradient of $\xi(n)$ is

$$\nabla\xi(n) = -e(n)\mathbf{x}^*(n) + \gamma\mathbf{w}_n$$

then the gradient descent algorithm becomes

$$\mathbf{w}_{n+1} = \mathbf{w}_n - \mu\nabla\xi(n) = \mathbf{w}_n - \mu\gamma\mathbf{w}_n + \mu e(n)\mathbf{x}^*(n)$$

which is the same as Eq. (9.53).

LMS Algorithms With Reduced Complexity. In spite of the computational efficiency of the LMS algorithm, additional simplifications may be necessary in some applications, such as high speed digital communications. There are several approaches that may be used to reduce the computational requirements of the LMS algorithm. One of these is the block LMS algorithm, which is identical to the LMS algorithm except that the filter coefficients are updated only once for each block of L samples [9]. In other words, the filter coefficients are held constant over each block of L samples, and the filter output $y(n)$ and the error $e(n)$ for each value of n within the block are calculated using the filter coefficients for that block. Then, at the end of each block, the coefficients are updated using an average of the L gradient estimates over the block. Thus, the update equation for the coefficients in the kth block is

$$\mathbf{w}_{(k+1)L} = \mathbf{w}_{kL} + \mu\frac{1}{L}\sum_{l=0}^{L-1} e(kL+l)\mathbf{x}^*(kL+l)$$

where the output for the kth block is

$$y(kL+l) = \mathbf{w}_{kL}^T\mathbf{x}(kL+l) \quad ; \quad l = 0, 1, \ldots, L-1$$

and the error is

$$e(kL+l) = d(kL+l) - \mathbf{w}_{kL}^T\mathbf{x}(kL+l) \quad ; \quad l = 0, 1, \ldots, L-1$$

Since $y(n)$ over each block of L values is the convolution of the weight vector $\mathbf{w}_{kL}$ with a block of input samples, the efficiency of the block LMS algorithm comes from using an FFT to perform this block convolution [12].

Another set of simplifications to the LMS algorithm are found in the *sign algorithms*. In these algorithms, the LMS coefficient update equation is modified by applying the sign operator to either the error $e(n)$, the data $x(n)$, or both the error and the data. For example, assuming that $x(n)$ and $d(n)$ are real-valued processes, the *sign-error* algorithm is

$$\boxed{\mathbf{w}_{n+1} = \mathbf{w}_n + \mu\ \mathrm{sgn}\{e(n)\}\mathbf{x}(n)} \qquad (9.55)$$

where

$$\mathrm{sgn}\{e(n)\} = \begin{cases} 1 & ; \quad e(n) > 0 \\ 0 & ; \quad e(n) = 0 \\ -1 & ; \quad e(n) < 0 \end{cases}$$

Note that the sign-error algorithm may be viewed as the result of applying a two-level quantizer to the error. The simplification in the sign-error algorithm comes when the step size is chosen to be a power of two, $\mu = 2^{-l}$. In this case, the coefficient update equation may be implemented using $p+1$ data shifts instead of $p+1$ multiplies. Compared to the LMS algorithm, the sign-error algorithm uses a noisy estimate of the gradient. Since replacing $e(n)$ with the sign of the error changes only the magnitude of the correction that is used to update $\mathbf{w}_n$ and does not alter the direction, the sign-error algorithm is equivalent to the LMS algorithm with a step size that is inversely proportional to the magnitude of the error.

It is interesting to note that the sign error algorithm may also be derived by using the LMS gradient descent algorithm to minimize the error

$$\xi(n) = |e(n)|$$

To show this, note that

$$\frac{\partial |e(n)|}{\partial w_n(k)} = \text{sgn}\{e(n)\}\frac{\partial e(n)}{\partial w_n(k)} = \text{sgn}\{e(n)\}x(n-k)$$

Therefore, the gradient vector is

$$\nabla\xi(n) = \text{sgn}\{e(n)\}\mathbf{x}(n)$$

and the result follows. Thus, the sign-error algorithm is sometimes referred to as the Least Mean Absolute Value (LMAV) algorithm [3].

Along the same lines, instead of using the sign of the error, the computational requirements of the LMS algorithm may be simplified by using the sign of the data as follows:

$$\boxed{\mathbf{w}_{n+1} = \mathbf{w}_n + \mu\, e(n)\text{sgn}\{\mathbf{x}(n)\}} \tag{9.56}$$

which is the *sign-data algorithm*. Note that, unlike the sign-error algorithm, the sign-data algorithm alters the direction of the update vector. As a result, the sign-data algorithm is generally less robust than the sign-error algorithm. In fact, examples have been found in which the coefficients diverge using the sign-data algorithm while converging using the LMS algorithm [28]. Note that the kth coefficient in the sign of the data vector may be written as follows:

$$\text{sgn}\{x(n-k)\} = \frac{x(n-k)}{|x(n-k)|}$$

Therefore, as in the normalized LMS algorithm where the weight vector is normalized by $\|\mathbf{x}(n)\|^2$, the sign-data algorithm individually normalizes each coefficient of the weight vector. Thus, the sign-data algorithm may be written as

$$w_{n+1}(k) = w_n(k) + \frac{\mu}{|x(n-k)|}\, e(n)x(n-k)$$

which is an LMS algorithm that has a different (time-varying) step size for each coefficient in the weight vector.

Finally, quantizing both the error and the data leads to the *sign-sign* algorithm, which has a coefficient update equation given by

$$\boxed{\mathbf{w}_{n+1} = \mathbf{w}_n + \mu\, \text{sgn}\{e(n)\}\text{sgn}\{\mathbf{x}(n)\}} \tag{9.57}$$

In this algorithm, the coefficients $w_n(k)$ are updated by either adding or subtracting a constant μ. For stability, a leakage term is often introduced into the sign-sign algorithm [3] giving an update equation of the form

$$\mathbf{w}_{n+1} = (1 - \mu\gamma)\mathbf{w}_n + \mu \text{ sgn}\{e(n)\}\text{sgn}\{\mathbf{x}(n)\}$$

Generally, the sign-sign algorithm is slower to converge than the LMS adaptive filter and has a larger excess mean-square error [5,11]. Nevertheless, the extreme simplicity of the update algorithm has made it a popular algorithm and has, in fact, been adopted in the international CCITT standard for 32kbps ADPCM [10].[5]

Variable Step-Size Algorithms. In selecting the step size μ in the LMS algorithm, there is a tradeoff between the rate of convergence, the amount of excess mean-square error, and the ability of the filter to track signals as their statistics change. Ideally, when the adaptation begins and $\mathbf{w}_n$ is far from the optimum solution, the step size should be large in order to move the weight vector rapidly toward the desired solution. Then, as the filter begins to converge in the mean to the steady-state solution, the step size should be decreased in order to reduce the excess mean-square error. This suggests the possibility of using a variable step size in the LMS adaptive filter. The difficulty, however, is in specifying a set of rules for changing the step size in such a way that the adaptive filter has a small excess mean-square error while, at the same time, maintaining the ability of the filter to respond quickly to changes in the signal statistics.

Assuming that $x(n)$ and $d(n)$ are real-valued processes, one approach that has been proposed for changing the step size, referred to as the *Variable Step* (VS) algorithm [19], is to use a coefficient update equation of the form

$$w_{n+1}(k) = w_n(k) + \mu_n(k)e(n)x(n-k)$$

where $\mu_n(k)$ are step sizes that are adjusted independently for each coefficient. The rules for adjusting $\mu_n(k)$ are tied to the rate at which the gradient estimate changes sign. With an estimated gradient given by

$$\nabla e^2(n) = -2e(n)x(n-k)$$

these rules are based on the premise that if the sign of $e(n)x(n-k)$ is changing frequently, then the coefficient $w_n(k)$ should be close to its optimum value where the gradient is equal to zero. On the other hand, if the sign is not changing very often, then the coefficient $w_n(k)$ is probably not close to its optimum value. Therefore, $\mu_n(k)$ is decreased by a constant, c_1, if m_1 successive sign changes are observed in $e(n)x(n-k)$, whereas $\mu_n(k)$ is increased by a constant, c_2, if $e(n)x(n-k)$ has the same sign for m_2 successive updates. In addition, hard limits are placed on the step size,

$$\mu_{\min} < \mu_n(k) < \mu_{\max}$$

in order to ensure that the VS algorithm converges in the mean with only a modest increase in computation, the VS algorithm may result in a considerable improvement in the convergence rate [19].

[5] See "32 kbit/s Adaptive differential pulse code modulation (ADPCM)," in the CCITT standard recommendation G.721, Melbourne 1988.

9.2.7 Gradient Adaptive Lattice Filter

Up to this point we have only considered FIR adaptive filters implemented in direct form. However, as discussed in Chapter 6, the lattice filter has some important advantages over a direct form structure. Therefore, in this section we consider the design of adaptive lattice filters. Our development begins with the derivation of the gradient adaptive lattice (GAL) for adaptive linear prediction. Then, in Section 9.2.8, we will consider the design of an adaptive lattice filter for estimating an arbitrary process $d(n)$.

As we saw in Chapter 6, an FIR lattice filter is parameterized in terms of its reflection coefficients, Γ_j. The outputs of the jth stage of the lattice filter are the jth-order forward and backward prediction errors,

$$\begin{aligned} e_j^+(n) &= x(n) + \sum_{k=1}^{j} a_j(k)x(n-k) \\ e_j^-(n) &= x(n-j) + \sum_{k=1}^{j} a_j^*(k)x(n-j+k) \end{aligned} \tag{9.58}$$

where the coefficients $a_j(k)$ are related to the reflection coefficients through the step-up and step-down recursions (see Section 5.2.4). In Section 6.6, we considered several different ways to design a lattice filter for linear prediction. These included the sequential minimization of either the mean-square forward prediction error, the mean-square backward prediction error, or the Burg error. For example, we found that sequentially minimizing the Burg error,

$$\xi_j^B(n) = E\left\{|e_j^+(n)|^2 + |e_j^-(n)|^2\right\}$$

results in a set of reflection coefficients given by

$$\Gamma_j^B = -2\frac{E\left\{e_{j-1}^+(n)\left[e_{j-1}^-(n-1)\right]^*\right\}}{E\left\{|e_{j-1}^+(n)|^2 + |e_{j-1}^-(n-1)|^2\right\}} \tag{9.59}$$

As an alternative to this solution, let us consider a gradient descent approach to minimize the Burg error using an update equation of the form[6]

$$\Gamma_j(n+1) = \Gamma_j(n) - \mu_j \frac{\partial \xi_j^B(n)}{\partial \Gamma_j^*} \tag{9.60}$$

From the lattice filter update equations

$$\begin{aligned} e_j^+(n) &= e_{j-1}^+(n) + \Gamma_j e_{j-1}^-(n-1) \\ e_j^-(n) &= e_{j-1}^-(n-1) + \Gamma_j^* e_{j-1}^+(n) \end{aligned} \tag{9.61}$$

it follows that the derivative in Eq. (9.60) is

$$\frac{\partial \xi_j^B(n)}{\partial \Gamma_j^*} = E\left\{e_j^+(n)\left[e_{j-1}^-(n-1)\right]^* + \left[e_j^-(n)\right]^* e_{j-1}^+(n)\right\}$$

[6]Note that we are allowing the step size to be different for each reflection coefficient.

Therefore, the update equation becomes

$$\Gamma_j(n+1) = \Gamma_j(n) - \mu_j E\Big\{e_j^+(n)\big[e_{j-1}^-(n-1)\big]^* + \big[e_j^-(n)\big]^* e_{j-1}^+(n)\Big\} \tag{9.62}$$

which is the *steepest descent adaptive lattice filter*. Since $e_j^+(n)$ and $\big[e_j^-(n)\big]^*$ depend on $\Gamma_j(n)$, substituting Eq. (9.61) into Eq. (9.62) it follows that $\Gamma_j(n)$ satisfies the first-order difference equation

$$\begin{aligned}\Gamma_j(n+1) = &\Big[1 - \mu_j E\Big\{|e_{j-1}^+(n)|^2 + |e_{j-1}^-(n-1)|^2\Big\}\Big]\Gamma_j(n)\\ &+ 2\mu_j E\Big\{e_{j-1}^+(n)\big[e_{j-1}^-(n-1)\big]^*\Big\}\end{aligned} \tag{9.63}$$

For stationary processes, $\Gamma_j(n)$ will converge to the Burg solution Γ_j^B in Eq. (9.59) provided the term multiplying $\Gamma_j(n)$ in Eq. (9.63) is less than one in magnitude,

$$\Big|1 - \mu_j E\Big\{|e_{j-1}^+(n)|^2 + |e_{j-1}^-(n-1)|^2\Big\}\Big| < 1$$

This, in turn, places the following constraint on the step size μ_j,

$$0 < \mu_j < \frac{2}{E\Big\{|e_{j-1}^+(n)|^2 + |e_{j-1}^-(n-1)|^2\Big\}} \tag{9.64}$$

Due to the expectation in Eq. (9.62), the steepest descent adaptive lattice filter is only of limited use since it requires that the second-order statistics of the forward and backward prediction errors be known. However, if we adopt the LMS approach of replacing these expected values with instantaneous values, then the update equation (9.62) becomes

$$\Gamma_j(n+1) = \Gamma_j(n) - \mu_j \Big\{e_j^+(n)\big[e_{j-1}^-(n-1)\big]^* + \big[e_j^-(n)\big]^* e_{j-1}^+(n)\Big\} \tag{9.65}$$

which is the *Gradient Adaptive Lattice* (GAL) filter [17,18]. With an analysis similar to that presented in Section 9.2.3 for the LMS adaptive filter, it follows that if $x(n)$ is wide-sense stationary, then the reflection coefficients $\Gamma_j(n)$ will converge in the mean to the values given in Eq. (9.59), provided that each step size μ_j satisfies the condition given in Eq. (9.64). Due to the fact that the lattice filter orthogonalizes the input process, thereby decoupling the estimation of the reflection coefficients, the convergence of the gradient adaptive lattice filter is typically faster than the direct form adaptive filter [21,27]. This orthogonalization also makes the convergence time essentially independent of the eigenvalue spread [17]. A drawback of the gradient adaptive lattice, however, is that it requires approximately twice as many operations per update than required for the direct form filter. The gradient adaptive lattice is summarized in Table 9.2.

Due to the difficulties involved in determining the step sizes necessary to satisfy the convergence constraint given in Eq. (9.64), the GAL algorithm is typically *normalized* using a time-varying step-size of the form

$$\mu_j(n) = \frac{\beta}{D_j(n)}$$

Table 9.2 ***The Gradient Adaptive Lattice Algorithm***

Parameters:	p = Filter order
	μ_j = step sizes, $j = 1, 2, \ldots, p$
Initialization:	$e_0^+(n) = e_0^-(n) = 0$
	$\Gamma_j(0) = 0$ for $j = 1, 2, \ldots, p$
Computation:	For $n = 0, 1, 2, \ldots$ and $j = 1, \ldots, p$ compute
	(a) $e_j^+(n) = e_{j-1}^+(n) + \Gamma_j(n)e_{j-1}^-(n-1)$
	(b) $e_j^-(n) = e_{j-1}^-(n-1) + \Gamma_j^*(n)e_{j-1}^+(n)$
	(c) $\Gamma_j(n+1) = \Gamma_j(n) - \mu_j\Big\{e_j^+(n)\big[e_j^-(n-1)\big]^* + \big[e_j^-(n)\big]^* e_{j-1}^+(n)\Big\}$

where β is a constant with $0 < \beta < 2$, and $D_j(n)$ is an estimate of the denominator in Eq. (9.64). An estimate that is frequently used is

$$D_j(n) = (1-\lambda)\sum_{k=0}^{n}\lambda^{n-k}\Big[|e_{j-1}^+(k)|^2 + |e_{j-1}^-(k-1)|^2\Big]$$

where λ is an exponential weighting factor with $0 < \lambda < 1$. An advantage of this estimate is that $D_j(n)$ may be evaluated efficiently with the recursion

$$D_j(n+1) = \lambda D_j(n) + (1-\lambda)\Big[|e_{j-1}^+(n+1)|^2 + |e_{j-1}^-(n)|^2\Big] \tag{9.66}$$

9.2.8 Joint Process Estimator

The gradient adaptive lattice developed in the previous section may be used for adaptive linear prediction or, equivalently, adaptive all-pole signal modeling. However, since it minimizes the mean-square forward and backward prediction errors, the gradient adaptive lattice is not directly applicable to other adaptive filtering applications such as system identification, channel equalization, or echo cancellation. On the other hand, recall that the *joint process estimator* developed in Section 7.2.4 may be used for the estimation of an arbitrary process $d(n)$. This joint process estimator may be made adaptive as shown in Fig. 9.18 and consists of two stages. The first stage is an adaptive lattice predictor as described in the previous section. The second stage produces an estimate of $d(n)$ by taking a linear combination of the backward prediction errors $e_j^-(n)$. Thus, we have an adaptive *lattice preprocessor* working in conjunction with an adaptive tapped delay line.

To analyze this adaptive joint process estimator, let us begin by assuming that the reflection coefficients $\Gamma_k(n)$ and the tapped delay line coefficients $b_n(k)$ are constant. With

$$e(n) = d(n) - \hat{d}(n) = d(n) - \mathbf{b}^T\mathbf{e}^-(n)$$

where $\mathbf{b} = \big[b(0),\ b(1),\ \ldots,\ b(p-1)\big]^T$ and $\mathbf{e}^-(n) = \big[e_0^-(n),\ e_1^-(n),\ \ldots,\ e_{p-1}^-(n)\big]^T$, we may find the coefficients $b(k)$ that minimize the mean-square error $\xi(n) = E\big\{|e(n)|^2\big\}$

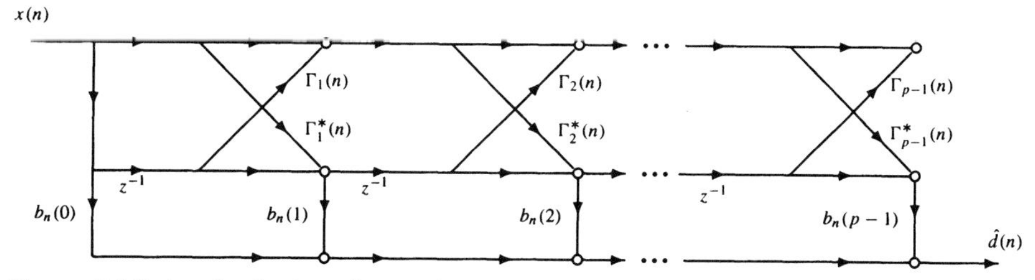

Figure 9.18 *An adaptive formulation of the lattice joint process estimator for estimating a process $d(n)$ from $x(n)$.*

by differentiating $\xi(n)$ with respect to $b^*(k)$ and setting the derivatives equal to zero. The result is a set of normal equations of the form

$$\mathbf{R}_e \mathbf{b} = \mathbf{r}_{de}$$

where

$$\mathbf{R}_e = E\{\mathbf{e}^-(n)[\mathbf{e}^-(n)]^H\}$$

is the autocorrelation matrix of the backward prediction error, and

$$\mathbf{r}_{de} = E\{d(n)[\mathbf{e}^-(n)]^H\}$$

is the cross-correlation between the desired process and the backward prediction error. These equations may be simplified considerably if we use the fact that the backward prediction errors are orthogonal (see Section 6.6),

$$E\{\mathbf{e}^-(n)\left[\mathbf{e}^-(n)\right]^H\} = \mathbf{D}_p = \text{diag}\{\epsilon_0,\ \epsilon_1,\ \ldots,\epsilon_p\}$$

This diagonal structure for $\mathbf{R}_e$ decouples the normal equations and allows us to solve for each coefficient $b(k)$ independently as follows:

$$b(k) = \frac{1}{\epsilon_k} E\{d(n)[e_k^-(n)]^*\} \tag{9.67}$$

Since this solution assumes that the processes are stationary and requires knowledge of the cross-correlation between $d(n)$ and $e_k^-(n)$, we may consider using a gradient descent approach to adaptively adjust the coefficients $b(k)$ as follows:

$$\mathbf{b}_{n+1} = \mathbf{b}_n - \mu \nabla \xi(n)$$

However, since the gradient involves an expectation,

$$\nabla \xi(n) = -E\{e(n)[\mathbf{e}^-(n)]^*\}$$

we may use the LMS approach of replacing the expected value with an instantaneous value. This results in the update equation

$$\boxed{\mathbf{b}_{n+1} = \mathbf{b}_n + \mu e(n)[\mathbf{e}^-(n)]^*} \tag{9.68}$$

which, in terms of the individual filter coefficients, is

$$b_{n+1}(k) = b_n(k) + \mu e(n)[e_k^-(n)]^*$$

In the overall system, there are two sets of parameters that are being updated in parallel: the reflection coefficients in the lattice predictor, $\Gamma_j(n)$, which are attempting to orthogonalize the input process $x(n)$, and the coefficients $b_n(k)$ that are used to form the estimate of $d(n)$. Due to the orthogonalization of $x(n)$, this filter tends to be less sensitive to the eigenvalue spread than the LMS adaptive filter. However, the backward prediction errors will not be orthogonal until the reflection coefficients have converged. Therefore, the backward prediction errors will be correlated and the behavior of the adaptive filter will depend on the eigenvalue spread until each $\Gamma_j(n)$ has converged. One interesting and important feature of this adaptive filter is that the orthogonalization of $x(n)$ does not depend on the desired signal $d(n)$. Therefore, if $x(n)$ is stationary, once the lattice predictor has converged, the backwards prediction error will remain orthogonal even if $d(n)$ is nonstationary.

9.2.9 Application: Channel Equalization

In this section, we will look briefly at the problem of adaptive channel equalization [20,26,27]. Adaptive equalizers are necessary for reliable communication of digital data across nonideal channels. The basic operation of a digital communication system may be described as follows. Let $d(n)$ be a digital sequence that is to be transmitted across a channel, with $d(n)$ having values of plus or minus one. This sequence is input to a pulse generator, which produces a pulse of amplitude A at time n if $d(n) = 1$ and a pulse of amplitude $-A$ if $d(n) = -1$. This sequences of pulses is then modulated and transmitted across a channel to a remote receiver. The receiver demodulates and samples the received waveform, which produces a discrete-time sequence $x(n)$. Although ideally $x(n)$ will be equal to $d(n)$, in practice this will rarely be the case. There are two reasons for this. First, since the channel is never ideal, it will introduce some distortion. One common type of distortion is *channel dispersion* that is the result of the nonlinear phase characteristics of the channel. This dispersion causes a distortion of the pulse shape, thereby causing neighboring pulses to interfere with each other, resulting in an effect known as *intersymbol interference* [24,26]. The second reason is that the received waveform will invariably contain noise. This noise may be introduced by the channel or it may be the result of nonideal elements in the transmitter and receiver. Assuming a linear dispersive channel, a model for the received sequence $x(n)$ is

$$x(n) = \sum_{k=-\infty}^{n} d(k)h(n-k) + v(n)$$

where $h(n)$ is the unit sample response of the channel and $v(n)$ is additive noise. Given the received sequence $x(n)$, the receiver then makes a decision as to whether a plus one or a minus one was transmitted at time n. This decision is typically made with a simple threshold device, i.e.,

$$\hat{d}(n) = \begin{cases} 1 & ; \quad x(n) \geq 0 \\ -1 & ; \quad x(n) < 0 \end{cases}$$

However, in order to reduce the chance of making an error, the receiver will often employ an equalizer in order to reduce the effects of channel distortion. Since the precise characteristics of the channel are unknown, and possibly time-varying, the equalizer is typically an adaptive filter. A simplified block diagram of the digital communication system is shown in Fig. 9.19.

As discussed in Section 9.1, one of the challenges in the design of an adaptive filter is generating the desired signal $d(n)$, which is required in order to compute the error $e(n)$.

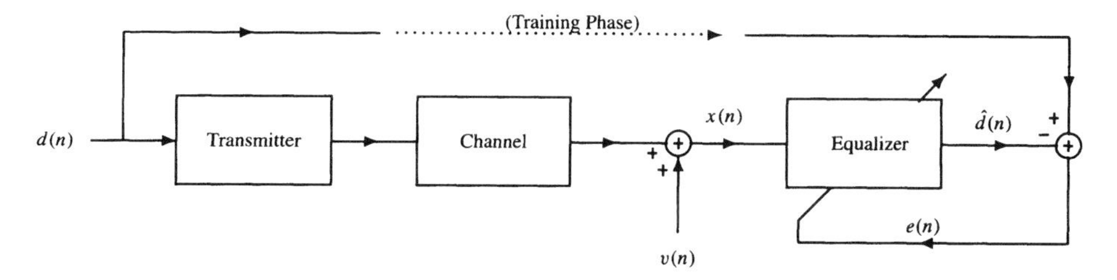

(a) Channel equalization using a training sequence.

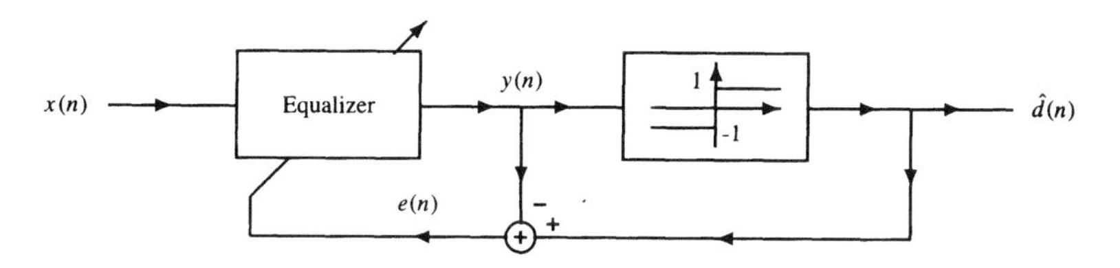

(b) The operation of the equalizer in decision-directed mode.

Figure 9.19 *A basic digital communication system with channel equalization.*

Without the error, a gradient descent algorithm will not work. For the channel equalizer, there are two methods that may be used to generate the error signal.

1. **Training method**. This method takes place during an initial training phase, which occurs when the transmitter and receiver first establish a connection. During this phase, the transmitter sends a sequence of pseudorandom digits that is known to the receiver. With knowledge of $d(n)$ the error sequence is easily determined and the tap weights of the equalizer may be *initialized.*
2. **Decision-directed method**. Once the training period has ended and data is being exchanged between the transmitter and receiver, the receiver has no a priori knowledge of what is being sent. However, there is a clever scheme that may be used to extract the error sequence from the output of the threshold device as illustrated in Fig. 9.19*b*. Specifically, note that if no errors are made at the output of the threshold device and $\hat{d}(n) = d(n)$, then the error sequence may be formed by taking the difference between the equalizer output, $y(n)$, and the output of the threshold device,

$$e(n) = y(n) - \hat{d}(n)$$

   This approach is said to be *decision directed* since it is based on the decisions made by the receiver. Although based on the threshold device making a correct decision, this approach will work even in the presence of errors, provided that they are infrequent enough. Of course, once the error rate exceeds a certain level, the inaccuracies in the error signal will cause the equalizer to diverge away from the correct solution thereby causing an increase in the error rate and eventually a loss of reliable communication. Before this happens, however, the receiver may request that the training sequence be retransmitted in order to re-initialize the equalizer.

To get a feeling for how well such an equalizer may work, we will look at the use of the LMS algorithm for the adaptive equalization of a linear dispersive channel that has a unit sample response given by [20,27]

$$h(n) = \begin{cases} \dfrac{1}{2} + \dfrac{1}{2}\cos\left[\dfrac{2\pi(n-2)}{W}\right] & ; \quad n = 1, 2, 3 \\ 0 & ; \quad \text{otherwise} \end{cases} \tag{9.69}$$

In addition to controlling the amount of distortion introduced by the channel, the parameter W affects the eigenvalues of the autocorrelation matrix $\mathbf{R}_x$ of the received signal $x(n)$. We will assume that the data is real-valued and that the input to the channel, $d(n)$, is a sequence of uncorrelated random variables with $d(n) = \pm 1$ with equal probability. Finally, we will assume that the additive noise $v(n)$ is zero mean white Gaussian noise with a variance $\sigma_v^2 = 0.005$ and that $v(n)$ is uncorrelated with $d(n)$.

We begin by finding the autocorrelation matrix for $x(n)$ and the corresponding eigenvalues. Recall that our model for $x(n)$ is

$$x(n) = d(n) * h(n) + v(n)$$

Since $d(n)$ is a sequence of uncorrelated random variables then $r_d(k) = \delta(k)$. Therefore, with $v(n)$ white noise with a variance σ_v^2, and with $v(n)$ being uncorrelated with $d(n)$, the autocorrelation sequence of $x(n)$ is

$$r_x(k) = h(k) * h(-k) + \sigma_v^2 \delta(k)$$

Since $h(n)$ is nonzero only for $n = 1, 2, 3$ we have

$$\begin{aligned} r_x(0) &= h^2(1) + h^2(2) + h^2(3) + \sigma_v^2 \\ r_x(1) &= h(1)h(2) + h(2)h(3) \\ r_x(2) &= h(1)h(3) \end{aligned}$$

and $r_x(k) = 0$ for all $k > 2$. Therefore, the autocorrelation matrix of $x(n)$ is a Toeplitz matrix of the form

$$\mathbf{R}_x = \text{Toep}\{r_x(0),\ r_x(1),\ r_x(2),\ 0, \ldots,\ 0\}$$

which is a *quintdiagonal* matrix, i.e., the only nonzero elements are along the main diagonal and the two diagonals above and below the main diagonal. The eigenvalues of $\mathbf{R}_x$, which affect the convergence of the adaptive filter, depend on the value of W as well as on the size, p, of the autocorrelation matrix $\mathbf{R}_x$, which is equal to the number of coefficients in the adaptive filter. Shown in Table 9.3 are the maximum and minimum eigenvalues of $\mathbf{R}_x$ along with the condition number χ for several different values of W and $p = 15$. From our discussions in Chapter 2 (see Property 7 on p. 48) we may upper bound the largest eigenvalue of $\mathbf{R}_x$ as the largest row sum which, for $p \geq 5$ is

$$\lambda_{\max} \leq r_x(0) + 2r_x(1) + 2r_x(2) \tag{9.70}$$

For example, for the given values of W listed in Table 9.3 we have

$$\lambda_{\max} \leq 1.8998\,(W = 2.8) \quad ; \quad \lambda_{\max} \leq 2.2550\,(W = 3.0) \quad ; \quad \lambda_{\max} \leq 2.6207\,(W = 3.2)$$

Note that these bounds are not tied to a specific value of p.

Let us now look at the performance of the adaptive equalizer using the normalized LMS algorithm to adjust the filter coefficients. With $\beta = 0.3$, $W = 3.0$, and $p = 15$, the equalizer is first initialized using a training sequence of length 500. Shown in Fig. 9.20*a*

Table 9.3 ***The Maximum and Minimum Eigenvalues for Three Different Channel Characteristics***

W	2.8	3.0	3.2
$r_x(0)$	1.0759	1.1300	1.1955
$r_x(1)$	0.3765	0.5000	0.6173
$r_x(2)$	0.0354	0.0625	0.0953
$\lambda_{\max}$	1.8803	2.2269	2.5835
$\lambda_{\min}$	0.4031	0.2652	0.1614
χ	4.6643	8.3966	16.006

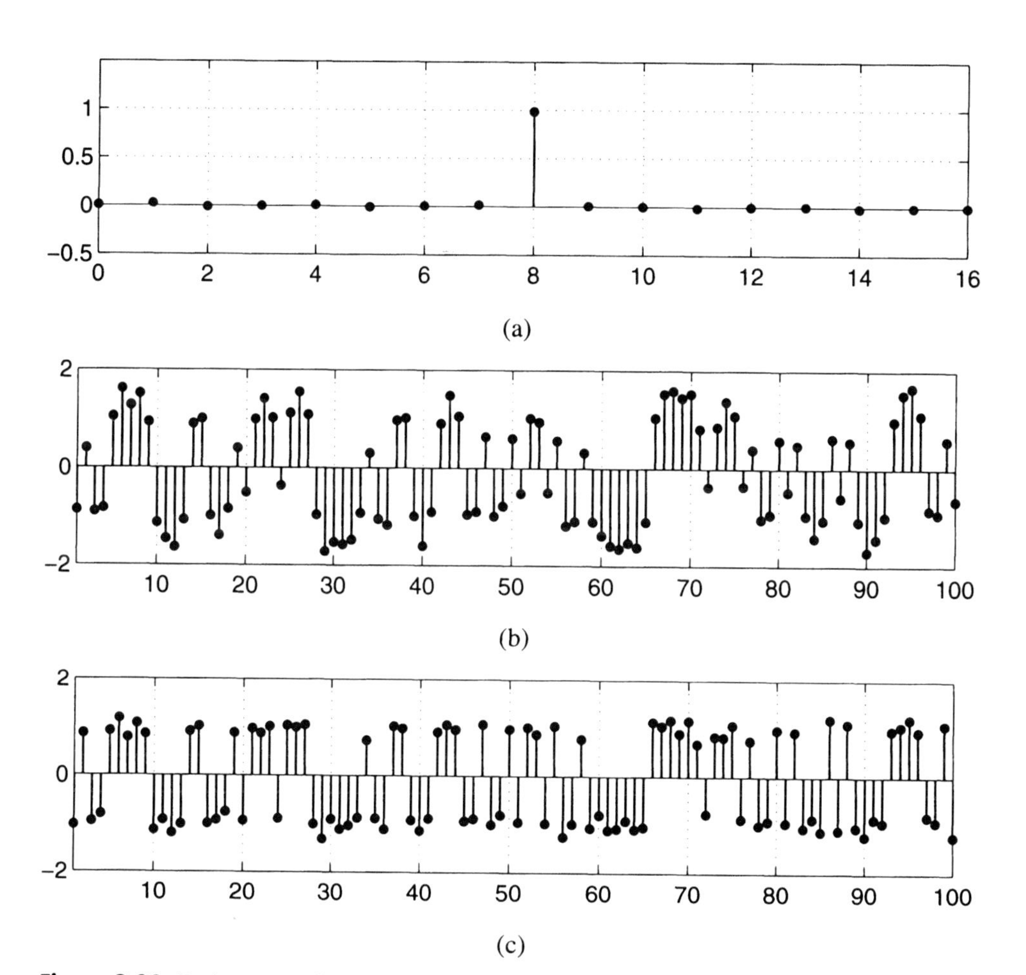

Figure 9.20 *Performance of the channel equalizer. (a) The convolution of the unit sample response of the channel with the equalizer at the end of the training sequence. (b) The received sequence after the training period has been completed. (c) The output of the equalizer.*

is the convolution of the unit sample response of the channel, $h(n)$, with the unit sample response of the equalizer $w(n)$ at the end of the training sequence. Since this convolution is approximately a unit sample at $n = 8$, the equalizer is a good approximation to the inverse of $h(n)$. However, after the training sequence, the equalizer no longer has knowledge of $d(n)$, and must rely on decision directed adaptation to generate the error sequence to drive the coefficient update equation. Shown in Fig. 9.20*b* is the received sequence, $x(n)$, for the first 100 values after the training period and in Fig. 9.20*c* is the output of adaptive equalizer. As we see, the equalizer is effective in producing a sequence $y(n)$ with $|y(n)| \approx 1$. Due to noise, however, and the fact that the equalizer is only an approximation to the inverse of $h(n)$, the equalized process is not exactly equal to $d(n)$. However, the equalizer output is accurate enough so that the threshold device produces an output $\hat{d}(n) = d(n)$.

9.3 ADAPTIVE RECURSIVE FILTERS

In the previous section, a variety of different methods were presented for designing an FIR (non-recursive) adaptive filter for producing the minimum mean-square estimation of a process $d(n)$ from measurements of a related process, $x(n)$. In this section, we turn our attention briefly to the problem of designing IIR (recursive) adaptive filters of the form

$$y(n) = \sum_{k=1}^{p} a_n(k) y(n-k) + \sum_{k=0}^{q} b_n(k) x(n-k) \tag{9.71}$$

where $a_n(k)$ and $b_n(k)$ are the coefficients of the adaptive filter at time n. Just as with linear shift-invariant (non-adaptive) filters, recursive adaptive filters potentially have an advantage over non-recursive filters in providing better performance for a given filter order (number of filter coefficients). However, with an adaptive recursive filter there are potential instability problems that may affect the convergence time as well as the general numerical sensitivity of the filter. High resonance poles, for example, may have time constants that are longer than the modes of the adaptive algorithm. In spite of these difficulties, there are many applications in which a recursive filter may be required. For example, suppose that a channel used to transmit a signal $d(n)$ introduces an echo so that the received signal is

$$x(n) = d(n) + \alpha d(n-N)$$

where $|\alpha| < 1$ and N is the delay associated with the echo. If both α and N are known, then the ideal *echo canceller* for recovering $d(n)$ from $x(n)$ is an IIR filter that has a system function given by

$$H(z) = \frac{1}{1 - \alpha z^{-N}}$$

However, since α and N are generally unknown and possibly time-varying, then it is more appropriate for the echo canceller to be an adaptive recursive filter. Although we could consider a nonrecursive adaptive filter, the order of the filter required for a sufficiently accurate estimate of $d(n)$ may be too large. For example, suppose that the inverse filter, $H(z)$, is expanded in a geometric series as follows:

$$H(z) = \frac{1}{1 - \alpha z^{-N}} = \sum_{k=0}^{\infty} \alpha^k z^{-Nk}$$

If p is large enough so that $|\alpha|^p \ll 1$, then we could form a finite-order approximation to

$H(z)$ as follows:

$$\hat{H}(z) = \sum_{k=0}^{p} \alpha^k z^{-Nk}$$

and consider an adaptive nonrecursive echo canceller of the form

$$\hat{d}(n) = \sum_{k=0}^{Np} b_n(k)x(n-k)$$

However, if $\alpha \approx 1$, which forces p to be large, or if $N \gg 1$, then the order of the adaptive filter, Np, required to produce a sufficiently accurate approximation to the inverse filter may be too large for this to be a viable solution.

We begin our discussion of adaptive recursive filters with a derivation of the conditions that are necessary in order for the coefficients of the recursive filter to minimize the mean-square error, $\xi(n) = E\{|e(n)|^2\}$, where $e(n) = d(n) - y(n)$ is the difference between the desired process $d(n)$ and the output of the adaptive filter, $y(n)$. We will initially assume that the filter is shift-invariant (non-adaptive) with

$$y(n) = \sum_{l=1}^{p} a(l)y(n-l) + \sum_{l=0}^{q} b(l)x(n-l) \tag{9.72}$$

If we let $\boldsymbol{\Theta}$ be the vector of filter coefficients,

$$\boldsymbol{\Theta} = \begin{bmatrix} \mathbf{a} \\ \mathbf{b} \end{bmatrix} = \big[a(1),\ a(2),\ \ldots,\ a(p),\ b(0),\ b(1),\ \ldots,\ b(q)\big]^T$$

and let $\mathbf{z}(n)$ denote the *aggregate data vector*

$$\begin{aligned} \mathbf{z}(n) &= \begin{bmatrix} \mathbf{y}(n-1) \\ \mathbf{x}(n) \end{bmatrix} \\ &= \big[y(n-1),\ y(n-2),\ \ldots,\ y(n-p),\ x(n),\ \ldots,\ x(n-q)\big]^T \end{aligned} \tag{9.73}$$

then the output of the filter at time n may be expressed in terms of $\boldsymbol{\Theta}$ and $\mathbf{z}(n)$ as follows:

$$y(n) = \mathbf{a}^T\mathbf{y}(n-1) + \mathbf{b}^T\mathbf{x}(n) = \boldsymbol{\Theta}^T\mathbf{z}(n) \tag{9.74}$$

In order for $\boldsymbol{\Theta}$ to minimize the mean-square error

$$\xi(n) = E\{|e(n)|^2\} = E\Big\{|d(n) - y(n)|^2\Big\}$$

it is necessary that the derivatives of $\xi(n)$ with respect to each of the coefficients, $a^*(k)$ and $b^*(k)$, be equal to zero, i.e., the gradient vector must be zero.[7] Since

$$\nabla\xi(n) = \nabla E\left\{|e(n)|^2\right\} = E\left\{\nabla|e(n)|^2\right\} = E\left\{e(n)\nabla e^*(n)\right\} \tag{9.75}$$

then the gradient will be zero when

$$E\left\{e(n)\nabla e^*(n)\right\} = 0 \tag{9.76}$$

[7] For an FIR filter, this is a necessary and sufficient condition since the mean-square error is a quadratic function of the coefficients $b(k)$. However, with the feedback coefficients $a(k)$, the mean-square error is no longer quadratic and $\xi(n)$ may have multiple local minima and maxima [31]. Therefore, this condition is only necessary and, in general, not sufficient.

Since $e(n) = d(n) - y(n)$ then

$$\nabla e^*(n) = -\nabla y^*(n)$$

and Eq. (9.76) becomes

$$E\left\{e(n)\nabla y^*(n)\right\} = 0 \tag{9.77}$$

The next step is to evaluate $\nabla y^*(n)$ by differentiating with respect to $a^*(k)$ and $b^*(k)$. In forming these derivatives, it must be kept in mind that $y(n)$ depends on $y(n-l)$ for $l = 1, 2, \ldots, p$, and each of these, in turn, depends on $a(k)$ and $b(k)$. Therefore, the partial derivatives of $y^*(n)$ with respect to $a^*(k)$ and $b^*(k)$ are

$$\begin{aligned}
\frac{\partial y^*(n)}{\partial a^*(k)} &= y^*(n-k) + \sum_{l=1}^{p} a^*(l)\frac{\partial y^*(n-l)}{\partial a^*(k)} \quad ; \quad k = 1, 2, \ldots, p \\
\frac{\partial y^*(n)}{\partial b^*(k)} &= x^*(n-k) + \sum_{l=1}^{p} a^*(l)\frac{\partial y^*(n-l)}{\partial b^*(k)} \quad ; \quad k = 0, 1, \ldots, q
\end{aligned} \tag{9.78}$$

Writing these equations in vector form leads to the following expression for the gradient vector:

$$\boxed{\nabla y^*(n) = \mathbf{z}^*(n) + \sum_{l=1}^{p} a^*(l)\nabla y^*(n-l)} \tag{9.79}$$

where $\mathbf{z}(n)$ is the aggregate data vector defined in Eq. (9.73). Finally, combining Eqs. (9.77) and (9.79) we obtain the following conditions that are necessary for $a(k)$ and $b(k)$ to minimize the mean-square error:

$$\begin{aligned}
E\left\{e(n)\left[y(n-k) + \sum_{l=1}^{p} a(l)\frac{\partial y(n-l)}{\partial a(k)}\right]^*\right\} &= 0 \quad ; \quad k = 1, 2, \ldots, p \\
E\left\{e(n)\left[x(n-k) + \sum_{l=1}^{p} a(l)\frac{\partial y(n-l)}{\partial b(k)}\right]^*\right\} &= 0 \quad ; \quad k = 0, 1, \ldots, q
\end{aligned} \tag{9.80}$$

Since these equations are nonlinear, they are difficult to solve for the optimum filter coefficients. This nonlinearity also implies that the solution to these equations may not be unique. Therefore, a solution may correspond to a local rather than a global minimum. As a result, these equations are generally not solved directly.

An alternative to solving Eq. (9.81) directly is to use a steepest descent approach to search for the solution iteratively. With $\boldsymbol{\Theta}_n$ the estimate of the solution at time n,

$$\boldsymbol{\Theta}_n = \begin{bmatrix} \mathbf{a}_n \\ \mathbf{b}_n \end{bmatrix} = \left[a_n(1),\ a_n(2),\ \ldots,\ a_n(p),\ b_n(0),\ \ldots,\ b_n(q)\right]^T$$

a steepest descent update for $\boldsymbol{\Theta}_n$ is given by

$$\boldsymbol{\Theta}_{n+1} = \boldsymbol{\Theta}_n - \mu\ \nabla\xi(n) \tag{9.81}$$

where μ is the step size and $\nabla\xi(n)$ is the gradient vector. The difficulty with this update equation is that the gradient vector involves an expectation

$$\nabla\xi(n) = -E\{e(n)\nabla y^*(n)\}$$

Therefore, unless the required ensemble averages are known, the steepest descent algorithm cannot be implemented. However, if we adopt the approach used in the LMS adaptive filter of replacing expected values with instantaneous values,

$$\hat{\nabla}\xi(n) = -e(n)\nabla y^*(n)$$

then the coefficient update equation (9.81) becomes

$$\boxed{\boldsymbol{\Theta}_{n+1} = \boldsymbol{\Theta}_n + \mu\, e(n)\nabla y^*(n)} \tag{9.82}$$

Expressed in terms of the filter coefficients $a_n(k)$ and $b_n(k)$, Eq. (9.82) becomes

$$\begin{aligned} a_{n+1}(k) &= a_n(k) + \mu\, e(n)\frac{\partial y^*(n)}{\partial a_n^*(k)} \\ b_{n+1}(k) &= b_n(k) + \mu\, e(n)\frac{\partial y^*(n)}{\partial b_n^*(k)} \end{aligned} \tag{9.83}$$

where the partial derivatives are evaluated using Eq. (9.79) with the filter coefficients $a(k)$ and $b(k)$ replaced by the time-varying coefficients $a_n(k)$ and $b_n(k)$,

$$\begin{aligned} \frac{\partial y^*(n)}{\partial a_n^*(k)} &= y^*(n-k) + \sum_{l=1}^{p} a_n^*(l)\frac{\partial y^*(n-l)}{\partial a_n^*(k)} \quad ; \quad k = 1, 2, \ldots, p \\ \frac{\partial y^*(n)}{\partial b_n^*(k)} &= x^*(n-k) + \sum_{l=1}^{p} a_n^*(l)\frac{\partial y^*(n-l)}{\partial b_n^*(k)} \quad ; \quad k = 0, 1, \ldots, q \end{aligned} \tag{9.84}$$

Equations (9.84) and (9.83) along with the filter output equation

$$y(n) = \mathbf{a}_n^T\mathbf{y}(n-1) + \mathbf{b}_n^T\mathbf{x}(n) = \boldsymbol{\Theta}_n^T\mathbf{z}(n)$$

form what is known as the *IIR LMS* algorithm [35]. Based on numerous simulations, it has been reported that, for sufficiently small step sizes, the IIR LMS algorithm generally converges to the filter coefficients that minimize the mean-square error [10,32]. However, compared to their FIR counterparts, convergence is generally much slower. In addition, due to the possibility of local minima in the mean-square error surface, convergence may not be to the global minimum.

We will now look at a couple of ways to simplify the IIR LMS adaptive filter. First, note that the derivatives of $y^*(n-l)$ in Eq. (9.84) are taken with respect to the *current* values of $a_n^*(k)$ and $b_n^*(k)$. Therefore, these expressions for the partial derivatives are not recursive and cannot be expressed in terms of a filtering operation. However, if the step size is small enough so that the coefficients are adapting slowly, then

$$\frac{\partial y^*(n-l)}{\partial a_n^*(k)} \approx \frac{\partial y^*(n-l)}{\partial a_{n-l}^*(k)} \quad ; \quad \frac{\partial y^*(n-l)}{\partial b_n^*(k)} \approx \frac{\partial y^*(n-l)}{\partial b_{n-l}^*(k)} \tag{9.85}$$

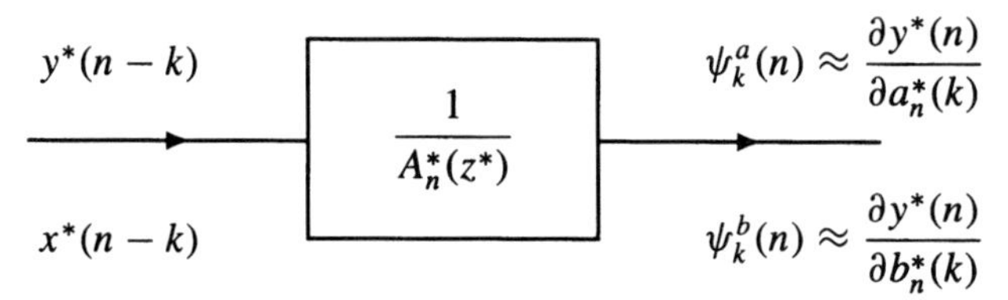

Figure 9.21 *The recursive evaluation of the gradient vector, formed by filtering $x^*(n-k)$ and $y^*(n-k)$ with the recursive filter $1/A_n^*(z^*)$.*

and the derivatives in Eq. (9.84) may be approximated as follows:

$$\begin{aligned}\frac{\partial y^*(n)}{\partial a_n^*(k)} &\approx y^*(n-k) + \sum_{l=1}^{p} a_n^*(l)\frac{\partial y^*(n-l)}{\partial a_{n-l}^*(k)}\\ \frac{\partial y^*(n)}{\partial b_n^*(k)} &\approx x^*(n-k) + \sum_{l=1}^{p} a_n^*(l)\frac{\partial y^*(n-l)}{\partial b_{n-l}^*(k)}\end{aligned} \tag{9.86}$$

Note that these expressions for the partial derivatives are now recursive since the partial derivatives in the sum correspond to delayed versions of the derivatives on the left. Therefore, let $\psi_k^a(n)$ and $\psi_k^b(n)$ be the approximations to the partial derivatives that are generated by these recursions, i.e.,

$$\begin{aligned}\psi_k^a(n) &= y^*(n-k) + \sum_{l=1}^{p} a_n^*(l)\psi_k^a(n-l)\\ \psi_k^b(n) &= x^*(n-k) + \sum_{l=1}^{p} a_n^*(l)\psi_k^b(n-l)\end{aligned} \tag{9.87}$$

As illustrated in Fig. 9.21, these approximations are generated by filtering $x^*(n-k)$ and $y^*(n-k)$ with the shift-varying recursive filter $1/A_n^*(z^*)$. A block diagram of this simplified IIR LMS adaptive filter is shown in Fig. 9.22. Note that there are $p+q+1$ recursive filters operating in parallel that produce the approximations $\psi_k^a(n)$ and $\psi_k^b(n)$ to the gradient vector. The outputs of these filters are then used to update the coefficients of the adaptive filter. The simplified IIR LMS algorithm is summarized in Table 9.4.

In spite of the simplification that results from the approximation made in Eq. (9.85), the implementation of $p+q+1$ shift-varying filters in parallel represents a significant computational load and requires a significant amount of storage. However, if we again assume that the step size μ is small, then the IIR LMS adaptive filter may be simplified further [22]. Specifically, note that the gradient estimate $\psi_k^a(n)$ is formed by delaying $y^*(n)$ and filtering this delayed process with the shift-varying filter $1/A_n^*(z^*)$ as illustrated in Fig. 9.23*a*. The estimate $\psi_k^b(n)$ is found in a similar fashion by filtering delayed versions of $x^*(n)$. If μ is small enough so that the coefficients $a_n(k)$ do not vary significantly over intervals of length p, then the filter $1/A_n^*(z^*)$ may be assumed to be shift-invariant and the cascade of the delay and the shift-varying filter may be interchanged as shown in Fig. 9.23*b*. As a result, $\psi_k^a(n)$ and $\psi_k^b(n)$ may be estimated by simply generating the filtered signals $y^f(n) = \psi_0^a(n)$ and $x^f(n) = \psi_0^b(n)$, and then delaying these signals. This method, referred to as the *filtered signal approach*, requires only two recursive filters to estimate the gradient vector as illustrated in Fig. 9.23*c*. Compared to the $p+q+1$ filters necessary for the simplified IIR LMS algorithm, this method is much more efficient and, for a sufficiently

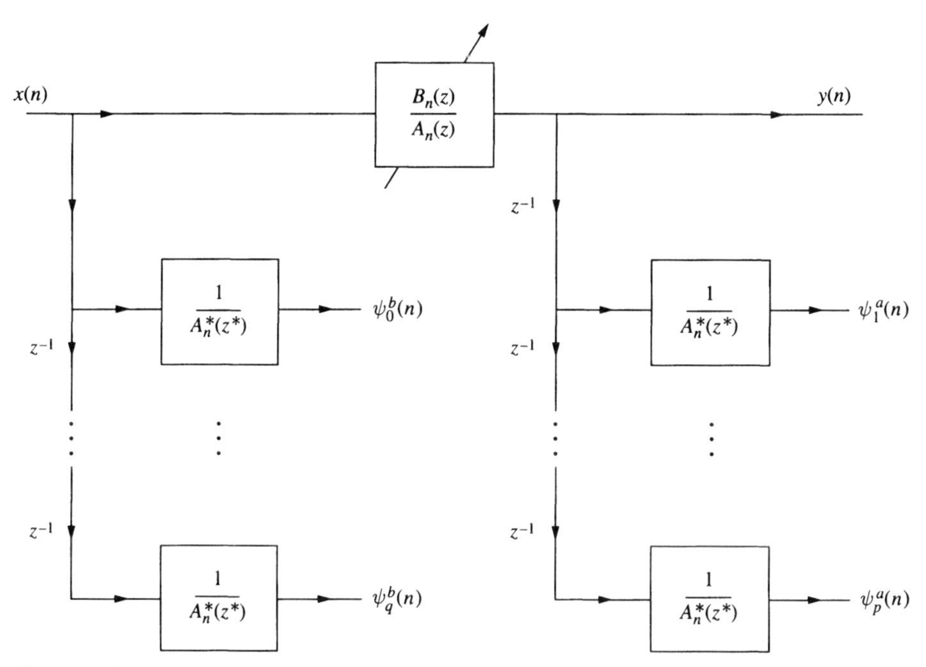

Figure 9.22 *A simplified IIR LMS adaptive filter that assumes that the step size is small enough so that an estimate of the gradient vector may be computed recursively. These estimates, $\psi_k^a(n)$ and $\psi_k^b(n)$, are used by the adaptive filter to update the filter coefficients. (Note that the inputs to the filters $1/A_n^*(z^*)$ must be conjugated prior to filtering.)*

Table 9.4 A Simplified IIR LMS Adaptive Filter

Parameters:	p, q = Filter order
	μ = step size
Initialization:	$\mathbf{a}_0 = \mathbf{b}_0 = \mathbf{0}$
Computation:	For $n = 0, 1, \ldots$ compute
	(a) $y(n) = \mathbf{a}_n^T \mathbf{y}(n-1) + \mathbf{b}_n^T \mathbf{x}(n)$
	(b) $e(n) = d(n) - y(n)$
	(c) For $k = 1, 2, \ldots, p$
	$\psi_k^a(n) = y^*(n-k) + \sum_{l=1}^{p} a_n^*(l)\psi_k^a(n-l)$
	$a_{n+1}(k) = a_n(k) + \mu e(n)\psi_k^a(n)$
	(d) For $k = 0, 1, \ldots, q$
	$\psi_k^b(n) = x^*(n-k) + \sum_{l=1}^{p} a_n^*(l)\psi_k^b(n-l)$
	$b_{n+1}(k) = b_n(k) + \mu e(n)\psi_k^b(n)$

small step size, has a performance that is similar to the IIR LMS algorithm [22]. The filtered signal adaptive recursive filter is summarized in Table 9.5. For a more complete treatment of adaptive recursive filters, see Reference [29].

Table 9.5 *The Equations for the Filtered Signal Approach to Adaptive Recursive Filtering*

Parameters:	p, q = Filter order μ = step size
Initialization:	$\mathbf{a}_0 = \mathbf{b}_0 = \mathbf{0}$
Computation:	For $n = 0, 1, \ldots$ compute
	(a) $y(n) = \mathbf{a}_n^T \mathbf{y}(n-1) + \mathbf{b}_n^T \mathbf{x}(n)$
	(b) $e(n) = d(n) - y(n)$
	(c) $y^f(n) = y(n) + \sum_{k=1}^{p} a_n^*(k) y^f(n-k)$ $x^f(n) = x(n) + \sum_{k=1}^{p} a_n^*(k) x^f(n-k)$
	(d) $a_{n+1}(k) = a_n(k) + \mu e(n) y^f(n-k)$ $b_{n+1}(k) = b_n(k) + \mu e(n) x^f(n-k)$

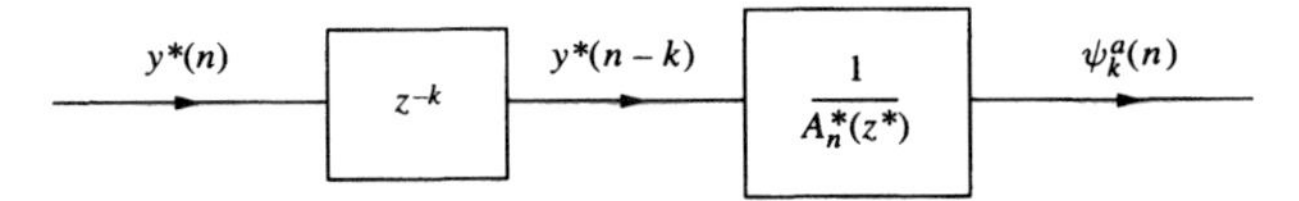

(a) The gradient approximation that is formed by filtering $y^*(n-k)$ with the shift-varying recursive filter $1/A_n^*(z^*)$.

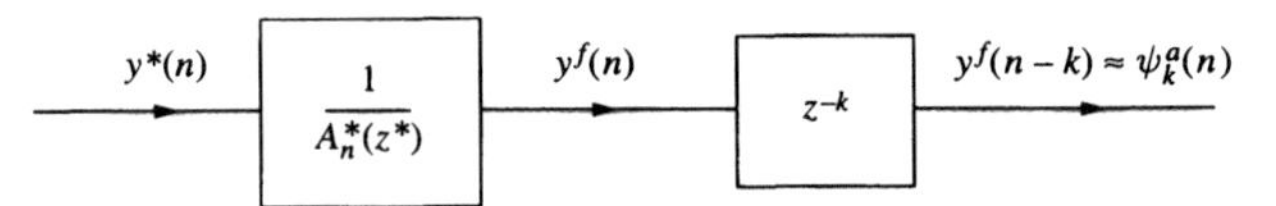

(b) Assuming that the filter coefficients are slowly varying, the delay and the recursive filter may be interchanged.

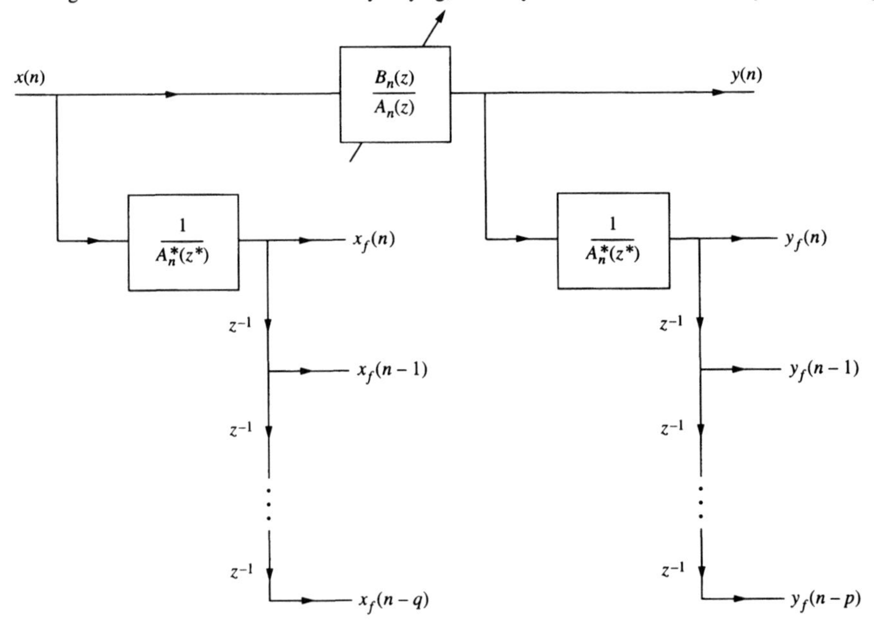

(c) The resulting simplification to the IIR LMS adaptive filter. (Note that $x(n)$ and $y(n)$ must be conjugated prior to filtering.)

Figure 9.23 *The filtered signal approach for IIR adaptive filtering.*

9.4 RECURSIVE LEAST SQUARES

In each of the adaptive filtering methods discussed so far, we have considered gradient descent algorithms for the minimization of the *mean-square error*

$$\xi(n) = E\{|e(n)|^2\}$$

The difficulty with these methods is that they all require knowledge of the autocorrelation of the input process, $E\{x(n)x^*(n-k)\}$, and the cross-correlation between the input and the desired output, $E\{d(n)x^*(n-k)\}$. When this statistical information is unknown, we have been forced to estimate these statistics from the data. In the LMS adaptive filter, for example, these ensemble averages are estimated using instantaneous values,

$$\hat{E}\{e(n)x^*(n-k)\} = e(n)x^*(n-k) \tag{9.88}$$

Although this approach may be adequate in some applications, in others this gradient estimate may not provide a sufficiently rapid rate of convergence or a sufficiently small excess mean-square error. An alternative, therefore, is to consider error measures that do not include expectations and that may be computed directly from the data. For example, a least squares error

$$\mathcal{E}(n) = \sum_{i=0}^{n} |e(i)|^2$$

requires no statistical information about $x(n)$ or $d(n)$, and may be evaluated directly from $x(n)$ and $d(n)$. There is an important philosophical difference, however, between minimizing the least squares error and the mean-square error. Minimizing the mean-square error produces the same set of filter coefficients for all sequences that have the same statistics. Therefore, the coefficients do not depend on the incoming data, only on their ensemble average. With the least squares error, on the other hand, we are minimizing a squared error that depends explicitly on the specific values of $x(n)$ and $d(n)$. Consequently, for different signals we get different filters. As a result, the filter coefficients that minimize the least squares error will be optimal for the given data rather than statistically optimal over a particular class of process. In other words, different realizations of $x(n)$ and $d(n)$ will lead to different solutions, even if the statistics of these sequences are the same. In this section, we will look at the filters that are derived by minimizing a weighted least squares error, and derive an efficient algorithm for performing this minimization known as *recursive least squares*.

9.4.1 Exponentially Weighted RLS

Let us reconsider the design of an FIR adaptive Wiener filter and find the filter coefficients

$$\mathbf{w}_n = \left[w_n(0),\ w_n(1),\ \ldots,\ w_n(p)\right]^T$$

that minimize, at time n, the weighted least squares error

$$\mathcal{E}(n) = \sum_{i=0}^{n} \lambda^{n-i} |e(i)|^2 \tag{9.89}$$

where $0 < \lambda \leq 1$ is an exponential weighting (forgetting) factor and

$$e(i) = d(i) - y(i) = d(i) - \mathbf{w}_n^T \mathbf{x}(i) \tag{9.90}$$

Note that $e(i)$ is the difference between the desired signal $d(i)$ and the filtered output at time i, using the *latest* set of filter coefficients, $w_n(k)$. Thus, in minimizing $\mathcal{E}(n)$ it is assumed that the weights $\mathbf{w}_n$ are held constant over the entire observation interval $[0, n]$. To find the coefficients that minimize $\mathcal{E}(n)$ we proceed exactly as we have done many times before by setting the derivative of $\mathcal{E}(n)$ with respect to $w_n^*(k)$ equal to zero for $k = 0, 1, \ldots, p$. Thus, we have

$$\frac{\partial \mathcal{E}(n)}{\partial w_n^*(k)} = \sum_{i=0}^{n} \lambda^{n-i} e(i) \frac{\partial e^*(i)}{\partial w_n^*(k)} = -\sum_{i=0}^{n} \lambda^{n-i} e(i) x^*(i-k) = 0 \tag{9.91}$$

for $k = 0, 1, \ldots, p$. Incorporating Eq. (9.90) into Eq. (9.91) yields

$$\sum_{i=0}^{n} \lambda^{n-i} \left\{ d(i) - \sum_{l=0}^{p} w_n(l) x(i-l) \right\} x^*(i-k) = 0$$

Interchanging the order of summation and rearranging terms we have

$$\sum_{l=0}^{p} w_n(l) \left[\sum_{i=0}^{n} \lambda^{n-i} x(i-l) x^*(i-k) \right] = \sum_{i=0}^{n} \lambda^{n-i} d(i) x^*(i-k) \tag{9.92}$$

We may express these equations concisely in matrix form as follows:

$$\boxed{\mathbf{R}_x(n) \mathbf{w}_n = \mathbf{r}_{dx}(n)} \tag{9.93}$$

where $\mathbf{R}_x(n)$ is a $(p+1) \times (p+1)$ exponentially weighted deterministic autocorrelation matrix for $x(n)$

$$\mathbf{R}_x(n) = \sum_{i=0}^{n} \lambda^{n-i} \mathbf{x}^*(i) \mathbf{x}^T(i)$$

with $\mathbf{x}(i)$ the data vector

$$\mathbf{x}(i) = \big[x(i),\ x(i-1),\ \ldots,\ x(i-p) \big]^T$$

and where $\mathbf{r}_{dx}(n)$ is the deterministic cross-correlation between $d(n)$ and $x(n)$,

$$\mathbf{r}_{dx}(n) = \sum_{i=0}^{n} \lambda^{n-i} d(i) \mathbf{x}^*(i) \tag{9.94}$$

Eq. (9.93) is referred to as the *deterministic normal equations*.

Having derived the equations that define the optimum filter coefficients, we may now evaluate the minimum squared error. With

$$\begin{aligned}
\mathcal{E}(n) &= \sum_{i=0}^{n} \lambda^{n-i} e(i) e^*(i) = \sum_{i=0}^{n} \lambda^{n-i} e(i) \left\{ d(i) - \sum_{l=0}^{p} w_n(l) x(i-l) \right\}^* \\
&= \sum_{i=0}^{n} \lambda^{n-i} e(i) d^*(i) - \sum_{l=0}^{p} w_n^*(l) \sum_{i=0}^{n} \lambda^{n-i} e(i) x^*(i-l)
\end{aligned}$$

note that if $w_n(l)$ are the coefficients that minimize the squared error, then it follows from Eq. (9.91) that the second term is zero and the minimum error is

$$\begin{aligned}
\{\mathcal{E}(n)\}_{\min} &= \sum_{i=0}^{n} \lambda^{n-i} e(i) d^*(i) \\
&= \sum_{i=0}^{n} \lambda^{n-i} \left\{ d(i) - \sum_{l=0}^{p} w_n(l) x(i-l) \right\} d^*(i) \\
&= \sum_{i=0}^{n} \lambda^{n-i} |d(i)|^2 - \sum_{l=0}^{p} w_n(l) \sum_{i=0}^{n} \lambda^{n-i} x(i-l) d^*(i)
\end{aligned}$$

Alternatively, we may express the minimum error in vector form as follows:

$$\left\{\mathcal{E}(n)\right\}_{\min} = \|\mathbf{d}(n)\|_\lambda^2 - \mathbf{r}_{dx}^H(n)\mathbf{w}_n \tag{9.95}$$

where $\|\mathbf{d}(n)\|_\lambda^2$ is the weighted norm of the vector $\mathbf{d}(n) = \left[d(n),\ d(n-1),\ \dots,\ d(0)\right]^T$.

Since $\mathbf{R}_x(n)$ and $\mathbf{r}_{dx}(n)$ both depend on n, instead of solving the deterministic normal equations directly, for each value of n, we will derive a recursive solution of the form

$$\mathbf{w}_n = \mathbf{w}_{n-1} + \Delta\mathbf{w}_{n-1}$$

where $\Delta\mathbf{w}_{n-1}$ is a correction that is applied to the solution at time $n-1$. Since

$$\mathbf{w}_n = \mathbf{R}_x^{-1}(n)\mathbf{r}_{dx}(n)$$

this recursion will be derived by first expressing $\mathbf{r}_{dx}(n)$ in terms of $\mathbf{r}_{dx}(n-1)$, and then deriving a recursion that allows us to evaluate $\mathbf{R}_x^{-1}(n)$ in terms of $\mathbf{R}_x^{-1}(n-1)$ and the new data vector $\mathbf{x}(n)$. From Eq. (9.94) it follows that the cross-correlation may be updated recursively as follows:

$$\mathbf{r}_{dx}(n) = \lambda\mathbf{r}_{dx}(n-1) + d(n)\mathbf{x}^*(n) \tag{9.96}$$

Similarly, the autocorrelation matrix may be updated from $\mathbf{R}_x(n-1)$ and the new data vector $\mathbf{x}(n)$ using the recursion

$$\mathbf{R}_x(n) = \lambda\mathbf{R}_x(n-1) + \mathbf{x}^*(n)\mathbf{x}^T(n) \tag{9.97}$$

However, since it is the inverse of $\mathbf{R}_x(n)$ that we are interested in, we may apply *Woodbury's Identity*, Eq. (2.29), to Eq. (9.97) to get the desired recursion. Woodbury's identity, repeated below for convenience, is

$$(\mathbf{A} + \mathbf{u}\mathbf{v}^H)^{-1} = \mathbf{A}^{-1} - \frac{\mathbf{A}^{-1}\mathbf{u}\mathbf{v}^H\mathbf{A}^{-1}}{1 + \mathbf{v}^H\mathbf{A}^{-1}\mathbf{u}}$$

Note that if we set $\mathbf{A} = \lambda\mathbf{R}_x(n-1)$ and $\mathbf{u} = \mathbf{v} = \mathbf{x}^*(n)$ then we obtain the following recursion for the inverse of $\mathbf{R}_x(n)$:

$$\mathbf{R}_x^{-1}(n) = \lambda^{-1}\mathbf{R}_x^{-1}(n-1) - \frac{\lambda^{-2}\mathbf{R}_x^{-1}(n-1)\mathbf{x}^*(n)\mathbf{x}^T(n)\mathbf{R}_x^{-1}(n-1)}{1 + \lambda^{-1}\mathbf{x}^T(n)\mathbf{R}_x^{-1}(n-1)\mathbf{x}^*(n)} \tag{9.98}$$

To simplify notation, we will let $\mathbf{P}(n)$ denote the inverse of the autocorrelation matrix at time n,

$$\mathbf{P}(n) = \mathbf{R}_x^{-1}(n)$$

and define what is referred to as the *gain vector*, $\mathbf{g}(n)$, as follows:

$$\mathbf{g}(n) = \frac{\lambda^{-1}\mathbf{P}(n-1)\mathbf{x}^*(n)}{1 + \lambda^{-1}\mathbf{x}^T(n)\mathbf{P}(n-1)\mathbf{x}^*(n)} \tag{9.99}$$

Incorporating these definitions into Eq. (9.98) we have

$$\mathbf{P}(n) = \lambda^{-1}\Big[\mathbf{P}(n-1) - \mathbf{g}(n)\mathbf{x}^T(n)\mathbf{P}(n-1)\Big] \tag{9.100}$$

An interesting and useful expression for the gain vector may be derived from Eq. (9.99) as follows. Cross-multiplying to eliminate the denominator on the right side of Eq. (9.99), we have

$$\mathbf{g}(n) + \lambda^{-1}\mathbf{g}(n)\mathbf{x}^T(n)\mathbf{P}(n-1)\mathbf{x}^*(n) = \lambda^{-1}\mathbf{P}(n-1)\mathbf{x}^*(n)$$

Next, bringing the second term on the left to the right side of the equation yields

$$\mathbf{g}(n) = \lambda^{-1}\Big[\mathbf{P}(n-1) - \mathbf{g}(n)\mathbf{x}^T(n)\mathbf{P}(n-1)\Big]\mathbf{x}^*(n)$$

Finally, from Eq. (9.100) we see that the term multiplying $\mathbf{x}^*(n)$ is $\mathbf{P}(n)$ and we have

$$\mathbf{g}(n) = \mathbf{P}(n)\mathbf{x}^*(n) \tag{9.101}$$

Thus, the gain vector is the solution to the linear equations

$$\boxed{\mathbf{R}_x(n)\mathbf{g}(n) = \mathbf{x}^*(n)} \tag{9.102}$$

which is the same as the deterministic normal equations for $\mathbf{w}_n$ in Eq. (9.93) except that the cross-correlation vector $\mathbf{r}_{dx}(n)$ has been replaced with the data vector $\mathbf{x}^*(n)$.

To complete the recursion, we must derive the time-update equation for the coefficient vector $\mathbf{w}_n$. With

$$\mathbf{w}_n = \mathbf{P}(n)\mathbf{r}_{dx}(n)$$

it follows from the update for $\mathbf{r}_{dx}(n)$ given in Eq. (9.94) that

$$\mathbf{w}_n = \lambda\mathbf{P}(n)\mathbf{r}_{dx}(n-1) + d(n)\mathbf{P}(n)\mathbf{x}^*(n) \tag{9.103}$$

Next, incorporating the update for $\mathbf{P}(n)$ given in Eq. (9.100) into the first term on the right side of Eq. (9.103), and setting $\mathbf{P}(n)\mathbf{x}^*(n) = \mathbf{g}(n)$ we have

$$\mathbf{w}_n = \Big[\mathbf{P}(n-1) - \mathbf{g}(n)\mathbf{x}^T(n)\mathbf{P}(n-1)\Big]\mathbf{r}_{dx}(n-1) + d(n)\mathbf{g}(n)$$

Finally, recognizing that $\mathbf{P}(n-1)\mathbf{r}_{dx}(n-1) = \mathbf{w}_{n-1}$ it follows that

$$\mathbf{w}_n = \mathbf{w}_{n-1} + \mathbf{g}(n)\Big[d(n) - \mathbf{w}_{n-1}^T\mathbf{x}(n)\Big]$$

which may be written as

$$\boxed{\mathbf{w}_n = \mathbf{w}_{n-1} + \alpha(n)\mathbf{g}(n)} \tag{9.104}$$

where

$$\alpha(n) = d(n) - \mathbf{w}_{n-1}^T\mathbf{x}(n) \tag{9.105}$$

is the difference between $d(n)$ and the estimate of $d(n)$ that is formed by applying the *previous* set of filter coefficients, $\mathbf{w}_{n-1}$, to the new data vector, $\mathbf{x}(n)$. This sequence, called the *a priori error*, is the error that would occur if the filter coefficients were not updated. The *a posteriori error*, on the other hand, is the error that occurs after the weight vector is updated,

$$e(n) = d(n) - \mathbf{w}_n^T\mathbf{x}(n)$$

Table 9.6 ***Recursive Least Squares***

Parameters:	$p =$ Filter order
	$\lambda =$ Exponential weighting factor
	$\delta =$ Value used to initialize $\mathbf{P}(0)$
Initialization:	$\mathbf{w}_0 = \mathbf{0}$
	$\mathbf{P}(0) = \delta^{-1}\mathbf{I}$
Computation:	For $n = 1, 2, \ldots$ compute
	$\mathbf{z}(n) = \mathbf{P}(n-1)\mathbf{x}^*(n)$
	$\mathbf{g}(n) = \dfrac{1}{\lambda + \mathbf{x}^T(n)\mathbf{z}(n)}\mathbf{z}(n)$
	$\alpha(n) = d(n) - \mathbf{w}_{n-1}^T\mathbf{x}(n)$
	$\mathbf{w}_n = \mathbf{w}_{n-1} + \alpha(n)\mathbf{g}(n)$
	$\mathbf{P}(n) = \dfrac{1}{\lambda}\left[\mathbf{P}(n-1) - \mathbf{g}(n)\mathbf{z}^H(n)\right]$

Note that when $\alpha(n)$ is small, the current set of filter coefficients are close to their optimal values (in a least squares sense), and only a small correction needs to be applied to the coefficients. On the other hand, when $\alpha(n)$ is large, the current set of filter coefficients are not performing well in estimating $d(n)$ and a large correction must be applied to update the coefficients.

One final simplification may be realized if we note that, in the evaluation of the gain vector $\mathbf{g}(n)$, and the inverse autocorrelation matrix $\mathbf{P}(n)$, it is necessary to compute the product

$$\boxed{\mathbf{z}(n) = \mathbf{P}(n-1)\mathbf{x}^*(n)} \tag{9.106}$$

Therefore, we may explicitly evaluate this *filtered information vector* and then use it in the calculation of both $\mathbf{g}(n)$ and $\mathbf{P}(n)$. Equations (9.99), (9.100), (9.104), (9.105), and (9.106) form what is known as the exponentially weighted *Recursive Least Squares* (RLS) algorithm, which is summarized in Table 9.6. The special case of $\lambda = 1$ is referred to as the *growing window RLS algorithm.*

The last issue that needs to be addressed concerns the initialization of the RLS algorithm. Since the RLS algorithm involves the recursive updating of the vector $\mathbf{w}_n$ and the inverse autocorrelation matrix $\mathbf{P}(n)$, initial conditions for both of these terms are required. There are two ways that this initialization is typically performed. The first is to build up the autocorrelation matrix recursively according to Eq. (9.97) until it is of full rank (typically $p+1$ input vectors), and then compute the inverse directly

$$\mathbf{P}(0) = \left[\sum_{i=-p}^{0} \lambda^{-i}\mathbf{x}^*(i)\mathbf{x}^T(i)\right]^{-1}$$

Evaluating the cross-correlation vector $\mathbf{r}_{dx}(0)$ in the same manner,

$$\mathbf{r}_{dx}(0) = \sum_{i=-p}^{0} \lambda^{-i} d(i)\mathbf{x}^*(i)$$

we may then initialize $\mathbf{w}_0$ by setting $\mathbf{w}_0 = \mathbf{P}(0)\mathbf{r}_{dx}(0)$. The advantage of this approach is that optimality is preserved at each step since the RLS algorithm is initialized at $n = 0$ with the vector $\mathbf{w}_0$ that minimizes the weighted least squares error $\mathcal{E}(0)$. A disadvantage, however, is that it requires the direct inversion of $\mathbf{R}_x(0)$, which requires on the order of $(p+1)^3$ operations. In addition, with this approach there will be a delay of $p+1$ samples before any updates are performed. Another approach that may be used is to initialize the autocorrelation matrix as follows:

$$\mathbf{R}_x(0) = \delta \mathbf{I}$$

where δ is a small positive constant, i.e., $\mathbf{P}(0) = \delta^{-1}\mathbf{I}$, and to initialize the weight vector to zero,

$$\mathbf{w}_0 = \mathbf{0}$$

The disadvantage of this approach is that it introduces a bias in the least squares solution [20]. However, with an exponential weighting factor, $\lambda < 1$, this bias goes to zero as n increases. A MATLAB program for the RLS algorithmis given in Fig. 9.24.

Unlike the LMS algorithm, which requires on the order of p multiplications and additions, the RLS algorithm requires on the order of p^2 operations. Specifically, the evaluation of $\mathbf{z}(n)$ requires $(p+1)^2$ multiplications, computing the gain vector $\mathbf{g}(n)$ requires $2(p+1)$ multiplications, finding the a priori error $\alpha(n)$ requires another $p+1$ multiplications, and the update of the inverse autocorrelation matrix $\mathbf{P}(n)$ requires $2(p+1)^2$ multiplications for a total of $3(p+1)^2 + 3(p+1)$. There is a similar number of additions. What is gained with this increase in computational complexity over the LMS algorithm, however, is an increase in performance. Generally, the RLS algorithm converges faster than the LMS algorithm and is less sensitive to eigenvalue disparities in the autocorrelation matrix of $x(n)$ for stationary

The RLS Algorithm

```
function [A,E] = rls(x,d,nord,lambda)
%
   delta=0.001;
   X=convm(x,nord);
   [M,N] = size(X);
   if nargin < 4,   lambda = 1.0;   end
   P=eye(N)/delta;
   W(1,:)=zeros(1,N);
   for k=2:M-nord+1;
       z=P*X(k,:)';
       g=z/(lambda+X(k,:)*z);
       alpha=d(k)-X(k,:)*W(k-1,:).';
       W(k,:)=W(k-1,:)+alpha*g.';
       P=(P-g*z.')/lambda;
   end;
```

Figure 9.24 *A MATLAB program for the RLS algorithm.*

processes. On the other hand, without exponential weighting, RLS does not perform very well in tracking nonstationary processes. This is due to the fact that, with $\lambda = 1$, all of the data is equally weighted in estimating the correlations. Although exponential weighting improves the tracking characteristics of RLS, it is not clear how to choose λ and, in some cases, the LMS algorithm may have better tracking properties [6,13] For a more complete discussion of the RLS algorithm, the reader is referred to Reference [20].

Example 9.4.1 *Linear Prediction Using RLS*

In this example, we compare LMS and RLS for the problem of linear prediction introduced in Example 9.2.1 (p. 509). With $x(n)$ generated according to the difference equation

$$x(n) = 1.2728x(n-1) - 0.81x(n-2) + v(n)$$

where $v(n)$ is unit variance white noise, we will consider a second-order RLS predictor of the form

$$\hat{x}(n) = w_n(1)x(n-1) + w_n(2)x(n-2)$$

Shown in Fig. 9.25a is a plot of the predictor coefficients $w_n(1)$ and $w_n(2)$ versus n using the growing window RLS algorithm ($\lambda = 1$). Compared to the LMS algorithm in Fig. 9.10,

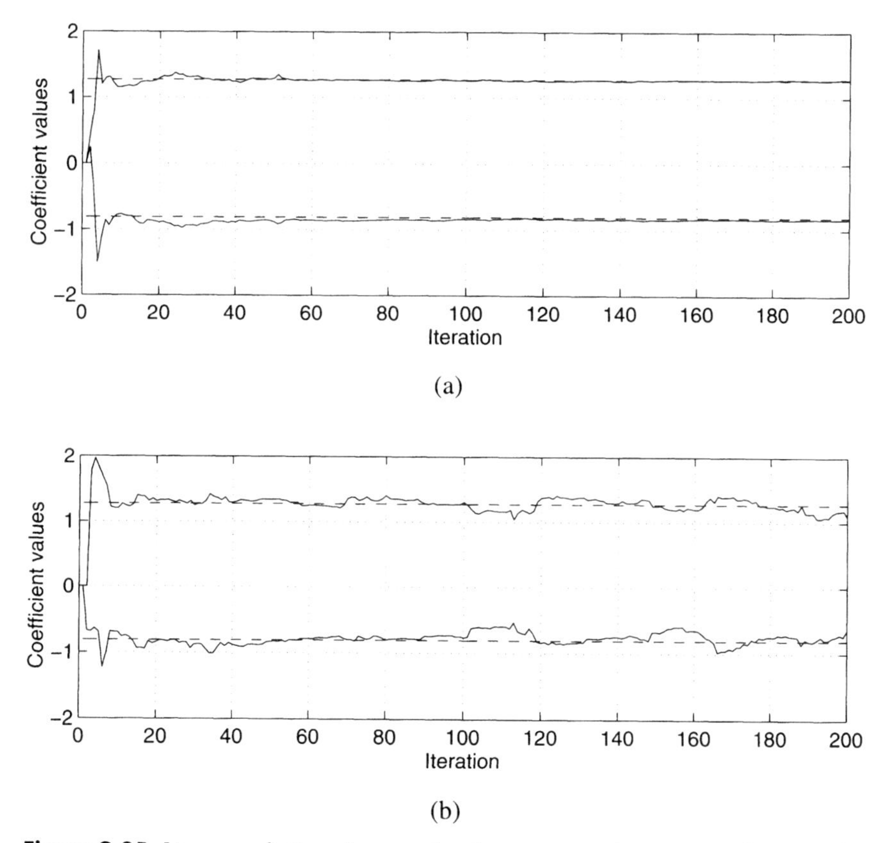

Figure 9.25 *Linear prediction of a second-order autoregressive process using recursive least squares. The dashed lines indicate the correct values for the coefficients. (a) Growing window RLS. (b) Exponentially weighted RLS with $\lambda = 0.95$.*

what is clearly evident is that the RLS algorithm has a much higher rate of convergence. Fig. 9.25*b*, on the other hand, shows the predictor coefficients using the exponential window RLS algorithm with $\lambda = 0.95$. What we observe is an increase in the fluctuations of the weights with the exponential window. This is due to the decrease in the effective length of the data window that is used to evaluate $\mathbf{R}_x(n)$. With a growing window RLS algorithm, if the signal properties are not changing in time then $\mathbf{R}_x(n)$ and the weights $\mathbf{w}_n$ become more stable as n increases.

As a final comparison, let us evaluate what happens if $x(n)$ is generated according to the time-varying difference equation

$$x(n) = a_n(1)x(n-1) - 0.81x(n-2) + v(n)$$

where $v(n)$ is unit variance white noise and $a_n(1)$ is a time-varying coefficient of the following form

$$a_n(1) = \begin{cases} 1.2728 & ; \quad 0 \le n < 100 \\ 0 & ; \quad 100 \le n \le 200 \end{cases}$$

This corresponds to a filter having a pair of poles at radius $r = 0.9$ and angle $\omega_0 = \pi/4$ when $a_n(1) = 1.2728$ and radius $r = 0.9$ and angle $\omega_0 = \pi/2$ when $a_n(1) = 0$. Thus, at time $n = 100$ the poles move from $\omega_0 = \pi/4$ to $\omega_0 = \pi/2$. Shown in Fig. 9.26*a* is a plot of $w_n(1)$ versus n for $\lambda = 1$. What we see is that the growing window RLS algorithm is not able to track the time-variations effectively due to the infinite length window (infinite memory). Shown in Fig. 9.26*b*, on the other hand, is a plot of $w_n(1)$ versus n for $\lambda = 0.9$. Note that with this decrease in the weighting factor, the RLS algorithm is able to more accurately track the true values. Finally, shown in Fig. 9.26*c* is a plot of the coefficients obtained with the LMS algorithm using a step size $\mu = 0.02$. As we see, the performance of the LMS algorithm in tracking nonstationary processes is similar to the exponentially weighted RLS algorithm.

9.4.2 Sliding Window RLS

The RLS algorithm minimizes the exponentially weighted least squares error $\mathcal{E}(n)$ given in Eq. (9.89). With a growing window RLS algorithm ($\lambda = 1$), each of the squared errors $|e(i)|^2$ from $i = 0$ to $i = n$ are equally weighted, whereas with an exponentially weighted RLS algorithm ($\lambda < 1$), the squared errors $|e(i)|^2$ become less important for values of i that are small compared to n. In both cases, however, the RLS algorithm has infinite memory in the sense that all of the data beginning from time $n = 0$ will affect the values of the coefficients $\mathbf{w}_n$. In some applications this may not be desirable. For example, for a nonstationary process whose statistics are changing rapidly in time, a small value for λ will be necessary in order to track the process. An alternative, however, is to minimize the sum of the squares of $e(i)$ over a finite window,[8]

$$\mathcal{E}_L(n) = \sum_{i=n-L}^{n} |e(i)|^2 \tag{9.107}$$

In addition to being able to more easily track nonstationary processes, the finite window allows for any data *outliers* to be *forgotten* after a finite number of iterations.

[8]Note that in the special case of $L = 0$, a gradient descent approach to minimize $\mathcal{E}_0(n)$ results in the LMS algorithm.

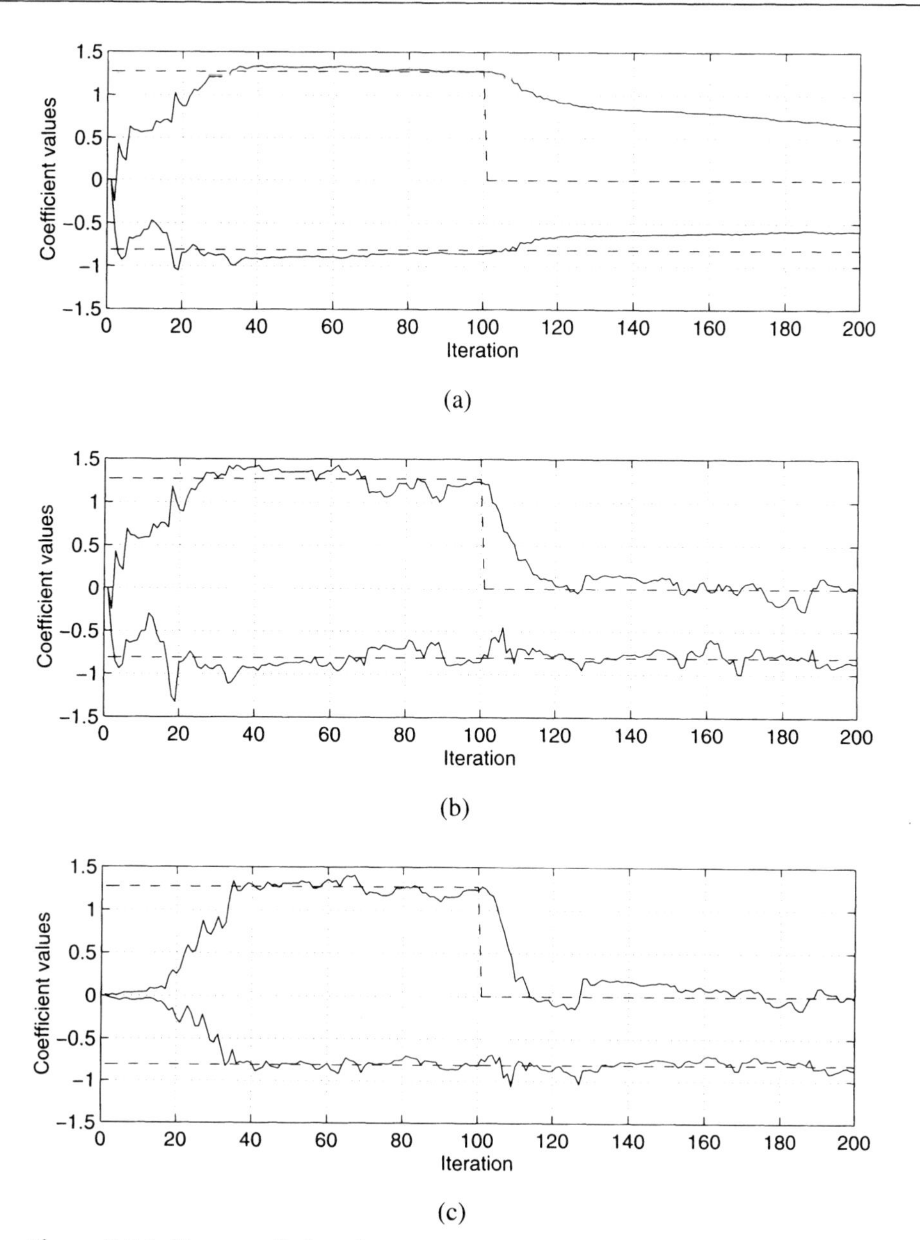

Figure 9.26 *Linear prediction of a nonstationary second-order autoregressive random process. The dashed lines indicate the correct values for the coefficients. (a) Growing window RLS. (b) Exponentially weighted RLS with $\lambda = 0.9$. (c) LMS algorithm with a step size $\mu = 0.02$.*

Following the same steps used to derive the deterministic normal equations for minimizing the exponentially weighted squared error, the filter coefficients $\mathbf{w}_n$ that minimize $\mathcal{E}_L(n)$ are found to be solutions to the equations

$$\mathbf{R}_x(n)\mathbf{w}_n = \mathbf{r}_{dx}(n) \tag{9.108}$$

where

$$\mathbf{R}_x(n) = \sum_{i=n-L}^{n} \mathbf{x}^*(i)\mathbf{x}^T(i) \tag{9.109}$$

and

$$\mathbf{r}_{dx}(n) = \sum_{i=n-L}^{n} d(i)\mathbf{x}^*(i)$$

Using a procedure similar to that used to derive the growing window RLS algorithm, a *sliding window* RLS algorithm may be developed that solves Eq. (9.108) recursively with a computational complexity on the order of p^2 operations. The sliding window RLS algorithm consists of the following two steps.

1. Given the solution $\mathbf{w}_{n-1}$ to Eq. (9.108) at time $n-1$, with the new data value $x(n)$, the weight vector $\widetilde{\mathbf{w}}_n$ is found that minimizes the error

$$\mathcal{E}_{L+1}(n) = \sum_{i=n-L-1}^{n} |e(i)|^2 \tag{9.110}$$

2. The weight vector $\mathbf{w}_n$ that minimizes $\mathcal{E}_L(n)$ is then determined by discarding the last data point, $x(n-L-1)$.

In the first step, note that we find the vector $\widetilde{\mathbf{w}}_n$ that minimizes the error $\mathcal{E}_{L+1}(n)$ that results from the addition of one additional data value. Therefore, we may use the growing window RLS algorithm to find $\widetilde{\mathbf{w}}_n$ from $\mathbf{w}_{n-1}$ as follows:

$$\begin{aligned}
\mathbf{g}(n+1) &= \frac{1}{1+\mathbf{x}^T(n)\mathbf{P}(n-1)\mathbf{x}(n)}\mathbf{P}(n-1)\mathbf{x}^*(n) \\
\widetilde{\mathbf{w}}_n &= \mathbf{w}_{n-1} + \mathbf{g}(n)[d(n) - \mathbf{w}_{n-1}\mathbf{x}^*(n)] \\
\widetilde{\mathbf{P}}(n) &= \mathbf{P}(n-1) - \mathbf{g}(n)\mathbf{x}^T(n)\mathbf{P}(n-1)
\end{aligned}$$

Note that $\widetilde{\mathbf{P}}(n)$ is the inverse of the matrix

$$\widetilde{\mathbf{R}}_x(n) = \sum_{k=n-L-1}^{n} \mathbf{x}^*(k)\mathbf{x}^T(k)$$

which is based on $L+2$ data values, and that $\widetilde{\mathbf{w}}_n$ is the solution to

$$\widetilde{\mathbf{R}}_x(n)\widetilde{\mathbf{w}}_n = \widetilde{\mathbf{r}}_{dx}(n)$$

where

$$\widetilde{\mathbf{r}}_{dx}(n) = \sum_{k=n-L-1}^{n+1} d(k)\mathbf{x}^*(k)$$

In the second step of the recursion, the last data point $x(n-L-1)$ is discarded to restore the $(L+1)$ point window. This is accomplished with a slight modification to the growing window RLS algorithm. Specifically, we begin with the matrix update

$$\mathbf{R}(n) = \widetilde{\mathbf{R}}(n) - \mathbf{x}^*(n-L)\mathbf{x}^T(n-L-1)$$

and the update of $\mathbf{r}_{dx}(n)$

$$\mathbf{r}_{dx}(n) = \widetilde{\mathbf{r}}_{dx}(n) - d(n-L-1)\mathbf{x}^*(n-L-1)$$

Finally, using the matrix inversion lemma and retracing the steps used to derive the RLS algorithm we obtain the following update equations

$$\begin{aligned}
\widetilde{\mathbf{g}}(n+1) &= \frac{1}{1-\mathbf{x}^T(n-L-1)\widetilde{\mathbf{P}}(n)\mathbf{x}^*(n-L-1)}\widetilde{\mathbf{P}}(n)\mathbf{x}^*(n-L-1) \\
\mathbf{w}_n &= \widetilde{\mathbf{w}}_n - \widetilde{\mathbf{g}}(n)[d(n-L-1) - \widetilde{\mathbf{w}}_n\mathbf{x}^*(n-L-1)] \\
\mathbf{P}(n) &= \widetilde{\mathbf{P}}(n) + \widetilde{\mathbf{g}}(n)\mathbf{x}^T(n-L-1)\widetilde{\mathbf{P}}(n)
\end{aligned}$$

which yields the desired solution and completes the derivation of the sliding window RLS algorithm. Compared to exponentially weighted RLS, the sliding window RLS requires about twice the number of multiplications and additions and, in addition, requires that $p+L$ values of $x(n)$ be stored. This storage requirement may potentially be a problem for long windows.

9.4.3 Summary

In this chapter, we have looked at a number of techniques for processing nonstationary signals using adaptive filters. These techniques have been extensively used in a variety of applications including system identification, signal modeling, spectrum estimation, noise cancellation, and adaptive equalization. Our treatment of adaptive filtering began with the design of an FIR filter to minimize the mean-square error $\xi(n) = E\{|e(n)|^2\}$ between a desired process $d(n)$ and an estimate of this process that is formed by filtering another process $x(n)$. First, we considered a steepest descent approach to minimize $\xi(n)$. However, since the gradient of $\xi(n)$ involves an expectation of $e(n)\mathbf{x}^*(n)$, this approach requires knowledge of the statistics of $x(n)$ and $d(n)$ and, therefore, is of limited use in practice. Next, we replaced the ensemble average $E\{|e(n)|^2\}$ with the instantaneous squared error $|e(n)|^2$. This led to the LMS algorithm, a simple and often effective algorithm that does not require any ensemble averages to be known. For wide-sense stationary processes, the LMS algorithm converges in the mean if the step size is positive and no larger than $2/\lambda_{\max}$, and it converges in the mean-square if the step size is positive and no larger than $2/\mathrm{tr}(\mathbf{R}_x)$. We then looked at several modifications of the LMS algorithm. The first was the normalized LMS algorithm, which simplifies the selection of the step size to ensure that the coefficients converge. Next was the leaky LMS algorithm, which is useful in overcoming the problems that occur when the autocorrelation matrix of the input process is singular. We then looked at the block LMS algorithm and the sign algorithms, which are designed to increase the efficiency of the LMS algorithm. In the block LMS algorithm, the filter coefficients are held constant over blocks of length L, which allows for the use of fast convolution algorithms to compute the filter output. The sign algorithms, on the other hand, achieve their simplicity by replacing $e(n)$ with $\mathrm{sgn}\{e(n)\}$, or $\mathbf{x}(n)$ with $\mathrm{sgn}\{\mathbf{x}(n)\}$, or both. Finally, we discussed how it would be beneficial to consider using a variable step size in the LMS algorithm. Ideally, when the filter coefficients are far from their optimum values, the step size should be large. Then, as the cofficients begin to converge to the optimum solution, the step size should be small in order to reduce the excess mean-square error. It is difficult, however, to define a set of rules for adapting the step size so that the adaptive filter will have a small excess mean-square error while, at the same time, maintaining the ability of the filter to respond quickly to changes in the signal statistics. Finally, we looked at the lattice filter as an alternative structure to use as an adaptive filter. Due to the orthogonalization of the input process, the gradient adaptive lattice filter converges more rapidly than the LMS adaptive filter, and tends to be less sensitive to the eigenvalue spread in the autocorrelation matrix of $x(n)$.

Following the FIR adaptive filters, we then turned our attention briefly to the design of adaptive recursive filters. These filters are much more complex than the FIR filters. In addition, since the mean-square error $\xi(n)$ is a nonlinear function of the filter coefficients, the adaptive filter may not converge or it may converge to a local rather than the global minimum. Also, since they are recursive, it is possible for these filters to become unstable. As a result, adaptive recursive filters are not as widely used in applications as their FIR counterparts.

This chapter concluded with a derivation of the recursive least squares algorithm to minimize a deterministic least squares error. Three different forms of RLS algorithm were derived: the growing window RLS algorithm, the exponentially weighted RLS algorithm, and the sliding window RLS algorithm. Although computationally more complex than the LMS adaptive filter, for wide-sense stationary processes the growing window RLS algorithm converges much more rapidly. However, in order to effectively track a nonstationary process, it is necessary to use either the exponentially weighted RLS algorithm or the sliding window RLS algorithm.

The treatment of adaptive filtering algorithms in this chapter is, by no means, complete. Many other approaches to FIR and IIR adaptive filtering have been developed and many papers have been published that analyze the performance of adaptive filtering algorithms and evaluate their effectiveness in applications [30]. A few of the more notable omissions include the fast RLS recursive algorithms [2,8,7], recursive least squares lattice filters [16,23], and nonlinear adaptive filters [25]. For a general and more comprehensive treatment of adaptive filters, one may consult any one of a number of excellent textbooks [1,3,10,20,37].

References

1. S. T. Alexander, *Adaptive Signal Processing: Theory and Applications*, Springer-Verlag, New York, 1986.
2. S. T. Alexander, "Fast adaptive filters: A geometric approach," *IEEE ASSP Mag.*, vol. 3, no. 4, pp. 18–28, October 1986.
3. M. G. Bellanger, *Adaptive Digital Filters and Signal Analysis*, Marcel Dekker, Inc., New York, 1987.
4. N. J. Bershad, "Analysis of the normalized LMS algorithm with Gaussian inputs," *IEEE Trans. Acoust., Speech, Sig. Proc.*, vol. ASSP-34, pp. 793–806, 1986.
5. N. J. Bershad, "On the optimum data non-linearity in LMS adaptation," *IEEE Trans. Acoust., Speech, Sig. Proc.*, vol. ASSP-34, pp. 69–76, February 1986.
6. N. J. Bershad and O. Macchi, "Comparison of RLS and LMS algorithms for tracking a chirped signal," *Proc. 1989 Int. Conf. Acoust., Speech, Sig. Proc.*, pp. 896–899, 1989.
7. G. Carayannis, D. G. Manolakis, and N. Kalouptsidis, "A fast sequential algorithm for least-squares filtering and prediction," *IEEE Trans. Acoust., Speech, Sig. Proc.*, vol. ASSP-31, no. 6, pp. 1394–1402, December 1983.
8. J. M. Cioffi and T. Kailath, "Fast recursive-least-squares transversal filters for adaptive filtering," *IEEE Trans. Acoust., Speech, Sig. Proc.*, vol. ASSP-32, no. 2, pp. 304–337, April 1984.
9. G. A. Clark, S. K. Mitra, and S. R. Parker, "Block implementation of adaptive digital filters," *IEEE Trans. Circuits and Systems*, vol. CAS-28, pp. 584–592, 1981.
10. P. M. Clarkson, *Optimal and Adaptive Signal Processing*, CRC Press, Boca Raton, FL, 1993.
11. T. Claasen and W. Mecklenbrauker, "Comparison of the convergence of two algorithms for adaptive FIR digital filters," *IEEE Trans. Acoust., Speech, Sig. Proc.*, vol. ASSP-29, pp. 670–678, June 1981.

12. C. F. N. Cowan and P. M. Grant, *Adaptive Filters*, Prentice-Hall, Englewood Cliffs, NJ, 1985.

13. E. Eleftheriou and D. Falconer, "Tracking properties and steady state performance of RLS adaptive filtering algorithms," *IEEE Trans. Acoust., Speech, Sig. Proc.*, vol. ASSP-34, no. 5, pp. 1097–1110, October 1986.

14. S. J. Elliott and P. A. Nelson, "Active noise control," *Sig. Proc. Magazine*, vol. 10, no. 4, October 1993, pp. 12–35.

15. P. L. Feintuch, "An adaptive recursive LMS filter," *Proc. IEEE*, pp. 1622–1624, November 1976.

16. B. Freidlander, "Lattice filters for adaptive processing," *Proc. IEEE*, vol. 70, no. 8, pp. 829–867, August 1982.

17. L. J. Griffiths, "A continuously-adaptive filter implemented as a lattice structure," *Proc. 1977 Int. Conf. Acoust., Speech, Sig. Proc.*, pp. 683–686, 1977.

18. L. J. Griffiths, "An adaptive lattice structure for noise cancellation," *Proc. 1978 Int. Conf. Acoust., Speech, Sig. Proc.*, pp. 87–90, 1978.

19. R. W. Harris, D. M. Chabries, and F. A. Bishop, "A variable step (VS) adaptive filter algorithm," *IEEE Trans. on Acoust., Speech, Sig. Proc.*, vol. ASSP-34, no. 2, pp. 309–316, April 1986.

20. S. Haykin, *Adaptive Filter Theory*, Prentice-Hall, Englewood-Cliffs, NJ, 1986.

21. M. L. Honig and D. G. Messerschmitt, "Convergence properties of an adaptive digital lattice filter," *IEEE Trans. Acoust., Speech, Sig. Proc.*, vol. ASSP-29, pp. 642–653, 1981.

22. S. Horvath Jr., "A new adaptive recursive LMS filter," in *Digital Signal Processing*, V. Cappellini and A.G. Constantinides, Eds., Academic Press, New York, pp. 21–26.

23. D. T. L. Lee, M. Morf, and B. Friedlander, "Recursive least squares ladder estimation algorithms," *IEEE Trans. Acoust., Speech, Sig. Proc.*, vol. ASSP-29, no. 3, pp. 467–481, June 1981.

24. R. W. Lucky, J. Salz, and E. J. Weldon Jr., *Principles of Data Communications*, McGraw-Hill, New York, 1968.

25. I. Pitas and A. N. Venetsanopoulos, *Nonlinear digital filters*, Kluwer Academic Publishers, Boston, 1990.

26. J. G. Proakis, *Digital Communication*, McGraw-Hill, New York, 1983.

27. E. H. Satorius and S. T. Alexander, "Channel equalization using adaptive lattice algorithms," *IEEE Trans. Comm.*, vol. COM-27, pp. 899–905, 1979.

28. W. A. Sethares, I. M. Y. Mareels, B. D. O. Anderson, C. R. Johnson, and R. R. Bitmead, "Excitation conditions for signed regressor least mean-squares adaptation," *IEEE Trans. Circuits Syst.*, vol. CAS-25, pp. 613–624, 1988.

29. J. Shynk, "Adaptive IIR filtering," *IEEE ASSP Magazine*, pp. 4–21, April 1989.

30. L. H. Sibul, Ed., *Adaptive Signal Processing*, IEEE Press, New York, 1987.

31. S. Stearns, "Error surfaces of recursive adaptive filters," *IEEE Trans. Acoust., Speech, Sig. Proc.*, vol. ASSP-29, pp. 763–766, 1981.

32. J. R. Treichler, "Adaptive algorithms for infinite impulse response filters," in *Adaptive Filters*, C. F. Cowan and P. M. Grant (Eds.), Prentice-Hall, Englewood Cliffs, NJ, 1985.

33. J. R. Treichler, C. R. Johnson Jr., and M. G. Larimore, *Theory and Design of Adaptive Filters*, John Wiley & Sons, 1987.

34. A. Weiss and D. Mitra, "Digital adaptive filters: Conditions for convergence, rates of convergence, effects of noise and errors arising from the implementation," *IEEE Trans. on Inf. Theory*, vol. IT-25, pp. 637–652.

35. S. White, "An adaptive recursive digital filter" *Proc. 9th Asilomar Conf. Circuits, Syst.*, pp. 21–25, 1975.

36. B. Widrow et. al., "Stationary and nonstationary learning characteristics of the LMS adaptive filter," *Proc. IEEE*, vol. 64, pp. 1151–1162, 1976.

37. B. Widrow and S. Stearns, *Adaptive Signal Processing*, Prentice-Hall, Englewood Cliffs, NJ, 1985.

9.5 PROBLEMS

9.1. In order for the steepest descent algorithm to converge, the step size must be in the range

$$0 < \mu < 2/\lambda_{\max}$$

However, in some cases it may be of interest to find the value of μ that gives the fastest rate of convergence. For a given step size μ, the rate of convergence for the weight vector $\mathbf{w}_n$ is dominated by the slowest converging mode in the expansion

$$\mathbf{w}_n = \mathbf{w} + \sum_{k=0}^{p} (1 - \mu\lambda_k)^n u_0(k)\mathbf{v}_k$$

(a) In terms of the eigenvalues λ_k, find the value for μ that maximizes the rate of convergence. In other words, find the value for μ that maximizes the rate of convergence of the slowest decaying mode.

(b) At what rate does the slowest mode decrease for the step size found in part (a)?

9.2. Suppose that the input to an adaptive linear predictor is white noise with an autocorrelation sequence $r_x(k) = \sigma_x^2\delta(k)$.

(a) Solve the normal equations and find the optimum pth-order one-step linear predictor, $\mathbf{w}$.

(b) Minimize the mean-square prediction error using the method of steepest descent with a step size $\mu = 1/(5\sigma_x^2)$ and an initial weight vector $\mathbf{w}_0 = [1, 1, \ldots, 1]^T$. Does the method of steepest descent converge to the solution found in part (a)?

9.3. Newton's method is an iterative algorithm that may be used to find the minimum of a nonlinear function. Applied to the minimization of the mean-square error

$$\xi(n) = E\{e^2(n)\}$$

where $e(n) = d(n) - \mathbf{w}^T\mathbf{x}(n)$, Newton's method is

$$\mathbf{w}_{n+1} = \mathbf{w}_n - \tfrac{1}{2}\mathbf{R}_x^{-1}\nabla\xi(n)$$

where $\mathbf{R}_x$ is the autocorrelation matrix of $x(n)$. Introducing a step-size parameter μ, Newton's method becomes

$$\mathbf{w}_{n+1} = \mathbf{w}_n - \tfrac{1}{2}\mu\mathbf{R}_x^{-1}\nabla\xi(n)$$

Comparing this to the steepest descent algorithm, we see that the step size μ is replaced with a matrix, $\mu\mathbf{R}_x^{-1}$, which alters the descent direction.

(a) For what values of μ is Newton's method stable, i.e., for what values of μ will $\mathbf{w}_n$ converge?

(b) What is the optimum value of μ, i.e., for what value of μ is the convergence the fastest?

(c) Suppose that we form an LMS version of Newton's method by replacing the gradient with a gradient estimate

$$\hat{\nabla}\xi(n) = \nabla e^2(n)$$

Derive the coefficient update equation that results from using this gradient estimate and describe how it differs from the LMS algorithm.

(d) Derive an expression that describes the time evolution of $E\{\mathbf{w}_n\}$ using the LMS Newton algorithm derived in part (c).

9.4. One way to derive the steepest descent algorithm for solving the normal equations $\mathbf{R}_x\mathbf{w} = \mathbf{r}_{dx}$ is to use a power series expansion for the inverse of $\mathbf{R}_x$. This expansion is

$$\mathbf{R}_x^{-1} = \mu \sum_{k=0}^{\infty} (\mathbf{I} - \mu \mathbf{R}_x)^k$$

where $\mathbf{I}$ is the identity matrix and μ is a positive constant. In order for this expansion to converge, $\mathbf{R}_x$ must be positive definite and the constant μ must lie in the range

$$0 < \mu < 2/\lambda_{\max}$$

where $\lambda_{\max}$ is the largest eigenvalue of $\mathbf{R}_x$.

(a) Let

$$\mathbf{R}_x^{-1}(n) = \mu \sum_{k=0}^{n} (\mathbf{I} - \mu \mathbf{R}_x)^k$$

be the nth-order approximation to $\mathbf{R}_x^{-1}$, and let

$$\mathbf{w}_n = \mathbf{R}_x^{-1}(n)\mathbf{r}_{dx}$$

be the nth-order approximation to the desired solution $\mathbf{w} = \mathbf{R}_x^{-1}\mathbf{r}_{dx}$. Express $\mathbf{R}_x^{-1}(n+1)$ in terms of $\mathbf{R}_x^{-1}(n)$, and show how this may be used to derive the steepest descent algorithm

$$\mathbf{w}_{n+1} = \mathbf{w}_n - \mu\big[\mathbf{R}_x\mathbf{w}_n - \mathbf{r}_{dx}\big]$$

(b) If the statistics of $x(n)$ are unknown, then $\mathbf{R}_x$ is unknown and the expansion for $\mathbf{R}_x^{-1}$ in part (a) cannot be evaluated. However, suppose that we approximate $\mathbf{R}_x = E\{\mathbf{x}(n)\mathbf{x}^T(n)\}$ at time n as follows

$$\hat{\mathbf{R}}_x(n) = \mathbf{x}(n)\mathbf{x}^T(n)$$

and use, as the nth-order approximation to $\mathbf{R}_x^{-1}$,

$$\hat{\mathbf{R}}_x^{-1}(n) = \mu \sum_{k=0}^{n} \Big[\mathbf{I} - \mu \mathbf{x}(k)\mathbf{x}^T(k)\Big]^k$$

Express $\hat{\mathbf{R}}_x^{-1}(n+1)$ in terms of $\hat{\mathbf{R}}_x^{-1}(n)$ and use this expression to derive a recursion for $\mathbf{w}_n$.

(c) Compare your recursion derived in part (b) to the LMS algorithm.

9.5. The convergence of a pth-order LMS adaptive filter depends on the eigenvalues of the autocorrelation matrix, $\mathbf{R}_x$, of the input process $x(n)$. These eigenvalues, in turn, depend upon the size, p, of $\mathbf{R}_x$. For example, it follows from the Bordering Theorem (see p. 48) that the maximum eigenvalue is a monotonically nondecreasing function of p, and the minimum eigenvalue is a monotonically nonincreasing function of p. In addition, it follows from the eigenvalue extremal property (see p. 97) that the maximum and minimum eigenvalues approach the maximum and minimum values of the power spectrum, $P_x(e^{j\omega})$, as $p \to \infty$,

$$\lambda_{\max} \to \max_{\omega} P_x(e^{j\omega}) \quad ; \quad \lambda_{\min} \to \min_{\omega} P_x(e^{j\omega})$$

Suppose that the input to an adaptive filter has an autocorrelation

$$r_x(k) = \alpha^{|k|} \quad ; \quad |\alpha| < 1$$

(a) Find the eigenvectors and eigenvalues of the 2×2 autocorrelation matrix $\mathbf{R}_x = \text{Toep}\{1, \alpha\}$. (Your answer will be in terms of α).

(b) Find the asymptotic values for the maximum and minimum eigenvalues of the $p \times p$ autocorrelation matrix $\mathbf{R}_x$ as $p \to \infty$.

(c) Find, as a function of α, the largest step size μ for convergence in the mean of the LMS algorithm, and find the slowest converging mode (assume that p is large).

9.6. The *condition number*, χ, of an autocorrelation matrix $\mathbf{R}_x$ may be bounded in terms of the power spectrum of the process $P_x(e^{j\omega})$ as follows (see Prob. 9.5),

$$\chi = \frac{\lambda_{\max}}{\lambda_{\min}} \le \frac{\max\limits_{\omega} P_x(e^{j\omega})}{\min\limits_{\omega} P_x(e^{j\omega})}$$

(a) Use this inequality to bound the condition number of the autocorrelation matrix for the moving average process

$$x(n) = w(n) + \alpha w(n-1)$$

where $w(n)$ is unit variance white noise.

(b) Repeat part (a) for the autoregressive process

$$x(n) = \alpha x(n-1) + w(n)$$

where $|\alpha| < 1$ and $w(n)$ is unit variance white noise.

(c) What does this bound imply about the performance of an adaptive filter when the input to the filter is a lowpass process with a power spectrum of the form

$$P_x(e^{j\omega}) = \begin{cases} 1 & ; \quad |\omega| < \omega_0 \\ 0 & ; \quad \omega_0 \le |\omega| \le \pi \end{cases}$$

9.7. Suppose that the input to an FIR LMS adaptive filter is a first-order autoregressive process with an autocorrelation

$$r_x(k) = c\alpha^{|k|}$$

where $c > 0$ and $0 < \alpha < 1$. Suppose that the step size μ is

$$\mu = \frac{1}{5\lambda_{\max}}$$

(a) How does the rate of convergence of the LMS algorithm depend upon the value of α?

(b) What effect does the value of c have on the rate of convergence?

(c) How does the rate of convergence of the LMS algorithm depend upon the desired signal $d(n)$?

9.8. The coefficient update equation for the LMS adaptive filter applies a correction to the weight $\mathbf{w}_n$ at time n as follows

$$\mathbf{w}_{n+1} = \mathbf{w}_n + \mu e(n)\mathbf{x}^*(n)$$

The *Block LMS* algorithm, on the other hand, accumulates these corrections for L samples, beginning at time n, while holding the weight vector $\mathbf{w}_n$ constant. A correction is then applied at the end of the block to form an update at time $n + L$ as follows

$$\mathbf{w}_{n+L} = \mathbf{w}_n + \mu \sum_{l=0}^{L-1} e(n+l)\mathbf{x}^*(n+l)$$

where

$$e(n+l) = d(n+l) - \mathbf{w}_n^T \mathbf{x}(n+l)$$

for $l = 0, 1, \ldots, L-1$.

(a) By evaluating the behavior of $E\{\mathbf{w}_n\}$ as a function of n, determine the conditions on the step size μ that are necessary for the block LMS algorithm to converge in the mean.

(b) Discuss the advantages and/or disadvantages of the block LMS algorithm compared to the standard LMS algorithm.

9.9. The first three autocorrelations of a process $x(n)$ are

$$r_x(0) = 1, \qquad r_x(1) = 0.5, \qquad r_x(2) = 0.5$$

Design a two-coefficient LMS adaptive linear predictor for $x(n)$ that has a misadjustment

$$\mathcal{M} = 0.05$$

and find the steady-state mean-square error.

9.10. Consider the single-weight adaptive filter shown in the figure below

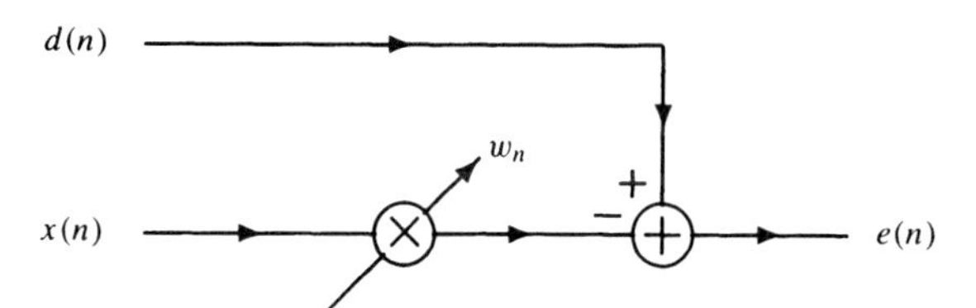

(a) Write down the LMS algorithm for updating the weight w.

(b) Suppose that $x(n)$ is a constant:

$$x(n) = \begin{cases} K & ; \quad n \geq 0 \\ 0 & ; \quad \text{otherwise} \end{cases}$$

Find the system function relating $d(n)$ to $e(n)$ using the LMS algorithm, i.e., find $H(z)$ in the figure below.

(c) Determine the range of values for μ for which $H(z)$ is stable.

9.11. The LMS adaptive filter minimizes the instantaneous squared error

$$\xi(n) = |e(n)|^2$$

Consider the modified functional

$$\xi'(n) = |e(n)|^2 + \beta \mathbf{w}_n^H \mathbf{w}_n$$

where $\beta > 0$.

(a) Derive the LMS coefficient update equation for $\mathbf{w}_n$ that minimizes $\xi'(n)$.

(b) Determine the condition on the step size μ that will ensure that $\mathbf{w}_n$ converges in the mean.

(c) If μ is small enough so that $\mathbf{w}_n$ converges in the mean, what does $\mathbf{w}_n$ converge to?

9.12. Show that the normalized LMS algorithm is equivalent to using the update equation

$$\mathbf{w}_{n+1} = \mathbf{w}_n + \mu e'(n)\mathbf{x}^*(n)$$

where $e'(n)$ is the error at time n that is based on the new filter coefficients $\mathbf{w}_{n+1}$,

$$e'(n) = d(n) - \mathbf{w}_{n+1}^T \mathbf{x}(n)$$

Discuss the relationship between μ and the parameter ϵ in the normalized LMS algorithm.

9.13. A process $x(n)$ is formed by passing white noise $w(n)$ through a filter that has a system function

$$H(z) = \frac{1}{1 - 0.08z^{-1} - 0.9z^{-2}}$$

The variance of the white noise is $\sigma_w^2 = (0.19)(0.18)$. The LMS algorithm with two coefficients is used to estimate a process $d(n)$ from $x(n)$.

(a) What is the maximum value for the step size, μ, in order for the LMS algorithm to converge in the mean? Hint: Use the inverse Levinson-Durbin recursion to find the autocorrelation sequence of $x(n)$.

(b) What is the time constant for convergence?

(c) What value for the step size would you use to maximize the rate of convergence of the weights?

(d) If the cross-correlation between $x(n)$ and $d(n)$ is zero,

$$E\{d(n)\mathbf{x}^*(n)\} = 0$$

what are the optimum filter coefficients $\mathbf{w} = [w(0), w(1)]^T$?

9.14. Griffiths developed an algorithm for adaptive beamforming referred to as the p-vector algorithm that eliminates the need for a reference signal $d(n)$ [17]. This algorithm may be derived as follows. Recall that the filter coefficient update equation for the LMS algorithm is

$$\mathbf{w}_{n+1} = \mathbf{w}_n + \mu e(n)\mathbf{x}^*(n) = \mathbf{w}_n + \mu d(n)\mathbf{x}^*(n) - \mu\big[\mathbf{w}_n^T \mathbf{x}(n)\big]\mathbf{x}^*(n)$$

Note that $d(n)$ is not explicitly needed in this update equation. Instead, what is required is the product $d(n)\mathbf{x}^*(n)$. Therefore, if $d(n)\mathbf{x}^*(n)$ is replaced with its expected value, $\mathbf{r}_{dx} = E\big\{d(n)\mathbf{x}^*(n)\big\}$, or by an estimate of this ensemble average, then we have an update equation that does not require knowledge of $d(n)$,

$$\mathbf{w}_{n+1} = \mathbf{w}_n + \mu \mathbf{r}_{dx} - \mu\big[\mathbf{w}_n^T \mathbf{x}(n)\big]\mathbf{x}^*(n)$$

This is the p-vector algorithm proposed by Griffiths.[9]

[9]The name p-vector algorithm comes from the fact that the vector $\mathbf{p}$ was used to denote the cross-correlation between $d(n)$ and $x(n)$.

(a) What constraints must be placed on the step size μ in order for $\mathbf{w}_n$ to converge in the mean?

(b) Develop a "leaky" p-vector algorithm and determine the range of values for μ for convergence in the mean. Assuming that μ is selected so that $\mathbf{w}_n$ converges in the mean, find $\lim_{n\to\infty} E\{\mathbf{w}_n\}$.

9.15. An FIR filter with system function $H(z)$ may be implemented using a frequency sampling structure as follows

$$H(z) = \frac{1 - z^{-M}}{M} \sum_{k=0}^{M-1} \frac{H(k)}{1 - e^{j2\pi k/M} z^{-1}} = H_1(z) H_2(z)$$

where $H_1(z)$ is a comb filter that has M zeroes equally spaced around the unit circle and $H_2(z)$ is a filterbank of resonators where the coefficients $H(k)$ are the DFT coefficients of $h(n)$, i.e.,

$$H(k) = \sum_{n=0}^{M-1} h(n) e^{-j2\pi kn/M}$$

Suppose that this structure is implemented as an adaptive filter using the LMS algorithm to adjust the filter (DFT) coefficients, $H(k)$. Derive the time-update equation for these coefficients and sketch the adaptive filter structure.

9.16. In many signal processing applications, it is important for a filter to have *linear phase*. This is particularly true in speech and image processing applications, where phase distortion produced by a filter may severely degrade the signal. Therefore, suppose that we would like to design an adaptive linear phase filter whose weights at time n satisfy the following symmetry constraint

$$w_n(k) = w_n(p-k) \quad ; \quad k = 0, 1, \ldots, p$$

For example, consider the two-coefficient linear phase adaptive filter shown in the figure below.

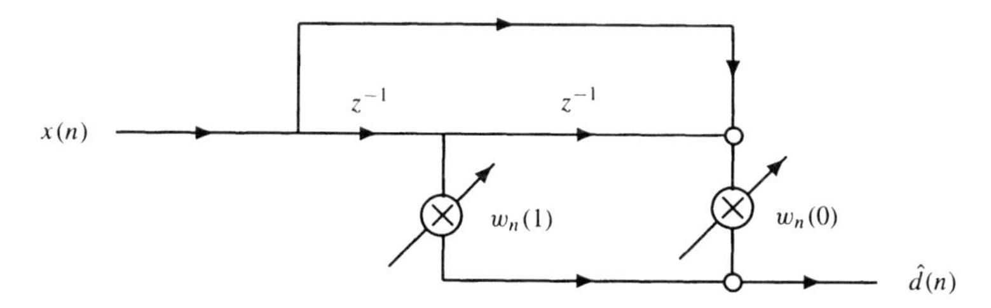

Note that this filter may be viewed as an FIR adaptive filter that is constrained to have the first coefficient equal to the third. As before, define the weight vector, $\mathbf{w}_n$, by

$$\mathbf{w}_n = [w_n(0), w_n(1)]^T$$

and the error sequence, $e(n)$, by

$$e(n) = d(n) - \hat{d}(n)$$

and assume that

$$r_x(0) = 1, \qquad r_x(1) = 0.5, \qquad r_x(2) = 0.1$$

(a) Derive the normal equations for the filter that minimizes the mean-square error

$$\xi = E\{[d(n) - \hat{d}(n)]^2\}$$

(b) Derive the LMS filter coefficient update equations for this constrained transversal filter.

(c) What values for the step size μ may be used if the weights are to converge in the *mean-square* sense?

(d) If we drop the linear phase constraint and use a three-coefficient LMS adaptive filter with $\mathbf{w} = [w(0), w(1), w(2)]^T$, what values of the step size μ may be used if the adaptive filter is to converge in the mean-square sense?

9.17. In some applications it is known that a given complex-valued process has a constant envelope, e.g. phase modulated signals. The *Constant Modulus Algorithm* (CMA) is an adaptive filtering technique that adjusts the filter coefficients in order to minimize the envelop variation. Given a complex signal $x(n)$ and a set of complex weights $w_n(k)$ at time n, the output of the adaptive filter is

$$y(n) = \mathbf{w}_n^H \mathbf{x}(n)$$

With the constant modulus algorithm, the weights are to be found that minimize the error

$$\xi(n) = \tfrac{1}{4} E\{(|y(n)|^2 - 1)^2\}$$

which is a non-negative measure of the average amount that the envelope of the filter output $y(n)$ deviates from a nominal level (unity in this case). Using Widrow's approach of estimating ensemble averages with one point sample averages, derive the CMA algorithm, which is an LMS version of a steepest descent algorithm to minimize the error $\xi(n)$ defined above.

9.18. The output $d(n)$ of an unknown system is given by

$$d(n) = \sum_{k=0}^{p} w(k)x(n-k) + v(n) = \mathbf{w}^T \mathbf{x}(n) + v(n)$$

where $w(k)$ are the unknown system parameters, $x(n)$ is the system input, and $v(n)$ is zero mean white noise with a variance of σ_v^2. The block diagram below shows an adaptive filter that is used to model the unknown system.

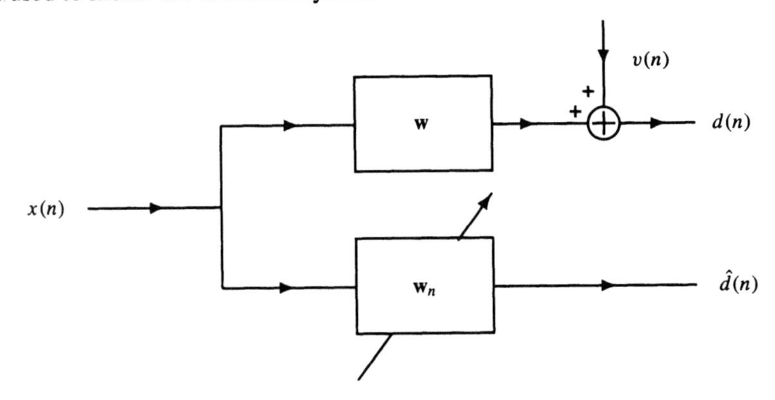

Assume that $x(n)$ is real-valued and suppose that we would like to find the weight vector $\mathbf{w}_n$ that minimizes the error

$$\xi(k) = E\{e^{2k}(n)\}$$

for some positive integer k where $e(n) = d(n) - \hat{d}(n)$.

(a) As in the LMS algorithm, use the instantaneous gradient vector and derive an LMS update equation for $\mathbf{w}_n$.

(b) Assuming that $v(n)$ is independent of $\mathbf{x}(n)$, and that the weight-error vector

$$\mathbf{c}_n = \mathbf{w}_n - \mathbf{w}$$

is close to zero, and that $c(n)$ is independent of $\mathbf{x}(n)$, show that

$$E\{\mathbf{c}_{n+1}\} = \left[\mathbf{I} - \mu k(2k-1)E\{v^{2k-2}(n)\}\mathbf{R}_x\right] E\{\mathbf{c}_n\}$$

where $\mathbf{R}_x$ is the autocorrelation matrix of $x(n)$.

(c) Show that the modified LMS algorithm derived in part (a) will converge in the mean if the step size μ satisfies the condition

$$0 < \mu < \frac{1}{k(2k-1)E\{v^{2(k-1)}(n)\}\lambda_{\max}}$$

where $\lambda_{\max}$ is the largest eigenvalue of the matrix $\mathbf{R}_x$.

(d) For $k = 1$, show that the results derived in parts (a), (b), and (c) reduce to those in the conventional LMS algorithm.

9.19. Adaptive filters are commonly used for linear prediction. Although harmonic signals such as sinusoids are perfectly predictable, measurement noise will degrade the performance of the predictor and add a bias to the coefficients, $\mathbf{w}$. For example, suppose that we want to design an adaptive linear predictor for a real-valued process $x(n)$ using the noisy measurements

$$y(n) = x(n) + v(n)$$

where $v(n)$ is zero mean white noise that is uncorrelated with $x(n)$. Assume that the variance of $v(n)$ is σ_v^2.

(a) Using the LMS algorithm

$$\mathbf{w}_{n+1} = \mathbf{w}_n + \mu e(n)\mathbf{y}(n)$$

find the range of values for μ for which the LMS algorithm converges in the mean and find

$$\lim_{n\to\infty} E\{\mathbf{w}_n\}$$

(b) The γ-LMS algorithm has been proposed as an adaptive filtering algorithm to combat the effect of measurement noise. Using the noisy observations, $y(n)$, this algorithm is

$$\mathbf{w}_{n+1} = \gamma\mathbf{w}_n + \mu e(n)\mathbf{y}(n)$$

where γ is a constant. Explain how the γ-LMS algorithm can be used to remove the bias in the steady-state solution of the LMS algorithm. Specifically, how would you select values for μ and γ?

9.20. In some applications, it may be necessary to delay the update of the filter coefficients for a short period of time. For example, in decision-directed feedback equalization, if a sophisticated algorithm such a Viterbi decoder is used to improve the decisions, then the desired signal and thus the error is not available until a number of samples later. Therefore, assume that $x(n)$ is real-valued, and consider the delayed LMS algorithm that has a filter coefficient update equation given by

$$\mathbf{w}_{n+1} = \mathbf{w}_n + \mu e(n-n_0)\mathbf{x}(n-n_0)$$

where

$$e(n - n_0) = d(n - n_0) - y(n - n_0)$$

Note that if the delay, n_0, is equal to zero then we have the conventional LMS algorithm.

(a) For $n_0 = 1$, determine the values of μ for which the delayed LMS algorithm converges in the mean.

(b) If $\lambda_k = 1$, for $k = 1, \ldots, p$ and if the step size $\mu = 0.1$, find the time constant, τ_0 for the LMS adaptive filter ($n_0 = 0$) and the time constant τ_1 for the delayed LMS adaptive filter with $n_0 = 1$.

9.21. In recent years, there has been an increasing interest in nonlinear digital filters. This interest has included the design of adaptive nonlinear filters. Volterra systems are an important class of nonlinear filters. Assuming that $x(n)$ is real-valued, a second-order Volterra digital filter has the form

$$y(n) = \sum_{k=0}^{K} a(k)x(n-k) + \sum_{k_1=0}^{K_1} \sum_{k_2=k_1}^{K_2} b(k_1, k_2)x(n-k_1)x(n-k_2)$$

Note that the output, $y(n)$, is formed from a linear combination of first-order (linear) terms $x(n-k)$, and a linear combination of second-order (nonlinear) terms $x(n-k_1)x(n-k_2)$. As a specific example, consider the following second-order digital Volterra filter with time-varying coefficients,

$$y(n) = a_n(0)x(n) + a_n(1)x(n-1) + b_n(0)x^2(n) + b_n(1)x(n)x(n-1)$$

Let $\mathbf{\Theta}_n$ be the coefficient vector

$$\mathbf{\Theta}_n = \left[a_n(0),\ a_n(1),\ b_n(0), b_n(1)\right]^T$$

and let $\mathbf{x}(n)$ be the data vector

$$\mathbf{x}(n) = \left[x(n),\ x(n-1),\ x^2(n),\ x(n)x(n-1)\right]^T$$

(a) Using the LMS update equation

$$\mathbf{\Theta}_{n+1} = \mathbf{\Theta}_n - \frac{1}{2}\mu \nabla e^2(n)$$

where $e(n) = d(n) - y(n)$, derive the coefficient update equations for $a_n(0)$, $a_n(1)$, $b_n(0)$, and $b_n(1)$.

(b) What condition must be placed on μ in order for the coefficient vector $\mathbf{\Theta}_n$ to converge in the mean?

(c) Describe what happens if the third-order statistics of $x(n)$ are zero, i.e.,

$$\begin{aligned} E\{x^3(n)\} &= 0 \\ E\{x^2(n)x(n-1)\} &= 0 \\ E\{x(n)x^2(n-1)\} &= 0 \end{aligned}$$

Discuss how you might improve the performance of the adaptive Volterra filter by having two step size parameters, μ_1 and μ_2, one for the linear terms and one for the nonlinear terms, and discuss how these parameters must be restricted in order for the filter to converge in the mean.

9.22. Consider the system identification problem shown in the figure below.

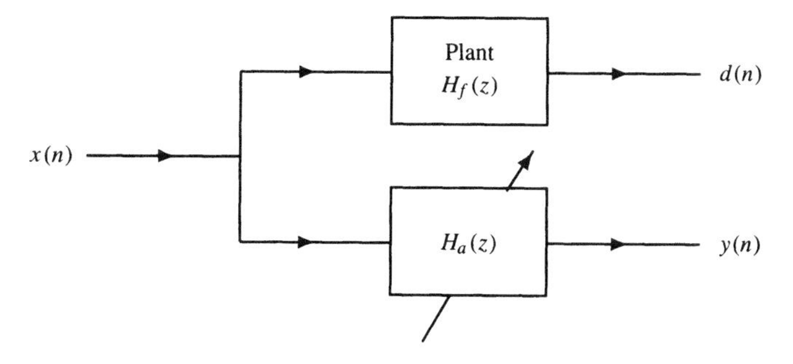

The plant has a rational system function of the form

$$H_f(z) = \frac{0.05 - 0.4z^{-1}}{1 - 1.1314z^{-1} + 0.25z^{-2}}$$

and the adaptive filter that is used to model $H_f(z)$ has two free parameters, a and b, as follows

$$H_a(z) = \frac{b}{1 - az^{-1}}$$

The input, $x(n)$, to both systems is unit variance white noise, and the goal is to find the values of a and b that minimize the mean-square error, $\xi = E\{|e(n)|^2\}$, where $e(n) = d(n) - y(n)$. This mean-square error is a bimodal function of a and b, having a global minimum at $(b, a) = (-0.311, 0.906)$ and a local minimum at $(b, a) = (0.114, -0.519)$.

(a) Write down the equations for the simplified IIR LMS adaptive filter.

(b) Repeat part (a) using the filtered signal approach.

(c) A simplification to the filtered signal approach that has been proposed by Feintuch [15] is to ignore the feedback terms in the equation for the gradient estimates $\psi_k^a(n)$ and $\psi_k^b(n)$. For real-valued signals, this simplification is

$$\begin{aligned}
\psi_k^a(n) &= y(n-k) + \sum_{l=1}^{p} a_n(l)\psi_k^a(n-l) \approx y(n-k) \\
\psi_k^b(n) &= x(n-k) + \sum_{l=1}^{p} a_n(l)\psi_k^b(n-l) \approx x(n-k)
\end{aligned}$$

Although more efficient than the filtered signal approach, this algorithm may converge to a false minimum, even when the simplified IIR LMS algorithm and the filtered signal approach converge to a minimum. Write down the adaptive filtering equations for $H_a(z)$ using Feintuch's algorithm.

(d) In order for the filter coefficients a_n and b_n in $H_a(z)$ to converge in the mean using Feintuch's algorithm, it is necessary that

$$\lim_{n\to\infty} E\{e(n)x(n)\} = 0$$

and

$$\lim_{n\to\infty} E\{e(n)y(n-1)\} = 0$$

Find the values of $E\{e(n)x(n)\}$ and $E\{e(n)y(n-1)\}$ at the global minimum of ξ. What does this imply about Feintuch's algorithm?

(e) Find the stationary point of the Feintuch adaptive filter, i.e., the value or values of a and b for which $E\big\{e(n)x(n)\big\} = 0$ and $E\big\{e(n)y(n-1)\big\} = 0$.

9.23. The Hyperstable Adaptive Recursive Filter (HARF) is an IIR adaptive filtering algorithm with known convergence properties. Due to its computational complexity, a simplified version of the HARF algorithm, known as SHARF, is often used. Although the convergence properties of HARF are not preserved in the SHARF algorithm, both algorithms are similar when the filters are adapting slowly (small μ). For real-valued signals, the coefficient update equations for the SHARF algorithm are

$$a_{n+1}(k) = a_n(k) + \mu_a y(n-k)v(n)$$
$$b_{n+1}(k) = b_n(k) + \mu_b x(n-k)v(n)$$

where

$$v(n) = e(n) + \sum_{k=1}^{K} c(k)e(n-k)$$

is the error signal that has been filtered with an FIR filter $C(z)$. Suppose that the coefficients of a second-order adaptive filter

$$H(z) = \frac{b(0) + b(1)z^{-1}}{1 + a(1)z^{-1}}$$

are updated using the SHARF algorithm with

$$C(z) = 1 + c(1)z^{-1} + c(2)z^{-2}$$

(a) Write down the SHARF adaptive filter equations for $H(z)$.

(b) If the SHARF algorithm converges in the mean, then $E\{v(n)\mathbf{x}(n)\}$ converges to zero where

$$\mathbf{x}(n) = [x(n),\ x(n-1),\ y(n-1)]^T$$

What does this imply about the relationship between the filter coefficients $c(1)$ and $c(2)$ and $E\{e(n)\mathbf{x}(n)\}$?

(c) What does the SHARF algorithm correspond to when $C(z) = 1$?

9.24. Modify the RLS algorithm so that the coefficients $w(k)$ satisfy the linear phase constraint, $w(k) = w(p-k)$. For example, with a five-coefficient filter, the coefficient vector $\mathbf{w}_n$ is of the form

$$\mathbf{w}_n = \big[w_n(0), w_n(1), w_n(2), w_n(1), w_n(0)\big]^T$$

9.25. There are many different ways that one may compare the performance of adaptive filtering algorithms. Suppose that we are interested in adaptive linear prediction and our measure of performance is the number of arithmetic operations required for the adaptive filter to converge. Let the time constant τ be used as the convergence time of the LMS algorithm. For the RLS algorithm, it is often stated that the rate of convergence is an order of magnitude faster than the LMS algorithm. Therefore, assume that the time constant for the RLS algorithm is one tenth that of the LMS algorithm.

(a) If the eigenvalues of the $p \times p$ autocorrelation matrix for $x(n)$ are

$$\lambda_1 = 1.0 \qquad \text{and} \qquad \lambda_2 = \cdots = \lambda_p = 0.01$$

and if we use a step size $\mu = 0.1$ for the LMS algorithm, for what order filter, p, are the RLS and LMS adaptive filters equal in terms of their computational requirements to reach convergence?

(b) For high order filters, $p >> 1$, the computational requirements of the RLS filter become large, and the LMS algorithm becomes an attractive alternative. For what reasons might you prefer to use the RLS algorithm in spite of its increased computational cost?

9.26. Let $\alpha(n)$ be the a priori error and $e(n)$ the a posteriori error in the RLS algorithm, and let

$$\mu(n) = \frac{1}{1 + \mathbf{x}^T(n)\mathbf{R}_x^{-1}(n-1)\mathbf{x}^*(n)}$$

be the scalar that is used in the calculation of the gain vector $\mathbf{g}(n)$.

(a) Show that $e(n)$ may be written in terms of $\alpha(n)$ and $\mu(n)$ by finding an explicit relation between $e(n)$, $\alpha(n)$, and $\mu(n)$. Hint: Begin with the RLS update equation for $\mathbf{w}_n$ and form the product $\mathbf{x}^T(n)\mathbf{w}_n$.

(b) Let $\mathbf{g}(n) = \mu(n)\mathbf{R}_x^{-1}(n-1)\mathbf{x}^*(n)$ be the gain vector in the RLS algorithm. Consider the time-varying filter that has coefficients $\mathbf{g}(n)$ and an input, $x(n)$, that is the same as that used in the RLS algorithm to compute the gain $\mathbf{g}(n)$, i.e.,

$$y(n) = \sum_{k=0}^{p-1} g_n(k)x(n-k)$$

For what signal, $d(n)$, will the difference between $d(n)$ and $y(n)$ be equal to $\mu(n)$ as illustrated in the figure below?

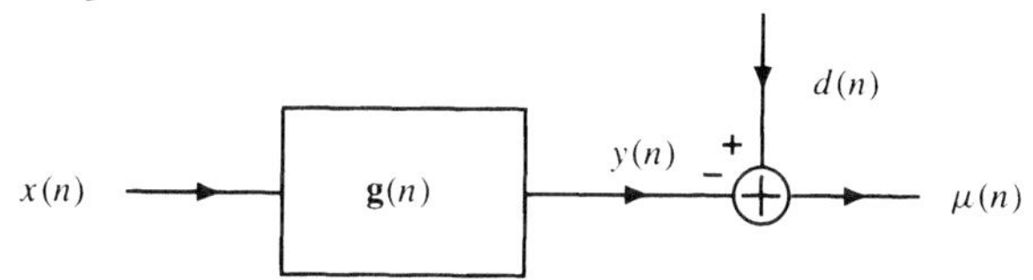

9.27. Suppose that the least-squares error used in the RLS algorithm is modified as follows

$$\mathcal{E}_a(n) = \sum_{i=0}^{n} \lambda^{n-i}|e(i)|^2 + \lambda^n \mathbf{w}_n^T \mathbf{w}_n$$

where $\mathbf{w}_n$ is the vector of filter coefficients at time n for a pth-order FIR adaptive filter and

$$e(i) = d(i) - \mathbf{w}_n^T \mathbf{x}(i)$$

Derive the equations for the optimal least squares filter $\mathbf{w}_n$ that minimizes $\mathcal{E}_a(n)$ for each n.

Computer Exercises

C9.1. In this exercise we consider the problem of adaptive linear prediction using the LMS algorithm. Let $x(n)$ be a second-order autoregressive process that is generated by the difference equation

$$x(n) + a(1)x(n-1) + a(2)x(n-2) = v(n)$$

where $v(n)$ is zero mean white noise with a variance σ_v^2. This process is the input to a two-coefficient LMS adaptive predictor.

(a) If $a(1) = -0.1$, $a(2) = -0.8$, and $\sigma_v^2 = 0.25$, for what values of the step size, μ, will the LMS algorithm converge in the mean? For what values of μ will the LMS algorithm converge in the mean-square?

(b) Implement an LMS adaptive predictor using $N = 500$ samples of $x(n)$ and make a plot of the squared prediction error, $e^2(n)$, versus n using step-sizes of $\mu = 0.05$ and $\mu = 0.01$. As described in Example 9.2.2, repeat this experiment for 100 different realizations of $x(n)$, and plot the learning curve by averaging the plots of $e^2(n)$ versus n. Estimate the corresponding values of the misadjustment by time-averaging over the final iterations of the ensemble-averaged learning curves. Compare these estimated values with theory.

(c) Estimate the steady-state values of the adaptive filter coefficients for step-sizes of $\mu = 0.05$ and $\mu = 0.01$. Note that you may do this by averaging the steady-state values of the coefficients (obtained at the final iteration of the LMS algorithm) over 100 independent trials of the experiment. Compare these estimated values with theory.

(d) Repeat the experiments in parts (b) and (c) with $a(1) = -0.1$, $a(2) = 0.8$, and $\sigma_v^2 = 0.25$.

(e) Repeat the above exercises using the sign-error, sign-data, and sign-sign algorithms.

C9.2. One of the many uses of adaptive filters is for system identification as shown in the figure below. In this configuration, the same input is applied to an adaptive filter and to an unknown system, and the coefficients of the adaptive filter are adjusted until the difference between the outputs of the two systems is as small as possible. Adaptive system identification may be used to model a system with slowly-varying parameters, provided both the input and output signals are available.

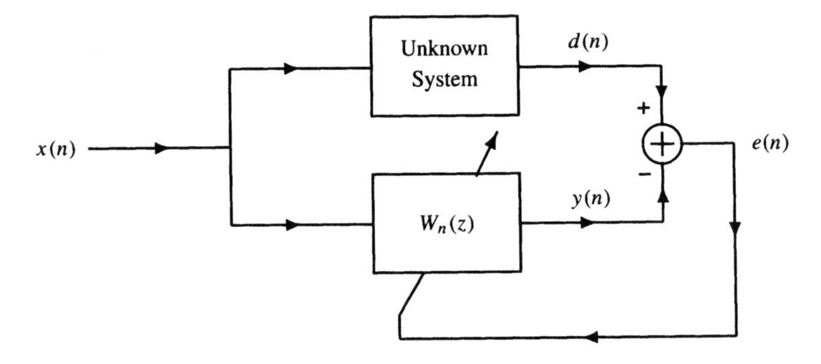

Let the unknown system that is to be identified be an FIR filter with unit sample response

$$g(n) = \delta(n) + 1.8\delta(n-1) + 0.81\delta(n-2)$$

With an input $x(n)$ consisting of 1000 samples of unit variance white Gaussian noise, create the reference signal $d(n)$.

(a) Determine the range of values for the step size μ in the LMS algorithm for convergence in the mean.

(b) Implement an adaptive filter of order $p = 4$ using the LMS algorithm. Set the initial weight vector equal to zero, and use a step size of $\mu = 0.1\mu_{\max}$, where $\mu_{\max}$ is the largest step size allowed for convergence in the mean. Let the adaptive filter adapt and record the final set of coefficients.

(c) Repeat part (b) using the normalized LMS algorithm with $\beta = 0.1$, and compare your results.

(d) As described in Example 9.2.2, make a plot of the learning curve by repeating the experiment described in part (b) for 100 different realizations of $d(n)$, and plotting the average of the plots of $e^2(n)$ versus n. How many iterations are necessary for the mean-square error to fall to 10% of its peak value? Calculate the theoretical value for the excess mean-square error and compare it to what you observe in your plot of the learning curve.

(e) Repeat parts (b) and (d) for $\mu = 0.01\mu_{\max}$ and $\mu = 0.2\mu_{\max}$.

(f) Repeat parts (b) and (d) for $p = 3$ and $p = 2$.

C9.3. In this exercise, we consider the system identification experiment described in Exercise C9.2, but this time, we look at what happens if the reference signal is corrupted by noise. Instead of $d(n)$, suppose that we observe

$$\widetilde{d}(n) = d(n) + \gamma v(n)$$

where $v(n)$ is unit variance white Gaussian noise.

(a) With an adaptive filter of order $p = 4$ and $\gamma = 0.1$, use the normalized LMS algorithm to model the system defined in Exercise C9.2. Use a step size of $\beta = 0.1$ and set the initial weight vector equal to zero. Let the filter adapt and record the final set of coefficients. Repeat your experiment using $p = 5$ and discuss your results.

(b) Repeat part (a) with $\gamma = 1$ and comment on how the accuracy of the model varies with the variance of the noise. Does the accuracy of your model depend on the step size?

C9.4. In the previous two exercises, we considered the problem of using an FIR adaptive filter to model an unknown FIR system. In this exercise, we consider system identification for an unknown IIR system. The arrangement for the filter is the same as that described in Exercise C9.2, except that the unknown system is an IIR filter with a system function

$$G(z) = \frac{1 + 0.5z^{-1}}{1 - 0.9z^{-1}}$$

Generate a 500-point sequence of unit variance white Gaussian noise, and use this process as the input to the unknown system and the adaptive filter.

(a) Use the normalized LMS algorithm with $p = 4$ and $\beta = 0.1$ to model $G(z)$. Record the final values of the filter coefficients and make a plot of the learning curve as described in Exercise C9.2. How many iterations are required for the filter to converge?

(b) Repeat part (a) for different values of p. What seams to be a "reasonable" value to use?

(c) Repeat parts (a) and (b) when the denominator of $G(z)$ is replaced with $A(z) = 1 - 0.2z^{-1}$. Explain your observations.

(d) Add white noise $v(n)$ to the output of the unknown system. Repeat part (a) with noise variances of $\sigma_v^2 = 0.01, 0.1$, and 1.0.

(e) Write a MATLAB m-file to implement the filtered signal approach for IIR adaptive filtering. With an adaptive recursive filter of the form

$$y(n) = a_n(1)y(n-1) + b_n(0)x(n) + b_n(1)x(n-1)$$

determine an appropriate value for the step size μ, and use your m-file to model the system $G(z)$. Discuss your results in terms of the convergence of the coefficients and the accuracy of the model. Describe what happens if the model order is increased, e.g., suppose you were to use two poles and two zeros in the adaptive filter.

(f) Add white noise $v(n)$ to the output of the unknown system. Repeat part (e) with noise variances of $\sigma_v^2 = 0.01, 0.1$, and 1.0.

C9.5. Modify the m-file for the normalized LMS algorithm, `nlms.m`, to take advantage of the computational simplification given in Eq. (9.51).

C9.6. In this problem we will compare LMS and RLS for adaptive linear prediction. As in Example 9.2.1, let $x(n)$ be a process that is generated according to the difference equation

$$x(n) = 1.2728x(n-1) - 0.81x(n-2) + v(n)$$

where $v(n)$ is unit variance white Gaussian noise. The adaptive linear predictor will be of the form

$$\hat{x}(n) = w_n(1)x(n-1) + w_n(2)x(n-2)$$

(a) Implement an RLS adaptive predictor with $\lambda = 1$ (growing window RLS) and plot $w_n(k)$ versus n for $k = 1, 2$. Compare the convergence of the coefficients $w_n(k)$ to those that are obtained using the LMS algorithm for several different values of the step size μ.

(b) Make a plot of the learning curve for RLS and compare it to the LMS learning curve (See Example 9.2.2 on how to plot learning curves). Comment on the excess mean-square error for RLS and discuss how it compares to that for LMS.

(c) Repeat part (b) for exponential weighting factors of $\lambda = 0.99, 0.95, 0.92, 0.90$ and discuss the trade-offs involved in the choice of λ.

(d) Modify the m-file for the RLS algorithm to implement a sliding window RLS algorithm.

(e) Let $x(n)$ be generated by the time-varying difference equation

$$x(n) = a_n(1)x(n-1) - 0.81x(n-2) + v(n)$$

where $v(n)$ is unit variance white noise and $a_n(1)$ is a time-varying coefficient given by

$$a_n(1) = \begin{cases} 1.2728 & ; \quad 0 \le n < 50 \\ 0 & ; \quad 50 \le n < 100 \\ 1.2728 & ; \quad 100 \le n \le 200 \end{cases}$$

Compare the effectiveness of the LMS, growing window RLS, exponentially weighted RLS, and sliding window RLS algorithms for adaptive linear prediction. What approach would you propose to use for linear prediction when the process has step changes in its parameters?

C9.7. Let $x(n)$ be a process that is generated according to the difference equation

$$x(n) = 0.8x(n-1) - 0.9x(n-2) + v(n)$$

where $v(n)$ is unit variance white Gaussian noise. Suppose that we would like to implement an adaptive linear predictor using the LMS algorithm,

$$\hat{x}(n) = w_n(1)x(n-1) + w_n(2)x(n-2)$$

(a) Generate a sequence $x(n)$ of length $N = 500$ using the difference equation above and determine a value for the LMS step size so that $\mathbf{w}_n$ converges in the mean to $\mathbf{w} = \begin{bmatrix} 0.8 & -0.9 \end{bmatrix}^T$ within about 200 iterations.

(b) Implement the LMS adaptive predictor and plot $w_n(k)$ versus n for $k = 1, 2$.

(c) Estimate the mean-square error $\xi(\infty)$ and compare it to the theoretical value.

(d) Write a MATLAB program or modify the m-file `lms.m` to implement the p-vector algorithm to estimate a process $d(n)$ from $x(n)$ (see Prob. 9.14),

$$\mathbf{w}_{n+1} = \mathbf{w}_n + \mu \mathbf{r}_{dx} - \mu\big[\mathbf{w}_n^T \mathbf{x}(n)\big]\mathbf{x}(n)$$

where $\mathbf{r}_{dx} = E\{d(n)\mathbf{x}(n)\}$.

(e) Find the vector, $\mathbf{r}_{dx} = E\{d(n)\mathbf{x}(n)\}$ for a one-step linear predictor, and implement the p-vector algorithm for $x(n)$. Plot $w_n(k)$ versus n and compare your results to the LMS adaptive linear predictor. Does the p-vector algorithm converge? If so, to what? What constraints are necessary on the step size for the p-vector algorithm to be well-behaved?

(f) Investigate the sensitivity of the p-vector algorithm to errors in the vector $\mathbf{r}_{dx}$ by making small changes in the values of $\mathbf{r}_{dx}$ used in part (e).

C9.8. The LMS algorithm is very popular due to its inherent simplicity. However, it suffers from relatively slow convergence properties, and its performance is severely affected when there is a large eigenvalue spread in the autocorrelation matrix of the input signal.

A number of different algorithms have been proposed to improve the convergence properties of the LMS algorithm that incorporate a variable step size parameter, $\mu(n)$. One approach that has been proposed is to use a monotonically decreasing step size of the form

$$\mu(n) = \frac{1}{c+n}$$

where c is some constant. Thus, the variable step size adaptive filter becomes

$$\mathbf{w}_{n+1} = \mathbf{w}_n + \mu(n)e(n)\mathbf{x}(n)$$

(a) Evaluate the effectiveness of this algorithm for linear prediction. What value should you choose for c? What limitations do you see that this algorithm might have?

(b) Another possible form for $\mu(n)$ would be to use

$$\mu(n) = \frac{1}{c_1 + c_2 n}$$

This would allow for better control in how fast the step size decreases to zero. Repeat part (a) for this variable step size adaptive filter.

C9.9. In this problem we look at the problem of adaptive noise cancellation without a reference (see Fig. 9.2*b*). Let

$$x(n) = d(n) + v(n)$$

where $d(n)$ is a unit amplitude sinusoid of frequency $\omega_0 = 0.01\pi$. Suppose that $v(n)$ is a moving average process that is generated as follows

$$v(n) = g(n) + 0.5g(n-2)$$

where $g(n)$ is unit variance white noise.

(a) What is the minimum value for the delay k_0 that may be used in the adaptive noise canceller?

(b) Write a MATLAB program for adaptive noise cancellation using the normalized LMS algorithm with $x(n-k_0)$ as the reference signal.

(c) Generate 1000 samples of $v(n)$ and $x(n)$ and use your MATLAB program to estimate $d(n)$ for filter orders $p = 5, 10, 15, 20$. Use values for k_0 that range from the minimum value determined in part (a) to $k_0 = 25$. What dependence on k_0, if any, do you observe? Explain. What is the effect of the filter order p? Given that the computational costs go up with p, what order do you think would be the best to use?

(d) Find the coefficients of an FIR Wiener filter to estimate and compare them with the steady-state values of the coefficients of the adaptive noise canceller.

APPENDIX MATLAB PROGRAMS

A.1 INTRODUCTION

One of the most effective ways for a student to gain a deep appreciation and understanding of signal processing is to process signals. There is a great deal of information to be gained by experimenting with algorithms, testing them on real signals and unusual sequences, developing "new approaches," and discovering at what point the theory begins to break down in practice. MATLAB provides an extremely powerful and versatile environment for performing signal processing experiments. As a result, we have included over thirty m-files in this book that cover everything from signal modeling and the Levinson recursion to spectrum estimation and adaptive filtering. In this appendix, we provide a description of each of these m-files, define the format of the inputs and outputs, give pointers to places where the user may encounter problems, and provide simple illustrative examples. Although it is assumed that the reader is familiar with MATLAB, the first-time user should be able to use the m-files in this book after reading this appendix. Anyone without MATLAB experience who wishes to go beyond what is presented here and begin writing their own m-files should consult the MATLAB Users Guide. The reader should keep in mind, however, that MATLAB has an online help facility that is accessible by typing `help` at the MATLAB prompt for general help, or by typing `help filename` for specific help on the m-file `filename`.

None of the m-files described in this appendix are particularly long and may be typed-in by hand without too much difficulty. However, all m-files are available by anonymous ftp from

```
ftp.ece.gatech.edu/pub/users/mhayes/stat_dsp
```

The reader may also want to browse the home page for this book on the Web at

```
http://www.ece.gatech.edu/users/mhayes/stat_dsp
```

for additional information such as new problems and m-files, errata and clarifications, and reader comments and feedback.

A.2 GENERAL INFORMATION

The m-files described in this appendix may be used to process discrete-time signals in a variety of different ways. Each m-file is written to accept either real or complex-valued signals. These signals are input as vectors and may be generated in a number of different ways. For example, a signal may be defined using a mathematical expression such as[1]

```
>>x = 2*cos([0:99]*pi/4);
```

which generates the signal $x(n) = 2\cos(n\pi/4)$ for $n = 0, 1, \ldots, 99$. Alternatively, a signal may be defined by specifying the signal values individually as in the command

```
>>x = [2 1 3 5 1];
```

which sets the first five values of $x(n)$ equal to 2, 1, 3, 5, 1, or with the command

```
>>x = [ones(1,32) zeros(1,32)];
```

which creates a *pulse* consisting of 32 values that are equal to 1 followed by 32 values that are equal to zero.

It is important to keep in mind that indices in MATLAB begin with one. For example, the signal

$$x(n) = \delta(n) + 2\delta(n-1) - 3\delta(n-2)$$

would be input using the command

```
>>x = [1 2 -3];
```

with the value of $x(0)$ being stored in the variable `x(1)`. Therefore, typing `x(1)` at the MATLAB prompt would return the value 1 whereas typing `x(0)` would result in the error message

```
>>x(0)
???  Index exceeds matrix dimensions.
```

Many of the m-files described in this appendix call other m-files. In particular, two m-files that are frequently called are `convm.m` and `covar.m`. The first, `convm.m`, is used to set up a convolution matrix. In other words, given a signal $x(n)$ of length N that is stored in the vector x, the command

```
>>X = convm(x,p);
```

produces a $(N + p - 1) \times p$ non-symmetric Toeplitz matrix $\mathbf{X}$ that has the N values of $x(n)$ in the first column (padded with zeros) and $\left[x(0),\ 0,\ \ldots,\ 0\right]$ in the first row.[2] For example,

```
>> x=[1 3 2];
>> X=convm(x,4)
```

[1]Note that >> represents the MATLAB command line prompt.

[2]Although common usage of the term Toeplitz matrix assumes that the matrix is square, here we use the term for a rectangular matrix to highlight the structure of the convolution matrix. Note that MATLAB allows the user to generate a rectangular Toeplitz matrix using the command `x=toeplitz(c,r)` where `c` and `r` are the first row and column, respectively, of the Toeplitz matrix, and `c` and `r` need not be the same length.

```
ans =

     1     0     0     0
     3     1     0     0
     2     3     1     0
     0     2     3     1
     0     0     2     3
     0     0     0     2
```

The second m-file, `covar.m`, forms a covariance matrix. Thus, if the sequence $x(n)$ is stored in the vector `x`, then

```
>> R=covar(x,p)
```

creates a $p \times p$ Hermitian Toeplitz matrix, `R`, of sample covariances. Listings of both of these m-files are given in Fig. A.1.

Many of the m-files in this book compute the DFT of a signal using the m-file `fft.m`. This m-file is called as follows:

```
>>X = fft(x,n);
```

where `x` is an input vector that contains the signal values that are to be transformed, and `X` is the output vector that contains the DFT coefficients. The second argument, `n`, is an optional integer that may be used to specify the length of the DFT. If `fft.m` is called without a second argument,

```
>>X = fft(x);
```

```
function X = convm(x,p)
%
% This function sets up a convolution matrix
%
N    = length(x)+2*p-2;
x    = x(:);
xpad = [zeros(p-1,1);x;zeros(p-1,1)];
for  i=1:p
     X(:,i)=xpad(p-i+1:N-i+1);
     end;


function R = covar(x,p)
%
% This function sets up a covariance matrix
%
x = x(:);
m = length(x);
x = x - ones(m,1)*(sum(x)/m);
R = convm(x,p+1)'*convm(x,p+1)/(m-1);
end;
```

Figure A.1 MATLAB *programs for generating a convolution matrix,* `convm.m`, *and a covariance matrix,* `covar.m`.

then the length of the DFT is equal to the length of the vector x. All of the m-files in this book that call `fft.m` set the value of n equal to 1024. Therefore, for signals longer than 1024-points, it may be necessary to modify the m-files that make a call to `fft.m`.

A.3 RANDOM PROCESSES

The computer exercises at the end of Chapter 3 require no special m-files and may be completed using only MATLAB functions and a few m-files from the Signal Processing Toolbox. Two of the m-files required in these exercises, `randn` and `rand`, are used to generate Gaussian and uniform white noise processes, respectively.[3] For example,

```
>>noise1=randn(1,100);
```

generates 100 samples of unit variance white Gaussian noise, and

```
>>noise2=2*rand(1,100)-1;
```

generates 100 samples of zero mean white noise that is uniformly distributed over the interval $[-1, 1]$. Once a white noise sequence has been created, it is straightforward to generate an autoregressive, moving average, or autoregressive moving average process. This may be done by filtering the white noise using the m-file `filter`. For example, the sequence of commands

```
>>a=[1, 0.5, 0.8];
>>b=[1, 2, 1];
>>x=filter(b,a,randn(1,100));
```

creates 100 samples of an ARMA(2,2) process that has a power spectrum

$$P_x(e^{j\omega}) = \frac{\left|1 + 2e^{-j\omega} + e^{-2j\omega}\right|^2}{\left|1 + 0.5e^{-j\omega} + 0.8e^{-2j\omega}\right|^2} = \frac{6 + 8\cos\omega + 2\cos 2\omega}{1.89 + 1.8\cos\omega + 1.6\cos 2\omega}$$

A.4 SIGNAL MODELING

In this section, we describe the m-files presented in Chapter 4, which are concerned primarily with signal modeling. These include m-files for the Padé approximation, Prony's method, Shanks' method, iterative prefiltering, the autocorrelation and covariance methods, and Durbin's method. In addition we describe the m-file to design a least squares inverse filter. In all cases, the signals that are used as inputs to these m-files may be either real-valued or complex-valued.

Padé Approximation. Given a signal $x(n)$, the Padé approximation finds the coefficients in the model

$$H(z) = \frac{B_q(z)}{A_p(z)} = \frac{\sum_{k=0}^{q} b_q(k)z^{-k}}{1 + \sum_{k=1}^{p} a_p(k)z^{-k}} \tag{A.1}$$

[3]See the MATLAB Users Manual or type `help randn` or `help rand` to see how to set the seed of the random number generator. This is useful when it is necessary to repeatedly generate the same noise process.

by matching the first $p+q+1$ signal values exactly, i.e., $x(n) = h(n)$ for $n = 0, 1, \ldots, p+q$. The m-file that is used to find the Padé approximation of a signal $x(n)$ is `pade.m`, and this m-file may be called as follows

```
>>[a,b]=pade(x,p,q);
```

There are three required inputs to this m-file. The first is the vector `x` that contains the values of the signal $x(n)$ that is to be modeled. The other two inputs, `p` and `q`, specify the model order, i.e., the order of the polynomials $A_p(z)$ and $B_q(z)$, respectively. The output consists of two vectors, `a` and `b`, that contain the model coefficients $a_p(k)$ and $b_q(k)$, respectively. As indicated in Eq. (A.1), the first coefficient of the vector `a` will always be equal to one.

Once the coefficients of the model have been found, the Padé approximation $\hat{x}(n)$ may be computed by evaluating the unit sample response of $H(z)$ as follows:

```
>>xhat=filter(b,a,[1 zeros(1,length(x)-1)]);
```

Alternatively, the DTFT of the Padé approximation may be plotted with the command

```
>>freqz(b,a)
```

Finally, the average squared error may be computed as follows

```
>>mse=norm(x-xhat);
```

where `xhat` is the Padé approximation that may be found as described above.

Special Notes: The model orders `p` and `q` must be non-negative integers and `p+q` must be less than the length of the vector `x`.

Prony's Method. As with the Padé approximation, Prony's method finds a model for a signal $x(n)$ of the form given in Eq. (A.1). Unlike the Padé approximation, however, the denominator coefficients $a_p(k)$ are found by minimizing the Prony error defined in Eq. (4.29). This requires solving the linear equations given in Eq. (4.35). Once the denominator coefficients have been determined, the numerator coefficients $b_q(k)$ are found using the Padé approach to match the signal exactly for the first $q + 1$ values of $x(n)$.

The m-file for Prony's method is `prony.m`, and the format of the command line is

```
>>[a,b,err]=prony(x,p,q);
```

As with `pade.m`, the inputs to `prony.m` include the vector `x` of signal values that are to be modeled, and the integers `p` and `q` that specify the model order. The outputs are the vectors `a` and `b` that contain the coefficients $a_p(k)$ and $b_q(k)$, respectively, and `err` which is the Prony error $\epsilon_{p,q}$ given in Eq. (4.44). The first coefficient of the vector `a` is always equal to one.

In the derivation of Prony's method, it is assumed that $x(n)$ is known for all $n \geq 0$. Therefore, there is an important practical issue that concerns how to address the problem of only being able to record and process a finite-length observation of $x(n)$. What is done in `prony.m` is to assume that $x(n)$ is zero for all values of n that are greater than the length of the input vector `x`. Therefore, `prony.m` uses the autocorrelation method to find the denominator coefficients, and the Padé approximation to find the numerator coefficients.

When using `prony.m` with `q=0` (an all-pole model) it is important to note that `prony.m` sets $b(0) = x(0)$. For an all-pole model, a better approach, in general, would be to set $b(0) = \sqrt{\epsilon_p}$ (an energy matching constraint) as follows,

```
>>[a,b,err]=prony(x,p,0);
>>b=sqrt(err);
```

Special Notes: The model orders p and q must be non-negative integers and p+q must be less than the length of the vector x.

Shanks' Method. Shanks' method provides an alternative to Prony's method of finding the numerator coefficients. Instead of forcing an exact fit to the data for the first $q+1$ values of n, Shanks' method performs a least squares minimization of the error

$$e'(n) = x(n) - \hat{x}(n)$$

The m-file for Shanks' method is `shanks.m` and the format used to call this m-file is

```
>>[a,b,err]=shanks(x,p,q);
```

where x is the vector that contains the signal that is to be modeled, p is the order of the denominator polynomial, and q is the order of the numerator polynomial. As with the previous two m-files, the output consists of two vectors, a and b, that contain the coefficients $a_p(k)$ and $b_q(k)$, respectively, and the model error `err`.

Special Notes: The model orders p and q must be non-negative integers and p+q must be less than the length of the vector x.

Iterative Prefiltering. Iterative prefiltering, or the method of Steiglitz and McBride, is an iterative algorithm to find a rational model for a signal $x(n)$ that minimizes the least squares error

$$\mathcal{E}_{LS} = \sum_{n=0}^{\infty} |x(n) - h(n)|^2$$

Although there is no general proof of convergence, iterative prefiltering often reaches an acceptable solution after 5 to 10 iterations. The m-file that implements the method of iterative prefiltering is `ipf.m` and may be called as follows

```
>>[a,b,err]=ipf(x,p,q,n,a0);
```

The inputs that are required are the signal vector, x, the number of poles in the model, p, the number of zeros, q, and the number of iterations n. In addition, there is an optional input vector, a0, that is used to initialize the recursion with a given set of denominator coefficients. If this input is left unspecified, then the initial condition is found using Prony's method. The outputs of the m-file are the model coefficients $a_p(k)$ and $b_q(k)$, which are stored in the vectors a and b, respectively, and the squared error, `err`.

The format of the m-file `ipf.m` allows one to continue the iteration after making an initial call to `ipf.m`. For example, typing

```
>>[a,b,err]=ipf(x,p,q,n1);
>>[a,b,err]=ipf(x,p,q,n2,a);
```

performs iterative prefiltering for n1 iterations, and then uses the vector a that is obtained after n1 iterations to initialize `ipf.m` for the next n2 iterations.

Special Notes: The model orders p and q must be non-negative integers and p+q must be less than the length of the vector x. The number of iterations, n, must be a positive integer, typically in the range from 5 to 10. The optional input vector a0 must be a non-zero vector of length p.

Autocorrelation Method. The autocorrelation method is an all-pole modeling technique that finds the all-pole coefficients $a_p(k)$ from the values of $x(n)$ for $n = 0, 1, \ldots, N$ by minimizing the Prony error

$$\mathcal{E}_p = \sum_{n=0}^{\infty} \left|e(n)\right|^2$$

where

$$e(n) = x(n) + \sum_{k=1}^{p} a_p(k)x(n-k)$$

Since $x(n)$ is assumed to be known only for $0 \le n \le N$, the error $e(n)$ may only be evaluated for $p \le n \le N$. Therefore, $\mathcal{E}_p$ cannot be minimized unless some assumptions are made about the values of $x(n)$ outside the interval $[0, N]$. With the autocorrelation method, $x(n)$ is assumed to be equal to zero for $n < 0$ and $n > N$, which is equivalent to applying a rectangular data window to $x(n)$. Although this window biases the solution, it ensures that the all-pole model will be stable. The m-file for the autocorrelation method is `acm.m` and may be called as follows

```
>>[a,err]=acm(x,p);
```

The input x is a vector that contains the signal values $x(n)$, and p is an integer that specifies the model order (number of poles). The output a is the vector of coefficients $a_p(k)$, and `err` is the modeling error, $\epsilon_p = \min\{\mathcal{E}_p\}$. The numerator coefficient, $b(0)$, of the all-pole model is typically selected to satisfy the energy matching constraint,

$$b(0) = \sqrt{\epsilon_p}$$

Another possibility, however, is

$$b(0) = x(0)$$

which forces the first value in the sequence to match the first value in the model.

<u>Special Notes</u>: The model order p must be a positive integer that is less than the length of the vector x.

Covariance Method. The covariance method is an alternative to the autocorrelation method for finding an all-pole model for a finite-length sequence $x(n)$, $n = 0, 1, \ldots, N$. Instead of assuming that the unknown values of $x(n)$ are equal to zero, the covariance method modifies the error that is to be minimized. The modification simply involves redefining the limits on the sum for $\mathcal{E}_p$ to begin at $n = p$ and end at $n = N$. Since the covariance method does not window the data, the model is generally more accurate than the autocorrelation method. However, the model is not guaranteed to be stable. The m-file for the covariance method is `covm.m` and it is called as follows:

```
>>[a,err]=covm(x,p);
```

As with the m-file `acm.m`, the inputs consist of the data vector x and the model order p, and the outputs are the model coefficients $a_p(k)$, which are stored in the vector a, and the error, which is stored in `err`.

<u>Special Notes</u>: The model order p must be a positive integer that is less than the length of the vector x.

Durbin's Method. Durbin's method is a procedure for finding a qth-order moving average model for a wide-sense stationary process $x(n)$. Thus, the model for $x(n)$ is of the form

$$x(n) = \sum_{k=0}^{q} b_q(k)w(n-k)$$

where $w(n)$ is unit variance white noise. This algorithm begins by finding a high-order ($p \gg q$) all-pole model for $x(n)$. Then, treating the model coefficients as the data, a qth-order all-pole model is found for these coefficients. The m-file for Durbin's method is `durbin.m`, which is called using a command of the form

```
>>b=durbin(x,p,q);
```

where `x` is the vector that contains the signal values, `p` is the order of the high-order all-pole model for $x(n)$, and `q` is the order of the moving average model. The output is the vector `b` of moving average coefficients.

Special Notes: The model orders `p` and `q` must be positive integers with `p` > `q`. It is generally recommended that the value of `p` be at least four times order of the moving average model, `q`. However, `p` must not exceed the length of the data vector `x`.

Least Squares Inverse Filter. Given a filter $g(n)$, the least squares inverse, $h(n)$, is the filter that produces the best approximation to a unit sample at time $n = n_0$ when convolved with $g(n)$,

$$g(n) * h(n) \approx \delta(n - n_0) \tag{A.2}$$

The m-file to find the least squares inverse filter is `spike.m`, and the format required to use this m-file is

```
>>[h,err]=spike(g,n0,n);
```

The inputs to this m-file include `g`, which is a vector that contains the coefficients of the filter that are used to design the least squares inverse, `n0`, which is the value of the delay n_0, and `n`, which is the order (number of coefficients) of the least squares inverse filter $h(n)$. The output `h` is the vector containing the coefficients of the least squares inverse filter, and `err` is the total squared error in the least squares inverse design.

Special Notes: The order of the inverse filter, `n`, must be a positive integer, and the delay, `n0`, must be a positive integer that is no larger than the length of the convolution $g(n)*h(n)$, i.e.,

```
n0 < length(g) + length(h) - 1
```

where `n = length(h)`.

A.5 LEVINSON RECURSION

In this section, we describe the m-files presented in Chapter 5, which are concerned with the Levinson and Levinson-Durbin recursions. Six of these m-files involve mappings between an autocorrelation sequence, the all-pole filter coefficients, and the reflection coefficients as illustrated in Fig. A.2. The seventh m-file is for the Levinson recursion, and may be used

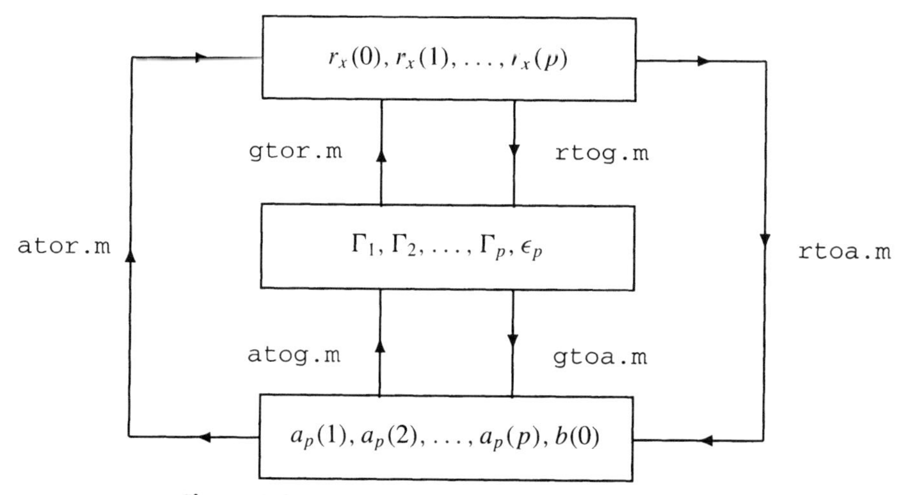

Figure A.2 *The m-files for converting between an autocorrelation sequence, the reflection coefficients, and the all-pole model.*

to find the solution to a general set of Toeplitz equations. Each of these m-files allow the inputs to be either real-valued or complex-valued vectors.

Levinson-Durbin Recursion. The Levinson-Durbin recursion is an efficient algorithm for solving a set of Hermitian Toeplitz equations of the form

$$\mathbf{R}_x \mathbf{a}_p = \epsilon_p \mathbf{u}_1$$

where $\mathbf{R}_x = \text{Toep}\{r_x(0), r_x(1), \ldots, r_x(p)\}$ is a Hermitian Toeplitz matrix and $\mathbf{u}_1$ is a unit vector with a one for the first coefficient. Since the Levinson-Durbin recursion is a mapping from a sequence of autocorrelations $r_x(k)$ to a set of model coefficients $a_p(k)$, the name of the m-file is `rtoa.m` (`r`-to-`a`).[4] The format of the command line for this m-file is

```
>>[a,epsilon]=rtoa(r);
```

and requires only one input, `r`, which is a vector that contains the sequence of autocorrelations, $r_x(k)$. The outputs include the vector `a`, the solution to the Toeplitz normal equations, and the scalar `epsilon`, which, in the case of all-pole signal modeling, represents the modeling error. It should be pointed out that `r` may be *any* complex vector, i.e., `r` need not be a valid autocorrelation sequence. However, if `r` is *not* a valid autocorrelation sequence, then $\mathbf{R}_x$ will not be positive semi-definite. In this case, `epsilon` may be negative and will not have any physical meaning in terms of a modeling error.

Although the Levinson-Durbin recursion is a mapping from a sequence of autocorrelations $r_x(k)$ to a set of model coefficients $a_p(k)$ and a set of reflection coefficients Γ_j, the m-file `rtoa.m` only finds the coefficients $a_p(k)$. A companion m-file is listed in Fig. A.3, `rtog.m`, that performs the mapping from $r_x(k)$ to the reflection coefficients Γ_j. The syntax for this m-file is

```
>>[gamma,epsilon]=rtog(r);
```

This m-file first calls `rtoa.m`, which produces the sequence $a_p(k)$, and then calls `atog.m`

[4]The inverse of this m-file is `ator.m` (`a`-to-`r`), which is described later.

```
function [gamma,epsilon] = rtog(r)
%
    [a,epsilon]=rtoa(r);
    gamma=atog(a);
    end;
```

Figure A.3 *An m-file for the Levinson-Durbin recursion that finds the set of reflection coefficients Γ_j from the autocorrelation sequence $r_x(k)$.*

(described below), which maps the coefficients $a_p(k)$ to the reflection coefficients Γ_j. The reflection coefficients may also be found from the sequence $r_x(k)$ as follows:

```
>>gamma=atog(rtoa(r));
```

Special Notes: None.

Step-up Recursion. The step-up recursion is an algorithm to find the direct-form coefficients $a_p(k)$ that correspond to a sequence of reflection coefficients Γ_j. The m-file for the step-up recursion is `gtoa.m` and may be called as follows:

```
>>a=gtoa(gamma);
```

The input to this m-file, `gamma`, is a vector that contains the reflection coefficients Γ_j and the output, a, is the vector of filter coefficients $a_p(k)$ with the leading coefficient equal to one. For example,

```
>>a=gtoa([.1, .2, .3]);
```

produces the vector `a = [1, 0.18, 0.236, 0.3]`.

Recall that the polynomial

$$A_p(z) = 1 + \sum_{k=1}^{p} a_p(k) z^{-k}$$

will be minimum phase (all of the roots inside the unit circle) if and only if the reflection coefficients are less than one in magnitude. Compare, for example, the results that are obtained with the following commands:

```
>>abs(roots(gtoa([.5, .6, .7])))
>>abs(roots(gtoa([.5, .6, 1.2])))
```

Special Notes: None.

Step-down Recursion. The step-down recursion is the inverse of the step-up recursion, producing a sequence of reflection coefficients from a set of filter coefficients. The m-file for the step-down recursion, `atog.m`, is called as follows:

```
>>g=atog(a);
```

where `a` is the vector that contains the filter coefficients $a_p(k)$, and `g` is the vector that contains the reflection coefficients Γ_j. The input `a` may be any vector of real or complex numbers, provided that the first coefficient is nonzero. Since `atog.m` normalizes the first coefficient of `a` to one, then the following two commands result in the same sequence of reflection coefficients,

```
>>atog([1, .9, .81]);
>>atog([2, 1.8, 1.62]);
```

As discussed in Section 5.2.4, the step-down recursion may be used to check the stability of a digital filter. For example, if $H(z) = B_q(e^{j\omega})/A_p(e^{j\omega})$ and the coefficients of the denominator polynomial $A_p(e^{j\omega})$ are stored in the vector `a`, then the command

```
>>max(abs(atog(a)))
```

will return a value that is less than one if the filter is stable, and a value that is greater than one if it is unstable.

Special Notes: The first coefficient of the vector `a` must be non-zero.

Inverse Levinson-Durbin Recursion. Given an autocorrelation sequence $r_x(k)$, the Levinson-Durbin recursion produces a sequence of reflection coefficients, Γ_j, and a sequence of direct-form filter coefficients $a_p(k)$. The inverse Levinson-Durbin recursion is a mapping that recovers the autocorrelation sequence from either Γ_j or $a_p(k)$. The two m-files that implement these recursions are `gtor.m` and `ator.m`. The first, `gtor.m`, finds the autocorrelation sequence from the reflection coefficients,

```
>>r=gtor(gamma,epsilon);
```

The only required input to this m-file is the vector `gamma` that contains the sequence of reflection coefficients. The second input, `epsilon`, is an optional scalar. With only one input, the command

```
>>r=gtor(gamma);
```

will return a normalized autocorrelation sequence with $r_x(0) = 1$. For example, with the reflection coefficient sequence $\Gamma = \left[0.2,\ -0.2,\ 0.2\right]^T$ we find

```
>> r=gtor([.2, -.2, .2])

r =

    1.0000   -0.2000    0.2320   -0.2614
```

Note that when `gamma` is a vector of length p, the output `r` will be a vector of length $p+1$. If, on the other hand, a value for `epsilon` is specified, then

```
>>r=gtor(gamma,epsilon);
```

will return an autocorrelation sequence that is scaled so that `gtor` is the inverse of `rtog`, i.e.,

```
>>r=gtor(rtog(r));
```

The second m-file, `ator.m`, finds the autocorrelation sequence from a set of filter coefficients, and may be called as follows:

```
>>r=ator(a,b);
```

The only required input to this m-file is the vector a of coefficients $a_p(k)$. The second input, b, is an optional scalar that corresponds to the value $b(0)$ in the all-pole model. With only one input,

```
>>r=ator(a);
```

returns a normalized autocorrelation sequence $r_x(k)$ with $r_x(0) = 1$. Note that if a is a vector of length p then the output r will also be a vector of length p. If a value for $b(0)$ is specified in the variable b, then

```
>>r=ator(a,b);
```

will return an autocorrelation sequence, properly scaled, so that r is the inverse of rtog.m, i.e.,

```
>>r=ator(rtoa(r));
```

Special Notes: None.

General Levinson Recursion. The general Levinson recursion is an efficient algorithm for solving a general set of Hermitian Toeplitz equations of the form

$$\mathbf{R}_p \mathbf{x}_p = \mathbf{b} \tag{A.3}$$

where $\mathbf{R}_p = \text{Toep}\{r_x(0), r_x(1), \ldots, r_x(p)\}$ is a Hermitian Toeplitz matrix and $\mathbf{b}$ is an arbitrary vector. These equations arise in a number of problems including the Padé approximation, Shanks' method, and FIR Wiener filtering. The m-file for the general Levinson recursion is glev.m and is called as follows

```
>>x=glev(r,b);
```

There are two required inputs to this m-file. The first is the vector r, which is the first column of the Toeplitz matrix $\mathbf{R}_p$. The second, b, is the vector of coefficients on the right-side of Eq. (A.3). The output is the vector x, which is the solution to the Toeplitz equations.

Special Notes: The vectors r and b must have the same length.

A.6 LATTICE FILTERS

In Chapter 6 we looked at all-pole models that are based on the sequential optimization of the reflection coefficients. These modeling techniques included the forward and backward covariance methods, and the Burg algorithm. In addition, we looked at the modified covariance method, which is a non-sequential version of the Burg algorithm. In this section, we describe the m-files for these methods.

The forward covariance method finds the reflection coefficients of an all-pole model for a signal $x(n)$ by sequentially minimizing the sum of the squares of the forward prediction error. A related modeling technique, the backward covariance method, sequentially minimizes the sum of the squares of the backward prediction error. For any sequence, either the forward covariance method or the backward covariance method will produce an unstable model. The Burg algorithm, however, which sequentially minimizes the sum of the squares of the forward and backward prediction errors, is guaranteed to give a stable model for any sequence $x(n)$. The modified covariance method, on the other hand, finds the all-pole model that minimizes the sum of the squares of the forward and backward prediction errors. Since this minimization is not performed sequentially, the modified covariance method finds the global minimum. However, the model that is found is not guaranteed to be stable.

The m-files for the lattice filter based signal models all have the same format. Each requires two inputs. The first is a vector `x` that contains the signal values that are to be modeled. The second is the order of the all-pole model, `p`. The outputs of each of these m-files include the vector `gamma`, which contains the reflection coefficients for the all-pole model, and the vector `err`, which contains the modeling errors for each model of order $k = 1$ to p.

The m-file for the forward covariance method is `fcov.m` and is called as follows:

```
>>[gamma,err] = fcov(x,p);
```

Although no m-file is given for the backward covariance method, one may be easily written by modifying `fcov.m`. Specifically, all that is required is to modify the line that computes the reflection coefficients (first line in the `for` loop) along with the line that computes the modeling error, `err(j)`, of the jth-order model (fourth line in the loop). The m-file for Burg's algorithm is `burg.m`, which has the following format:

```
>>[gamma,err] = burg(x,p);
```

The m-file for the modified covariance method on the other hand, is `mcov.m`, which may be called as follows:

```
>>[gamma,err] = mcov(x,p);
```

Once the reflection coefficients of the all-pole model have been found, the approximation $\hat{x}(n)$ of $x(n)$ may be computed as follows:

```
>>xhat=filter(1,gtoa(gamma),[1, zeros(length(x)-1)]);
```

This command first converts the reflection coefficients into the direct form filter coefficients. These coefficients are then used to find the unit sample response of the filter that is used to model $x(n)$.

Special Notes: For each m-file, the model order `p` must be a positive integer that is less than the length of the vector `x`.

A.7 OPTIMUM FILTERS

In Chapter 7 we looked at the problem of designing a filter that would produce the minimum mean-square estimate of a process $d(n)$ from a related process $x(n)$. Although no m-files were given in Chapter 7, many of the filters may be designed and implemented using standard MATLAB functions. As an example, here we will illustrate how to design an FIR Wiener filter for noise cancellation as described Sect. 7.2.3. Specifically, we will look at the filtering problem considered in Example 7.2.6.

Let $x(n) = d(n) + v_1(n)$ where $d(n)$, the desired signal, is a sinusoid of frequency $\omega = 0.05\pi$ that is observed in the presence of additive noise, $v_1(n)$. Given a related noise process, $v_2(n)$, that is correlated with $v_1(n)$, a Wiener noise cancellation filter to estimate

$d(n)$ may be designed as follows. As defined in Example 7.2.6, let

$$v_1(n) = 0.8v_1(n-1) + g(n)$$
$$v_2(n) = -0.6v_2(n-1) + g(n)$$

where $g(n)$ is a zero-mean, unit variance white Gaussian noise process. We may generate 200 samples of each of these processes with the following sequence of commands:

```
>>d=sin(0.05*pi*[1:200] + 2*pi*rand);
>>g=randn(1,200);
>>v1=filter(1,[1 -0.8],g);
>>v2=filter(1,[1 0.6],g);
>>x=d+v1;
```

The coefficients of the Wiener filter to estimate $v_1(n)$ from $v_2(n)$ are found by solving the linear equations $\mathbf{R}_{v_2}\mathbf{w} = \mathbf{r}_{xv_2}$. First, however, we must estimate the autocorrelation matrix of $v_2(n)$, and the cross-correlation between $x(n)$ and $v_2(n)$. Given $x(n)$ and $v_2(n)$ these may be estimated as follows:

```
>>Rv2=covar(v2,4);
>>rxv2=convm(x,4)'*convm(v2,4)/(length(x)-1);
```

Next, the linear equations are solved with the command

```
>>w=rxv2(1,:)/Rv2;
```

and the estimate of the additive noise $v_1(n)$ generated by filtering $v_2(n)$ as follows:

```
>>v1hat=filter(w,1,v2);
```

Finally, the estimate of $d(n)$ is the difference between $x(n)$ and $\hat{v}_1(n)$,

```
>>dhat=x-v1hat;
```

A.8 SPECTRUM ESTIMATION

In Chapter 8 we considered the problem of estimating the power spectrum $P_x(e^{j\omega})$ of a wide-sense stationary process $x(n)$ from a finite sequence of observations. In this section, we describe the m-files for spectrum estimation that are found in Chapter 8. The inputs to these m-files vary, depending on what parameters need to be specified in the spectrum estimation technique. Each, however, requires an input vector x, which may be either real-valued or complex-valued, that contains the values of the process $x(n)$ that is to be analyzed. For the classical spectrum estimation m-files, the output is a vector Px that contains 1024 samples of the estimated spectrum, $\hat{P}_x(e^{j\omega})$, over the frequency range $[0, 2\pi]$. For the remaining m-files (except the Pisarenko harmonic decomposition), the vector Px contains 1024 values of the spectrum estimate in dB, i.e., $10\log_{10}\hat{P}_x(e^{j\omega})$.

Once $\hat{P}_x(e^{j\omega})$ has been computed, making a plot of the spectrum estimate is straightforward. For example, the periodogram may be plotted as follows:

```
>>Px=periodogram(x);
>>plot(Px)
```

or, more simply, by typing

```
>>plot(periodogram(x))
```

Alternatively, to plot the spectrum in dB we may use the command

```
>>plot(10*log10(Px))
```

To label the frequency axis in Hertz, we set FS equal to the sampling frequency, f_s, and type the commands

```
>>f=FS*[0:1023]/1024;
>>plot(f,Px)
```

Note that if FS=2, then the axis will be labeled in units of π. Finally, it should be pointed out that if $x(n)$ is real, then the power spectrum is symmetric,

$$\hat{P}_x(e^{j\omega}) = \hat{P}_x(e^{-j\omega})$$

In this case, it is only necessary to plot the first half of the array Px,

```
>>plot(Px(1:512))
```

or,

```
>>plot(f(1:512),Px(1:512))
```

One of the experiments described in Chapter 8 is to generate an overlay plot of the spectrum estimates that are generated from an ensemble of realizations of a given process. These plots are useful in studying the variability or *stability* of a spectrum estimation technique, and in estimating the expected value of $\hat{P}_x(e^{j\omega})$. To generate such a plot, it is necessary to write an m-file similar to the one given in Fig. A.4, which generates num periodograms of a process consisting of a sum of sinusoids in white noise. The output of this m-file, Px, is a matrix having num columns where each column is a periodogram of one of the sequences in the ensemble. For example, the command

```
>>Px=overlay(40,[0.4*pi,0.45*pi],[5,5],1,50);
```

generates 50 periodograms using 40 samples of the process

$$x(n) = 5\sin(0.4\pi n + \phi_1) + 5\sin(0.45\pi n + \phi_2) + v(n)$$

where ϕ_1 and ϕ_2 are random variables that are uniformly distributed over the interval $[-\pi, \pi]$, and $v(n)$ is unit variance white Gaussian noise. From the matrix Px, an overlay plot of the periodograms may be displayed with the command

```
>>plot(10*log10(Px))
```

and the average of the spectrum estimates may be plotted as follows:

```
>>plot(10*log10(sum(Px')/num))
```

In the following sections we describe the m-files for spectrum estimation, beginning with the classical methods in Section A.8.1. In Section A.8.2, we then describe the m-files for the nonclassical methods, which include the minimum variance method, the maximum entropy method, and the model-based methods. Finally, the m-files for frequency estimation are presented in Section A.8.3.

Generating Multiple Spectrum Estimates

```
function Px=overlay(N,omega,A,sigma,num)
% Calculates the periodogram using an ensemble of realizations of a
% process consisting of a sum of complex exponentials in white noise.
%
% Px=overlay(N,omega,A,sigma,num)
%       N     :  length of the signal
%       omega :  vector containing the sinusoid frequencies
%       A     :  vector containing the sinusoid amplitudes
%       sigma :  variance of the white noise
%       num   :  size of the ensemble (number of periodograms)
%
   jj=length(omega);
   n=1:N;
   for i=1:num
        x=sigma*randn(1,N);
        for j=1:jj
              phi=2*pi*rand(1);
              x=x+A(j)*sin(omega(j)*n+phi);
              end;
        Px(:,i)=periodogram(x);     % This command may be replaced
                                    % with another m-file.
        end;
```

Figure A.4 *An m-file to find the periodogram of an ensemble of realizations of a process consisting of a sum of complex exponentials in white noise. These periodograms may then be used to generate overlay plots or an estimate of the expected value of the periodogram.*

A.8.1 Classical Methods

In this section we describe the m-files for the classical approaches to spectrum estimation, which include the periodogram, the modified periodogram, Bartlett's method, Welch's method, and the Blackman-Tukey method. We begin with the m-file for the periodogram, which is used in each of the other m-files described in this section.

The Periodogram. The periodogram of $x(n)$, using samples from $n = n_1$ to $n = n_2$, is defined by

$$\hat{P}_{per}(e^{j\omega}) = \frac{1}{n_2 - n_1 + 1}\left|\sum_{n=n_1}^{n_2} x(n)e^{-jn\omega}\right|^2$$

The m-file to compute the periodogram is `periodogram.m`, and the format of the command line for this m-file is

```
>>Px=periodogram(x,n1,n2);
```

The only required input is the vector x, which contains the signal values $x(n)$. The inputs `n1` and `n2` are optional integers that indicate the values of $x(n)$ within the vector x that are to be used, i.e., `x(n1)` to `x(n2)`. If these integers are left unspecified, as in the command

```
>>Px=periodogram(x);
```

then the periodogram is computed using all of the values in the vector `x`. The output `Px` is a vector that contains 1024 samples of the periodogram from $\omega = 0$ to $\omega = 2\pi$.

Special Notes: The inputs `n1` and `n2` must be positive integers with `n1` < `n2` ≤ `length(x)`. This program calls the m-file `fft.m` and, by default, an FFT of length 1024 is used. Therefore, if `n2−n1+1` is greater than 1024, then `periodogram.m` must be modified to accommodate a longer FFT.

The Modified Periodogram. The modified periodogram allows the user to window $x(n)$ prior to computing the periodogram,

$$\hat{P}_M(e^{j\omega}) = \frac{1}{n_2 - n_1 + 1} \left| \sum_{n=n_1}^{n_2} w(n)x(n)e^{-jn\omega} \right|^2$$

The window $w(n)$ allows one to trade-off spectral resolution, which is determined by the width of the main lobe of $W(e^{j\omega})$, and spectral masking, which is due to the sidelobes of $W(e^{j\omega})$. The m-file for the modified periodogram, `mper.m`, is called as follows:

```
>>Px = mper(x,win,n1,n2);
```

The only required inputs are `x`, the vector of signal values $x(n)$, and `win`, an integer that specifies the type of window that is to be used. The correspondence between values of `win` and the type of window is given in Fig. A.5. As with the periodogram, the optional inputs `n1` and `n2` may be used to select a subset of the values of $x(n)$ within the array `x`. If `n1` and `n2` are left unspecified, then all of the values in the vector `x` will be used.

Special Notes: Same as for the m-file `periodogram.m`. In addition, `win` should be an integer within the range from `win=1` to `win=5`.

Bartlett's Method. Bartlett's method produces a spectrum estimate by averaging the periodograms that are formed from nonoverlapping subsequences of $x(n)$. Periodogram averaging allows one to trade spectral resolution for a reduction in the variance of the estimate. The m-file for Bartlett's method is `bart.m`, and the format of the command line is

```
>>Px=bart(x,nsect);
```

where `x` is a vector of signal values and `nsect` is the number of subsequences that are to be extracted from $x(n)$. As with the periodogram, the output `Px` is a vector of length 1024 that contains the values of the estimated spectrum from $\omega = 0$ to $\omega = 2\pi$.

Special Notes: The number of subsequences, `nsect`, must be a positive integer. Although `nsect` is allowed to be as large as `length(x)`, a much smaller value is typically used in order to ensure that the Bartlett estimate has sufficient resolution.

```
win=1 : Rectangular window
win=2 : Hamming window
win=3 : Hanning window
win=4 : Bartlett window
win=5 : Blackman window
```

Figure A.5 *The types of windows and the corresponding values for the parameter* `win` *in the m-files for the modified periodogram, Welch's method, and the Blackman-Tukey method.*

Welch's Method. Welch's method is a generalization of Bartlett's method that allows a window to be applied to $x(n)$, thereby replacing the periodograms in Bartlett's method with modified periodograms. In addition, Welch's method allows the subsequences that are extracted from $x(n)$ to overlap. The m-file for Welch's method is `welch.m` and the format of the command line is

```
Px = welch(x,L,over,win);
```

The inputs to this m-file include the vector `x`, which contains the signal values $x(n)$, an integer `L`, which defines the length of the sections that are to be extracted from the vector `x`, the parameter `over`, which is a real number in the range $0 \leq$ `over` < 1 that defines the amount of overlap between the subsequences, and `win`, an integer that specifies the type of window that is to be used according to the list given in Fig. A.5. The overlap that is specified by `over` is given as a percentage. For example, if `over=0`, then there is no overlap, and if `over=0.5`, then there will be a 50% overlap between subsequences. If the value of `win` is unspecified, then a rectangular window is used. If, in addition, the value of `over` is undefined, then `over` is set to zero, and

```
Px = welch(x,L);
```

is equivalent to Bartlett's method. However, this way of implementing Batlett's method differs from the command

```
Px = bartlett(x,nsect);
```

in that it allows the user to specify the section length rather than the number of sections. Finally, if `welch.m` is used with only one input as in the command

```
Px = welch(x);
```

then a rectangular window is used, the overlap is set to zero, and the section length is set equal to the length of the vector `x`. Thus, this command is equivalent to the command

```
Px = periodogram(x);
```

for computing the periodogram.

Special Notes: Same as for the m-file `periodogram.m`. In addition, the value for the overlap must satisfy $0 \leq$ `over` < 1. Overlaps of 50% (`over=.5`) and 75% (`over=.75`) are commonly used.

Blackman-Tukey Method. The Blackman-Tukey method of periodogram smoothing involves multiplying an estimated autocorrelation sequence, $\hat{r}_x(k)$, with a window, $w(k)$, and then taking the discrete-time Fourier transform,

$$\hat{P}_{BT}(e^{j\omega}) = \sum_{k=-M}^{M} \hat{r}_x(k)w(k)e^{-jk\omega}$$

The m-file for the Blackman-Tukey method is `per_smooth.m` and the format of the command line is

```
Px = per_smooth(x,win,M,n1,n2);
```

The only required inputs for this m-file are `x`, `win`, and `M`. As with the previous m-files, `x` is a vector that contains the signal values, `win` is an integer that defines the type of window that is to be used as given in Fig. A.5, and `M` is the length of the window. The inputs `n1` and n2 are optional integers that indicate the values of $x(n)$ within the vector x that are to be used. If these integers are left unspecified then the entire data record is used.

Special Notes: Same as for the m-file for the modified periodogram, `mper.m`. In addition, the length of the window, M, must be in the range $0 < \texttt{M} \leq \texttt{length(x)}$.

A.8.2 Nonclassical Methods

In this section we describe the nonclassical spectrum estimation techniques, which include the minimum variance method, the maximum entropy method, and the model-based methods.

Minimum Variance Method. For a wide-sense stationary process $x(n)$, the minimum variance estimate of the power spectrum is

$$\hat{P}_{MV}(e^{j\omega}) = \frac{p}{\mathbf{e}^H \mathbf{R}_x^{-1} \mathbf{e}} \tag{A.4}$$

where $\mathbf{R}_x$ is the $p \times p$ autocorrelation matrix. The m-file that is used to compute the minimum variance estimate is `minvar.m`, and the format of the command line is

```
>>Px = minvar(x,p);
```

where x is a vector that contains the values $x(n)$, and p is the order of the minimum variance estimate. The order p is equal to the order of the bandpass filters used in the minimum variance filterbank, and is equal to the size of the autocorrelation matrix, $\mathbf{R}_x$. The output, Px, is a vector that contains 1024 samples of the minimum variance estimate (in dB) from $\omega = 0$ to $\omega = 2\pi$. Instead of evaluating Eq. (A.4) directly, `minvar.m` finds the minimum variance estimate using the expansion

$$\hat{P}_{MV}(e^{j\omega}) = \frac{p}{\displaystyle\sum_{i=1}^{p} \frac{1}{\lambda_i} |\mathbf{e}^H \mathbf{v}_i|^2}$$

where λ_i and $\mathbf{v}_i$ are the eigenvalues and eigenvectors, respectively, of $\mathbf{R}_x$. Note that each term in the denominator, $\mathbf{e}^H \mathbf{v}_i$, is the DTFT of the eigenvector $\mathbf{v}_i$. These terms are evaluated in `minvar.m` using a 1024-point FFT.

Special Notes: The order p must be a positive integer. Typically, $\texttt{p} \approx \texttt{N}/3$, where N is the length of the data sequence (assuming that this value is not too large to perform an eigenvalue decomposition of $\mathbf{R}_x$).

Maximum Entropy Method. The maximum entropy method of spectrum estimation finds an all-pole model for a process using the autocorrelation method, and then uses the model parameters $a_p(k)$ to estimate the spectrum as follows:

$$\hat{P}_{mem}(e^{j\omega}) = \frac{\epsilon_p}{\left|1 + \displaystyle\sum_{k=1}^{p} a_p(k) e^{-jk\omega}\right|^2}$$

The m-file for the maximum entropy method is `mem.m`, and the format of the command line is

```
>>function Px = mem(x,p)
```

where x is the vector that contains the signal values, p is the model order, and Px is a vector that contains 1024 samples of the maximum entropy spectrum (in dB) from $\omega = 0$ to $\omega = 2\pi$.

Special Notes: The order p must be a positive integer that does not exceed the length of the vector x.

Model-based Methods. A model-based approach to spectrum estimation involves the selection of a model, the estimation of the model parameters, and the evaluation of the spectrum using the estimated parameters. The spectrum of a process $x(n)$ may be estimated using any of the signal modeling m-files described in Sect. A.4 and Sect. A.6. For example, an autoregressive spectrum estimate using Burg's method may be found as follows:

```
>>[a,err] = burg(x,p);
>>Px=10*log10(err)-20*log10(abs(fft(a,1024)));
```

A.8.3 Frequency Estimation

In this section we describe the m-files that are used for processes that consist of a sum of complex exponentials in white noise. We begin with the noise subspace methods, which are frequency estimation algorithms and include the Pisarenko harmonic decomposition, the MUSIC algorithm, the eigenvector method, and the minimum norm algorithm. Then we discuss the m-files that use a signal subspace approach, which involves a principal components analysis of the autocorrelation matrix $\mathbf{R}_x$.

Noise Subspace Methods. A noise subspace method to estimate the frequencies of p complex exponentials in noise involves the use of a frequency estimation function of the form

$$\hat{P}_x(e^{j\omega}) = \frac{1}{\sum_{i=p+1}^{M} \alpha_i |\mathbf{e}^H \mathbf{v}_i|^2} \tag{A.5}$$

where $\mathbf{v}_i$ are vectors that lie in the noise subspace of $\mathbf{R}_x$, and α_i are constants. In the Pisarenko harmonic decomposition, the frequency estimation function is

$$\hat{P}_{PHD}(e^{j\omega}) = \frac{1}{|\mathbf{e}^H \mathbf{v}_{\min}|^2}$$

where $\mathbf{v}_{\min}$ is the eigenvector of $\mathbf{R}_x$ that has the minimum eigenvalue. The m-file for Pisarenko's method is phd.m, and the format of the command line to use this m-file is

```
>>[vmin,sigma]=phd(x,p);
```

where x is a vector that contains the values of $x(n)$, and p is an integer that defines the number of complex exponentials in $x(n)$.[5] The output of this m-file is the eigenvector having the smallest eigenvalue, vmin, along with the minimum eigenvalue, sigma, which may be used as an estimate of the white noise variance. The frequencies of the complex exponentials may be estimated from vmin by finding the angles of the eigenfilter $V_{\min}(z)$ as follows:

```
>>freq=angle(roots(vmin));
```

[5]For sinusoids in noise, p is equal to twice the number of sinusoids.

Alternatively, the frequencies may be estimated from the peaks of the eigenspectrum, which may be plotted using the command

```
>>plot(abs(fft(vmin,1024)))
```

The MUSIC algorithm is another frequency estimation technique of the form given in Eq. (A.5). However, with the MUSIC algorithm, the constants α_i are equal to one and $\mathbf{v}_i$ are the $M - p$ eigenvectors of $\mathbf{R}_x$ that have the smallest eigenvalues. The m-file for the MUSIC algorithm is `music.m`, and the format of the command line is

```
>>Px = music(x,p,M);
```

where `x` is a vector that contains the signal values, `p` is a positive integer that specifies the number of complex exponentials in $x(n)$, and `M` is an integer that specifies the size of the autocorrelation matrix and, consequently, defines the number of eigenvectors, `M-p`, that are to be used to form the frequency estimation function $\hat{P}_x(e^{j\omega})$. The integer `M` must be larger than `p` and, in the special case of `M=p+1`, the MUSIC algorithm is equivalent to the Pisarenko harmonic decomposition. The output `Px` is a vector that contains 1024 equally spaced samples of the frequency estimation function (in dB) from $\omega = 0$ to $\omega = 2\pi$.

The eigenvector method is a frequency estimation algorithm that is similar to the MUSIC algorithm. As with the MUSIC algorithm, the vectors $\mathbf{v}_i$ for $i = p + 1$ to M are the eigenvectors that have the smallest eigenvalues. The constants α_i, however, are equal to the inverse of the eigenvalues, $\alpha_i = 1/\lambda_i$. The m-file for the eigenvector method is `ev.m` and the format to use this m-file is

```
>>Px = ev(x,p,M);
```

The inputs, `x`, `p`, and `M` and the output `Px` are as described for the m-file `music.m`.

The last frequency estimation algorithm is the minimum norm method, which uses a frequency estimation function of the same form as that used for the Pisarenko harmonic decomposition,

$$\hat{P}(e^{j\omega})_{MN} = \frac{1}{|\mathbf{e}^H\mathbf{a}|^2}$$

However, instead of using the eigenvector having the smallest eigenvalue, the minimum norm method uses the vector $\mathbf{a}$ in the noise subspace that has the minimum norm. The m-file for the minimum norm algorithm is `min_norm.m` and the format of the command line is

```
>>Px = min_norm(x,p,M);
```

where `x`, `p`, and `M` are as described for the MUSIC algorithm, and `Px` contains the samples of the frequency estimation function in dB.

Special Notes: For each of the noise subspace m-files, `M` and `p` must be positive integers with `M` $\geq$ `p+1`. Since the m-files are frequency estimation algorithms, the vector `Px` is not an estimate of the power spectrum, and should only be used to extract estimates of the complex exponential frequencies, ω_k.

Principal Components Spectrum Estimation. For a wide-sense stationary process consisting of p complex exponential in noise, a principal components method of spectrum estimation finds a reduced rank approximation to the autocorrelation matrix using the p

principal eigenvectors,

$$\hat{\mathbf{R}}_s = \sum_{i=1}^{p} \lambda_i \mathbf{v}_i \mathbf{v}_i^H$$

and then estimates the power spectrum from $\hat{\mathbf{R}}_x$. For example, a principal components approach using the Blackman-Tukey method and a Bartlett window yields the estimate

$$\hat{P}_{PC-BT}(e^{j\omega}) = \frac{1}{M}\mathbf{e}^H \hat{\mathbf{R}}_s \mathbf{e}$$

An m-file to find $\hat{P}_{PC-BT}(e^{j\omega})$ is `bt_pc.m`, which has a command line for the form

```
>>Px = bt_pc(x,p,M);
```

where `x` is the vector containing the signal values, `p` is the number of complex exponentials, and `M` is the size of the autocorrelation matrix $\mathbf{R}_x$ (length of the Bartlett window). The output `Px` is a vector containing 1024 samples of the estimated power spectrum (in dB) from $\omega = 0$ to $\omega = 2\pi$.

Although `bt_pc.m` is the only m-file that was given in Chapter 8 for principal components spectrum estimation, using this m-file as a model, m-files for the other principal components methods, such as the minimum variance method and the maximum entropy method, may be easily written.

A.9 ADAPTIVE FILTERING

In Chapter 9, we considered the problem of estimating a process $d(n)$ from the values of a related process $x(n)$. In this section we describe three m-files that may be used to solve this problem. One of these m-files is for the LMS algorithm, and another is for the normalized LMS algorithm. Both of these algorithms estimate $d(n)$ by filtering $x(n)$ with a time-varying tapped delay line,

$$\hat{d}(n) = \sum_{k=0}^{p-1} w_n(k)x(n-k)$$

where the coefficients $w_n(k)$ are determined using an update equation of the form

$$\mathbf{w}_{n+1} = \mathbf{w}_n + \mu(n)e(n)\mathbf{x}^*(n)$$

The third m-file is for the RLS algorithm, which finds the least squares estimate of $d(n)$. Although many other types of adaptive filtering algorithms were discussed in Chapter 9, in most cases, m-files for these algorithms may be easily implemented with only a slight modifications to one of the m-files described in this section.

LMS. With an FIR LMS adaptive filter, the coefficient update equation has the form

$$\mathbf{w}_{n+1} = \mathbf{w}_n + \mu e(n)\mathbf{x}^*(n)$$

The m-file for the LMS adaptive filter is `lms.m` and may be called as follows:

```
>> [W,E] = lms(x,d,mu,nord,a0);
```

To use this m-file, four inputs are required; the fifth is optional. The first input is the vector x, which contains the values of $x(n)$ that are to be used to estimate $d(n)$. Second is the vector d of desired signal values, $d(n)$. Third is the step size mu, which, for convergence in the mean, must be non-negative and less than $1/\lambda_{\max}$ where $\lambda_{\max}$ is the largest eigenvalue of the autocorrelation matrix of $x(n)$. Finally, nord is an integer that defines the order of the adaptive filter (number of filter coefficients). The optional input a0 allows one to define the initial conditions of the adaptive filter. The output W is a matrix that has the coefficients of the adaptive filter at time n stored in the nth row. Thus, a plot of each filter coefficient, $w_n(k)$, as a function of n may be plotted as in Fig. 9.10 with the command

```
>> plot(W)
```

The optional output E is a vector that contains the values of the error, $e(n) = \hat{d}(n) - d(n)$, at time n.

Special Notes: The length of the vector d, the desired signal, must be at least as long as the length of the input vector x. The step size mu must be non-negative and it generally should be much smaller than $1/\lambda_{\max}$ where $\lambda_{\max}$ is the largest eigenvalue of the autocorrelation matrix of $x(n)$. The order of the filter, nord, must be a non-negative integer and smaller than the length of the vector x. Finally, the input a0 must be a vector with a length equal to the value of nord.

Normalized LMS. The normalized LMS algorithm, which has a coefficient update equation of the form

$$\mathbf{w}_{n+1} = \mathbf{w}_n + \beta \frac{\mathbf{x}^*(n)}{\epsilon + \|\mathbf{x}(n)\|^2} e(n)$$

is simpler to use than the LMS algorithm in that it allows the step size to be selected without having to know the largest eigenvalue of the autocorrelation matrix of $x(n)$. The normalized LMS algorithm has two parameters. The first, β, is the normalized step size, and the second, ϵ, is a parameter that stabilizes the algorithm in case $\|\mathbf{x}(n)\|^2$ becomes small. The m-file for the normalized LMS algorithm is nlms.m, which may be called as follows:

```
>> [W,E] = nlms(x,d,beta,nord,a0);
```

The inputs and outputs are the same as those described for the LMS algorithm. Specifically, the input is stored in the vector x, the desired signal is stored in the vector d, the step size β is set by the value of beta, the filter order is specified by nord, and the optional initial conditions for the adaptive filter may be specified with a0. The step size β should be positive and no larger than one, $0 < \beta < 1$. In nlms.m, the parameter ϵ is set equal to 0.0001. Therefore, if another value is desired, the m-file would need to be modified accordingly. As with the LMS algorithm, the output W is a matrix with the nth row containing the adaptive filter coefficients at time n, and E is a vector that contains the errors, $e(n) = \hat{d}(n) - d(n)$.

Special Notes: Same as for the m-file lms.m except that the step size must be in the range 0 < beta < 1.

Recursive Least Squares. The recursive least squares algorithm produces the set of filter coefficients $w_n(k)$ at time n that minimize the weighted least squares error

$$\mathcal{E}(n) = \sum_{i=0}^{n} \lambda^{n-i} |e(i)|^2$$

The m-file for the RLS algorithm is `rls.m`, and is called using a command of the following form:

```
>> [W,E] = rls(x,d,lambda,nord);
```

The inputs to this m-file include the input signal vector `x`, the desired signal vector `d`, the exponential weighting factor `lambda`, and the filter order `nord`. The outputs of `rls.m` are the same as those for `lms.m` and `nlms.m`.

Special Notes: The length of the vector `d`, the desired signal, must be at least as long as the length of the input vector `x`. The exponential weighting factor, `lambda`, must be greater than zero and less than or equal to one. If `lambda` is equal to one, then the m-file implements the growing window RLS algorithm. The order of the filter, `nord`, must be a non-negative integer that is smaller than the length of the vector `x`.

SYMBOLS

OPERATIONS

$\mathbf{a}^T$, $\mathbf{A}^T$	Transpose of a vector $\mathbf{a}$ or matrix $\mathbf{A}$ (p. 21, 27)
$\mathbf{a}^H$, $\mathbf{A}^H$	Hermitian transpose of a vector $\mathbf{a}$ or matrix $\mathbf{A}$ (p. 21, 27)
$\mathbf{a}^*$, $\mathbf{A}^*$	Complex conjugate of a vector $\mathbf{a}$ or matrix $\mathbf{A}$ (p. 21, 27)
$\mathbf{a}_j^R$, $A_j^R(z)$	The reciprocal vector and its z-transform (p. 223, 224)
$\mathbf{A}^{-1}$	Inverse of $\mathbf{A}$ (p. 28)
$\mathbf{A}^{+}$	Pseudoinverse of $\mathbf{A}$ (p. 32, 34)
$\mathrm{Cov}(x, y)$	Covariance between x and y (p. 66)
$\det(\mathbf{A})$	Determinant of $\mathbf{A}$ (p. 29)
$E\{x\}$	Expected value of x (p. 63)
$\Pr\{A\}$	The probability of event A (p. 58)
$\mathrm{Toep}\{\mathbf{a}\}$	Hermitian Toeplitz matrix with $\mathbf{a}$ as the first column (p. 38)
$\mathrm{tr}(\mathbf{A})$	Trace of $\mathbf{A}$ (p. 30)
$\mathrm{Var}\{x\}$	Variance of x (p. 64)
∇_x	Gradient vector (p. 49)
$\Vert\mathbf{x}\Vert$	Norm of the vector $\mathbf{x}$ (p. 22)
$\langle\mathbf{a}, \mathbf{b}\rangle$	Inner product of $\mathbf{a}$ and $\mathbf{b}$ (p. 22)
$[P(z)]_+$	The causal (positive-time) part of $P(z)$ (p. 191)
$[P(z)]_-$	The anti-causal (negative-time) part of $P(z)$ (p. 191)

PRINCIPAL SYMBOLS

$\mathbf{a}_p$	Augmented vector of all-pole model coefficients (p. 148)
$\bar{\mathbf{a}}_p$	Vector of all-pole model coefficients (p. 136)
$\mathbf{b}_q$	Vector of numerator coefficients $b_q(k)$ (p. 137)
$c_x(k)$	The autocovariance of a WSS process $x(n)$ (p. 77)
$c_x(k, l)$	The autocovariance of $x(n)$ (p. 77)
c_{xy}	The covariance between x and y (p. 66)
$c_{xy}(k, l)$	The cross-covariance between $x(n)$ and $y(n)$ (p. 79)
$\mathbf{c}_n$	Weight error vector (p. 501)
$\mathbf{C}_x$	Autocovariance matrix (p. 85)
$e_p^+(n)$, $E_p^+(z)$	Forward prediction error and its z-transform (p. 290)
$e_p^-(n)$, $E_p^-(z)$	Backward prediction error and its z-transform (p. 291)
$\mathbf{e}_i$	Vector of complex exponentials (p. 427)
$\mathcal{E}(n)$	Least squares error for RLS (p. 541)
$\mathcal{E}_p$	All-pole modeling error (p. 161)
$\mathcal{E}_{p,q}$	Prony error (p. 145)
$\mathcal{E}_p^+$	The squared p-th order forward prediction error (p. 290, 308)
$\mathcal{E}_p^-$	The squared p-th order backward prediction error (p. 292, 313)
$\mathcal{E}_p^B$	Burg error of order p (p. 317)
$\mathcal{E}_p^C$	Covariance error of order p (p. 182)

$\mathcal{E}_p^M$	Error for the modified covariance method (p. 322)
$\mathcal{E}_{LS}$	Least squares error (p. 132)
$\mathcal{E}_S$	Shanks' error (p. 156)
$f_x(\alpha)$	Probability density function (p. 61)
$F_x(\alpha)$	Probability distribution function (p. 61)
$h(n)$	Unit sample response (p. 11)
$H(e^{j\omega})$	Frequency response (p. 13)
$H(z)$	System function (p. 15)
$\mathbf{I}$	Identity matrix (p. 28)
$\mathbf{J}$	Exchange matrix (p. 36)
$\mathbf{m}_x$	The mean vector (p. 85)
$m_x(n)$	Mean of a random process $x(n)$ (p. 77)
m_x	The mean of the random variable x (p. 66) or WSS process $x(n)$ (p. 82)
$\mathcal{M}$	Figure of merit (p. 424) or level of misadjustment (p. 513)
P_A	Projection matrix (p. 34)
$\hat{P}_{AR}(e^{j\omega})$	Autoregressive spectrum estimate (p. 442)
$\hat{P}_{ARMA}(e^{j\omega})$	Autoregressive moving average spectrum estimate (p. 450)
$\hat{P}_{B}(e^{j\omega})$	Spectrum estimate using Bartlett's method (p. 413)
$\hat{P}_{BT}(e^{j\omega})$	Blackman-Tukey estimate (p. 421)
$\hat{P}_{EV}(e^{j\omega})$	Frequency estimation function, eigenvector method (p. 465)
$\hat{P}_{M}(e^{j\omega})$	Modified periodogram (p. 410)
$\hat{P}_{MA}(e^{j\omega})$	Moving average spectrum estimate (p. 448)
$\hat{P}_{mem}(e^{j\omega})$	Spectrum estimate using the maximum entropy method (p. 436)
$\hat{P}_{MN}(e^{j\omega})$	Frequency estimation function, minimum norm method (p. 466)
$\hat{P}_{MU}(e^{j\omega})$	Frequency estimation function, MUSIC algorithm (p. 464)
$\hat{P}_{MV}(e^{j\omega})$	Minimum variance spectrum estimate (p. 429)
$\hat{P}_{PC-AR}(e^{j\omega})$	Principal components autoregressive spectrum estimate (p. 472)
$\hat{P}_{PC-BT}(e^{j\omega})$	Principal components Blackman-Tukey estimate (p. 471)
$\hat{P}_{PC-MV}(e^{j\omega})$	Principal components, minimum variance estimate (p. 471)
$\hat{P}_{per}(e^{j\omega})$	Periodogram (p. 394)
$\hat{P}_{phd}(e^{j\omega})$	Pisarenko harmonic decomposition (p. 459)
$\hat{P}_{W}(e^{j\omega})$	Spectrum estimate using Welch's method (p. 418)
$P_x(e^{j\omega}),\ P_x(z)$	Power spectrum of $x(n)$ (p. 95)
$\mathrm{Res}\big[\hat{P}_x(e^{j\omega})\big]$	Resolution of $\hat{P}_x(e^{j\omega})$ (p. 402)
$r_x(k)$	The autocorrelation of a WSS process $x(n)$ (p. 83)
$r_x(k,l)$	The autocorrelation of $x(n)$ (p. 77)
r_{xy}	The correlation between x and y (p. 66)
$r_{xy}(k,l)$	The cross-correlation between $x(n)$ and $y(n)$ (p. 79)
$\mathbf{R}_x$	Autocorrelation matrix (p. 85)
$\mathbf{s}_j,\ \mathbf{s}_j^{\star}$	Singular predictor vectors (p. 270)

$S_j(z)$, $S_j^*(z)$	Singular predictor polynomials (p. 269)
$u(n)$	Unit step (p. 8)
$u_0(\omega)$	Unit impulse (p. 13)
$\mathbf{u}_1$	Unit vector with one as its first element (p. 25)
$\mathbf{v}_{\min}$	Eigenvector having the minimum eigenvalue (p. 459)
$V_{\min}(z)$	Eigenfilter corresponding to the minimum eigenvalue (p. 459)
$\mathbf{V}$	Matrix of eigenvectors (p. 44)
$\mathcal{V}$	Variability (p. 424)
$w_R(n)$	Rectangular window (p. 178)
$W_R(e^{j\omega})$	DTFT of a rectangular window (p. 409)
$w_B(n)$	Bartlett window (p. 398)
$W_B(e^{j\omega})$	DTFT of a Bartlett window (p. 399)
$x^f(n)$, $y^f(n)$	Filtered signals used in an adaptive recursive filter (p. 538)
$x_N(n)$	Windowed signal (p. 394)
β	Normalized step size (p. 514)
γ_j	Parameter used in the Levinson-Durbin recursion (p. 217)
Γ_j	Reflection coefficient (p. 218)
$\boldsymbol{\Gamma}_p$	Reflection coefficient vector (p. 222)
Γ_j^B	Reflection coefficients, Burg's method (p. 317)
Γ_j^I	Reflection coefficients, Itakura's method (p. 316)
Γ_j^+	Reflection coefficients, forward covariance method (p. 308)
Γ_j^-	Reflection coefficients, backward covariance method (p. 313)
δ_j	Levinson recursion parameter (p. 265)
$\delta(n)$	Unit sample (p. 8)
ϵ_p	Minimum all-pole modeling error (p. 163)
$\epsilon_{p,q}$	Minimum Prony modeling error (p. 148)
$\varepsilon_j(n)$	Singular forward prediction error (p. 298)
λ_i	Eigenvalue (p. 41)
$\lambda_{\max}$	Maximum eigenvalue (p. 47)
$\boldsymbol{\Lambda}$	Diagonal matrix of eigenvalues (p. 44)
μ	Adaptive filter step size (p. 500)
ξ	Mean-square error (p. 68)
$\xi(n)$	Mean-square error at time n (p. 499)
$\xi_{ex}(n)$	Excess mean-square error (p. 511)
$\rho(\mathbf{A})$	Rank of the matrix $\mathbf{A}$ (p. 28)
ρ_{xy}	Correlation coefficient (p. 66)
σ_x^2	Variance of the random variable x (p. 64) or WSS process $x(n)$ (p. 82)
$\sigma_x^2(n)$	Variance of the random process $x(n)$ (p. 77)
τ	Adaptive filter time constant (p. 503)
τ_j, τ_j^*	Parameters that appear in the split Levinson recursion (p. 271)
χ	Condition number of a matrix (p. 503)
$\psi_k^a(n)$, $\psi_k^b(n)$	Gradient estimate used in an adaptive recursive filter (p. 538)
Ω	Sample space (p. 58)

Index

CPSIA information can be obtained at www.ICGtesting.com
Printed in the USA
BVOW06s1815260713

326932BV00010B/317/A

9 780471 594314